Mathematics, Informatics, Science and Education International Conference (MISEIC 2019)

Journal of Physics: Conference Series Volume 1417

Surabaya, Indonesia
28 September 2019

ISBN: 978-1-7138-1322-4
ISSN: 1742-6588

Printed from e-media with permission by:

Curran Associates, Inc.
57 Morehouse Lane
Red Hook, NY 12571

proceedings.com

Some format issues inherent in the e-media version may also appear in this print version.

Institute of Physics
Dirac House, Temple Back
Bristol BS1 6BE UK

Phone: 44 1 17 929 7481
Fax: 44 1 17 920 0979

techtracking@iop.org

Additional copies of this publication are available from:

Curran Associates, Inc.
57 Morehouse Lane
Red Hook, NY 12571 USA
Phone: 845-758-0400
Fax: 845-758-2633
Email: curran@proceedings.com
Web: www.proceedings.com

TABLE OF CONTENTS

PHYSICS

PURE AND APPLIED MATHEMATICS

COMPUTER SCIENCE

CHEMISTRY

BIOLOGY

COMPUTER SCIENCE EDUCATION

MATHEMATICS EDUCATION

SCIENCE EDUCATION

Journal of Physics: Conference Series **1417** (2019) 011001 doi:10.1088/1742-6596/1417/1/011001

Organizer

Mathematics, Informatics, Science, and Education International Conference (MISEIC) 2019

Universitas Negeri Surabaya, Surabaya, Indonesia

September 28th, 2019

Welcome speech by Rector of Universitas Negeri Surabaya

Assalamu'alaikum wr wb.

The honorable Keynote speakers: Prof. Hernando Ombao, Ph.D, Assoc. Prof. Boon Chuan Low, Prof. Tomonori Ichinose, and Dr. Yusuf Fuad, M.App.Sc.

Dearest guests, presenters, and participants of the Mathematics, Informatics, Science, and Education International Conference (MISEIC) 2019.

Praise be to Allah SWT for His blessing moment, hence we can attend this valuable moment to share some ideas in this conference.
It is my pleasure, as the Rector and on behalf of Universitas Negeri Surabaya to welcome all of you, from abroad and within Indonesia, to participate in the Mathematics, Informatics, Science, and Education International Conference (MISEIC) 2019.

Ladies and Gentlemen,
The Sustainable Development Goals (SDGs), also known as the Global Goals, were adopted by all United Nations Member, including Indonesia, as a universal call to action to end poverty, protect the planet and ensure that all people enjoy peace and prosperity by 2030. The Global Goals address the global challenges we face, including those related to poverty, inequality, climate, environmental degradation, prosperity, and peace and justice, which can be described in 17 Sustainable Development Goals (SDGs). We can play important role and do significant action to support the achievement of these goals as we are educator and researchers as well.

Ladies and Gentlemen,
Inspired by the spirit of Sustainable Development Goals, the MISEIC 2019 brings the theme "Trends, Advancement, and Innovation in Mathematics, Informatics, Science and Education Toward Sustainable Development Goals". We do hope that this conference will be the home for the researchers, the lecturers, and the practitioners in Mathematics, Informatics, Science and Education to communicate their original scientific ideas based on their updated research. Their contribution will be valuable input for the development of these fields.

Ladies and Gentlemen,
We are grateful to all institutions, sponsors, all organizing and scientific committee members, and all participants for making this conference possible and successful. We wish you all have a pleasant stay here in Surabaya, the Heroic City.
May you have a very successful conference.
Thank you.

Prof. Dr. Nur Hasan, M.Kes

Journal of Physics: Conference Series **1417** (2019) 011001 doi:10.1088/1742-6596/1417/1/011001

Welcome speech by Chairman of the Mathemathics, Informatics, Science, and Education International Conference (MISEIC) 2019

Assalamu'alaikum wr wb.
The honorable Rector of Universitas Negeri Surabaya,
The honorable Keynote speakers: Prof. Hernando Ombao, Ph.D, Assoc. Prof. Boon Chuan Low, Prof. Tomonori Ichinose, and Dr. Yusuf Fuad, M.App.Sc.

Dearest guests, presenters, and participants of the Mathematics, Informatics, Science, and Education International Conference (MISEIC) 2019.

Welcome to Surabaya, welcome to Universitas Negeri Surabaya, welcome to the venue. We are proud to host the Mathematics, Informatics, Science, and Education International Conference (MISEIC) 2019.

Ladies and Gentlemen,
MISEIC 2019 is the 3rd Mathematics, Informatics, Science and Education International Conference (MISEIC) organized by Faculty of Mathematics and Natural Sciences Universitas Negeri Surabaya. This international conference aims to bridge the scientists, education experts and practitioners, and students in the scientific forum through sharing ideas and issues about theoretical and practical knowledge in mathematics, informatics, science and STEM education. The theme for this year conference is "Trends, Advancement, and Innovation in Mathematics, Informatics, Science and Education Toward Sustainable Development Goals" as one of our contribution to success the achievement of The Sustainable Development Goals (SDGs).

Ladies and Gentlemen,
This conference is a great event for the scientists, education experts and practitioners to share their ideas and research finding. MISEIC 2019 is attended by presenters who come from overseas and Indonesia. After the plenary session, the presenters will present their articles in the parallel session based on their research field. Hopefully, this session will be a chance for scientific sharing as well as building future networking.

Ladies and Gentlemen,
Last but not least, we have to acknowledge that this conference will not be possible without tremendous supports and help from those who really give their all-out efforts and hardworking. I am very grateful to all members of the organizing committee and scientific committee for their great work to support this conference

We wish you all have a memorable moment in Surabaya and have a fruitful and very successful conference.

Thank You,
A'yunin Sofro, Ph.D.

Abstract of STEACH Keynote Speakers

BCH Domain as a Versatile Scaffold Module in Cell Signaling and Mechanobiology

Assoc. Prof. Boon Chuan Low
Mechanobiology Institute Singapore, Department of Biological Sciences and University Scholars Programme, National University of Singapore

Protein-protein interactions define specificity in cell signaling that requires precise timing and placement of interacting partners. Deregulation of such intricate networks often lead to pathologies. My laboratory pioneers the characterization of the BNIP-2 and Cdc42GAP Homology (BCH) domain as a versatile scaffold protein domain that regulates small GTPases, kinases and metabolic signaling at specific locales. Specifically, BCH domains control Ras, Rho and Rac GTPases signaling by engaging them with their immediate activators, the guanine nucleotide exchanger factors (GEFs) or their inactivators, the GTPase-activating Proteins (GAPs). This leads to dynamic actin cytoskeletal rearrangement and also activation of endosomal Ras/MAPK signaling for cell morphogenesis, motility, growth and differentiation. One BCH-containing protein BNIP-H/Caytaxin also facilitates shuttling of ATP citrate lyase (ACL) on the kinesin motor from cell body to neurite termini where it produces acetylcholine at the synapses. In this talk, I will discuss the plasticity of BCH domains in the context of its molecular evolution, unique structure, versatile functions and their links to cancer and neurological disorder such as ataxia. The significance of BCH domain as a concerted scaffold device that bridges mechanical cues to cell dynamics, will be explored in the context of the emerging field of mechanobiology.

Journal of Physics: Conference Series **1417** (2019) 011001 doi:10.1088/1742-6596/1417/1/011001

Current Challenges in the Analysis of High Dimensional Brain Signals

Prof. Hernando Ombao, Ph.D
King Abdullah University of Science and Technology, Saudi Arabia

Advances in imaging technology has given unprecedented access for neuroscientists to examine various facets of how the brain "works". Brain activity is complex. A full understanding of brain activity requires careful study of its multi-scale spatial-temporal organization (from neurons to regions of interest; and from transient events to long-term temporal dynamics). It is also multi-faceted and cannot be fully characterized by a single data modality. To fully appreciate brain processes, one must integrate various data that probe into both the anatomical structure and specific functionality such as electrical, metabolic and hemodynamic.

There are many challenges to analyzing brain data. First, brain data is massive - these are recordings across many location and over long recording times. Second, it has a complex structure with non-stationary properties that evolve over space and time. Third, brain data is often dominated by noise. Thus, this environment has provided big opportunities for data scientists to develop new tools and models for addressing the current research in the neuroscience community. This paper will highlight these challenges confronting in this interface between neuroscience and the data science (computational science, statistical learning and modeling).

Journal of Physics: Conference Series **1417** (2019) 011001 doi:10.1088/1742-6596/1417/1/011001

The Effectiveness of the Methods and Approaches of ESD for 2030 Sustainable Development Goals

Prof. Tomonori Ichinose
Director of Research Center for International Understanding in Education/Miyagi University of Education, Japan

ESD is a key element of quality education. Its cross-cutting competencies in cognitive, socio-emotional and behavioural dimensions of learning bear relevance to all areas of education. (UNESCO Executive Board 20 February 2019). This study, the result of four years of research in 279 primary and secondary schools in Japan, reveals that in integrated study periods, and with the engagement of local human resources, an increase of activities across different classrooms and grades are a clear result of enhanced ESD and students' collaboration, cooperativeness, cooperative attitude, and ability to communicate are interconnected. From a pluralistic perspective of ESD, teaching students to confront tensions and conflicts and tackling issues with no single answer seems to be difficult. Since the ESD curriculum has been constructed through whole-school educational activities in Japan by linking it with other subjects and areas, utilizing integrated study periods, it is important to examine approaches to understanding the effectiveness of a whole-school approach of ESD.

Key words
SDGs, Holism, Pluralism, Japanese UNESCO-Associated Schools, Whole-school approach

Uniqueness and Stability of Stationary Periodic Behavior of PWM DC-DC Converters Using Multifrequency Modelling

Dr. Yusuf Fuad, M.App.Sc.
Department of Mathematics, Faculty of Math. and Nat. Sci.
Universitas Negeri Surabaya, Indonesia.

It is well known that the power conversion is commonly achieved using switching elements. The study of switched power converters is an area of active research. As the voltage and current ratings and switching characteristics of power semiconductors devices keep improving, the range of application continues to expand in areas such as communications and data handling systems, portable battery-operated equipment and uninterrupted power source. This study highlights the theory of PWM DC-DC converters for the standard Buck, Boost, Buck-Boost and Ćuk converters. Some aspects concerning PWM DC-DC converters will be discussed based on a multifrequency averaging, which involves Fourier series, as a generalization of the state space averaging techniques. It provides an efficient way to find the stationary state responses which are nonsinusoidal. The stationary periodic behavior plays an important field of study of PWM DC-DC converters. Moreover, it has been shown that a stabilizing controller for the state space averaging of PWM DC-DC standard converters does automatically imply a stabilizing controller for the converters itself. Utilizing the multifrequency modelling, the stationary periodic signals converter can be determined by solving a set of linear equations. Then, the existence and uniqueness of the stationary periodic signals, as well as the stability aspect of the converters, can be analyzed for open and closed loop situations. Simulations are indeed demonstrated, using Boost and Ćuk converters, in open and closed loop situations.

Keywords Multifrequency modelling, pulse-width modulated (PWM), periodic behavior, stability.

MISEIC 2019 Rundown

Time	Activity
07.00-08.00	Registration
08.00-08.45	Opening Ceremony of MISEIC Welcome Dance Singing Indonesia National Anthem & Mars Unesa Chairman Speech Rector Speech & Opening Praying Token of Appreciation
08.45-09.00	Photo Session
PLENARY SESSION 1	
09.00-09.30	**Boon Chuan Low** (*Mechanobiology Institute Singapore, Department of Biological Sciences and University Scholars Programme, National University of Singapore*) with the title "BCH Domain as a Versatile Scaffold Module in Cell Signaling and Mechanobiology"
09.30-10.00	**Hernando Ombao** (*Biostatistics Group, King Abdullah University of Science and Technology, Saudi Arabia*) with the title "Current Challenges in the Analysis of High Dimensional Brain Signals"
10.00-10.30	Discussion
PLENARY SESSION 2	
10.30-11.00	**Tomonori Ichinose** (*Research Center for International Understanding in Education/Miyagi University of Education, Japan*) with the title "The Effectiveness of the Methods and Approaches of ESD for 2030 Sustainable Development Goals"
11.00-11.30	**Yusuf Fuad** (*Applied Mathematics Research Group, Department of Mathematics, Universitas Negeri Surabaya*) with the title "Uniqueness and Stability of Stationary Periodic Behavior of PWM DC-DC Converters Using Multifrequency

Journal of Physics: Conference Series **1417** (2019) 011001 doi:10.1088/1742-6596/1417/1/011001

	Modelling"
11.30-12.00	Discussion
12.00-13.00	Lunch Break
13.00-15.00	Parallel Session I
15.00-15.15	Coffee Break
15.15-16.00	Parallel Session II

Mathematics, Informatics, Science, and Education International Conference

(MISEIC) 2019

Universitas Negeri Surabaya, Surabaya, Indonesia

September 28th, 2019

Chairperson	: A'yunin Sofro, M.Si., Ph. D.
Vice Chairperson	: Dr. Eko Hariyono, M.Pd.
Secretary	: Affiati Oktaviarina, S.Si., M.Sc.
	Dini Kinati Fardah, S.Pd., M.Pd.
Treasurer	: Ika Kurniasari, S.Pd., M.Pd.
	Laily Rosdiana, S.Pd., M.Pd.
Organising Committee	: Utama Alan Deta, MPd, M.Si.
	Setya Chendra Wibawa, M.T.
	Reni Ambarwati, M.Sc.

Advisory Board and Scientific Committee:

Prof. Pitoyo Hartono (中京大学 Chukyo University, Japan)
Prof. Fou-Lai Lin 林福來 (國立臺灣師範大學 National Taiwan Normal University, Taiwan)
Dr.ir. Wim T. van Horssen (TU Delft, The Netherlands)
Husam Al-Najar, PhD (The Islamic University of Gaza, Palestine)
Prof. Hernando Ombao, PhD (King Abdullah University of Science and Technology, Saudi Arabia)
Prof. Tomonori Ichinose 市瀬 智紀 (宮城教育大学 Miyagi University of Education, Japan)
Dr Dieky Adzkiya, MSi (Sepuluh Nopember Institute of Technology, Indonesia)
Prof. Dr Endang Susantini, MPd (Universitas Negeri Surabaya, Indonesia)
Prof. Dr Dwi Juniati, MSi (Universitas Negeri Surabaya, Indonesia)
Nadi Suprapto, PhD (Universitas Negeri Surabaya, Indonesia)
Arie Wardhono, PhD (Universitas Negeri Surabaya, Indonesia)
Dr I. G. P. Asto Buditjahjanto, MT (Universitas Negeri Surabaya, Indonesia)
Dr Abadi, MSc (Universitas Negeri Surabaya, Indonesia)
Dr Elly Matul Imah, MKom (Universitas Negeri Surabaya, Indonesia)
Rooselyna Ekawati, PhD (Universitas Negeri Surabaya, Indonesia)

Journal of Physics: Conference Series **1417** (2019) 011003 doi:10.1088/1742-6596/1417/1/011003

Organizer
Mathematics, Informatics, Science, and Education International Conference (MISEIC) 2019
Universitas Negeri Surabaya, Surabaya, Indonesia
September 28th, 2019

The conference is organized by **Universitas Negeri Surabaya**

Website: http://www.unesa.ac.id/

Journal of Physics: Conference Series **1417** (2019) 011004 doi:10.1088/1742-6596/1417/1/011004

Photograph

Mathematics, Informatics, Science, and Education International Conference (MISEIC) 2019

Universitas Negeri Surabaya, Surabaya, Indonesia

September 28th, 2019

Figure 1. Welcoming adress from the Chaiman of MISEIC 2019

Figure 2. Keynote Speakers and the Dean and the Staff of Mathematics and Natural Sciences Faculty. From the left to right (Dr. Masriyah, M.Pd, Prof. Madlazim, Prof. Boon Chuan Low, Prof. Hernando Ombao).

Journal of Physics: Conference Series **1417** (2019) 011004 doi:10.1088/1742-6596/1417/1/011004

Figure 3. Dr. Yusuf Fuad, M.App.Sc. from Universitas Negeri Surabaya, Indonesia (Keynote Speaker MISEIC 2019)

Figure 4. Prof. Hernando Ombao from KAUST Kingdom of Saudi Arabia (Keynote Speaker MISEIC 2019)

Figure 5. Participants at Parallel Session

Figure 6. Parallel Session

Peer review statement

All papers published in this volume of *Journal of Physics: Conference Series* have been peer reviewed through processes administered by the proceedings Editors. Reviews were conducted by expert referees to the professional and scientific standards expected of a proceedings journal published by IOP Publishing.

Journal of Physics: Conference Series **1417** (2019) 012001 doi:10.1088/1742-6596/1417/1/012001

Identification of Phase $CaCO_3/MgO$ in Bangkalan Dolomite Sand as An Antibacterial Substance

L Rohmawati[1*], S P Sholicha[1], S Holisa SP[1] and W Setyarsih[1]

[1]Department of Physics, Universitas Negeri Surabaya, Surabaya, Indonesia

*e-mail: lydiarohmawati@unesa.ac.id

Abstract. Dolomite sand has several ingredients such as $CaCO_3$ and MgO which have the potential as an antibacterial substance. The making of this antibacterial substance uses the heating/calcination method of dolomite powder with a variation of holding time from 0.5 to 2.5 hours at a temperature of 700°C. The nanoparticles produced from the calcination process were then characterized by XRD to identify the formed phase. This study aimed to determine the effect of holding time on phase formation on dolomite. The results of this study are identified as forming four phases namely calcite ($CaCO_3$), periclase (MgO), calcium hydroxide ($Ca(OH)_2$) and lime (CaO) with different percentages in each variation of holding time. The $Ca(OH)_2$ and CaO phases include the impurity phase with a lower percentage than the $CaCO_3$ and MgO phase percentages. The most dominant phase is $CaCO_3$ and the MgO phase formed optimally at 1 hour holding time with a percentage of 47.1% $CaCO_3$ and 35.9% MgO.

1. Introduction

Dolomite is limestone which is mostly found in Indonesia, especially in the regions of Java, such as Tuban, Lamongan and Bangkalan. Each region has different dolomite content. The dolomite content of Tuban include: 92.40% Ca, 6.5% Mg, 0.45% Fe, 0.055% Mn, 0.54% Yb and 0.073% Cu [1]. Dolomite from the Lamongan area has several elements as 35.7% CaO, 16.7% MgO, 2.19% SiO2, 0.02% Al_2O_3 and 0.01% Fe_2O_3 [2]. The content of Bangkalan Madura dolomite as 63.42% CaO, 26.39% MgO, 5.93% Na2O, 1.20% SiO_2, 0.86% Al_2O_3 and 0.74% Fe_2O_3 [3]. Both of CaO and MgO elements are optimal content of dolomite, especially in Tuban and Bangkalan areas. But the MgO element from Bangkalan is greater compared to the MgO from Tuban. MgO has an important role as an antibacterial agent that is good and safe to use in humans [4] [5]. So, in this study, the dolomite used is from Bangkalan Madura. $CaCO_3/MgO$ composite in dolomite can be formed in a simple calcination process.

Calcination of dolomite at 700°C for one hour produced optimum $CaCO_3/MgO$ phases using analysis on match software of XRD characterization and obtained a percentage of 42.4% MgO (periclase), 55% $CaCO_3$ (calcite) and 2.6% CaO (lime) [6]. In that research, dolomite was calcined at 600°C, 700°C, and 800°C for one hour. Yamamoto reported that heating dolomite at 800°C formed $CaCO_3/MgO$ composites which can be known from the results of XRD characterization. $CaCO_3/MgO$ composites have the good antibacterial potential for the bacteria Staphylococcus aureus and Escherichia coli [7]. Jannah about $CaCO_3$ composites from shellfish and MgO with a composition of 80 wt% $CaCO_3$ and 20 wt% MgO have the best inhibitory properties as an antibacterial material in S.

aureus and E. coli bacteria [8]. Utilization of dolomite as an antibacterial has not been widely studied, it is generally used as a building material, fertilizer, cement mixture and insulating material [9].

Beside of $CaCO_3/MgO$ composites can be used as an abrasive but can also be used as an antibacterial material [7]. Antibacterial in this composite is due to the presence of metal elements and active oxygen groups such as O_2- superoxide anions on the surface of antibacterial agents [7], [10]. MgO acts as an inhibitor of bacterial growth and $CaCO_3$ can remove stains and plaque on teeth [7]. These bacteria are Staphylococcus aureus, *Escherichia coli* and *Streptococcus* mutants [11], [12]. These bacteria are usually in the oral cavity so that it can cause several diseases of the oral cavity and teeth such as dental caries and periodontitis and cause plaque on the teeth that cause inflammation in the oral cavity [13], [14].

Based on several studies above, the role of dolomite as an antibacterial is very important, especially the presence of $CaCO_3/MgO$. Therefore, in this study dolomite extraction was carried out with a simple calcination method at a temperature of 700°C [6] with a holding time variation of 0.5 to 2.5 hours. The calcination results were characterized by XRD and the results were analysed using match software to identify the optimum $CaCO_3/MgO$ phase, which can later be used as a reference for application as an antibacterial material.

2. Materials and Methode

2.1. Materials and tools

The material used in this study is dolomite lime from Jaddih Hill, Bangkalan-Madura. The tools used for dolomite extraction include mortar and pestle, 200 mesh sieve and furnace with type L5 / 11 / B170. The characterization tool used was XRD Bragg-Brentano Philips X'pert Diffractometer with Cu-Kα radiation, range 2θ = 15-90°, data step = 0.02°/minute, detector zero position = 0.023. Match software is used to analyse data resulting from XRD characterization.

2.2. Preparation of $CaCO_3/MgO$

Referring to the research of Sholicha et al. (2019) [6] preparation of dolomite powder containing $CaCO_3/MgO$ was carried out by smoothing dolomite powder which had an average particle size of about 7μm using mortar and pestle. Then dolomite sieved using a 200-mesh sieve, then weighed using a digital balance sheet of 5 grams and calcined at a temperature of 700°C at a heating rate of 10°C / minute with a holding time variation of 0.5 hours, 1 hour, 1.5 hours, 2 hours and 2.5 hours. After reaching 700°C, the powder is cooled to room temperature. Furthermore, the XRD characterization is performed to identify the optimum phase from the holding time variations in the calcination of dolomite.

3. Result and Discussions

XRD characterization results show the phases formed with some data obtained, namely diffraction intensity (I), diffraction peak position data (2θ) and distance between fields (d) or miller indices. The data obtained is used to determine the optimum phase formed in the sample. The sample tested was Bangkalan dolomite powder. Analysis of XRD characterization is done using software Match! that is to match the ICSD database data with the XRD characterization data. Dolomite was calcined at 700°C with a variation of holding time 0.5 to 2.5 hours then XRD was tested which showed four phases formed namely phase CaCO3, MgO, $Ca(OH)_2$ and CaO (Figure 1).

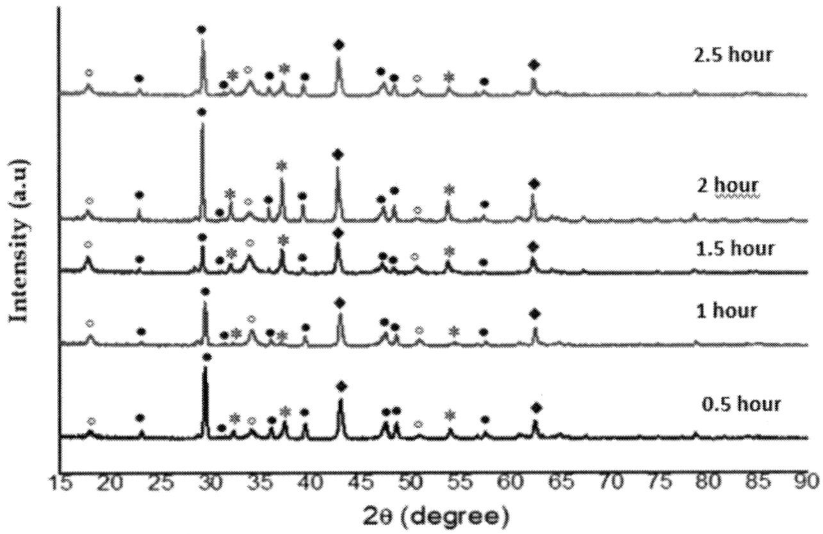

Figure 1. Identification of the Dolomite Powder Phase at temperature 700°C.
(○) $Ca(OH)_2$, (●) $CaCO_3$, (♦) MgO, (*) CaO

Phases formed in dolomite powder have different percentages (Table 1). The phase of $CaCO_3$ (calcite) is formed at the highest diffraction intensity which is at $2\theta = 29.42°$ with the miller index (104), according to [15], which revealed that the phase of $CaCO_3$ was formed at $2\theta = 29°$ with the miller index (104). Other diffraction peaks for the $CaCO_3$ phase are formed at the miller index (012), (006), (110), (113), (202), (024), (018), (116), (211), (122), (010), and others that are in accordance with PDF cards number 96-900-9669. The diffraction peak in the MgO (periclase) phase is formed at $2\theta = 42.96°$ with the miller index (200). Other MgO diffraction peaks are formed at the miller index (111), (202), (311) and (222), according to PDF cards number 96-101-1118. At a holding time of 1.5 hours, the intensity of the $CaCO_3$ phase is lower when compared to other holding times, such as at an angle of $2\theta = 57.44°$ which shows the percentage of $CaCO_3$ phase of 25.8% (Table 1). Besides that, the intensity for the MgO phase is also formed minimum in samples with holding time 0.5 and 2 hours, which is at an angle of 37.38°, 42.94°, 62.29°, 74.66° and 78.60°.

The other two phases other than the phase of $CaCO_3$ and MgO are the phases of CaO and $Ca(OH)_2$ which are impurities. Lime phase (CaO) is formed at an angle of $2\theta = 32.26°$, 37.40°, 54.31°, 64.72° and 67.37° with the miller index (111), (200), (202), (311), (222), (400) and (331) according to PDF cards number 96-101-1096. Phase Ca (OH) 2 is formed at an angle of $2\theta = 18.05°$, 28.70°, 34.11°, 36.97°, 47.46°, 50.83°, 54.31°, 56.61°, 62.35°, 63.80°, 64.72° 71.82°, 77.08°, 79.04°, 81.59°, 84.83° and 86.29° with miller indices (001), (100), (011), (102), (012), (110), (111), (003), (200), (201), (013), (103), (022), (004), (113), (121), (104) and (023) which are suitable for PDF cards numbers 96-100-8782. The two impurities were formed at holding time 0.5 to 2.5 hours with different percentages according to Table 1, but at holding time 1 hour the phase of CaO and $Ca(OH)_2$ was formed with a low intensity at an angle of $2\theta = 32.26°$ and 37,40° for the CaO phase and angles $2\theta = 63.80°$ and 79.04°.

The highest percentage of $CaCO_3$ phase is formed at 0.5 hour holding time, while the highest MgO phase is formed at 1 hour holding time (Table 1). In this case, the highest percentage MgO holding time at 1 hour was chosen compared to the highest $CaCO_3$ composition at 0.5-hour detention because the MgO phase can be used to inhibit antibacterial activity while $CaCO_3$ is only used to remove stains and plaque on teeth [16], [17]. With a low percentage of the phase of CaO and $Ca(OH)_2$ at 1 hour holding time, so the phase of $CaCO_3$ and MgO becomes the dominant phase with a percentage of 47.1% $CaCO_3$ and 35.9% MgO.

The presence of impurities such as CaO and $Ca(OH)_2$ provides its benefits, which have certain functions namely the $Ca(OH)_2$ phase can be used as an antimicrobial and anti-fungal which can kill bacteria because it has a strong alkaline pH (12.3), where bacteria cannot live in an alkaline environment [18], [19], [20]. Calcium oxide (CaO) can be used as an antimicrobial because it has the same superoxide and alkaline pH as $Ca(OH)_2$ which can kill bacteria [21]. The pH value is very influential on antibacterial material in inhibiting the activity of bacteria or microorganisms. The higher the pH value, the antibacterial activity increases as a solvent with alkaline properties mixed into the antibacterial material will increase the pH value and the antibacterial activity [10].

Table 1. Percentage of phase formed in $CaCO_3$/MgO composite using Analysis Match! software

Holding Time (hour)	Percentage of Phase (%)			
	$CaCO_3$	MgO	CaO	$Ca(OH)_2$
0.5	56.3	29.4	8.9	5.3
1	47.1	35.9	0.9	16.0
1.5	25.8	35.8	17.8	20.6
2	49.9	29.5	17.1	3.5
2.5	47.4	34.2	7.0	11.5

The phase of $CaCO_3$ and MgO in dolomite which is the dominant and formed phase at 1 hour holding time is used as a candidate for antibacterial material because it has a high percentage compared to other holding times. According to research by [14], the longer holding time cause the fewer phase to form, but in this research with variations in the holding time, forming phase with different percentages and isn't constant that which the fewer or more formed of phase at dolomite.

4. Conclusion

The optimum phase with the highest percentage of MgO and a low percentage of impurity phases namely is CaO and $Ca(OH)_2$ which are formed at a holding time of 1 hour with a percentage of $CaCO_3$ of 47.1% and MgO of 35.9%. This shows that Indonesia has natural materials such as dolomite which can be used as good antibacterial material, namely at calcination of 700°C with a holding time of 1 hour.

Acknowledgement

Thanks to the Direktorat Riset Pengabdian Masyarakat Direktorat Jendral Penguatan Riset dan Pengembangan Kementerian Riset, Teknologi, dan Pendidikan Tinggi multi-year at UNESA Rector's decision number 507/UN38/HK/LT/2019. This study has been supported by the Department of Physics Universitas Negeri Surabaya. The study process was being done at the Materials Physics Laboratory of Universitas Negeri Surabaya, Laboratory of Materials and Metallurgy of Institut Teknologi Sepuluh Nopember in Surabaya.

References

[1] Arifin Z, Apriliani NF, Zainuri M and Darminto 2017 Characterization of Precipitated $CaCO_3$ Synthesized from Dolomite *IOP Conf. Series: Materials Science and Engineering* 196

[2] Febriana E 2011 *Kalsinasi Dolomit Lamongan untuk Pembuatan Kalsium-Magnesium Oksida sebagai Bahan Baku Kalsium dan Magnesium Karbonat Presipitat* (Depok: Universitas Indonesia)

[3] Solihin S, Arini T and Febriana E 2012 Synthesis of Ultra Fine Grain Magnesium Carbonate Part 1. Calcination Behaviour of Indonesian Dolomite *J. Metalurgi* **27** (3) 173-178

[4] Stoimenov P K, Rosalyn L K, George L M and Kenneth J K 2002 Metal Oxide Nanoparticles as Bactericidal Agents *Langmuir* **18** (17) 6679-6686
[5] Sundrarajan M, Suresh J and Gandhi R R 2012 A Comparative Study on Antibacterial Properties of MgO Nanoparticle Prepared Under Different Calcination Temperature *Digest J. Nanomaterials and Biostructures* **7** (3) 983-989
[6] Sholicha S P, Sabrina G J, Rohmawati L and Setyarsih W 2019 Preparation of $CaCO_3$/MgO From Bangkalan's Dolomite for Raw Biomaterial *IOP Conf. Series: Materials Science and Engineering* **1171**
[7] Yamamoto O, Ohira T, Alvarez K and Fukuda M 2010 Antibacterial Characteristics of $CaCO_3$-MgO Composites *Mater. Sci. Eng. B: Solid-State Materials for Advanced Technology* **173** 208-212
[8] Jannah Z, Mubarok H, Syamsiyah F, Putri A A H and Rohmawati L 2018 Preparation of Calcium Carbonate (from Shellfish)/Magnesium Oxide Composites as an Antibacterial Agent. *IOP Conf. Series: Materials Science and Engineering* **367**
[9] Royani A, Sulistiyono E and Sufiandi D 2016 Pengaruh Suhu Kalsinasi pada Proses Dekomposisi Dolomit *Jurnal Sains Materi Indonesia* **18** (1) 41-46
[10] Ohira T and Yamamoto O 2012 Correlation Between Antibacterial Activity and Crystallite Size On Ceramics *Chemical Engineering Science* **68** (1) 355-361
[11] Aas A, Jorn A, Paster B J, Stokes L N, Olsen I and Dewhirst F E 2005 Defining the Normal Bacterial Flora of the Oral Cavity *J. Clinical Microbiology* **43** 5721-5732
[12] Sharma A and Somani R 2009 Dermatoglypic Interpretation of Dental Caries and its Correlation to Salivary Bacteria Interactions: An in Vivo Study *J. Indian Soc. Pedod. Prevent Dent.* **27** (1) 17-21
[13] Sabir A 2005 Aktivitas Antibakteri Flavonoid Propolis Trigona sp terhadap Bakteri Streptococcus mutans (in vitro) *Dent. J.* **38** (3) 135-141
[14] Islam M R, Ogura Y, Asadulghani M, Ooka T, Murase K, Gotoh Y and Hayashi T 2012 A Sensitive and Simple Plaque Formation Method for the Stx2 Phage of Eschericia Coli 0157:H7, which does not Form Plaques in The Standard Plating Procedure *Plasmid* **67** (3) 227-235
[15] Rahman M A, Halfar J and Shinjo R 2013 X-Ray Diffraction is A Promising Tool to Characterize Coral Skeletons *J. Sci: Advances in Material Physics and Chemistry* **3** 120-125
[16] Yamamoto O, Ohira T, Mohan DJ, Fukuda M, Ozkal B, Sawai J and Nakagawa Z 2008 Antibacterial Characteristics of Carbon Coated $CaCO_3$/MgO Powder Led by The Pyrolysis of Poly (Vinyl Alcohol) - Dolomite Mixture *Carbon* **46** 77-81
[17] Sawai J, Kojima H and Igarashi H 2000 Antibacterial Characteristics of Magnesium Oxide Powder *World J. Microbilogy and Biothecnology* **16** (2) 187-194
[18] Farhad A and Mohammadi Z 2005 Calcium Hidroxide: A Review *Int. Dent. J.* **55** (5) 293-301
[19] Chai W L, Hamimah H, Cheng A C and Sallam A A 2007 Susceptibility of Enterococcus Faecalis Biofilm to Antibiotics and Calcium Hydroxide *J. Oral Sci.* **49** 161-166
[20] Hapsari C A, Putra L A, Rahma R R N and Irawan R S 2012 Pengaruh Kelembapan, Temperatur, dan pH pada Proses Bioremediasi Menggunakan Bakteri Bacillus sp. Bulking Agent Serabut Buah Bintaro *J. Polusi Tanah dan Air Tanah* **1** (1) 1-9
[21] Roy A, Gauri S S, Bhattacharya M and Bhattacharya J 2013 Antimicrobial Activity of CaO Nanoparticles *J. Biomedical Nanotechnology* **9** 1570-1578
[22] Efhana D P and Zainuri M 2014 Pengaruh Variasi Waktu Penahanan Proses Kalsinasi Terhadap Prekursor Bahan Katoda Lithium Ferrophospate (LFP) *J. Sains dan Seni Pomits* **3** 2337-3520

Solvent Effect on Viscoelastic Behaviour and Morphology of Polyaniline Coating at QCM Sensor

N P Putri[1], D W Pravitasari[2], F Al Aziz[2], Maulidiah[2], D J D H Santjojo[2], Masruroh[2*], and S P Sakti[2]

[1]Physics Department, Universitas Negeri Surabaya, Surabaya, Indonesia
[2]Physics Department, Brawijaya University, Malang, Indonesia

*e-mail: ruroh@ub.ac.id

Abstract. The effect of different solvents and solution concentration on viscoelastic behaviour and morphology of polyaniline (PANi) coating was investigated in this study. The PANi coating on a QCM sensor was created by spin coating with various solvents, such as NMP, m-cresol, and DMF with varying concentrations of 1-4% by weight of PANi. The viscoelastic behaviour was observed by impedance analyser and morphology of the PANi coating was observed via optical microscope and SEM/EDX. The viscoelasticity was measured to determine whether the coating produces an acoustic load (loading effect). The observations showed that the different solvents give a different effect on the impedance value of the PANi coating. With m-cresol solvent, the impedance value in the series resonant frequency was so large, indicating the PANi coating was not rigid and produce a loading effect on QCM sensor. From morphology observation showed that homogeneity of the polyaniline coating is still poor. There is an agglomeration of polyaniline powder which is insoluble in the solvent used. From the viscoelastic properties of the coating, the optimum value was obtained in PANi coating with NMP solvent.

1. Introduction

Quartz crystal microbalance (QCM) is one of the piezoelectric sensors that allow selective detection of biological events in real-time. This sensor has a working principle based on changes in the oscillation frequency which is proportional to the changes in the deposited mass. QCM is the most widely used acoustic resonator device because it allows the study of viscoelastic properties of materials, molecular adsorbs and living cell mobility. The Saurbrey equation assumes that the mass or film deposited on the QCM surface will follow the QCM vibration and therefore the coated QCM will only behave as if it were thicker. The effective wavelength of the crystal will increase and cause the resonance frequency to decrease. This model assumes that there is no energy dissipation and thus only applies to thin, rigid and homogeneous films and has the same acoustic properties as QCM materials. QCM can function as a sensor if the deposited mass does not have a viscoelastic effect so that QCM does not have a damping experience. The sensitivity and selectivity of QCM can be adjusted to the selection of the right chemo sensitive film.

Conductive polymers for example, polypyrrole, politiophene, and polyaniline, have been used as sensors for hydrocarbon gas sensors, CO, NOx, and organic vapors [1]. With low operating temperatures, conductive polymers are better candidates for sensor materials than metal oxide-based sensors. Among the types of conductive polymers, polyaniline (PANi) is one of the materials that are in great demand because it has several superior features, including being easily polymerized, has good redox reversibility, and has good environmental resistance. Polyaniline has been proven to be an active gas sensor that can detect a variety of gases for example, alcohol [2], chloroform [3], free radicals [4], hydrocarbons [5] and scents [6].

To form a polyaniline thin film on the QCM surface, several deposition methods can be carried out, one of which is the spin coating method. Spin coating is a procedure used to apply a uniform thin film to flat substrates. This method is very easy and has been proven to form a homogeneous thin film. Several factors influence the quality of the thin film formed using the spin coating method, including the spin coater rotational speed, solvent selection and the solution's concentration [7]. Spin coater rotational speed will affect the deposited film thickness which will also affect the thin film sensitivity. However, a thick film will give the loading effect on QCM, so that the optimal rotating speed is needed where the thin film formed will not give the loading effect. Several solvents can be used to dissolve polyaniline. The physical and chemical properties of each solvent can be used as a reference for choosing a solvent that is adjusted to the use of thin film. The concentration of the deposited solution on the surface of QCM will influence the morphology of thin film formed. The thin film formed is expected to have porous morphology so that it is suitable when functioning as a sensor.

In this research, a PANi deposition was carried out on the QCM substrate using the spin coating method. Spin coating deposition parameter variations were performed, namely variations in the type of solvent, solution concentration, and rotational speed. The effect of deposition factors of spin coating method on viscoelastic properties and morphology of the PANi thin film deposited on QCM surface was explored. The viscoelastic properties of thin film were analysed from changes in the impedance values due to variations in rotational speed, type of solvent, and solution's concentration, while the morphology and microstructure of the film were analysed based on the results of Optical Microscope and Scanning Electron Microscope (SEM).

2. Experimental Procedures

2.1 *Synthesis of polyaniline solution*

To make a polyaniline solution, polyaniline (PANi) powder must be synthesized first. The synthesis of PANi uses the oxidation method as has been done in previous studies [8]. The oxidized polyaniline powder was then dried and crushed with mortar until a fine powder was obtained with granules that could escape in the 325-mesh filter. The powder passed through the filter was then dissolved with a variety of solvents, namely N-Methylpyrolidine (NMP), m-cresol, and N.N-dimethylformamide (DMF). The concentration of the solution is also varied, which is 1 - 4% by weight of PANi. Making the solution using a magnetic stirrer with a speed of 1500 rpm for 2 hours, then continued with ultra-sonication for 2 hours to obtain a homogeneous solution. After obtaining a homogeneous solution, the solution is then dropped over the surface of the QCM and rotated with a variety of rotational speed 1000- 4000 rpm for 60 seconds. The coated QCM then dried in an oven at temperature 100°C for 10 minutes.

2.2 *Characterization*

Before and after coated with PANi, the impedance value of QCM was measured using an impedance analyser to see changes in the QCM viscoelastic properties due to PANi deposition. The morphology of the PANi coating was analysed using an optical microscope and Scanning Electron Microscope (SEM). Energy Dispersive X-Ray Spectroscopy (EDX) is used to identify elements in the polyaniline film deposited above QCM.

3. Result

3.1 *Variation of solvent*

The deposition process of PANi on the QCM surface by spin coating method requires the presence of a homogeneous solution, so the role of the solvent is very important in this method. Because PANi is one of the polymers that have low solubility in organic solvents, the selection of the right solvent must be done first by considering their solubility parameters. The solubility parameter can be used to predicting polymer solubility. The basis of these so-called Hansen solubility parameters (HSP) is that the total energy of the vaporization of a liquid consists of several individual parts. These arise from dispersion forces (d), permanent dipole forces (p) and molecular hydrogen bonding (h) [9]. The overall solubility parameter can be calculated by:

$$\delta = \sqrt{\delta_d^2 + \delta_p^2 + \delta_h^2} \qquad (1)$$

From the above equation, it was found that the solubility parameter value of m-cresol was 11.11 $MPA^{1/2}$, NMP was 21.48 $MPA^{1/2}$, and DMF was 12.14 $MPA^{1/2}$ [10]. Thus, PANi will dissolve faster in NMP solvents because the solubility coefficient is the greatest among other solvents. The solubility parameter (δ) is directly proportional to the solvent evaporation energy (ΔE_{vap}) and inversely proportional to the mean molar volume (V_m). Evaporation energy will correlate with solvents vapor pressure which will affect the morphology of the deposited polymer on the QCM surface.

Quartz Crystal Microbalance response measurement can be done using a frequency measurement, dissipation signal and also impedance measurement. Among many methods, the electrical impedance is one of the preferred analyses. The impedance measurement can lead us to investigate the material properties, such as complex shear modulus, viscosity and loading effect caused by coatings on QCM's surface [11]. In this study, impedance measurements were carried out on QCM before and after PANi coated to analyse loading effect obtained by QCM due to the solvents used. It was found that the deposition of PANi with various solvents gave a different loading effect on QCM (Figure 1).

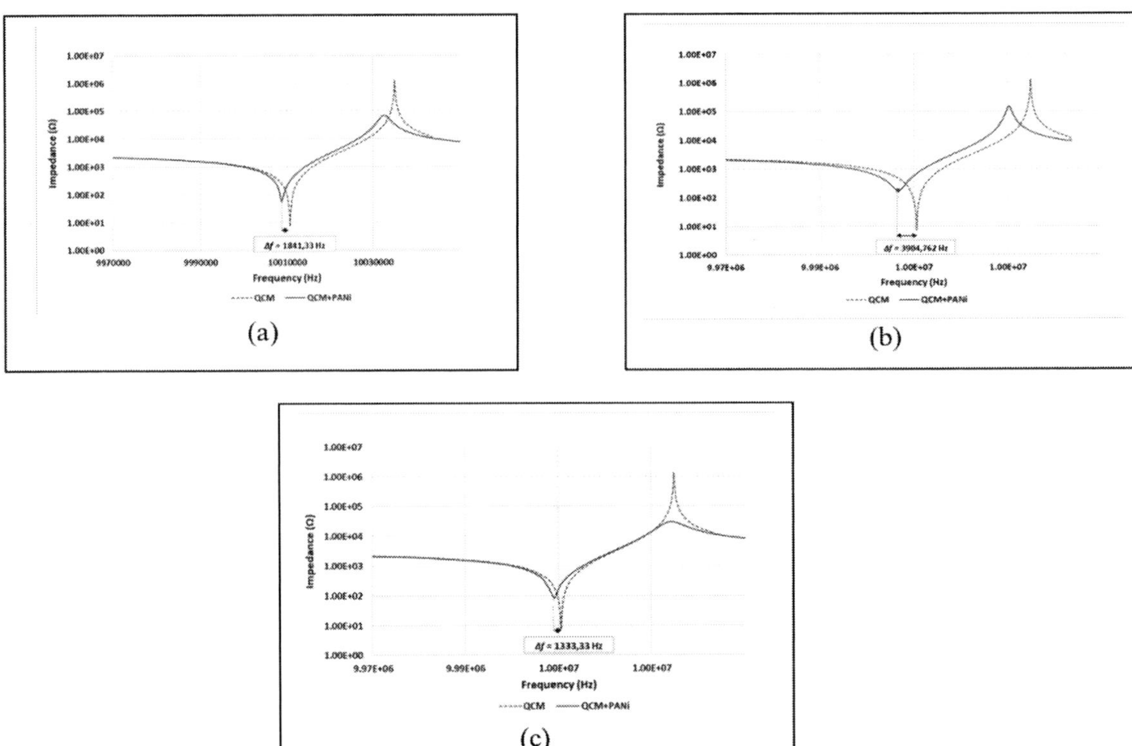

Figure 1. Frequency changes in QCM due to PANi coating with various solvents (a) NMP, (b) m-cresol, and (c) DMF

From the impedance-frequency graph in Figure 1, a minimum impedance value identifies the series resonant frequency, and the maximum impedance value identifies a parallel resonant frequency [12]. The measurement result showed that the series resonant increased, indicating a change in impedance value. From the shift, we can see that PANi coating shows different viscoelasticity, which causes significant differences in the impedance value as a result of various solvent. Solution or analyte sorption not only increases the mass of the coating (mass effect) but also changes the shear parameters of PANi (viscoelastic effect) [13]. The biggest frequency shift value is owned by m-cresol solvent, whereas for NMP and DMF solvents the frequency shift that occurs is not too significant (Table 1). M-cresol is a PANi solvent which can increase PANi conductivity by secondary doping. M-cresol is also hygroscopic, so it can react with water vapor in the environment. The interaction between PANi and m-cresol and water vapor causes an increase in mass deposited on QCM's surface. Different solvents can provide different viscoelastic effect, this can be seen from the change in measured impedance value (increase in the minimum curve peak). From the impedance change data in Table 1, it can be seen that m-cresol also gives a large viscoelastic effect on PANi. Changes in the viscoelastic properties of PANi due to m-cresol solvents cause QCM to get a loading effect so that it cannot function as a good sensor.

Table 1. Changes in frequency and impedance values due to variations in PANi solvent

Solvent	Frequency changes (Hz)	Impedance changes (Ω)
NMP	1841,333	49,119
m-cresol	3904,762	169,018
DMF	1333,333	76,716

3.2 Variation of the solution's concentration

Impedance measurement data with variations in solvent and concentration can be seen in the graph in Figure 2. A good impedance value is a small one, which is related to the viscoelastic nature of the thin film formed above QCM. If the PANi film impedance value is too large (> 100Ω), then it can be said that the PANi film formed is rubbery, otherwise, if the PANi film's impedance value <100Ω, then the film is still rigid. It can be seen that the optimum impedance value is obtained at the PANi film derived from NMP and DMF solutions at 2% concentration. In all samples, films with concentrations other than 2% have a greater impedance value. In a solution with a concentration of 1%, there is a small amount of PANi powder, so that when the spin coating process is carried out, there will be an interaction between the solvent (which has a greater concentration) with QCM which results in a loading effect. QCM drying process after spin coating cannot completely remove the solvent (trapping process occurs). On the other hand, in solutions with concentrations greater than 2%, more dissolved PANi powder, so that it will add deposited mass to the QCM surface which results in an added loading effect on QCM.

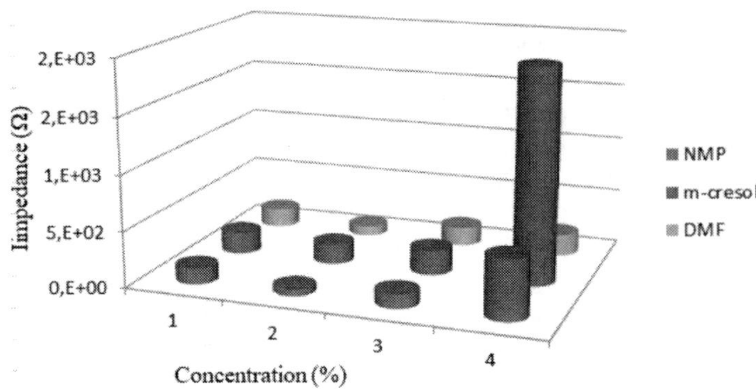

Figure 2. The impedance value of the PANi coating with various solvent and solution concentration

The surface morphology of PANi coating is shown in Figure 3. All the images correspond to coating with PANi in the NMP solvent. There is a difference in dispersion between the PANi films at different solution concentrations. Coating derived from 2% PANi solution in NMP solvents (Figure 3.a) have better morphology, showing a more even dispersion compared to the second film (Figure 3.b). This morphology is directly related to solution concentration. As explained above, that with different concentrations of solution, the amount of PANi powder dispersed in the solution is also different. So that the solution with a concentration of 4% has more PANi powder content when compared to a solution with a concentration of 2%. With the increasing number of PANi powder in the solution will reduce its solubility, so that during the spin coating process, many PANi powders will clot and form agglomeration on the QCM surface. This will obviously affect the PANi layer impedance value. From the impedance measurement data (Figure 2), it can be seen that the value of the impedance of the coating coming from the solution with a concentration of 4% has a very large impedance. So, it can be concluded that the agglomeration of PANi powder gives a greater loading effect on QCM. PANi prepared by oxidation with HCl doping has properties that are not easily soluble to organic solvents, so the right solution concentration is needed so that PANi particles can be dispersed properly.

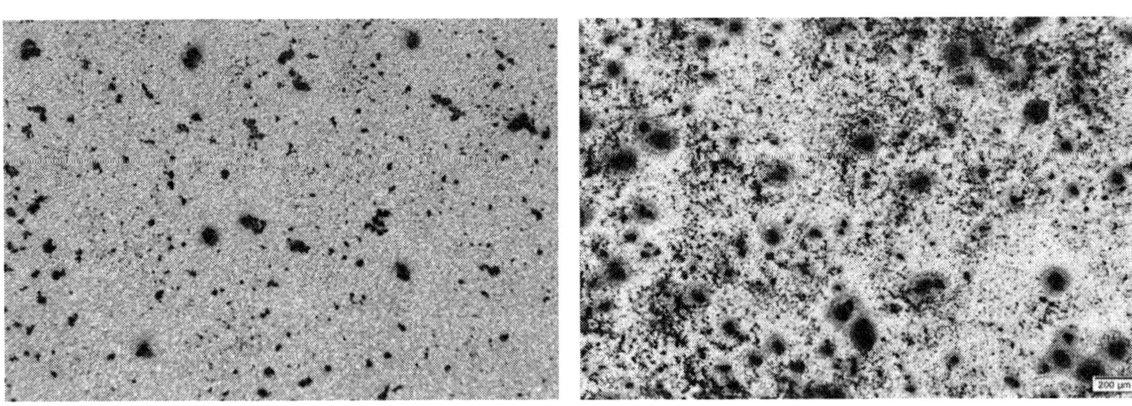

(a) (b)

Figure 3. Observation of PANi layer morphology using an optical microscope (100x magnification) with variations in the concentration (a) 2% and (b) 4%

3.3 Variation of rotational speed

The effect of spin coating rotational speed on the viscoelastic properties of the PANi coating can be seen in Figure 4. It can be seen that the spin coater's rotational speed greatly affects the thickness of the film which has an impact on the magnitude of the PANi coating impedance value. The optimum value is obtained in the sample rotated at 4000 rpm. When the rotating speed is low, thicker films will be produced compared to high rotating speeds. With a high rotating speed, the films formed will be thinner and not burden QCM. PANi solution with 2% NMP solvent dripped on the QCM surface will be distributed evenly because of the centrifugal force so that the film will experience thinning during the spinning process.

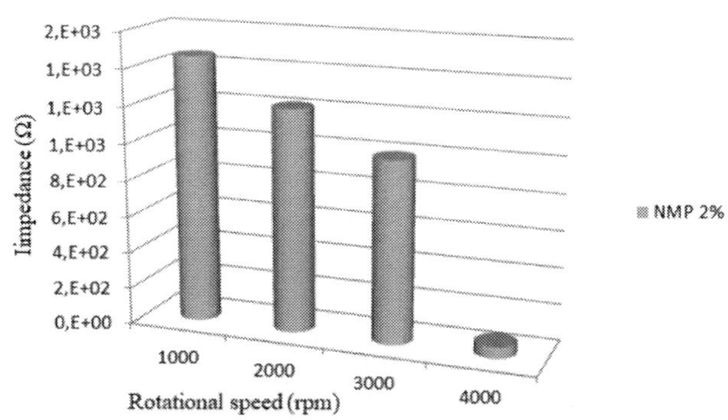

Figure 4. The impedance value of the PANi coating with variations in rotational speed

Rotational speed variations also affect the morphology of the PANi coating formed (Figure 5). At 1000 rpm rotation speed, there are islands of PANi particles in the formed coating. This can occur because when the rotating speed used is low, the process of spreading the solution is slow so that the formed film is not uniform and agglomeration occurs. On the contrary, with a high rotational speed, the centrifugal force on the spin coating turns greater and forcing the film to spread faster, this causes fewer agglomerations.

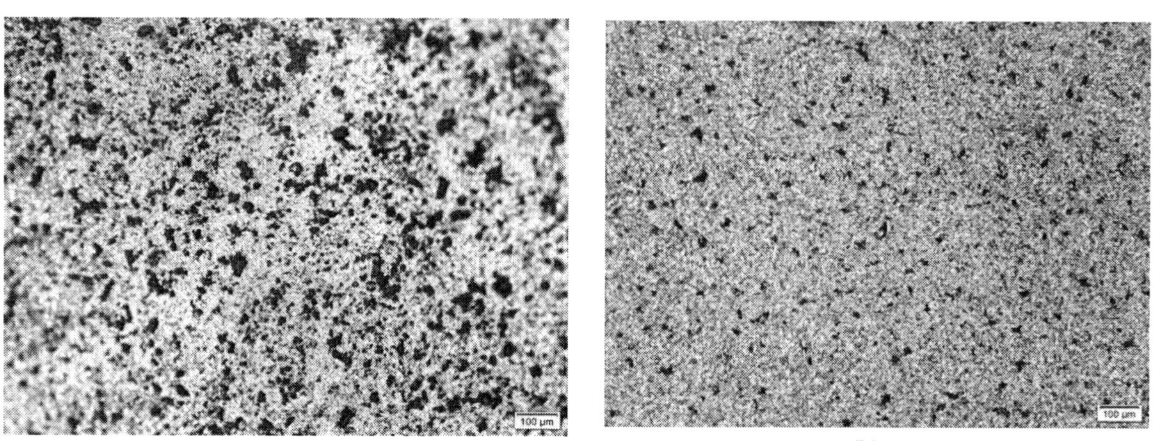

(a) (b)

Figure 5. Morphology of PANi coating in 2% NMP with the rotational speed of (a) 1000 rpm and (b) 4000 rpm

3.4 *Microstructure*

The morphology of the PANi coating with DMF 2% solvent can be seen in Figure 6. The effect of the solvent on the film morphology can be explained by the value of the vapor pressure of each solvent, as mentioned in the initial part of the discussion. Vapor pressure value at 20°C for NMP solvent is 31.6 Pa, while m-cresol is 18.6 Pa and DMF has a vapor pressure of 380 Pa. From the value of the vapour pressure of each solvent it can be seen that DMF is very volatile, much faster than the other two solvents. Solvents that rapidly evaporate can result in the dissolution of the dissolved PANi, as can be seen in Figure 6. The morphology of the PANi coating above the QCM surface is still not homogeneous, there are still many PANi powder which is agglomerated and do not spread evenly. This can result in an uneven distribution of stress, which increases the viscoelasticity of QCM.

PANi powder which is deposited on the QCM surface has a spherical shape and nanometre dimension, it can be seen that the powder diameter is in the range of 70-80 nm. PANi powder dissolving process in each solvent has several stages, namely, starting with grinding to pass the filter until stirred with a magnetic stirer for 2 hours and ultrasonied for 2 hours as well. This process can reduce the size of the PANi powder to nano dimensions.

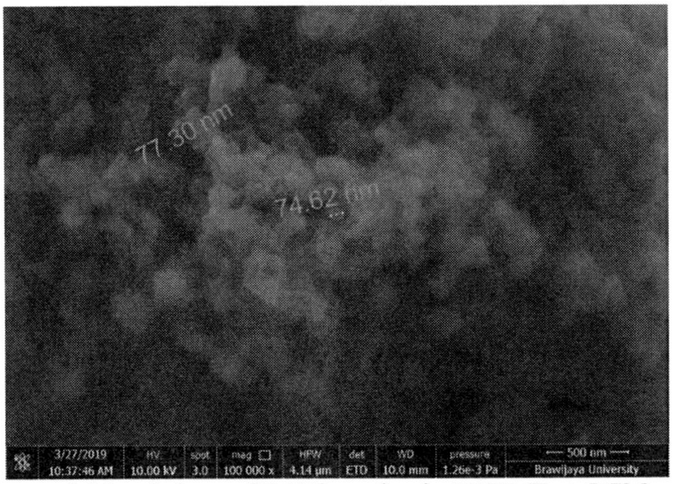

Figure 6. Morphology of PANi coating in 2%DMF at QCM surface

From the EDS analysis (Figure 7) it can be seen that the most deposited element on the QCM surface is carbon (C) of 77.50 %wt and oxygen (O) of 13.58%wt. This proves the existence of PANi which has a molecular formula $(C_6H_4NH)n$ on the QCM surface. The presence of Cl element (1.51 %wt) is due to the polymerization process of PANi which uses HCl, while the Ag element is the electrode on QCM. Ag elements still detected in EDS spectra show that the silver electrodes in QCM have not all been coated by PANi.

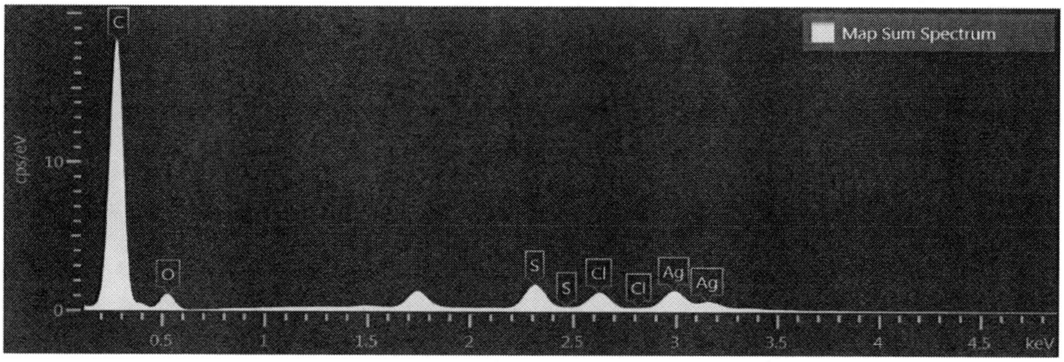

Figure 7. EDS spectrum of PANi coating at QCM surface

4. Conclusion

From this study it can be concluded that to form a thin film of PANi on the surface of QCM using the spin coating method, the recommended solvent is NMP with a solution concentration of 2% and rotational speed of spin coating of 4000 rpm. It has been investigated that with a 2% PANi-NMP solution the film formed does not give a loading effect on QCM so that it can still be used as a sensor.

References

[1] Ram M K and Bhethanabotla V R 2010 *Sensors for Chemical and Biological Applications*

[2] Ayad M M, El-Hefnawey G, and Torad N L 2009 A Sensor of Alcohol Vapours Based on Thin Polyaniline Base Film and Quartz Crystal Microbalance, *J. Hazard. Mater.* **168** (1) 85–88

[3] Shinen M H, Essa F O, and Naji A S 2014 Study the Sensitivity of Quartz Crystal Microbalance (QCM) Sensor Coated with Different Thickness of Polyaniline for Determination Vapors of Ether, Chloroform , Carbon tetrachloride and Ethyl acetate, *Chem. Mater. Res.* **6** (3) 7-12

[4] Banerjee S, Konwar D, and Kumar A 2012 Chemical Polyaniline Nanofiber Reinforced Nanocomposite Coated Quartz Crystal Microbalance Based Highly Sensitive Free Radical Sensors *Sensors Actuators B. Chem.* **171–172** 924–931

[5] Ayad M M, El-Hefnawey G, and Torad N L 2008 Quartz Crystal Microbalance Sensor Coated with Polyaniline Emeraldine Base for Determination of Chlorinated Aliphatic Hydrocarbons *Sensors Actuators, B Chem.* **134** (2) 887–894

[6] Tiggemann L, *et al.* 2016 Low-cost Gas Sensors with Polyaniline Film for Aroma Detection *J. Food Eng.* **180** 16–21

[7] Tyona M D 2013 A Theoritical Study on Spin Coating Technique *Advances Mater. Res.* **2** (4) 195-208

[8] Putri N P, Kusumawati D H, Widiyanti N and Munasir 2018 Synthesis of Polyaniline/Cellulose Composite as Humidity Sensor *J. Phys.: Conf. Series* **997**

[9] Nasir R, Mukhtar H and Man Z, Miscibility Studies of Polyethersulfone (PES), N-Methyl-2-Pyrrolidone (NMP) and Alkanolamines on the Basis of Hansen Solubility Parameters, *Int. J. Scientific Eng. Technol.* **3** (5) 450-453

[10] Hansen C M 1967 The Three Dimensional Solubility Parameter and Solvent Diffusion Coefficient (Copenhagen: Danish Technical Press)

[11] Sakti S P, Akbar E R N, Kamasi D D and Naba A 2019 Impedance Measurement of the Quartz Crystal Microbalance using Phase Gain Detector and Digital Storage Oscilloscope *Mater. Sci. Eng.* **546**

[12] Masruroh, Djoko D J D H, Rahayu S and Sakti S P 2016 Viscoelastic and Morphological Behaviour of Stearic Acid Film on Top of Polystyrene as Immobilisation Matrix for QCM Sensor *Mater. Sci. Forum* **848** 757-762

[13] Behling C, Lucklum R and Hautmann P 1998 Response of Quartz-crystal Resonators to Gas and Liquid Analyte Exposure *Sens. and Actuators A* **68** 388-398

Discriminating a Nuclear Blast from A Natural Seismic Event Using A Full Moment-Tensor Inversion Method

T Prastowo[1,2*], Madlazim[1,2] and Supardiyono[1,2]

[1]Physics Department, Universitas Negeri Surabaya, Surabaya, Indonesia
[2]Center for Earth Science Studies, Universitas Negeri Surabaya, Surabaya, Indonesia

*e-mail: tjiptoprastowo@unesa.ac.id

Abstract. Identification of seismic signals if they are generated by tectonic or volcanic origin or underground nuclear tests is of primary importance for impacts on environmental issues. The recent blast on September 3, 2017 has been in question due to its potential destruction on the global environment. Here, we applied a method of full moment-tensor inversions to analyse broadband waveforms generated by such an explosion. The signals were recorded by a network of 4 regional seismic stations (MAJO, INCN, MDJ, BJT), situated between 360 and 1,100 km away from the explosion epicenter. The results showed that a variance reduction of 72.6%, providing high confidence for all values obtained from the method used: a seismic moment of $M_o = 6.25 \cdot 10^{24}$ dyne cm, corresponding to a moment magnitude of $M_w = 5.8$, with a dominant volumetric component (ISO) of 50% relatively compared to a 6.8% double-couple (DC) part and a 43.2% compensated linear vector dipole (CLVD) part, and a centroid-depth of 1.0 km. The relative dominance of the ISO seismic component over the other two components, along with the shallow source extracted from the cross-correlation and source mechanism analysis, suggests that the September 3, 2017 explosion was induced by a suspected nuclear test.

1. Introduction

Routine monitoring and accurate identification of seismic signals generated by natural earthquakes or underground nuclear detonations have long been performed for reasons associated with safety issues. Detection of nuclear explosions using a combined model of source mechanisms was proposed by [1] in response to inconsistencies of each particular model with field data obtained from the explosions. Discriminants of artificial explosions from natural events are important and can possibly be obtained from analysis of broadband seismic waveforms recorded by regional stations [2, 3] or observed as teleseismic sources [4]. Relevant work on direct comparisons between human-induced explosions, including a nuclear test, and earthquakes of either tectonic or volcanic origin was explored by [5] using both numerical modelling and surface wave data analysis.

As detonations produce seismic signals propagating beneath the surface to a large area of coverage, experts could then examine propagating waves back in time to the exact source location and determine the corresponding source mechanism. In regard to the analysis of surface waves, a common method, the so-called P/S ratio, describing comparison between P- and S-wave amplitudes hence their energies, was explored by [6] to separate artificial signals from those of earthquakes. Using this direct method, the P/S ratio from nuclear explosions is found to be larger than that from naturally occurring events, suggesting the clear dominance of P-wave energy over energies from other seismic wave components. In contrast, seismic energy induced by earthquakes is distributed over both body and surface waves, and in some cases most of the energy is taken away from the source by S-wave. However, [7] showed the *P*/*S* discriminant dependence on frequency, making it applicable with caution as the generation of *S*-disturbances in underground blasts, such as a nuclear test, may be possible, leading to a failure in using the discriminant [8].

As a nuclear weapon brings potential destruction on its surroundings, a watch for the development of nuclear weapons testing around the globe is in demand. The North Korean government is suspected to perform nuclear tests in a series of years since 2006 to date with increasing magnitudes measured. These were the single-blast explosion detected on October 9, 2006 with $M = 4.3$ [9], the second test observed on May 25, 2009 with $M = 4.7$ [10], the third test on February 12, 2013 with $M = 5.1$ [11], the fourth on January 6, 2016 with $M = 5.1$ [12]. All of these previous studies clearly distinguished underground explosions from natural earthquakes recorded by surrounding regional observatories (complete information can be found at https://ds.iris.edu/ds/nodes/dms/specialevents/).

In this study, we extend the methodology of discriminating explosions from earthquakes based on real-time monitoring of a nuclear explosion, including travel time inversion for depth estimates [13] to a full moment-tensor inversion method using broadband waveforms acquired from regional stations. We examine and analysis the North Korean's nuclear blast on September 3, 2017 at 03:30:08 UTC with a magnitude of $M = 6.3$ by GEOFON-GFZ at http://geofon.gfz-potsdam.de/eqinfo/special/korea. We then use the recent blast as a case study and compare it to earlier tests by the same government. The main aim of the present case is thus to determine whether the latest detonation is a nuclear bomb. In addition, the method developed for this case is proposed with an expected accuracy in prediction of explosion-earthquake discriminant better than that of relevant work, as discussed in [13].

2. Method

The case considered in this study is the latest explosion in North Korea with a measured magnitude of $M = 6.3$ on September 3, 2017 at 03:30:08 UTC. The epicenter of the explosion is located at 41.33° N and 129.03° E. Datasets in the form of broadband seismic digital data were obtained from a network of regional stations managed by The International Federation of Digital Seismograph Networks (FDSN), routinely supervising IU: Global Seismograph Network (GSN-IRIS/USGS), which is freely accessible at http://www.fdsn.org/networks/detail/IU/ and another IC: New China Digital Seismograph Network (NCDSN) available at http://www.fdsn.org/networks/detail/IC/. These provided 4 regional stations used in the current study, namely IU.MAJO, IU.INCN, IC.MDJ, and IC.BJT, where the corresponding locations relative to the source location are listed in table 1. For comparison, all the datasets used were also collected from https://ds.iris.edu/wilber3/find_event.

Table 1. Regional stations used in this study with the corresponding epicentral distances.

FDSN code	Station	Latitude	Longitude	Epicentral Distance	
				in degree	in km
IU	MAJO	36.55° N	138.20° E	8.67°	954
	INCN	37.48° N	126.62° E	4.33°	476
IC	MDJ	44.62° N	129.59° E	3.35°	368
	BJT	40.02° N	116.17° E	9.94°	1093

Figure 1 provides a geographical map of the September 3, 2017 explosion, including all stations used for characterizing the explosion. The yellow star indicates the epicenter whereas the red triangles represent the 4 stations listed in table 1; each station is positioned within a radius of 10° (one degree is about 110 km) away from the epicenter. Green-triangle stations are not used for nuclear test detection as some of them are located outside the radius of 10° permitted or positioned at the perimeter but giving unclear waveforms.

The method of full moment-tensor seismic inversions used in the present work was detailed in [14]. We here provide the saline points of the method using broadband seismograms of seismic signals filtered at low frequency ranges. A full moment-tensor is usually decomposed into two components, that is, an isotropic (ISO) component and a deviatoric tensor. The latter is then further transformed into a double-couple (DC) part, commonly believed to be responsible for most of tectonic earthquakes, and a compensated-linear vector dipole (CLVD) part, normally associated with volcanic activities. For determination of source mechanisms exploded on September 3, 2017, we have used a MTINV 3.05 software (with permission from Dr. Ichinose). Analysis of broadband data was performed by inverting 3 local seismic waveforms (abbreviated as **Z** = vertical, **R** = radial, and **T** = tangential components) directly obtained from a network of the 4 regional stations in table 1 to extract the source parameters and their corresponding source mechanisms.

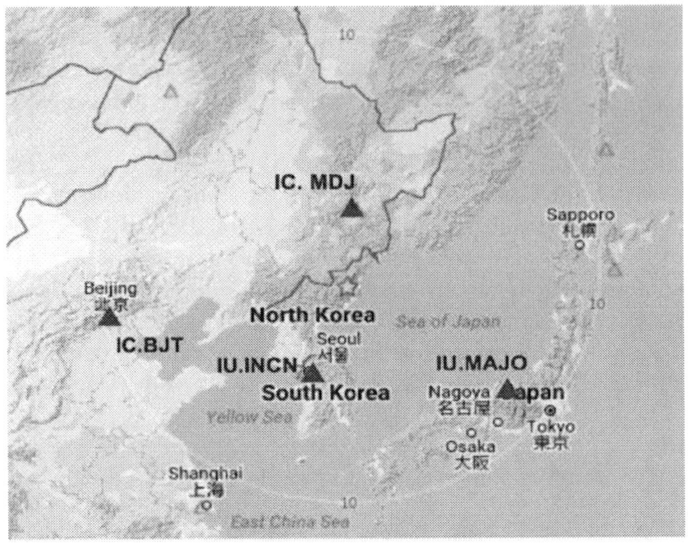

Figure 1. The source location and regional stations from https://ds.iris.edu/wilber3/find_event

3. Results and Discussions

Following the methodology in [14], we have performed the full moment-tensor inversion method to analyse broadband seismic waveform signals recorded from the recent blast on September 3, 2017 initiated by the North Korean government, as investigated by [13] but with different techniques and procedures, in order to determine source parameters and mechanisms of the detonation. The results are given in two correlated Figures (Figure 2 and Figure 3), from which relevant discussions are provided. Note that these Figures contain all about seismic information required for characterizing the explosion. The following paragraphs will provide such information concerning with the source parameters and their corresponding source mechanisms.

Figure 2 describes waveform fittings from field observations and numerical simulations completed with all 3 local components of seismic waveforms for recorded micro-displacements (symbolized here as **Z** = vertical, **R** = radial, and **T** = tangential components, respectively). Different from teleseismic observational datasets used by [4], for the datasets in all the 4 regional stations used in the present case (the detailed stations are provided in table 1), black-colored signals represent waveforms from surveys and corresponding red-colored ones show synthetic waveforms. The results of the full moment-tensor seismic inversions for all broadband seismograms recorded by the 4 stations can be seen in Figure 2. Numerical values characterizing the blast-event are featured, including the origin time of the explosion at 03:30:08 UTC, the source epicenter measured in longitude and latitude degrees (41.33° N and 129.03° E, respectively), the source centroid-depth at 1 km below the ground, the station geographical locations in radial distances and azimuthal positions with respect to the source epicenter (see table 1 for complete information), nodal planes (NP1 and NP2) quantifying types of strike, dip, and rake for each nodal plane, the event seismic moment of $M_o = 6.25 \cdot 10^{24}$ dyne cm, which directly corresponds to the moment magnitude of $M_w = 5.80$, as well as 6 major elements of the fully decomposed moment-tensors ($M_{xx} = 1.84$, $M_{xy} = 0.07$, $M_{xz} = 1.87$, $M_{yy} = 1.58$, $M_{yz} = 0.51$, $M_{zz} = 5.95$) and the following sets of paired moment-tensors $M_{yx} = M_{xy}$, $M_{zx} = M_{xz}$, and $M_{zy} = M_{yz}$ as they are symmetric tensors.

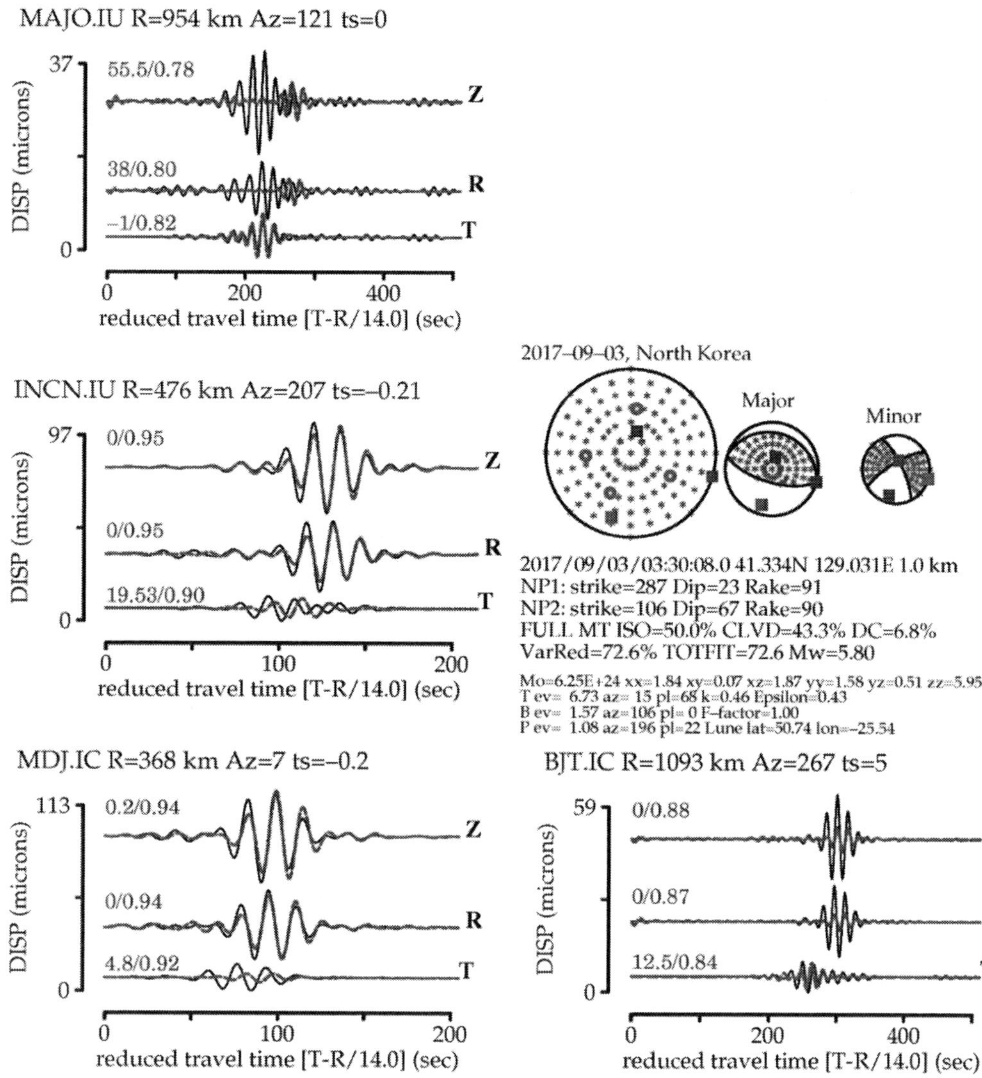

Figure 2. Broadband data with local **Z**, **R**, and **T** waveforms from the September 3, 2017 blast-event, showing fittings of observed (black) and simulated (red) waveforms from the 4 regional stations used (MAJO and INCN provided by IU network, MDJ and BJT provided by IC network).

From a statistical analysis perspective, the resulting seismic inversions give a value of 72.6% for the variance reduction, providing a high level of confidence for all values of the parameters examined. From a physical ground, the seismic inversion method is different from the *P*/*S* ratio analysis in that this method does not rely on direct comparison between the surface wave amplitudes or energies [6]. Instead, the full moment-tensors are decomposed into two parts, isotropic and deviatoric components. The volumetric (ISO) part of 50.0% was found to be dominant, indicating that the case considered in this study is likely to distribute its energy radially outward. This type of radiant energy distribution in nature is similar to properties of a compressional-longitudinal wave [7]. This is a supporting evidence for the explosion in question, prompting that it was detonated by a nuclear weapon test.

Following standardized decomposition for the full seismic moment-tensors, as discussed in [3, 5], the deviatoric component is further decomposed into the double-couple (DC) and compensated linear vector dipole (CLVD) tensors. Here, the DC mechanism was measured as a relatively small proportion of only 6.8%, suggesting that the source mechanism is unlikely due to shear-dislocation movement of combined compressional and extensional waves, usually found for seismic activities of tectonic origin. The 43.2% CLVD component completed the full moment-tensor decomposition, for which the energy is rapidly released from the source location via vertically spatial energy distribution through soil layers beneath the surface reaching up to a remarkable distance of tens of kilometers above the zero point. Along with the depth reported, these findings resolve the September 3, 2017 blast as a nuclear bomb, similar to those from earlier nuclear tests previously claimed by [9, 10, 11, 12].

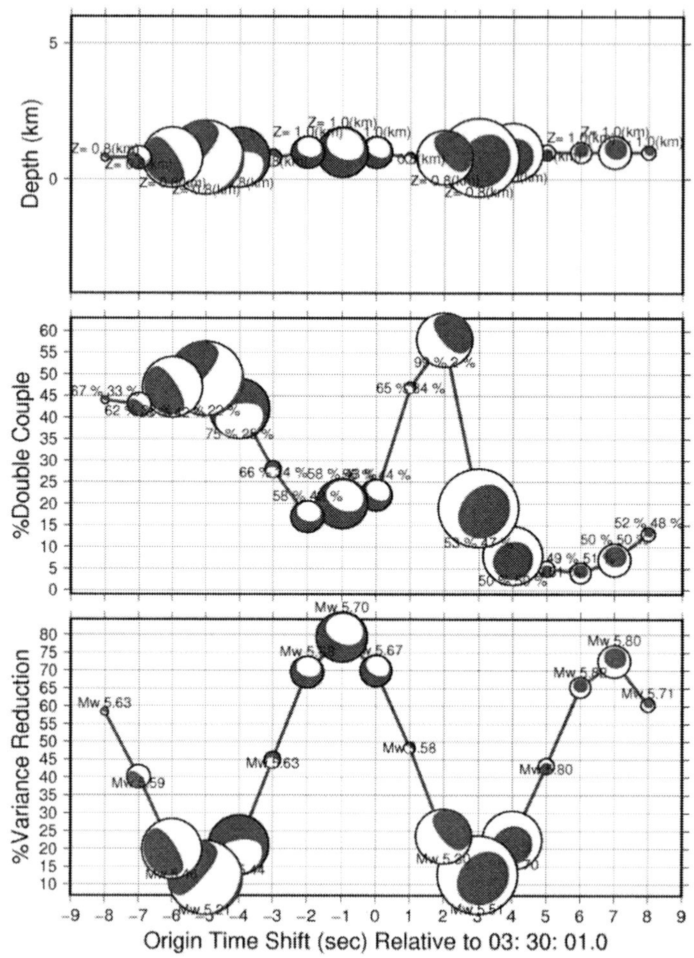

Figure 3. Graphics, showing the cross-correlation of the variance reduction, DC part, and the centroid-depth with respect to time taken for the moment-tensor inversions as the horizontal axis. The variance reduction achieves its peak at 73% corresponding to a seismic moment of M_w = 5.8 (bottom panel), which correlates to about 7% DC component (middle panel) and a 1.0 km centroid-depth (top panel).

Plotted in Figure 3, graphics as a result of the full moment-tensor inversions for the case examined. Three plots show a cross-correlation of variance reduction, the DC and its counterpart, the ISO (both in per cent), and the centroid-depth (given in km). With respect to the time shift estimated for the first explosion at 03:30:08 UTC, the September 3, 2017 blast rapidly released its huge amount of energy only within seconds. The energy, measured here as seismic moment, matches with an observed peak of approximately 73% statistical variance reduction, corresponding to a measured moment magnitude of M_w = 5.8, followed by earthquakes with a relatively small portion of about 7% for a DC mechanism at the moment the ground explosion was detected at 03:30:08 UTC to only a maximum value of 15% several seconds later. These support for a claim that the energy is not associated with tectonic release. The dominance of the ISO component over the other, particularly the DC, can thus be a good indicator for explosion-earthquake discriminant [3, 14], in addition to other seismic discriminants, for example, the ratio of body to surface wave amplitudes or energies in [4], the *P*/*S* ratio analysis in [6, 7, 8], and the direction of the first vertical micro-displacement in [13].

Regarding all the above findings, from this stage we can make a definitive statement that owing to the full moment-tensor inversions developed by [14], the event in question had a significant isotropic component of 50% compared to only a 7% double-couple with an adequately high level of confidence. This, as in most similar cases, places the relative proportion of the ISO component in a major role in monitoring of broadband seismic signals generated by nuclear tests. Together with the results derived from the full moment-tensor inversions detailed in Figure 2, these parameters of the cross-correlation in Figure 3 yield the same centroid-depth of the source, estimated to be only 1 km below the surface. This finding of the source centroid-depth is in good agreement with that reported by GEOFON-GFZ, which released the depth to be 1 km (accessed at http://geofon.gfz-potsdam.de/eqinfo/special/korea). But using the Real Time Monitoring Nuclear Explosion (RTMNE) method that included determination of the source location or hypocenter, [13] reported a depth of up to 2 km deep beneath the ground. However, this method gave no clue about source magnitude and arguably provided imprecise source depth determination, as claimed by [15]. In regard to the difference in source depth estimates, we here to some degree believe that the calculated results from the current method, including the source depth, are obtained with a better resolution and hence a sufficiently high level of accuracy.

In addition, characterization of seismic signals requires a common measurement of energy release, which is, for practical purpose, best represented by a magnitude scale of the signal. Regarding this, GEOFON-GFZ reported a magnitude of $m_b = 6.3$ for the September 3, 2017 suspected event based on measurements of body seismic waves whereas this work measures the magnitude of $M_w = 5.8$ based on the centroid full moment-tensors. In part, the main reason for this magnitude difference is that most of the seismic energy release is taken away by the propagation of surface waves rather than body waves. However, within uncertainties in magnitude measurements both scales arguably have no influence on the determination of whether the blast was underground nuclear detonation or generated by a common natural earthquake. Another method of magnitude determination of a particular seismic signal using earthquake rupture duration as a local parameter recorded by a local network of seismic stations was discussed by [16] in terms of 'duration magnitude' but the methodology presented is not relevant to the present study. Considering possibly increasing scales of any artificial explosions in coming years, it seems that the source depth is likely to be more important for the explosion-earthquake discriminant than the magnitude. While the hypocentre of natural events is usually found from tens of kilometres to a greater depth in some cases, the common source depth of underground nuclear detonation is found to be relatively shallow to only within 1-2 km deep for technical reasons.

4. Conclusions

We have utilized the full moment-tensor inversions to distinguish signals generated by a nuclear test, blasted on September 3, 2017 by the North Korean government from those induced by a natural event. The signals were recorded by a network of 4 regional stations, namely MAJO and INCN (IU network) and MDJ and BJT (IC network), in which these stations were within a radius of 1,100 km away from the epicenter. The results give a value of 72.6% for the variance reduction, providing a high level of confidence for all calculations, including a seismic moment of $6.25 \cdot 10^{24}$ dyne cm, corresponding to a moment magnitude of M_w 5.8, with the dominant ISO component of 50%, and the less important parts of the 6.8% DC and the 43.2% CLVD parts, and the centroid-depth of only 1 km. The clear dominance of the ISO component over the DC part, along with the relatively shallow source depth extracted from careful analysis of the cross-correlation and source mechanisms, suggests that the explosion in quest was generated by a nuclear weapon test. This finding is in good agreement with the results officially announced by some institutions around the world, suggesting that the methodology in the present work is therefore reliable and sensible for monitoring of broadband seismic signals with a better resolution in discriminating nuclear blasts from natural events.

Acknowledgments
The authors personally acknowledge Dr. Gene A. Ichinose for firmly allowing us to use MTINV 3.05, a programming package for seismic data inversions as well as The Incorporated Research Institute for Seismology (IRIS) network and The International Federation of Digital Seismograph Networks (FDSN) for the data available for use in this work.

References

[1] Massé R P 1981 *Bull. Seismo. Soc. Am.* **71** (4) 1249-1268
[2] Song X J, Helmberger D V and Zhao L 1996 *Geophys. J. Int.* **125** (1) 15-29
[3] Dreger D and Woods B 2002 *Tectonophys.* **356** (1-3) 139-156
[4] Dahy A S and Hassib H G 2009 *Euro. J. Appl. Sci.* **1** (4) 47-52
[5] Cho K-H 2014 *Appl. Geophys.* **11** (4) 429-436
[6] Fisk M D 2006 *Bull. Seismo. Soc. Am.* **96** (6) 2348-2367
[7] Walter W R, Matzel E, Pasyanos M E, Harris D B, Gok R and Ford S R 2007 *Proc. 29th Monitor. Res. Rev.* (Denver: Colorado, US) 684-93
[8] Houng S E 2018 *Bull. Seismo. Soc. Am.* **108** (1) 218-229
[9] Zhao L-F, Xie X-B, Wang W-M, Hao J-L and Yao Z-X 2008 *Bull. Seismo. Soc. Am.* **98** (6) 2571- 2589
[10] Shin J S, Sheen D-H and Kim G 2010 *Geophys. J. Int.* **180** (1) 243-250
[11] Zhao L-F, Xie X-B, Wang W-M, Hao J-L and Yao Z-X 2014 *Seismo. Res. Lett.* **85** (1) 130-134
[12] Zhao L-F, Xie X-B, Wang W-M, Hao J-L and Yao Z-X 2016 *Geophys. J. Int.* **206** (1) 1487-1491
[13] Prastowo T and Madlazim 2018 *J. Phys. Conf.: Series* 953 012211
[14] Ichinose G A, Anderson J G, Smith K D and Zeng Y 2003 *Bull. Seismo. Soc. Am.* **93** (1) 61-84
[15] Prastowo T 2019 *J. Phys. Conf.: Series* **1171** 012002
[16] Cholifah L, Puspitasari D A, Mahanani B M P and Prastowo T 2019 *J. Phys. Conf.: Series* **1153** 012018

Simple Glucose Measurement System Based on Uncladded Fiber Bragg Grating Etched with Nitric Acid

Asnawi[1,3*], E Rahmawati[1], N Qomariyah[2], N A Siddiq[3], S K Goshals[4], Y H Pramono[3]

[1]Department of Physics, Universitas Negeri Surabaya, Surabaya, Indonesia
[2]Department of Biology, Universitas Negeri Surabaya, Surabaya, Indonesia
[3]Department of Physics, Institut Teknologi Sepuluh Nopember, Surabaya, Indonesia
[4]Department of Physics, University Technology Malaysia, Johor Bahru, Malaysia

*e-mail: asnawi@unesa.ac.id

Abstract. The demand for a non-invasive, fast, and accurate glucose detection for diabetic patients is increasing nowadays. It is predicted that diabetic patient will increase from 415 million people at 2018 to 640 million people at 2050. Glucose is also present in saliva, therefore glucose concentration measurement in saliva can be an alternative way during diabetics monitoring. In this study, we measure 0-60% glucose concentration through optical systems. The optical system consists of an erbium-doped fibber laser, uncladded fiber bragg gratings that have been etched with nitric acid (HNO_3), and optical spectrum analyser (OSA). The measured parameter is a Bragg wavelength shift due to changes in glucose solution concentration. The results show that the wavelength shifts to lower value (blue shift) as the glucose concentration increases. The sensitivity obtained from this measurement system is 1.2 pm/% in glucose solution. This result is a preliminary study to detect glucose concentration in saliva based on Fibber Bragg Grating which has advantages for diabetic patients such as simple setup, non-invasive, fast, and low cost.

1. Introduction

It is estimated that the number of adults suffering from diabetes mellitus in the world at 2018 reaches 415 million people [1]. The number of diabetic patients will continue to increase, in 2040 it is predicted that more than 640 million people will suffer from a disease called "the silent killer" [2]. Called as the silent killer because diabetes mellitus can affect all organs of the body and trigger various other diseases such as wounds that difficult to heal, lung infections, and strokes [3].

A method that widely used for early detection and monitoring of diabetes is through measuring glucose levels in the blood. However, in practice, measurement of glucose levels must be done by piercing the patient's finger several times a day (invasive) which can cause temporary discomfort, bruising, and infection [4]. Beside blood, biomarker containing glucose are saliva [5,6]. Saliva can be easily obtained from patients (non-invasive) and will reduce discomfort during glucose levels checking. Some well-established technologies used in measuring glucose levels in saliva such as Liquid Chromatography - Mass Spectrometry (LC-MS) [7] and UV-VIS Spectrophotometry [8]. Measurements of saliva glucose levels using these two technologies are relatively complex and complicated, therefore they can only be used in the laboratory. Furthermore, the process requires considerable time, expensive reagents, and need highly trained operators. As a result, this technology cannot be used to monitor glucose levels at home and portable.

In terms of measuring glucose levels, optical sensors are superior to electrical sensors because their sensitivity is higher, lighter, and smaller [9]. Fiber Bragg Grating (FBG) has been used as an optical glucose-based sensor and shows high sensitivity, fast response time, and a wide measurement range [10,11]. However, the structure of the FBG that has been used must be in the form of a D-shaped fiber so that it requires complex fabrication and is very prone to broken. In this study, the fabrication and characterization of uncladded standard FBG was etched by nitric acid solution. The performance of the proposed sensor was analysed in terms of sensitivity and stability. All the experiments were performed in room temperature.

2. Theory

The single mode FBG is designed to work in telecommunication wavelengths, where a single wavelength termed by center wavelength or Bragg wavelength is reflected back [12]. The center wavelength of the FBG's reflection band ($\lambda_B \lambda_B$) given in Eq. (1) where $\Lambda\Lambda$ is period of the Bragg grating and $n_{eff} n_{eff}$ is the effective refractive index. The effective refractive index is the combination of the refractive index of the core and cladding.

$$\lambda_B = 2\, n_{eff}\, \Lambda \lambda_B = 2\, n_{eff}\, \Lambda \tag{1}$$

When the cladding was removed and introduced to medium (analytes), the effective index will experience some discrepancy, causing a shift in the reflected wavelength. Due to total internal reflection between the cladding and the surrounding medium, the light could not escape out of the fiber, though is guided back to the core of the fiber. Total internal reflection (TIR) between the cladding and the surrounding produces an evanescent wave that be the basis for the sensing potential of a Fiber Bragg Grating based sensor. The penetration depth ($d_p d_p$) of the evanescent wave at the core-analytes interface is given in Eq. (2)[13].

$$d_p = \frac{\lambda}{2\pi(n_{eff}^2 \sin\theta^2 - n_{analyte}^2)^{1/2}} d_p = \frac{\lambda}{2\pi(n_{eff}^2 \sin\theta^2 - n_{analyte}^2)^{1/2}} \tag{2}$$

The bragg grating affect the effective index, therefore change penetration depth of the evanescent wave into the surrounding.

3. Experimental Details

3.1. Preparation of Sensing Element

Standard FBG with 1554 nm central wavelength and reflectivity > 90% purchased from Beijing Samyon Instruments were used as the main part of glucose sensing device. D(+) Glucose anhydrous, deionized water, isopropyl alcohol, and nitric acid purchased from SAP chemicals, Indonesia. The polymer coating of FBG was removed using nitric acid (HNO_3) 65% by immersing the Bragg grating part of the fiber into a glass beaker for 15 minutes until polymer coating look slightly yellowed. It should be noted that extra care must be taken because bragg grating part is very fragile. The FBG was lifted from the acid then FBG cladding removed manually with a fiber stripper. The exposed part over the FBG is cleaned with isopropyl alcohol.

3.2. Characterization

The experimental arrangement used for the glucose concentration sensing is shown in Fig 1. Homemade erbium doped fiber laser operated at central wavelength 1550 nm with laser diode (THORLABS, 980 nm) pumping was used for the light source. The ends of the prepared FBG were attached to the light source and Optical Spectrum Analyzer (OSA, Anritsu MS9740A). Light from source was coupled into one end of the prepared FBG and the other end was coupled with an OSA to record the wavelength shift.

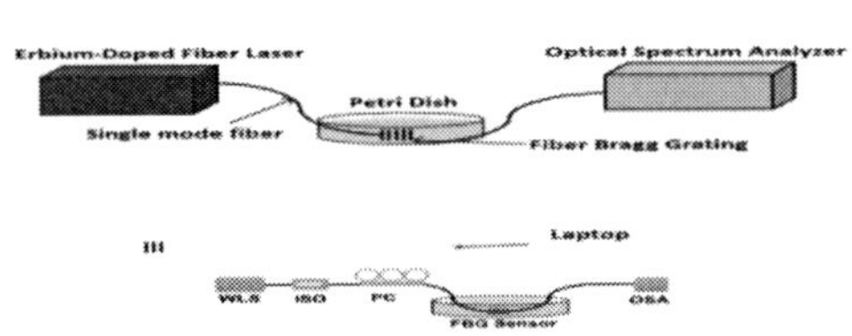

Figure 1. Experimental setup for sugar concentration detection based on Fiber Bragg Grating

During each measurement, petri dish contains 15 ml volume of various glucose concentration that dissolved in deionized water (wt%) is introduced into the FBG-sensing portion and real-time wavelength shift were monitored. Wavelength shift were acquired after 100 s of exposure to each glucose solution [14]. The wavelength shift was recorded for each concentration of glucose in solution. Aqueous samples of glucose with concentration varying from 0% to 60% (prepared in our laboratory) in increments of 10% sugar concentration were used to study the response of sensor toward glucose.

4. Results and Discussion

4.1. Sensing Response

Figure 2. illustrates the wavelength shift of glucose solution at various concentration measured using uncladded FBG after etched with nitric acid. As the glucose concentration increase, the center wavelength starts shifting to the lower wavelength (blue shift). Meanwhile the normal FBG (with cladding) did not show any wavelength shift. At 0% glucose concentration (only deionized water), the center of the wavelength is 1550.0708 nm. Figure 2 (b) shows that the proposed sensor response to glucose concentration is linear in the range 0 - 60% and has a line equation $y = -0.12x + 1{,}550.07 y = -0.12x + 1{,}550.07$ and the value of $R^2 = 0.99 R^2 = 0.99$. Based on this performance, the sensitivity of the proposed sensor can be calculated which is equal to 1.2 pm/% in the measurement range of 0 - 60% glucose concentration.

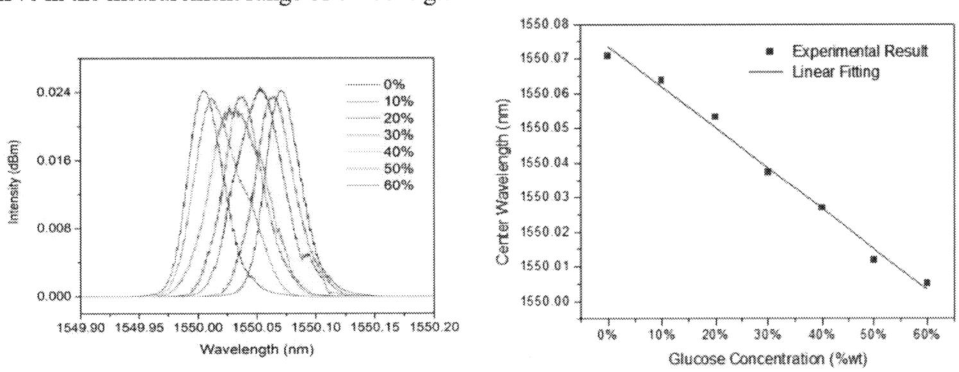

(a) (b)

Figure 2. Result of glucose measurement using the proposed sensor (a) Wavelength spectrum analyzed by Optical Spectrum Analyzer (b) Center wavelength shifted based on different glucose concentration

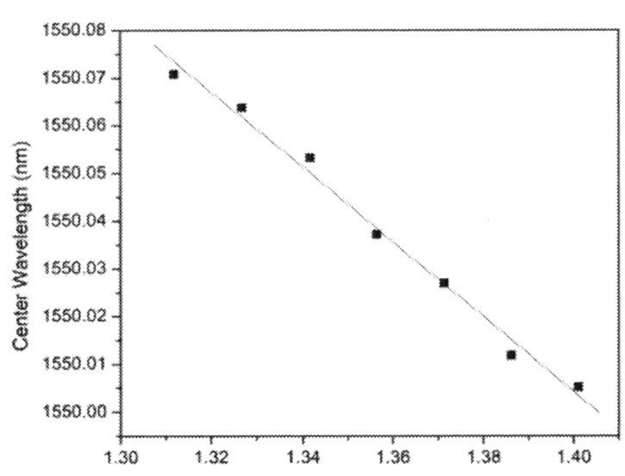

Figure 3. Plot of glucose refractive index to the transmitted wavelength center of FBG

Shifting the center of the wavelength to a smaller value is caused by changes in the refractive index of various glucose solutions. Based on study done by [6], the higher the concentration of glucose, the greater the refractive index. The refractive index change will change the effective refractive index of the FBG and cause changes in the center of the wavelength as formulated by Eq. (1). The greater the effective refractive index, the bigger the center value of the reflection wavelength and the smaller the value of the transmission wavelength. The plot of the effect of the refractive index of glucose at different concentration provided by using the data in [15] to the shift in the transmitted center wavelength is shown in Figure 3.

4.2. *Sensor Stability*

The cost of the sensor is decided by manufacturing expenses and its repeated use. The repeated use may decrease the cost per use. The stability of the proposed sensor was tested by measuring response toward glucose of concentration 30% for 4 days. The response of the sensor is shown in Fig. 4. Sensor has shown approximately similar response for 4 days consecutively. The results obtained indicate that the sensor has good stability in measurement.

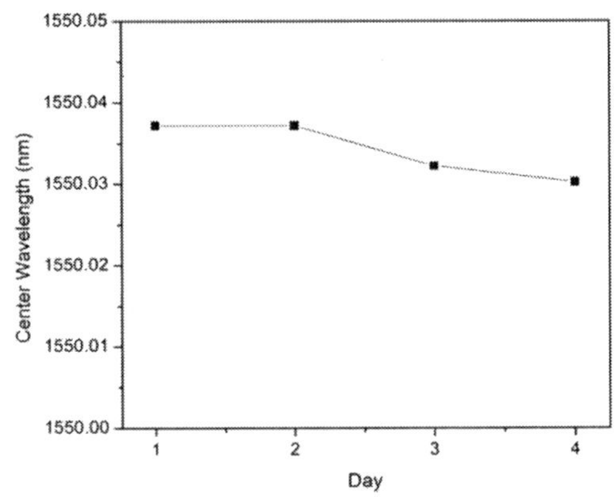

Figure 4. Stability curve of proposed sensor measured for 30% sugar concentration

5. Conclusion

In this report, the fabrication and characterization of glucose sensor based etched Fiber Bragg Grating has been presented. The sensor response was studied by exploring the wavelength shift properties at various glucose solution concentration (wt %). The FBG was etched with nitric acid solution, hence providing simple sensor fabrication. The proposed sensor showed good results for both high and low concentrations of sugar and has sensitivity 1.2 pm/% in glucose solution due to refractive index change of glucose solution. The stability of the sensor for 4 days is adequate. This result is a preliminary study to detect glucose concentration in saliva based on Fiber Bragg Grating which has advantages for diabetic patients such as simple setup, non-invasive, fast, and low cost.

References

[1] Wiviott S D, Raz I, Bonaca M P, Mosenzon O, Kato E T, Cahn A, Silverman M G, Zelniker T A, Kuder J F, Murphy S A and Bhatt D L 2019 Dapagliflozin and Cardiovascular Outcomes in Type 2 Diabetes. *New England J. Med.*, **380** (4) 347

[2] Ogurtsova K, da Rocha Fernandes J D, Huang Y, Linnenkamp U, Guariguata L, Cho N H, Cavan D, Shaw J E and Makaroff L E 2017 IDF Diabetes Atlas: Global Estimates for the Prevalence of Diabetes for 2015 and 2040. *Diabetes Res.Clinical Pract.*, **128** 40

[3] Fatimah R N 2015 Diabetes Melitus Tipe 2. *J. Majority*, **4** (5)

[4] Yusof M I B M, Renaldi A, Septiandi I, Kurniawan B and Fauzi A 2016 Mengenal Alat Pemantau Gula Darah dari Air Liur dengan "On-Chip Electrochemical Sensing" *Majalah Farmasetika* **1** (1) 3

[5] Yamaguchi M, Mitsumori M and Kano Y 1998 Development of Non Invasive Procedure for Monitoring Blood Glucose Levels Using Saliva *Proceedings of the 20th Annual International Conference of the IEEE Engineering in Medicine and Biology Society, Biomedical Engineering Towards the Year 2000 and Beyond (Cat. No. 98CH36286)* **20** 1763

[6] Panchbhai A S, Degwekar S S and Bhowte R R 2010 Estimation of Salivary Glucose, Salivary Amylase, Salivary Total Protein and Salivary Flow Rate in Diabetics in India *J. Oral Sci.* **52** (3) 359

[7] Jurysta C, Bulur N, Oguzhan B, Satman I, Yilmaz T M, Malaisse W J and Sener A 2009 Salivary Glucose Concentration and Excretion in Normal and Diabetic Subjects. *BioMed Res. Int.*

[8] Enejder A M, Scecina T G, Oh J, Hunter M, Shih W, Sasic S, Horowitz G L and Feld M S 2005 Raman Spectroscopy for Non Invasive Glucose Measurements. *J. Biomedical Optics* **10** (3) 031114

[9] El-Sherif M, Bansal L and Yuan J 2007 Fiber Optic Sensors for Detection of Toxic and Biological Threats *Sensors* **7** (12) 3100

[10] Zhou K, Chen X, Zhang L and Bennion I 2006 Implementation of Optical Chemsensors based on HF-etched Fibre Bragg Grating Structures *Measurement Sci. Technol.* **17** (5) 1140

[11] Zhou K, Chen X, Zhang L and Bennion I 2004 High-sensitivity Optical Chemsensor based on etched D-fibre Bragg Gratings *Electronics Lett.* **40** (4) 232

[12] Broadway C, Rui M, Junior A G L, Marques C and Caucheteur C 2018 Towards Commercial Polymer Fiber Bragg Grating Sensors: Review and Applications *J. Lightwave Technol.*

[13] Arasu P T, Al-Qazwini Y, Onn B I and Noor A S M 2012 Fiber Bragg Grating based Surface Plasmon Resonance Sensor Utilizing FDTD for Alcohol Detection Applications *2012 IEEE 3rd International Conference on Photonics* 93

[14] Pahurkar V G, Tamgadge Y S, Gambhire A B and Muley G G 2015 Evanescent Wave Absorption based Polyaniline Cladding Modified Fiber Optic Intrinsic Biosensor for Glucose Sensing Application. *Measurement* **61** 9

[15] Weast R C and Astle M J 1981 CRC *Handbook of Chemistry and Physics 61st edition.* (Chemical Rubber Co)

Journal of Physics: Conference Series 1417 (2019) 012005 doi:10.1088/1742-6596/1417/1/012005

Development of Optical Fiber Sensor Performance as a Tool to Determine Cholesterol Concentration

M Budiyanto[1], A R Purnomo[1], W B Sabtiawan[1], Suhariningsih[2], M Yasin[2]
[1]Department of Natural Science, Faculty of Mathematics and Natural Science, Universitas Negeri Surabaya, Surabaya 60231, Indonesia
[2]Department of Physics, Faculty of Science and Technology, Universitas Airlangga, Surabaya 60115, Indonesia

e-mail: mohammadbudiyanto@unesa.ac.id

Abstract. The purpose of this study was to determine cholesterol concentration by developing the performance of optical sensors. The use of this optical sensor is equipped with an optical fiber bundle to guide the wave in the hope of not reducing the intensity of the wave source. This sensor is simpler and inexpensive to use to determine the concentration of a material. The research method for determining cholesterol levels can be determined through the character of the laser beam to the material through a fiber optic sensor. The optical fiber sensor method is simpler and cheaper to use to determine the concentration of a material. Based on this description, research on fiber optic sensors will be developed as an instrument model to determine cholesterol concentration using a fiber optic bundle type sensor. The method used is the modulation of laser beam intensity. A beam of light that enters the optical fiber of the receiver and is processed by an optical detector into an electrical signal that will be displayed on the computer. The experimental results will obtain a linear relationship between the sensor output peak voltage as a function of variations in cholesterol concentration and parameters and sensor performance will be measured including sensitivity and linearity. The results showed that maximum voltage detection showed a linear decrease with increasing concentration with a sensitivity of 0.15 mV/ppm and a linearity of more than 97%. high sensitivity, simplicity of design, and low cost of promoting the developed sensor results in good performance detecting cholesterol concentration.

1. Introduction

Cholesterol is a fat molecule in cells divided into LDL, HDL, total cholesterol and triglycerides. Cholesterol is one of the nutrients that are needed by the body in addition to other nutrients such as carbohydrates, proteins, vitamins, and minerals. Cholesterol is present in the bloodstream or cells of the body that are needed for the formation of cell walls and as a raw material for several hormones. However, if excessive levels of cholesterol in the blood can cause illness, including coronary heart disease and stroke. Normal cholesterol should be under 200 mg/dl. Cholesterol is also a basic ingredient in the formation of steroid hormones. Excessive cholesterol in the body will be buried in the walls of blood vessels and cause a condition called atherosclerosis, which is narrowing or hardening of the arteries. This condition is the forerunner to the occurrence of heart disease and stroke.

The detection of cholesterol levels has been carried out by researchers such as sensitive and selective electrochemical sensors using silver nanoparticles modified glass carbon electrodes for the determination of cholesterol in serum bovine [1]. Electrochemical sensors are based on molecular film printing on carbon nanoparticles that are modified electrodes for cholesterol determination [2]. The layer of carbon nanotubes and bienzyme biosensor-based gold nanoparticles for cholesterol detection [3]. High-performance electrochemical biosensors for the detection of total cholesterol [4]. Cholesterol detection with amperometry using the electrochemical oxidation method [5]. electrochemical and optical polymers by using biosensors to detect cholesterol [6]. The weaknesses of the sensors are sensor performance optimization such as sensitivity, linear range, and detection limits are still not optimal. Several ways and techniques for calculating solution concentrations such as cholesterol have been carried out. Among them, UV-Vis spectrophotometer is used to determine the blood cholesterol level of patients experiencing hypercholesterolemia. Development of biosensors as analytical instruments for determining cholesterol in traditional foods using UV-Vis spectrophotometer. UV-Vis spectrophotometers require a stable complexing to increase the adsorption sensitivity. Another obstacle is the analysis with UV-Vis spectrophotometer which has a detection limit.

There are many ways to determine the results of cholesterol concentration through laser beam propagation in certain solutions by using sensors through fiber optic waveguides. Sensors using optical fiber are simpler to be used to determine the concentration of a material. Sensors using optical fibers are also used to determine the concentration of the salt solution in a material with high sensitivity measurement results [7]. From this brief description of optical sensors that use fiber developed by researchers to make optical fiber sensors as an instrument to determine cholesterol concentrations by using optical fiber sensor types of optical fiber bundles. This study is to determine the profile of the intensity of the beam output of the receiving optical fiber (RF) detected by the photodetector of the object shift. Besides, to determine the performance of the sensor by using a fiber optic sensor bundle to determine cholesterol concentration. This research expects to get an experimental model (setup) of a fiber optic sensor system bundle to determine cholesterol concentration. Describe the intensity profile of the receiving fiber (RF) beam output which is detected by the photodetector against the object shift. With the final goal of the study is to determine the performance of the session by using two types of optical fibers to determine cholesterol concentrations. The development of this instrument is based on the concept of the interaction of the He-Ne laser beam to a material that takes into account physical quantities.

Laser light includes examples of visible light electromagnetic waves. A laser is a device that emits light through a process called stimulated emission. The laser is a continuation of LASER (Light Amplification by Stimulated Emission of Radiation). Lasers are devices that use quantum mechanical effects, induced or stimulating emissions, to produce coherent light rays. Lasers also have the character of emitting photons in coherent light. Laser light is usually monochromatic, for example has a single wavelength or color, and is emitted in a subtle beam. This is different from ordinary light sources that emit photons that can be seen in all directions, usually encompassing the wavelengths of the broad electromagnetic spectrum. Lasers can be understood through the use of theories of quantum mechanics and thermodynamics.

Laser light is unique and is within the range of the electromagnetic wave spectrum. The light that comes out of the laser with a variety of wavelengths and variations in intensity is applied in various fields, including telecommunications, meteorology, metrology, biology, chemistry, military, cybernetics, industry, and medicine. Laser in this study will propagate using optical fiber. Optical fiber is a kind of cable transmission channel made of glass or plastic that is very smooth and can be used to transmit light signals from one place to another. The light source used is usually a laser. Optical fiber with a diameter of approximately 120 micrometers. The light in the optical fiber does not come out because the refractive index of the glass is greater than the refractive index of the air because the laser has a very narrow spectrum. In principle, optical fiber reflects and refracts the amount of light that propagates in it. The efficiency of optical fibers is determined by the purity of the building blocks of glass. The purer the glass, the less light is absorbed by optical fibers.

2. Method

The principle and method of this research are the basic principle of the properties contained in the He-Ne laser beam which involves a solution. By analyzing the physical quantities of the laser beam through the solution and transmitted through the solution including the intensity and output voltage that is read by the detection device. Detection devices convert wave energy into electrical energy using optical detectors. The tools and materials used in research to determine this cholesterol concentration are Laser Helium-Neon, standard cholesterol, fiber optic type bundle, silicon detector, digital voltmeter. Design of optical fiber sensor design with optical fiber type as shown in Figure 1.

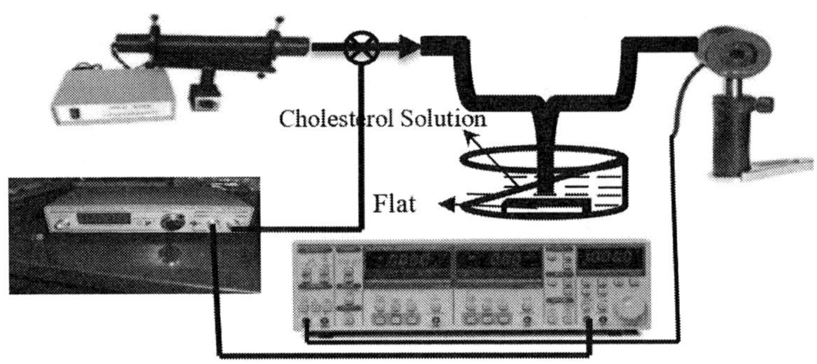

Figure 1. Optical fiber sensors determine the concentration of cholesterol by using a fiber bundle

Research Procedure by characterizing the shift sensor to determine the shifting of the fiber bundle sensing channel to the mirror. Measurement of physical magnitude data when the fiber bundle is placed coincides with a flat and concave mirror i.e. at shift $z = 0$. The optical fiber reflection bundle is placed on the micrometer position by sliding every 50 µm. At each shift read the measured output voltage so that the data obtained in the form of a detector output voltage as a function of fiber bundle shift. Detection is carried out on several variations of concentration, among others 0 ppm, 50 ppm, 100 ppm, 150 ppm, 200 ppm, 250 ppm, 300 ppm.

Data Analysis in this experiment, research variables can be made including the dependent variable is the peak voltage of the sensor output (in mV units) on optical fiber bundles, the independent variable is the concentration of cholesterol (in units of percent volume). The next step in the experimental test, the data obtained is in the form of the peak voltage value of the sensor output as a function of cholesterol concentration variation. From these functions, it can be determined the front linear region (front slope) and the back linear region (back slope). Linearity test of the front slope and back slope area is used a linear regression test. With the acquisition of peak voltage value data as a function of cholesterol concentration. From these results, sensor performance will be obtained which includes resolution, sensitivity, linearity, dynamic range and linear range of cholesterol sensors from three types of optical fibers.

3. Results and Discussion

The amount of electric voltage read on an optical detector as a function of shifting the bundle of fibers for each concentration from 0 ppm to 300 ppm. Experimental data shows that there is a relationship between the variation of cholesterol concentration changes to the maximum output voltage at each concentration. The higher the cholesterol concentration the lower the measured detector output voltage. Detection data of the detector output voltage starts when the bundle fibers placed coincide with a mirror ie at shift $z = 0$. The bundle fibers placed on the micrometer position are shifted every 50 µm. At each shift position, the detector output voltage is measured to obtain the maximum output voltage from each variation of cholesterol concentration. The maximum output voltage value from the variation of cholesterol concentration as in table 1.

Table 1. Data on maximum electrical voltage for seven cholesterol concentration samples

No	Concentration (ppm)	Electric voltage output (mV)
1	0	194.4
2	50	182.3
3	100	179.2
4	150	168.1
5	200	159.7
6	250	151.8
7	300	150.1

Figure 2. shows that the maximum voltage appears to decrease linearly with increasing concentrations of cholesterol solutions.

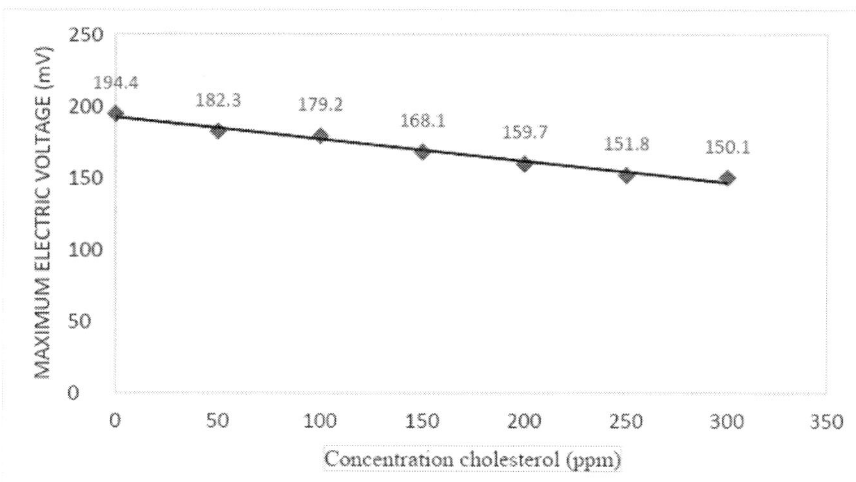

Figure 2. Maximum voltage for variations in cholesterol concentration

Detection analysis results show that the maximum output voltage shows a linear decrease in the increase in the concentration of cholesterol solution with a sensitivity of 0.15 mV/ppm and linearity of 97.65%. Based on the results of experiments determining cholesterol concentrations using SSO quasi Gaussian beam shows that the optical fiber output power profile will increase with increasing distance of the object shift from the optical fiber bundle, after reaching the peak output power will decrease exponentially with increasing distance in each variation of cholesterol concentration.

The results of peak stress analysis of variations in cholesterol concentration using SSO flat mirror reflecting sensitivity 0.15 mV/ppm and linearity 97.65%. The results of this sensitivity indicate an approach to the value of sensitivity using a quasi Gaussian file and in accordance with the analysis obtained theoretically [8].

4. Conclusion

The conclusion that can be made based on the results and discussion of this study is that the output power profile of the results of experiments using a Gaussian quasi beam shows that the optical power output will increase with increasing object shifts from optical fiber, after reaching the peak the output power will decrease exponentially with increasing distance. The performance of optical fiber sensors to determine cholesterol concentration reaches high sensitivity, linearity, and accuracy with a sensitivity value of 0.15 mV/ppm, linearity of more than 97.65%, linear range of 0 - 300 ppm.

References

[1] Nantaphola S, Chailapakul O, and Weena S 2015 Sensitive and selective electrochemical sensor using silver nanoparticles modified glassy carbon electrode for determination of cholesterol in bovine serum *Sensors and Actuators B* **207** pp 193–198

[2] Ji J, Zhou Z, Zhao X, Sun J, and Sun X 2015 Electrochemical sensor based on the molecularly imprinted film at Au nanoparticles-carbon nanotubes modified electrode for determination of cholesterol *Biosensors and Bioelectronics* **66** pp 590–595

[3] Cai X, Xia Gaoa L W, and Lin X 2013 A layer-by-layer assembled and carbon nanotubes/gold nanoparticles-based bienzyme biosensor for cholesterol detection *Sensors and Actuators B* **181** pp 575– 583

[4] Ahmadalinezhad A and Chen A 2011 High-performance electrochemical biosensor for the detection of total cholesterol *Biosensors and Bioelectronics* **26** pp 4508-4513

[5] Soylemez S, Yesemin A U, Cevher Gundo H, Ergun Y, and Levent T 2015 Electrochemical and optical properties of a conducting polymer and its use in a novel biosensor for the detection of cholesterol *Sensors and Actuators B* **212** pp 425–433

[6] Faria J B 1998 A theoretical analysis of the bifurcated fiber bundle displacement sensor *IEEE Trans on Inst and Meas* **47** pp 742-747

[7] Yasin M, Ahmad H, Thambiratnam K, Jasim A A, Phang S W, and Harun S W 2013 Design of Multimode Tapered Fibre Sensor for Glucose Detection *Optoelectronics and Advanced Materials* **7**(5– 6) pp 371-376

[8] Rahman H A, Harun S W, Saidin N, Yasin M, and Ahmad H 2012 Fiber Optic Displacement Sensor for Temperature Measurement *IEEE Sensors Journal* **12**(5) pp 1361 - 1364

A Water Surface Cleaning Robot

E Rahmawati[*1], I Sucahyo[1], A Asnawi[1], M Faris[1], M A Taqwim[1], D Mahendra[1]

[1]Physics Department, Universitas Negeri Surabaya, Indonesia

*e-mail: endahrahmawati@unesa.ac.id

Abstract. This paper describes the design of a robot for cleaning rubbish floating on the water surface. Three important issues for designing the aquatic robots are a cost-effective solution along with robustness and durability. Due to the nature of the cleaning work, we designed the vehicle structure that can provide high stability, good ability in maneuver and can easily collect all the waste flowing in between. A pontoon shaped hull works best for this case and fulfils all the hydrostatic, structural stability criteria. For removal and collection of surface waste, a motor-driven collecting-arm system has been designed for collecting the wastes and redeploying it into a rectangular basket on the hull. This design provides simple and effective waste removal and accommodates large amounts of waste within a small space. For the prototype, the hulls were made up of styrofoam which is wrapped by fibre and then coated with waterproof and resin. It supported together by aluminum. This light and tough structure support the total weight of the system. The propulsion system based on a differential drive mechanism has been designed, which allows the robots to take a 360 turn on the spot and provides high thrust. Electronic circuit and motors have been placing inside the hull, in order to protect them from water. The robot is manually controlled by remote control based on Xbee Pro wireless modules. The testing of the robot prototype proved to be effective in waste collecting and removal. The maximum trash loads that robot can bear is up 16 kg.

1. Introduction

Waste is an environmental problem that always arises from year to year and still cannot be resolved entirely. We frequently found garbage from various places dumped into rivers, waterways, or reservoirs. The rubbish can clog the flow of water, causing water to become dirty and smelly so that it often overflows and causes disasters, including flooding. How to clean waste from water areas requires extensive resources, for example, by cleaning staff and using excavators [1,2].

This study aims to provide an alternative solution to the problem of waste in water areas by developing robotics technology capable of operating in water areas. The proposed applied research is expected to be an alternative solution to prevent disasters, especially floods. Robotics technology developed in the form of eco-robot with the main task of collecting waste. The robot is designed to be controlled r manually by remote control. The development method of this research refers to ADDIE. This method including of analysing of the robotic cleaning system, designing the robot, developing the robot, implementing robot to clean waste in limited water areas, evaluating the effectiveness of robot in cleaning up trash for the more extensive area. This article is focusing on the design and development of the robot.

Previously some robots have been built for cleaning water surface [3, 4]. A few results are available in the open references, which have discussed the development of such particular purpose robot [5, 6]. This work aims at developing a more versatile and efficient system by the usage of the aqua robot. This robot has an excellent opportunity to expand its functions in the future. Such conducting activities for removing algae, leaves and twigs; spraying of chemicals at the appropriate locations; checking the water quality as well as deploying payload are planned to be done autonomously. These functions may save a lot of human effort and provide a sustainable solution to the pervasive problem. For navigation and trash cleaning on the ground, many well-designed algorithms have been developed earlier for both single robotic systems as well as for swarms [7]. However, because of the difference in the dynamic environment, propulsion system, and the difficulty to accurately determine the current position based on relative velocity and acceleration, these algorithms cannot be directly used on aquatic surfaces. Also, the navigation algorithms developed earlier for autonomous aquatic robots have not been designed with cleaning as an integral part of them [8,9].

2. Robot Design

The three crucial consideration for designing the robot is a cost-effective solution along with robustness, and durability. Consider the function of cleaning work, the robot structure is designed in order to provide high stability, excellent manoeuvrability and can easily collect all the waste flowing around. A pontoon shaped hull works best for this case and fulfils all the hydrostatic, seakeeping, and structural stability criteria [10]. For removal and collection of surface waste, a motor-driven arm-manipulator system has been designed for collecting the wastes in a section covered with a net. This design provides simple and effective waste removal and accommodates large amounts of trash within a small space.

Furthermore, as the collected waste floats on the water surface, the robot does not need to support its weight. The lower section of the belt is placed below the water level to take out the floating weeds easily. For the prototype (Fig. 1), the pontoon shaped hulls were made up of two shaped styrofoam. The styrofoam was wrapped by fibreglass matt and then was coated by aqua proof and resin. This design result in a light and strong pontoon hull, minimum risk of a leak. Two pontoon hulls supported together with the help of an aluminium truss. This light and robust structure support the total weight of the system. The design parameters of this prototype are given in Table 1. To navigate autonomously in small and constrained spaces, a propulsion system that can provide high mobility and short turning radius is needed. Due to its flat front and the central waste collection system, the robot is also required to overcome high form and skin drag given by:

$$F_D = C_D \cdot v^2 \quad (1)$$

which F_D is drag force in Newton, C_D is drag coefficient and v is the speed of the robot.

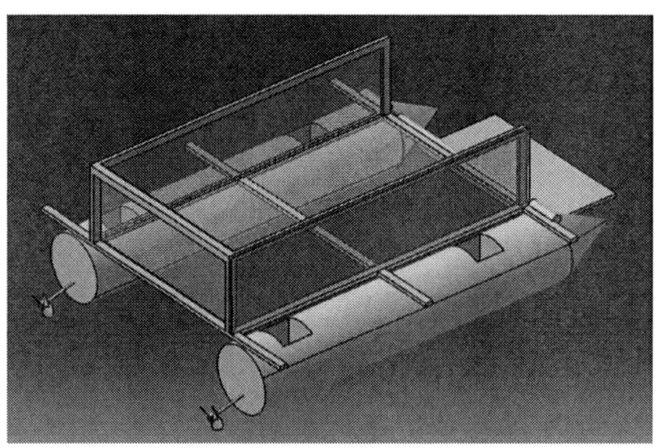

Figure 1. Robot prototype

Figure 2. Dimension of robot

Table 1. Robot specification

Design parameter	*Value*
Dimensions (overall)	1200mm x 800mm x 500mm
Net weight (only robot)	8 kg
Maximum weight (with load)	28 kg
Average battery while cleaning	1 hour
Propeller type	three blades
Control system	manual (remote control)
Drive motors	Motor Driver BTS7960 43A
Wireless module	Xbee Pro S2C ZigBeeBP24CZ7WIT-004

Placing the electronic circuits, drive motors, batteries, and motors inside the hull, in order to protects them from water leakage inside its casing. As the robot is designed to traverse manually, it is remote controlled based on Xbee Pro S2C modules. The electronic circuit is relatively simple. It consists of an Arduino based microcontroller, three motor drivers, two DC motors which were attached to the flexible shaft and propellers, one power window DC motor. Power window motor has been used for arm manipulator which collects trash and raise it to basket. The basket has been attached on the hull.

Robot has been controlled by remote control through Arduino UNO which send data (command) to transmitter. The data from transmitter has been received by receiver. Receiver send data to Arduino UNO. Arduino UNO send data to motor DC, then motor DC will act refer to data or instruction.

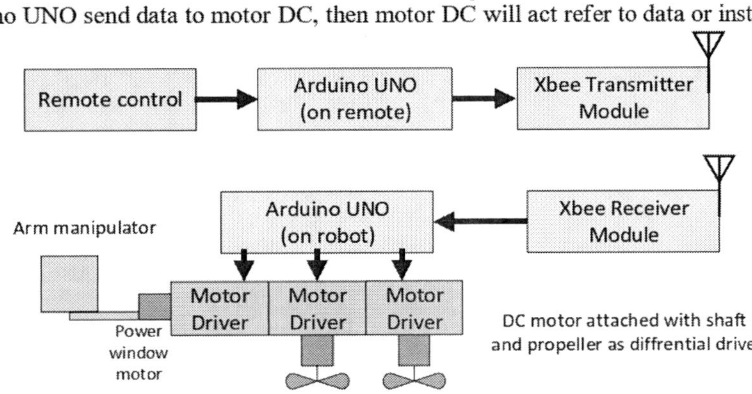

Figure 3. Control system for differential drive of robot

3. Results

Consider designing the hull of the robot; there were several technical matters that are geometric design, reducing hydrodynamic resistance, minimizing weight and preventing drown because of leakage. The primary requirement of the hull design is to generate a sufficient force suspending the robot in water. To overcome the problem, the robot developed by us uses two pontoons hulls to suspend the robot. The payload of the two pontoons is 20 kg, and their weight is 8kg. How to achieve static balance in the water is an essential issue in designing the robot. For increasing the payload, previous research adds more pontoons [10]. While the waste container is full or empty, there must be no difficulties for a robot Iie in water steadily. The robot is made symmetric along the moving direction. The two pontoons are placed so that the centre of the floating force is located at the symmetric axis. The weights are placed at the centre of mass of the robot.

The application of the control system using XBee wireless to the robot motion system has been successfully carried out by utilizing the control method by wireless communication with the serial data transmission format. The control system is examined by giving instructions on the remote that are well received by the robot with the appropriate motion response with the commands written on the program. The robot is controlled s moves supposed moving with the trajectory as in figure 3 and collects garbage along the path. The robot can operate well and able to accommodate garbage loads up to 20 kg.

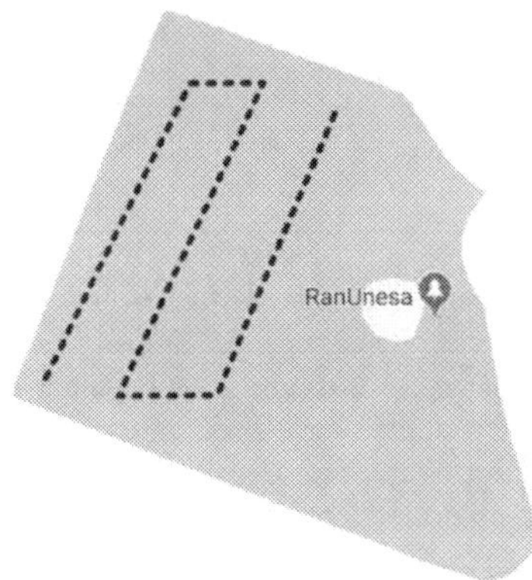

Figure 4. Path of Robot motion during testing

4. Conclusion

A water surface cleaning Robot was a relative success, the team has created an outline for future improvement in terms of research and theory, and implementation. On a high level, more research should have been done regarding the interfacing between module so the robot not only has single task but also multi task such as selecting trash refer to its material and also monitoring water quality. Consider designing the hull of the robot; there were several technical matters that are geometric design, reducing hydrodynamic resistance, minimizing weight and preventing drown because of leakage. The primary requirement of the hull design is to generate a sufficient force suspending the robot in water. To overcome the problem, the robot developed by us uses two pontoons hulls to suspend the robot. The payload of the two pontoons is 20 kg, and their weight is 8kg. How to achieve static balance in the water is an essential issue in designing the robot. For increasing the payload, previous research adds more pontoons [10,11]. While the waste container is full or empty, there must be no difficulties for a robot Iie in water steadily. The robot is made symmetric along the moving direction. The two pontoons are placed so that the centre of the floating force is located at the symmetric axis. The weights are placed at the centre of mass of the robot.

The application of the control system using XBee wireless to the robot motion system has been successfully carried out by utilizing the control method by wireless communication with the serial data transmission format. The control system is examined by giving instructions on the remote that are well received by the robot with the appropriate motion response with the commands written on the program.

The robot is controlled s moves supposed moving with the trajectory as in Figure 3 and collects garbage along the path. The robot can operate well and able to accommodate garbage loads up to 20 kg.

References

[1]. Savitri P A, Purba W S and Zulkifli M 2018 Pengelolaan Sampah di Indonesia *Statistik Lingkungan Hidup Indonesia 2018* Badan Pusat Statistik

[2]. Griffiths G 2003 *Technology and Applications of Autonomous Under- water Vehicles* (UK:

Taylor and Francis)
[3]. Ramos P, Cruz N, Matos A, Nevesand M V, Pereira F L 2001 *Monitoring an Ocean Outfall Using an AUV. Proc. MTS/IEEE Oceans'01* Honolulu
[4]. Saravana K G, Kumar S, Ragavan R, Balakrishnan M 2016 *IJSRP* **6** (4) 525
[5]. Dandan K, Albitar H, Ananiev A and Kalakov I 2015 *Int. J. Adv. Robot Syst.* **12** (12) 1
[6]. Nurlansa O, Istiqomah D A, and Pawitra M A S 2014 *IJFCC* **3** (5) 367
[7]. Yuyi Z, Yu Z, Huanxin L, Yunjia L and Liang L 2013 Cleaning Robot. *Int. J. Adv. Robot Syst.* **10** (5) 1
[8]. Attamimi M, Araki T, Nakamura T and Nagai T 2013. Visual Recognition System for Cleaning Tasks by Humanoid Robots *Int. J. Adv. Robot Syst.* **10** (11) 1
[9]. Ruangpayoongsak N, Sumroengrit J, Leanglum M 2017 A Floating Waste Scooper Robot On Water Surface *17th International Conference on Control, Automation and Systems (ICCAS 2017)*
[10]. Wang Z, Liu Y, Yip H W, Peng B, Qiao S, and He S 2008 Design and Hydrodynamic Modeling ofA Lake Surface Cleaning Robot *Proceedings of the 2008 IEEE/ASME International Conference on Advanced Intelligent Mechatronics* China
[11]. Agrawal P and Bhattacharaya B 2013 Aquatic Multi-Robot System for Lake Cleaning *Nature-Inspired Mobile Robotics* **171**

Spectral Approach to Modeling Dependence in Multivariate Time Series

Hernando Ombao

Statistics Program, King Abdullah University of Science and Technology (KAUST), Saudi Arabia

E-mail: hernando.ombao@kaust.edu.sa

Abstract. Consider a multivariate time series such as prices of stocks from various sectors, amount of rainfall in many geographical locations, and brain signals from many different locations on the scalp. The goal of this paper is to present the spectral approach to modeling dependence between components of the multivariate time series. There are many measures of dependence - the most popular being cross-correlation or partial cross-correlation. This measure is easy to compute and easy to understand but it coarse in a sense that it is not able to identify the underlying frequencies that are responsible for driving the dependence. In the stock price example, two stocks may be highly correlated but it would be helpful to see if this correlation is driven by the daily fluctuations or by millisecond-level fluctuations. In the neuroscience example, when two brain regions exhibit a high level of correlation, it will be important if this synchronicity is due to low-frequency oscillations or high-frequency oscillations. Here we present an overview of the underlying principles through specific spectral models which decompose the signals into oscillations of various frequencies and then model lead-lag dependence via these oscillations.

1. Introduction

Consider a multivariate time time series $\mathbf{X}_t = [X_1(t), \ldots, X_D(t)]'$ which are time series recordings from D locations in space (i.e., D electroencephalogram (EEG) channels on the scalp or D sectors as in the stock price example). The primary question that we will address in this paper is to assess the nature in which activity in channel d_1 is associated with activity in channel d_2.

In more general settings, one asks the additional question of whether or not dependence between channels d_1 and d_2 are driven by the *other* channels. The goal is to present an approach to modeling dependence between components of this D-dimensional network through the oscillatory components of each univariate time series $X_d(t)$.

There are many approaches to modeling dependence between components of a multivariate time series. The most common is pairwise cross-correlation which measures

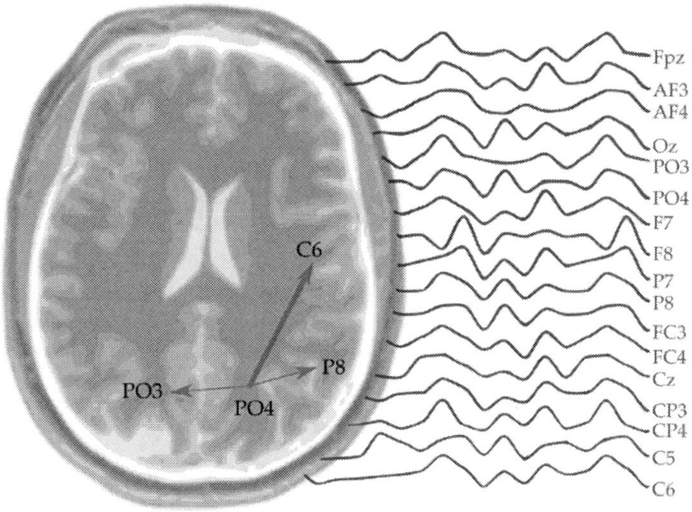

Figure 1. Modeling dependence in brain activity through the EEG channels.

the strength of linear dependence between two components. Another common measure is partial cross-correlation which gives a measure of strength of the direct linear link between two components (after having removed the linear effect of all the other components in the network). Here, we illustrate how one can dig deeper than the classical correlation (or covariance) in order to understand more complex dependence structure between components. For ease in exposition, we shall develop a model only for $D = 2$ keeping in mind that these ideas are generalizable for higher dimensions.

We now motivate our approach of decomposing observed time series in terms of various oscillations. Under weak stationarity, the Cramér representation of a bivariate time series $\mathbf{X}(t) = [X_1(t), X_2(t)]'$ is given by

$$\begin{pmatrix} X_1(t) \\ X_2(t) \end{pmatrix} = \int_{-0.5}^{+0.5} \exp(i2\pi\omega t) \begin{pmatrix} dZ_1(\omega) \\ dZ_2(\omega) \end{pmatrix}$$

which expresses the d-th time series $X_d(t)$ (for $d = 1, 2$) as a sum of infinitely many sinusoidal oscillations $\exp(i2\pi\omega t)dZ_p(\omega)$ where the frequency ω lies in $(-0.5, 0.5)$. In this representation, the basis consists of the Fourier waveforms $\exp(i2\pi\omega t)$ with random zero-mean coefficients $[dZ_1(\omega), dZ_2(\omega)]$. Under weak stationarity, the random coefficients are uncorrelated across different frequencies, i.e., $\mathbb{C}\text{orr}(dZ_{d_1}(\omega), dZ_{d_2}(\lambda)) = 0$ for $\omega \neq \lambda$.

One most common measure of dependence, called coherency, gives an indication of synchrony between $X_1(t)$ and $X_2(t)$ for each frequency oscillation ω, i.e.,

$$\rho_{12}(\omega) = \mathbb{C}\text{orr}(\exp(i2\pi\omega t)dZ_1(\omega), \exp(i2\pi\omega t)dZ_2(\omega)) = \mathbb{C}\text{orr}(dZ_1(\omega), dZ_2(\omega)).$$

Journal of Physics: Conference Series **1417** (2019) 012007 doi:10.1088/1742-6596/1417/1/012007

Here, coherency is complex-valued and thus its phase captures the lead-lag relationship between $X_(t)$ and $X_2(t)$. But its squared-modulus, called coherence, lies in the unit interval, i.e.,

$$\eta_{12}(\omega) := |\rho_{12}(\omega)|^2 \in [0, 1].$$

The above motivates an intuitive method in [1] for estimating the true value of coherence via linear filtering which we briefly describe below. Define $\{\psi_j^{\Omega}\}$ be absolutely summable filter coefficients with frequency response function with support mainly in the interval $\Omega \in (0, 0.5)$. In EEG analyses, we measure dependence over various disjoint intervals that correspond the standard bands which are delta $(0-4)$ Hertz, theta $(4-8)$ Hertz, alpha $(8-12)$ Hertz, beta $(12-30)$ Hertz and gamma $(30-50)$ Hertz.

Next, for observed time series $\{[X_1(t), X_2(t)]', \ t = 1, \ldots, T\}$, denote the *$\Omega$-filtered* time series to be, respectively,

$$X_{1,\Omega}(t) = \sum_j \psi_j^{\Omega} X_1(t-j) \text{ and } X_{2,\Omega}(t) = \sum_j \psi_j^{\Omega} X_2(t-j)$$

so that the filtered series $X_{1,\Omega}(t)$ and $X_{2,\Omega}(t)$ will each have spectra that is concentrated on the frequency band Ω. Without loss of generality, assume that the mean of the filtered time series are both zero, i.e.,

$$\frac{1}{T}\sum_{t=1}^{T} X_{d,\Omega}(t) = 0 \quad \text{for } d = 1, 2.$$

In [1], the estimate of coherence at frequency band Ω is obtained by shifting one of the filtered time series to obtain the maximum absolute cross-correlation. In the following sections, we develop the proposed spectral approach and then compare the various spectral-based approaches to modeling dependence in multivariate time series.

2. The spectral decomposition of multivariate time series

In this section, we now define the discretized Cramér representation of a weakly stationary time series. First, segment the frequency axis $(0, 0.5)$ into non-overlapping bands Ω_k, $k = 1, \ldots, K$ that completely cover $(0, 0.5)$. That is,

(a.) $\Omega_k \cap \Omega_{k'} = \varnothing$ for $k \neq k'$

(b.) $\cup_k \Omega_k = (0, 0.5)$.

Recall that the notation for Ω_k-filtered version of time series $X_d(t)$ to be $X_{d,\Omega}(t)$. Then, the discretized Cramér representation of $[X_1(t), X_2(t)]'$ is given by

$$\begin{pmatrix} X_1(t) \\ X_2(t) \end{pmatrix} \approx \begin{pmatrix} \sum_k X_{1,\Omega_k}(t) \\ \sum_k X_{2,\Omega_k}(t). \end{pmatrix} \tag{1}$$

Here, each $X_d(t), d = 1, 2$ is decomposed into different oscillations (at specific frequency bands). As noted, under weak stationarity, the oscillations at different frequency bands are uncorrelated, i.e., for any d_1 and d_2, any t and s, and $k \neq k'$,

$$\mathrm{Cov}(X_{d_1,\Omega_k}(t), X_{d_2,\Omega_{k'}}(s)) = 0. \tag{2}$$

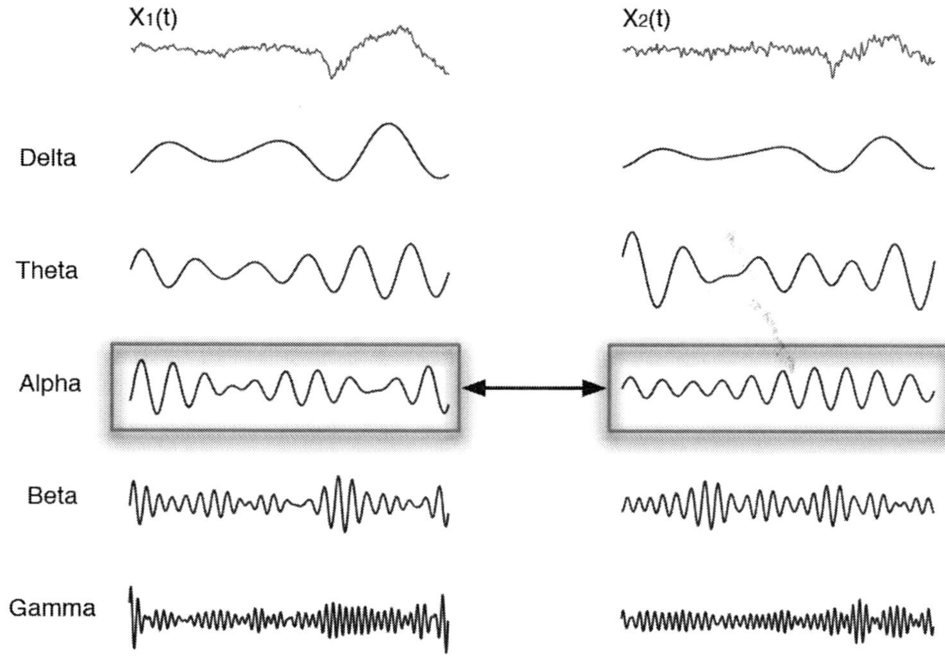

Figure 2. Each time series $X_1(t)$ and $X_2(t)$ are decomposed into oscillations at various frequency bands: delta (0-4) Hertz, theta (4-8) Hertz, alpha (8-12) Hertz, beta (12-30) Hertz, gamma (30-50) Hertz.

See Figure 2 where each time series $X_1(t)$ and $X_2(t)$ are decomposed into oscillations at various frequency bands. We now illustrate that covariance, as a measure of dependence, can be decomposed into the covariance between the different frequency components. From Equations 1 and 2,

$$\mathbb{C}\text{ov}(X_1(t), X_2(s)) \approx \sum_k \mathbb{C}\text{ov}(X_{1,\Omega_k}(t), X_{2,\Omega_k}(s)).$$

Remarks. From the decomposition of covariance above, one can identify the frequency band Ω_{k^*} that drives the linear association between $X_1(t)$ and $X_2(t)$.

3. Different measures of spectral dependence

As already noted, the most common measures of frequency band-specific synchrony is coherence. In practice, we estimate coherence between $X_1(t)$ and $X_2(t)$ at each of the frequency bands by computing the squared sample cross-correlations of each of the filtered components (see [1]). Other approaches to estimating coherence involves computing the Fourier transform and the Fourier periodogram matrix and then smoothing the auto- and cross-periodograms (see [2], [3], [4], [?]).

For non-stationary time series, Loéve [5] developed the harmonizable process representation that allowed oscillations at different frequencies to be correlated. Thus,

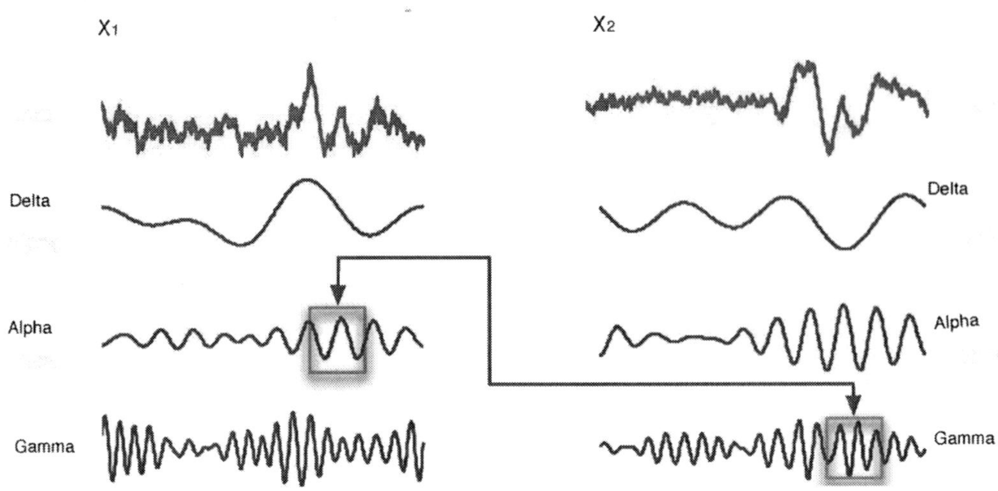

Figure 3. Each time series $X_1(t)$ and $X_2(t)$ are decomposed into oscillations at various frequency bands. Dual-frequency between the alpha and gamma bands that is evolving across time [10].

under this process, the dual-frequency (band) coherence is defined as

$$\rho_{12}(\Omega_1, \Omega_2) = |\,\mathbb{C}\mathrm{orr}(X_{1,\Omega_1}(t), X_{2,\Omega_2}(t))|^2.$$

For further details on these processes, see [6], [7], [8], [9].

Under the harmonizable process framework, dual-frequency coherence captures the *overall* dependence across the entire observation period. This is a limitation since this dependence could vary across time (see Figure 3). Thus, in [10], the evolutionary dual-frequency coherence concept was developed within the context of the discretized harmonizable process.

All the aforementioned coherence measures do not directly measure lead-lag dependence. Here, we propose a spectral cross-oscillatory model below to examine how the oscillatory activity at the Ω_1 band for the first time series may depend on the oscillatory activity at all bands $\Omega_1, \ldots, \Omega_K$ for the second time series:

$$X_{1,\Omega_1}(t) = \sum_{\ell=1}^{L} \sum_{k=1}^{K} X_{2,\Omega_k}(t-\ell) + epsilon_{1,\Omega_1}(t) \tag{3}$$

where $epsilon_{1,\Omega_1}(t)$ is white noise with mean 0 and variance σ^2. The optimal lag L can be selected using information criteria such as the Akaike information criterion or the Bayesian Schwartz information criterion.

4. Modeling dependence via latent sources

One way to account for dependence between time series $X_1(t)$ and $X_2(t)$ is to model these as functions of latent processes. Define the latent processes as follows: $Z_\delta(t)$ be a latent process with power on the delta band; $Z_\theta(t)$ as the theta process; $Z_\alpha(t)$ as the alpha process; $Z_{beta}(t)$ as the beta process and $Z_\gamma(t)$ as the gamma process. Each of these processes can be approximated by second-order autoregressive processes. In [11], the observed time series $X_1(t)$ and $X_2(t)$ are considered to be linear mixtures of these latent processes:

$$\begin{pmatrix} X_1(t) \\ X_2(t) \end{pmatrix} = \begin{pmatrix} q_{1\delta} & q_{1\theta} & q_{1alpha} & q_{1beta} & q_{1\gamma} \\ q_{2\delta} & q_{2\theta} & q_{2alpha} & q_{2beta} & q_{2\gamma} \end{pmatrix} \begin{pmatrix} Z_\delta(t) \\ Z_\theta(t) \\ Z_\alpha(t) \\ Z_{beta}(t) \\ Z_\gamma(t) \end{pmatrix}.$$

Thus, $X_1(t)$ and $X_2(t)$ are coherent at, say the alpha band, if both $q_{1\alpha} \neq 0$ and $q_{2\alpha} \neq 0$, i.e., both time series contain the alpha latent process.

In the representation above, the latent processes are assumed to be independent. However, for some time series, it is possible for some latent processes to be dependent. Moreover, the relationship could be non-linear. In [12], general models are being studied where the theta and gamma latent processes are coupled in some non-linear way

$$Z_\gamma(t) = A(Z_\theta(t - \ell|)Z_\gamma(t - \ell).$$

Here, the amplitude of the gamma latent process at time t may depend on the *phase* of the theta process at time $t - \ell$ via the function $A(.)$ and in the amplitude of the past gamma process value.

5. Summary

In this paper, we demonstrate different approaches to modeling dependence through oscillatory processes via the Cramér and Loéve representations and also via the mixtures (possibly non-linear) of oscillatory latent processes. Through these representations, we formally define various measures of spectral dependence. One advantage of these measures is that they are easily interpretable and that the estimation can be readily implemented via the fast Fourier transform and linear filtering. These measures are demonstrated to be highly informative in analyzing brain signals.

References

[1] Ombao H and Van Bellegem S. 2008. Coherence Analysis: A Linear Filtering Point Of View *IEEE Transactions on Signal Processing* vol. **56(6)** pp 2259-2266.
[2] Shumway R and Stoffer D. 2017. Time Series Analysis and Its Applications With R Examples. *Springer: New York.*
[3] Ombao H, Raz J, von Sachs R and Malow B. 2001. Automatic Statistical Analysis of Bivariate Non-Stationary Time Series, *Journal of the American Statistical Association*, **96**, pp 543-560.
[4] Ombao H, von Sachs R and Guo W. 2005. SLEX Analysis of Multivariate Non- Stationary Time Series, *Journal of the American Statistical Association*, **100,** pp 519-531.
[5] Loéve M. 1955. Probability Theory. *The University series in higher Mathematics.*

[6] Napolitano A. 2012. Generalizations of Cyclostationary Signal Processing: Spectral Analysis and Applications. *John Wiley & Sons, Ltd. IEEE Press.*

[7] Dudek A. 2016. First and second order analysis for periodic random arrays using block bootstrap methods. *Electronic Journal of Statistics*, **10(2)**, pp 2561-2583.

[8] Lii K and Rosenblatt M. 2002. Spectral analysis for harmonizable processes. *Annals of Statistics*, **30**, pp 258-297.

[9] Olhede S and Ombao H. 2013. Covariance of Replicated Modulated Cyclical Time Series. *IEEE Transactions on Signal Processing*, **61**, pp 1944-1957.

[10] Gorrostieta C, Ombao H and von Sachs R. 2018. Time-Dependent Dual Frequency Coherence in Multivariate Non-Stationary Time Series. *Journal of Time Series Analysis*, **40**, 3-22.

[11] Gao X, Shahbaba B, Fortin N and Ombao H. 2019. Evolutionary State-Space Models With Applications to Time-Frequency Analysis of Local Field Potentials. *Statistica Sinica,* Accepted for publication.

[12] Kazimierska K, Dudek A and Ombao H. 2019. Nonlinear coupling between oscillatory processes. *Working paper.*

[13] Cho Y and Bridgeman B 2012 Relationship of TOEFL iBT Scores to Academic Performance : Some Evidence from American Universities *Language Testing* vol. **29 (3)** pp 421-442

[14] Berger D E 2017 Introduction to Binary Logistic Regression and Propensity Score Analysis *Working paper*

Hypothesis testing of Geographically weighted bivariate logistic regression

M Fathurahman[1,2], Purhadi[3], Sutikno[4], and V Ratnasari[5]

[1] Department of Statistics, Mulawarman University, Samarinda, Indonesia
[2] Doctoral Student at the Department of Statistics, Institut Teknologi Sepuluh Nopember, Surabaya, Indonesia
[3,4,5] Department of Statistics, Institut Teknologi Sepuluh Nopember, Surabaya, Indonesia

e-mail: fathur@fmipa.unmul.ac.id

Abstract. In this study, the hypothesis testing of geographically weighted bivariate logistic regression (GWBLR) procedure is proposed. The GWBLR model is a bivariate logistic regression (BLR) model which all of the regression parameters depend on the geographical location in the study area. The geographical location is expressed as a point coordinate in two-dimensional geographic space (longitude and latitude). The response variables of BLR model are constructed from a (2 □ 2) contingency table and it follows the multinomial distribution. The purpose of this study is to test the GWBLR model parameters. There are three hypothesis tests. The first is a parameters similarity test using the Vuong test method. The test is to obtain a significant difference between GWBLR and BLR. The second is a simultaneous test using the likelihood ratio test method. The simultaneous test is to obtain the simultaneous significance of the regression parameters. The last is a partial test using Wald test method. The result showed that the Vuong statistic and Wald statistic have an asymptotic standard normal distribution, whereas the likelihood ratio statistic has an asymptotic chi-squared distribution.

1. Introduction

Geographically weighted regression (GWR) is an effective technique for modelling spatial non-stationary data [1, 2]. Hypothesis testing is one of the statistical inference tools that have an important contributed in GWR modelling and it was developed [3, 4, 5]. The response variable of the GWR model in previous studies is quantitative data and normally distributed. However, in fact, not only the response variable is quantitative data and normally distributed, but also is qualitative (categorical) data and follows other distributions, such as Bernoulli, binomial, and multinomial.

Recently, the GWR models with response variable are categorical data have been proposed. GWR and a logistic regression model are combined to form a geographically weighted logistic regression (GWLR) [6]. The proposed model has two categories of the response variable which follows Bernoulli distribution and it was applied to model the spatial variation in the relation between erosion (presence or absence) and several controlling variables for the Afon Dyfi in West Wales. Furthermore, a geographically weighted logistic model (GWLM) was employed and it has shown that the GWLM is an efficient tool to account for spatial heterogeneity, showing better fit and residual properties compared to the logistic regression model (LRM) and logistic mixed model (LMM) in modeling the occurrence of cloud cover, using the spatial data derived from satellite imagery and GIS [7]. On the other hand, the GWLM and global logistic model (GLM) were used to model and analyze the spatial variation in the human factors associated with forest fires [8]. GWLM techniques have shown a high predictive potential for human-caused wildfire occurrence modelling, surpassing classical regression techniques like GLM and allowing the detection of non-stationary relationships between response and independent variables. However, the number of categories in the response variable of the GWR model for categorical data is also more than two categories. There are several studies have been developed.

The geographically weighted multinomial logistic regression (GWMLR) model was introduced and it has the number of category of the response variable is more than two categories [9, 10]. In the GWMLR model, each category is unordered and follows the multinomial distribution. As a follow-up, previous studies, the geographically weighted ordinal logistic regression (GWOLR) and geographically weighted ordinal logistic regression semiparametric (GWOLRS) model were proposed [11, 12]. Like the GWMLR model, the response variable of GWOLR and GWOLRS model follows the multinomial distribution and has more than two categories, but each category is ordered. In the GWOLRS model, some of the independent variables are global and the other variables are local [12].

The previous studies extend the GWR models for categorical data that have only one response variable (univariate). However, in many fields of research, several cases have two response variables (bivariate). Therefore, a new spatial regression model is proposed in this study, namely the geographically weighted bivariate logistic regression (GWBLR). The GWBLR model is to focus on two variable responses which each variable has two categories. This study aims to test the GWBLR model parameters. There are three kinds of hypothesis tests. The first test is a parameters similarity test by using the Vuong test method. This test is used to obtain if there is a significant difference between GWBLR model and BLR model. The second test is a simultaneous test by using a likelihood ratio test (LRT) method. This test aimed to obtain the simultaneous significance of the regression parameters. The last test is a partial test by using the Wald test method. This test is used to obtain the significance of each parameter in the regression model.

2. Bivariate logistic regression

The bivariate logistic regression (BLR) is an extension of the univariate logistic regression when there are two categorical data of response variables and they are correlated to each other. In this study, each of the response variables has two categories. Let Y_1 and Y_2 are response variables of the BLR model. The response variables can be shown in Table 1 and the joint probability of the response variables in Table 1 is presented in Table 2.

Table 1. The (2×2) contingency table of the response variables.

Y_1	Y_2		Total
	$Y_2 = 1$	$Y_2 = 0$	
$Y_1 = 1$	Y_{11}	Y_{10}	Y_{1+}
$Y_1 = 0$	Y_{01}	Y_{00}	$Y_{0+} = n - Y_{1+}$
Total	Y_{+1}	$Y_{+0} = n - Y_{+1}$	$Y_{++} = n$

Table 2. The joint probability of the response variables.

Y_1	Y_2		Total
	$Y_2 = 1$	$Y_2 = 0$	
$Y_1 = 1$	p_{11}	p_{10}	p_{1+}
$Y_1 = 0$	p_{01}	p_{00}	$p_{0+} = 1 - p_{1+}$
Total	p_{+1}	$p_{+0} = 1 - p_{+1}$	$p_{++} = 1$

Based on Table 1 and Table 2, the random variables Y_{11}, Y_{10}, Y_{01}, and Y_{00} are follows the multinomial distribution with joint probability function defined by

$$P(Y_{11} = y_{11}, Y_{10} = y_{10}, Y_{01} = y_{01}) = \prod_{g=0}^{1}\prod_{h=0}^{1}\frac{p_{gh}^{y_{gh}}}{y_{gh}!}, 0 < p_{gh} < 1,$$

where $g, h = 0,1$; $y_{gh} = 0,1$; $y_{00} = 1 - y_{11} - y_{10} - y_{01}$; and $p_{00} = 1 - p_{11} - p_{10} - p_{01}$.

Furthermore, the BLR model can be written as follows

$$g_1(x) = logit\ (p_1(x)) = \beta_1^T x,$$
$$g_2(x) = logit\ (p_2(x)) = \beta_2^T x,$$
$$g_3(x) = ln\ ln\ \psi_1(x) \ = \beta_3^T x,$$

where $x = [1\ x_1\ x_2 \cdots x_k]^T$ is a vector of the covariate, $\beta_1^T = [\beta_{01}\ \beta_{11}\ \beta_{21} \cdots \beta_{k1}]$, $\beta_2^T = [\beta_{02}\ \beta_{12}\ \beta_{22} \cdots \beta_{k2}]$, and $\beta_3^T = [\beta_{03}\ \beta_{13}\ \beta_{23} \cdots \beta_{k3}]$ are the parameters, $p_1(x)$ is the marginal probability function of Y_1, $p_2(x)$ is the marginal probability function of Y_2, and $\psi_1(x)$ is the odds ratio that is shown an association between Y_1 and Y_2. The marginal probability function and the odds ratio are defined by

$$p_1(x) = p_{1+}(x) = \frac{e^{\beta_1^T x}}{1+e^{\beta_1^T x}},$$
$$p_2(x) = p_{+1}(x) = \frac{e^{\beta_2^T x}}{1+e^{\beta_2^T x}},$$
$$\psi_1(x) = \frac{p_{11}(x)p_{00}(x)}{p_{10}(x)p_{01}(x)}.$$

According to [13], the joint probability of $p_{11}(x)$ in Equation can be obtained as follows

$$p_{11}(x) = \{\frac{a_1 - \sqrt{a_1^2 + b_1}}{2(\psi_1(x)-1)}, \psi_1(x) \neq 1\ \ p_1(x)p_2(x), \psi_1(x) = 1$$

where $a_1 = 1 + (\psi_1(x) - 1)(p_1(x) + p_2(x))$, $b_1 = -4\psi_1(x)(\psi_1(x) - 1)p_1(x)p_2(x)$ with $p_1(x)$ and $p_2(x)$ in Equations and. Furthermore, based on Table 2 and Equation, the joint probabilities of $p_{10}(x)$, $p_{01}(x)$, and $p_{00}(x)$ are obtained as follows

$$p_{10}(x) = p_1(x) - p_{11}(x),$$
$$p_{01}(x) = p_2(x) - p_{11}(x),$$
$$p_{00}(x) = 1 - p_{11}(x) - p_{10}(x) - p_{01}(x)$$
$$= 1 - p_1(x) - p_2(x) + p_{11}(x).$$

3. Geographically weighted bivariate logistic regression

The geographically weighted bivariate logistic regression (GWBLR) is developed of the BLR model which all of the regression parameters depend on the geographical location in the study area. The geographical location is expressed as a point coordinate in two-dimensional geographic space (longitude and latitude) [14].

Let $u_i = (u_{1i}, u_{2i})$ denotes a vector of point coordinate for i^{th} location where the data is observed for $i = 1,2,\ldots,n$, with u_{1i} is latitude and u_{2i} is longitude, then from BLR model in Equations -, it can be developed to the new BLR model with all of the parameters depend on the geographical location which is called GWBLR model. Here, the parameters are assumed to be functions of the location on which the observations are obtained. Suppose, all of the parameters of the BLR model in Equations - depend on the geographical location, then GWBLR model at i^{th} location with coordinate u_i has an expression as follows

$$h_1(x_i) = logit\ (\pi_1(x_i)) = \beta_1^T(u_i)x_i, i = 1,2,\ldots,n$$
$$h_2(x_i) = logit\ (\pi_2(x_i)) = \beta_2^T(u_i)x_i,$$
$$h_3(x_i) = ln\ ln\ \psi_2(x_i) \ = \beta_3^T(u_i)x_i,$$

where

$x_i = [1\ x_{1i}\ x_{2i} \cdots x_{ki}]^T$ is a vector of independent variables at i^{th} location,
$\beta_1^T(u_i) = [\beta_{01}(u_i)\ \beta_{11}(u_i)\ \beta_{21}(u_i) \cdots \beta_{k1}(u_i)]$, $\beta_2^T(u_i) = [\beta_{02}(u_i)\ \beta_{12}(u_i)\ \beta_{22}(u_i) \cdots \beta_{k2}(u_i)]$, and $\beta_3^T(u_i) = [\beta_{03}(u_i)\ \beta_{13}(u_i)\ \beta_{23}(u_i) \cdots \beta_{k3}(u_i)]$ are the parameters at i^{th} location,
$\pi_1(x_i)$ is the marginal probability function of Y_1 at i^{th} location, $\pi_2(x_i)$ is the marginal probability function of Y_2 at i^{th} location, and $\psi_2(x_i)$ is the odds ratio that is shown an association between Y_1 and

Y_2 at i^{th} location. The marginal probability function of response variables and the odds ratio at i^{th} location is defined by

$$\pi_1(x_i) = \pi_{1+}(x_i) = \frac{expexp\,(\beta_1^T(u_i)x_i)}{1+expexp\,(\beta_1^T(u_i)x_i)},$$
$$\pi_2(x_i) = \pi_{+1}(x_i) = \frac{expexp\,(\beta_2^T(u_i)x_i)}{1+expexp\,(\beta_2^T(u_i)x_i)},$$
$$\psi_2(x_i) = \frac{\pi_{11}(x_i)\pi_{00}(x_i)}{\pi_{10}(x_i)\pi_{01}(x_i)}.$$

Based on the joint probability function in Equation, the joint probability of $\pi_{11}(x_i)$ in Equation can be determined by

$$\pi_{11}(x_i) = \{\frac{a_2-\sqrt{a_2^2+b_2}}{2(\psi_2(x_i)-1)}, \psi_2(x_i) \neq 1 \quad \pi_1(x_i)\pi_2(x_i), \psi_2(x_i) = 1$$

where $a_2 = 1 + (\psi_2(x_i) - 1)(\pi_1(x_i) + \pi_2(x_i))$, $b_2 = -4\psi_2(x_i)(\psi_2(x_i) - 1)\pi_1(x_i)\pi_2(x_i)$ with $\pi_1(x_i)$, $\pi_2(x_i)$, and $\psi_2(x_i)$ in Equations -. The joint probabilities of $\pi_{10}(x_i)$, $\pi_{01}(x_i)$, and $\pi_{00}(x_i)$ are obtained as follows

$$\pi_{10}(x_i) = \pi_1(x_i) - \pi_{11}(x_i),$$
$$\pi_{01}(x_i) = \pi_2(x_i) - \pi_{11}(x_i),$$
$$\pi_{00}(x_i) = 1 - \pi_1(x_i) - \pi_2(x_i) + \pi_{11}(x_i).$$

4. Hypothesis testing of the GWBLR model parameters

Hypothesis testing consists of three kinds of tests. The first test is a parameter similarity test between GWBLR and BLR. The second test is a simultaneous test of GWBLR parameters. The last test is a partial test of GWBLR parameters.

4.1. Parameter similarity test between GWBLR and BLR

The aim of this test is used to obtain a significant difference between GWBLR model and BLR model. Parameter similarity test is also called the goodness of fit test [15]. The null hypothesis (H_0) and the alternative hypothesis (H_1) are as follows

$$H_0: \beta_{rs}(u_i) = \beta_{rs}, \quad r = 1, 2, \dots, k; s = 1, 2, 3; \; i = 1, 2, \dots, n,$$
H_1: at least one of $\beta_{rs}(u_i) \neq \beta_{rs}$.

where r is the index of the covariate, k is the number of covariates, s is the index of parameters, i is the index of research sample (the study area), and n is the number of the study area.

The test statistic for hypotheses in Equation can be determined by using the Vuong test method. The part of this method can be determined by using the likelihood ratio test (LRT) procedure [15]. Suppose the parameters set under H_0 is $\omega = \{\beta_{rs}, r = 0, 1, 2, \dots, k; s = 1, 2, 3\}$. Thus, to determine the likelihood function under H_0

$$L(\omega) = \prod_{i=1}^{n}\left[p_{11i}^{y_{11i}}(x_i)p_{10i}^{y_{10i}}(x_i)p_{01i}^{y_{01i}}(x_i)p_{00i}^{y_{00i}}(x_i)\right].$$

Let $p_{ghi}(x_i) = p_{ghi}$, for $g, h = 0{,}1$ and $i = 1, 2, \dots, n$. The likelihood function in Equation can be rewritten as

$$L(\omega) = \prod_{i=1}^{n}\left[p_{11i}^{y_{11i}}p_{10i}^{y_{10i}}p_{01i}^{y_{01i}}p_{00i}^{y_{00i}}\right].$$

Based on Equation to determine maximum log-likelihood function under H_0

$$ln\,ln\,L(\hat{\omega}) = \sum_{i=1}^{n}(y_{11i}\,ln\,ln\,\hat{p}_{11i} + y_{10i}\,ln\,ln\,\hat{p}_{10i} + y_{01i}\,ln\,ln\,\hat{p}_{01i} + y_{00i}\,ln\,ln\,\hat{p}_{00i}),$$

where $\hat{p}_{11i}$, $\hat{p}_{10i}$, $\hat{p}_{01i}$, and $\hat{p}_{00i}$ for $i = 1, 2, \dots, n$ are formulated as follows

$$\hat{p}_{11i} = \{\frac{a_3 - \sqrt{a_3^2 + b_3}}{2(\hat{\psi}_3 - 1)}, \hat{\psi}_3 \neq 1 \quad \hat{p}_{1i}\hat{p}_{2i}, \hat{\psi}_3 = 1$$

where $a_3 = 1 + (\hat{\psi}_3 - 1)(\hat{p}_{1i} + \hat{p}_{2i})$, $b_3 = -4\hat{\psi}_3(\hat{\psi}_3 - 1)\hat{p}_{1i}\hat{p}_{2i}$, $\hat{p}_{1i} = \frac{e^{\hat{\beta}_1^T x_i}}{(1+e^{\hat{\beta}_1^T x_i})}$),

$\hat{p}_{2i} = \frac{e^{\hat{\beta}_2^T x_i}}{(1+e^{\hat{\beta}_2^T x_i})}$), and $\hat{\psi}_3 = \frac{\hat{p}_{11i}\hat{p}_{00i}}{\hat{p}_{10i}\hat{p}_{01i}}$, with $\hat{\beta}_1^T = [1\; \hat{\beta}_{01}\; \hat{\beta}_{11}\; \hat{\beta}_{21} \cdots \hat{\beta}_{k1}]$

and $\hat{\beta}_1^T = [1\ \hat{\beta}_{01}\ \hat{\beta}_{11}\ \hat{\beta}_{21} \cdots \hat{\beta}_{k1}]$. Thus, $\hat{\beta}_1^T$ and $\hat{\beta}_2^T$ are maximum likelihood estimators (MLEs) for β_1 and β_2.

$$\hat{p}_{10i} = \hat{p}_{1i} - \hat{p}_{11i}.$$
$$\hat{p}_{01i} = \hat{p}_{2i} - \hat{p}_{11i}.$$
$$\hat{p}_{00i} = 1 - \hat{p}_{1i} - \hat{p}_{2i} + \hat{p}_{11i}.$$

Furthermore, to obtain parameters set under population is

$$\Omega = \{\beta_{rs}(u_i), r = 0,1,2,\dots,k; s = 1,2,3; i = 1,2,\dots,n\}.$$

Thus, the likelihood function and maximum log-likelihood function under population are given by

$$L(\Omega) = \prod_{i=1}^{n}\left[\pi_{11i}^{y_{11i}}\pi_{10i}^{y_{10i}}\pi_{01i}^{y_{01i}}\pi_{00i}^{y_{00i}}\right],$$
$$ln\ ln\ L(\hat{\Omega}) = \sum_{i=1}^{n}(y_{11i}\ ln\ ln\ \hat{p}_{11i} + y_{10i}\ ln\ ln\ \hat{p}_{10i} + y_{01i}\ ln\ ln\ \hat{p}_{01i} + y_{00i}\ ln\ ln\ \hat{p}_{00i}),$$

where $\hat{\pi}_{11i}$, $\hat{\pi}_{10i}$, $\hat{\pi}_{01i}$, and $\hat{\pi}_{00i}$ for $i = 1,2,\dots,n$ are formulated as follows

$$\hat{\pi}_{11i} = \{\frac{a_4 - \sqrt{a_4^2 + b_4}}{2(\hat{\psi}_4 - 1)}, \hat{\psi}_4 \neq 1\quad \hat{\pi}_{1i}\hat{\pi}_{2i}, \hat{\psi}_4 = 1$$

where $a_4 = 1 + (\hat{\psi}_4 - 1)(\hat{\pi}_{1i} + \hat{\pi}_{2i})$, $b_4 = -4\hat{\psi}_4(\hat{\psi}_4 - 1)\hat{\pi}_{1i}\hat{\pi}_{2i}$, $\hat{\pi}_{1i} = \frac{e^{\hat{\beta}_1^T(u_i)x_i}}{[1+e^{\hat{\beta}_1^T(u_i)x_i}]}$, $\hat{\pi}_{2i} = \frac{e^{\hat{\beta}_2^T(u_i)x_i}}{[1+e^{\hat{\beta}_2^T(u_i)x_i}]}$, $\hat{\beta}_1^T(u_i) = [1\ \hat{\beta}_{01}(u_i)\ \hat{\beta}_{11}(u_i)\ \hat{\beta}_{21}(u_i) \cdots \hat{\beta}_{k1}(u_i)]$, $\hat{\beta}_2^T(u_i) = [1\ \hat{\beta}_{02}(u_i)\ \hat{\beta}_{12}(u_i)\ \hat{\beta}_{22}(u_i) \cdots \hat{\beta}_{k2}(u_i)]$, and $\hat{\psi}_4 = \frac{\hat{\pi}_{11i}\hat{\pi}_{00i}}{\hat{\pi}_{10i}\hat{\pi}_{01i}}$ with $\hat{\beta}_1^T(u_i)$ and $\hat{\beta}_2^T(u_i)$ are maximum likelihood estimators (MLEs) for $\beta_1(u_i)$ and $\beta_2(u_i)$.

$$\hat{\pi}_{10i} = \hat{\pi}_{1i} - \hat{\pi}_{11i}.$$
$$\hat{\pi}_{01i} = \hat{\pi}_{2i} - \hat{\pi}_{11i}.$$
$$\hat{\pi}_{00i} = 1 - \hat{\pi}_{1i} - \hat{\pi}_{2i} + \hat{\pi}_{11i}.$$

The test statistic for testing hypotheses in Equation can be obtained by

$$V = \frac{\sqrt{n}\left(\frac{1}{n}\sum_{i=1}^{n} m_i\right)}{\sqrt{\sqrt{n}\left(\frac{1}{n}\sum_{i=1}^{n}(m_i - \underline{m})^2\right)}}$$

where $m_i = ln\ ln\ L(\hat{\Omega}) - ln\ ln\ L(\hat{\omega})$, $\underline{m} = \frac{1}{n}\sum_{i=1}^{n} m_i$, and n is sample size. The $ln\ ln\ L(\hat{\omega})$ and $ln\ ln\ L(\hat{\Omega})$ are obtained in Equations and.

The Vuong statistic in Equation was asymptotically standard normal distributed. Therefore, reject H_0 if $|V| > Z_{1-\alpha/2}$, where the critical value $Z_{1-\alpha/2}$ is $1 - \alpha/2$ quantile from a standard normal distribution.

4.2. Simultaneous test

This test is used to determine the simultaneous significance of the regression parameters. Consider the hypotheses

$$H_0: \beta_{1s}(u_i) = \beta_{2s}(u_i) = \cdots = \beta_{ks}(u_i) = 0, s = 1,2,3;\ i = 1,2,\dots,n$$
$$H_1: \text{at least one of the } \beta_{rs}(u_i) \neq 0, r = 1,2,\dots,k.$$

The test statistic for hypotheses in Equation can be determined by using the LRT method [15]. Suppose the parameters set under H_0 is $\omega^* = \{\beta_{01}(u_i), \beta_{02}(u_i), \beta_{03}(u_i), i = 1,2,\dots,n\}$. Thus, to determine the likelihood function under H_0

$$L(\omega^*) = \prod_{i=1}^{n}\left[\left(\pi_{11i}^*(x_i)\right)^{y_{11i}}\left(\pi_{10i}^*(x_i)\right)^{y_{10i}}\left(\pi_{01i}^*(x_i)\right)^{y_{01i}}\left(\pi_{00i}^*(x_i)\right)^{y_{00i}}\right].$$

Let $\pi_{ghi}^*(x_i) = \pi_{ghi}^*$, for $g,h = 0,1$ and $i = 1,2,\dots,n$. The likelihood function in Equation can be rewritten as

$$L(\omega^*) = \prod_{i=1}^{n}[(\pi_{11i}^*)^{y_{11i}}(\pi_{10i}^*)^{y_{10i}}(\pi_{01i}^*)^{y_{01i}}(\pi_{00i}^*)^{y_{00i}}].$$

Based on Equation to determine maximum log-likelihood function under H_0

$$ln\ ln\ L(\hat{\omega}^*) = \sum_{i=1}^{n}(y_{11i}\ ln\ ln\ \hat{\pi}_{11i}^* + y_{10i}\ ln\ ln\ \hat{\pi}_{11i}^* + y_{01i}\ ln\ ln\ \hat{\pi}_{11i}^* + y_{00i}\ ln\ ln\ \hat{\pi}_{11i}^*),$$

where $\hat{\pi}_{11i}^*$, $\hat{\pi}_{10i}^*$, $\hat{\pi}_{01i}^*$, and $\hat{\pi}_{00i}^*$ for $i = 1,2,\dots,n$ are formulated as follows

$$\hat{\pi}_{11i}^{*} = \{\frac{a_5 - \sqrt{a_5^2 + b_5}}{2(\hat{\psi}_5 - 1)}, \hat{\psi}_5 \neq 1 \ \ \hat{\pi}_{1i}^{*}\hat{\pi}_{2i}^{*}, \hat{\psi}_5 = 1$$

where $a_5 = 1 + (\hat{\psi}_5 - 1)(\hat{\pi}_{1i}^{*} + \hat{\pi}_{2i}^{*})$, $b_5 = -4\hat{\psi}_5(\hat{\psi}_5 - 1)\hat{\pi}_{1i}^{*}\hat{\pi}_{2i}^{*}$, $\hat{\pi}_{1i}^{*} = \frac{e^{\hat{\beta}_{01}(u_i)x_i}}{[1+e^{\hat{\beta}_{01}(u_i)x_i}]}$, $\hat{\pi}_{2i}^{*} = \frac{e^{\hat{\beta}_{02}(u_i)x_i}}{[1+e^{\hat{\beta}_{02}(u_i)x_i}]}$, and $\hat{\psi}_5 = \frac{\hat{\pi}_{11i}^{*}\hat{\pi}_{00i}^{*}}{\hat{\pi}_{10i}^{*}\hat{\pi}_{01i}^{*}}$ with $\hat{\beta}_{01}(u_i)$ and $\hat{\beta}_{02}(u_i)$ are maximum likelihood estimators (MLEs) for $\beta_{01}(u_i)$ and $\beta_{02}(u_i)$.

$$\hat{\pi}_{10i}^{*} = \hat{\pi}_{1i}^{*} - \hat{\pi}_{11i}^{*}.$$
$$\hat{\pi}_{01i}^{*} = \hat{\pi}_{2i}^{*} - \hat{\pi}_{11i}^{*}.$$
$$\hat{\pi}_{00i}^{*} = 1 - \hat{\pi}_{1i}^{*} - \hat{\pi}_{2i}^{*} + \hat{\pi}_{11i}^{*}.$$

Furthermore, to obtain parameters set under population. The parameters set underpopulation are similar to the parameter similarity test in Section 4.1. Therefore, the likelihood function and maximum log-likelihood function underpopulation are obtained by Equations and.

The test statistic for hypotheses in Equation is given by

$$G^2 = -2[ln\ ln\ L(\hat{\omega}^*)\ - ln\ ln\ L(\hat{\Omega})\ ],$$

where $L(\hat{\Omega})$ and $L(\hat{\omega}^*)$ are obtained in Equations and the test statistic in Equation can be rewritten as follows

$$G^2 = 2[\sum_{i=1}^{n}(y_{11i}\ ln\ ln\ \hat{\pi}_{11i}\ + y_{10i}\ ln\ ln\ \hat{\pi}_{10i}\ + y_{01i}\ ln\ ln\ \hat{\pi}_{01i}\ + y_{00i}\ ln\ ln\ \hat{\pi}_{00i}\ ) + \\ - \sum_{i=1}^{n}(y_{11i}\ ln\ ln\ \hat{\pi}_{11i}^{*}\ + y_{10i}\ ln\ ln\ \hat{\pi}_{10i}^{*}\ + y_{01i}\ ln\ ln\ \hat{\pi}_{01i}^{*}\ + y_{00i}\ ln\ ln\ \hat{\pi}_{00i}^{*}\ )].$$

The likelihood ratio (LR) statistic is also referred to as the *Wilk's lambda* statistic [16]. The LR statistic in Equation has an asymptotic chi-squared distribution with v degree of freedom, where v is the difference between the effective number of parameters in GWBLR model without independent variables (the reduced model) and GWBLR model with independent variables (the full model). Therefore, reject H_0 when $G^2 > \chi^2_{(v,1-\alpha)}$, where $\chi^2_{(v,1-\alpha)}$ is the $1-\alpha$ quantile from a chi-square distribution (χ^2) with v degree of freedom.

4.3. *Partial test*

The last test on hypothesis testing of the GWBLR model parameters is a partial test. This test is used to determine the significance of each parameter in the regression model. The form of hypotheses can be expressed as

$$H_0: \beta_{rs}(u_i) = 0, r = 1, 2, \ldots, k; s = 1, 2, 3;\ i = 1, 2, \ldots, n,$$
$$H_1: \beta_{rs}(u_i) \neq 0.$$

The test statistic for hypotheses in Equation is done by using the Wald test method. The basic idea of this method is following properties of the MLEs, particularly asymptotically normally distributed. Therefore, the Wald statistic is formulated by

$$W = \frac{\hat{\beta}_{rs}(u_i)}{\sqrt{\widehat{Var}(\hat{\beta}_{rs}(u_i))}} \sim N(0,1),$$

where $\widehat{Var}(\hat{\beta}_{rs}(u_i))$ is the diagonal elements of the matrix $[I(\beta(u_i))]^{-1}$ and $I(\beta(u_i))$ is a Fisher information matrix [17]. The $[I(\beta(u_i))]^{-1}$ the matrix can be determined by

$$[I(\beta(u_i))]^{-1} = \left(-E\left[\frac{\partial^2 ln ln\ L(\beta(u_i))}{\partial^2 \beta(u_i)}\right]\right)^{-1} = \\ \left(E\left[\left(\frac{\partial ln ln\ L(\beta(u_i))}{\partial \beta(u_i)}\right)\left(\frac{\partial ln ln\ L(\beta(u_i))}{\partial \beta(u_i)}\right)^T\right]\right)^{-1}.$$

The Wald statistic in Equation was asymptotically standard normal distributed [17]. Thus, reject H_0 if $|W| > Z_{1-\alpha/2}$, where the critical value $Z_{1-\alpha/2}$ is $1-\alpha/2$ quantile from a standard normal distribution.

5. Conclusion

GWBLR model is a local form of BLR which all of the regression parameters depend on the geographical location. Hypothesis testing on the GWBLR model is based on the GWR method, that is, hypothesis testing is done locally, dependent on the spatial weighting function in the study area. There are three kinds of hypothesis tests. The first test is a parameter similarity test by using the Vuong test method. The test is used to determine a significant difference between GWBLR and BLR. The Vuong test statistic was asymptotically standard normal distributed. The second test is a simultaneous test using the LRT method. The simultaneous test is used to determine the simultaneous significance of the regression parameters. The LR statistic has an asymptotic chi-squared distribution with the degree of freedom is the difference between the effective number of parameters in the reduced model and the full model. The last test is a partial test using the Wald test method and it was applied to determine the significance of each parameter in the regression model. The Wald test statistic was asymptotically standard normal distributed.

References

[1] Brunsdon C, Fotheringham A S and Charlton M 1996 Geographically weighted regression: a method for exploring spatial nonstationarity *Geographical Analysis* **28**(4) 281-298
[2] Brunsdon C, Fotheringham S and Charlton M 1998 Geographically weighted regression: modelling spatial non-stationarity *Journal of the Royal Statistical Society Series D (The Statistician)* **47**(3) 431-443
[3] Brunsdon C, Fotheringham A S and Charlton M 1999 Some notes on parametric significance tests for geographically weighted regression *Journal of Regional Science* **39**(3) 497-524
[4] Leung Y, Mei C-L and Zhang W-X 2000a Statistical tests for spatial nonstationarity based on the geographically weighted regression model *Environment and Planning A* **32** 9-32
[5] Leung Y, Mei C-L and Zhang W-X 2000b Testing for spatial autocorrelation among the residuals of the geographically weighted regression *Environment and Planning A* **32** 871-890
[6] Atkinson P M, German S E, Sear D A and Clark M J 2003 Exploring the relations between riverbank erosion and geomorphological controls using geographically weighted logistic regression *Geographical Analysis* **35**(1) 58-82
[7] Wu W and Zhang L 2013 Comparison of spatial and non-spatial logistic regression models for modeling the occurrence of cloud cover in North-eastern Puerto Rico *Applied Geography* **37** 52-62
[8] Rodrigues M, de la Riva J and Fotheringham S 2014 Modeling the spatial variation of the explanatory factors of human-caused wildfires in spain using geographically weighted logistic regression *Applied Geography* **48** 52-63
[9] Luo J and Kanala N K 2008 Modeling urban growth with geographically weighted multinomial logistic regression *Proc. of SPIE* **7144** 71440M-1-71440M-2
[10] Wang Y, Kockelman K M and Wang X 2011 Anticipation of land use change through use of geographically weighted regression models for discrete response *Transportation Research Record: Journal of the Transportation Research Board* No. 2245 111-123
[11] Purhadi, Rifada M and Wulandari S P 2012 Geographically weighted ordinal logistic regression model *International Journal of Mathematics and Computation* **16**(3) 116-126
[12] Wardhani N W S, Pramoedyo H and Dianati Y N 2014 Food security and vulnerability modeling of East Java Province based on Geographically Weighted Ordinal Logistic Regression Semiparametric (GWOLRS) Model *Journal of Degraded and Mining Lands Management* **2**(1) 231-234
[13] El-Sayed A M, Islam M A and Alzaid A A 2013 Estimation and test of measures of association for correlated binary data *Bulletin of the Malaysian Mathematical Sciences Society* **36**(4) 985-1008
[14] Fotheringham A S, Charlton M and Brunsdon C 2002 *Geographically Weighted Regression: The analysis of spatially varying relationships* (Chichester: John Wiley & Sons)

[15] Triyanto, Purhadi, Otok B W and Purnami S W 2016 Hypothesis testing of geographically weighted multivariate poisson regression *Far East Journal of Mathematical Sciences* **100**(5) 747-762
[16] Harini S, Purhadi, Mashuri M and Sunaryo S 2012 Statistical test for multivariate geographically weighted regression model using The Method of Maximum Likelihood Ratio Test *International Journal of Applied Mathematics & Statistics* **29**(5) 110-115
[17] Greene W H 2003 *Econometric Analysis* (New Jersey: Pearson Education)

The Optimal Implementation of Chlorination, Treatment, and Education Controls in Reducing the Transmission of Cholera Disease

I Fitria[1*], S M Hasanah[1], Subchan[1], A P Pratama[2]
[1]Department of Mathematics, Institut Teknologi Kalimantan, Balikpapan
[2]Department of Mathematics, Institut Teknologi Sepuluh November, Surabaya

e-mail: *irma.fitria@lecture.itk.ac.id

Abstract. Cholera is an acute disease caused by the bacteria Vibrio Cholerae. Cholera is a contagious and deadly disease that requires immediate medical treatment. The mathematical studies are conducted to create models of cholera transmission. The cholera transmission models are carried out by implementing the optimal controls, namely the chlorination for bacteria, control treatment for quarantined individuals and education campaigns for susceptible and infected individuals. The optimal control efforts are given to minimize the spreading of cholera disease. The method used to solve this optimal control problem is Pontryagin Minimum Principle. Based on the simulation, the provision of optimal controls in forms of education and treatment in individuals and chlorination can minimize the concentration of bacteria and the number of infected humans.

1. Introduction

Cholera is an infectious disease caused by an infection bacteria Vibrio Cholerae with diarrhea symptoms accompanied by acute vomiting due to enterotoxins produced by bacteria [1]. Cholera can spread as an endemic, epidemic or pandemic disease. Since 1917, there are seven pandemics that spread across Europe. Vibrio Cholerae bacteria first appeared in Sulawesi, Indonesia and caused a cholera epidemic [2]. Cholera can kill around 95,000 people and affect 2.9 million more every year [3, 4]. In 2016, overall 54% of cases were reported from Africa, 13% from Asia and 32% from Hispaniola. Cholera is still a major public health problem and affects the developing world population, especially countries without sanitation resources [3].

Vibrio Cholerae bacteria into the body through foods or drinks that have been contaminated. The bacterium secretes enterotoxins (poisons) in the intestinal tract so that diarrhea is accompanied by acute vomiting [1]. Someone who has been infected with the Vibrio Cholerae bacterium within a few days will lose a lot of body fluids and enter dehydration [2]. Bacteria can only grow at a very high pH, which is between 8.5 - 9.5 and very good growth at pH 7.0 [5]. Cholera is currently still a concern in the medical world. Research on cholera has also been carried out by [6-8]. Various efforts were made to prevent the spread of cholera such as vaccination, clean water sanitation [9] and various other treatments [10-11]. On the other hand, treatment is the most important thing to eradicate disease [10-11]. Several alternative treatments for preventing cholera have been investigated, for example by [12-15].

These problems in the field of mathematics have been studied with mathematical modeling approaches [16]. Research on cholera has been studied by [17] through the optimum control approach of infectious diseases with its control variables namely of education and chlorination, [18] conducting research on optimal control treatment of humans infected with cholera, by providing treatment to infected individuals and [19] conducted a study on the model of the spread of cholera by giving control in the form of medication and intervention through improved sanitation, education and quarantine. In addition, similar studies have also been studied by [20-31].

Furthermore, in this study, a mathematical model of the spread of cholera was reconstructed using a model that had been developed by [19] with optimal control variables in the form of chlorination in bacteria and control treatment in patients infected with the disease in the form of treatment and education of susceptible individuals. The purpose of giving control is to minimize the number of individuals infected cholera and reduce the concentration of Vibrio Cholerae bacteria.

2. Model Formulation

The model developed in this study using a model developed in the research of [19] for cholera, namely the SEIQRB (Susceptible-Educated-Infected-Quarantined-Recovered-Bacteria) model with the subpopulation of infected individuals divided into two, namely asymptomatic and symptomatic. The model construction divided the human subpopulation into six classes $N(t)$, namely individuals susceptible $S(t)$, educated human $E(t)$, individuals infected cholera with asymptomatic $I_A(t)$, individuals infected cholera with symptomatic $I_S(t)$, the human that is under treatment through quarantine $Q(t)$, individuals recovered $R(t)$. In addition, the bacterial concentration is expressed in $B(t)$. In this model, control variables are used in the form of chlorination (u_1), quarantine treatment (u_2) and education for susceptible individuals (u_3).

Mathematically, the following are the results of a dynamic system model reconstruction with optimal control

$$\dot{S} = \Lambda + vR + \epsilon E - \mu S - u_3\psi S - \beta\frac{B}{k+B}S, \tag{1}$$
$$\dot{E} = u_3\psi S - \epsilon E - \mu E - \gamma E, \tag{2}$$
$$\dot{I}_A = p\beta\frac{B}{k+B}S + p\gamma E - \mu I_A - \alpha_2 I_A, \tag{3}$$
$$\dot{I}_S = (1-p)\beta\frac{B}{k+B}S + (1-p)\gamma E - \mu I_S - \mu_S I_S - u_2\delta I_S, \tag{4}$$
$$\dot{Q} = u_2\delta I_S - \mu Q - \mu_Q Q - \alpha_1 Q, \tag{5}$$
$$\dot{R} = \alpha_1 Q + \alpha_2 I_A - \mu R - vR, \tag{6}$$
$$\dot{B} = \eta I_A + \eta I_S - u_1 B - dB, \tag{7}$$

with u_1, u_2 and u_3 is the level of giving optimal control efforts with limits

$$0 \le u = (u_1, u_2, u_3) \le 1.$$

Here are some parameter used Λ is constant human recruitment rate, μ is natural mortality rate, β is rate consumption bacterial through contaminated sources, k is half the constant saturation of the bacterial concentration, the degree of movement from subpopulation susceptible to infected cholera due contact between infected individuals and contaminated sources namely $\beta\frac{B}{k+B}$, ψ is rate subpopulation of educated susceptible individuals, ϵ is rate subpopulation educated stop following education campaign, γ is rate subpopulation educated infected cholera (very unlikely), p is rate subpopulation with infected asymptomatic, α_1 and α_2 are rate recovering subpopulation infected with symptomatic and asymptomatic, δ is the length of time an individual was quarantined, μ_S and μ_Q are death rate of individuals symptomatic and quarantined, v is individuals recovered so that becomes susceptible human, η is concentration bacterial growth because individuals infected release the Vibrio Cholerae into the aquatic environment and d is natural death rate of Vibrio Cholerae cause decrease concentration bacteria.

3. Optimal Control Problems

The problem of optimal control in this study was solved using the Pontryagin Minimum Principle [32-33]. Next, the dynamic system model is as follows

$$\dot{x}(t) = f(x(t), u(t), t),$$

so that the set of state variables according to Equations (1) - (7) can be written as follows

$$x(t) = (S(t), E(t), I_A(t), I_s(t), Q(t), R(t), B(t))^T,$$

with the initial state $x(t_0) = x_0$.

The purpose of this research problem is to obtain optimal control by minimizing the infected subpopulation, the number of bacteria and the costs that will be incurred on chlorine, education, and treatment. Mathematically, the purpose of optimal control can be expressed in the form of the objective function as follows

$$J = \frac{1}{2}\int_{t_0}^{t_f} [C_1 I_A{}^2(t) + C_2 I_S{}^2(t) + C_3 B^2(t) + C_4 u_1{}^2(t) + C_5 u_2{}^2(t) + C_6 u_3{}^2(t)]\, dt, \quad (8)$$

where t_0 is a start time and t_f is a final time specified, as well as $C_i > 0$ for each $i = 1,2,3,4,5,6$ is the parameter weights or price coefficient issued at each control during the day period. The type of problem optimal control of this research is fixed-final time (t_f) and free-final state $\left(x(t_f)\right)$ with conditions $0 < t < t_f$. *State* at the start $\left(x(t_0)\right)$ already set and state in finish $\left(x(t_f)\right)$ not set. So, every existing system use the state boundary conditions at the initial time as follows,

$$S(t_0) = S_0, E(t_0) = E_0, I_A(t_0) = I_{A_0}, I_S(t_0) = I_{S_0}, Q(t_0) = Q_0, R(t_0) = R_0, B(t_0) = B_0.$$

The boundary condition for type fixed-final time and the free-final state as following

$$x(t_0) = x_0, \lambda^*(t_f) = \left(\frac{\partial \phi}{\partial x}\right)_{*t_f}, \quad (9)$$

where $x_0 = \{S_0, E_0, I_{A_0}, I_{S_0}, Q_0, R_0, B_0\}$.

4. Result and Discussion

Based on Section 3, it is known that the dynamic system model for the spread of cholera with optimal control, objective function, and boundary conditions. Furthermore, the problem of optimal control can be solved by several steps, namely first forming the Pontryagin function [32-33]. Based on Equation (8) the problems in this research can be arranged into the following equation

$$\begin{aligned} H(x(t), u(t), \lambda(t), t) &= L(x(t), u(t), t) + \lambda'(t) f(x(t), u(t), t), \\ &= \frac{1}{2}(C_1 I_A{}^2(t) + C_2 I_S{}^2(t) + C_3 B^2(t) + C_4 u_1{}^2(t) + \\ &+ C_5 u_2{}^2(t) + C_6 u_3{}^2(t)) + \lambda_S\left(\Lambda + vR + \epsilon E - \mu S - u_3\psi S - \beta\frac{B}{k+B}S\right) + \\ &+ \lambda_E(u_3\psi S - \epsilon E - \mu E - \gamma E) + \lambda_{I_A}\left(p\beta\frac{B}{k+B}S + p\gamma E - \mu I_A - \alpha_2 I_A\right) + \\ &+ \lambda_{I_S}\left((1-p)\beta\frac{B}{k+B}S + (1-p)\gamma E - \mu I_S - \mu_S I_S - u_2\delta I_S\right) + \\ &+ \lambda_Q\left(u_2\delta I_S - \mu Q - \mu_Q Q - \alpha_1 Q\right) + \lambda_R(\alpha_1 Q + \alpha_2 I_A - \mu R - vR) + \\ &+ \lambda_B(\eta I_A + \eta I_S - u_1 B - dB) \end{aligned} \quad (10)$$

Next, minimize H to u. The optimal control variables namely chlorination (u_1), quarantine treatment (u_2), and education (u_3), so that it is obtained

$$\hat{u}^*{}_1 = \frac{\lambda_B B}{C_4}, \quad (11)$$

$$\hat{u}^*{}_2 = \frac{\lambda_{I_S}\delta I_S - \lambda_Q \delta I_S}{C_5}, \quad (12)$$

$$\hat{u}^*{}_3 = \frac{\lambda_S\psi S - \lambda_E\psi S}{C_6}. \quad (13)$$

The optimal control equation (11) - (13) is a proportion, so it is defined by $0 \le u_1 \le 1,\ 0 \le u_2 \le 1$ and $0 \le u_3 \le 1$. Then, the optimal control value can be stated as follows,

$$u^*{}_i = \left\{\{0, \hat{u}^*{}_i\}, 1\right\}, \quad (14)$$

where $i = 1,2,3$. The next step is to form a state equation. The equation of state in optimal conditions is obtained by reducing the optimal function of Pontryagin H to λ like the following

$$\dot{x}^* = \frac{\partial H^*}{\partial \lambda}. \quad (15)$$

Based on Equation (15) so that the state equation is obtained for each subpopulation. Next is the state equation in the susceptible subpopulation.

$$\dot{S}^* = \Lambda + vR + \epsilon E - \mu S - u^*_3\psi S - \beta\frac{B}{k+B}S, \quad (16)$$

$$\dot{E}^* = u^*_3\psi S - \epsilon E - \mu E - \gamma E, \quad (17)$$

$$\dot{I_A}^* = p\beta\frac{B}{k+B}S + p\gamma E - \mu I_A - \alpha_2 I_A, \quad (18)$$

$$\dot{I_S}^* = (1-p)\beta\frac{B}{k+B}S + (1-p)\gamma E - \mu I_S - \mu_S I_S - u^*_2\delta I_S, \quad (19)$$

$$\dot{Q}^* = u^*_2\delta I_S - \mu Q - \mu_Q Q - \alpha_1 Q, \quad (20)$$

$$\dot{R}^* = \alpha_1 Q + \alpha_2 I_A - \mu R - vR, \quad (21)$$

$$\dot{B}^* = \eta I_A + \eta I_S - u^*_1 B - dB. \quad (22)$$

Next, form a costate equation by lowering the Pontryagin H function negatively to the state as follows.

$$\dot{\lambda}^* = -\frac{\partial H^*}{\partial x}. \quad (23)$$

Equation (23) is used to form the costate equation in each subpopulation. The following is the costate equation of the susceptible subpopulation.

$$\dot{\lambda_S}^* = \mu\lambda_S + u^*_3\psi\lambda_S + \beta\frac{B}{k+B}\lambda_S - u^*_3\psi\lambda_E - \lambda_{I_A}p\beta\frac{B}{k+B} - \lambda_{I_S}(1-p)\frac{B}{k+B}, \quad (24)$$

$$\dot{\lambda_E}^* = -\epsilon\lambda_S + \epsilon\lambda_E + \mu\lambda_E + \gamma\lambda_E - p\gamma\lambda_{I_A} - (1-p)\gamma\lambda_{I_S}, \quad (25)$$

$$\dot{\lambda}_{I_A}^* = -C_1 I_A + \mu\lambda_{I_A} + \alpha_2\lambda_{I_A} + \alpha_2\lambda_R - \eta\lambda_B. \quad (26)$$

$$\dot{\lambda}_{I_S}^* = -C_2 I_S + \mu\lambda_{I_S} + \mu_S\lambda_{I_S} + u^*_2\delta\lambda_{I_S} - u^*_2\delta\lambda_Q - \eta\lambda_B \quad (27)$$

$$\dot{\lambda_Q}^* = \mu_Q\lambda_Q + \mu\lambda_Q + \alpha_1\lambda_Q - \alpha_1\lambda_R \quad (28)$$

$$\dot{\lambda_R}^* = \mu\lambda_R - v\lambda_S + v\lambda_R \quad (29)$$

$$\dot{\lambda_B}^* = -C_3 B + \frac{S\beta k\lambda_S}{(k+B)^2} - \frac{Sp\beta k\lambda_{I_A}}{(k+B)^2} - \frac{S(1-p)\beta k\lambda_{I_S}}{(k+B)^2} + u^*_1\lambda_B + d\lambda_B \quad (30)$$

In the next step, numerical simulation is carried out on the model of spreading cholera with optimal control [34-35]. In this section, the results of the simulation have been explained. The following are the parameter values shown in Table 1. The parameter values in Table 1 are used to obtain numerical simulation results. Simulations with optimal control using the Forward-Backward method Sweep Runge-Kutta Order 4. The purpose of the numerical simulation is to determine the effect of the effectiveness of giving optimal control in each subpopulation.

Table 1. Parameter Value

Parameter	Value	Reference	Parameter	Parameter	Reference
Λ	$\frac{24.4N(0)}{365000}$	[9]	d	$1/30$	[9]
μ	$2.2493.10^{-5}$	[9]	$S(0)$	5750	[9]
β	0.08	[9]	$E(0)$	0	[15]
k	10^6	[9]	$I_A(0)$	1000	[15]
v	0.4/365	[9]	$I_S(0)$	700	[15]
δ	0.05	[9]	$Q(0)$	0	[9]
ϵ	0.003	[15]	$R(0)$	0	[9]
ψ	0.008	[15]	$B(0)$	275000	[9]
γ	0.005	[15]	t_f	100 days	[15]
p	0.78	[15]	α_2	0.15	[15]
α_1	0.2	[9]	μ_S	0.00127	[15]
μ_Q	0.0001	[9]	η	50	[15]

It is reviewed from the comparison of the number of individuals at the beginning time with the number of individuals at the end or at certain times through comparison before and after giving optimal control in each subpopulation. The following are the simulation results of each subpopulation.

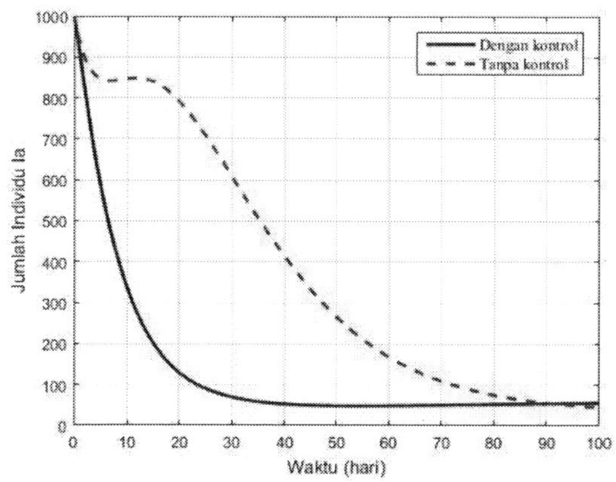

Figure 1. Infectious Individuals Asymptomatic

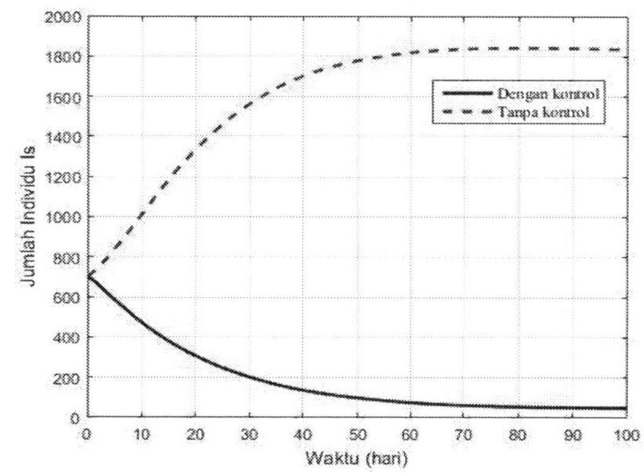

Figure 2. Infectious Individuals Symptomatic

Based on Figure 1, it is known that the number of individuals I_A with optimal control, every day decreases compared to without optimal control. Decreasing the graph with optimal control significantly at t = 10 days. The graph decreases rapidly compared to without optimal control. It is assumed that the number of subpopulations at the beginning is 1000 individuals. The number of individuals without optimal control when t = 10 days has increased, which is to 850 individuals while the number of individuals with optimal control decreases to 330 individuals. Therefore, education of cholera for susceptible and educated subpopulations is very important. So, in this case, the level of effectiveness of giving optimal control in the Infected Asymptomatic subpopulation in the form of education is 22%.

Individuals symptomatic (can see Figure 2) if not treated immediately will increase contamination of the environment. Thus, the number of individuals affected by cholera has increased, given the rapid spread of bacteria. However, if the Infected Symptomatic subpopulation is given optimal control in the form of treatment during quarantine, then the number of individuals infected with the disease will decrease significantly. At the end of the number of subpopulations with optimal control as many as 50 individuals. This means that the level of effectiveness in providing optimal control in the form of treatment has an effect of 97.27%. So, in this case, the goal is to minimize the number of individuals infected with symptomatic can be fulfilled. Next, we can see the rate of change in the number of bacterial concentration in Figures 4.

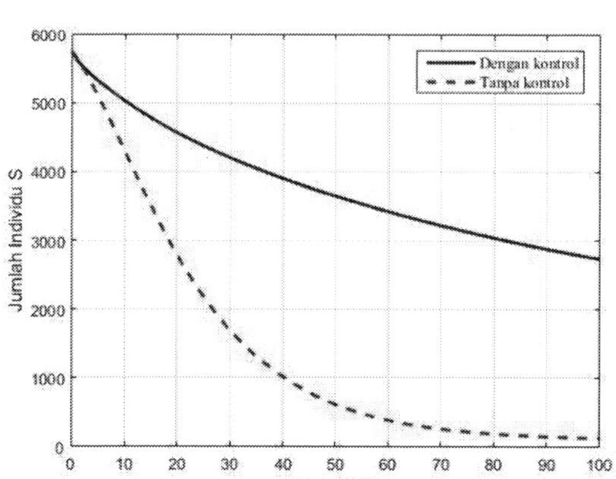

Figure 3. Individuals Susceptible

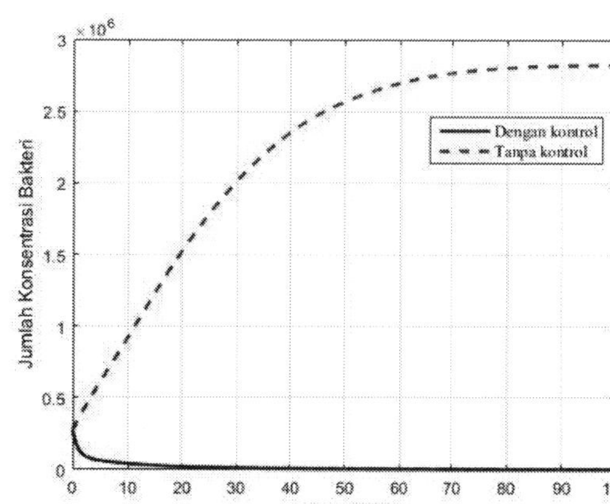

Figure 4. Bacteria Concentration

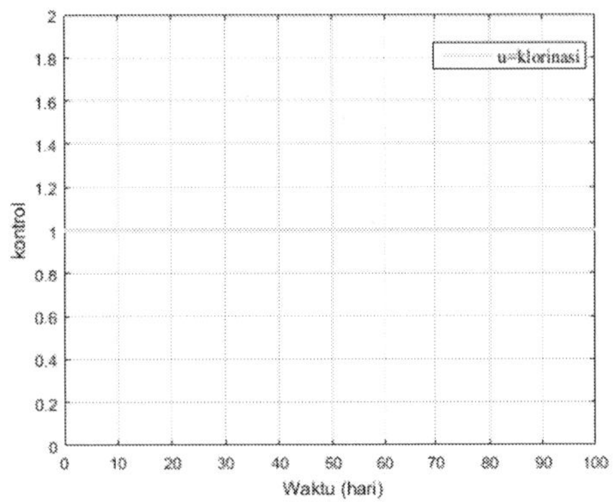

(a)

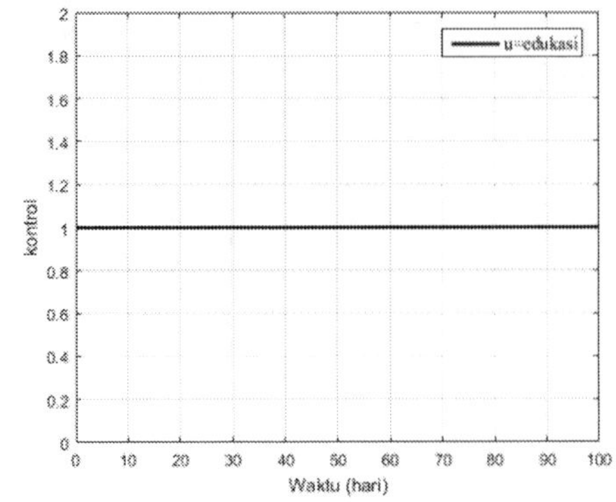

(b)

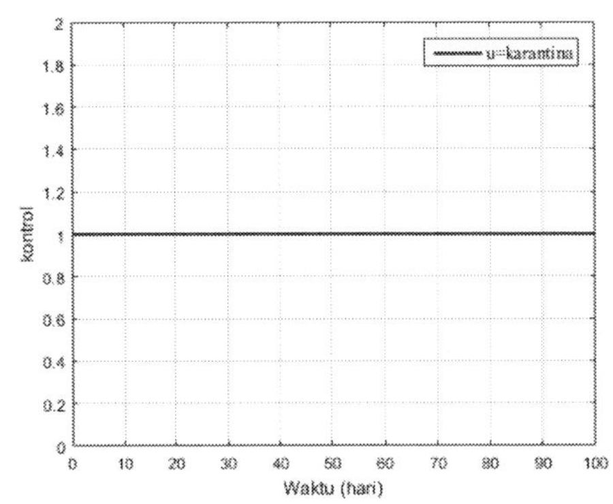

(c)

Figure 5. The Optimal Control in (a) Chlorination, (b) Education, (c) Quarantine

Based on Figure 3 it is known that susceptible individuals decrease when given optimal control. The number of individuals susceptible to optimal control at the end time is 2733.28 or around 2733 individuals while without optimal control the number of individuals at the end is 134 individuals. Therefore, susceptible individuals need to be given control in the form of education. So, if the educated individual increases, the number of individuals infected with the disease will decrease. When given control in the form of education, the susceptible subpopulation will decrease until the end time. This means that the effectiveness of providing education to individuals susceptible has an effect of 95%. Furthermore, it can be seen the rate of change in the number of individuals infected with asymptomatic and symptomatic.

The concentration level at the initial time was 275000 cell/ml. Figure 4 shows a graph of the ratio of changes in the number of bacterial concentrations with and without optimal control. Without optimal control, the rate of development of bacterial concentrations increases significantly from day 10 to $9{,}5 \times 10^5$, whereas when given optimal control the level of bacterial concentration decreases on day 10 to 4×10^4 and continues to decline to 5144 at the end. The level of bacterial concentration without control at the end time is $2{,}8 \times 10^6$. So, in this case, the goal is to minimize bacterial concentration can be attained. In more detail, the optimal control value is shown in Figure 5.

5. Conclusion

Based on the results and discussion in this study, it can be concluded that after reconstructing the model by adding optimal control variables in the form of education and treatment in individuals and chlorination can minimize the concentration of bacteria and the number of infected humans.

References

[1] Merrel D S, Butler S M, Qadri F, Dolganov, N A, Alam A, Cohen M B, Calderwood S B, Schoolnik G K, Camili A 2002 *Nature* **417** (6889) 642-645
[2] Setiati S, Idrus A, *et al* 2014 *Ilmu Penyakit Dalam* (Jakarta: InternaPublishing)
[3] World Health Organization 2017 Cholera annual report 2016, [online]: https://www.who.int/wer/2017/wer9236/en/
[4] World Health Organization 2018 Cholera, [online]: http://www.who. int/news-room/fact-sheets/detail/cholera
[5] Amelia S 2005 *Vibrio Cholerae* (Medan: Universitas Sumatera Utara)
[6] Guli M M 2016 *J. Biocelebes* **10** (2) 18-24
[7] Guntina R K and Sri A F K 2017 *J. Farmaka Suplemen* **15** (1) 92-104
[8] I Wayan Y W, Retno K and Anak A G R D 2015 *J. Metamorfosa,* **11** (1) 16-22
[9] Unicef Indonesia 2012 Ringkasan Kajian Air Bersih, Sanitasi, dan Kebersihan, [online]: http://www.unicef.org
[10] World Health Organization 2018 Prevention of Cholerae, [online] tersedia di : http://www.who.int/en/news-room/feature-stories/detail/prevention-for-a-cholera-free-world
[11] World Health Organization 2017 Prevention for a cholera free world. [online] tersedia di : https://www.who.int/en/news-room/feature-stories/detail/prevention-for-a-cholera-free-world
[12] Christy A S 2012 *J. Ilmu Kesehatan dan Kedokteran Keluarga* **8** (2) 97-102
[13] Karmila 2016 *Skripsi Daya Hambat Ekstrak Daun Mengkudu (Morinda Citrifolia L.) terhadap Pertumbuhan Bakteri Penyebab Diare* (Makassar: Universitas Alauddin Makassar)
[14] Sugiarti L, Dwi S and Sofiyatul N J 2019 *J. Pengabdian Kesehatan* **2** (1) 63-77
[15] Utami P R and Don A R 2018 *Seminar Nasional Pelestarian Lingkungan (SENPLING)* (Riau: Universitas Riau) 586
[16] Cahyono 2013 *Pemodelan Matematika Edisi Pertama* (Yogyakarta: Graha Ilmu)
[17] Bakhtiar T 2015 *Peran Edukasi dan Klorinasi dalam Pengendalian Penyakit Menular: Sebuah Pendekatan Kontrol Optimum Proc Seminar dan Rapat Tahunan Bidang MIPA BKS-PTN Barat* (Pontianak: Universitas Tanjungpura) 430-440
[18] Lemos-Paiao A P, Christina J S and Delfim F M T 2016 *J. of Computational and App Math* **318** 168-180
[19] Syafi'I A M, Irma F, *et al* 2018 *Conf. Nasional Matematika XIX* (Malang: Universitas Brawijaya)
[20] Adaniyah A 2016 *Skripsi Analisis Kestabilan dan Kontrol Optimal pada Model Matematika Epidemi Kolera* (Surabaya: Universitas Airlangga)
[21] Asfarina G T, Syafruddin S and Sukarna 2016 *Skripsi Analisis Kestabilan Penyebaran Penyakit Kolera Menggunakan Model SEIRS dengan Vaksinasi dan Faktor Treatment* (Makassar: Universitas Negeri Makassar)
[22] Dany A R 2016 *Skripsi Model Matematika Penyebaran Kolera dengan Mempertimbangkan Intervensi Kesehatan Masyarakat* (Surabaya: Universitas Airlangga)
[23] Fitrianah A 2015 *Skripsi Analisis Dinamika Model Penyebaran Penyakit Kolera* (Bogor: Institut Pertanian Bogor)
[24] Fitriyah A 2015 *Thesis Suatu Analisis dari Model Matematika Penyakit Kolera* (Yogyakarta: Universitas Gajah Mada)
[25] Maisura H and Sumarno P S 2018 *J. of Math and Its App* **17** (1) 33-46
[26] Mahmudah Y, Fatmawati and Yayuk W 2013 *J. Matematika* **2** (1) 73-80
[27] Nirwani N, V H Badshah and R Khandelwal 2015 *Advances in Applied Science Research*, **6** (6) 181-186
[28] Renny 2009 *Proc. Seminar Nasional Matematika dan Pendidikan Matematika*, Jurusan Pendidikan Matematika
[29] Stephen E and Nkuba N 2015 *Applied and Computational Math* **4** (2) 53-63
[30] Wang J and Chairat M 2011 *Canadian Applied Mathematics Quartely* **19** (3)

[31] Zulaikha T and Intan F 2017 *Proc. Seminar Nasional Integrasi Matematika dan Nilai Islami* **1** (1) 41-51
[32] Naidu D S 2002 *Optimal Control System* (New York: Idaho State University)
[33] Subchan and Zbikowski R 2009 *Computational Optimal Control Tools and Practice* (United Kingdom: John Wiley & Sons Ltd)
[34] Burden R L and Faires J D 2011 *Numerical Analysis Ninth Edition* (Camda: Brooks/Cole Cengage Learning)
[35] Lenhart S and John T W 2007 *Optimal Control Applied to Biological Models* (New York: Chapman & Hall/CRC Taylor and Francis Group)

Rainfall Prediction in East Java Using Spatial Extreme Value Theory

A A Artha[1] and A Sofro[1]
[1]Mathematics Department, Universitas Negeri Surabaya, Ketintang, Surabaya 60231, East Java, Indonesia

E-mail: ayuninsofro@unesa.ac.id

Abstract. Extreme rainfall is a phenomenon that occurs because of the high intensity of rainfall which causes many negative impacts. For example, is flooding that can ruin crops for farmers. To minimize this hazard, rainfall prediction is carried out. One methods that can be used is spatial extreme value theory using max stable approach and copula approach. From it, we will get the return level (predictive value) rainfall by paying attention to the location elements in it. This study discusses the max-stable approach using the Smith model and the copula approach using the Copula Gaussian model. Parameter estimation used is Maximum Likelihood Estimation (MLE) method. Because it takes into spatial elements, the trend surface model is also used. At the end of the study, Akaike Information Criterion (AIC) is used to compare and select the best model that can be used in predicting rainfall. Spatial extreme value theory will be applied to rainfall data in East Java which consists of nine rainfall observation stations. The results of spatial parameter estimation show that the highest of rainfall intensity value will occur around rainfall observation station namely Sawahan which is located in Nganjuk district in a period of two years. From AIC, it can be said that the model using copula Gaussian is the best model compare with max-stable approach as it provides the smallest AIC values.

1. Introduction

In this time, the condition of the earth became unstable. There are many disasters that causes many casualties, road damage, floods, and the other material losses. Natural disasters that occur come from unusual earth activities. As an example of the occurrence of floods, landslides, or rainstorms in an area caused by rainfall with high intensity. To anticipate this problem, predictions of extreme rainfall can be made. Natural events such as high rainfall, high tides, high winds, and etc occur because they are influenced by conditions from a place (spatial). Therefore, rainfall in one area will depend on rainfall in other areas which are close together due to spatial dependence [1]. R. Jane in 2016 used Copula Dependency on Radio Waves, Davison and Gholamrezaee in 2010 used max-stable application for temperature data and Mc.Neil in 2013 used modelling dependencies with copula. So to predict extreme rainfall, spatial extreme value theory can be used with the max-stable approach and the copula approach because the approaches enable to resolve dependencies in many locations [1]. The max-stable approach and the copula approach are statistical methods used to analyze extreme events that occur on earth, such as rainfall, wind speed, humidity, and so on. Then forecasting and modeling extreme rainfall can be done by using the parameter estimation that is maximum likelihood estimation (MLE) [2]. This study takes the issue of rainfall in East Java. East Java is one of the provinces that calculated in the contributing of rice production nationally. Around 17% of national rice production comes from East Java (BPS, 2016). Meanwhile, based on the East Java Agriculture Service in the first quarter of 2010 rice fields were damaged due to the impact of flooding reaching a significant value of 6,972.49 ha. So the rainfall in East Java is an important thing to be examined how rainfall patterns occur. The important thing in the studying of EVT is determine the return level which is the maximum rainfall threshold value [3]. So, the return level is an extreme rainfall prediction value that may occur in the future so that it can be used as a source of information that can be used to anticipate and minimize the impact.

2. Literature Review

2.1 *Extreme Value Theory*

Extreme Value Theory (EVT) is a theory that studies the occurrence of extreme values that occur in the earth. Identification of extreme values with EVT can be done with two methods that known as the Block Maxima (BM) method and the Peaks Over Threshold (POT) method. The BM method is a method which takes the maximum value in a period which called block and the POT method is a method where a value will pass through a certain threshold [4]. So to determine the extreme value from a data, BM method and POT method can be used [5]. In this study, BM method will be used because rainfall data is a seasonal data type where data patterns are influenced by season. BM method is a method that can identify extreme values based on the highest values of observation data. The block maxima method applies the Fisher and Tippet (1928) theorem in Gilli and Kellezi (2006), where in the theorem states that sample data of extreme values taken by the BM method will follow the Generalized Extreme Value (GEV) distribution function. Suppose there are $X_1, X_2, \ldots, X_n$ for each X_n is a random variable that follows the distribution function F. where $M_n = Max\{X_1, X_2, \ldots, X_n\}$ is the maximum value of n observations, and M_n has a distribution function F_n. So the M_n distribution follows the cdf from GEV like the following equation:

$$Pr\left(\frac{M_n - b_n}{a_n} \le x\right) = F^n(a_n x + b_n) \rightarrow F(\mu, \sigma, \xi), \text{ as } n \rightarrow \infty,$$

$$M_n \sim F(\mu, \sigma, \xi).$$

where GEV has the cumulative distribution function (cdf) as follows:

$$F(\mu, \sigma, \xi) = \begin{cases} \exp\left\{-[1 + \xi\left(\frac{x-\mu}{\sigma}\right)]^{-\frac{1}{\xi}}\right\}, & \xi \neq 0. \\ \exp\left\{-\exp\left(-\frac{x-\mu}{\sigma}\right)\right\}, & \xi = 0. \end{cases} \quad (2.1)$$

and have a probability distribution function (pdf) like the following equation:

$$f(\mu, \sigma, \xi) = \begin{cases} \frac{1}{\sigma}\{1 + \xi\left(\frac{x-\mu}{\sigma}\right)\}^{-\frac{1}{\xi}-1} \exp\{-[1 + \xi\left(\frac{x-\mu}{\sigma}\right)]^{-\frac{1}{\xi}}]\}, & \xi \neq 0. \\ \frac{1}{\sigma}\exp\left(-\frac{x-\mu}{\sigma}\right)\exp\{-\exp(\frac{x-\mu}{\sigma})\} & ,\xi = 0. \end{cases} \quad (2.2)$$

where x is the extreme value obtained from the BM method.. GEV has three distribution parameters, that known as location parameters/μ, scale parameters/σ, and shape parameters/ ξ. Form of the parameters follow 3 distributions, namely Rev. Weibull ($\xi < 0$), Gumbel ($\xi = 0$), and Frechet ($\xi > 0$) [6]. The difference of the distribution can be found in the tip of the three shape parameters which makes the distribution difficult to observe so the GEV distribution serves to combine the three by following the GEV distribution form.

2.2 *Maximum Likelihood Estimation (MLE)*

The GEV estimation parameter can be done by using the Maximum Likelihood Estimation (MLE) method. The following are parameter estimates for the parameters $\hat{\mu}, \hat{\sigma}, \hat{\xi}$ with MLE using pdf from GEV like equation (2.2). Then the likelihood function with $\xi \neq 0$ is:

$$L(\mu, \sigma, \xi) = \prod_{i=1}^{n} \frac{1}{\sigma}\{1 + \xi\left(\frac{x_i - \mu}{\sigma}\right)\}^{-\frac{1}{\xi}-1} \exp\{-[1 + \xi\left(\frac{x_i - \mu}{\sigma}\right)]^{-\frac{1}{\xi}}]\} \quad (2.3)$$

From the results of the first derivative (2.3), if the form of the likelihood function is not explicit, a numerical approach is needed, namely the Newton Raphson method.

2.3 Dependencies
In statistical theory, modelling EVT in a location is considered insufficient to deal with various extreme events because extreme events involve many adjacent locations that are influential factors. So the extreme events observed in two or more different locations require an approach that is the max-stable approach and the copula approach [7].

2.3.1 Max-Stable Approach
This method used to stabilize data by transforming extreme value distribution (GEV) into the Frechet distribution. This is needed because the three types of GEV distribution, the most heavy-tailed distribution ($\xi > 0$) is the Frechet distribution. Therefore, the process of stabilizing data from extreme value distribution into the Frechet distribution be required that known as the max-stable process. The transformation process of cdf GEV into Frechet pdf using the following equation:

$$z(x) = \left\{1 + \frac{\hat{\xi}(x_{ij}-\hat{\mu})}{\hat{\sigma}}\right\}^{\frac{1}{\hat{\xi}}} \tag{2.4}$$

where z is a max-stable process with a Frechet distribution, $\hat{\mu}, \hat{\sigma}, \hat{\xi}$ are parameter estimates from GEV, and x_{ij} is the extreme value of the first observation at the observation station j. Based on equation (2.5), there are several models that formed of max-stable processes. One model that usually used is the Smith model. This model is considered the simplest model by involving normal distribution [7]. This model has the following PDF:

$$f(z_j(x), z_{j+1}(x)) = \exp\left\{-\frac{1}{z_{j+1}}\Phi\left(\frac{h}{2}+\frac{1}{h}\log\left[\frac{z_{j+1}}{z_j}\right]\right) - \frac{1}{z_{j+1}}\Phi\left(\frac{h}{2}+\frac{1}{h}\log\left[\frac{z_j}{z_{j+1}}\right]\right)\right\} \tag{2.5}$$

where h is the euclid distance from 2 locations j and j + 1, z is the max-stable process with equation (2.5), and Φ is a cdf from the normal distribution. To get the spatial dependencies, the trend surface model can be used. This model is able to explain dependencies between locations using the coordinates of the location of rainfall measurements, so that the trend surface model follows the following equation [2] :

$$\begin{aligned} \mu_j &= \beta_{1,0} + \beta_{1,1}lon_j + \beta_{1,2}lat_j \\ \sigma_j &= \beta_{2,0} + \beta_{2,1}lon_j + \beta_{2,2}lat_j \\ \xi_j &= \beta_{3,0} \end{aligned} \tag{2.6}$$

where lon_j is the longitudinal coordinate and lat_j is the latitude coordinate in the j-rainfall station. So to get the $\hat{\beta}$ parameter estimation in a number of observation stations j, the maximum likelihood estimation method is used. So the parameter estimation of $\hat{\beta}$ is:

$$\hat{\beta} = \arg\max L(\beta, z)$$

However, the parameter estimates for max-stable approach are considered if applied in many locations. This means that the more locations used, the greater error value generated by the AIC method [8]. So that another approach is carried out, namely using the Copula approach which can connect the number of locations [9].

2.3.2 Copula Approach
Copula was introduced by Abe Sklar in 1959 through the sklar theorem. According to the sklar theorem, copula is a function that connects multivariate distribution functions with marginal distribution [9]. Copula can be explore and characterize dependency structures between random variables through marginal distribution function. Copula is divided into two types of families, namely elliptical copula and archimedian copula. For the case of Spatial Extreme the copula model that can be used is copula elliptical. Copula which is included in the elliptical copula is Copula Gaussian and Student's t-copula [2]. In this study Gaussian copula was used because this copula model is the most popular model used and is

a simple model. In general, copula can be used universally with uniform standard functions. So the combined distribution function of the random variable x that follows the uniform standard distribution is:

$$C(u_1, \dots, u_j) = F(F'(u_1), \dots, F'(u_j))$$

where F is a combined pdf of random variable x. So that it can be defined that:

$$f(x_1, \dots, x_j) = c(F_1(x_1), \dots, F_1(x_1))$$

In this study, extreme value copula is used as a spatial approach. Extreme value copula is more suitable for heavy-tailed data, when the parameter $\xi > 0$. From the three types of GEV distribution, namely reversed weibull, gumbel, and Frechet, the distribution that is the most heavytail is the Frechet distribution. So the parameter estimation results from GEV will follow the Frechet distribution where the parameter ξ for the Frechet distribution is $\xi > 0$ [10]. In the Copula Gaussian for the extreme spatial case the transformation process uses the GEV cdf with the transformation equation defined in the following equation:

$$u_j = F_j(x_{ij}) \tag{2.7}$$

where F_j is cdf GEV and x_{ij} is j-th and i observation data. So copula cdf follows the following equation:

$$C(u_1, \dots, u_j) = \Phi(\Phi'(u_1), \dots, \Phi'(u_j)) \tag{2.8}$$

where Φ is cdf from the normal distribution because the copula used is copula gaussian. So according to the Sklar theorem, copula has a opportunity joint function as follows:

$$f(x_{1,1}, x_{1,2}, \dots, x_{i,j}) = f(x_{1,1}).f(x_{1,2}). \dots, f(x_{i,j}).c(u_1, \dots, u_j) \tag{2.9}$$

where f is pdf GEV because the distribution used is the GEV distribution.

After transformation into copula, the spatial GEV estimation parameter of copula will be performed using MLE. So the results will be obtaining from the parameter estimation copula using trend surface models like equation (2.9) [11]. So it is necessary to estimate the parameter $\hat{\beta}$ using the joint opportunity function with copula. So the form of the log-likelihood from the pdf above is:

$$\ln(\hat{\beta}) = \sum_{i=1}^{n} \sum_{j=1}^{m} \ln f(x_{ij}).c(u_1, \dots, u_j)$$

where x_{ij} is the i-extreme value at the jth observation station.

2.4 Return Level

To examine extreme events, determine the return value is also an important thing. Return value is the maximum value that is expected to occur in the coming period. The concept used in the return value is to predict rare events that will occur such as extreme rainfall that has a negative impact. Return Level is also an application of GEV distribution. From the cdf on GEV, the probability of extreme rainfall is 1-p, where p is the return period with $p = \frac{1}{k}$, where k is the number of blocks formed so that [12]:

$$\hat{R}_j = \hat{\mu}_j - \frac{\hat{\sigma}_j}{\hat{\xi}_j}[1 - (-\ln(1-p))^{-\hat{\xi}_j} \tag{2.10}$$

Which means that the return value from extreme events on the jth observation station can be determined using the equation. Where $\hat{\mu}_j$ is the estimation of the jth location parameter, $\hat{\sigma}_j$ is the estimation of the jth scale parameter, $\hat{\xi}_j$ is the estimation of the jth shape parameter, and p is return period.

2.5 Akaike's Information Criterion (AIC)

Akaike's Information Criterion (AIC) is the best model selection method. So this method is used to determine or select the best model that will be produced from this study. In the certain contexts, choosing a simple model is better than choosing a complex model. According to Ligas and Banasick [13] AIC is defined by the following equation:

$$AIC = -2 \log\log L(\beta) + 2q \tag{2.11}$$

Where $\log L(\beta)$ is the log-likelihood function of each approach and q is the number of parameters estimated. A lower AIC value indicates the best model that can be used.

2.6 *Extreme Rainfall*
Rainfall is the amount of water that falls on a flat land surface during a certain period measured in millimeters (mm). Extreme rainfall has parameters with rainfall intensity>100 millimeters per day, while rainfall with intensities > 50 millimeters per day is included in rainfall with heavy intensity. (BMKG, 2016). Rainfall is data that classified into spatial data. So, spatial extreme rainfall is extreme rainfall that occurs in various locations and has spatial data elements such as coordinates and the others [14].

3. Result and Discussion
In this study, we used secondary data that obtained directly from the website of the Meteorology and Geophysics Agency (BMKG), namely dataonline.bmkg.go.id. The data used is rainfall data in East Java which consists 9 of rainfall observation stations starting in 2009 until 2018.

The first step is make 4 periods (blocks) of rainfall in one year which in one block contains 3 months rainfall data based on the moonsun rainfall pattern, namely the December-Janurai-February block (DJF), March-April-May block (MAM), June-July-August block (JJA), and September-October-November block (SON). After forming many blocks, the maximum value of each block will be taken, called the Block Maxima (BM) method. So in the intervals from 2009-2018, 40 blocks will be formed. From the results of the block maxima method, a correlation between the locations obtained will be checked. So it can be concluded that all locations have a correlation or interconnection between locations with each other. Because the distribution of extreme values taken using the BM method will follow the GEV distribution, GEV parameter estimation will be carried out univariately using maximum likelihood estimation with 3 parameters, namely location parameters, scale parameters, and form parameters. Because the data used contains many locations, it is indicated that there are spatial elements that affect extreme rainfall that occurs. So the approach that can be used is the max-stable approach and the copula approach [15]. The first step to calculate dependencies using the max-stable approach is to transform the extreme value distribution into the Frechet distribution. This process is called the max-stable process. This transformation uses equation (2.8). To obtain spatial parameter estimates with this approach, the Smith model is used, followed by a trend surface model. As for the copula approach, it uses the copula unit transformation so that it forms a gaussian copula model. From the copula gaussian model, parameter estimates can be obtained which are also followed by the trend surface model. The parameter estimates used by both approaches are MLE.

After obtaining spatial parameter estimates, the rainfall prediction values obtained from each approach are used. So that in the following tables of return value with a 2-years return period at each station can be shown in the following table:

Table 1. Return level of max-stable approach

Station	Year		
	2019-2020	2021-2022	2023-2024
Banyuwangi	274.55	338.84	368.78
Juanda	527.54	660.56	722.5
Kalianget	327.55	410.37	448.94
Karangkates	590	735.93	803.89
Malang	563.13	703.17	768.39
Perak 1	535.88	671.73	734.98
Sangkapura	534	675.07	740.77
Sawahan	699.44	875.34	957.25
Tretes	666.08	693.87	758.5

Table 2. Return level of copula approach

Station	Year		
	2019-2020	2021-2022	2023-2024
Banyuwangi	293.12	379.9	423.19
Juanda	562.06	737.87	825.59
Kalianget	334.79	438.16	489.73
Karangkates	642.81	842.64	942.33
Malang	609.26	798.92	893.54
Perak 1	568.95	747.43	836.47
Sangkapura	546.53	721.47	808.75
Sawahan	758.33	996.55	1115.39
Tretes	597.41	783.7	876.75

From table 1 and 2, it can be concluded that the highest intensity rainfall occurs around the Sawahan rainfall observation station both with the max-stable approach and the copula approach respectively. Whereas to determine the best results from the two approaches used, the Akaike Information Criterion (AIC) method is used. Based on this method, according to Ligas and Banasick [13] an approach that has the lowest AIC value is the best approach that can be used. So that the results of the AIC method are as follows:

Table 3. AIC value

Approach	AIC Value
Max-stable approach	37764.95
Copula Gaussian approach	4549.22

Therefore, from Table 3 it can be said that the model using copula Gaussian is the best model compare with max-stable approach as it provides the smallest AIC values [2].

4. Conclusion

The return level for extreme rainfall events in East Java with two approach give the result that the highest intensity of rainfall will occur in the same location which is around the Sawahan rainfall observation station located in Nganjuk district. But it has different rainfall intensity. The best model produced based on the AIC is the copula approach with an AIC value of 4549.22

References

[1] C. E. R. Cooley, "A Survey of spatial extremes :Measuring Spatial Dependence and Modelling Spatial Effect," *Stat. J.*, 2012.
[2] S. P. M. R. Davison A.C, "Statistical modelling of spatial extremes," *Stat. Sci.*, 2012.
[3] H. T. S. A. Oktaviarina, A. Sofro, "Generalized Extreme Value for Climate Change Analysis," *Atl. Highlights Eng.*, 2018.
[4] L. F. E. McNeil, "Modelling dependence with copulas and applications to risk management," *Elsevier*, 2003.
[5] Gili M. Kellezi, "An application of extreme value theory for measuring financial risk," *Comput. Econ.*, 2006.
[6] Schlather, "Models for stationary max-stable random," 2002.
[7] R. A. Davis, "Statistical inference for max-stable processes in space and time," *J. R. Stat. Soc.*, 2013.
[8] F. Ramadhani, "Extreme Value Theory (EVT) Application On Estimating The Distribution Of Maxima," in *Conference Proceedings*, 2017.
[9] Nelsen, "An Introduction to Copula," *Springer Ser. Stat.*, 2006.
[10] M. Ribatet, "Spatial Extremes Package." 2018.

[11] J. Tawn, "Modelling spatial extreme events with environmental applications," *Spat. Stat.*, 2017.
[12] A. Rudvik, "Dependence Structures in Stable Mixture Models with an Application to Extreme Precipitation," 2012.
[13] Ligas M. and Banasik P., "Local height transformation through polynominal regression," *J. Geod. Cartogr.*, 2012.
[14] L. D. Haan, "A spectral representation for max-stable processes," *Ann. Probab.*, 1984.
[15] WeiB, "Copula Parameter Estimation: Numerical Considerations and Implications for Risk Management," *J. Risk*, 2010.

Forecasting Surabaya Rainfall in 2017 Using First Order Autoregressive Model through INLA Bayesian Approach

Hery Tri Sutanto[1]
[1]Mathematics Department, Universitas Negeri Surabaya, C8 Building, Ketintang Surabaya, Indonesia

E-mail: herysutanto@unesa.ac.id,

Abstract. Water is one of the natural factors which affects the life of organisms on earth. If in one area, there is a sufficient amount of rain, then it will support the life of organisms in that area. However, if an area has an excessive amounts of rain, flooding will occur and it will damage the infrastructure, the agricultural land and some existing ecosystems. One way to reduce unwanted floods is by predicting accurate flood discharge. It can be used not only to design water buildings but also to manage water flow. Flood discharge can be predicted using the time series method, namely the First Order Autoregressive model, also known as the AR model (1). Estimating parameters in the AR model (1) can be done using the Moment, least Square and Maximum Likelihood methods. However, because the sample size used in this study is small then the Bayesian method can be used to estimate the AR model parameters (1). It has some advantages such as: 1. It can produce logical interpretations to conclude statistically in time series analysis 2. The output can always be updated based on the latest information. The Markov Chain Monte Carlo method is one method that can be used for Bayesian inference calculations, specifically the Gibbs sampler and Metropolis-Hastings algorithms. Both of these algorithms are more flexible and easily adapted to estimate complex models with many parameters or characterized by posterior nonstandard distributions. The MCMC theory states that the convergence distribution of the simulation value to the posterior distribution has a number of iterations to infinity, the posterior distribution will not be feasible. So that the processing of data consume a lot of time to reach convergence, then if it is not stopped then it will give suggestions about how some MCMC iterations become convergent. If the model is more complex then the MCMC algorithm is slower and the calculation becomes not feasible. Therefore, Rue et al. (2009) proposed a fast and accurate algorithm called the Integrated Nested Laplace Approximation Algorithm (INLA).

1. Introduction

Water is one of the factors that influence the survival of living things on the earth. If an area has a sufficient amount of rain, then it can supports all life in the area. Conversely, if an area has an excessive amounts of rain that will exceed the carrying capacity of the environment, then the presence of water will harm agriculture, plantations, housing and other living things. It will cause damage to infrastructure that has been built by the government, damage to agricultural land, and damage to some existing ecosystems [1].

Regard to the flood, efforts are needed to overcome it. Rainfall data recorded by BMKG (Meteorology and Geophysics Agency) can be used to predict the amount of rainfall that will occur in the future. This allows the government to be able to plan, improve development and anticipate future floods. In order to reduce an unwanted flooding, we can just predict the incidence of it. Furthermore, it can be used to design water buildings, rivers, gutters, irrigation, water seepage, dams and water management [2].

Geographically, the city of Surabaya has an area of 326.81 km^2 with an average rainfall of 1,321 mm per year. From May to October in Surabaya, there is a dry season. In November to April the rainy season occurs. If the rainfall is too many and the carrying capacity of the environment (the place of seepage, rivers, ditches, rice fields, lakes) is inadequate then there will be a big flood in Surabaya [3].

There are several studies on rainfall in Indonesia using univariate time series. One of them is an autoregressive (AR1) model which is assumed to be a stationary rain condition while the rain is not stationary. Then the AR (1) model has developed using the ARMA (Autoregressive Moving -Average) model which is assumed to be a stationary rain process and lognormal distribution. To estimate parameters in the AR (1) model: 1). Moment Method, 2). Least Square Method, and 3). Maximum Likelihood Method [4,5].

It is because the sample size used in this study is small then to estimate the parameters of the AR model (1), we use Bayesian method. The reason for involving Bayesian analysis in time series analysis are, 1). In this approach, it can produce a logical interpretation to conclude statistically in time series analysis, 2). The output can always be updated based on the latest information [5].

The Monte Carlo Markov Chain method is a supporter of calculations for Bayesian inference. In particular the Gibbs sampler and Metropolis-Hastings algorithm. They are more flexible and easily adapted to estimate complex models that use many parameters or characteristics by a nonstandard posterior distribution. The MCMC theory states that the distribution of the convergence of the simulation value to the target density (posterior distribution) if the number of iterations is infinite then the posterior distribution is not feasible to run an infinite Markov chain. Hence, in order to process the data, it takes time too long to reach convergence, if it is not stopped, it will give a suggestion how some MCMC iterations need convergence [6,5].
The main key in drawing conclusions with MCMC is greater caution (a lot of time) to decide and monitor the conference stage in determining the best setting (parameterization, prior distribution, initial value and distribution of Martin-Hasting proposals) which results in a reliable and accurate MCMC output (Brooks et al., 2011).

Several methods for producing existing posterior distributions have their advantages and disadvantages. Rue et al. (2009) have developed a method for estimating parameters that is better and more efficient than the methods described earlier. This method is called Integrated Nested Laplace Approximation (INLA) [7].

Regard to the background above, here the following problem in this study. a) how to use the Order One autoregressive model to predict 2017 rainfall in Surabaya using 2016 rainfall data, b) How to use the Bayesian Order One Autoregressive Model in INLA to predict 2017 rainfall in Surabaya using 2016 rainfall data.

2. Study Literature

2.1. Time Series

The main purpose of time series analysis is to produce predictions of future values based on current data. Time series analysis is a method used to study theoretically and application. According to Cryer and Chan (2008), time series data is a series of observational data arranged according to time sequence. Time series forecasting is a model to predict future time values based on data (events) that have occurred [4,6].

2.2. Stationary Data

If a time series data is plotted then there is no evidence of change in the mean value from time to time then it can be said that the data series is stationary at its mean. If a time series data is plotted then it does not show any change in variance from time to time then it is said that the data series is stationary in its variants. If a time series data is plotted then it shows the change in mean (some trend-cycle) over time then it is said the data series is not stationary at its mean. If a time series data is plotted, then it does not show any change in mean and variance from time to time then it is said that the data series is not stationary at the mean and its variants [4,8].

Journal of Physics: Conference Series **1417** (2019) 012011 doi:10.1088/1742-6596/1417/1/012011

2.3. *Stationary Average Data*

According to (Hanke J E, Ritse A., and Witchen, 1999) the autocorrelation values on stationary data to the average will not be significant after the second or third lag time, while the non-stationary data has a significant autocorrelation value after the third lag time or several periods. Stationary to the average can be known graphically by looking at the ACF plot, the data is said to be stationary to the average if at most the first 3 lags on the ACF plot do not exceed the ± $\pm\frac{2}{\sqrt{n}}$. If the data is not stationary to the average, it is necessary to do a differentiation process until the data is stationary to the average [4,8]. The first form of differentiation can be stated as follows.

$$\nabla Y_t = Y_t - Y_{t-1}$$

Statistic used

$$t_{hit} = \frac{\varphi^*}{Se(\varphi^*)}$$

where,
$\hat{\varphi}^*$: Estimated value of the Autoregressive (AR) parameter
$Se(\hat{\varphi}^*)$: Error standar of $\hat{\varphi}^*$

2.4. *Forecasting Fist Order Autoregressive Model or AR Model (1)*

The simplest model of AR (1) with zero on its mean [9]

$$Y_t = C + \rho\ Y_{t-1} + e_t \qquad \text{where } -1 < \rho < 1$$
$$t = 1, 2, \ldots, n$$

It is a stationary AR (1) process (Box et al, 1970)
Y_t is t^{th} observation time series
$\{e_t\}$ is a process
C is an arbitrary constant
ρ is a parameter autoregressive

2.5. *Laplace Approximation*

The INLA approach was first proposed by Rue et al. It is to determine a Bayesian inference algorithm based on integrated nested Laplace approximations (INLA). INLA was designed specifically for Gaussian latent models that are quite flexible which are used for different applications. See Rue et al. [10] who review some of the examples used in the R-INLA package program.
Suppose $f(x)$ is probability density function for random variable x, then

$$\int f(x)\,dx = \int exp(\log f(x))\,dx \qquad (1)$$

The value of $\log f(x)$ can be obtained by using Taylor series and take $x = x_0$ as follows

$$\log(f(x)) \approx \log f(x_0) + (x - x_0)\frac{\partial \log f(x)}{\partial x}\Big]_{x=x_0} + \frac{(x-x_0)^2}{2}\frac{\partial^2 \log f(x)}{\partial x^2}\Big]x = x_0 \qquad (2)$$

If $x_0 = x^* = argmax\ \log f(x)$ then $(x - x_0)\frac{\partial log f(x)}{\partial x}\Big]x = x_0 \;= 0$

Equation (2) will be

$$\log(f(x)) \approx \log f(x^*) + \frac{(x-x^*)^2}{2}\frac{\partial^2 \log f(x)}{\partial x^2}\Big]x = x^* \qquad (3)$$

Substitute (3) into (1) then

$$\int f(x)dx \approx \int exp\left(\log f(x^*) + \frac{(x-x^*)^2}{2}\frac{\partial^2 \log f(x)}{\partial x^2}\Big]x = x^*\right) dx$$
$$= \exp(log\ f(x^*)\ \int exp\left(\frac{(x-x^*)^2}{2}\frac{\partial^2 \log f(x)}{\partial x^2}\Big]x = x^*\right) dx \qquad (4)$$

Where the integrand relates to the density of normal distribution.
Choose

$$\sigma^{2*} = -\frac{1}{\frac{\partial^2 \log f(x)}{\partial x^2}\Big]x=x^*}$$

Such that

$$\int f(x)dx \approx exp(\log f(x^*)) \int exp\left(-\frac{(x-x^*)^2}{2\sigma^{2*}}\right) dx \quad (5)$$

where $exp\left(\frac{(x-x^*)^2}{2}\frac{\partial^2 \log f(x)}{\partial x^2}\Big| x = x^*\right)$ is a kernel of normal distribution (x^*, σ^{2*})

If the integral above has boundary at domain (α, β) then (5) becomes

$$\int_\alpha^\beta f(x)\, dx \approx f(x^*) \sqrt{2\pi\, \sigma^{2*}} \; (\phi(\beta) - \phi(\alpha)) \quad (6)$$

where $\phi(.)$ is cumulative density function of normal distribution.

2.6. *Latent Gaussian Model*

INLA is designed with the Gaussian latent model, where the observed (response) variable y_i is assumed to belong to a distribution family (not necessarily from an exponential family) [7]. Some parameters of the $\emptyset_i$ family are links from the additive predictor structure of η_i through a link function $g(.)$, such that $g(\emptyset_i) = \eta_i$.

Additive linear predictor $\eta_i$is defined as follows

$$\eta_i = \beta_0 + \sum_{m=1}^{m=M} \beta_m\, x_{mi} + \sum_{l=1}^{L} f_l(Z_{li})$$

where:

β_0 is a scalar for intercept

$\beta = \{\beta_1, \beta_2, \dots, \beta_M\}$ is a coeficient of linear effect quantity from severalcovariates $x = (x_1, \; x_2, \dots, x_M)$ in response and $f = \{f_1\,(.), \dots, f_L(.)\}$ is a collection of function which is defined from a covariate $Z = (z_1, \dots, z_L)$. In $f_L(.)$, it can be assumed some different form such that smoothness, non linear effect covariate, time series, seasonal effect, intercept and random slope (temporal random effect or spatial random effect). A group of latent gaussian model is flexibel and able to accommodate a wide range of models ranging from generalize and dynamic linear models to spatial and spatio-temporal (Martins et al., 2013).

A group of spatio-temporal model is usually constructed by hierarchical bayesian with three step. First step is construct an observation mode $\pi(y|x)$ where $\pi(.|.)$ is density and y is an observation vector. If it is assumed y_i independent, latent component (cannot be observed) x and hyperparameter vector θ then the distribution of N observation is likelihood:

$$\pi(y|x,\theta) = \prod_{i=1}^{i=N} \pi(y_i|x_i,\theta)$$

Second step is latent gaussian field $\pi(x|\theta)$ where a prior of multivariate gaussian with mean is equal to zero and precision matrix ϱ for x. It depends on hyperparameter θ (third step) which is no need to be Gaussian. Hence, $x \sim N\left(0, \varrho^{-1}(\theta)\right)$ with density function as follows:

$$\pi(x|\theta) = (2\pi)^{-N/2} \; |\varrho(\theta)|^{1/2}\, exp\left(-\frac{1}{2}x^T\varrho(\theta)x\right)$$

Suppose, a components of latent gaussian field x is conditionally independent such that $\varrho(\theta)$ is sparse precision matrix [7].

3. Discussion

3.1. *AR Model (1) Identification*

The rainfall data is identify as follows. Find the scatter plot of the data from day one to last day in the same year.

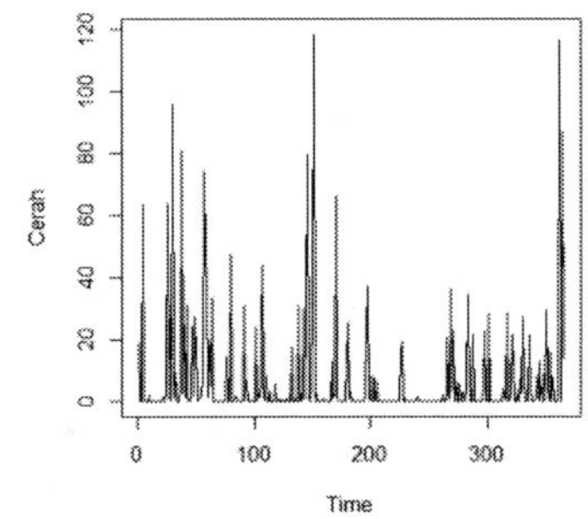

The results of the plot turned out to be not stationary data. Therefore data needs to be transformed with first differencing. Unfortunately, because on the first differencing, the data remained not stationary on the 210th day up to the 366th day, then it requires to differencing in the second order to obtain stationary data.

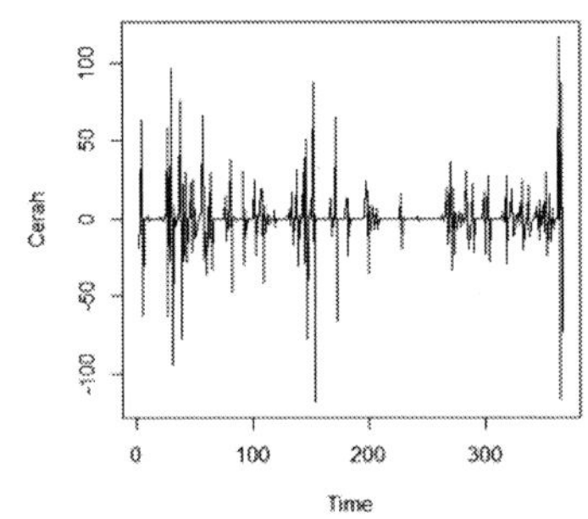

3.2. AR Model (1) and test parameters

Regard to the identification above, there are some possiblities model such as:

i. Model AR(1) atau model ARMA(1,0) atau Model ARIMA(1,0,0)
ii. Model Arma (0,1) atau Arima(0,0,1)
iii. Model Arma(1,1) atau Arima(1,0,1)
iv. Model Arima(1,2,0)
v. Model Arima(0,2,1)

By using R software and forecast library, we get

```
> library(forecast)
> auto.arima(sbydat)
Series: sbydat
ARIMA(1,0,1) with non-zero mean
Coefficients:
         ar1      ma1    mean
      0.8633  -0.7966  6.5592
s.e.  0.1347   0.1591  1.2537
sigma^2 estimated as 260.5:  log likelihood=-1535.84
AIC=3079.67   AICc=3079.78   BIC=3095.28
```

So, Arma (1,1) or Arima (1,0,1) had been choosen then it can be used to predict the rainfall for next two months or next 60 days based on the data with program as follow.

```
> sby_arma_c<-forecast(arma_c,h=60,level=c(99.5))
> sby_arma_c
```

Therefore, the result can be plotted like follows

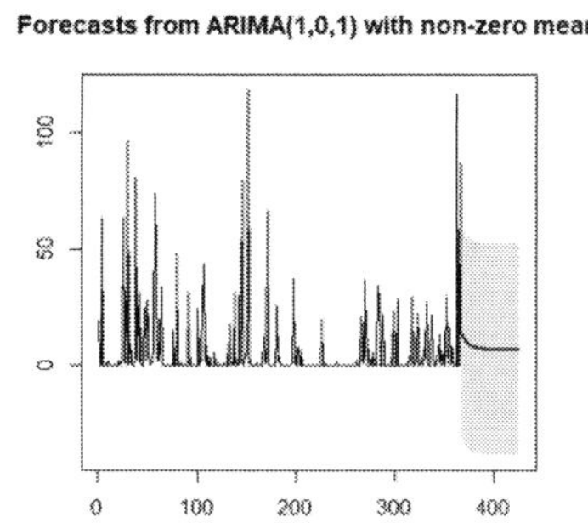

3.3. *AR Model (1) and Bayesian INLA*

```
library(INLA)
temp = read.csv("D:/BAYESIANINLA2019/Dataiinla/Tugas AR1 INLA/Olahdatahujansby2016.csv", header=F)
n = nrow(temp)
data = data.frame(y = temp[,1], t=1:n)
dates <- data[,2]
plot(dates, data$y, lwd=2, xlab='harian', ylab='Data Curah Hujan')
lines(dates,data$y)
```

It gave this result

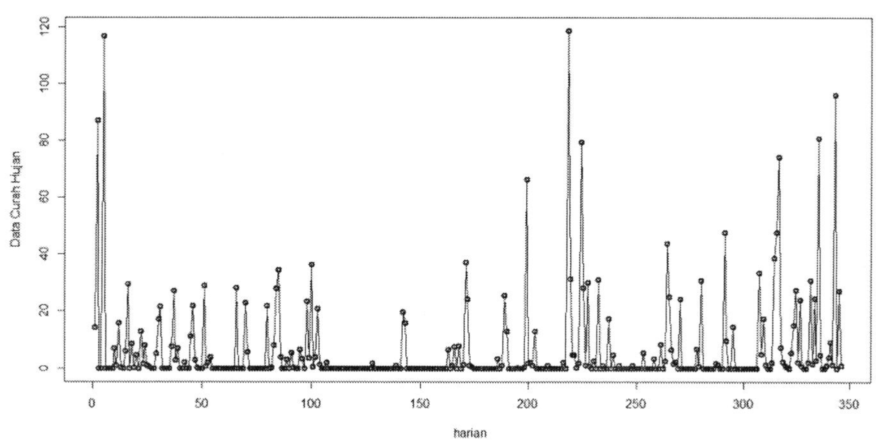

By observing the likelihood and AR(1) model, we can get this result

```
family <- "gaussian"
formula1 <- y~ f(t,model='ar1')
```

Calculate the posterior distribution $\pi(\theta|y)$

where $\theta = (\theta_1, \theta_2)$

θ_1 is hyperparameter with $\theta_1 = \log(K)$

θ_2 is hyperparameter with $\theta_2 = \log(1 + \rho_1 - \rho)$

By using transformation, it will give $\pi(K|y)$ and $\pi(\rho|y)$

where $K = \tau(1 - \rho_2)$

By using this program

```
res1 <- inla(formula=formula1,data=data,family=family)
summary(res1)
```

```
Call:
   "inla(formula = formula1, family = family, data = data)"
Time used:
    Pre = 2.83, Running = 1.96, Post = 0.183, Total = 4.97
Fixed effects:
              mean    sd 0.025quant 0.5quant 0.975quant  mode kld
(Intercept) 6.393 0.847      4.729    6.393      8.056 6.393   0
Random effects:
  Name    Model
    t AR1 model
Model hyperparameters:
                                        mean       sd    0.025quant
Precision for the Gaussian observations 4.00e-03 0.00e+00     0.003
Precision for t                         1.91e+04 1.86e+04  1270.650
Rho for t                               2.00e-03 6.98e-01    -0.987
                                        0.5quant  0.975quant    mode
Precision for the Gaussian observations  4.00e-03   4.00e-03   0.004
Precision for t                          1.36e+04   6.83e+04 3432.27
Rho for t                                4.00e-03   9.88e-01  -0.997
Expected number of effective parameters(stdev): 1.00(0.00)
Number of equivalent replicates : 365.94
Marginal log-Likelihood:  -1555.17
```

4. Conclusion

Before processing, it takes 2.83 and then it runs (running) = 1.96. Furthermore, it takes 0.183 (Post), hence it takes 4.97 in total. In summary, the output above shows the distribution of β_0 (intercept), distribution τ (precision for Gaussian observation), K distribution (precision for t) and distribution ρ (Rho for t) which have 95% of the shortest Highest Posterior Density Region (HPDR) is τ (precision for Gaussian observations). From the results of the fitting, we can also see two types of points that almost overlap everywhere, which means that the results of the model and real data are almost the same.

References

[1] Tim D 2007 *Fundamentals of Hydrology, Routledge Fundamentals of Physical geography* (New Zealands)

[2] Hidayah E, Anwar N E, and Iriawan N 2010 Evaluating Error of Temporal Disaggregation from Daily into hourly Rainfall using Heytos Model at sampean catchments Area IPTEK *The Journal of Technology and Science* **21**(1)

[3] Kristianta F and Fithrisani K 2016 Predict rainfall in East Surabaya with Wector Autoregressive Neural Network *Journal Sains and Art ITS* **5**(2)

[4] Agung N G I 2009 *Time Series Data Analysis Using EViews* (Asia: John Wiley & Sons Pte Ltd.)

[5] Maoumi H, Niekerk J, and Staden P 2010 *Bayesian Analysis of AR(1) model* (South Africa: University of Pretoria)

[6] Detting M 2016 *Applied Time Series Analysis, Institute for Data Analysis and Process Design* (Winterthur: Zurich University of Applied Sciences)

[7] Blangiardo M and Cameletti M 2015 *Spatial and Spatio-temporal Bayesian Models with R_INLA* (United Kingdom: John Wiley & Sons, Ltd)

[8] Rosadi D 2011 *Ekonometrika & Analisis Runtun Waktu Terapan dengan Eviews Aplikasi untuk bidang ekonomi, bisnis dan keuangan* (Yogyakarta)

[9] Zheng R and Bakka H 2018 *Analysing unemployment data with the AR1 Model*

[10] Rue H and Martino 2009 *Approximation Bayesian Inference for Latent Gaussian Models Using INLA*

[11] Chib S 1991 *Bayes regression with autoregressive errors A Gibbs sampling approach* (USA: Washington University)

[12] Gomes-Rubio V and Rue H *2017 Markov Chain Monte Carlo with the Integrated Nested Laplace Approximation* (Spain)

[13] Otieno E 2013 *Bayesian Spatial and Spatiotemporal Modeling (Applied to precipitation data set).*

Multinomial Logistic Regression and Support Vector Machine for Osteoarthritis Classification

C Aroef[1*], R P Yuda[1], Z Rustam[1], and J Pandelaki[2]

[1]Department of Mathematics, University of Indonesia, Depok 16426, Indonesia
[2]Department of Radiology, RSUPN dr. Cipto Mangunkusumo, Jakarta 10430, Indonesia

e-mail: *chelvian.aroef@ui.ac.id, rivan.pratama@sci.ui.ac.id, rustam@ui.ac.id

Abstract. Everyone joints go through a cycle of damage and repair during their lifetime, but sometimes the body's process to repair our joints can cause changes in their shape or structure. When these changes happen, it's known as osteoarthritis. Osteoarthritis is the most common form of arthritis, affecting millions of people worldwide. Osteoarthritis causes pain, swelling, stiffness in the areas, and decreased the ability to move for the sufferers. Therefore it requires accurate method of classification. Many methods have been used to classify osteoarthritis, but this study will apply Multinomial Logistic Regression and Super Vector Machine (SVM) as the machine learning methods. We used CT scan result data from RSUPN dr. Cipto Mangunkusumo, Central Jakarta. The results show the SVM provides better results than Multinomial Logistic Regression in terms of classification accuracy. The highest accuracy of SVM reaches around 85%, while Multinomial Logistic Regression only 71%.

1. Introduction

All joints go through a cycle of damage and repair during their lifetime, but sometimes the process of repairing the joints may cause changes in their shape or even structure. When these changes happen in one or more of the joints in the knee, it's known as osteoarthritis (OA) [1]. In osteoarthritis, the cartilage in the knee joint gradually wears away [2]. As the cartilage wears away, it becomes frayed and rough, and the protective space between the bones decreases [2]. As the cartilage getting rougher, the knee doesn't move as smoothly as it should, and it might feel painful and stiff [1].

Osteoarthritis usually happens to older people because of the ability of cartilage to heal decreases [3]. As the person getting heavier, the risk of getting OA is also getting higher, because the joints will get more pressure from the body [3]. Every pound of weight gain adds 3 to 4 pounds of extra weight on the knees [3]. OA may also happen to an athlete, cause the overuse of the knee [3]. If you feel the pain when you are active, but gets a little better with the rest, maybe you should check your knee because it's one of the symptoms of OA [3]. The other symptoms are the joints become stiff and swollen, making it difficult to bend and straighten the knee [1]. OA impacts most of your activity because it damages the knee which mostly used for activities. There are several treatment options available for OA, one of them is exercise [3]. Exercise will strengthen the muscle around the knee and makes the joint more stable and decrease pain [3].

The other option is losing some weight, losing even a small amount of weight can significantly decrease knee pain from OA.

The previous study on OA classification used AdaBoost Support Vector Machine [4]. This study will apply another method of machine learning which never been used before for Osteoarthritis classification, such as Multinomial Logistic Regression (MLR) and Support Vector Machine (SVM). Multinomial Logistic Regression is a development method of Logistic Regression when the dependent variable is more than two classes. Super Vector Machine is known for its classification method for two classes or more. Both methods are applied because the OA is going to be classified into 4 classes. We will these methods in terms of accuracy. By using both methods, it is expected to help the health sector to be able to classify osteoarthritis more efficiently.

2. Methods

2.1. *Data*

The data for this study is obtained from dr. RSUPN Cipto Mangunkusumo (RSCM), Central Jakarta, by the result of Cartigram scan, which composed of six features, which are: Gender, Age, Femur Lateral, Tibia Lateral, Femur Medial, and Tibia Medial, and one prediction class. The data includes 42 observations which classify into four classes.

Table 1. Results of Cartigram of Osteoarthritis Data from the Department of Radiology, RSCM.

Number of Patient	Gender	Age	Femur Lateral			Tibia Lateral			Femur Medial			Tibia Medial			Grade OA
			1	2	3	4	5	6	1	2	3	4	5	6	
1	F	74	46.5	50.5	36.5	53.5	40	32	49	65.5	58	46	50	49.5	3
2	F	62	46	40.5	71.5	53	34	36.5	54	54	47	43.5	33	34	1
3	M	65	50.5	48.5	46	27	23.3	29	45	72.5	35	35	34.5	71	2
4	F	74	41	36.5	53	39	34	46	53	34	45.5	52.5	37	24	4

2.2. *Logistic Regression*

Logistic Regression in many ways is the natural complement of ordinary linear regression whenever the target variable is categorized.

For a target (dependent) variable Y with two class and predictor (independent) variable X, let $g(x) = Pr\ (X = x) = 1 - Pr(X = x)$, the logistic regression model has a linear form for Logit with probability as follows:

$$Logit[g(x)] = log\left(\frac{g(x)}{1 - g(x)}\right) = \alpha + \beta x, \qquad where\ the\ odds = \frac{g(x)}{1 - g(x)} \tag{1}$$

The Logit has a form of linear approximation, and logit is equaled with the logarithm of the odds. The parameter β is the rate of increase or decrease of the S-shaped curve of $g(x)$.

2.3. *Multinomial Logistic Regression*

The logistic regression can be extended to models with multiple predictor variables. Let *u* define the number of predictors for a binary response Y by $x_1, x_2, \dots, x_u$. The model for log odds as follow:

$$Logit[Pr\ Pr\ (Y = 1)\ ] = \alpha + \beta_1 x_1 + \beta_2 x_2 + \cdots + \beta_u x_u \tag{2}$$

and the alternative formula, directly specifying $h(x)$ is

$$h(x) = \frac{exp(\alpha + \beta_1 x_1 + \beta_2 x_2 + \cdots + \beta_u x_u)}{1 - exp(\alpha + \beta_1 x_1 + \beta_2 x_2 + \cdots + \beta_u x_u)} \tag{3}$$

The parameter β_i, for $1 < i \leq u$ and $1 < j \leq u$ and $i \neq j$, refers to the effect of x_i on the log odds that $Y = 1$, controlling other x_j, for instance, $exp(\beta_i)$ is the multiplicative effect on the odds of a one-unit increase x_i, at fixed level of other x_j.

If we have n independent observations with u-predictor variables, and the target variable has q categories, to build the logits in the multinomial case, one of the categories must be considered as the base level and all the logits functions are constructed relative to it. There are no conditions to turn the category into the base level, so we can choose randomly, we will turn category z as the base level. Since there is no order, it is possible that any category could be labeled z. Let H_j denote the multinomial probability of observation for the jth category, to find the relationship between this probability and the u-predictor variables, $X_{1,}X_2, \dots, X_u$, the multinomial logistic regression model is:

$$log \left[\frac{H_j(x_i)}{H_z(x_i)}\right] = \alpha_{01} + \beta_{1j}x_{1i} + \beta_{2j}x_{2i} + \cdots + \beta_{uj}x_{ui} \quad (4)$$

where $j = 1,2, \dots, z - 1$ and $i = 1,2, \dots ,n$. Since all the H's adds to unity this reduces to

$$log\ log\ \left(H_j(x_i)\right) = \frac{exp(\alpha_{01} + \beta_{1j}x_{1i} + \beta_{2j}x_{2i} + \cdots + \beta_{uj}x_{ui})}{1 + \sum_{j=1}^{z-1}\ \alpha_{01} + \beta_{1j}x_{1i} + \beta_{2j}x_{2i} + \cdots + \beta_{uj}x_{ui}} \quad (5)$$

for $j = 1,2, \dots, z - 1$. The model parameters are determined by the method of multinomial linear.

2.4. *Support Vector Machine*

Support Vector Machine (SVM) is one of a machine learning algorithms usually used for classification and regression introduced by Russian Scientist, Vapnik in 1992. For the first time of SVM appearance, SVM was used only for binary classification, but now it could be used for more than two-class classification, usually called multiclass classification. SVM takes the form of mapping input space into higher dimensional space to support nonlinear classification problems where the maximum separation of the hyperplane is constructed. The hyperplane is a linear pattern (depends on the kernel) that maximizes the margin resulting maximum value between classification classes.

2.4.1. Characteristic of SVM [6]. Given dataset $\{x_m, y_m\}_{m=1}^N$ where N is the number of samples, $x_m \in R^D$ is a feature vector from sample-m, where D is the number of features (dimension), and y_m is the class label. For binary classification $y_m \in \{-1, +1\}$, but for more than two (multiclass) classification problem $y_m \in \{1,2, \dots, u\}$ where u is the number of classes. The main goal of using Support Vector Machine is to find the best hyperplane for the data:

$$q \cdot x + p = 0 \quad (6)$$

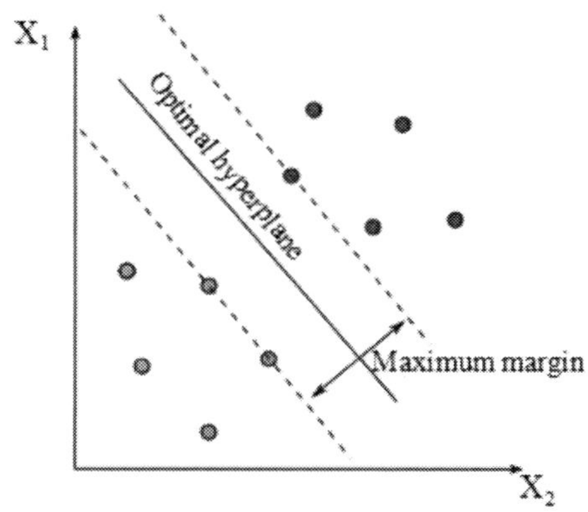

Figure 1. SVM trying to find the best hyperplane to separate data into two classes.

The problem of SVM optimization can be summarized as follows:

$$\frac{1}{2}\ ||q||^2 \tag{7}$$

$$s.t. y_i(q^T \cdot x_i + p) \geq 1, \forall m = 1, \ldots, N \tag{8}$$

The objective function (7) to find $q \in R^n$ and $p \in R^n$ subject to (8), where q is the weights vector and p is bias. By solving the equation (7) and (8), we get the equation of q and p as follows:

$$q = \sum_{m=1}^{N} a_m y_m x_m \tag{9}$$

$$p = \frac{1}{N_s} \sum_{m \in s} \left(y_m - \sum_{m \in s} a_m y_m x_m \right) \tag{10}$$

By the equation we get the decision function as follows:

$$f(x) = q \cdot x + p \tag{11}$$

which could maximize the margins.

3. Experimental Result

This study used software R Studio version 1.1.463 for both Multinomial Logistic Regression and Support Vector Machine.

3.1. Osteoarthritis Classification using Multinomial Logistic Regression

Table 2. Results of Accuracy Osteoarthritis Classification using Multinomial Logistic Regression.

Data Training (%)	Accuracy (%)
10	50
20	54.54
30	51.85
40	66.67
50	60

60	62.50
70	72.72
80	71.14
90	50

As a result of the table, the data training at 70% has the best accuracy among the rest with around 72.72%. The worst accuracy is recorded at data training 10% and 90% with 50%.

3.2. *Osteoarthritis Classification using Support Vector Machine*

Table 3. Results of Accuracy Osteoarthritis Classification using Support Vector Machine with Radial Basis Kernel Function, parameter C=4, and $\sigma = 0.1259338$.

Data Training (%)	Accuracy (%)
10	67.56
20	54.54
30	62.06
40	66.67
50	60
60	62.50
70	72.72
80	85.71
90	67.67

As a result on the table, the data training at 80% has the best accuracy among the rest with around 85.71%. The worst accuracy is recorded at data training 20% with 54.54%.

4. Conclusion

This study applied two different classifiers methods, which are Multinomial Logistic Regression (MLR) and Support Vector Machine (SVM) to compare the accuracy. From the simulation for SVM with Radial Basis Function (RBF) kernel, we get the best parameter which is $C = 4$ *and* $\sigma = 0.1259338$. From the experiment, we get the best accuracy is around 72.72% for Multinomial Logistic Regression from 70% training data. Support Vector Machine has better accuracy for 85.71% from 80% training data. These results show the performance of Support Vector Machine with RBF kernel is better than Multinomial Logistic Regression in terms of accuracy.

Acknowledgment

This research is supported financially by the University of Indonesia, with DRPM PIT-9 2019 research grant scheme, ID number NKB-0039/UN2.R3.1/HKP05.00/2019

References

[1] Versusarthritis accessed 14 July 2019 see https://www.versusarthritis.org/about-arthritis/conditions/osteoarthritis-of-the-knee/

[2] Orthoinfo accessed 14 July 2019 see https://orthoinfo.aaos.org/en/diseases--conditions/arthritis-of-the-knee/

[3] Webmd accessed 14 July 2019 see https://www.webmd.com/osteoarthritis/ostearthritis-of-the-knee-degenerative-arthritis-of-the-knee#1

[4] Adyalam T R, Rustam Z 2018 Classification of Osteoarthritis Disease Severity Using Adaboost Support Vector Machine *J. Phys. Conf. Ser.* **1108** (1) 012062

[5] Abdalla M E 2012 An Application on Multinomial Logistic Regression Model *Pakistan Journal of*

Statistics and Operation Research
[6] Cristianini N and Taylor J S 2000 *An Introduction to Support Vector Machines and Other Kernel-based Learning Methods* Cambridge University Press

Journal of Physics: Conference Series **1417** (2019) 012013 doi:10.1088/1742-6596/1417/1/012013

Subsampling Method for Big Data in Poisson Regression

A N I Pradana and A Sofro*

Mathematics Department, Universitas Negeri Surabaya, Ketintang, Surabaya 60231, East Java, Indonesia

* Email: ayuninsofro@unesa.ac.id

Abstract. Big Data is very large, too fast and complex that cannot be managed with traditional tools; and must use new methods and tools to process it. Subsampling method is one method that can be used to solve big data problems. This paper focuses on estimation in Poisson Regression using Maximum Likelihood Estimation (MLE). In this study, it will be discussed about the implementation of the Subsampling method of The Demand for Medical Care data. To understand the performance of subsampling method, the paper provided a scenario with different sample sizes. The results of this study obtained parameter estimates and Poisson regression models on each subsample. Based on the Akaike Information Criterion, the best model is providing the smallest value. From the scenario, the sample size of the subsampling which is close to sampling tend to have smaller value of AIC.

1. Introduction

Big data is a collection of data that is too large so that there is a lot of information that is almost unlimited in number and requires large computational costs in analyzing it. The simplest way to reduce the computational cost of a procedure is to subsample data before doing anything else. With the development of science and technology, the statistical strategy that can solve the big data problem is using the subsampling method which provides a more effective. The approach is taking subsamples from the original data set and using the subsamples to estimate models, predictions and statistical inferences. There are several variants of the subsampling algorithm, for example to solve Ordinary Least Square (OLS) in linear regression for big data [1] and Leverage techniques are used for more informative samples from complete data subsets for linear regression [2].

The subsampling technique can be applied in most of the statistical analysis approach. If we want to predict or find out the relationship between variables, then regression analysis can be used which has two types of variables, namely the response (Y) variable and predictor variable (X). In this paper, we will focus on Poisson regression as one method to describe the relationship between several independent variables and the dependent variable in the form of discrete data [3].

One of the count data is about the demand for medical data which is concerned with the number of requests medical care for individuals aged 66 years. The total observation is 4406 covered by Medicare public insurance obtained from the United States National Medical Expenditure Survey (NMES) [4]. If all data is used to model, it is very inefficient and takes a long time. How to analyze the data efficiently is becoming increasingly important. Therefore, we need a more powerful approach, one of them is using the subsampling method which can make it easier to analyze. It also can overcome computational problems because of large data [5].

2. Method

2.1 Big Data
Big data is data that exceeds the capacity process of a conventional database system. Data is too large, moving too fast or not following the existing database architecture structure. Big data is voluminous and complex data sets; and therefore, it is difficult to be analyzed [6]. To get the value of the data, we must choose an alternative way to process it [7]. There are three main dimensions in big data, namely 3V: Volume showing the amount of data, Velocity shows the speed at which data is generated and analyzed, Variety shows on many forms of data taken, both structured and unstructured [8].

2.2 *Subsampling Method*
The first Subsampling method was introduced by Carlstein (1986) as a tool for estimating the parameters of statistical sampling distribution calculated from the sample [9]. The subsampling method is a method for dealing with big data by taking a subsample from the original data set concerning the probability distribution and using this sample as a substitute of the original data to estimate the model, predictions and statistical inference. In subsample retrieval for parameter estimation can use simple random sampling [5].

2.3 Poisson Regression
Poisson regression is one of the nonlinear regression methods. Poisson regression analysis is one of the statistical analytical tools. Poisson regression aims to describe the relationship between response / dependent variables in the form of discrete data and several predictor / independent variables [3]. In Poisson must meet the assumption of equidispersion, which means that the variance value of the response variable Y given by $X = x$ must be equal to the mean value ie $Var(x) = E(x) = \mu$. Suppose there is a response variable that $y_i$is the number of occurrences of an event and is a measure of observation, then the probability function y is:

$$f(y_i) = \frac{e^{-\mu_i}\mu_i{}^{y_i}}{y_i!}, y_i = 1,2,3, \dots n, \qquad (1)$$

or can be written $y_i \sim Poi(\mu_i), i = 1,2, \dots, n$.

where are the parameters $\mu_i > 0$. If Y Poisson distribution has the same mean and variance, it will be shown as follows[10]:

$$E(y) = \mu \text{ and } var(y) = \mu.$$

Furthermore, the Poisson regression model can be written as

$$y_i = exp\ exp\ (x_i^T \beta)\ . \qquad (2)$$

2.4 Maximum Likelihood Estimation (MLE)
The Maximum Likelihood Estimation (MLE) method is one method for estimating parameters. This method was first introduced by RA Fisher in 1912. This estimation method can be applied to most problems and has a strong intuitive appeal, and often produces a good predictor of parameters β.

Let $y_1, y_2, \dots, y_n$ be a continuous random variable of size n with a probability function $f(y_1, \beta)$ and β is an unknown parameter.

In Poisson regression, the likelihood function is as follows:

$$L(y,\beta) = \prod_{i=1}^{n} f(y_i, \beta)$$
$$= \prod_{i=1}^{n} \left[\frac{e^{-\mu_i}\mu_i{}^{y_i}}{y_i!}\right] = \frac{\prod_{i=1}^{n} \mu_i{}^{y_i} exp\,(-\sum_{i=1}^{n} \mu_i)}{\prod_{i=1}^{n} y_i!}, \qquad (3)$$

And the log-likelihood function can be written as follows:

$$l(\beta) = log\ log\ L(y\ ;\ \beta)\ = log\ log\ \prod_{i=1}^{n}\ [\frac{e^{-\mu_i}\mu_i^{y_i}}{y_i!}]$$
$$= \sum_{i=1}^{n}\ y_i(x_i^T\beta) - \sum_{i=1}^{n}\ exp(x_i^T\beta) - log \sum_{i=1}^{n}\ y_i!. \quad (4)$$

Parameter values β get by maximizing the opportunity function. This is done by looking for the first partial derivative of the likelihood function for each parameter and then zero. So, MLE $\hat{\beta}$ is the solution to the following equation:

$$\frac{\partial loglog\ L(y\ ;\ \beta)}{\partial\beta} = 0.$$

2.5 *Subsampling Methods in Poisson Regression*

In the first sampling approach, which is choosing a small subset of complete data called the subsampling step, then using the sample to estimate the model parameters.
General Subsampling algorithm:

- Subsampling
  Taking a random subsample with a size $r > p$ where p is the number of predictor variables of all data using sampling probability distribution $\{\pi_i\}_{i=1}^{n}$. Record the data selected $\{x_i^*, y_i^*\}_{i=1}^{r}$ as predictor variables and response variables, as well as π_i^*, to $i = 1,2, \ldots, r$.
- Estimation
  Maximize the log-likelihood function to get estimates β based on the subsample.
  Based on the log-likelihood function of Poisson regression in equation (4), the log-likelihood function in the subsample can be written as follows:
  $$l(\beta) = \frac{1}{r}\sum_{i=1}^{r}\ [y_i^*(x_i^{*T}\beta) - exp(x_i^{*T}\beta)]. \quad (5)$$
  Maximization can be applied with the Newton Rapshon method, with some iterations to converge.
  In the subsampling method, the subsample number is getting closer to the sample, the smaller the MSE value means that the formed model is better if the subsample number is close to the number of samples [11].

2.6 *The Demand for Medical Care*

Demand for Medical Care is a health care request for the elderly population covered by Medicare, a public insurance program that offers substantial protection against health care costs. The request for treatment includes the number of visits to doctors / non-doctors in the office and the number of visits to doctors / non-doctors in outpatient hospitals. To model the demand for medical care, the number of doctor visits as a response variable was used and there were some factors including the number of hospitalizations at the hospital, self-perceived health status, number of patients in chronic conditions, gender, number of years of education, private insurance indicators.

Data were obtained from the United States National Medical Expenditure Survey (NMES) conducted in 1987 and 1988 to provide an overview of how Americans use and pay for health services. NMES is based on representation, national probability samples of civilians, non-institutionalized populations, and individuals who received long-term care facilities during 1987. Under the NMES survey, more than 38,000 people in 15,000 households throughout the United States were interviewed every three months about insurance coverage health, services used, and the costs and sources of payment.

2.7 *Akaike's Information Criterion (AIC)*

Akaike's Information Criterion (AIC) is the best selection criterion among some models. The best model is a model that has a minimum AIC value among all other models.
Akaike's Information Criterion (AIC) is defined as follows [12]:

$$AIC = -2\ ln\ ln\ L(\beta) + 2k\ , (6)$$

where $L\ (\beta)$ is the likelihood value and k is the number of parameters.

Journal of Physics: Conference Series **1417** (2019) 012013 doi:10.1088/1742-6596/1417/1/012013

3. Results and Discussion

3.1 Subsampling Method for Poisson Regression

3.1.1 Taking Subsample

The first step is taking subsample at random or random sampling without returning, each member has the same opportunity to be chosen as a subsample member. The subsample is chosen randomly without regard to levels using R software. Sampling is used because the sample size is too large.Subsample taken based on the subsampling algorithm is as much as $r > p$ where p is the number of predictor variables and r is the number of subsamples taken. In sample data, y as a response variable and x as a predictor variable while in subsample data, the response variable is expressed as y^* and the predictor variable is expressed as x^*. In this study used 4406 observationsand 7 variables and selected subsample numbers $r = 400$and $r > p = 6$.

3.1.2 Parameter Estimation of the Poisson Regression Model

In the next step after taking subsample, parameter estimation is done by using Maximum Likelihood Estimation (MLE) on r subsample data as the original data by maximizing equation (7) as follows:

$$l(\beta) = \frac{1}{r}\sum_{i=1}^{r} \left[y_i^*\left(x_i^{*T}\beta\right) - exp\left(x_i^{*T}\beta\right)\right]. \qquad (7)$$

The results of the parameter estimation can be seen in table 1.

Table 1. Estimation of parameters of the Poisson Regression Model on Subsample data with $r = 400$

rameter	Independent Variables	stimation
0	(*intercept*)	1.22569
1	Number of hospital stay (x_1)	0.31707
2	Health status (x_2)	-0.42079
3	Number of chronic diseases(x_3)	0.16563
4	Gender (x_4)	-0.16948
5	umber of years of education(x_5)	0.02108
6	Insurance indicator (x_6)	-0.05957

From the results of the parameter estimation above, the Poisson regression model formed is:

$$y_i = e^{(1.22569 + 0.31707x_1 - 0.42079x_2 + 0.16563x_3 - 0.16948x_4 + 0.02108x_5 - 0.05957x_6)}$$

Based on the Poisson regression model above, each increase in the number of hospitalizations increases the number of physician visits by exp (0.31707) if the other variables are not included in the model. Each increase in the number of chronic illnesses suffered by a person increases the number of physician visits by exp (0.16563) if other variables are not included in the model. Each increase in one year of education increases the number of doctor visits by as much as exp (0.02108) if the other variables are not included in the model.

3.2 Scenario

In this section, we will discuss some subsample scenarios to investigate approach performance. The scenario is taking several subsampling with different sample sizes, to determine differences in the estimated results of subsample data parameters. It is also to find the best model among the several subsamples. The amount of r taken based on the algorithm is $r > p$. Table 2 is about the results of parameter estimation for several subsamples.

Table 2. Parameter estimation of Poisson Regression model on subsample data $r = 600$, $r = 800, r = 1000$

Parameter	Independent Variables	Estimation		
		600	800	1000
0	*(intercept)*	1.17868	1.15744	1.15574
1	Number of hospital stay (x_1)	0.21098	0.20994	0.18079
2	Health status (x_2)	−0.32223	−0.26647	−0.46835
3	Number of chronic diseases(x_3)	0.20214	0.23141	0.16359
4	Gender (x_4)	−0.16222	−0.07386	−0.13272
5	umber of years of education(x_5)	0.00701	0.00980	0.02679
6	Insurance indicator (x_6)	0.24328	0.09469	0.07794

Based on the table, the Poisson regression model formed on each subsample is as follows :

For $r = 600 : y_i = e^{(1.17868 + 0,21098x_1 - 0,32223\,x_2 + 0,20214\,x_3 - 0,16222\,x_4 + 0,00701\,x_5 + 0,24328x_6)}$

For $r = 800 : y_i = e^{(1,15744 + 0,20994x_1 - 0,26647x_2 + 0,23141x_3 - 0,07386x_4 + 0,00980x_5 + 0,09469x_6)}$

For $r = 1000 : y_i = e^{(1,15574 + 0,18079x_1 - 0,46835x_2 + 0,16359x_3 - 0,13272x_4 + 0,02679x_5 + 0,07794x_6)}$

3.3 Selection of the Best Model

Based on the results of processing data from some subsamples used in this research, the AIC value will be obtained to find out the best model. The AIC value is obtained based on the log-likelihood value. The value of the log-likelihood for each subsample (r) are as follows, for $r = 400$ is -4.730267, $r = 600$ is *8.002661*, $r = 800$ *is 5.192549* and $r = 1000$ is *4.988598*. The AIC calculation results can be seen in Table 3 below:

Table 3. AIC results on each subsample model

Subsample (r)	
	24.67021
	24.3851
	23.9772

It can be seen from Table 3 that at r (the number of subsamples) 1000 has the smallest AIC values. It means that at 1000 number of subsample, the model provides the best among all of the others.

4. Conclusion

It can be concluded that the subsampling method is one approach that can deal with big data by taking a subsample from completed data. Meanwhile, when the number of subsamples tends to the number of samples, thus the value of AIC is getting smaller.

References

[1] Drineas P, Mahoney M W, and Sarlós S M T 2011 *Numer. Math.* **117** 219
[2] Ma P and Sun X 2014 *Wire's Comp. Stat.* **7** 70
[3] Cameron A C and Trivedi P K 1998 *Regression Analysis of Cound Data* (Cambridge: Cambridge University Press)
[4] Deb P and Trivedi P K 1998 J. Appl. Econom. 12 313
[5] Meng C, Wang Y, Zhang X , Mandal A , and Ma P 2017 Effective Statistical Methods for Big Data Analytics (USA: IGI Global)

[6] Shofiyah F and Sofro A 2018 *J. Phy IOP Conf. Ser.* **1108** 012107
[7] Dumbill E 2012 Planning for Big Data (New York: O'Reilly)
[8] Phillips-Wren G and Hoskisson A 2014 Frontiers in Artificial Intelligence and Applications (IOS Press Ebooks)
[9] Fukuchi J I 1999 Biom. 86 591
[10] Myers R H, Montgomery D C, Vining G G, and Robinson T J 2010 Generalized Linear Models with Applications in Engineering and the Sciences (New Jersey: John Wiley)
[11] Wang H, Zhu R, and Ma P 2017 J. of the Am. Stat. Assoc. 113 829
[12] Akaike H 1974 *IEEE* **19** 716

Oscillation Phenomena of *Rayleigh* Equation Disturbed by Superposition of Two Harmonic Excitation

A Alfan and N A C Dewi

Matematika, Universitas Billfath Lamongan, Jawa Timur, Indonesia.

*Email: arisalfan@gmail.com, alivianihaya@gmail.com

Abstract. The *Rayleigh* Nonlinear Oscillation Equation Phenomenon disturbed by two external harmonic forces. One of which is small, will be discussed. Using the *Multiple Scales* method to obtain analytical solutions, several cases of resonance are determined. Steady-state solutions and their stabilities will be investigated. Depending on the value of amplitudes and frequencies of the external forces given, for a non-trivial steady-state solution, there are two possibilities, namely a periodic solution, and a non-periodic solution that raises the beating phenomenon.

1. Introduction

Disturbed system phenomena are often found in everyday life considering that there is almost no ideal system without any outside force interference. Therefore, it is necessary to pay attention to the forces that affect the system, even though they are quite small. Some ways that can be done to overcome this problem is to approach the solution using analytical, numeric or frequently used methods, namely a combination of the two. One analytical method commonly used is asymptotic expansion commonly known as perturbation method and more specifically that will be used in writing this article, namely the *Multiple Scales* method. The method is suitable for analyzing the influence of parameters that are of little value to the system.

The equation that will be discussed in this article is *Rayleigh* equation. *Rayleigh* equation is an equation that has a stable solution at a certain limit cycle for $t \rightarrow \infty$. One particular example of the popular *Rayleigh* equation is the *van der pol* equation. In general, a one-dimensional homogeneous *Rayleigh* equation is,

$$\ddot{u} + F(u, \dot{u})\dot{u} + u = 0$$

The discussion of nonlinear system solutions is widely discussed in [3] [4]. Many applications in engineering, medicine and the like are found in [5]. The discussion will begin with completing *Rayleigh* equation which has been given two different dimensionally sized external forces using *Multiple Scales* method. Using this method leads to different cases which will be discussed in the next section. In each case a discussion of the solution and equilibrium was carried out.

2. Multiple Scale

Given *Rayleigh* equation which has been done non-dimensionless process

$$\ddot{u} + u = \epsilon\left(\dot{u} - \frac{1}{3}\dot{u}^3\right)$$

By adding two harmonic external forces, an equation will be obtained,

$$\ddot{u} + u = \epsilon\left(\dot{u} - \frac{1}{3}\dot{u}^3\right) + k_1 \cos\cos\left(\Omega_1 t + \theta_1\right) + \epsilon k_2 \cos\cos\left(\Omega_2 t + \theta_2\right) \tag{2.1}$$

Using the Multiple Scales method means we assume that u is a functional solution that depends not only on t and ϵ, but also on $\epsilon t, \epsilon^2 t; \epsilon^3 t, \ldots$ so that it can be written,

$$u = \hat{u}(T_0, T_1, T_2, \ldots ; \epsilon)$$

where $T_n = \epsilon^n t$. This assumption means that the solution can be written as a combination of several different time scales ranging from fast movements to slow movements after getting a solution in each order. For example, clockwork on a wall clock (not digital), with $\epsilon = \frac{1}{60}$, T_0 is a time scale so that a fast solution movement can be observed properly which in this case is indicated by a second hand motion. T_1 represents the time scale in which the solution movement can be seen when observing the solution until $t = \epsilon^{-1}$ in this case can be indicated by minute motion, as well as T_2 which shows the time scale in which the solution movement can be observed while observing the movement of the solution to $t = \epsilon^{-2}$ in this case it can be represented by a hours motion. Thus T_0 represents the fast-moving time scale of the solution, T_n is a time scale for measuring movements that are slower ϵ^n times than T_0 (Nayfeh et al, 1995). By dividing the time scale t into $T_0, T_1, T_2, \ldots$ it means changing the equation that was originally in the form of an ordinary differential equation into a partial differential equation so it is necessary to redefine $\frac{d}{dt}$ and $\frac{d^2}{dt^2}$. Using the chain rule is obtained,

$$\frac{d}{dt} = D_0 + \epsilon D_1 + \epsilon^2 D_2 + \cdots$$

$$\frac{d^2}{dt^2} = D_0^2 + 2\epsilon D_0 D_1 + \epsilon^2 (D_1^2 + D_0 D_2) + \cdots$$

with $D_n = \frac{\partial}{\partial T_n}$ dan $D_n^2 = \frac{\partial^2}{\partial T_n^2}$. Next expand the solution u,

$$u = u_0(T_0, T_1, T_2, \ldots) + \epsilon u_1(T_0, T_1, T_2, \ldots) + \cdots \quad (2.2)$$

Substitute the results of the expansion to Equation (2.1) and ignore the order ϵ^2 or higher terms obtained,

Order ϵ^0 : $D_0^2 u_0 + u_0 = k_1 \cos\cos(\Omega_1 T_0 + \theta_1)$

Order ϵ^1 : $D_0^2 u_1 + u_1 = -2D_0 D_1 u_0 + D_0 u_0 - \frac{1}{3}(D_0 u_0)^3 + k_2 \cos\cos(\Omega_2 T_0 + \theta_2)$

The solution to the equation ϵ^0 can be expressed as,

$$u_0 = a \cos\cos(T_0 + \beta) + \frac{k_1}{1-\Omega_1^2} \cos\cos(\Omega_1 T_0 + \theta_1) \quad (2.3)$$

with a and β functions in T_1 as well as $\Omega_1 \neq 1$. In other forms, the solution can be written,

$$u_0 = \frac{a}{2}\left(e^{i(T_0+\beta)} + e^{-i(T_0+\beta)}\right) + \frac{k_1}{2(1-\Omega_1^2)}\left(e^{i(\Omega_1 T_0+\theta_1)} + e^{-i(\Omega_1 T_0+\theta_1)}\right)$$

$$= Ae^{iT_0} + Be^{i(\Omega_1 T_0)} + \bar{A}e^{-iT_0} + \bar{B}e^{-i(\Omega_1 T_0)}$$

where $A = \frac{a}{2}e^{i\beta}$ dan $B = \frac{1}{2(1-\Omega_1^2)}e^{i\theta}$. Substitute u_0 in the equation order ϵ,

$$D_0^2 u_1 + u_1 = (-2\dot{A} + A)ie^{iT_0} - \frac{1}{3}(-A^3 ie^{3iT_0} - B^3 i\Omega_1^3 e^{3i\Omega_1 T_0} -$$
$$3A^2 Bi\Omega_1 e^{(2+\Omega_1)iT_0} + 3A^2\bar{A}ie^{iT_0} - 3\bar{A}^2 Bi\Omega_1 e^{(\Omega_1-2)iT_0} - 3AB^2 i\Omega_1^2 e^{(2\Omega_1+1)iT_0} +$$
$$3\bar{A}B^2 i\Omega_1^2 e^{(2\Omega_1-1)iT_0} + 3B^2\bar{B}i\Omega_1^3 e^{i\Omega_1 T_0} + 6A\bar{A}Bi\Omega_1 e^{i\Omega_1 T_0} + 6AB\bar{B}i\Omega_1^2 e^{iT_0}) + \frac{k_2}{2}e^{i(\Omega_2 T_0+\theta_2)} + cc \quad (2.4)$$

with $\dot{A}$ declaring $\frac{\partial A}{\partial T_1}$ and cc declares complex conjugate term of the right side. Equation (2.4) contains a secular term that is a term that depends on T_0 so that if solution u_1 is applied to Equation (2.3), u_1 which should have an order smaller than u_0, when $t \geq \epsilon^{-1}$ the order of u_1 will be equal or more height of u_0. This causes the solution Equation (2.3) to be infinite or cause the phenomenon of resonance, namely the amplitude of the solution continues to increase for $t \to \infty$. Therefore, in order to obtain a uniform expansion of solutions, secular term must be eliminated. Secular term are indicated by a term containing e^{iT_0}. To eliminate secular terms Equation (2.4), it must be,

$$-2\dot{A} + A - A^2\bar{A} - 2AB\bar{B}\Omega_1^2 = 0$$

Paying attention to equality (2.4), even though only a small external force is added, but if $\Omega_2 \approx 1$ there will be an addition to the secular term. This shows that one of the causes of the addition of the secular term is accumulation of the external force frequency to the frequency of the cause of resonance. Because the two external forces work together on the system, there is a possibility of resonance originating from the two external forces. From Equation (2.4), the superposition of the external forces causes resonance and possible small dividers namely,

1. $\Omega_1 \approx 1$ dan $\Omega_2 \approx 1$
2. $\Omega_1 \approx 3$ dan $\Omega_2 \approx 1$
3. $3\Omega_1 \approx 1$ dan $\Omega_2 \approx 1$

For the first case because of the value of $\Omega_1 \approx \Omega_2$ then both can be written as a unit with provisions such as, Nayfeh 1995, so that the case is similar to the problem of one external force. The following will be investigated for cases $\Omega_2 \approx 1$ with Ω_1 far from 3 or $\frac{1}{3}$ before discussing the combination of Ω_1 and Ω_2.

3. Case $\Omega_2 \approx 1$

3.1. *Approximate Solutions*

For σ_2 is a detuning parameter that states the closeness of Ω_2 with number one. Write,

$$\Omega_2 = 1 + \epsilon\sigma_2$$

Subtitles in Equation (2.4) then there will be additional secular term so to eliminate them,

$$-2\dot{A} + A - A^2\bar{A} - 2AB\bar{B}\Omega_1^2 - i\frac{k_2}{2}e^{i(\sigma_2 T_1+\theta_2)} = 0$$

for $A = \frac{a}{2}e^{i\beta}$ and $B = \frac{k_1}{2(1-\Omega_1^2)}e^{i\theta}$ we get,

$$-\dot{a} - ai\dot{\beta} + \frac{a}{2} - \frac{a^3}{8} - a\Lambda^2 - i\frac{k_2}{2}e^{i(\sigma_2 T_1+\theta_2-\beta)} = 0 \quad (3.1)$$

Separating the real and imaginary parts from Equation (3.1) will be obtained,

$$\dot{a} = \frac{a}{2} - \frac{a^3}{8} - a\Lambda^2 + \frac{k_2}{2}sin(\sigma_2 T_0 + \theta_2 - \beta), \qquad a\dot{\beta} = -\frac{k_2}{2}cos(\sigma_2 T_1 + \theta_2 - \beta) \quad (3.2)$$

As the initial approach a uniform solution can be written,

$$u = acos(T_0 + \beta) + \frac{k_1}{1-\Omega_1^2}cos(\Omega_1 T_0 + \theta_1) + O(\epsilon)$$

where $O(\epsilon)$ states the order solution ϵ, a and β are obtained from system (3.2). In order not to depend explicitly on time, a new variable $\gamma = \sigma_2 T_1 + \theta_2 - \beta$ is introduced so that equation (3.2) becomes,

$$\dot{a} = \frac{a}{2} \quad \frac{a^3}{8} \quad a\Lambda^2 + \frac{k_2}{2}sin(\gamma), \qquad a\dot{\gamma} = a\sigma_2 + \frac{k_2}{2}cos(\gamma) \quad (3.3)$$

Completing Equation (3.3) will be obtained a and γ. Therefore solutions can be written as,

$$u = acos(\Omega_2 t - \gamma + \theta_2) + \frac{k_1}{1-\Omega_1^2}cos(\Omega_1 T_0 + \theta_1) + O(\epsilon)$$

Lease follow these instructions as carefully as possible so all articles within a conference have the same style to the title page. This paragraph follows a section title so it should not be indented.

3.2. *Steady State Solution*

Steady state solution are obtained when $\dot{a} = \dot{\gamma} = 0$ so that they meet,

$$\frac{a}{2} - \frac{a^3}{8} - a\Lambda^2 = -\frac{k_2}{2}sin(\gamma) \quad (3.4)$$

$$a\sigma_2 = -\frac{k_2}{2}cos(\gamma) \quad (3.5)$$

Squaring the two fields of Equation (3.4) and (3.5) then adding them together will be obtained,

$$\left(\frac{a}{2} - \frac{a^3}{8} - a\Lambda^2\right)^2 + (a\sigma_2)^2 = \left(\frac{k_2}{2}\right)^2 \quad (3.6)$$

Equation (3.6) is commonly called the response frequency equation. Figure 1 shows the response to changes in amplitude due to changes in k_2 by specifying the detuning parameter and also the response to changes in amplitude due to changes in detuning parameters by setting k_2. Further observed, it will be known that for $0 < k_2 < 0.5284$ there are three equilibrium points in some interval and for the rest interval there is only one equilibrium point. For $0.5284 < k_2 < 1$ there will be some interval so that there is only one equilibrium point, than followed by some interval that have three equilibrium points than followed by interval with one equilibrium point depending on the selection of k_2. For $k_2 > 1$, it will only have one equilibrium point for each real σ_2 value with a maximum amplitude that decreases for the larger k_2 as shown in Figure 1. Next, the stability of the equilibrium point will be investigated.

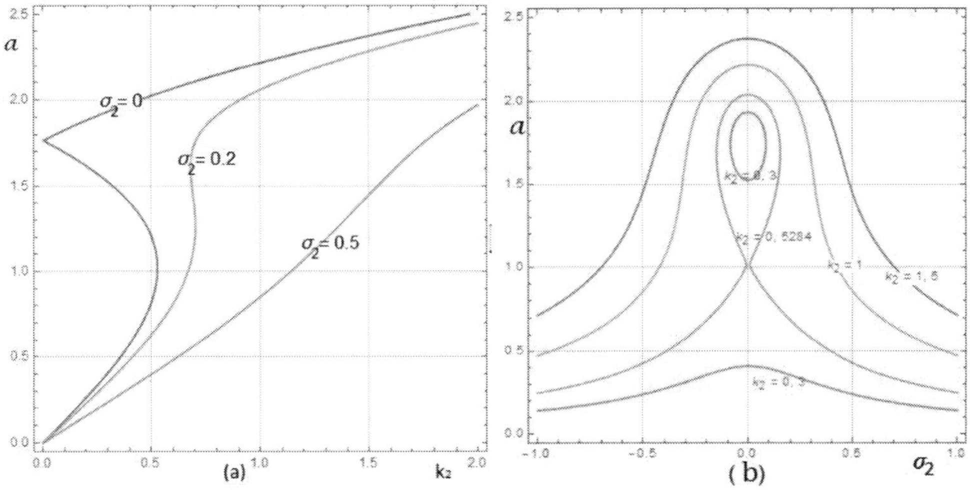

Figure 1. With $\Omega_1 = 2, k_1 = 1$ changes in amplitude to (a) k_2 with some σ_2, (b) Detuning parameters with some values of k_2

3.3. Equilibrium Points Stability

Stability analysis can be done by investigating the movement of points around the equilibrium point. For that to be defined,

$$a = a^* + a_0 \, \gamma = \gamma^* + \gamma_0 \tag{3.7}$$

with a^*, γ^* is a small parameter and a_0, γ_0 is the equilibrium point so that it satisfies Equation (3.4) and (3.5). By substituting (3.7) to Equation (3.3) and maintaining the linearity of a^* and γ^* there will be a coefficient matrix,

$$\begin{vmatrix} \left(\frac{1}{2} - \frac{3a_0^2}{8} - \Lambda^2\right) - \lambda & \left(\frac{k_2}{2}\cos\gamma_0\right) \\ -\left(\frac{k_2}{2a_0^2}\cos\gamma_0\right) & -\left(\frac{k_2}{2a_0}\sin\gamma_0\right) - \lambda \end{vmatrix} = 0 \tag{3.8}$$

So we get,

$$\lambda^2 - \left(1 - 2\Lambda^2 - \frac{a_0^2}{2}\right)\lambda + \Delta = 0 \tag{3.9}$$

where $\Delta = \sigma^2 + \left(\frac{1}{2} - \Lambda^2 - \frac{3a_0^2}{8}\right)\left(\frac{1}{2} - \Lambda^2 - \frac{a_0^2}{8}\right)$.

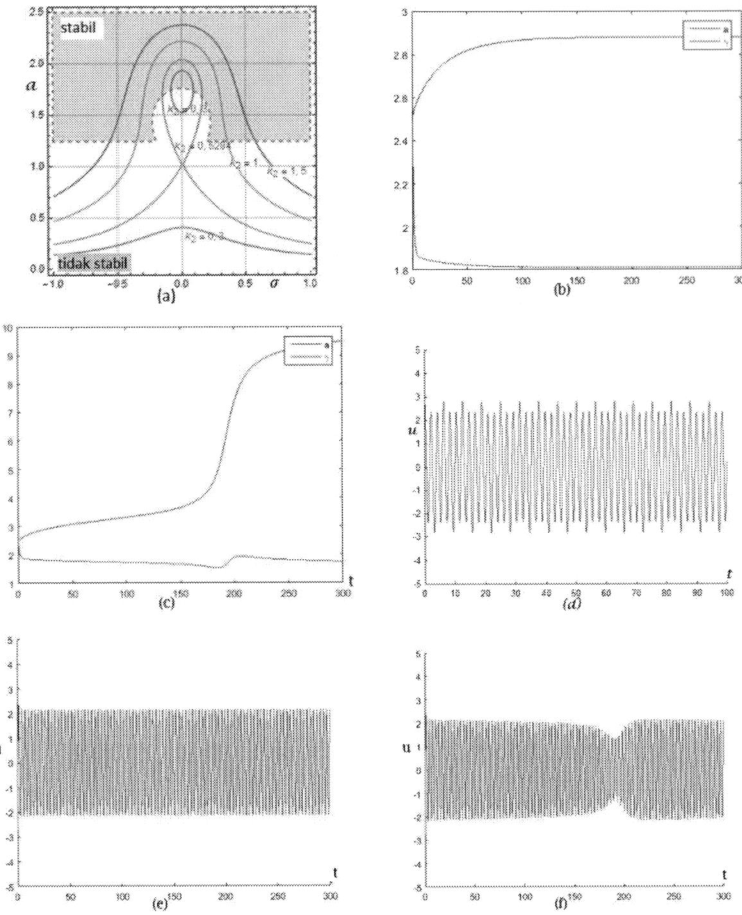

Figure 2. With $\epsilon = 0.1$ and $\Omega_1 = 2$ (a) The equilibrium point curve for $k_1 = 1$ and the stability area (b) Change a and γ for $\sigma_2 = 0.08$ (c) Change a and γ for $\sigma_2 = 0.09$ (d) external force (e) Plot solution for $\sigma_2 = 0.08$ (f) Plot solution for $\sigma_2 = 0.09$.

A stable periodic solution will be obtained if $\Delta > 0$ and $\frac{a_0^2}{2} > 1 - 2\Lambda^2$ by set k_1, ϵ and Ω_1 so that the equilibrium stability area will not depend on the amplitude of the external force k_2 as shown in Figure 2 (a) . The equilibrium points in the shading area are stable while those outside the shading area are unstable. If given $k_1 = 1$, $\epsilon = 0.1$, $\Omega_1 = 2$ and $k_2 = 0.3$ there will be a stable equilibrium point producing the periodic solution when $|\sigma_2| < 0.0855$ which means for a long time, the solution's amplitude will be a constant amplitude a_0. Whereas if chosen $|\sigma_2| > 0.0855$ there will be a beating phenomenon, namely a solution whose amplitude oscillates as shown in Figures 2 (c) and (f).

4. Case $\Omega_1 \approx 3$ and $\Omega_2 \approx 1$

4.1. Approximate Solution

For σ_2 is a detuning parameter that states the closeness of Ω_2 with number one. Write,

$$\boldsymbol{\Omega_1 = 3 + \epsilon\sigma_1, \qquad \Omega_2 = 1 + \epsilon\sigma_2}$$

Subtitles in Equation (2.4) then there will be additional secular term so to eliminate them,

$$-2\dot{A} + A - A^2\bar{A} - 2AB\bar{B}\Omega_1^2 + \bar{A}^2 B\Omega_1 e^{\sigma_1 iT_1} - \frac{ik_2}{2} e^{i(\sigma_2 T_1 + \theta_2)} = 0$$

with $\boldsymbol{A = \frac{a}{2} e^{i\beta}}$ and $\boldsymbol{B = \frac{k_1}{2(1-\Omega_1^2)} e^{i\theta}}$ we get a new system,

$$\begin{aligned} \dot{a} &= \frac{a}{2} - \frac{a^3}{8} - a\Lambda^2 + \frac{a^2}{4}\Lambda\cos(\sigma_1 T_1 + \theta_1 - 3\beta) + \frac{k_2}{2}\sin(\sigma_2 T_1 + \theta_2 - \beta) \\ a\dot{\beta} &= \frac{a^2}{4}\Lambda\sin(\sigma_1 T_1 + \theta_1 - 3\beta) - \frac{k_2}{2}\cos(\sigma_2 T_1 + \theta_2 - \beta) \end{aligned} \tag{4.1}$$

In order not to depend explicitly on time, a new variable $\gamma = \boldsymbol{\sigma T_1} - \boldsymbol{3\beta}$ is introduced where $\sigma_1 = \boldsymbol{3\sigma_2} = \boldsymbol{\sigma}$ so that equation (4.1) becomes,

$$\dot{a} = \frac{a}{2} - \frac{a^3}{8} - a\Lambda^2 + \frac{a^2}{4}\Lambda\cos(\gamma + \theta_1) + \frac{k_2}{2}\sin\left(\frac{1}{3}\gamma + \theta_2\right)$$
$$\frac{a}{3}\dot{\gamma} = \frac{a\sigma}{3} - \frac{a^2}{4}\Lambda\sin(\gamma + \theta_1) + \frac{k_2}{2}\cos\left(\frac{1}{3}\gamma + \theta_2\right) \quad (4.2)$$

Completing Equation (4.2) will be obtained $\boldsymbol{a}$ and γ. Therefore solutions can be written as,

$$\boldsymbol{u = a\cos\left(\frac{1}{3}(\Omega_1 t \quad \gamma)\right) + \frac{k_1}{1-\Omega_1^2}\cos(\Omega_1 T_0 + \theta_1) + O(\epsilon)} \quad (4.3)$$

4.2. Steady State Solution

Steady state solution are obtained when $\dot{a} = \dot{\gamma} = 0$ so that they meet,

$$\frac{a}{2} - \frac{a^3}{8} - a\Lambda^2 = -\left(\frac{a^2}{4}\Lambda\cos(\gamma + \theta_1) + \frac{k_2}{2}\sin\left(\frac{1}{3}\gamma + \theta_2\right)\right)$$
$$\frac{a\sigma}{3} = \frac{a^2}{4}\Lambda\sin(\gamma + \theta_1) - \frac{k_2}{2}\cos\left(\frac{1}{3}\gamma + \theta_2\right) \quad (4.4)$$

Squaring the two fields of Equation (4.4) then adding them together will be obtained frequency response equation,

$$\left(\frac{a}{2} - \frac{a^3}{8} - a\Lambda^2\right)^2 + \left(\frac{a\sigma}{3}\right)^2 = \left(\frac{a^2}{4}\Lambda\right)^2 + \left(\frac{k_2}{2}\right)^2 - \frac{a^2k_2}{4}\Lambda\sin\left(-\frac{2}{3}\gamma + \theta_2 - \theta_1\right) \quad (4.5)$$

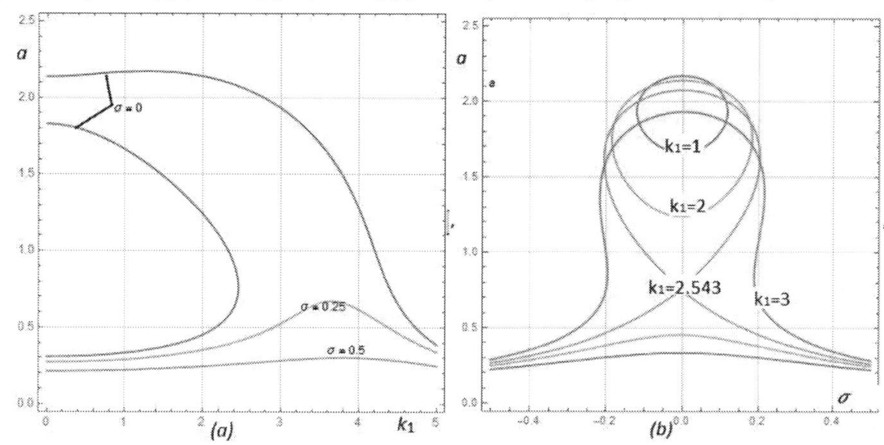

Figure 3. With $\epsilon = 0.1, k_2 = 0.3$ (a) Change in amplitude versus k_1 with several σ, (b) Change in amplitude versus detuning parameter with some values of k_1.

For $0 < k_1 < 2.545$ there are three equilibrium points around the detuning parameter σ zero then for σ which is large enough there is one equilibrium point. Whereas for $2.545 < k_1 < 3.43$ there are σ regions around zero which have one equilibrium point, also there are regions with three equilibrium points then for σ is large enough there is only one equilibrium point. For $k_1 > 3.43$ there is only one equilibrium point for each σ. Then the stability of the equilibrium point will be investigated.

4.3. Equilibrium Points Stability

Using the same way as the previous section applied to the system (4.2) the coefficient matrix is fulfilled,

$$\begin{vmatrix} J_1 - \lambda & J_2 \\ J_3 & J_4 - \lambda \end{vmatrix} = 0 \quad (4.6)$$

where,

$$J_1 = \frac{1}{2} - \frac{3a_0^2}{8} - \Lambda^2 + \frac{a_0}{2}\Lambda\cos(\gamma_0 + \theta_1); \quad J_2 = -\frac{a_0^2}{4}\sin(\gamma_0 + \theta_1) + \frac{k_2}{6}\cos\left(\frac{\gamma_0}{3} + \theta_2\right)$$
$$J_3 = -\frac{3}{4}\Lambda\sin(\gamma_0 + \theta_1) - \frac{3k_2}{2a_0^2}\cos\left(\frac{\gamma_0}{3} + \theta_2\right); \; J_4 = -\frac{3a_0}{4}\Lambda\cos(\gamma_0 + \theta_1) - \frac{k_2}{2a_0}\sin\left(\frac{\gamma_0}{3} + \theta_2\right)$$

so we get,

$$\lambda^2 - \left(1 - 2\Lambda^2 - \frac{a_0^2}{2}\right)\lambda + \Delta = 0 \tag{4.7}$$

with $\Delta = \left(\frac{1}{2} - \frac{a_0^2}{8} - \Lambda^2\right)^2 - \frac{a_0^2}{4}\Lambda^2 + \left(\frac{\sigma}{3}\right)^2 - \frac{a_0^2}{4}\left(\frac{1}{2} - \frac{a_0^2}{8} - \Lambda^2 - \frac{a_0^2}{2}\Lambda\cos(\gamma_0 + \theta_1)\right)$.

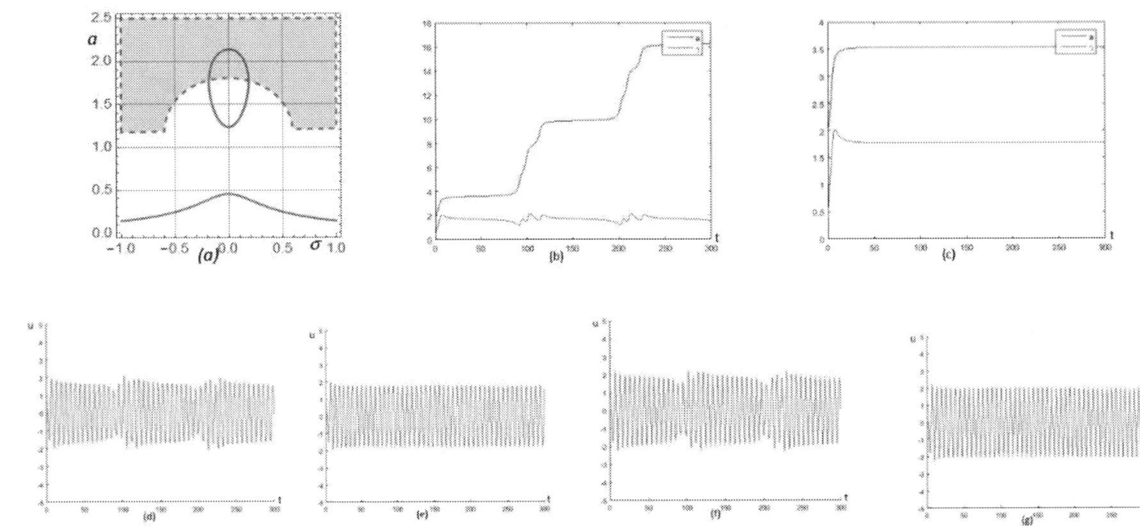

Figure 4. With $\epsilon = 0.1$ and $k_2 = 0.3$ (a) The equilibrium curve for $k_1 = 2$ and the stability area (b) Change a and γ for $\sigma = 0.235$ (c) Change a and γ for $\sigma = 0.23$ (d) Homogeneous Solution $\sigma = 0.235$ (e) Homogeneous solution for $\sigma = 0.23$ (f) Plot solution $\sigma = 0.235$ (g) Plot the solution when $\sigma = 0.23$.

From Equation (4.7), there will be a stable equilibrium point producing the periodic solution if $\frac{a_0^2}{2} > 1 - 2\Lambda^2$ and $\Delta > 0$. It can be seen that the conditions obtained by the periodic solution are stable depending on the selection of k_1, which means that the condition of the periodic solution stability changes with changes in k_1. The equilibrium point in the shading area in Figure 4 (a) is stable while the other is unstable. For example for $\epsilon = 0.1$, $k_2 = 0.3$ selected $k_1 = 2$, a stable equilibrium solution for periodic solutions will be obtained when $|\sigma| < 0.233$ as Figure 4 (a). The selection of σ outside the area will cause a homogeneous solution to produce a beating phenomenon, namely the amplitude of the periodic oscillating solution as shown in Figures 4 (d) and (f).

5. Case $3\Omega_1 \approx 1$ and $\Omega_2 \approx 1$

5.1. *Approximate Solution*

A For σ_2 is a detuning parameter that states the closeness of Ω_2 with number one. Write,

$$3\Omega_1 = 1 + \epsilon\sigma_1; \qquad \Omega_2 = 1 + \epsilon\sigma_2$$

Subtitles in Equation (2.4) then there will be additional secular term so to eliminate them,

$$-2\dot{A} + A - A^2\bar{A} - 2AB\bar{B}\Omega_1^2 + \frac{1}{3}B^3\Omega_1^3 e^{\sigma_1 iT_1} - \frac{ik_2}{2}e^{i(\sigma_2 T_1 + \theta_2)} = 0$$

with $A = \frac{a}{2}e^{i\beta}$ and $B = \frac{k_1}{2(1-\Omega_1^2)}e^{i\theta}$ we get a new system,

$$\begin{aligned} \dot{a} &= \frac{a}{2} - \frac{a^3}{8} - a\Lambda^2 + \frac{1}{3}\Lambda^3\cos(\sigma_1 T_1 + 3\theta_1 - \beta) + \frac{k_2}{2}\sin(\sigma_2 T_1 + \theta_2 - \beta) \\ a\dot{\beta} &= \frac{1}{3}\Lambda^3\sin(\sigma_1 T_1 + 3\theta_1 - \beta) - \frac{k_2}{2}\cos(\sigma_2 T_1 + \theta_2 - \beta) \end{aligned} \tag{5.1}$$

In order not to depend explicitly on time, a new variable $\gamma = \sigma T_1 - \beta$ is introduced where $\sigma_1 = \sigma_2 = \sigma$ so that equation (5.1) becomes,

$$\dot{a} = \frac{a}{2} - \frac{a^3}{8} - a\Lambda^2 + \frac{1}{3}\Lambda^3\cos(\gamma + 3\theta_1) + \frac{k_2}{2}\sin(\gamma + \theta_2)$$
$$a\dot{\gamma} = a\sigma - \frac{1}{3}\Lambda^3\sin(\gamma + 3\theta_1) + \frac{k_2}{2}\cos(\gamma + \theta_2) \quad (5.2)$$

Completing Equation (5.2) will be obtained a and γ. Therefore solutions can be written as,

$$u = a\cos(3\Omega_1 t - \gamma) + \frac{k_1}{1\ \Omega_1^2}\cos(\Omega_1 T_0 + \theta_1) + O(\epsilon) \quad (5.3)$$

5.2. *Steady State Solutions*

Steady state solution are obtained when $\dot{a} = \dot{\gamma} = 0$ so that they meet,

$$\frac{a}{2} - \frac{a^3}{8} - a\Lambda^2 = -(\frac{1}{3}\Lambda^3 cos(\gamma + 3\theta_1) + \frac{k_2}{2} sin(\gamma + \theta_2)) \quad (5.4)$$
$$a\sigma = \frac{1}{3}\Lambda^3 sin(\gamma + 3\theta_1) - \frac{k_2}{2} cos(\gamma + \theta_2) \quad (5.5)$$

Squaring the two fields of Equation (5.4) and (5.5) then adding them together will be obtained frequency response equation,

$$(\frac{a}{2} - \frac{a^3}{8} - a\Lambda^2)^2 + (a\sigma)^2 = (\frac{1}{3}\Lambda^3)^2 + (\frac{k_2}{2})^2 + \frac{\Lambda^3 k_2}{3} sin(\theta_2 - 3\theta_1)$$

For $0 < k_1 < 2{,}546$ there are three equilibrium points around $\sigma = 0$ then for σ which is quite large there is one equilibrium point. Whereas for $2{,}546 < k_1 < 2.83$, there are σ regions around zero which have one equilibrium point, there are also regions with three equilibrium points then for σ large enough there is only one equilibrium point. For $k_1 > 2.83$ there is only one equilibrium point for each σ. Then the stability of the equilibrium point will be investigated.

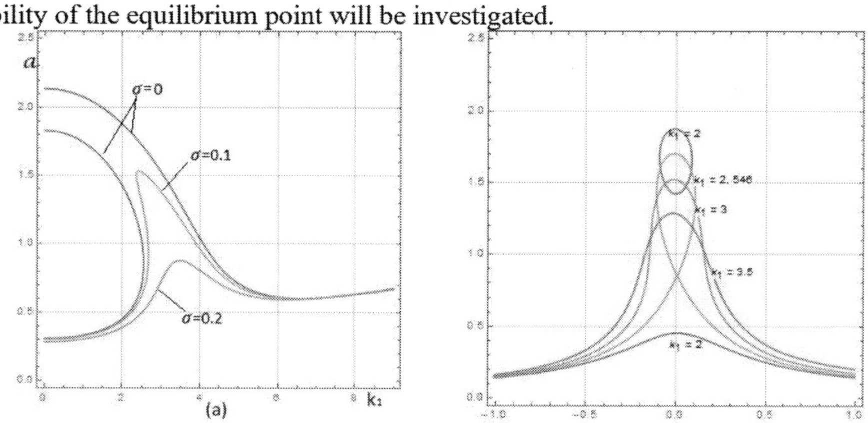

Figure 5. (a) Change a to k_1 with some σ, (b) Change in amplitude to detuning parameter with some values of k_1.

5.3. *Equilibrium Points Stability*

Using the same method as the previous section applied to the system (5.2), the coefficient matrix is fulfilled,

$$\begin{vmatrix} J_1 - \lambda & J_2 \\ J_3 & J_4 - \lambda \end{vmatrix} = 0$$

with,

$$J_1 = \frac{1}{2} - \frac{3a_0^2}{8} - \Lambda^2; \qquad J_2 = \frac{k_2}{2} cos(\gamma_0 + \theta_2) - \frac{1}{3}\Lambda^3 sin(\gamma_0 + 3\theta_1)$$
$$J_3 = \frac{1}{3a_0^2}\Lambda^3 sin(\gamma_0 + 3\theta_1) - \frac{k_2}{2a_0^2} cos(\gamma_0 + \theta_2); \quad J_4 = -(\frac{1}{3a_0}\Lambda^2 cos(\gamma_0 + 3\theta_1) + \frac{k_2}{2a_0} sin(\gamma_0 + \theta_2))$$

So we get,

$$\lambda^2 - (1 - 2\Lambda^2 - \frac{a_0^2}{2})\lambda + \Delta = 0 \quad (5.6)$$

with $\Delta = (\frac{1}{2} - \frac{3a_0^2}{8} - \Lambda^2)(\frac{1}{2} - \frac{a_0^2}{8} - \Lambda^2) + \sigma^2$.

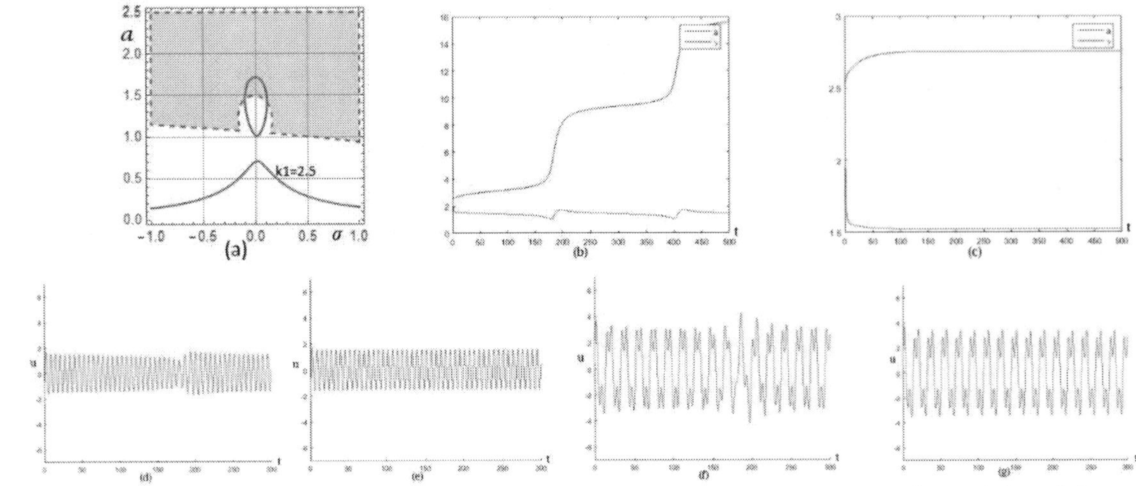

Figure 6. With $\epsilon = 0.1$ and $k_2 = 0.3$ (a) The equilibrium point curve for $k_1 = 2.5$ and the stability area (b) Change a and γ for $\sigma = 0.11$ (c) Change a and γ for $\sigma = 0.1$ (d) Homogeneous Solution $\sigma = 0.11$ (e) Homogeneous Solution for $\sigma = 0.1$ (f) Plot solution when $\sigma = 0.11$ (g) Plot the solution when $\sigma = 0.1$.

From Equation (5.6), a periodic solution point will be stable when $\frac{a_0^2}{2} > 1 - 2\Lambda^2$ and $\Delta > 0$. It can be seen that the conditions obtained by the periodic solution are stable depending on the selection of k_1, which means that the condition of the periodic solution stability changes with changes in k_1. The shaded area in Figure 6 (a) shows the area that meets the stability requirements for $\epsilon = 0.1, k_2 = 0.3$ and $k_1 = 2.5$ is in $|\sigma| < 0.1$ in other words the equilibrium point in the area is stable while the other is unstable . The choice of σ in the area that has a stable equilibrium point causes the solution to converge at a constant amplitude the point remains stable a_0. The selection of σ outside the area will cause the amplitude of the homogeneous solution to oscillate to beating as shown in Figures 6 (b), (d) and (f).

6. Conclusion

The magnitude of the frequency and amplitude of the external force given can affect the stability of the solution. Stable solution leads to a constant amplitude and phase so that a solution is obtained in the form of a harmonic oscillation wave. Unstable solution leads to an amplitude or phase which changes over time. Solutions that have periodic amplitude fluctuating over time will give rise to a *beating* phenomenon.

7. References

[1] Bolanov A, Janson N, Postnov D and Sosnovtseva O 2008 *Synchronization: From Simple to Complex* (German:Springer)
[2] Elnaggar A M and El-Basyouny A F 1993 *Acta Mech. Sinica* **9** 61
[3] Nayfeh A H 1993 *Introduction to Perturbation Techniques* (USA: John Willey and Sons)
[4] Nayfeh A H and Mook D T 1995 *Nonlinear Oscillations* (USA: John Willey and Sons)
[5] Pikovsky A, Rosenblum M and Kurths J 2001 *Synchronization A Universal Concept in Nonlinear Sciences* (Cambridge: Cambridge University Press)
[6] Cousteix J and Mauss J 2007 *Asymptotic analysis and boundary layers* (Germany: Springer)
[7] Hazewinkel M 1991 *Encyclopaedia of Mathematics Orbit — Rayleigh Equation* (Netherlands: Springer)

Multiclass Classification of Brain Cancer with Multiple Multiclass Artificial Bee Colony Feature Selection and Support Vector Machine

S A A Kharis*, I Hadi and K A Hasanah

Department of Mathematics, Faculty of Mathematics and Science, Universitas Negeri Jakarta, Jl. Rawamangun Muka, Jakarta, 13220, Indonesia

* E-mail: sellyanas@yahoo.com

Abstract. A World Health Organization reported that the mortality rate due to brain cancer is the highest in the Asian continent. It is critical importance that brain cancer can be detected earlier so that the treatment process can be carried out more precisely and will be able to extend the life expectancy of brain cancer patients. Taking advantage of microarray data, machine learning methods can be applied to help brain cancer prediction according to its type. This problem can be referred to as a multiclass classification problem. Using the one versus one approach, there will be as many as $\frac{k(k-1)}{2}$ two-class problems, where k indicates the number of classes. In this paper, Multiple Multiclass Artificial Bee Colony (MMABC) implemented as a feature selection method and Support Vector Machine (SVM) as a classification method. ABC algorithm proved successful in solving optimisation problems with high dimensionality, and SVM can produce accurate and robust classification results. The data obtained from Broad Institute data. The data consist of 7129 features and 42 samples. From the experiment, the accuracy of Multiple SVM using a feature selection based MMABC method reached 95.24% accuracy in usage 300 best features; this percentage slightly more superior than SVM method without feature selection.

1. Introduction

Cancer is a group of diseases characterised by the uncontrolled growth and spread of abnormal cells [1]. If it happens to the brain, then it is called 'brain cancer.' Brain cancer is commonly attacking young generation, from children to adults below 40 years old [2]. More than two-thirds of adults diagnosed with brain cancer will die within two years of diagnosis [3, 4]. Based on other data from National Cancer Institute, there were about 17.760 deaths in 2019 which were caused by brain cancer and 23.820 new cases [5]. Besides, A World Health Organization reported that mortality rate due to brain cancer was and is still the highest in Asia continent [6]. However, the attention and help which was given by the government to this brain cancer case were not as much as they did to the other types of cancer. Therefore, there should be researches in detecting and treating it.

On the other side, gens in certain bodies can be interpreted numerically, which is usually called microarray data. In detecting data, microarray data can be analysed by seeing the combination of some of the gens in patient's body. The characteristics from the medical data have many features (gens), that the amount of them could reach hundreds or even thousands, and little number of samples, which were

Journal of Physics: Conference Series **1417** (2019) 012015 doi:10.1088/1742-6596/1417/1/012015

fewer than one hundred, made it need a method that could help to detect cancer easier and in objective way. In the machine learning context, the problem of detecting cancer using microarray data is one of the classification problems. In a simple case, the classification problem is detecting whether cancer or not cancer (normal), so there will be two classes. However, this classification problem can be analysed further, which is detecting the types of cancer. Detecting the type of cancer is not restricted to two classes, but it can be more than that.

Some researchers have examined the multiclass problem. In 2015, Zhang developed the Multiple Support Vector Machine Recursive Feature Elimination algorithm [7]. This method is a development from the Multiple Support Vector Machine Recursive Feature Elimination (MSVM-RFE) concept, which was stated by Duan and Multiclass Support Vector Machine Recursive Feature Elimination (MCSVM-RFE) which was stated by Zhou. In Zhou's research, the *one-versus-rest* method was used to form the two-classes problems from multiclass problems. In this research, another alternative is used, which is *one-versus-one.* Another research done by Azminuddin and his colleagues with the multi-class svm model uses 1V1 strategy to classify the wall-following robot navigation data [8]. Besides, the research of multiclass classification was done by Thirumalaimuthu Thirumalaiappan, using the SVM design based on multi knowledge-based system (SMK) [9]. Other than the usage of SVM, another classifier can be applied to the multiclass classification, just like the one which was applied by Juntao Li in his research, by using multinomial regression [10]. In this research, multiclass classification technique was submitted with the optimal artificial bee colony (ABC) method. The usage of ABC is chosen based on the previous research: ABC algorithm performs better than Differential Evolution (DE), Particle Swarm Optimization (PSO), and an Evolutionary Algorithm (EA) for multi-dimensional and multimodal numeric problems with high dimensionality. Besides that, ABC algorithm is flexible and straightforward to use and robust the optimisation of the algorithm [11]. SVM as a classifier is chosen because it can produce accurate and robust classification results [12].

2. Materials and Method

2.1. Materials

The data used in this research from Broad Institute (http://www.broadinstitute.org/cgi-bin/cancer/datasets.cgi). It was actually obtained in .gct format. By using the R programming language, data was converted into .csv format to make it easier to be observed. The data consisted of four types of brain cancers, such as Medulloblastoma (MD), Malignant Glioblastoma (MG), Primitive Neuroectodermal Tumors (PNET) and Atypical Teratoid Rhabdoid Tumor (ATRT). The data also consisted of 7129 features and 42 samples with ten samples for MD, ten samples for MG, ten samples for ATRT, eight samples for PNET, and four samples for Normal. In brain cancer microarray data, the features are the gens and the labels are the types of brain cancer.

2.2. Method

2.2.1. Multiple Multiclass Artificial Bee Colony (MMABC)

MMABC is a concept development of *Artificial Bee Colony* for the multiclass problems. In this research, the approach used is *one versus one.* With this type of approach, each of probabilities of the pair's counts, so there will be $\frac{k(k-1)}{2}$ types of ordering features on MMABC and classifier model, which k indicates the number of classes. The classification problem of brain cancer from Broad Institute is multiclass, so there will be ten types of binary class classification. Those pairs are MD and MG, MD and normal, MD and PNET, MD and ATRT, MG and normal, MG and PNET, MG and ATRT, normal and PNET, Normal and ATRT, PNET and ATRT. On each problem of the mention binary class problem, change the label of first-class into 1 and the second class into -1. Then, pick the number of features to be applied. Make limits to each of the binary class problems with the chosen features. After that, train the classifier for each of those binary class problems. The process above shown in Figure 1.

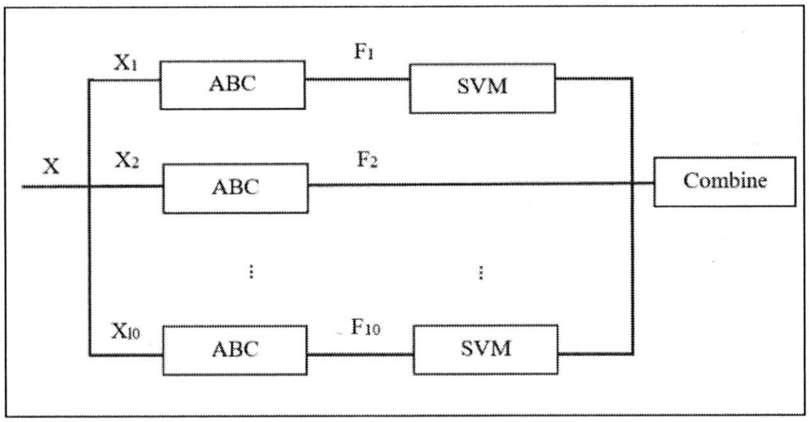

Figure 1. MMABC process

The MMABC algorithm with *one versus one* (OVO) approach, which k class of the data shown in Figure 2. First, divide the multiclass problems by using the OVO approach, then we will get $\frac{k(k-1)}{2}$ of binary class problems. Second, initialise done by *scout bee* by using this equation:

$$X_i^j = X_{min}^j + \omega(X_{max}^j - X_{min}^j),\ 0 \leq \ \omega \leq 1. \tag{1}$$

Annotation:
X_i^j : Food source, $i = 1,2, \dots, SM$
X_{min}^j : Minimum features index, $j = 1,2, \dots, D$
X_{max}^j : Maximum features index, $j = 1,2, \dots, D$

On ABC method as the selection feature, the control parameter of X is $X_{min} = 1$ and $X_{max} =$ maximum features indexX_{max}. After the initialisation has been done, count the value of fitness food source X_i, $i = 1, \dots, SM$, based on this equation (2).

$$fit(X_i) = akurasi(X_i) \tag{2}$$

Third, the bees colony will be divided into three sections. They are employed bee, onlooker bee, and scout bee. The employed bee comes to the location of food source and looks for new food source which has the nectar more than the surroundings have in the same area. The searching for a new food source (V_i), is using this equation (3).

$$V_i^j = X_i^j + \Phi(X_i^j - X_k^j), \tag{3}$$

X_k is the chosen food source which is chosen randomly, $k = 1,2, \dots, SM, k \neq i$, and $-1 \leq \Phi \leq 1 -$ $1 \leq \Phi \leq 1$. Because the food sources consist of features index (with the value of 1 until the total of features from one matrix data), so the choice of population will be applied with this requirement:

If $V_i^j < X_{min} V_i^j < X_{min}$, then

$$V_i^j = (V_i^j) mod(Total\ Fitur) + 1 \tag{4}$$

If $V_i^j > X_{max}$, so

$$V_i^j = (V_i^j) mod(Total\ Fitur) + 1 \tag{5}$$

After that, count fitness value of each food sources and make the greedy selection: If the value of fitness V_i is higher than *fitness* X_i, then V_i will replace X_i, If the value of fitness X_i is higher than *fitness* V_i, then X_i will stay, and abandon food was added into 1.

After the greedy selection made, find the probabilities to find out which food source will be replaced on the next phase, by using equation (6).

$$p_i = \frac{fitness_i}{\sum_{n=1}^{SM} \; fitness_i}, \tag{6}$$

which $fitness_i$ is a quality of the food source towards-i on the employed bee phase.

If the searching process of food sources has finished, then the employed bee will share the information about the quality of the food sources and the probability to the onlooker bee. Onlooker bee chooses the new food source based on the given probabilities of the mentioned food. In this process, onlooker bee increases the probabilities of the new food sources based on the equation (3), and the value of fitness will be found out by using the equation (2). Then, do the greedy selection process.

```
Input:
    Data

    The chosen of total features which will be put into classifier (d)
     Output:
     The chosen features of each data subset of data () and classifier model
              Initialization:
                   QUOTE  subsets:
              For
                   Initialize the population, following the equation (1)
                   Find the fitness value of each food sources, based on the equation (2)
                   Iteration = 1
                   While Iteration < Maximum Iteration
                             Produce the new food sources on the employed bee phase, based on the
                equation (3)
                             Find the fitness value of each food sources
                             Do the greedy selection
                             Find the probabilities of fitness value
                             Produce the new food sources on the onlooker bee phase, based on the
                equation (3)
                             Find the fitness value of each food sources
                             Do the greedy selection
                             If the fitness value is not increasing during the iteration, and Abandon
                food is out of the limit, then produce the food sources based on the equation
                             Collecting quality food sources that have been obtained
                   Iteration = 1 + Iteration
```

Figure 2. MMABC algorithm

2.2.2. The procedure to solve the multiclass problem of brain cancer with MMABC and SVM

In this research, the usage of SVM as the classification method is combined with MMABC as the feature selection method to solve the multiclass classification problem of brain cancer. The learning process to solve the multiclass classification problem is shown in Figure 3.

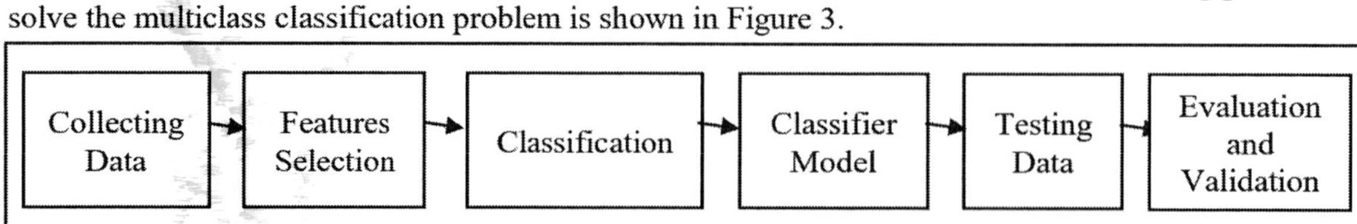

Figure 3. Feature Selection Flowchart

The result of classification is running time and accuracy with or without using feature selection. Accuracy is one metric for evaluating classification models [13]. The accuracy value in percentage is between 0 and 100. The accuracy is measured by using matrix confusion. Table 1 shows the matrix confusion for the multiclass problem with the amount of k class of the examples [14].

Table 1. Confusion Matrix Form for Multiclass Classification

		Predicted Class			
		Class 1	**Class 2**	**…**	**Class k**
Actual Class	Class 1	f_{11}	f_{12}	…	f_{1k}
	Class 2	f_{21}	f_{22}	…	f_{2k}
	⋮	⋮	⋮	⋮	⋮
	Class k	f_{k1}	f_{k2}	…	f_{kk}

The accuracy of the multiclass problem can be measured by equation 7 [15].

$$\text{Accuracy} = \frac{\textit{The number of true predictions}}{\textit{The number of predictions}} = \frac{\textit{The diagonal values on matrix}}{\textit{All values on confusion matrix}} = \frac{\sum_{j=1}^{m} f_{jj}}{\sum_{k=1}^{m}\sum_{j=1}^{m} f_{jk}} \tag{7}$$

3. Results and Discussion

In this paper, there are four types of simulations done. They are brain cancer classification using SVM and linear kernel without feature selection, brain cancer classification using SVM and RBF kernel without feature selection, brain cancer classification using MMABC, and SVM with linear kernel, and brain cancer classification using MMABC and SVM with RBF kernel simulation. The simulation was done by using 3-cross validation, the data was divided into three parts, which each part of the data will be the test data and the average will be obtained. Based on the SVM, count the perpendicular distance of data with each of the hyperplane. If the distance of one sample on the test data is closer to hyperplane for class -1 than the hyperplane to the class 1, the sample is predicted in class -1. By doing this, on every binary class case, the prediction will be obtained from every sample of the test data.

To unite the results of the mentioned predictions, the class label for each of the binary class cases will be changed back into the multiclass label. With the max-wins principal, class, which commonly shows up is the last prediction of a sample on the test data. After all the samples on the data test have been successfully predicted, then make the confusion matrix based on Table 1. The next one is operating the accuracy with the equation (7). The first simulation, which will be done is doing the classification process using SVM with the linear kernel without feature selection. Table 2 shows the confusion matrix of the classification simulation's results of brain cancer using SVM and linear kernel without feature selection.

Table 2. Confusion matrix's types of brain cancer classification using SVM and linear kernel without feature selection

		Predicted Class				
		MD	**MG**	**ATRT**	**Normal**	**PNET**
Actual Class	MD	8	1			1
	MG	1	5	3		1
	ATRT			10		
	Normal				4	
	PNET	1	2			5

Based on the mentioned confusion matrix, the accuracy for overall classes is 76.19 %, which consist of 32 samples classified as true, and 10 samples which had error classifications, such as 1 data of MD classified as MG, 1 data of MD classified as PNET, 1 data of MG classified as MD, 3 data of MG classified as ATRT, 1 data of MG classified as PNET, 1 data of PNET classified as MD, 2 data of PNET classified as MG. The running time for simulation 1 was 20.488 seconds. After that, do the simulation 2 (classification process using SVM and RBF Kernel without feature selection). Table 3 shows the confusion matrix of the classification simulation's results of brain cancer using SVM with RBF kernel without feature selection.

Table 3. Confusion matrix's types of brain cancer classification using SVM and RBF kernel without feature selection

		Predicted Class				
		MD	**MG**	**ATRT**	**Normal**	**PNET**
Actual Class	MD	9		1		
	MG	1	9			
	ATRT			10		
	Normal		1		1	2
	PNET		1			7

Based on the mentioned confusion matrix, the accuracy for overall classes is 85.71%, which consist 36 samples classified as true, and 6 samples which had error classifications, such as 1 data of MD classified as ATRT, 1 data of MG classified of MD, 1 normal data classified as MG, 2 normal data classified as PNET, 1 data of PNET classified as MG. The running time for simulation 2 was 9.8114 seconds. Simulation 3 and 4 are the types of brain cancer classifications using MMABC and SVM with linear kernel and RBF, respectively. The features used were 10, 50, 100, 200, 300, 400, 500, 600, 700, 800, 900, and 1000. Table 4 shows the confusion matrix of the classification simulation's results of brain cancer using MMABC and SVM with linear and RBF kernel.

Table 4. The accuracy result of brain cancer classifications using MMABC and SVM with linear kernel and RBF

The number of features used	Linear Kernel		RBF Kernel	
	The average accuracy	**The average running time (seconds)**	**The average accuracy**	**The average running time (seconds)**
10	88.05%	0.2669	85.71%	0.3716
50	85.71%	0.2741	88.10%	0.3782
100	85.71%	0.3371	92.86%	0.3865
200	90.48%	0.3788	80.48%	0.4003
300	92.86%	0.3898	95.24%	0.4152
400	85.71%	0.3984	92.86%	0.4180
500	88.10%	0.4343	92.86%	0.4228
600	90.48%	0.4652	88.10%	0.4373
700	80.95%	0.5091	90.48%	0.4457
800	76.19%	0.5305	85.71%	0.4505
900	90.48%	0.5489	80.95%	0.4536
1000	85.71%	0.5764	90.48%	0.4790
Mean	86.70%	0.4258	88.65%	0.4216

The accuracy above can also be expressed in Figure 4. Based on the simulation result, the number of the most optimum of the usage of the linear kernel on MMABC-SVM is 300 features with the accuracy

92.86% meanwhile the number of the most optimum features for using RBF Kernel on MMABC-SVM is 300 features with 95.24%. The running time is shown in Figure 5.

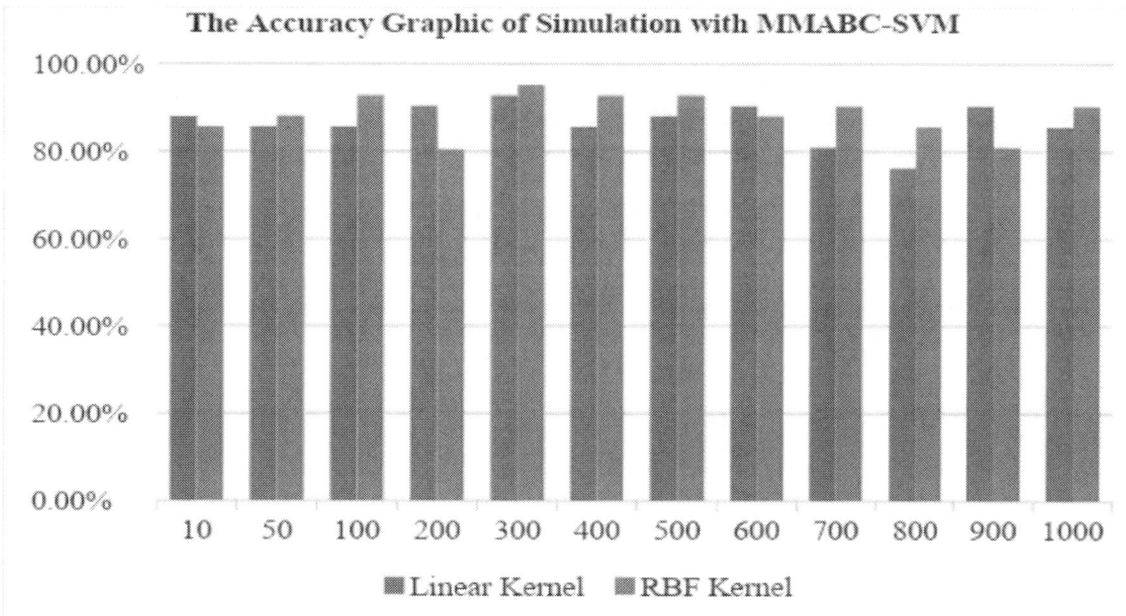

Figure 4. The accuracy graphic of simulation with Linear Kernel and RBF for MMABC-SVM

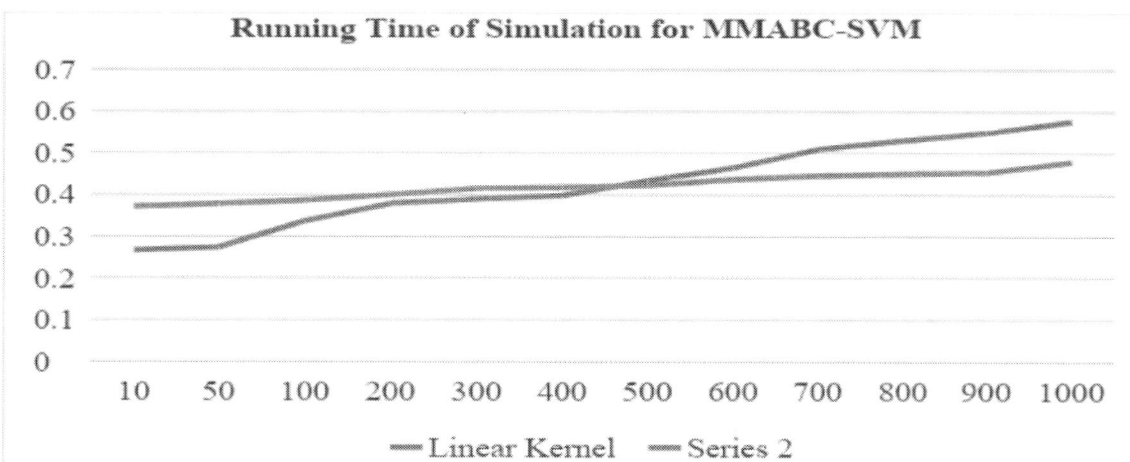

Figure 5. The running time graphic of simulation with Linear Kernel and RBF for MMABC-SVM

From four simulations, the highest accuracy is 95.24% by using MMABC-SVM with RBF kernel with a 0.4152 seconds amount of 300 features. The average of the accuracy using MMABC is higher if it is compared with the classification without the selection feature, for both linear kernel and RBF. On the graphic of the running time, it can be seen generally that the more features used, the length of running time needed. It also can be assumed that the importance of the selection feature process is to increase the accuracy and to reduce the running time.

4. Conclusion

The classification of brain cancer based on the data from Broad Institute can be solved by using machine learning. This problem is a multiclass classification problem. In this paper, the approach used is one versus one, so 1 multiclass classification problem became ten problems of binary class classification. The classification method used is SVM. 4 simulations have been done on this research. In the first simulation, the usage of SVM and linear kernel without feature selection resulted the accuracy 76.19%, with the running time 20.488 seconds. On the second simulation, the usage of SVM and RBF Kernel without feature selection resulted the accuracy 85.71 % with the running time 9.8114 seconds. In the third simulation, the usage of SVM and MMABC with linear kernel resulted in the highest accuracy 92.86 % and the running time 0.3898 with the number of usage 300 features. In the fourth simulation, the usage of SVM and MMABC with RBF Kernel resulted in the highest accuracy 95.24 % and 0.4152 of the running time with the number of usage 300 features. From all of those four simulations, the MMABC-SVM method with RBF Kernel is a method that reduces the highest average of the accuracy and the fastest of the running time.

Acknowledgements

This research is fully funded and supported by *Lembaga Penelitian dan Pengabdian Masyarakat* (Institute of Research and Community Services) Universitas Negeri Jakarta.

References

[1] American Cancer Society 2019 *Cancer Facts & Figures* (Atlanta: American Cancer Society) 1-72.
[2] Australian Institute of Health and Welfare 2017 *Australian Cancer Incidence and Mortality (ACIM)* (Canberra: AIHW 7)
[3] Gilbert M R, Dignam J J, Armstrong T S, Wefel J S, Blumenthal D T, Vogelbaum M A, Colman H, Chakravarti A, Pugh S, Won M, Jeraj R, Brown P D, Jaeckle K A, Schiff D, Stieber V W, Brachman D G, Werner-Wasik M, Tremont-Lukats I W, Sulman E P, Aldape K D, Walter J. Curran J and Mehta M P 2014 *N. Engl. J.Med* **370** 699-708
[4] Chinot O L, Wick W, Mason W, Henriksson R, Saran F, Nishikawa R, Carpentier A F, Hoang-Xuan K, Kavan P, Cernea D, Brandes A A, Hilton M, Abrey L and Cloughesy T 2014 *N.Engl. J.Med* **370** 709-722
[5] National Cancer Institute 2019 *Cancer Stat Facts: Brain and Other Nervous System Cancer* (Rockville: National Cancer Institute)
[6] S Tandel G, Biswas M, Kakde O, Tiwari A, Suri H, Turk M 2019 *Cancers*, **1** 111
[7] Zhang L and Huang X 2015 *International Joint Conference on Neural Networks (IJCNN)* 1-6
[8] Azis A I S, Suhartono V and Himawan H 2017 *J. Teknol. Inf. CyberKU* **13** 170
[9] Ramanathan T T and Sharma D 2017 *Procedia Comput. Sci.* **115** 307
[10] Li J, Wang Y, Song X and Xiao H 2018 *Comput. Biol. Med.* **100** 1
[11] Karaboga D and Basturk B 2007 *J. Glob. Optim.* 39 459
[12] Auria L, Moro R A 2008 *DIW Discussion Papers*, No. 811, (Berlin: Deutsches Institut für Wirtschaftsforschung)
[13] Hossin, M and Sulaiman, M N 2015 *Int. J. Data Min. Knowl. Manag. Process* **5** 1
[14] Manliguez, C 2016 *Generalized confusion matrix for multiple classes*
[15] Santra, A K and Christy, C J 2012 *Int. J. Comput. Sci. Issues* **9** 322

Journal of Physics: Conference Series 1417 (2019) 012016 doi:10.1088/1742-6596/1417/1/012016

Hierarchical Clustering Algorithm Based on Density Peaks using Kernel Function for Thalassemia Classification

S Hartini[1] **and Z Rustam**[1]
[1]Department of Mathematics, University of Indonesia, Depok 16424, Indonesia,

Email: sri.hartini@sci.ui.ac.id, rustam@ui.ac.id

Abstract. Thalassemia is an inherited blood disorder and one of the most common genetic diseases in the world, with its classification due to the processes involved in patients' treatment. Therefore, it is essential to obtain a precise diagnosis due to an appropriate treatment receives a higher lifetime value. A new method based on the kernel, modified from hierarchical clustering based on density peaks (HCDP) was proposed in this research. Furthermore, the concept of *k*-nearest neighbor and hierarchical clustering, consists of three steps were utilized. These steps include local density calculation, hierarchy representation, and optimal cluster extraction. The modification of this technique is therefore based on the polynomial kernel function, which was expected to be more accurate in separating data that cannot be detached linearly. Data utilized was obtained from Harapan Kita Hospital, West Jakarta, Indonesia, and it consists of 82 thalassemia and 68 non-thalassemia samples. Using the proposed method, the performance of HCDP with/without kernel function in this paper were examined using 10-fold cross-validation and compared using the confusion matrix by calculating its F1-Score. The results concluded that hierarchical clustering based on density peaks gives approximately 67.77 percent F1-Score, while 70.06 percent is obtained when the method is combined with the kernel function.

1. Introduction

Thalassemia, which is derived from two Greek words *thalassa* (sea) and *haima* (blood), is the most common genetic disorders that are inherited [1]. This disease is prevalent in most countries of the world such as the Mediterranean, Central Asia, Middle East, Southern China, India, Far East, and also the countries along the northern coast of South America and Africa [2]. The total incidence of patients with this disease each year is estimated at 1:100,000 in every part of the world, and 1:10,000 people in the European Union [3].

According to the Thalassemia International Federation, it is estimated that 7% of the population in the world is a carrier of a severe hemoglobin disorder including thalassemia and 300-500,000 children are born with it each year. Furthermore, there were also reports that only 200,000 people living with this disease are registered and receiving regular treatment around the world [4]. Thalassemia classification, therefore, aims to give advance research in clinical care for precise diagnosis. This classification enables the delivery of appropriate treatment capable of improving patients' quality of life and increasing their survival rate.

Numerous classification studies have previously been carried out, involving various machine learning techniques. Paokanta et al. [5] utilized some methods such as *k*-Nearest Neighbors (KNN), Bayesian Networks (BNs), Multi-Layer Perceptron (MLP), Multinomial Logistic Regression, and also Naïve Bayes with the accuracy of 80 percent. Some neural networks and decision trees also used red blood cells, reticulocytes, and platelets characteristics [6]. Using the Complete Blood Count (CBC) test and artificial intelligence algorithms, Yousefian et al. [7] found that red blood cells, hemoglobin, mean corpuscular volume, and hematocrit are essential characteristics in classifying thalassemia.

Hierarchical clustering based on density peaks has proven more accurate for several complex datasets [8]. Therefore, in this study, hierarchical clustering based on density peaks using kernel function was adopted and expected to be more accurate in the classification.

2. Material and Methods

2.1. *Material*

Thalassemia dataset was obtained from Harapan Kita Hospital in West Jakarta, Indonesia. This dataset consists of 82 thalassemia and 68 non-thalassemia samples with numerical characteristics described by ten attributes, as shown in Table 1.

Table 1. Thalassemia dataset attributes

Attribute Name	Description
HGB	The number of hemoglobin in grams per deciliter of blood
HCT	Hematocrit or the volume percentage of red blood cells
WBC	The number of white blood cells per microliter of blood
Basophils	The basophils percentage in the blood
Eosinophils	The eosinophils percentage in the blood
Rod neutrophils	The bands (almost mature) neutrophils percentage with rod-shaped
Segmented neutrophils	The segmented (fully mature) neutrophils percentage in the blood
Lymphocytes	The lymphocytes percentage in the blood
Monocytes	The monocytes percentage in the blood
PLT	The number of platelet count per microliter of blood

2.2. *Methods*

This study uses a hierarchical clustering method based on density peaks with/without kernel function. The performances of these two methods are evaluated using 10-fold cross-validation and later compared.

2.2.1. *Kernel Function.* The kernel function for every $x, y \in R^n$ is defined as below [9]

$$K(x,y) = (\phi(x), \phi(y)) \tag{1}$$

where ϕ is the function for mapping the input data to the feature space. Meanwhile, the distance between two points is defined as follows [9]

$$\text{dist}(x,y) = \sqrt{K(x,x) - 2K(x,y) + K(y,y)} \tag{2}$$

This kernel function representation maps the data into a high dimensional feature space useful in solving complex real-world applications which cannot be solved by linear functions [10]. The kernel function can be linear, polynomial, and Gaussian radian basis kernel function [11]. However, polynomial kernel function with the formula expressed in equation (3) is used in this research. The polynomial degree used can be any value, but for simplicity, only the first ten polynomial with degree p is examined.

$$\mathrm{K}(x,y) = (x \, . \, y + 1)^p \tag{3}$$

2.2.2. *Hierarchical Clustering Based on Density Peaks.* There are three steps in this method: k-nearest density calculations, hierarchy illustration in a graph, and cluster optimization, which is used to obtain the final clusters. k-nearest density calculation uses k-nearest neighbors whose i's density defined as below.

$$\rho_i = (i, j) \tag{4}$$

where $kNN(i) = \{j \in D \mid d(i,j) \leq dist(i, NN_k(i)\}$and $NN_k(i)$ is the kth nearest point to i by dist(.) computed using standard Euclidean distance [8]. In order to implement k-nearest neighbors, KD tree was used to compute the distance efficiently. It is defined as the multidimensional binary search trees in which data structure is divided into k partition search space [12].

After computing k-nearest density ρ_i, its distance δ_i from points with a higher density is calculated as below [8].

$$\delta_i = dist(i, j) \tag{5}$$

The hierarchical clustering was used based on the density peaks to obtain a tree which gives distance and density information. It was conducted by treating data as vertices in a graph. However, each vertex not related to the maximal density was connected with an edge weight δ_i to its nearest neighbor of higher density [8]. The edge was then sorted dan iteratively deleted in decreasing order of weight. After that, a clustering tree containing all partitions was obtained.

Finally, the problem was handled as an optimization issue to maximize the total stabilities of the clusters shown below [8].

$$S = \sum_{i=1}^{K} S(C_i) \tag{6}$$

where K is defined as the number of clusters and $S(C_i)$ as [8].

$$S(C_i) = \sum_{x_j \in C_i} \left(\frac{1}{\delta_{exclude}(x_j, C_i)} - \frac{1}{\delta_{emerge}(C_i)} \right) \tag{7}$$

where $\delta_{exclude}$ is the maximum weight with deleting edges when excluding point x_j from the cluster and δ_{emerge} being the maximum weight whose cluster C_i was separated from the previous one [8].

2.2.3. *Hierarchical Clustering Based on Density Peaks using Kernel Function.* When the kernel function was used, there was not much difference with the original. There were also three steps in hierarchical clustering based on density peaks using kernel function. These include k-nearest density calculations, hierarchy representation in a graph, and cluster optimization. The difference exists due to the substitution of the distance formula with kernel function, thereby, making it possible to compare the performance of these two methods in the classification. k-nearest density calculation uses k-nearest neighbors i's density is defined as

$$\rho_i = \sqrt{K(i,i) - 2K(i,j) + K(j,j)} \tag{8}$$

where $kNN(i) = \{j \in D \mid K(i,j) \leq K(i, NN_k(i))\}$ and $NN_k(i)$ is the kth nearest point to i by dist(.). After computing its k-nearest density ρ_i, its distance δ_i from points with a higher density is calculated as follows:

$$\delta_i = \sqrt{K(i,i) - 2K(i,j) + K(j,j)} \tag{9}$$

The hierarchical clustering based on density peaks using kernel function was used to obtain a tree conducted by treating data as vertices in a graph. Each vertex besides the one for maximum density was connected with an edge weight δ_i to its nearest neighbor of higher density. The edge was then sorted dan iteratively removed in decreasing order of weight, and the clustering problem was transformed into a sub-tree partition problem. Finally, the optimal cluster extraction was carried out in the same way as the hierarchical clustering based on density peaks without using kernel function.

2.2.4. Performance Measure. F1-score was used in this paper to evaluate the methods based on the confusion matrix. It has a balance interest in precision and sensitivity, which distinguishes the correct classification of labels within different classes [13]. They are the measures used to measure how well the method obtained relevant information [14]. The formula of F1-Score is shown below.

$$\text{F1-Score} = \frac{2\ x\ sensitivity\ x\ precision}{sensitivity + precision}\ x\ 100\% \tag{10}$$

with the sensitivity and precision defined below.

$$\text{Sensitivity} = \frac{TP}{TP+FN}\ x\ 100\% \tag{11}$$

$$\text{Precision} = \frac{TP}{TP+FP}\ x\ 100\% \tag{12}$$

in which TP is the number of thalassemia samples correctly diagnosed, FN is the number of thalassemia samples incorrectly diagnosed, and FP is the number of non-thalassemia samples incorrectly diagnosed [15].

3. Results and Discussions

The performance of hierarchical clustering based on density peaks with/without kernel function was examined using 10-fold cross-validation. In this, the dataset was split randomly into ten mutually exclusive folds of approximately equal size [16]. Furthermore, each fold was taken as the validation data for testing the classification model, and this process was repeated ten times, which was advantageous to the repeated random sub-sampling [17]. The performance of hierarchical clustering based on density peaks is given in Table 2 below.

Table 2. Performance measurements of hierarchical clustering based on density peaks

Fold	F1-Score (%)	Running Time (s)
1	69.43	5.59
2	67.65	1.11
3	69.57	1.26
4	68.06	0.98
5	68.75	0.83
6	66.32	1.00
7	66.67	1.18
8	69.43	0.95
9	63.44	1.01
10	68.42	1.36

The average value is calculated, with the F1-Score and running time of the hierarchical clustering based on density peaks method acquired is 67.77 percent and 1.53 seconds, respectively. The first ten polynomial degrees are examined with the kernel using the same 10-fold cross-validation. The performance of hierarchical clustering based on density peaks using the polynomial kernel is given in Table 3.

Table 3. Performance measurements of hierarchical clustering based on density peaks using a polynomial kernel function

Polynomial Degree	Average of F1-Score (%)	Average of Running Time (s)
1	67.77	1.26
2	55.32	1.17
3	68.29	1.17
4	68.37	1.25
5	69.55	1.25
6	69.55	1.27
7	69.55	1.20
8	70.06	1.29
9	70.06	1.28
10	70.06	1.16

According to Table 3, we can see that the average of F1-Score and also the running time varied according to the polynomial degree used, as illustrated in Figure 1 and Figure 2 below.

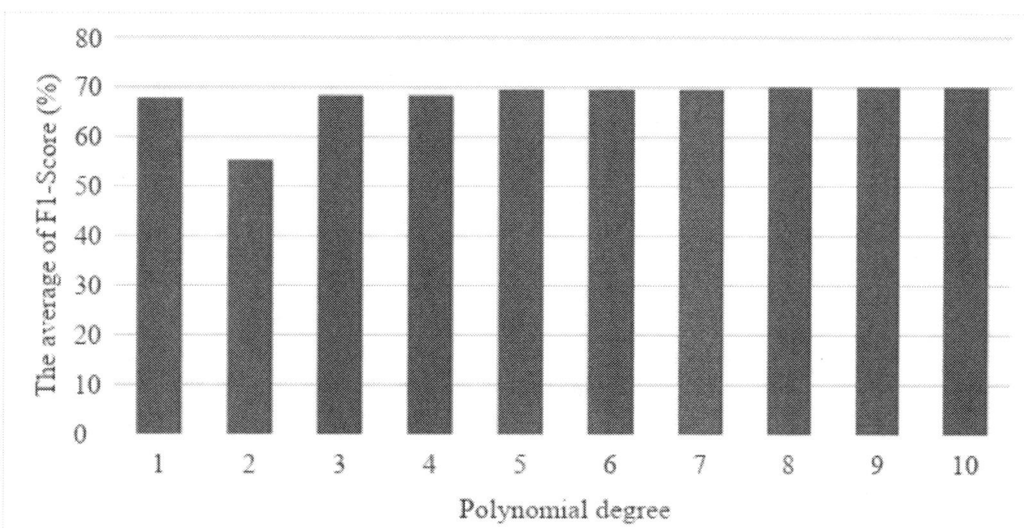

Figure 1. The average of F1-Score of hierarchical clustering based on density peaks using polynomial kernel functions

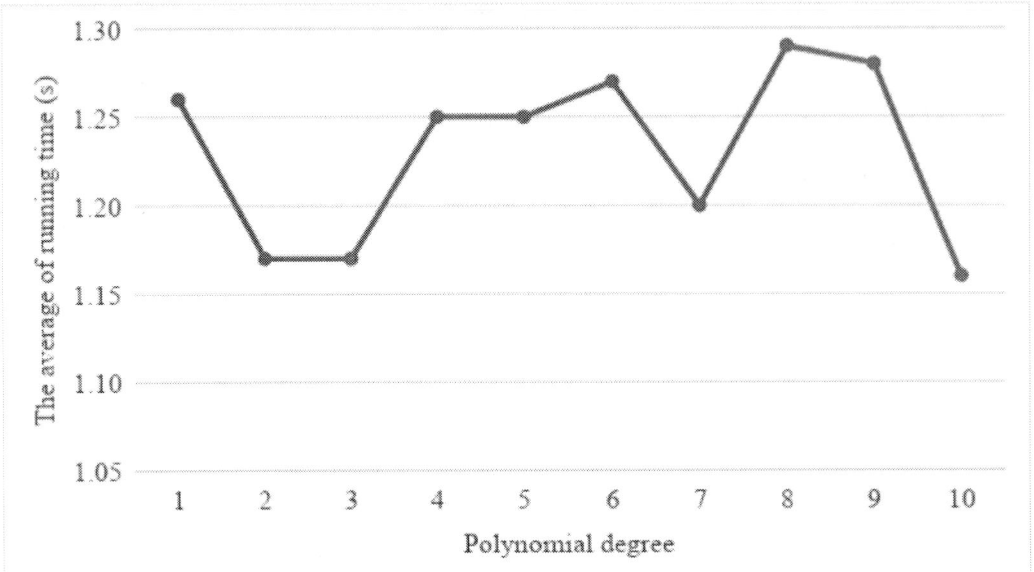

Figure 2. The average of the running time of hierarchical clustering based on density peaks using polynomial kernel functions

The highest average value of F1-Score is 70.06 percent, which was obtained using the eighth, ninth, tenth-degree polynomial kernel. Furthermore, these F1-Scores were obtained after 1.29 seconds, 1.28 seconds, and 1.16 seconds. Therefore, tenth-degree polynomial kernel gives better performance in average running time. The performance of hierarchical clustering based on density peaks using tenth-degree polynomial is given in Table 4.

Table 4. Performance measurements of hierarchical clustering based on density peaks using a tenth-degree polynomial kernel function

Fold	F1-Score (%)	Running Time (s)
1	71.43	1.14
2	68.29	1.16
3	70.19	1.16
4	69.57	1.14
5	70.19	1.16
6	68.29	1.24
7	69.57	1.14
8	72.04	1.16
9	70.19	1.17
10	70.81	1.15

In this table, the F1-Score reaches up to 70 percent in several folds using this polynomial kernel function, with 72.04 percent as the highest F1-Score obtained.

4. Conclusion

Thalassemia, as an inherited blood disorder, is the most common genetic disorders in the world. It is necessary to classify it since the initial process of executing appropriate treatment is capable of improving patients' quality of life and increase their survival rate. The dataset was obtained from Harapan Kita Hospital, West Jakarta, Indonesia. This dataset consists of 82 thalassemia and 68 non-thalassemia samples which were used to evaluate the performance of hierarchical clustering based on density peaks with/without the kernel function.

These two methods were examined using 10-fold cross-validation. From the experiments, hierarchical clustering based on density peaks gives approximately 67.77 percent F1-Score in 1.53 seconds. However, 70.06 percent of F1-Score is obtained when combined with a polynomial kernel function. These values are relatively good, but still, need some improvement, although both methods had good running time, which is less than 2 seconds for almost all experiments.

It is required that further research be carried out using the new model in order to improve their performance. Other kernel parameters can also be used for the same purpose, with the possibility to observe other performance measurements. Higher F1-Score is then expected to give more accurate and precise classification to obtain a better diagnosis.

Acknowledgments
This research was financially supported with a PDUPT 2019 research grant scheme (ID number NKB-1621/UN2.R3.1/HKP.05.00/2019) by the Indonesian Ministry of Research and Higher Education. The authors are grateful to the Harapan Kita Hospital in West Jakarta, Indonesia, for their kindness in providing the thalassemia dataset. Also, the authors are thankful for all reviewers involved in the process of improvement of this article.

References
[1] Galanello R and Origa R 2010 Beta-thalassemia *Orphanet J. of Rare Diseases* **5**(11) pp 1–15
[2] Vincent O, Oluwaseyi B, James B and Saidat L 2016 Coinheritance of B-Thalassemia and Sickle Cell Anaemia in Southwestern Nigeria Ethiop *J Health Sci* **26**(6) pp 517–522
[3] Nigam N, Nigam S, Agarwal M, and Singh P K 2017 B-thalassemia: from clinical symptoms to the management *Int. J. of Contemporary Med. Research* **5**(5) pp 1066–1070
[4] Thalassemia International Federation https://thalassaemia.org.cy/ Accessed 23 June 2019
[5] Paokanta P, Harnpornchai N, Srichairatanakool S and Ceccarelli M 2011 The knowledge discovery of B-Thalassemia using principal components analysis: PCA and machine learning techniques *Int. J. of e-Education, e-Business, e-Management and e-Learning* **1**(2) pp 169–174
[6] Wongseree W, Chaiyaratana N, Vichittumaros K, Winichagoon P, and Fucharoen S 2007 Thalassemia classification by neural networks and genetic programming *Inf. Sci.* **177** pp 771–786
[7] Yousefian F, Banirostam T and Azarkeivan A 2017 Prediction thalassemia based on artificial intelligence techniques: a survey *Int. J. of Adv. Research in Comp. and Com. Eng* **6(8)** pp 281–287
[8] Rong Z, Yong Z, Shengzhong F and Nurbol L 2018 A novel hierarchical clustering algorithm based on density peaks for complex datasets *Hindawi: Complexity* **2018** pp 1-8
[9] Stitson M O, Weston J A E, Gammerman A, Vovk V, Vapnik V 1996 Theory of support vector machines. Technical Report CSD-TR-96-17, Royal Holloway, University of London
[10] Bezdek J C 1981 *Pattern recognition with fuzzy objective function algorithms* (New York: Plenum Press)
[11] Liu L, Shen B and Wang X 2014 Research on Kernel Function of Support Vector Machine *In: Huang YM., Chao HC., Deng DJ., Park J. (eds) Advanced Technologies, Embedded and Multimedia for Human-centric Computing. Lecture Notes in Electrical Engineering* (Springer, Dordrecht)
[12] Bentley J L 1975 Multidimensional binary search trees used for associative searching. *Com. of the ACM* **18(9)** pp. 509-517
[13] Sokolova M, Japkowicz N, and Szpakowicz S 2006 Beyond accuracy, F-score and ROC: A family of discriminant measures for performance evaluation. *Adv. in Artificial Intelligence, Lecture Notes in Comp. Sci.* 4304

[14] Ting K M 2011 Precision and Recall *In: Sammut C., Webb G.I. (eds) Encyclopedia of Machine Learning* (Springer, Boston, MA)
[15] Rahideh A and Shaheed M H 2011 Cancer classification using clustering-based gene selection and artificial neural networks. *Proc. 2nd Int. Conf. on Control, Inst., and Automation.* (Shiraz, New Jersey: IEEE) pp 1175–1180
[16] Kohavi R 1995 A study of cross-validation and bootstrap for accuracy estimation and model selection *Int. Joint Conf. on Artificial Intelligence*
[17] Raju S K and Murty M R 2018 Support Vector Machine with k-fold cross validation model for software fault prediction *Int. J. of Pure and Applied Maths* pp 321-334

Comparison of Confidence Intervals for the TG Estimator in Capture-recapture Data

Orasa Anan[1*], Preedaporn Kanjanasamranwong[1], W. Chantarangsri[2], and A'yunin Sofro [3],

[1]Department of Mathematics and Statistics, Faculty of Science, Thaksin University, Phatthalung, Thailand
[2]Department of Mathematics,Faculty of Sciences and Technology,Nakhon Pathom Rajabhat University, Nakhon Pathom, Thailand
[3]Mathematics Department, Universitas Negeri Surabaya, Indinesia

E-mail: *aorasa@tsu.ac.th

Abstract. Capture-recapture techniques are very powerful tool and widely used for estimating an elusive target population size. Capture-recapture count data is presented in form of frequencies of frequencies data. They consist of the frequency of unites detected exactly once, twice, and so on, and the frequency of undetected unites is unknown. As consequence, the resulting distribution is a zero-truncated count distribution. The binomial distribution is selected as a simple model if the maximum number of counting occasions is known. It counting occasions are not known in advance, the series of frequencies assumed to be the Poisson distribution. In fact, the target population might be heterogeneous because it has different characteristics, resulting in over or under dispersion based on the basic models. The mixed Poisson, which is the exponential-Poisson mixture model, have been widely used to construct population size estimator for capture-recapture data. The original Turing estimator provides a good performance under the Poisson distribution. Additionally, an extension of Turing estimator, called the Turing-based geometric distribution with non-parametric approach was proposed (TG) for the heterogeneous population. It gives an easy way to estimate the target population size. In this work, we derived uncertainty measures for the TG estimator by considering two sources of variance (M1), and the second way is using only one source of variance (M2). It is emphasised that although the analytic approaches to compute uncertainty measures can be easily used in practice, there are valid asymptotically and requires a large sample size. Therefore, re-sampling approaches, true bootstraps (M3), imputed bootstrap (M4) and reduced bootstrap (M5), are proposed as alternative methods to get uncertainty measures. The study compares performance of variance and confidence interval of paralytics and re-sampling methods by using a simulation study. Overall, the imputed bootstrap is the best choice for estimating variance and constructing confidence interval for the TG estimator. The analytic approach with two sources of variance remains successful to estimate variance and calculate confidence interval in the case of large. It is very clear that the reduced bootstrap and the analytic approach with one source of variance are not appropriate in all situations. For the true bootstrap, the true value of population size is often unknown in nature; therefore, it quite useless for capture-recapture study.

Keywords : capture-recapture; uncertainty; population size estimator

1. Introduction

Capture-recapture techniques have been originally used in ecological science with the object to estimate the elusive wildlife population size.[1]. However, this technique has been extended in many areas aim to deal with under report problem such as epidemiology and public health [2],[3] social science [4] computer system [[5]] marketing [6]. In this paper aims to investigate the uncertainty in the TG estimator that was proposed in [7] . This non-parametric estimator is suitable for unobserved heterogeneity based geometric distribution in capture-recapture data. In this work we compare five uncertainty measurements from analytic and bootstrap approaches. A large-scale of simulation study is provided to lean about behaviour of variance estimation methods. Their comparisons also show how often the confidence interval covers the true value of population sizes. The article is organized as follows: we introduce the Poisson mixture model and the TG estimator in the section 2. In section 3, we show detail of variance estimation methods. Following by the simulation results in the section 4. In section 5, we present some real data examples. Finally, we summarize and point out some remarks in section 6.

2. The Poisson mixture model

Let x_i, $i = 1, 2, 3, 4, ..., N$ denote the number of time individual i is observed covering the m occasions, and let $g_x = Pr(X = x)$, f_x is the frequency of individuals observed exactly x times so that $x = 0, 1, 2, 3, ..., m$. Note that f_0 is missing and it can be estimated by $\widehat{N}\widehat{p}$. The Poisson distribution often assumed to be a basic model in capture-recapture study; however, the unobserved heterogeneity usually occurs as the variance is greater/lower than mean. The Poisson mixture model is more realistic than the original model. Refer to [8] suggested a mixed Poisson model with arbitrary mixing $h(x)$ to model the form of heterogeneity as

$$g_x = \int_0^\infty \frac{exp(-\lambda)\lambda^x}{x!} h(\lambda;\theta)d\lambda. \tag{2.1}$$

The exponential distribution was selected as the arbitrary mixing density; $h(x) = \frac{1}{\theta}exp\left(-\frac{\lambda}{\theta}\right)$ in the mixed Poisson model. The marginal of x achieved that

$$g(x) = \int_0^\infty \frac{exp(-\lambda)\lambda^x}{x!} exp(-\frac{\lambda}{\theta}) = \frac{1}{\theta^x}\int_0^\infty \lambda^x exp{-}\lambda(1+\theta^{-1})d\lambda = \frac{1}{1+\theta}\left(\frac{\theta}{1+\theta}\right)^x = p(1-p)^x, \tag{2.2}$$

where $p = \frac{p}{1+\theta}$ is the probability of success based on the geometric distribution. The geometric distribution is potentially flexible candidate for unobserved heterogeneity in capture recapture study, it can be applied to many real life situations. The population size estimator based-geometric distribution was proposed by [7] to deal with an violation assumption of Poisson model, and [9] reconstructed in 2.1

2.1. A Turing based-geometric distribution

The Turing based-geometric distribution estimator was proposed by [7], let $X \sim Geo(p)$, $g_0 = p$ and $E(X) = \frac{1-p}{p}$,

$$g_0 = p = \sqrt{p^2} = \sqrt{\frac{(1-p)p^2}{(1-p)}} = \sqrt{\frac{1-p}{(1-p)/p}} = \sqrt{\frac{g_1}{E(X)}}. \tag{2.3}$$

In practice, the g_x can be estimated by the $\frac{f_x}{n}$, leading to

$$\hat{g} = \sqrt{\frac{f_1/n}{S/n}} = \sqrt{\frac{f_1}{S}}, \tag{2.4}$$

where $S = \sum_{x=0}^{m} x f_x = \sum_{x=1}^{m} x f_x$. As a consequence, the Turing based geometric distribution estimator (TG) was given as

$$\widehat{N}_{TG} = \frac{n}{1 - \sqrt{\frac{f_1}{S}}}. \quad (2.5)$$

The interesting points of the TG estimator are that the formula is not complicated for users in practice, and it includes all of information of sample, by using S.

Theorem 2.1. The TG estimator is the asymptotically unbiased estimator under the geometric distribution

$$\lim_{N \to \infty} \frac{E(\widehat{N}_{TG})}{N}. \to 1.$$

See [7] for more detail.

3. Variance estimation techniques

3.1. Analytical approaches

$M1$:Analytic approaches with two sources of variances. According to [10], the variance of TG estimator was derived under a conditional technique mixed with the delta method. As the $\widehat{N}_{TG} = \frac{n}{1-\sqrt{\frac{f_1}{S}}}.$, the variance of $\widehat{N}_{TG}$ arise from two sources; the first one is influenced by the random variable n, and the second one come from the predictive value $\hat{g}_0$ based on the observed n individuals. Thus

$$Var(\widehat{N}) = Var_n\left\{E(\widehat{N}|n)\right\} + E_n\left\{Var(\widehat{N}|n)\right\}. \quad (3.1)$$

From the first term in (3.1), the approximation $E(\widehat{N}|n)$ can be justified by the delta-method, so that $E(\widehat{N}|n) \approx \frac{n}{1-g_0}$,

$$\widehat{Var}_n\left\{E(\widehat{N}|n)\right\} = \frac{1}{(1-g_0)^2} Var(n) = \frac{N(1-g_0)g_0}{(1-g_0)^2} = \frac{n\sqrt{\frac{f_1}{S}}}{\left(1-\sqrt{\frac{f_1}{S}}\right)^2}. \quad (3.2)$$

Since $E(n) = N(1-g_0)$ and $\widehat{g}_{0(TG)} = \sqrt{\frac{f_1}{S}}$. Additionally, for the second term of (3.1) is estimated as

$$E_n\left\{Var(\widehat{N}|n)\right\} = E_n\left\{Var\left(\frac{n}{1-\sqrt{\frac{f_1}{S}}}|n\right)\right\} = E_n\left\{n^2 Var\left(\frac{1}{1-\sqrt{\frac{f_1}{S}}}\right)\right\} = n^2\left\{\frac{S+f_1}{4S^2\left(1-\sqrt{\frac{f_1}{S}}\right)^4}\right\} \quad (3.3)$$

Finally, the variance of TG estimator is given as

$$\widehat{Var}(\widehat{N}_{TG}) = \frac{n\sqrt{\frac{f_1}{S}}}{\left(1-\sqrt{\frac{f_1}{S}}\right)^2} + n^2\left\{\frac{S+f_1}{4S^2\left(1-\sqrt{\frac{f_1}{S}}\right)^4}\right\}. \quad (3.4)$$

$M2$: Analytic approach with one source of variance. The second way of variance estimation, we treat n as fixed therefore the variance of $\widehat{N}_{TG}$ come from only one source as following;

$$\widehat{Var}(\widehat{N}_{TG}) = n^2\widehat{Var}\frac{1}{1-\hat{g}_0} = n^2\left\{\frac{S+f_1}{4S^2\left(1-\sqrt{\frac{f_1}{S}}\right)^4}\right\}. \tag{3.5}$$

This approach is expected more realistic for capture-recapture data in practice since it avoids a computational software system error [9],[11].

3.2. Resampling approaches

The resampling techniques was proposed as alternative choices in the literature [12] to estimate variance and construct confidence interval in capture-recapture framework. Since the variance of the TG estimation by the normalized approximation method is valid asymptotically, large sample sizes are required as a basic assumption. In reality, a target population size can be small and this might lead to underestimation or overestimation of the variance. Additionally, model uncertainty in the analytic approaches usually occur due to the model structure depends on data from only one sample. These limitations of analytic variance approach might lead to invalid confidence intervals for the true population size and achieve an unsatisfactory coverage probability. This study, the resampling approaches; True Bootstrap (M3); Imputed Bootstrap (M4); Reduced Bootstrap (M5) are suggested as an alternative choices to measure variance and construct confidence intervals. The advantages of using the bootstrap approaches are not reverent to the consideration model and straightforward to implement by using computer programming. The basic assumption of those approaches is that the capture-recapture history can be defined by the multinomial likelihood as $\binom{N}{f_0, f_1, f_2, f_3, ..., f_m} p_0^{f_0}p_1^{f_1}p_2^{f_2}...p_m^{f_m}$ The algorithms for variance estimation are given as follows:

Step 1: Estimating population size: $\widehat{N}_{TG} = \frac{n}{1-\sqrt{\frac{f_1}{S}}}$.

Step 2: For T samples consist of individuals from the ordinary data as:

(i) True Bootstrap ($M3$)
Let $\widehat{p}_{M3} = \left\{\frac{f_0}{N}, \frac{f_1}{N}, \frac{f_2}{N}..., \frac{f_m}{N}\right\}$. This method can be use to estimate variance of population size estimator only if the population size is known in advance. Each individual is drown from the multinomial distribution with parameter N and $\widehat{p}_{M3}$.

(ii) Imputed Bootstrap ($M4$)
Let $\widehat{p}_{M4} = \left\{\frac{\widehat{f}_0}{\widehat{N}_{TG}}, \frac{\widehat{f}_0}{\widehat{N}_{TG}}, \frac{\widehat{f}_0}{\widehat{N}_{TG}}, ..., \frac{\widehat{f}_0}{\widehat{N}_{TG}}\right\}$. Under the M4, each individual is drown from the multinomial distribution with parameter $\widehat{N}_{TG}$ and $\widehat{p}_{M4}$

(iii) Reduced Bootstrap ($M5$)
Let $\widehat{p}_{M5} = \left\{\frac{\widehat{f}_1}{n}, \frac{\widehat{f}_2}{n}..., \frac{\widehat{f}_m}{n}\right\}$. Under the M5, each individual is drown from the multinomial distribution with parameter n and $\widehat{p}_{M4}$

Step 3 For each sample, $\widehat{N}_{TG}$ is estimated based on the geometric distribution model.
Step 4 Estimate the variance of $\widehat{N}_{TG}$ under the T bootstrapped samples as

$$Var(\widehat{N}) = \frac{1}{T-1}\sum_{t=1}^{T}\left(\widehat{N}_{TG,t} - E(\widehat{N}_{TG})\right) \tag{3.6}$$

4. Simulation

The simulation study is taken following the geometric distribution with different level of heterogeneity. The objective of this is to investigate the performance of proposed methodologies. Then $X \sim Geo(p)$ where $p = 0.1$, 0.2, 0.3, 0.4, 0.5, 0.6, 0.7 and 0.8. The population size N is set at $N = 50, 80$ and 100 for small sizes, $N = 250, 500$ and 750 for medium sizes, and $N = 1,000, 5,000, 10,000$ and $50,000$ for large sizes. It is 1,000 samples from the true model.

4.1. Bias and Variance of TG estimator

Figure 1 shows the behaviour of TG estimator under different setting. The performance of estimator is evaluated in term of relative bias; $RBias = \frac{1}{N}\{E(\widehat{N} - N)\}$ and relative variance; $\frac{1}{N}\widehat{Var(N)}$ As can be seen that the TG estimator performs very well under the $Geo(p)$ generated

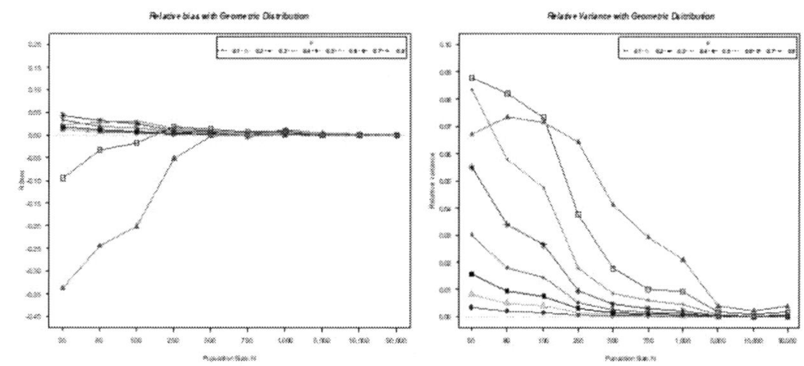

Figure 1. Bias and variance of TG estimator

data. Increasing population sizes lead to a better performance of estimation in term of both variance and bias. Hence, it can be confirmed that the TG estimator has an asymptotic unbiased property with respect to population size. Another interesting point is that the TG estimator are likely to be identical with the parametric estimator under the geometric distribution, namely MLEGeo [9]. The simulation results were shown that they are the best choices for estimating population size in CR data if and only if data follow the geometric distribution.

4.2. Comparing variance estimation

The objective of this section is to compare the performance of five variance approximation methods. For the convenience, the ratio of standard error of estimation, which is defined as the estimated standard error from each approach divided by the true standard error. The simulation result is represented in Figure 2. The findings suggest that imputed bootstrap might be the best choice for estimating variance of TG estimator, especially when the population size is greater than 250. The analytic approach with two sources of variances remains realistic and valid method. The analytic method with one source of variance and reduced bootstrap underestimate in all situations. We do not suggest both of them to estimate the variance of the TG estimator. The true bootstrap tends to identical estimation with the imputed bootstrap but it can be use when the population size is known.

4.3. Comparing confidence intervals

The performance of a confidence interval (CI) method is assessed by how often the confidence interval covers the true value of N. The coverage probability is used for comparing the

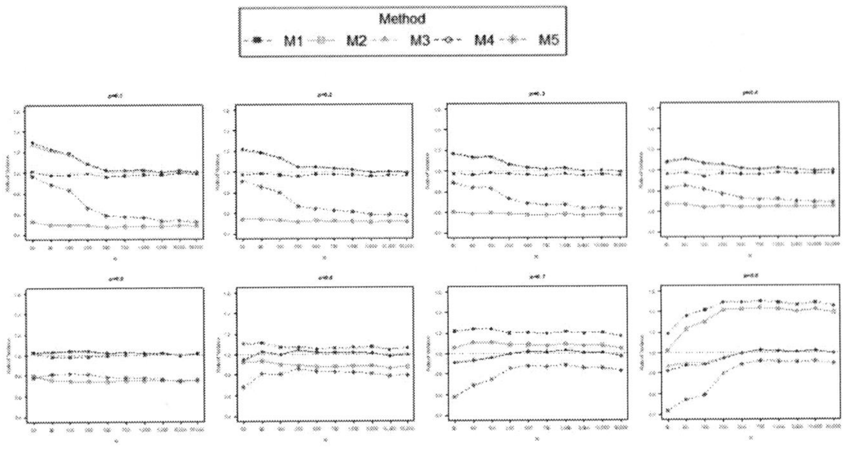

Figure 2. Ratio of standard errors from five methods to the true standard error when data are generated under a geometric distribution; M1: Analytic approach with two source of variance, M2:Analytic approach with one source of variance M3: True bootstrap, M4: Imputed bootstrap and M5: Reduced bootstrap

performance of five variance estimation methods with the nominal coverage probability, setting at 0.95 or 95%. A common procedure of 95% CI is construed by $\widehat{N} \pm Z_{0.975}\sqrt{\widehat{Var}(\widehat{N})}$. For the bootstrap methods, considering all of $\widehat{N}_t, t = 1, 2, 3, ..., 1000$, the standard error can be computed by taking the sample standard deviation from the bootstrap samples results. The approximate of 95% CI of $\widehat{N}$ can be achieved by using the percentile methods. Since the simulation has R replication runs, the coverage probability (Cov) can be calculated as $Cov = (\frac{1}{R}\sum_{r=1}^{R} A_t) \times 100$, where A_t equal to 1 if the true value N is in the target confidence interval, and 0 otherwise.

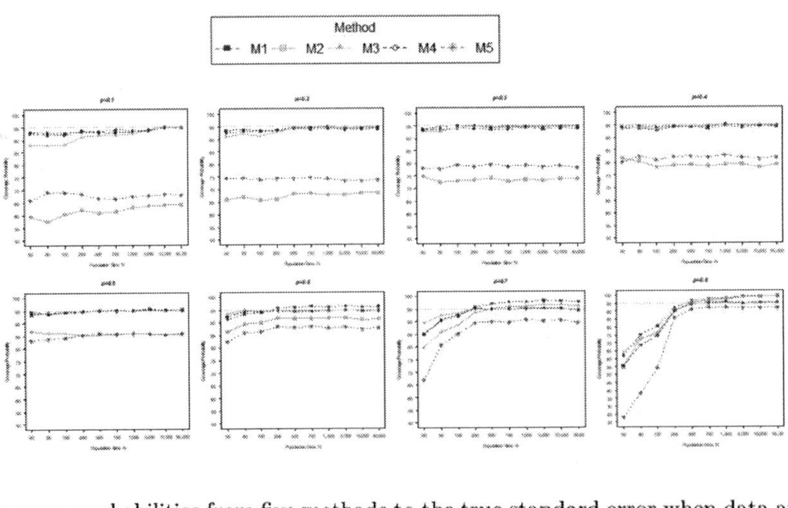

Figure 3. Coverage probabilities from five methods to the true standard error when data are generated under a geometric distribution; M1: Analytic approach with two source of variance, M2:Analytic approach with one source of variance M3: True bootstrap, M4: Imputed bootstrap and M5: Reduced bootstrap

We compare five methods to get confidence interval of the TG estimator. The simulation results as Figure 3 suggest that the imputed bootstrap performs the best to construct the confidence intervals of the TG estimator. Overall, this method provides coverage probabilities close to the nominal 95%. For the analytic approach with one source of variance and reduce bootstrap methods suffer in providing a good performance of coverages for all situations. The confidence intervals of analytic approach with two sources of variances might affect the coverage probability for a small population size in some cases; however, it remains useful in practice. Since it is not only has a good performance but also easy to compute and do not require computational programming method. We recommend to use the imputed bootstrap and analytic approach with two sources of variances for investigating uncertainty of the TG estimator.

5. Real data examples

5.1. Golf tees data

The first example is the golf tees data, $N = 250$ groups of golf tees were placed in survey regain at the University of St Andrews, see [13]. This example is useful for comparing the performance of several variance estimation methods in the case of a true value of population size is known. A total of $n = 162$ groups of golf tees were observed, so the number of hidden in the experiment need to be estimated. The frequency distribution is given as $(f_0, f_1, f_2, ..., f_m) = (88, 46, 28, 21, 13, 23, 14, 6, 11)$. We use the ratio plots [14] for investigating the suitable models as Figure 4, we expect that the TG estimator would perform well. The TG estimator provides $\widehat{N} = 228$. Interestingly, variance estimation from M4 is slightly greater than M 1 and M4. As we expect M2 and M5 are not appropriated for constructing confidence intervals, they do not cover the true value N because of underestimate of variance of estimation. We suggest to use M1 or M4 to estimate variance and construct the confidence interval for golf tees data.

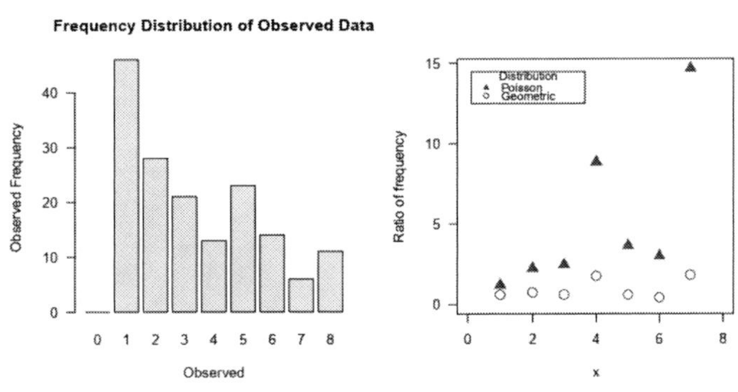

Figure 4. Frequency distribution (left) and the ratio plot of Poisson and Geometric models

Table 1. Comparison of five methods of variance estimation for golf tees data

Method	$(\widehat{N})_{Mean}$	$(\widehat{N})_{Med}$	$\widehat{S.E.}(\widehat{N})$	$95\%CI$	Length
M 1	228	-	12.03	(204 – 252)	48
M 2	228	-	7.165	(214 – 242)	28
M 3	228	228	13.20	(203 – 255)	52
M 4	228	228	12.35	(204 – 253)	49
M 5	229	228	7.77	(214 – 245)	31

5.2. *Heroin users in Bangkok data*

A further example based upon the geometric distribution is the heroin drug user in Bangkok data which was analysed in [15] using the Poisson mixture model to estimate the number of heroin drug users in Bangkok in 2003. This data set is collected from 61 health treatment centres between 1 October and 31 December 2001. From Figure 5, the log ratios suggest to select the population size estimator based on the geometric model. The resulting estimated population size by using the TG estimator is $\widehat{N} = 12,175$. Since the true value of the number of drug users is unknown in advance, so that M3 can not use for estimating variance and produce confidence interval in this case. M1 and M3 are more realistic than M2 and M5, which often underestimate of variance, supporting by the simulation study.

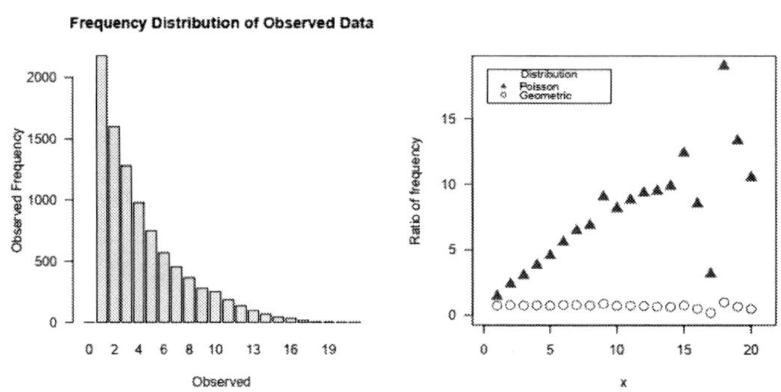

Figure 5. Frequency distribution (left) and the ratio plot of Poisson and geometric models

Table 2. Comparison of five methods of variance estimation for heroin users in Bangkok data

Method	$(\widehat{N})_{Mean}$	$(\widehat{N})_{Med}$	$\widehat{S.E.}(\widehat{N})$	$95\%CI$	Length
M 1	$12,175$	-	73.99	$(12,030 - 12,320)$	290
M 2	$12,175$	-	41.41	$(12,094 - 12,256)$	162
M 3	–	–	–	–	–
M 4	$12,176$	$12,176$	76.48	$(12,026 - 12,325)$	299
M 5	$12,175$	$12,175$	45.57	$(12,087 - 12,266)$	179

5.3. *Dengue per village in Chaiang Mai in 2013*

A case study on dengue fever in Chaiang Mai was analysed in [16]. The data were collected from the laboratory to confirm dengue patients in 2013. To estimate total of patients by using the TG estimator that is confirmed by the ratio plots as Figure 6, Finding that an estimate of dengue patients is $1,617$, which is larger than using the MLEGeo estimator as original analysis. We suggest $M3$ and $M4$ to investigate the uncertainty for this data that was supporting by the simulation study.

6. Conclusion

Capture-recapture approaches are widely us to estimate the hidden population size in may areas, and the TG estimator was proposed as an alternative choice to solve an unobserved heterogeneity in CR data. The TG estimator is mentioned as an asymptotically unbiased estimator with respect to the population size under the geometric distribution. This work we compare an analytic approaches and bootstrap techniques to estimate uncertainty of population

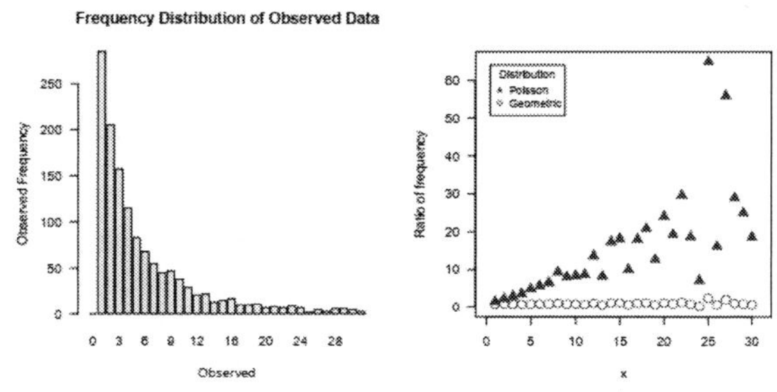

Figure 6. Frequency distribution (left) and the ratio plot of Poisson and geometric models

Table 3. Comparison of five methods of variance estimation for Dengue per village in Chaiang Mai in 2013

Method	$(\widehat{N})_{Mean}$	$(\widehat{N})_{Med}$	$\widehat{S.E.}(\widehat{N})$	95%CI	Length
M 1	1,617	-	19.88	(1,578 − 1,656)	78
M 2	1,617	-	9.27	(1,599 − 1,635)	36
M 3	–	–	–	–	–
M 4	1,663	1,663	25.82	(1,615 − 1,715)	100
M 5	1,663	1,661	17.28	(1,632 − 1,700)	68

size in the capture-recapture data. The further result is that the analytic approach with one source of variance, which was often suggested for investigating uncertainty in capture-recapture study [9], [17], presents a poor coverage probability. The analytic approach and the reduce bootstrap should discontinue in capture-recapture study. The imputed bootstrap is the most accurate but it require a computer intensive programming. Overall, the analytic approach with two sources of variance behaves close to the imputed bootstrap. Therefore, we suggest to use the analytic approach with two sources of variance to estimate variance and construct confidence interval in practice.

Acknowledgement(s) : I would like to thank the referees for his comments and suggestions on the manuscript. This work was supported by Thaksin University.

[1] Darroch, J. N., and D. Ratcliff, A note on capture-recapture estimation, Biometrics (1980) 149-153.
[2] Wesson, Paul, Richard Lechtenberg, Arthur Reingold, Willi McFarland, and Neena Murgai, Evaluating the Completeness of HIV Surveillance Using Capture–Recapture Models, Alameda County, California, AIDS and Behavior. 22 (7) (2018) 2248-2257.
[3] Baujat, Geneviève, Rémy Choquet, Stéphane Bouée, Viviane Jeanbat, Laurène Courouve, Amélie Ruel, Caroline Michot et al, Prevalence of fibrodysplasia ossificans progressiva (FOP) in France: an estimate based on a record linkage of two national databases, Orphanet journal of rare diseases. 12 (1) (2017) 123.
[4] Van Der Heijden, Peter GM, Maarten Cruyff, and Hans C. Van Houwelingen, Estimating the size of a criminal population from police records using the truncated Poisson regression model, Statistica Neerlandica. 57(3) (2003) 289-304.
[5] Accettura, Nicola, Giovanni Neglia, and Luigi Alfredo Grieco, The Capture-Recapture approach for population estimation in computer networks, Computer Networks. 89 (2015) 107-122.
[6] Yauck, Mamadou, Louis-Paul Rivest, and Greg Rothman, Capture-Recapture Methods for Data on the Activation of Applications on Mobile Phones, Journal of the American Statistical Association. (2018) 1-33.
[7] Anan, Orasa, Dankmar Böhning, and Antonello Maruotti, On the Turing estimator in capture–recapture count data under the geometric distribution, Metrika. (2018) 1-24.
[8] Chao, Anne, Estimating the population size for capture-recapture data with unequal catchability, Biometrics. (1987) 783-791.
[9] Niwitpong, Sa-aat, Dankmar Böhning, Peter GM van der Heijden, and Heinz Holling, Capture–recapture

estimation based upon the geometric distribution allowing for heterogeneity, Metrika. 76 (4) (2013) 495-519.
[10] Böhning, Dankmar, A simple variance formula for population size estimators by conditioning, Statistical Methodology. 5 (5) (2008) 410-423.
[11] Böhning, Dankmar, Ratio Plot and Ratio Regression with Applications to Social and Medical Sciences, Statistical Science. 31 (2016) 205-218.
[12] Anan, Orasa,Böhning, Dankmar, and Antonello Maruotti, Uncertainty estimation in heterogeneous capture–recapture count data, Journal of Statistical Computation and Simulation. 87 (10) (2017) 2094-2114.
[13] Anan, Orasa, Böhning, Dankmar, and Antonello Maruotti, Population size estimation and heterogeneity in capture–recapture data: a linear regression estimator based on the Conway–Maxwell–Poisson distribution, Statistical Methods and Applications. 26 (1) (2017) 49-79.
[14] Böhning, Dankmar, Ratio plot and ratio regression with applications to social and medical sciences, Statistical Science. 31 (2) (2016) 205-218.
[15] Böhning, Dankmar, Busaba Suppawattanabodee, Wilai Kusolvisitkul, and Chukiat Viwatwongkasem, Estimating the number of drug users in Bangkok 2001: A capture–recapture approach using repeated entries in one list, European Journal of Epidemiology. 19(12) (2004) 1075.
[16] Böhning, Dankmar, Peter GM van der Heijden, and John Bunge, eds. Capture-recapture Methods for the Social and Medical Sciences. CRC Press, 2017.
[17] Böhning, Dankmar, Power series mixtures and the ratio plot with applications to zero-truncated count distribution modelling, Metron. 73 (2) (2015) 201-216.

A Split Plot Design for an Optimal Mixture Process Variable Design of a Baking Experiment

Mika Alvionita Sitinjak[1], Utami Dyah Syafitri[1,2], Erfiani[1]
[1] Departement of Statistics, IPB University, Bogor, Indonesia

[2]Corresponding author: utamids@apps.ipb.ac.id

Abstract. A mixture process variable (MPV) design consists of mixture design and process variable(s). The problem in MPV experiment is the number of experiment runs will be larger if the process variable increases. An optimal design can be a solution to produce a good design with a certain criterion and a limited number of runs. In practice, the compositions of the mixture design are running on each level of the process variable(s). It has a consequence that the randomization is restricted. A split-plot design can be an alternative to overcome the problem. In this research, the whole plots of the split-plot design were the levels of the process variable(s) and the subplots were the compositions of the mixture experiments. In addition, two optimality criteria were used: D-optimality and I-optimality criterion. The D-optimal design is searching a design by minimizing covariance of model parameters meanwhile the I-optimal design is seeking a design by minimizes average of the prediction variance. The study case was a baking experiment in which consisted of three flours and a process variable. It is surprised that the D-optimal design out performed compared to the I-optimal design in terms of the variance prediction in this case.

Keywords: Mixture experiment, Process Variable, D-Optimal design, I-Optimal design

1. Introduction

Many industries need experiments to get good results or products. However, a good experiment needs more runs in order to get a good model. This has a consequence on budget and time. To solve this problem, a good experimental design is needed. The experimental design is a series of tests for desired changes that come from the input variables of a process or system so that the reasons for the change in the response output can be observed and identified [1].

A {q,m} simplex-lattice design for a mixture experiment was proposed first time by [2] . The mixture experiment is combination of two or more components in order to make a product. The special features of mixture experiments are the components between 0 and 1, and the sum of all components are unity. Hence, the experimental region is a simplex. It is meaning that the components are dependence.

In many cases, the response does not only depend on proportions of component, but also depends on the process variable [3,4,5]. Combination between the process variable(s) and the mixture design is called a mixture process variable (MPV) design. The mixture process variable experiment has limitation when the number of process variables increase then the number of experimental runs also increase. In addition, the process variable is usually not flexible to change from one level to another level. On other side, the mixture design is easier to make than changing the level of the process variable(s). This situation brings to restricted randomization. One of the experiments of restricted randomization is a split plot experiment [5]. Further literature about MPV with split plot approach can be found in [6,7,8,9].

The split plot experiment consists of whole plots and sub plots. In this case, the process variable is a whole plot factor and the mixture compositions are sub plot factors. There are two randomizations in the split plot experiment. The first randomization is how to allocate the level of the process variable on the whole plots meanwhile the second randomization is how to put on the mixture compositions as the sub plots on a certain whole plot.

An optimal design is a branch of experimental design in which the researcher has flexibility to construct the design based on the real condition [10,11]. The optimal design is searching a good design based on the certain criterion. The common criterion used is the D-optimality criterion which is based on parameter prediction. As the mixture experiments are the special case of response surface methodology, so the I-optimality criterion is suggested to use [11,12]. Hence, in this paper, the D-optimality and the I-optimality criterion were used to find the optimal design.

2. Materials

The case study consisted of three mixture components and a process variable. The proportion of the first, the second, and the third ingredient were x_1 , x_2 and x_3, respectively. The constraints of each ingredient are shown in Table 1. The levels of the process variable were -1, 0, 1. The code -1 represents the lowest value, the code 0 represents the middle level, and the code 1 represents the highest value.

Table 1. The constraints of the ingredients

Ingredient	Constraints
1st	$x_1 \geq 0.1$
2nd	$x_2 \geq 0.1$
3rd	$x_3 \geq 0.6$

The experimental region of the mixture experiment is shown in Figure 1. The experimental region is a part of a whole simplex (the white area). Furthermore, the design points of the mixture experiments would be run on each level of the whole plot factor.

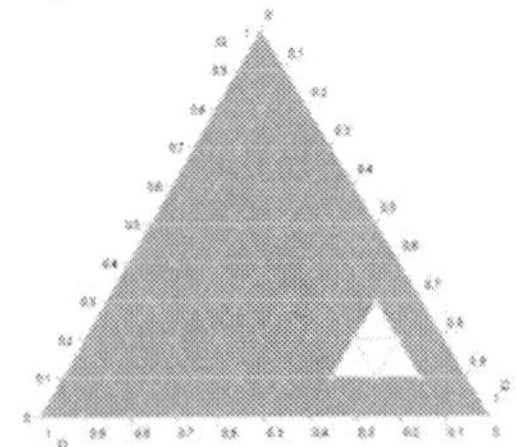

Figure 1. The experimental region of the mixture experiment

3. Methods

To construct the optimal design, a coordinate exchange algorithm was used. The steps of generating the designs were:

Step 1. Determine model of the split-plot design. The model which written in matrix notation is shown below:

$$\boldsymbol{y} = \boldsymbol{X\beta} + \boldsymbol{Z\gamma} + \boldsymbol{\varepsilon}$$

where $\boldsymbol{X}$ represents model matrix $n \times p$ containing all whole plot and subplot. The matrix $\boldsymbol{Z}$ represents model matrix $n \times b$ of matrix zeroes and ones in the whole plot. The vector $\boldsymbol{\gamma}$

represents random effects of the b whole plot and the vector $\boldsymbol{\varepsilon}$ represents random errors. The assumptions of the model are

$$\boldsymbol{\gamma} \sim N(\mathbf{0}_b, \sigma_\gamma^2 \boldsymbol{I}_b)$$
$$\boldsymbol{\varepsilon} \sim N(\mathbf{0}_n, \sigma_\varepsilon^2 \boldsymbol{I}_n)$$
$$cov(\varepsilon, \gamma) = 0_{n \times b}$$

Based on the assumptions, then covariance matrix of $\boldsymbol{y}$ is

$$\boldsymbol{V} = (\sigma_\varepsilon^2 \boldsymbol{I}_n + \sigma_\gamma^2 \boldsymbol{I}_b \boldsymbol{ZZ}') = \sigma_\varepsilon^2 (\boldsymbol{I}_n + \eta \boldsymbol{ZZ}')$$

where $\eta = \frac{\sigma_\gamma^2}{\sigma_\varepsilon^2}$. The generalized least squares estimator of maximum likelihood estimator for parameter vector $\boldsymbol{\beta}$ is

$$\widehat{\boldsymbol{\beta}} = (\boldsymbol{X}'\boldsymbol{V}^{-1}\boldsymbol{X})^{-1}\boldsymbol{X}'\boldsymbol{V}^{-1}\boldsymbol{Y}$$

and covariance matrix is

$$cov(\widehat{\boldsymbol{\beta}}) = (\boldsymbol{X}'\boldsymbol{V}^{-1}\boldsymbol{X})^{-1} = \sigma_\varepsilon^2 \, (\boldsymbol{X}'(\boldsymbol{I}_n + \eta \boldsymbol{ZZ}')^{-1}\boldsymbol{X})^{-1}$$

Furthermore, the information matrix is written as

$$\boldsymbol{M} = \boldsymbol{X}'\boldsymbol{V}^{-1}\boldsymbol{X} = \sigma_\varepsilon^2 \, (\boldsymbol{X}'(\boldsymbol{I}_n + \eta \boldsymbol{ZZ}')^{-1}\boldsymbol{X})$$

Step 2. Determine the value of $\eta = 1, 5$, and 10. The larger η is meaning that the variance of the whole plot is larger.

Step 3. Determine the mixture process variable (MPV) model. The mixture model used in this paper was a second degree Scheffe model. The combination between the mixture model and the process variable model can be written as:

$$E(\boldsymbol{Y}) = \beta_1 x_1 + \beta_2 x_2 + \beta_3 x_3 + \beta_{12} x_1 x_2 + \beta_{13} x_1 x_3 + \beta_{23} x_2 x_3 + \delta_1 x_1 w + \delta_2 x_2 w + \delta_3 x_3 w + \alpha_1 w + \alpha_2 w^2$$

The model involves six parameters of the mixture experiments, the parameters of the interaction between the mixture components and the process variable, and two parameters of the process variable. In total, there are 11 parameters.

Step 4. Determine the number of experimental runs. Considering the number of parameters, the 21 runs were defined.

Step 5. Determine D-optimal design and the I-optimal design by the coordinate exchange algorithm. JMP software was used to generate the designs.

Step 6. Compare the designs using the Fraction of Design Space (FDS). The better design is the design which has lower prediction variance entire the experimental region than another design.

4. Results and Discussion

In this case study, the size of whole plot was 7 and the size of sub plot was 3. Hence, there were 21 runs in total. The final D-optimal design of different value of η is shown in Table 2 meanwhile the final I-optimal design of different value of η is shown in Table 3.

The largest D-optimality criterion of the D-optimal design was achieved when $\eta = 1$. However, this design had the largest I-optimality criterion. This is meaning that the design was optimal in terms of parameter precision but not in terms of the average of variance prediction. Vice versa, the design which optimal in terms of the average of variance prediction was the D-optimal design when $\eta = 5$. The I-efficiency of the D-optimal design when $\eta = 5$ compared to the design when $\eta = 1$ was 1.9401. the D-efficiency of the D-optimal design when $\eta = 5$ compared to the design when $\eta = 1$ was 0.9623.

The third design of the D-optimality design was the design when $\eta = 10$. This design was the worst design in terms of the parameter precision. Nevertheless, this design was similar as the D-optimal design when $\eta = 5$ in terms of variance prediction. The I-efficiency of the design was 0.9902.

Table 2. D-optimal designs in various value of η

Whole Plots	$\eta = 1$				$\eta = 5$				$\eta = 10$			
	x_1	x_2	x_3	w	x_1	x_2	x_3	w	x_1	x_2	x_3	w
1	0.2	0.1	0.7	-1	0.1	0.3	0.6	-1	0.3	0.1	0.6	-1
1	0.1	0.3	0.6	-1	0.3	0.1	0.6	-1	0.1	0.1	0.8	-1
1	0.1	0.1	0.8	-1	0.1	0.1	0.8	-1	0.1	0.3	0.6	-1
2	0.1	0.3	0.6	1	0.1	0.3	0.6	1	0.1	0.1	0.8	1
2	0.3	0.1	0.6	1	0.1	0.2	0.7	1	0.1	0.2	0.7	1
2	0.1	0.2	0.7	1	0.3	0.1	0.6	1	0.3	0.1	0.6	1
3	0.1	0.2	0.7	0	0.2	0.1	0.7	-1	0.2	0.1	0.7	1
3	0.2	0.1	0.7	0	0.2	0.2	0.6	-1	0.3	0.1	0.6	1
3	0.2	0.2	0.6	0	0.1	0.2	0.7	-1	0.1	0.3	0.6	1
4	0.3	0.1	0.6	1	0.1	0.3	0.6	1	0.1	0.3	0.6	-1
4	0.2	0.2	0.6	1	0.2	0.1	0.7	1	0.3	0.1	0.6	-1
4	0.1	0.1	0.8	1	0.1	0.1	0.8	1	0.1	0.1	0.8	-1
5	0.2	0.1	0.7	1	0.1	0.1	0.8	-1	0.1	0.2	0.7	-1
5	0.1	0.1	0.8	1	0.1	0.3	0.6	-1	0.2	0.1	0.7	-1
5	0.1	0.3	0.6	1	0.3	0.1	0.6	-1	0.2	0.2	0.6	-1
6	0.1	0.1	0.8	-1	0.1	0.2	0.7	1	0.1	0.3	0.6	1
6	0.2	0.2	0.6	-1	0.1	0.1	0.8	1	0.1	0.2	0.7	1
6	0.3	0.1	0.6	-1	0.2	0.2	0.6	1	0.2	0.2	0.6	1
7	0.1	0.3	0.6	-1	0.3	0.1	0.6	1	0.2	0.1	0.7	1
7	0.3	0.1	0.6	-1	0.2	0.2	0.6	1	0.2	0.2	0.6	1
7	0.1	0.2	0.7	-1	0.2	0.1	0.7	1	0.1	0.1	0.8	1
The D-optimal criterion	42.1418				27.6157				1.4308			
The I-optimal criterion	0.9658				0.4978				0.5027			

The characteristic of the I-optimal designs was different with the D-optimal designs. The optimal design in terms of the average variance prediction was the I-optimal design when $\eta = 1$ and $\eta = 5$. The I-optimal design with $\eta = 10$ was the worst design in terms of the prediction of variance. However, this design had similar with the I-optimal design when $\eta = 1$ in terms of the D-optimality criterion. The D-efficiency and the I-efficiency of the I-optimal design when $\eta = 10$ compared to the I-optimal design when $\eta = 1$ were 1.005 and 0.5864, respectively. This result was opposite with the characteristic of the D-optimal designs.

Table 3. I-optimal designs in various value of η

Whole Plots	$\eta = 1$				$\eta = 5$				$\eta = 10$			
	x_1	x_2	x_3	w	x_1	x_2	x_3	w	x_1	x_2	x_3	w
1	0.1	0.3	0.6	1	0.2	0.1	0.7	-1	0.1	0.2	0.7	-1
1	0.2	0.1	0.7	1	0.1	0.3	0.6	-1	0.2	0.2	0.6	-1
1	0.1	0.2	0.7	1	0.2	0.2	0.6	-1	0.3	0.1	0.6	-1
2	0.1	0.1	0.8	1	0.2	0.1	0.7	0	0.1	0.2	0.7	1
2	0.2	0.2	0.6	1	0.2	0.2	0.6	0	0.1	0.3	0.6	1
2	0.3	0.1	0.6	1	0.1	0.2	0.7	0	0.2	0.1	0.7	1
3	0.2	0.2	0.6	0	0.2	0.1	0.7	0	0.2	0.1	0.7	-1
3	0.2	0.1	0.7	0	0.1	0.2	0.7	0	0.1	0.2	0.7	-1
3	0.1	0.2	0.7	0	0.2	0.2	0.6	0	0.2	0.2	0.6	-1
4	0.2	0.1	0.7	0	0.1	0.1	0.8	1	0.2	0.2	0.7	1
4	0.1	0.2	0.7	0	0.2	0.2	0.6	1	0.1	0.1	0.8	1
4	0.2	0.2	0.6	0	0.1	0.2	0.7	1	0.2	0.2	0.6	1
5	0.2	0.2	0.6	-1	0.1	0.2	0.7	0	0.2	0.1	0.7	0
5	0.3	0.1	0.6	-1	0.2	0.1	0.7	0	0.2	0.2	0.6	0
5	0.1	0.2	0.7	-1	0.2	0.2	0.6	0	0.1	0.2	0.7	0
6	0.2	0.1	0.7	-1	0.1	0.2	0.7	-1	0.2	0.1	0.7	-1
6	0.1	0.3	0.6	-1	0.3	0.1	0.6	-1	0.1	0.3	0.6	-1
6	0.1	0.1	0.8	-1	0.1	0.1	0.8	-1	0.1	0.1	0.8	-1
7	0.2	0.1	0.7	0	0.3	0.1	0.6	1	0.1	0.2	0.7	1
7	0.2	0.2	0.6	0	0.1	0.3	0.6	1	0.3	0.1	0.6	1
7	0.1	0.2	0.7	0	0.2	0.1	0.7	1	0.2	0.1	0.7	1
The D-optimal criterion	0.2185				0.0179				0.2301			
The I-optimal criterion	0.5629				0.5659				0.9599			

To compare the D-optimal designs and the I-optimal designs, the FDS plots were used. Figure 2 shows the FDS plots of the D-optimal design and the I-optimal design in various values of η. In general, the I-optimal design when $\eta = 1$, 5, and 10 had lower prediction variance among the experimental region (more than 95% of the experimental region). This is meaning that the I-optimal designs were better than the D-optimal designs in terms of the variance prediction.

In contrary, the results of the FDS plot for $\eta = 5$ and 10 were different from the I-efficiencies of the I-optimal designs compared to the D-optimal designs. The I-efficiencies of the I-optimal design when η= 5 was 0.87975 meanwhile the I-optimal design when $\eta = 10$ was 0.5237. Only the I-optimal design when $\eta = 1$, the result of the FDS plot was in line with the I-efficiency (1,7158).

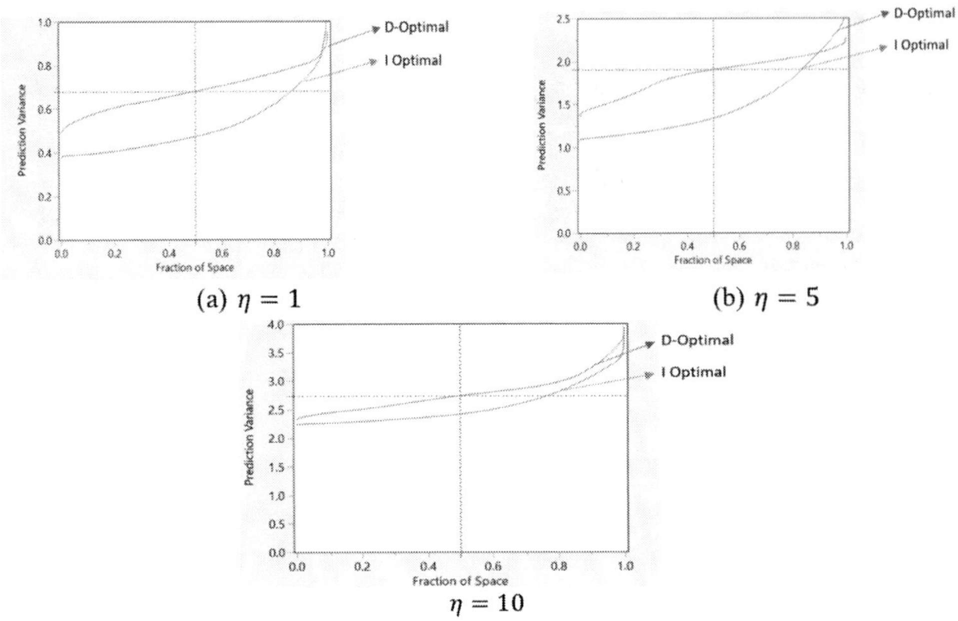

Figure 2. The FDS plots of the D-optimal and the I-optimal design on the various values of η

5. Conclusion

The two optimal designs with two different criteria were presented in this paper. The optimal designs depended on the value of η and the optimality criterion. In terms of parameter precision, the D-optimal designs were better than the I-optimal designs. Furthermore, in terms of variance prediction, the I-optimal designs were better than the D-optimal designs based on the FDS plots. However, some results of the FDS plots did not coincide with the I-efficiency.

6. Acknowledgement

We are especially grateful to the Ministry of Research, Technology and Higher Education of the Republic of Indonesia and partnership company for funding and collaboration.

References

[1] Montgomery D C 2013 *Design and Analysis of Experiments* 8th ed. (New York: John Wiley & Sons).

[2] Schefee, H 1958 Experiments with mixtures *Journal of the Royal Statistical Society Series B* **20(2)** 344-360

[3] Cornell J A 1988 Analyzing Data from Mixture Experiment Containing Process Variables: A Split Plot Approach *Journal of Quality Technology* **20(1)** 2-23.

[4] Goldfarb H B, Borror C M, Mongomery D C 2003 Mixture Process Variable Experiment with Noise Variables. *Journal of Quality Technology* **35(4)** 393-405.

[5] Kowalski S M, Cornell J, Vining G G 2000 A new model and class of designs for mixture experiments with process variables *Journal of communications in statistics, theory and methods* **29** 2255-2280.

[6] Goos P, Donev A N 2007 Tailor Made Split Plot Designs for Mixture and Process Variables *Journal of Quality Technology* **39** 326-339.

[7] Goos P, Vanderbroek M 2003 D-optimal Split-Plot Designs with Given Numbers and Sizes of Whole Plots *Journal of Technometrics* **45(3)** 235-245.

[8] Goos P, Vanderbroek M 2001 Optimal Split Plot Designs *Journal of Quality Technology* **33(4)** 436-450
[9] Njoroge GG, Simbauni JA, Koske JA; 2017; An Optimal Split Plot Design for Performing a Mixture Process Experiment. Science Journal of Applied Mathematics and Statistics; 5:1; 15-23.
[10] Atkinson A, Donev A N, Tobias R D 2007 *Optimum Experimental Designs with SAS* (Oxford: Oxford University Press).
[11] Goos P and Jones B 2011 *Optimal Design of Experiments* (Cornwall: John Wiley & Sons)
[12] Goos P and Syafitri U 2016 I-optimal design of mixture experiments Journal of the American Statistical Association **111**(514) 899-911

Analysis between Temperature and Wind Speed in East Java using Bivariate Extreme Value Theory

A Oktaviarina and *A Sofro
Mathematics Department, Universitas Negeri Surabaya, Indonesia

*Corresponding author: ayuninsofro@unesa.ac.id

Abstract. Extreme weather now is the biggest effect of climate change. It is also a major issue over the past decade. High heat and hurricanes are some of its indicators. Both natural disasters have a big impact not only on humans but also on the presence of animals and the environment. Therefore, modelling of the disaster, especially for prediction extreme events is the basic need to minimize the damage and the losses. Both of the hazards come from extreme temperatures and wind speeds, respectively. Hence, this paper will analyze the issues using extreme value theory (EVT) approach. We will investigate it separately and then analyze both two cases simultaneously. The results show that using peaks over threshold univariate or separately approach is the best model among other approaches since providing the smallest AIC value. Meanwhile, the value of AIC for the block maxima is smaller than peaks over threshold using the bivariate approach.

1. Introduction

Drought and lack of clean water usually occurs in the long dry season due to extreme weather. In this season, the wind blows with a tendency to be high and air temperatures increase, see[1]. To anticipate the issue, further analysis needs to be done such as the prediction of maximum temperature and maximum wind speed. Using early warning information, we believe that the community can prepare and minimize the effect of the hazards.

One of the appropriate approaches for extreme phenomena is the extremes value theory (EVT). Extreme value theory is a branch of a statistical method for conducting extraordinary cases. Block maxima and peak over threshold are some methods in EVT, see [2]. The researchers have been interested in the research such as application in passing maneuvers, see [3]. Tanarthe has predicted the return level of hot days in the Persian Gulf using the Peak Over Threshold method, see [4].

In this case, between temperature and speed wind has correlation. Therefore, the most appropriate method is using bivariate extremes value theory. Some studies about these have been developed. Chen Wang has developed to obtain joint crash probability, see [5]. Application in different traffic conflict indicators for road safety estimation and actual crash data can also be found in [6].

In this paper, we will focus on modeling temperature and wind speed. We will provide the models and the return level.

2. Method

2.1 Extreme Value Theory

Extreme value theory concerns with phenomena of extreme data. The method has two approaches which are block maximums and peaks over threshold. EVT uses a single process is called univariate EVT, see [7].

2.2 Univariate Generalized Extreme Value (UGEV) Distribution

Let $X_1, X_2, \ldots, X_n$ is a independent random variables sequence with a common distribution function $F(x)$, and let $M_n = max\{X_1, X_2, \ldots, X_n\}$ with $P\{M_n \leq y\} = F^n(y)$ correspondents to Block Maximum. When $n \to \infty, M_n$ will converge to a univariate generalized extreme value (UGEV) distribution:

$$F(\mu,\sigma,\xi) = exp\left[-\left(1+\xi\left(\frac{y-\mu}{\sigma}\right)\right)^{-1/\xi}\right], \xi \neq 0$$

(2.1)

and probability density function for UGEV as follows

$$f(\mu,\sigma,\xi) = \frac{1}{\sigma}\left[1+\xi\left(\frac{y-\mu}{\sigma}\right)\right]^{-\frac{1}{\xi}-1} exp\left[-\left(1+\xi\left(\frac{y-\mu}{\sigma}\right)\right)^{-1/\xi}\right], \xi \neq 0$$

(2.2)

where (μ, σ, ξ) are the location, scale and shape parameters respectively. The Frechet case is obtained when $\xi > 0$, the negative Weibull when $\xi < 0$. The Gumbel case is defined by continuity when $\xi \to 0$, see [8].

The univariate GEV parameters can be estimated using maximum likelihood estimation (MLE), see [10]. The process of MLE for univariate GEV using equation (2.2) as follows

$$L(\mu,\sigma,\xi|x_1,x_2,\ldots,x_n) = \prod_{i=1}^{n} f(x_i;\mu,\sigma,\xi)$$

$$L(\mu,\sigma,\xi) = \prod_{i=1}^{n}\frac{1}{\sigma}\left[1+\xi\left(\frac{y_i-\mu}{\sigma}\right)\right]^{-\frac{1}{\xi}-1} exp\left[-\left(1+\xi\left(\frac{y_i-\mu}{\sigma}\right)\right)^{-1/\xi}\right]$$

$$ln[L(\mu,\sigma,\xi)] = \ln\left[\prod_{i=1}^{n}\frac{1}{\sigma}\left[1+\xi\left(\frac{y_i-\mu}{\sigma}\right)\right]^{-\frac{1}{\xi}-1} exp\left[-\left(1+\xi\left(\frac{y_i-\mu}{\sigma}\right)\right)^{-\frac{1}{\xi}}\right]\right].$$

2.3 Univariate Generalized Pareto (UGP) Distribution

The univariate generalized Pareto distribution for fitting peak over threshold method which has the function

$$F(z;\sigma_u,\xi) = 1-\left(1+\xi\left(\frac{z-u}{\sigma_u}\right)\right)^{-\frac{1}{\xi}}, \xi \neq 0$$

(2.3)

and probability distribution function for UGPD is

$$f(z;\sigma_u,\xi) = \frac{1}{\sigma_u}\left(1+\xi\left(\frac{z-u}{\sigma_u}\right)\right)^{-1-\frac{1}{\xi}}, \xi \neq 0$$

(2.4)

where u is a high threshold, see [9].

The univariate GEV parameters can be estimated using MLE, see [10]. The procedure of MLE for univariate GPD using equation (2.4) is

$$L(\sigma,\xi|x_1,x_2,\dots,x_n)=\prod_{i=1}^{n} f(x_i;\mu,\sigma,\xi)$$

$$L(\sigma,\xi)=\prod_{i=1}^{n}\frac{1}{\sigma_u}\left(1+\xi\left(\frac{z_i-u}{\sigma_u}\right)\right)^{-1-\frac{1}{\xi}}$$

$$ln[L(\mu,\sigma,\xi)]=\ln\left[\prod_{i=1}^{n}\frac{1}{\sigma_u}\left(1+\xi\left(\frac{z_i-u}{\sigma_u}\right)\right)^{-1-\frac{1}{\xi}}\right]$$

2.4 Bivariate Generalized Extreme Value (BGEV)

Bivariate Generalized Extreme Value distribution function is

$$G(x,y)=\exp[-V(\tilde{x},\tilde{y})],\quad \tilde{x}>0,\ \tilde{y}>0 \tag{2.5}$$

where $V(\tilde{x},\tilde{y})=2\int_0^1 max\left(\frac{x}{\tilde{x}},\frac{1-w}{\tilde{y}}\right)dH(w)$ and H is a distribution function on $w\epsilon[0,1]$ satisfying the mean constraint $\int_0^1 wdH(w)=1/2$ [7].

The BGEV parameters can be estimated using MLE. The procedure can be found in [7].

2.5 Bivariate Generalized Pareto (BGP)

Bivariate Generalized Extreme Value distribution function is

$$G(x,y)=\exp[-V(\tilde{x},\tilde{y})],\ x>u_x,y>u_y \tag{2.6}$$

where $V(\tilde{x},\tilde{y})=2\int_0^1 max\left(\frac{x}{\tilde{x}},\frac{1-w}{\tilde{y}}\right)dH(w)$ and H is a distribution function on $w\epsilon[0,1]$ satisfying the mean constraint $\int_0^1 wdH(w)=1/2$, see [7].

Censored likelihood estimation function for Bivariate GPD is obtained as

$$L(\theta)=\prod_{i=1}^{n}\psi(\theta,(x_i,y_i))$$

where θ denotes the parameters of G and

$$\psi(\theta,(x_i,y_i))=\begin{cases}\frac{\partial^2 F}{\partial x\partial y}|_{(x,y)}, if\ (x,y)\epsilon R_{1,1}\\ \frac{\partial G}{\partial x}|_{(x,u_y)}, if\ (x,y)\epsilon R_{1,0}\\ \frac{\partial G}{\partial y}|_{(u_x,y)}, if\ (x,y)\epsilon R_{0,1}\end{cases}$$

otherwise $\psi(\theta,(x_i,y_i))=G(u_x,u_y)\ if\ (x,y)\epsilon R_{0,0}$, see [12].

3. Results and Discussion

In this paper, data are daily of the temperature maximum and wind speed maximum in Surabaya. It is measured at the Perak I Meteorological Station in July 2018 to June 2019 which can download from the official website, i.e. bmkg.go.id.

The procedure here is describe the data, we start to analyze using univariate extreme value theory using block maxima and peak over threshold. Then, we analyze using bivariate extreme value theory using block maxima and peak over threshold. The last step is to determine the best model for data based on AIC value.

The data consist of temperature maximum and wind speed maximum. The total observation is 365 daily data. The lowest wind speed data is 2 m/s and the highest is 9 m/s. The lowest temperature data is 24^0 C and the highest one is $37{,}2^0$ C.

As we mentioned previous, the data is modelled using the block maxima univariate method. Estimated parameters of the block maxima method are given as follow

Table 1. Temperature & Wind speed Parameter Estimation of Block Maxima Univariate

No	Coefficients	Temperature		Wind Speed	
		Estimate	Std Error	Estimate	Std Error
1	Location (μ)	33.29	0.08	4.0	0.08
2	Scale (σ)	1.45	0.05	1.33	0.05
3	Shape (ξ)	-0.35	0.017	-0.17	0.03

Table.1 shows that data follow generalize extreme value (GEV) distribution, especially Weibull distribution because all the shape parameter have negative value , see [13].

The univariate GEV distribution result from Table 1 is

$$f(y, \mu = 33.29, \sigma = 1.45, \xi = -0.35) = \frac{1}{1.45}\left[1 - 0.35\left(\frac{y - 33.29}{1.45}\right)\right]^{\frac{1}{0.35}-1}, \xi \neq 0$$

Then, the data is modelled with peaks over threshold method with threshold=33 for temperature and threshold=3 for wind speed. In Table 2, it shows that data has generalized Pareto distribution. The detail of estimated parameters is shown in Table 2.

Table 2. Temperature and wind speed estimated parameters of peaks over threshold univariate

No	Coefficients	Temperature		Wind Speed	
		Estimate	Std Error	Estimate	Std Error
1	Location (μ)	-	-	-	-
2	Scale (σ)	1.90	0.13	3.04	0.19
3	Shape (ξ)	-0.43	0.04	-0.49	0.036

The peaks over threshold univariate model can be served as follows

$$f(z; \sigma_u = 1.9, \xi = -0.43) = \frac{1}{1.9}\left(1 - 0.43\left(\frac{z - u}{1.9}\right)\right)^{-1+\frac{1}{0.43}}, \quad \xi \neq 0$$

Data is modeled using the bivariate block maxima methods. Table 3 that the data have generalized extreme value and e Weibull distribution because the shape parameter had a negative sign.

Table 3. Temperature and Wind speed Parameter Estimation of Block Maxima Bivariate

No	Coefficient	Estimate	Std Errors	Coefficient	Estimate	Std Errors
1	Location (μ)	4.02	0.08	Location (μ)	33.29	0.08
2	Scale (σ) 1	1.34	0.05	Scale (σ) 2	1.46	0.05
3	Shape (ξ) 1	-0.15	0.04	Shape (ξ) 2	-0.35	0.02

The block maxima bivariate model follows bivariate generalized extreme value (BGEV) distribution, so the distribution is

$$G(x,y) = \exp[-V(\tilde{x},\tilde{y})],\ x > u_x, y > u_y$$

$$\text{where } V(\tilde{x},\tilde{y}) = \left(1 - 1.59\left(\frac{x}{5.57}\right)\right)^{1/1.59}\left(1 - 0.79\left(\frac{y}{2.43}\right)\right)^{.1/0.79}$$

Using threshold=3. 33 for modeling data with peaks over threshold method. The result is given in Tble 4 below:

Table 4. Temperature and Wind speed Parameter Estimation of Peaks Over Threshold Bivariate

No	Parameter	Estimate	Parameter	Estimate
1	Location (μ) 1	-	Location (μ) 2	-
2	Scale (σ) 1	5.57	Scale (σ) 2	2.43
3	Shape (ξ) 1	-1.59	Shape (ξ) 2	-0.79

The model of bivariate generalized pareto distribution based on in table 4 is

$$G(x,y) = \exp[-V(\tilde{x},\tilde{y})],\ x > u_x, y > u_y$$

$$\text{where } V(\tilde{x},\tilde{y}) = \left(1 - 1.59\left(\frac{z-3.33}{5.57}\right)\right)^{-1+1/1.59}\left(1 - 0.79\left(\frac{y-3.3}{2.43}\right)\right)^{.-1+1/0.79}$$

The last step is determining the best model. From Table 5, we can see that block maxima bivariate model is better than peaks over threshold which has AIC=2541.71.

Although, peaks over threshold univariate is the best model among all of the approaches with AIC=1521.65.

Table 5. AIC values of different Models

No	Model	AIC
1	Univariate Block Maxima	2552.28
2	Univariate POT	1521.65
3	Bivariate Block Maxima	2541.71
4	Bivariate POT	2000010

4. Conclusion

Temperature and wind speed data are modelled by univariate and bivariate model. The peaks over threshold univariate is the best model due to providing the smallest AIC value.

5. Acknowledgements

We would like to thank Universitas Negeri Surabaya, Indonesia, which give us a grant and opportunity to disseminate an intellectual of our research.

References

[1]. Gamage A U, Pearson D and Hanna F 2016 *South East Asia J. of Public Health* **6** 3
[2]. Coles S G 2001 An introduction to Statistical Modeling of Extreme Values (Switzerland: Springer)
[3]. Farah H and Azevedo L 2017 *IATSS Res.* **41** 12
[4]. Tanarhte M, Hadjinicolaou P, and Lelieveld J 2015 *Clim. Res.* **63** 99
[5]. Wang C, Xu C, and Dai Y 2019 *Accid. Anal. Prev.* **123** 365
[6]. Zheng L, Sayed T, and Essa M 2018 *Acc. Anal. & Prev.* **123** 314
[7]. Zheng L and Sayed T 2019 *Transp. Res.* **103** 211
[8]. Ribatet M 2011 A User' s Guide to the POT Package (Version 1. 4) An Introduction to the EVT the univariate case
[9]. Gilleland E and Katz R W 2016 *J. Stat. Softw.* **72**
[10]. Oktaviarina A, Sofro A, and Sutanto H 2018 *Proc. of the Int. Conf. on Sci. and Tech.* **18.2018.170**
[11]. Sciences H, Journal J, June H, Sandoval C E and Raynal-villaseñor J 2008 *Hydro. Sci. J.* **53**
[12]. Zheng L, Ismail K, Sayed T, and Fatema T, 2018 *Accid. Anal. Prev.* **120** 83
[13]. Kim Y, Jang S C, and Lim T J 2015 *Nucl. Eng. Technol.* **1** 59

The New Convergence Definition for Sequence of *k*-Dimensional Subspaces of an Inner Product Space

Manuharawati[1], D N Yunianti[1], M Jakfar[1]

[1]Mathematics Department, Universitas Negeri Surabaya, Surabaya, 60321, Indonesia

Emails: [1]manuharawati@unesa.ac.id

Abstract. We discuss the convergence for sequence of subspaces of an inner product space. This paper is an extension of the work by Manuharawati et al [10 and 11]. In this paper, we present a concept of convergence of sequence of k-dimensional subspaces of an inner product space. The properties of the concept are established. Moreover, we also study its connection with angles in an inner product space.

Mathematics Subject Classification 2000: Primary 40A05; Secondary 14M15, 15A63, 46B20

1. Introduction

The concept of convergence of sequences of points has been extended by several authors (see [1–9]) to the concept of convergence of sequences of sets. The one of these such extensions that we will consider in this paper is Wijsman convergence. We will define convergence for sequences of subspaces and establish some basic results regarding these notions.

In [10,11], Manuharawati et al. have introduced a more geometrical definition of the convergence of a one-dimensional sequence of a one-dimensional subspace of a normed space and explained its connection with angle between two k-dimensional subspaces of a innerproduct space. Their definition of the convergence of a sequence of 1-dimensional subspace, however, is based on a definition convergence of a sequence on vector which we found a gap which we have generalized it to 2-dimensional subspace. The purpose of this note is to generalize their definition for k-dimensional subspaces and at the same time to explain the relationship between convergence of sequence with the angle between two k-dimensional subspaces of real inner product space.

Let us start with fundamental definitions from the literature. Let $(X, \langle\cdot,\cdot\rangle)$ be an inner product space.

Distance between two vectors u and v in X.

$$d(u, v) = \|u - v\|$$

Distance between a vector u and subspaces V of X.

$$d(u, V) = \inf\{\|u - v\| : v \in V\} = \|u - u^V\|$$

Distance or gap between two subspaces of X (see [12, pp. 197-199])

$$\begin{aligned} d(U,V) &= \sup\{\inf\{\|u-v\| : v\in V,\} : u\in U, \|u\|=1\} \\ &= \sup\{\|u-\mathrm{proj}_V\, u\| : u\in U, \|u\|=1\} \\ &= \sup_{\substack{u\in U \\ \|u\|=1}} \|u-\mathrm{proj}_V\, u\| \end{aligned}$$

Inspired from the distances, we can define the limit of a sequence of k-dimentional subspaces in the space.

In the following sections, an explicit definition for convergence of sequence of k-dimentional subspaces of a normed space will presented. As a consequence of our definition, the continuity property of angle between two k-dimensional subspaces of inner product spaces is obtained.

2. Results

Given an inner product space $(X,\langle\cdot,\cdot\rangle)$ of dimension k or higher (may be infinite). Let U_n and V be k-dimensional subspaces of X for every $n\in\mathbb{N}$.

Definition 2.1. We shall say that the sequence (U_n) converges to the subspace V, written $U_n\to V$ if

$$d(U_n,V) = \sup_{\substack{u_n\in U \\ \|u_n\|=1}} \|u_n-\mathrm{proj}_V\, u_n\| \to 0$$

as $n\to\infty$.

Let $U_n = span\{u_{1_n}, u_{2_n}, \dots, u_{k_n}\}$ and $V = span\{v_1, v_2, \dots, v_k\}$ be k-dimensional subspaces of X, where $\{u_{1_n}, u_{2_n}, \dots, u_{k_n}\}$ and $\{v_1,\dots,v_k\}$ are orthonormal for every $n\in\mathbb{N}$. The following theorems are some of the properties of the convergence concept for sequence of k −dimensional subspace of X.

Theorem 2.2. *$U_n\to V$ if and only if $\|u_{i_n}-\mathrm{proj}_V\, u_{i_n}\|\to 0$ for every $i=1,2,\dots,k$.*

Proof.

Given $U_n\to V$, then by Definition 2.1, $\sup_{\substack{u_n\in U \\ \|u_n\|=1}} \|u_n-\mathrm{proj}_V\, u_n\|\to 0$. It follows that for every $i = 1,2,\dots,k$ $u_{i_n}\in U_n$ with $\|u_{i_n}\|=1$ then

$$\|u_{i_n}-\mathrm{proj}_V\, u_{i_n}\| \le \sup_{\substack{u_n\in U_n \\ \|u_n\|=1}} \|u_n-\mathrm{proj}_V\, u_n\|\to 0.$$

Conversely, we know that $\mathrm{proj}_V\, u_{i_n} = \sum_{j=1}^{k}\langle u_{i_n}, v_j\rangle v_j$ and for every $i=1,2,\dots,k$

$$\begin{aligned} &\sup_{\substack{u_n\in U_n \\ \|u_n\|=1}} \|u_n-\mathrm{proj}_V\, u_n\| \\ &= \sup_{\substack{u_n\in U_n \\ \|u_n\|=1}} \left\|(\alpha_1u_{1_n}+\alpha_2u_{2_n}+\dots+\alpha_ku_{k_n}) - \mathrm{proj}_V\,(\alpha_1u_{1_n}+\alpha_2u_{2_n}+\dots+\alpha_ku_{k_n})\right\| \\ &\le \sup_{\substack{u_n\in U_n \\ \|u_n\|=1}} \sum_{i=1}^{k}\|\alpha_iu_{i_n}-\mathrm{proj}_V(\alpha_iu_{i_n})\| \\ &= \sum_{i=1}^{k}|\alpha_i|\|u_{i_n}-\mathrm{proj}_V(u_{i_n})\|\to 0. \end{aligned}$$

■

Theorem 2.3. *$U_n\to V$ if and only If $\sum_{j=1}^{k}\langle u_{i_n}, v_j\rangle^2\to 1$ for every $i=1,2,\dots,k$.*

Proof.

We know that

$$\operatorname{proj}_V u_{i_n} = \sum_{j=1}^{k} \langle u_{i_n}, v_j \rangle v_j.$$

Since $\{v_1, v_2, \dots, v_k\}$ is orthonormal, we can get

$$\begin{aligned}
\left\|u_{i_n} - \operatorname{proj}_V u_{i_n}\right\|^2 &= \langle u_{i_n} - \sum_{j=1}^{k} \langle u_{i_n}, v_j \rangle v_j \,, u_{i_n} - \sum_{j=1}^{k} \langle u_{i_n}, v_j \rangle v_j \rangle \\
&= \langle u_{i_n}, u_{i_n} \rangle - 2 \langle u_{i_n}, \sum_{j=1}^{k} \langle u_{i_n}, v_j \rangle v_j \rangle + \langle \sum_{j=1}^{k} \langle u_{i_n}, v_j \rangle v_j \,, \sum_{l=1}^{k} \langle u_{i_n}, v_l \rangle v_l \rangle \\
&= 1 - 2 \sum_{j=1}^{k} \langle u_{i_n}, v_j \rangle \langle u_{i_n}, v_j \rangle + \sum_{j=1}^{k} \langle u_{i_n}, v_j \rangle \left(\sum_{l=1}^{k} \langle u_{i_n}, v_l \rangle \langle v_j, v_l \rangle \right) \\
&= 1 - \sum_{j=1}^{k} \langle u_{i_n}, v_j \rangle^2.
\end{aligned}$$

So, $\left\|u_{i_n} - \operatorname{proj}_V u_{i_n}\right\| \to 0$ if and only if $\sum_{j=1}^{k} \langle u_{i_n}, v_j \rangle^2 \to 1$ as $n \to \infty$. By Theorem 2.2, we get $U_n \to V$ if and only if $\sum_{j=1}^{k} \langle u_{i_n}, v_j \rangle^2 \to 1$ as $n \to \infty$.

■

Theorem 2.4. $U_n \to V$ *if and only if* $\left\|\operatorname{proj}_V u_{i_n}\right\| \to 1$ *for every* $i = 1,2, \dots, k$.
Proof.
Since $\{v_1, v_2, \dots, v_k\}$ is orthonormal, we can get

$$\begin{aligned}
\left\|\operatorname{proj}_V u_{i_n}\right\|^2 &= \langle \sum_{j=1}^{k} \langle u_{i_n}, v_j \rangle v_j \,, \sum_{l=1}^{k} \langle u_{i_n}, v_l \rangle v_l \rangle \\
&= \sum_{j=1}^{k} \langle u_{i_n}, v_j \rangle \left(\sum_{l=1}^{k} \langle u_{i_n}, v_l \rangle \langle v_j, v_l \rangle \right) \\
&= \sum_{j=1}^{k} \langle u_{i_n}, v_j \rangle^2
\end{aligned}$$

So, $\sum_{j=1}^{k} \langle u_{i_n}, v_j \rangle^2 \to 1$ if and only if $\left\|\operatorname{proj}_V u_{i_n}\right\| \to 1$ as $n \to \infty$. By Theorem 2.2, we get $U_n \to V$ if and only if $\left\|\operatorname{proj}_V u_{i_n}\right\|$ as $n \to \infty$.

■

Hereafter we shall employ the standard n-norm $\|\cdot,\cdot, \dots, \cdot\|$ on X, given by

$$\|u_1, u_2, \dots, u_n\| := \begin{vmatrix} \langle u_1, u_1 \rangle & \langle u_1, u_2 \rangle & \dots & \langle u_1, u_n \rangle \\ \langle u_2, u_1 \rangle & \langle u_2, u_2 \rangle & \dots & \langle u_2, u_n \rangle \\ \vdots & \vdots & \ddots & \vdots \\ \langle u_1, u_n \rangle & \langle u_2, u_3 \rangle & \dots & \langle u_n, u_n \rangle \end{vmatrix}^{\frac{1}{2}}$$

Here assume $n > 2$ (see {6} and {10}). Note particularly that $\|u_1, u_2, \dots, u_n\| = \sqrt{\det[\langle u_i, u_j \rangle]}$ is nothing but the Gram's determinant generated by $u_1, u_2, \dots, u_n$ (see [13] or [14]). Geometrically, being the square root of the Gram's determinant, $\|u_1, u_2, \dots, u_n\|$ represents the volume of the n-dimensional parallelepiped spanned by $u_1, u_2, \dots, u_n$.

The following theorems are properties of convergent sequence related standard n-norm.

Theorem 2.5. $U_n \to V$ *if and only if* $\left\|u_{i_n}, \mathrm{proj}_V\, u_{i_n}\right\| \to 0$ *as* $n \to \infty$ *for every* $i = 1,2,\dots,k$.
Proof.
We know that

$$\mathrm{proj}_V\, u_{i_n} = \sum_{j=1}^{k} \langle u_{i_n}, v_j\rangle v_j.$$

Since $\{v_1, v_2, \dots, v_k\}$ is orthonormal, we can observe that

$$\begin{aligned}
\left\|u_{i_n}, \mathrm{proj}_V\, u_{i_n}\right\|^2 &= \begin{vmatrix} \langle u_{i_n}, u_{i_n}\rangle & \langle u_{i_n}, \mathrm{proj}_V\, u_{i_n}\rangle \\ \langle \mathrm{proj}_V\, u_{i_n}, u_{i_n}\rangle & \langle \mathrm{proj}_V\, u_{i_n}, \mathrm{proj}_V\, u_{i_n}\rangle \end{vmatrix} \\
&= \langle u_{i_n}, u_{i_n}\rangle\langle \mathrm{proj}_V\, u_{i_n}, \mathrm{proj}_V\, u_{i_n}\rangle - \langle u_{i_n}, \mathrm{proj}_V\, u_{i_n}\rangle^2 \\
&= \langle u_{i_n}, u_{i_n}\rangle \left\langle \sum_{j=1}^{k} \langle u_{i_n}, v_j\rangle v_j, \sum_{l=1}^{k} \langle u_{i_n}, v_l\rangle v_l \right\rangle - \left(\left\langle u_{i_n}, \sum_{j=1}^{k} \langle u_{i_n}, v_j\rangle v_j\right\rangle\right)^2 \\
&= \langle u_{i_n}, u_{i_n}\rangle \sum_{j=1}^{k} \langle u_{i_n}, v_j\rangle \left\langle v_j, \sum_{l=1}^{k} \langle u_{i_n}, v_l\rangle v_l \right\rangle - \left(\sum_{j=1}^{k} \langle u_{i_n}, v_j\rangle^2\right)^2 \\
&= \sum_{j=1}^{k} \langle u_{i_n}, v_j\rangle^2 - \left(\sum_{j=1}^{k} \langle u_{i_n}, v_j\rangle^2\right)^2 \\
&= \left(1 - \sum_{j=1}^{k} \langle u_{i_n}, v_j\rangle^2\right) \sum_{j=1}^{k} \langle u_{i_n}, v_j\rangle^2.
\end{aligned}$$

As for inner product, we have Cauchy-Schwarz inequality, then $\left|\sum_{j=1}^{k} \langle u_{i_n}, v_j\rangle^2\right| \le \sum_{j=1}^{k} \left\|u_{i_n}\right\|^2 \left\|v_j\right\|^2 = k$. Consequently, $\left\|u_{i_n}, \mathrm{proj}_V\, u_{i_n}\right\| \to 0$ if and only if $\sum_{j=1}^{k} \langle u_{i_n}, v_j\rangle^2 \to 1$ as $n \to \infty$. By Theorem 2.3, we get $U_n \to V$ if and only if $\left\|u_{i_n}, \mathrm{proj}_V\, u_{i_n}\right\| \to 0$ as $n \to \infty$.

■

Theorem 2.6. $U_n \to V$ *if and only if* $\left\|u_{i_n}, v_1, v_2, \dots, v_n\right\| \to 0$ *as* $n \to \infty$ *for every* $i = 1,2,\dots,k$.
Proof.
Since $\{v_1, v_2, \dots, v_k\}$ is orthonormal, we can observe that

$$\begin{aligned}
\left\|u_{i_n}, v_1, v_2, \dots, v_k\right\|^2 &= \begin{vmatrix} \langle u_{i_n}, u_{i_n}\rangle & \langle u_{i_n}, v_1\rangle & \langle u_{i_n}, v_2\rangle & \dots & \langle u_{i_n}, v_k\rangle \\ \langle v_1, u_{i_n}\rangle & \langle v_1, v_1\rangle & \langle v_1, v_2\rangle & \dots & \langle v_1, v_k\rangle \\ \langle v_2, u_{i_n}\rangle & \langle v_2, v_1\rangle & \langle v_2, v_2\rangle & \dots & \langle u_2, v_k\rangle \\ \vdots & \vdots & \vdots & \ddots & \vdots \\ \langle v_k, u_{i_n}\rangle & \langle v_k, v_1\rangle & \langle v_k, v_2\rangle & \dots & \langle v_k, v_k\rangle \end{vmatrix} \\
&= \begin{vmatrix} 1 & \langle u_{i_n}, v_1\rangle & \langle u_{i_n}, v_2\rangle & \dots & \langle u_{i_n}, v_k\rangle \\ \langle v_1, u_{i_n}\rangle & 1 & 0 & \dots & 0 \\ \langle u_{i_n}, v_2\rangle & 0 & 1 & \dots & 0 \\ \vdots & \vdots & \vdots & \ddots & \vdots \\ \langle v_k, u_{i_n}\rangle & 0 & 0 & \dots & 1 \end{vmatrix} \\
&= 1 - \langle u_{i_n}, v_1\rangle^2 + \langle u_{i_n}, v_2\rangle^2 + \dots + \langle u_{i_n}, v_k\rangle^2
\end{aligned}$$

$$= 1 - \sum_{j=1}^{k} \langle u_{i_n}, v_j \rangle^2$$

So, $\sum_{j=1}^{k} \langle u_{i_n}, v_j \rangle^2 \to 1$ if and only if $\|u_{i_n}, v_1, v_2, \dots, v_k\| \to 0$ as $n \to \infty$. By Theorem 2.3, we get $U_n \to V$ if and only if $\|u_{i_n}, v_1, v_2, \dots, v_k\| \to 0$ as $n \to \infty$.

■

Remark 2.7. Theorem 2.2 until Theorem 2.6 said that the following statements are equivalent:

(i) $U_n \to V$
(ii) $\|u_{i_n} - \text{proj}_V\, u_{i_n}\| \to 0$ *for every* $i = 1,2, \dots, k$.
(iii) $\sum_{j=1}^{k} \langle u_{i_n}, v_j \rangle^2 \to 1$ *for every* $i = 1,2, \dots, k$.
(iv) $\|\text{proj}_V\, u_{i_n}\| \to 1$ *for every* $i = 1,2, \dots, k$.
(v) $\|u_{i_n}, \text{proj}_V\, u_{i_n}\| \to 0$ *as* $n \to \infty$ *for every* $i = 1,2, \dots, k$.
(vi) $\|u_{i_n}, v_1, v_2, \dots, v_n\| \to 0$ *as* $n \to \infty$ *for every* $i = 1,2, \dots, k$.

Theorem 2.8. (Homogeneity Property) *The Definition 2.1 satisfy homogeneity property. It means, the definition is dependent of the choice of orthonormal bases for V and* U_n *for every* $n \in \mathbb{N}$.

Proof.

First note that the statement of

$$\|u_{i_n}, v_1, v_2, \dots, v_n\| \to 0, \forall i = 1,2$$

is independent of the choice of basis for U_n and V for every $n \in \mathbb{N}$. Further, since the value is also invariant under any change of basis for U_n and V. Indeed, the value is unchanged if we

(a) swap u_{1_n} and u_{2_n},
(b) swap v_1 and v_2,
(c) replace u_{1_n} by $\frac{\alpha u_{1_n} + \beta u_{2_n}}{\|\alpha u_{1_n} + \beta u_{2_n}\|}$,
(d) replace v_1 by $\frac{\alpha v_1 + \beta v_2}{\|\alpha v_1 + \beta v_2\|}$, with $\alpha \neq 0$ or $\beta \neq 0$.

■

Corollary 2.9. $U_n \to V$ if and only if $\sum_{l=1}^{k} \langle u_{i_n} + u_{j_n}, v_l \rangle^2 \to 2$ *for every* $i = 1,2, \dots, k$.

Proof.

Let U_n be k-dimensional subspace with an orthonormal basis $u_{1_n}, u_{2_n}, \dots, u_{k_n}$. Using Gram-Schmidt orthonormal process, we can construct another orthonormal basis, say $u'_{1_n}, u'_{2_n}, \dots, u'_{k_n}$, with $u'_1 = \frac{u_{i_n} + u_{j_n}}{\sqrt{2}}$ for sum $i, j \in \{1,2, \dots, k\}$. According to Homogeneity Property and Theorem 2.3, we get

$$0 \leftarrow \sum_{l=1}^{k} \langle \frac{u_{i_n} + u_{j_n}}{\sqrt{2}}, v_l \rangle^2 = \sum_{l=1}^{k} \frac{1}{2} \langle u_{i_n} + u_{j_n}, v_l \rangle^2 = \frac{1}{2} \sum_{l=1}^{k} \langle u_{i_n} + u_{j_n}, v_l \rangle^2$$

as $n \to \infty$.

■

Corollary 2.10. $U_n \to V$ if and only if $\sum_{l=1}^{k} \langle u_{i_n}, v_l \rangle \langle u_{j_n}, v_l \rangle \to 0$ *where* $i \neq j$ *for every* $i, j = 1,2, \dots, k$.

Proof.

Observe that

$$\langle u_{i_n} + u_{j_n}, v_l \rangle^2 = \left(\langle u_{i_n}, v_l \rangle + \langle u_{j_n}, v_l \rangle \right)^2 = \langle u_{i_n}, v_l \rangle^2 + \langle u_{j_n}, v_l \rangle^2 + 2 \langle u_{i_n}, v_l \rangle \langle u_{j_n}, v_l \rangle.$$

Thus, we have

$$\langle u_{i_n}, v_l \rangle \langle u_{j_n}, v_l \rangle = \frac{1}{2} \left(\langle u_{i_n} + u_{j_n}, v_l \rangle^2 - \langle u_{i_n}, v_l \rangle^2 - \langle u_{j_n}, v_l \rangle^2 \right).$$

Consequently, if $i \neq j$ then

$$\sum_{l=1}^{k}\langle u_{i_n},v_l\rangle\langle u_{j_n},v_l\rangle=\frac{1}{2}\left(\sum_{l=1}^{k}\langle u_{i_n}+u_{j_n},v_l\rangle^2-\sum_{l=1}^{k}\langle u_{i_n},v_l\rangle^2-\sum_{l=1}^{k}\langle u_{j_n},v_l\rangle^2\right).$$

Using Theorem 2.3 and Corollary 2.9, we get $U_n\to V$ if and only if $\sum_{l=1}^{k}\langle u_{i_n},v_l\rangle\langle u_{j_n},v_l\rangle\to 0$ as $n\to\infty$.

■

The following theorem stated that the limit of sequences of k-dimensional subspaces is unique.

Theorem 2.11. (Uniqueness Property) *If $U_n\to V$ and $U_n\to W$ as $n\to\infty$, then $V=W$.*

Proof.
Let $U_n\to V$ as $n\to\infty$. By Definition 3.1, we have

$$\|u_{i_n},v_1,v_2,\dots,v_k\|\to 0,\forall i=1,2$$

as $n\to\infty$.
Since $U_n\to W$ as $n\to\infty$ then we also have

$$\|u_{i_n},w_1,w_2,\dots,w_k\|\to 0,\forall i=1,2$$

as $n\to\infty$.
By using Triangle Inequality, we get

$$\begin{aligned}\|w_i,v_i,v_2,\dots,v_k\|&=\|w_i-u_{i_n}+u_{i_n},v_1,v_2,\dots,v_k\|\\&\le\|w_1-u_{1_n},v_1,v_2,\dots,v_k\|+\|u_{1_n},v_1,v_2,\dots,v_k\|.\end{aligned}$$

■

Example 2.12. *If $U_n=span\{u_1,u_2,\dots,u_n\}=U$ then $U_n\to U$ as $n\to\infty$.* For every $i=1,2$

$$\|u_{i_n},u_1,u_2\dots,u_n\|=0\to 0.$$

3. Furthermore Results

Let $(X,\langle\cdot,\cdot\rangle)$ be a real inner product space. If $U=span\,\{u1,u_2,\dots,u_k\}$ and $V=span\,\{v_1,v_2,v_3,..,v_k\}$ are k-dimensional subspaces of X, then the angle between subspaces U and V is defined by $\theta(U,V)$ with $0\le\theta(U,V)<\frac{\pi}{2}$ and $\cos\theta(U,V)=\|\mathrm{proj}_V\,u_1,\mathrm{proj}_V\,u_2,\dots,\mathrm{proj}_V\,u_k\|$ (see [15,16,17]).

Theorem 3.3. Let $U_n=span\{u_{1_n},u_{2_n},\dots,u_{k_n}\}$ and $V=span\{v_1,v_2,\dots,v_k\}$ be k-dimensional subspaces of X, where $\{u_{1_n},u_{2_n},\dots,u_{k_n}\}$ and $\{v_1,\dots,v_k\}$ are orthonormal for every $n\in\mathbb{N}$. $U_n\to V$ *if and only if $\theta(U_{i_n},V)\to 0$ as $n\to\infty$.*
Proof.
Let

$$\begin{aligned}\|\mathrm{proj}_V\,u_1,\mathrm{proj}_V\,u_2,\dots,\mathrm{proj}_V\,u_k\|^2&=\det(\langle\mathrm{proj}_V\,u_i,\mathrm{proj}_V\,u_j\rangle)\\&=\det\left(\langle\sum_{l=1}^{k}\langle u_{i_n},v_l\rangle v_l,\sum_{m=1}^{k}\langle u_{j_n},v_m\rangle v_m\rangle\right)\\&=\det\left(\sum_{l=1}^{k}\langle u_{i_n},v_l\rangle\langle v_l,\sum_{m=1}^{k}\langle u_{j_n},v_m\rangle v_m\rangle\right)\end{aligned}$$

$$= \det\left(\sum_{l=1}^{k}\langle u_{i_n}, v_l\rangle\left(\sum_{m=1}^{k}\langle u_{j_n}, v_m\rangle\langle u_j, v_m\rangle\right)\right)$$

$$= \det\left(\sum_{l=1}^{k}\langle u_{i_n}, v_l\rangle\,\langle u_{j_n}, v_m\rangle\right)$$

According to Corollary 2.10, If $U_n \to V$, then $\sum_{l=1}^{k}\langle u_{i_n}, v_l\rangle\,\langle u_{j_n}, v_m\rangle \to 0$ where $i \neq j$ for every $i, j = 1,2,\ldots,k$. Because the determinant function is continuous, then we have

$$\|\mathrm{proj}_V\, u_1, \mathrm{proj}_V\, u_2, \ldots, \mathrm{proj}_V\, u_k\|^2 \to 1.$$

Then the arccosinus function is continuous, the we get

$$\theta\left(U_{i_n}, V\right) \to \arccos 1 = 0$$

as $n \to \infty$.

■

Acknowledgment

I would like to thank DIPA State University of Surabaya for financial support.

References

[1] Aubin J P and Frankowska H 1990 *Set-Valued Analysis*, vol. 2 (Boston, USA: Birkhauser)

[2] Baronti M and Papini P L 1986 Convergence of sequences of sets in *Methods of Functional Analysis in Approximation Teory* **76** pp 135–155 (Basel, Switzerland: Birkhauser)

[3] Beer G 1989 Convergence of continuous linear functionals and their level sets *Archiv der Mathematik* **52**(5) pp 482– 491

[4] Beer G 1985 On convergence of closed sets in a metric space and distance functions *Bulletin of the Australian Mathematical Society* **31**(3) pp 421–432

[5] Borwein J and Vanderwerff J 1994 Dual Kadec-Klee norms and the relationships between Wijsman, slice, and Mosco convergence *Te Michigan Mathematical Journal* **41**(2) pp 371–387

[6] Kisi O and Nuray F 2013 On S_λ^L (I)-asymptotically statistical equivalence of sequences of sets *Mathematical Analysis* **2013** Article ID 602963

[7] Sonntag Y and Zalinescu C 1993 Set convergences: an attempt of classifcation *Transactions of the American Mathematical Society* **340**(1) pp 199–226

[8] Wijsman R A 1964 Convergence of sequences of convex sets, cones and functions *Bulletin of the American Mathematical Society* **70** pp 186–188

[9] Wijsman R A 1966 Convergence of sequences of convex sets, cones and functions. II *Transactions of the American Mathematical Society* **123** pp 32–45

[10] Manuharawati, Yunianti D N, Jakfar M 2018 A Sequence Convergence of 1 −Dimensional Subspace in a Normed Space *Journal of Physics: Conference Series* **1108** Article ID 012085

[11] Manuharawati, Yunianti D N, Jakfar M 2019 The Concept of Convergence for 2-Dimensional Subspaces Sequence in Normed Spaces *The Australians Journal of Mathematical Analysis and Application* **16**(2) pp 1-11 (Accepted)

[12] Kato T 1995 *Perturbation Theory for Linear Operators, Classics Math.* (Berlin: Springer-Verlag)

[13] Gantmacher F R 1998 *The Theory of Matrices*. Vol. I, (New York: Chelsea Publ. Co., New York 1960, 247–256. cf. Reprint of the 1959 translation)

[14] Mitrinovic D S, Pecaric J E, Fink A M 1993 *Classical and New Inequalities in Analysis* (Dordrecht: Kluwer Academic Publishers) pp 595–603.

[15] Gunawan H and Neswan O 2005 On Angles Between Subspaces of Inner Product Spaces *J. Indo. Math. Soc.* **11**
[16] Gunawan H, Neswan O, and Setya-Budhi W 2005 A Formula for Angles between Two Subspaces of Inner Product Spaces *Beitr. zur Algebra und Geometrie Contributions to Algebra and Geometry* **46(**2) pp 311-320
[17] Manuharawati and Jakfar M 2017 Thy-angle in an n-Normed Space *Far East Journal of Mathematical Sciences* **102**(5) pp 979-994

Two-Level Handling Model of Tuberculosis

Yusuf Fuad, Dimas Maulana, Yuliani Puji Astuti
Mathematics Department, Universitas Negeri Surabaya
E-mail: yusuffuad@unesa.ac.id

Abstract. Tuberculosis (TB) is an infectious disease caused by the Mycobacterium tuberculosis which most commonly attacks the lungs. In 2017, ten million people contracted TB and 1.6 million people died of TB. It is estimated that one million children suffering from TB and 230,000 children die due to TB, including children who die both TB and HIV. Global estimation show that there are ten million TB cases which are equivalent to 133 cases per hundred thousand population numbers, the country with the highest case is India by 27%, China by 9% and Indonesia by 8%. Some research related to the mathematical model of the spread of TB, for example the S-E-I/L model by Soyoung, the S-L-I/J-T model by Zhang, and the S-L-I-T model by Trauer. In this research proposal we will reconstruct a two-level model of the spread of tuberculosis infection, which means that this tuberculosis infection attacks two populations simultaneously, namely adult and individual children. The model used is the S-E-I-R/S-E-I-R model. The first individual who is attacked, in this case is an adult individual, then infects the individual child. Later, we obtain the disease free equilibrium which is $\left(\frac{\alpha N_a}{\mu}, 0, 0, 0, \frac{\alpha N_c}{\mu}, 0, 0, 0\right)$.

1. Introduction

Tuberculosis (TB) is an infectious disease caused by the bacterium Mycobacterium tuberculosis which most often attacks the lungs. In 2017, ten million people contracted TB and 1.6 million died from TB (including three hundred thousand people who suffered from TB along with HIV). Also in 2017, there are an estimated one million children who suffer from TB and 230,000 children die from TB, including children who have died from TB and HIV. [1].

TB, a chronic infectious disease caused by Mycobacterium tuberculosis, has become a major public health problem. One-third of the world's population is infected with this bacterium, reported by WHO that three million people die every year (World Health Organization, 2018a). Although almost 95% of TB infections occur in developing countries, the emergence of HIV and drug resistance to TB bacteria have changed the dynamics of infection worldwide. [2]. Other factors that cause the TB epidemic are the elimination of TB control programs, such as poor drug consumption, poverty, low rates of early TB detection, HIV infection, diabetes and immigration. [3].

In 2017, global estimates show that there are ten million TB cases (around nine to eleven million) which are equivalent to 133 cases (range 120-148 cases) per hundred thousand population. Of these cases, the highest prevalence was found in the Southeast Asian region by 44%, Africa by 25%, West Pacific by 18% and several other cases in the Mediterranean region by 7.7%, North and South America by 2.7% and cases that were the least was found in the European region, which was 2.7%. Whereas the countries with the highest cases were India sequentially by 27%, China by 9% and Indonesia by 8% [1].

Children in the first five years of age are vulnerable to fatal and serious TB infection, even more so without the BCG vaccine. Accurate data regarding the incidence and prevalence of TB are not available because of the difficulty of TB diagnosis in children. However, the few available data sources indicate that the prevalence of TB in children aged 0-14 years is estimated to be 0.3% of cases found based on radiological examinations, and 0.15% of bacterial cases [4].

Research related to the mathematical model of TB spread, for example the mathematical model by Kim So-Young discusses the spread of TB and intervention strategies for TB mitigation using the S-E-I/L model, where the latent compartment in the study is divided into two, namely latent with high risk (E) and low risk (I) [5]. However, there was no recovered compartment in the model used by Kim. This was later refined by Jinhui Zhang. The model used by Zhang is similar to the model used by Kim, but there is a recovered or treatment compartment, so the model used is S-L-I/J-T. In Zhang's study, there was only one latent compartment, but the infection compartment was divided into two, namely infected and treated in hospital (J) and infected but not treated in hospital (I) [6]. While the model used by Trauer is similar to the model used by Young. In the infection compartment, there is only compartment I, so the model used is S-L-I-T [7]. In this paper we would like to construct a S-E-I-R/S-E-I-R model.

2. Mathematical Model

Let subscripts a denote the adult population, and c denote the children. Also, let S, E, I and R be the Susceptible, Exposed, Infectious, and Recovered compartment respectively. The TB started both in either adult or children. The adult population is N_a while children population is N_c. The ratio of change in susceptible population is denoted by α. The natural death is denoted by μ. While β, λ, γ, and $\frac{I}{N}$ is the ratio of TB infection, ratio of adult population, population contact and the percentage of infected population and the whole population, respectively. The death ratio due to TB is d, the recovery ratio of TB is r, and k is the infection ratio of TB infection. We assumed that there is no reinfection in this case and the propagation of TB is slow. We can construct a compartmental model as follow:

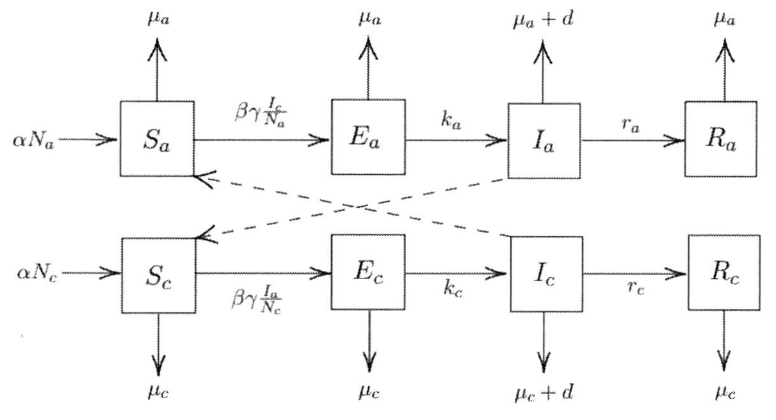

Figure 1. The compartment diagram of the TB

The infection begis in both adult and children population. The change ratio of susceptible plays an important role in determining the susceptible population in both adult and children. In this paper, we assume that both ratio in adult and children are the same, then the cross-infection occurs. The infected adult infect the susceptible children and vice versa, and contibute a ratio of either $\frac{I_c}{N_a}$ or $\frac{I_a}{N_c}$. We assume that the bacteria propagates slowly, the susceptible compartment the susceptible will transfer to exposed compartment. The exposed population has a k variable which donates the infection ratio, differs in adult and children population. We obtain the mathematical model:

$$\frac{dS_a}{dt} = \alpha N_a + \beta\gamma S_a \frac{I_c}{N_a} - \mu_a S_a \tag{1a}$$
$$\frac{dE_a}{dt} = \beta\gamma S_a \frac{I_c}{N_a} - \mu_a E_a - k_a E_a \tag{1b}$$
$$\frac{dI_a}{dt} = k_a E_a - (\mu_a + d) I_a - r_a I_a \tag{1c}$$
$$\frac{dR_a}{dt} = r_a I_a - \mu_a R_a \tag{1d}$$
$$\frac{dS_c}{dt} = \alpha N_c + \beta\gamma S_c \frac{I_a}{N_c} - \mu_c S_c \tag{1e}$$
$$\frac{dE_c}{dt} = \beta\gamma S_c \frac{I_a}{N_c} - \mu_c E_c - k_c E_c \tag{1f}$$
$$\frac{dI_c}{dt} = k_c E_c - (\mu_c + d) I_c - r I_c \tag{1g}$$
$$\frac{dR_c}{dt} = r_c I_c - \mu_c R_c \tag{1h}$$

By setting the right-hand side of equations 1 to zero and solving them, we obtained to equilibria, that is the disease-free equilibirum:

$$T_1 = \left(\frac{\alpha N_a}{\mu}, 0, 0, 0, \frac{\alpha N_c}{\mu}, 0, 0, 0 \right) \tag{2}$$

It is not easy to obtain the endemic equilibrium due to the complexity of the equation, however, we can still obtain and show the endemic equilibrium numerically. The endemic equilibrium:

$$T_2 = (S_a^*, E_a^*, I_a^*, R_a^*, S_c^*, E_c^*, I_c^*, R_c^*) \tag{3}$$

where:

$$S_a^* = \frac{(\gamma\alpha\beta k_a (\mu_c + k_c)(\mu_c + d + r_c) N_a + \mu_c N_c (\mu_c + k_c)(\mu_c + d + r_c)(\mu_a + d + r_a)(\mu_a + k_a))}{(\mu_a (\mu_c + k_c)(\mu_c + d + r_c) N_a + N_c \alpha\beta\gamma k_c)\gamma\beta k_a}$$
$$E_a^* = -\frac{1}{k_a (\mu_a + k_a)\beta\gamma (\mu_a (\mu_c + k_c)(\mu_c + d + r_c) N_a + N_c \alpha\beta\gamma k_c)}$$
$$I_a^* = -\frac{1}{\beta\gamma (\mu_a (\mu_c + k_c)(\mu_c + d + r_c) N_a + N_c \alpha\beta\gamma k_c)(\mu_a + k_a)(\mu_a + d + r_a)}$$
$$R_a^* = -\frac{1}{\beta\gamma (\mu_a (\mu_c + k_c)(\mu_c + d + r_c) N_a + N_c \alpha\beta\gamma k_c)(\mu_a + k_a)(\mu_a + d + r_a)\mu_a}$$
$$S_c^* = \frac{(\mu_a (\mu_c + k_c)(\mu_c + d + r_c) N_a + N_c \alpha\beta\gamma k_c)(\mu_a + k_a)(\mu_a + d + r_a) N_c}{k_c \beta\gamma (\gamma\alpha\beta k_a N_a + \mu_c N_c (\mu_a + k_a)(\mu_a + d + r_a))}$$
$$E_c^* = -\frac{1}{k_c \beta\gamma (\gamma\alpha\beta k_a (\mu_c + k_c) N_a + \mu_c N_c (\mu_a + k_a)(\mu_a + d + k_a)(\mu_c + k_c))}$$
$$I_c^* = -\frac{1}{(\gamma\alpha\beta k_a (\mu_c + k_c)(\mu_c + d + r_c) N_a + \mu_c N_c (\mu_c + k_c)(\mu_c + d + r_c)(\mu_a + d + r_a)(\mu_a + k_a))\gamma\beta}$$
$$R_c^* = -\frac{1}{(\gamma\alpha\beta k_a (\mu_c + k_c)(\mu_c + d + r_c) N_a + \mu_c N_c (\mu_c + k_c)(\mu_c + d + r_c)(\mu_a + d + r_a)(\mu_a + k_a))\gamma\beta\mu_c}$$

3. Conclusion

In this paper, a two-level handling of TB model has been constructed. We can obtain two equilibrium point, i.e., the disease-free equilibrium point and the endemic equilibrium point. We can easily obtain the disease-free equilibrium but the endemic equilibrium is not easy to obtain analitically. By setting some values to the parameters, we can obtain the equilibria numerically.

References

[1] World Health Organization 2018 Global Tuberculosis Report 2018 Tech. rep. World Health Organization Geneva
[2] Okuonghae D 2013 *Appl. Math. Model.* **37** 6786–6808 ISSN 0307-904X URL `https://www.sciencedirect.com/science/article/pii/S0307904X13000929`
[3] Malik M, Larasati M and Aldila D 2018 *J. Phys. Conf. Ser.* **1108** 012061 ISSN 1742-6588 URL `http://stacks.iop.org/1742-6596/1108/i=1/a=012061?key=crossref.3a457ab6d368206e0ce8f07f08e66e1c`
[4] Mishra B K and Srivastava J 2014 *J. Egypt. Math. Soc.* **22** 311–316 ISSN 1110-256X URL `https://www.sciencedirect.com/science/article/pii/S1110256X13000965`
[5] Kim S, de los Reyes A A and Jung E 2018 *J. Theor. Biol.* **443** 100–112 ISSN 0022-5193 URL `https://www.sciencedirect.com/science/article/pii/S0022519318300420`
[6] Zhang J, Li Y and Zhang X 2015 *J. Theor. Biol.* **365** 159–163 ISSN 0022-5193 URL `https://www.sciencedirect.com/science/article/pii/S0022519314006080`
[7] Trauer J M, Denholm J T and McBryde E S 2014 *J. Theor. Biol.* **358** 74–84 ISSN 0022-5193 URL `https://www.sciencedirect.com/science/article/pii/S0022519314002963`

Analysis between GPA and TOEFL Score For Postgraduate Student of Sports Education using Bivariate Logistic Regression

A Sofro†[3], A Oktaviarina†, E Mintarto ‡ and O Wiriawan‡

† Mathematics Department, Universitas Negeri Surabaya, East Java, Indonesia
‡ Sport Science Faculty, Universitas Negeri Surabaya, East Java, Indonesia

E-mail: ayuninsofro@unesa.ac.id

Abstract. Due to the globalization era, there is the challenge of competition in all fields including finding work in sports education. To meet the stakeholder's needs, it is necessary to have some criteria for prospective workers and also a well-planned process in academic. Grade Point Average (GPA), is one of the most important achievements during the postgraduate programme. Meanwhile, another academic component is English as a Foreign Language Test Score (TOEFL) which also has a very significant impact on seeing the job.

This paper will analyze valuable factors that influence GPA and TOEFL using a bivariate approach. In this case, the GPA will be defined as two levels, i.e low and high. TOEFL is also able to define as two categories, i.e more and less than 450. Hence, bivariate logistic regression is the most appropriate approach. The result shows that the coefficient of age has strong evidence to reject the null hypothesis at a significant level of 0.05. It means that the variable of age has a huge contribution to influence the score of the TOEFL for the postgraduate programme in sports education.

1. Introduction

GPA is a reflection of the level of academic success of students, which contains the personal nature of students such as discipline, perseverance in learning, hard work and mindset development. GPA is one indicator that is considered by a company or institution in recruiting employees. There are also other factors such as aptitude tests, interviews, and work experience. Another academic component is TOEFL which is considered by companies for recruiting employees.

Since the last decade, Indonesia has launched the era of globalization and free markets. This era requires expertise to communicate internationally. The competition is in all fields including for seeking a job. To meet the criteria of the shareholders, qualified human resources is needed. From an academic point of view, a fairly long and planned process is needed to realize the criteria for prospective workers.

Sports education in Universitas Negeri Surabaya is the only postgraduate programme in East Java. The goal is to produce graduates who are of high quality and are capable of carrying out

[3] To whom correspondence should be addressed.

development tasks in the field of sports for the community. The postgraduate students in sports education come from various circumstances, i.e. original, high school status. To provide a high quality of postgraduate students in the programme, further investigation needs to be done such as analysis for GPA and the score TOEFL as basic information for making decisions.

The study between score TOEFL and GPA has been developing massively. Although the TOEFL scores can not be used to predict the academic performance, the student who gets higher TOEFL iBT tend to earn a higher GPA as academic performance, see [1] and [2]. Meanwhile, some evidence shows that there is a strong correlation between TOEFL iBT and GPA which can be found in [3].

The GPA can be categorized as high if it exceeds 3.5 and low for others. Similar to the previous measurement, the TOEFL score, it will be high if it is above 450 and low for others. Therefore, the GPA and TOEFL scores follow the binomial bivariate distribution. One method for finding relationships where response variables have bivariate categorically (high and low), with factors of one or more independent variables is using bivariate binomial regression.

The logistic regression is very well known method to accommodate categorized response, see [4], [5] and [6]. Meanwhile, the application of logistic regression in education also can be found in [7] and in public health see [8] and [9]. Many data including education become more than one response variable and complex. This issue has been accommodated by some developing logistic regression theories, say bivariate approach see [10]. The application of the epidemiology model can also be seen in [11]. The approach has been developed under VGAM package for user friendly, see [12], [13] and [14]. Finally, the package has been improved in 2008 by Zelig package, see [15] and [16].

In this paper, we will focus on analyzing the GPA and the score of TOEFL for Postgraduate students of sports education using bivariate logistic regression. The result of the models will be also discussed.

2. Method

In this section, to understand a bivariate logistic regression, we start with a simple approach. i.e logistic regression.

2.1. Logistic Regression Model

Suppose that y_i is categorical response variable with the value of either 0 or 1 and $i = 1, ..., n$. We can define with value 1 if it is success and 0 for failure. The covariate variables are defined as x_i, where also $i = 1, .., n$ and n is the sample size. The model is following a binomial distribution with logit as the link function. Hence, the model of logistic bivariate is following

$$\begin{aligned} y_i|\pi_i &\sim Bin(1, \pi_i), \\ \log\left(\frac{\pi_i}{1-\pi_i}\right) &= \mathbf{x}_i^T\boldsymbol{\beta}. \end{aligned} \tag{1}$$

The logit function can be rewritten down as following

$$\frac{\pi_i}{1-\pi_i} = \exp(\mathbf{x}_i^T\boldsymbol{\beta}) \tag{2}$$

$$\pi_i + \pi_i \exp(\mathbf{x}_i^T\boldsymbol{\beta}) = \exp(\mathbf{x}_i^T\boldsymbol{\beta})$$

$$\pi_i = \frac{\exp(\mathbf{x}_i^T\boldsymbol{\beta})}{1+\exp(\mathbf{x}_i^T\boldsymbol{\beta})}. \tag{3}$$

Since each y_i is defined as a binomial distribution, hence the probability of y_i can be represented as follows

$$P(y_i|\boldsymbol{\beta}) = \frac{n_i!}{y_i!(n_i-y_i)!}\pi_i^{y_i}(1-\pi_i)^{n_i-y_i}, \quad i = 1, ..., n.$$

In order to estimate the parameters ($\boldsymbol{\beta}$), we will use maximum likelihood estimator approach. The first procedure is to construct the likelihood function of the logistic regression model. The likelihood function can be written down as follows

$$\begin{aligned} L(\boldsymbol{\beta}|y_i)) &= \prod_{i=1}^{n} P(y_i|\boldsymbol{\beta}) \\ &= \prod_{i=1}^{n} \left(\pi^{y_i}(1-\pi_i)^{1-y_i}\right) \\ &= \prod_{i=1}^{n} \left(\frac{\pi_i}{1-\pi_i}\right)^{y_i} (1-\pi_i) \\ &= \prod_{i=1}^{n} \left(\exp(\mathbf{x}_i^T\boldsymbol{\beta})\right)^{y_i} \left(1 - \frac{\exp(\mathbf{x}_i^T\boldsymbol{\beta})}{1+\exp(\mathbf{x}_i^T\boldsymbol{\beta})}\right). \end{aligned} \tag{4}$$

Then, we involve a natural log to the likelihood function for convenient, hence the log likelihood function can be defined as follows

$$l(\boldsymbol{\beta}) = \sum_{i=1}^{n} \left(y_i(\mathbf{x}_i^T\boldsymbol{\beta}) - \log(1+\exp(\mathbf{x}_i^T\boldsymbol{\beta}))\right). \tag{5}$$

Finally, we maximize the equation 5 to obtain the parameter model. One of the method to optimize is using Newton Rahpson method.

2.2. Bivariate Logistic Regression

In this model, there are two categorical (binary) response variables which can be defined as y_1 and y_2. To construct the joint probability of the model, we can start from the marginal of dependent variables and the odds ratio which measures the association between the two categorical response variables. We also define as well that y_{vw} are the random bivariate variable which can contain two values, either 1 or 0. Hence, in this case we will involve four random bivariate variables, i.e y_{11}, y_{10}, y_{01} and y_{00}. The probabiility of the random bivariate variables are $\pi_{11}, \pi_{10}, \pi_{01}$ and π_{00} respectively. Thus, the model with the logit link function can be defined as follows

$$\begin{aligned} y_{11} &\sim Bin(y_{11}|\pi_{11}), \\ y_{10} &\sim Bin(y_{10}|\pi_{10}), \\ y_{01} &\sim Bin(y_{01}|\pi_{01}). \end{aligned} \tag{6}$$

and the odd ratio as the association measurement of two dependent variables are defined as

$$\Psi = \frac{\pi_{00}\pi_{01}}{\pi_{10}\pi_{11}} \tag{7}$$

where

$$\begin{aligned} \pi_{11} &= P(y_1 = 1, y_2 = 1) \\ \pi_{10} &= P(y_1 = 1, y_2 = 0) \\ \pi_{01} &= P(y_1 = 0, y_2 = 1) \\ \pi_{00} &= P(y_1 = 0, y_2 = 0). \end{aligned}$$

Meanwhile the logit link function for marginal probability for each dependent variables are follows

$$\begin{aligned} \log\left(\frac{\pi_j}{1-\pi_j}\right) &= \eta_j, \quad j = 1,2 \\ \log(\Psi(\mathbf{x})) &= \eta_3 \end{aligned} \tag{8}$$

where $\eta_j = \beta_j^k \mathbf{x}$. The marginal probability of binary response variables with p independent variables can be rewritten as following

$$\begin{aligned} \pi_1 &= P(y=1) = \frac{\exp(\beta_{01} + \beta_{11}x_1 + \beta_{21}x_2 + ... + \beta_{p1}x_p)}{1 + (\exp(\beta_{01} + \beta_{11}x_1 + \beta_{12}x_2 + ... + \beta_{p1}x_p)} \\ \pi_2 &= P(y=2) = \frac{\exp(\beta_{02} + \beta_{12}x_1 + \beta_{22}x_2 + ... + \beta_{p2}x_p)}{1 + (\exp(\beta_{02} + \beta_{12}x_1 + \beta_{22}x_2 + ... + \beta_{p2}x_p)} \\ \Psi &= \exp(\gamma_0 + \gamma_1 x_1 + \gamma_2 x_2 + ... + \gamma_p x_p) \end{aligned}$$

The other quantities can be calculated using formula as following

$$\begin{aligned} \pi_{11} &= \begin{cases} \frac{1}{2}(\Psi - 1)^{-1} - a - \sqrt{a^2 + b} & ; \Psi \neq 1 \\ \pi_1 \pi_2 & ; \Psi = 1 \end{cases} \\ \pi_{10} &= \pi_1 - \pi_{11} \\ \pi_{01} &= \pi_2 - \pi_{11} \\ \pi_{00} &= 1 - \pi_{10} - \pi_{01} - \pi_{11} \end{aligned}$$

where $a = 1 + (\pi_1 + \pi_2)(\Psi - 1)$ and $b = -4\Psi(\Psi - 1)\pi_1\pi_2$.

Let the two binary bivariate response variables are denoted as (y_{1i}, y_{2i}) with the values either 1 or 0, where $i = 1, .., n$, n is the number of observation. We have discussed in the beginning sub section, we can recall that the model has defined some binary bivariate random variables independently as , i.e y_{11}, y_{10}, y_{01} and y_{00} and the probability are $\pi_{11}, \pi_{10}, \pi_{01}$ and π_{00} respectively. Hence, the probability density function of the model can be written as follows

$$\begin{aligned} P(\mathbf{y}) &= \pi_{vw}^{y_{vw}}; \quad v = 0,1 \quad w = 0,1. \\ &= \pi_{11}^{y_{11}} \pi_{10}^{y_{10}} \pi_{01}^{y_{01}} \pi_{00}^{y_{00}}; \end{aligned}$$

Then the likelihood function of the model can be denoted as

$$L(\boldsymbol{\beta}) = \prod P(\mathbf{y}) = \prod \pi_{11}^{y_{11}} \pi_{10}^{y_{10}} \pi_{01}^{y_{01}} \pi_{00}^{y_{00}}. \tag{9}$$

The log likelihood function of equation 9 are written down as

$$l(\boldsymbol{\beta}) = \sum (y_{11} \ln \pi_{10} + y_{10} \ln \pi_{01} + y_{01} \ln \pi_{11} + y_{11} \ln \pi_{11}). \tag{10}$$

We need to maximize the equation 10 to estimate the parameter of model ($\boldsymbol{\beta}$).

2.3. Data and Variables

In this paper, we have surveyed postgraduate students in sports education programme, Universitas Negeri Surabaya in 2018 using a random sampling technique. At the end of the survey, we have got 31 participants. Some of the dependent variables are GPA (y_1) and TOEFL score (y_2), whilst independent variables are GPA and TOEFL score when an undergraduate programme, age, and employment status.

Variable	Max	Min	Mean
GPA Undergraduate	3.5	2.98	3.32
GPA Postgraduate	3.88	3.28	3.72
TOEFL Score (Undergraduate)	470	384	412.93
TOEFL Score (Postgraduate)	507	333	430.12

Table 1. Descriptive statistics for some variables

3. Result and Discussion

Table 1 shows the statistics descriptive for some variables, i.e. GPA and TOEFL scores. Meanwhile, the statistics descriptive of the independent variables such as age and employee status is provided in Table 2.

Variable	Category	Frequency	Percentage ()
Age	20-25	13	41.9
	25-30	18	58.1
Employee Status	Employee	19	61.3
	Unemployee	12	38.7

Table 2. Descriptive statistics for some independent variables

In this paper, we start to analyze the data using the simple approach, i.e logistic regression model. In this case, the dependent variable is defined as the TOEFL scores in the postgraduate programme, whilst the independent variables are GPA and TOEFL scores in the undergraduate program, age, and employee status. The estimated parameter using the approach can be found in Table 3. From Table 3 shows that the coefficient of the age variable has strong evidence

Coefficients	Estimate	Std Error	Z value	P-value
(Intercept)	0.300	4.054	0.074	0.941
GPA undergraduate (x_1)	0.384	0.475	0.809	0.418
TOEFL undergraduate (x_2)	0.059	0.520	0.115	0.908
Employee status (x_3)	1.212	0.983	1.233	0.217
Age (x_4)	-2.280	0.938	-2.430	0.015

Table 3. Estimated parameters TOEFL data using Logistic Regression Model

to reject the null hypothesis at $\alpha = 0.05$. It means that the variable is a significant factor to influence the TOEFL scores. Meanwhile, we also can express that the coefficient for age is -2.280 which refers to the log odd ratio between age (20-25) group and age (25-30) group since we denote that the younger group is null. The odd ratio equals 0.1022842 which means that the odd for age (25-30) is about 10% higher than the odd for age (20-25) to have a good TOEFL score. The model using logistic regression can be written down as follows

$$\pi = \frac{\exp(0.300 + 0.384x_1 + 0.059x_2 + 1.212x_3 - 2.280x_4)}{1 + \exp(0.300 + 0.384x_1 + 0.059x_2 + 1.212x_3 - 2.280x_4)}.$$

Based on the [2], a student who earns a higher TOEFL score tend to have a good GPA as well. Then, we provide analyzing the GPA and TOEFL score simultaneously using the bivariate approach. The two dependent variables are GPA and TOEFL score in the postgraduate programme, whilst the GPA and TOEFL in undergraduate, employee status and age are independent variables. The estimated parameters using bivariate logistic regression can be found in Table 4.

Journal of Physics: Conference Series **1417** (2019) 012022 doi:10.1088/1742-6596/1417/1/012022

Coefficients	Estimate	Std Error	Z value	P-value
(Intercept):1	2.976	3.853	0.773	0.439
(Intercept):2	-0.572	3.887	-0.147	0.882
(Intercept):3	1.627	0.990	1.643	0.100
GPA undergraduate :1	-0.073	0.418	-0.176	0.860
GPA undergraduate :2	0.417	0.460	0.908	0.364
TOEFL undergraduate:1	-0.459	0.512	-0.897	0.369
TOEFL undergraduate:2	0.127	0.496	0.258	0.796
Employee status :1	0.620	0.874	0.709	0.478
Employee status :2	1.203	0.951	1.265	0.206
Age :1	-0.361	0.655	-0.551	0.581
Age :2	-1.962	0.877	-2.235	0.025

Table 4. Estimated parameters using the Bivariate Logistic Regression

From Table 4, we can see that the coefficient of age variable for logit function in the second response variable (TOEFL scores) has strong evidence to reject the null hypothesis at $\alpha = 0.05$. It means that the age variable is a significant aspect to contribute to the TOEFL score which is similar to the logistic regression approach. From Table 4, we also can express that the coefficient of age for the second response variable is -1.962 referring to the log odd ratio between age (20-25) group and age (25-30) group. It means that the odd of age (25-30) group is 14% higher that odd of age (20-25) group to earn a high TOEFL score due to the older group signed as 1. The model using the bivariate approach from Table 4 can be written down as follows

$$\begin{aligned}\pi_1 &= \frac{\exp(2.976 - 0.073x_1 - 0.459x_2 + 0.620x_3 - 0.361x_4)}{1 + \exp(2.976 - 0.073x_1 - 0.459x_2 + 0.620x_3 - 0.361x_4)} \\ \pi_2 &= \frac{\exp(-0.572 + 0.417x_1 0.127x_2 + 1.203x_3 - 1.962x_4)}{1 + \exp(-0.572 + 0.417x_1 0.127x_2 + 1.203x_3 - 1.962x_4)}\end{aligned}$$

and the odd ratio is $\Psi = \exp(1.627) = 5.0923$.

For the goodness of fit model, we provide Akaike Criterion Information (AIC) to evaluate models in Table 5. From the Table 5 shows that bivariate approach provides the smallest AIC

Model	AIC
Logistic Regression	41.76
Bivariate Logistic Regression	40.79

Table 5. AIC value for two models

values. It can be said that the bivariate logistic model is the best model for analyzing GPA and TOEFL scores.

4. Conclusion

Age is a significant variable to influence TOEFL scores using univariate and bivariate logistic regression. The bivariate approach has provided the best model for the data as producing the smallest value of AIC.

References

[1] Cho Y and Bridgeman B 2012 Relationship of TOEFL iBT Scores to Academic Performance : Some Evidence from American Universities *Language Testing* vol. **29 (3)** pp 421-442

[2] Vu L T and Vu P H 2013 Is the TOEFL Score A reliable Indicator of International Graduate Student's Academic Achievement in American Higher Education? *International Journal on Studies in English Language and Literature* vol. **1**
[3] Ginther A and Yan X 2017 Interpreting the relationship between TOEFL iBT Scores and GPA: Language Proficiency, Policy and Profiles *Language Testing* vol. **35(2)** pp 271-295
[4] Berger D E 2017 Introduction to Binary Logistic Regression and Propensity Score Analysis *Working paper*
[5] Tranmer M and Elliot M Binary Logistic Regression *Canthie Marsh Centre for Census and Survey Research*
[6] Sprerandei S 2014 Understanding Logistic Regression Analysis *Biochemia Medica* vol. **24 (1)** pp 12-18
[7] Peng C Y J, Lee K L and Ingersoll G M 2002 An Introduction to Logistic Regression Analysis and Reporting *The Journal of Education Research* vol. **96** pp 3-14
[8] Park H A 2013 An Introduction to Logistic Regression : From Basic concept to interpretation with particular Attention to Nursing Domain *Journal Korean Academic Nursing* vol. **43** pp 164 - 164
[9] Ruspriyanty D I and Sofro A 2018 Analysis of Hypertension Disease using Logistic and Probit Regression *Journal of Physics Conference Series* vol. **1108(1)** : 012054
[10] McCullagh P and Nelder J A 1983 Generalized Linear Models, Chapman and Hall
[11] Fitzmaunce G M, Laird N M, Zahner G E P, and Daskalakis C 1995 Bivariate Logistic Regression of Childhood Psychopathology Ratings Using Multiple Informants *American Journal of Epidemiology* vol. **142** pp 1194 -1203
[12] Yee T W The VGAM Package for Categorical Data Analysis
[13] Yee T W and Hastie T J 2003 Reduced Rank Vector Generalized Linear Models *Statistical Modelling* col. **3** pp 15 -41
[14] Yee T W 2008 VGAM family functions for Bivariate Binomial Responses
[15] Imai K, King G, and Lau O 2008 Zelig: Everyone's Statistical Software
[16] Imai K, King G and Lau O 2008 Toward a common Framework for Statistical Analysis and Development *Journal of computational and Graphical Statistics* vol. **17** pp 892 - 913

Metrics on a *G*-metric Space

Muhammad Jakfar[1], Manuharawati[1], Dwi Nur Yunianti[1], Mey Dita Kumala[1]
[1]Department of Mathematics, Universitas Negeri Surabaya, Indonesia

E-mail: muhammadjakfar@unesa.ac.id

Abstrak. Given a G-metric space, we offer a simple way to derived metrics from G-metric and realize that any G-metric space is a metric space. We also show that, in certain cases, the metric can be derived from G-metric in such a way that the convergence and completeness in the G-metric is equivalent to those in the derived metric.

1 Introduction

Since the concept of metric space was introduced by Frechet in 1906 [1], it has a very important basic role in mathematics and its application. Many mathematicians conduct research on this space because there are many mathematical concepts that can be discussed in this space, for example Manuharawati, et al. in [2]. Next many researchers try to generalize this space.

The first attempt to generalize the ordinary distance function to a distance of three points was introduced by Gahler [3, 4] in 1993. K. S. Ha, et al. [5] showed that a 2-metric is not a generalization of the usual notion of a metric. It was mentioned by Gahler [3] that the notion of a 2-metric is an extension of an idea of ordinary metric and geometrically $d(x, y, z)$ represents the area of a triangle formed by the points x, y, and z in X as its vertices. But this is not always true. Sharma ([6]) showed that $d(x, y, z) = 0$ for any three distinct points $x, y, z \in \mathbb{R}^2$. B. C. Dhage in 1992 introduced a new class of generalized metrics called D-metrics [7]. However, several errors for fundamental topological properties in a D-metric space were found by Mustafa and Sims [8]. Due to these considerations, Mustafa and Sims [9] proposed a more appropriate notion of a generalized metric space, called G-metric space. Many mathematicians (for example M. Abbas, et al. ([10])) get some fixed point theorems in G-metric space.

In this paper, we shall show that we can define some metrics on G-metric space and realize that any G-metric space is a metric space. We also show that, in certain cases, the metric can be derived from G-metric in such a way that the convergence and completeness in the G-metric is equivalent to those in the derived metric. Finally, we develop a fixed point theorem for G-metric space.

Throughout this paper, we need the following concepts.

Definition 1.1. (in [1]) Given a set $X \neq \emptyset$. A Function $d: X \times X \to \mathbb{R}$ is called metric on X if for every $x, y, z \in X$ satisfy this following conditions:

(D1) $d(x,y) \geq 0$, and $d(x,y) = 0$ if and only if $x = y$.

(D2) $d(x,y) = d(y,x)$

(D3) $d(x,y) \leq d(x,z) + d(z,y)$.

The Set X which equipped by metric is called metric space, and denoted by (X, d).

Definition 1.2. (Mustafa and Sims in [9]). Given a set $X \neq \emptyset$. A Function $G: X \times X \times X \to \mathbb{R}$ is called metric-G on X if for every $x, y, z \in X$ satisfy this following conditions:

(i) $G(x,y,z) = 0$ jika dan hanya jika $x = y = z$.

(ii) $G(x,x,y) > 0$ jika $x \neq y$

(iii) $G(x,y,z) = G(y,x,z) = G(x,z,y) = \cdots$ (simetris terhadap semua permutasi ketiga variable)

(iv) $G(x,x,y) \leq G(x,y,z)$ untuk semua $x, y, z \in X$ dengan $y \neq z$.

(v) $G(x,y,z) \leq G(x,a,a) + G(a,y,z)$ untuk semua $x, y, z. a \in X$.

The Set X which equipped by G-metric on X is called G-metric space, and denoted by (X, G).

Preposition 1.3. (in [9]) Let (X, G) be a G-metric space, then for any x, y, z, and a in X it follows that:

(1) $G(x,y,z) \leq G(x,x,y) + G(x,x,z)$

(2) $G(x,y,y) \leq 2G(y,x,x)$

Theorem 1.4. (in [9]) For any metric space (X, d), the following function is a G-metric on X

$$G_d(x_1, x_2, x_3) = maks\{d(x_1,x_2), d(x_2,x_3), d(x_3,x_1)\}.$$

Corollary 1.5. Every metric space is a G-metric space.

2 Preliminary Results

Suppose hereafter that (X, G) is a G-metric space. Define the following functions d_{G_∞}, d_{G_1} and d_{G_p} on X^2 by

$$d_{G_\infty}(x,y) := \max\{G(x,y,y), G(y,x,x)\}, \quad (2.1)$$

$$d_{G_1}(x,y) := G(x,y,y) + G(y,x,x), \quad (2.2)$$

and

$$d_{G_p}(x,y) := (G(x,y,y)^p + G(y,x,x)^p)^{\frac{1}{p}}, \text{for } p > 1 \quad (2.3)$$

Then we have the following results.

Theorem 2.1. *The function d_{G_∞} defines a metric on X.*
Proof.
We will verify that d_{G_∞} satisfies the three properties of metric.
(1) Clearly, $d_{G_\infty}(x,y) = \max\{G(x,y,y), G(y,x,x)\} \geq 0$. And also, if $x = y$ then $G(x,y,y) = 0$ and $G(y,x,x) = 0$. Hence, $d_{G_\infty}(x,y) = 0$. Conversely, if $d_{G_\infty}(x,y) = 0$ then $G(x,y,y) = 0$ and $G(y,x,x) = 0$. It means $x = y$.
(2) $d_{G_\infty}(x,y) = \max\{G(x,y,y), G(y,x,x)\} = \max\{G(x,y,y), G(y,x,x)\} = d_{G_\infty}(y,x)$.
(3) Observe that

$$\begin{aligned} d_{G_\infty}(x,y) &= \max\{G(x,y,y), G(y,x,x)\} \\ &\leq \max\{G(x,z,z) + G(z,y,y), G(y,z,z) + G(z,x,x)\} \\ &\leq \max\{G(x,z,z), G(z,x,x)\} + \max\{G(z,y,y), G(y,z,z)\} \\ &= d_{G_\infty}(x,z) + d_{G_\infty}(z,y). \end{aligned}$$

■

Theorem 2.2. *The function d_{G_1} defines a metric on X.*
Proof.
We will verify that d_{G_1} satisfies the three properties of metric.
(1) Clearly, $d_{G_1}(x,y) = G(x,y,y) + G(y,x,x) \geq 0$. And also, if $x = y$ then $G(x,y,y) = 0$ and $G(y,x,x) = 0$. Hence, $d_{G_1}(x,y) = 0$. Conversely, if $d_{G_1}(x,y) = 0$ then $G(x,y,y) = 0$ and $G(y,x,x) = 0$. It means $x = y$.
(2) $d_{G_1}(x,y) = G(x,y,y) + G(y,x,x) = G(y,x,x) + G(x,y,y) = d_{G_1}(y,x)$.
(3) Observe that

$$\begin{aligned} d_{G_1}(x,y) = G(x,y,y) + G(y,x,x) &\leq \big(G(x,z,z) + G(z,y,y)\big) + \big(G(y,z,z) + G(z,x,x)\big) \\ &\leq \big(G(x,z,z) + G(z,x,x)\big) + \big(G(y,z,z) + G(z,y,y)\big) = d_{G_1}(x,z) + d_{G_1}(z,y). \end{aligned}$$

■

Theorem 2.3. *The function d_{G_p} defines a metric on X for $p > 1$.*
Proof.
We will verify that d_G satisfies the three properties of metric.
(1) Clearly, $d_{G_p}(x,y) = (G(x,y,y)^p + G(y,x,x)^p)^{\frac{1}{p}} \geq 0$. And also, if $x = y$ then $G(x,y,y) = 0$ and $G(y,x,x) = 0$. Hence, $d_{G_p}(x,y) = 0$. Conversely, if $d_{G_p}(x,y) = 0$ then $G(x,y,y) = 0$ and $G(y,x,x) = 0$. It means $x = y$.
(2) $d_{G_p}(x,y) = (G(x,y,y)^p + G(y,x,x)^p)^{\frac{1}{p}} = (G(x,y,y)^p + G(y,x,x)^p)^{\frac{1}{p}} = d_{G_p}(y,x)$.
(3) Using Minkowski's inequality, Observe that

$$\begin{aligned} d_{G_p}(x,y) = (G(x,y,y)^p + G(y,x,x)^p)^{\frac{1}{p}} &\leq \Big(\big(G(x,z,z) + G(z,x,x)\big)^p + \big(G(y,x,x) + G(z,y,y)\big)^p\Big)^{\frac{1}{p}} \\ &\leq (G(x,z,z)^p + G(z,x,x)^p)^{\frac{1}{p}} + (G(y,x,x)^p + G(z,y,y)^p)^{\frac{1}{p}} = d_G(x,z) + d_G(z,y). \end{aligned}$$

■

***Corollary* 2.4.** Every G-metric space is a metric space.

Definition 2.5. Let (X, G) be a G-metric space, then for $x_0 \in X$, $r > 0$, the G-ball with center x_0 and radius r is

$$B_G(x_0, r) = \{y \in X : G(x_0, y, y) < r\}$$

Proposition 2.6. Let (X, G) be a G-metric space, then for any $x_0 \in X$ and $r > 0$, we have,

(1) If $G(x_0,x,y) < r$ then $x,y \in B_G(x_0,r)$,
(2) If $y \in B(x_0,r)$ then there exists a $\delta > 0$ such that $B_G(y,\delta) \subset B_G(x_0,r)$.
Proof. (1) follow directly from (G3), while, (2) follows from (G5) with $\delta = r - G(x_0,y,y)$.

■

Proposition 2.7. Let (X,G) be a G-metric space, then for any $x_0 \in X$ and $r > 0$, we have,

$$B_G\left(x_0,\frac{1}{2}r\right) \subset B_{d_{G\infty}}(x_0,r) \subset B_G(x_0,r),$$

$$B_G\left(x_0,\frac{1}{3}r\right) \subset B_{d_{G_1}}(x_0,r) \subset B_G\left(x_0,\frac{2}{3}r\right),$$

and

$$B_G\left(x_0,\left(\frac{1}{2^p+1}\right)^{\frac{1}{p}} r\right) \subset B_{d_{G_1}}(x_0,r) \subset B_G\left(x_0,\left(\frac{2^p}{2^p+1}\right)^{\frac{1}{p}} r\right).$$

Proof.

(i) Let $x \in B_{G_1}\left(x_0,\frac{1}{2}r\right)$. Then $G(x_0,x,x) < \frac{r}{2}$. It follows that

$$d_{G_\infty}(x_0,x) = \max\{G(x_0,x,x),G(x,x_0,x_0)\} \le \max\{G(x_0,x,x),2G(x_0,x,x)\} \le 2G(x_0,x,x) < r.$$

So, $x \in B_{d_{G\infty}}(x_0,r)$.

Let $x \in B_{d_{G\infty}}(x_0,r)$. Then $d_{G_\infty}(x_0,r) = \max\{G(x_0,x,x),G(x,x_0,x_0)\} < r$. Since $G(x_0,x,x) < 2G(x,x_0,x_0)$, it follows that

$$G(x_0,x,x) \le \max\{G(x_0,x,x),G(x,x_0,x_0)\} < r.$$

Thus, $G(x_0,x,x) < r$. So, $x \in B_G(x_0,r)$.

(ii) Let $x \in B_{G_1}\left(x_0,\frac{1}{3}r\right)$. Then $G(x_0,x,x) < \frac{r}{3}$. It follows that

$$d_{G_1}(x_0,x) = G(x_0,x,x) + G(x,x_0,x_0) \le G(x_0,x,x) + 2G(x_0,x,x) \le 3G(x_0,x,x) < r.$$

So, $x \in B_{d_{G_1}}(x_0,r)$.

Let $x \in B_{d_{G_1}}(x_0,r)$. Then $d_{G_1}(x_0,r) = G(x_0,x,x) + G(x,x_0,x_0) < r$. Since $G(x_0,x,x) < 2G(x,x_0,x_0)$, it follows that

$$\frac{3}{2}G(x_0,x,x) \le G(x_0,x,x) + G(x,x_0,x_0) < r.$$

Thus, $G(x_0,x,x) < \frac{2}{3}r$. So, $x \in B_G(x_0,\frac{2}{3}r)$.

(iii) Let $x \in B_{G_p}\left(x_0,\left(\frac{1}{2^p+1}\right)^{\frac{1}{p}} r\right)$. Then $G(x_0,x,x) < \frac{r}{(2^p+1)^{\frac{1}{p}}}$. It follows that

$$d_{G_p}(x_0,x) = (G(x_0,x,x)^p + G(x,x_0,x_0)^p)^{\frac{1}{p}} \le \left(G(x_0,x,x)^p + \left(2G(x,x_0,x_0)\right)^p G(x_0,x,x)^p\right)^{\frac{1}{p}}$$

$$\le (2^p+1)^{\frac{1}{p}} G(x_0,x,x) < r.$$

So, $x \in B_{d_{G_p}}(x_0,r)$.

Let $x \in B_{d_{G_p}}(x_0,r)$. Then $d_{G_p}(x_0,r) = (G(x_0,x,x)^p + G(x,x_0,x_0)^p)^{\frac{1}{p}} < r$. Since $G(x_0,x,x) < 2G(x,x_0,x_0)$, it follows that

$$\left(\frac{2^p+1}{2^p}\right)^{\frac{1}{p}} G(x_0,x,x) \le (G(x_0,x,x)^p + G(x,x_0,x_0)^p)^{\frac{1}{p}} < r.$$

Thus, $G(x_0,x,x) < \left(\frac{2^p}{2^p+1}\right)^{\frac{1}{p}} r$. So, $x \in B_G\left(x_0,\left(\frac{2^p}{2^p+1}\right)^{\frac{1}{p}} r\right)$.

■

3 Convergence and Completeness

Definition 3.1. Let (X, G) be a G-metric space. The sequence $(x_n) \subset X$ is convergent to x if it converges to x in the G-metric topology, $\tau(G)$.

Proposition 3.2. *Let (X, G) be a G-metric space, then for a sequence $(x_n) \subset X$ and point $x \in X$ the following are equivalent.*

(1) (x_n) *is G-Convergent*
(2) $d_{G_{infty}}(x_n, x) \to 0$, *as* $n \to 0$.
(3) $d_{G_1}(x_n, x) \to 0$, *as* $n \to 0$.
(4) $d_{G_p}(x_n, x) \to 0$, *as* $n \to 0$.
(5) $G(x_n, x_n, x) \to 0$, *as* $n \to 0$.
(6) $G(x_n, x, x) \to 0$, as $n \to 0$.
(7) $G(x_m, x_n, x) \to 0$, as $m, n \to 0$.

Proof.
The equivalent of (1) and (2), (1) and (3), and also (1) and (4) follow from Proposition 2.7. That (3) implies (5) (and (6)) follow from Equations 2.2, the definition d_{G_1}. (5) implies (6) is a consequence of (5) of Proposition 1.3, while (6) entails (7) follows from (3) of proposition 1.3. Finally, that (7) implies (3) follows from (2.2) and (5) of Proposition 1.3.

■

Definition 3.3. Let (X, G) be a G-metric Space, the the sequence $(x_n) \subset X$ is said to be G-Cauchy if for every $\varepsilon > 0$, there exists $N \in \mathbb{N}$ such that $G(x_n, x_m, x_l) < \varepsilon$ for all $n, m, l \geq N$.

The next proposition follows directly from the definitions.

Proposition 3.4. *In a G-metric space, the following are equivalent.*

(1) The sequence (x_n) *is G-Cauchy.*
(2) (x_n) *is Cauchy sequence in metric space* (X, d_{G_∞}).
(3) (x_n) *is Cauchy sequence in metric space* (X, d_{G_1}).
(4) (x_n) *is Cauchy sequence in metric space* (X, d_{G_p}).

***Corollary** 3.5. Every G-Convergence sequence in a G-metric space is G-Cauchy.*

Definition 3.6. A G-metric space (x, G) is said to be G-complete if every G-Cauchy sequence in (X, G) is G-convergent.

***Proposition** 3.7. In a G-metric space, the following are equivalent.*

(1) (X, G) *is G-complete.*
(2) (X, d_{G_∞}) *is complete.*
(3) (X, d_{G_1}) *is complete.*
(4) (X, d_p) *is complete.*

4 Concluding Remarks

We have shown that a *G*-metric space is a metric space and that the metric can be derived from the *G*-metric in such a way that the convergence and completeness in the *G*-metric is equivalent to those in the derived metrics.

5 References

[1] Frechet M 1906 Sur quelques points du calcul fonctionnel *Rendiconti del Circolo Matematico di Palermo* **22**(1) pp 1–74

[2] Manuharawati, Yunianti D N, and Jakfar M 2018 Some Basic Properties of Uniformly Symmetrically Continuous (Real Valued) Functions on Metric Spaces *Journal of Physics: Conference Series* **947**(1) 012006

[3] Gahler S 1963 2-metrische raume und ihre topologische struktur *Math. Nachr.* **26** pp 115-148

[4] Gahler S 1966 Zur geometric 2-metrische raume, Rev. *Roum. Math. Pures et Appl.* **11** pp 664-669

[5] Ha K S, Cho Y J, and White A 1988 Strictly convex and 2-convex 2-normed spaces, *Math. Japonica* **33**(3) pp 375-384

[6] Sharma A K 1980 A note on fixed points in 2-metric spaces *Indian J. Pure Appl. Math.* **11**(2) pp 1580-1583

[7] Dhage B C 1984 *A study of some fixed point theorem, Ph.D. Thesis* (Marathwada Univ. Aurangabad)

[8] Mustafa Z and Sims B 2003 Some remarks concerning D-metric spaces *Proceedings of the International Conferences on Fixed Point Theory and Applications*, Valencia (Spain), pp 189-198

[9] Mustafa Z and Sims B 2006 A new approach to generalized metric spaces *Journal of Nonlinear and Convex Analysis* **7**(2) pp 289-297

[10] Abbas M, Sintunavarat W, Kumam P 2012 Coupled fixed point of generalized contractive mappings on partially ordered G-metric spaces *Fixed Point Theory and Appl*ications **2012**(31)

Modelling the Influence of Unhealthy Human Behaviour on the Spread of Zoonosis Disease that May Cause a Possible Future Pandemic

WA Sidik[1]
[1]Department of Mathematics, Jenderal Soedirman University

E-mail: wuryatmo.sidik@gmail.com

Abstract. This paper addresses some strategies to improve the quality of management system in the laboratories by presenting models and methods for analyzing the effect of un-healthy human behavior on zoonosis disease spread and possible future pandemic. The possible future pandemic is modelled by a hypothetical mutated virus as a result of virus recombination. The variability of seven disease states among anima and humans are described as deterministic processes and modeled in the form of a well-defined initial value problem. The epidemic model has six disease state equilibria of which four are globally asymptotically stable and the others are locally asymptotically stable. Analyses show that: (i) to control the spread of disease means to control the effective rate of disease transmissions (ii) containing the disease in the animal world does not stop the spread of the implicated diseases in the human world. (iii) the spread of the mutant virus has a bigger magnitude than original in terms of the proportion of individuals acquiring the disease. (iv) reducing the contact among the animals will reduce the spread of the disease in the animal and human population but not stop the spread of the mutant virus. (v) social distancing programs reduce the number of human casualties.

1.Introduction

Most pandemics originate in animals, are caused by viruses, and are driven to emerge by ecological, behavioral, or socioeconomic changes. Acquired immunodeficiency syndrome (AIDS) of humans is caused by two lentiviruses, human immunodeficiency virus types 1 and 2 (HIV-1 and HIV-2). Here, we describe the origins and evolution of these viruses, and the circumstances that led to the AIDS pandemic. Both HIVs are the result of multiple cross-species transmissions of simian immunodeficiency viruses (SIVs) naturally infecting African primates [1]. In 2003, severe acute respiratory syndrome coronavirus (SARS-CoV) emerged and caused over 8,000 human cases of infection and more than 700 deaths worldwide. Zoonotic SARS-CoV likely evolved to infect humans by a series of transmission events between humans and animals for sale in China [2].

Humans can be infected with zoonotic influenza viruses, such as avian influenza virus subtype A(H5N1), A(H7N9), and A(H9N2) and swine influenza virus subtype A(H1N1), A(H1N2) and A(H3N2). Human infections are primarily acquired through direct contact with infected animals or contaminated environments, some these viruses have not acquired the ability of sustained transmission among humans. The primary concern surrounding the diseases is its ability to infect humans and their mutations. Simultaneous infection of humans by the zoonotic influenza viruses could theoretically generate novel influenza viruses with pandemic potential [3], [4], [5]. A pandemic caused by such a virus could become one of the worst in history because of the potential that the virus would be highly pathogenic such as H5N1 [6], [7].

Many epidemiology models have been developed for understanding the spread and control of the diseases. Various issues have been considered including the risk of the future pandemic among human populations. Most of the existing models, however, are based on the underlying assumption that the disease is already communicable among humans such as 1918 (Spanish) flu [8], [9], swine-origin influenza-A (H1N1) [10], [11] or 1957 Asian flu [12].

Studies based on molecular-biological and genomic analysis have been carried out to track possible virus mutation and recombination of the zoonotic influenza viruses [13], [14]. A virus mutation process could be in the form of antigenic shift or antigenic drift. Antigenic shift is the process by which two or more different strains of a virus combine to form a new subtype having a mixture of the surface antigens of the original strains. Antigenic drift describes small and gradual changes in the surface proteins (antigens) of the virus through random mutational processes [15], [16].

To the best of our knowledge, only a few published studies model the threat of an emerging pandemic due to H5N1 virus mutation by antigenic drift [17] or antigenic shift [18], [19] mechanisms. In animal-human cross-species virus transmission models, such as avian-human [18] and swine-human [20], a hypothetical mutant virus was viewed as a shift virus mutation of the virus. Later, a model that incorporates both drift and shift as evolution mechanisms of influenza was proposed in [19]. As in [13], the drift evolution of influenza was modeled by the total number of the amino-acid substitutions during the strain circulating time.

The exiting antigenic shift models assume that the avian influenza virus in a human is re-assorted and then mutated in "vivo" [18], [19]. The problem with these models is that there is no natural method for defining the mutation probability. It is assumed in this study that a pandemic threat is posed by a virus mutation as a result of recombination between a zoonotic influenza virus and a human influenza virus.

2. Material and methods

2.1 Assumptions

Human infection of a zoonotic virus (*ZV*) happens by means of direct virus transmission from infected animals to humans [21]. The existing models such as in [18],[19], consider one infected disease state only. In this study, the infected humans with *ZV* were classified as symptomatic and asymptomatic. This partition is important, since individuals in these groups are treated differently. Symptomatic individuals are very sick and are usually isolated while asymptomatic individuals are usually still active [22], [23].

It is assumed that only the asymptomatic individuals have the potential to host a virus mutation and that double virus co-infection happens only when an infected but asymptomatic human with *ZV* is subsequently infected by a human virus (*HV*). It was assumed that the virus re-combination of *ZV* and *HV* mutate in 'vivo' into a new virus strain. The virus was referred to as a mutant virus (*MV*). It was assumed that the mutant virus can transmit between humans.

Thus, this study considers three influenza viruses that are spreading among linked populations of animals (or animals for short) and humans. The first virus is *ZV*. It is transmissible between animals, transmissible from animals to humans, but not between humans. The virus is highly virulent among animals and humans. The second virus is *HV*. It spreads and is communicable among humans but does not infect animals. It is not extremely virulent among humans. The third virus is the hypothetical *MV* described above. This virus is assumed to have the ability to spread among humans easily with much greater virulence and thus posing the threat of a pandemic.

2.2 *The Model*

Individuals in the interrelated population of animals and humans are assigned to seven compartments, each representing a specific disease state. Let $Z(t)$ be the vector of disease state variables

$$Z(t) = (X(t), Y(t), H(t), S(t), A(t), B(t), M(t))$$

where $X(t)$ is the proportion of susceptible animals, $Y(t)$ is the proportion of infected animals with *ZV*, $H(t)$ is the proportion of susceptible humans, $S(t)$ is the proportion of infectious humans with *HV*, $A(t)$ is the proportion of asymptomatic humans with *ZV*, $B(t)$ is the proportion of symptomatic humans with *ZV* and $M(t)$ is the proportion of infectious humans with *MV*. Each subpopulation is assumed to be homogeneous in the sense that its individuals have the same infectious periods, immunity periods and contact rates with individuals in other subpopulations.

Starting from an initial disease state

$$Z(0) = Z_0, Z_0 = (X_0, Y_0, H_0, S_0, A_0, B_0, M_0) \in \Omega,$$

the dynamic of the disease state $Z(t) \in \Omega$ is described by an initial value problem

$$\dot{X} = \eta_x - \delta_x X - \alpha_{yx}\kappa_{yx}YX + r_y Y$$
$$\dot{Y} = \alpha_{yx}\kappa_{yx}YX - (\delta_x + m_y + r_y)Y$$
$$\dot{H} = \eta_h - \delta_h H - \alpha_{yh}\kappa_{yh}YH - \alpha_{sh}\kappa_{sh}SH - \alpha_{mh}\kappa_{mh}MH + r_s S$$
$$\dot{S} = \alpha_{sh}\kappa_{sh}SH - (\delta_h + m_s + r_s)S$$
$$\dot{A} = \gamma_a\alpha_{yh}\kappa_{yh}YH - \mu\alpha_{sa}\kappa_{sa}SA - (\delta_h + m_a)A$$
$$\dot{B} = (1-\gamma_a)\alpha_{yh}\kappa_{yh}YH - (\delta_h + m_b)B$$
$$\dot{M} = \mu\alpha_{sa}\kappa_{sa}SA + \alpha_{mh}\kappa_{mh}MH - (\delta_h + m_m)M, \qquad (1)$$
$$Z(0) = Z_0, Z_0 \in \Omega, \qquad (2)$$

where $\Omega \subseteq R_+^7$ is the set of all disease states,

$$\Omega = \{Z(t) | 0 \le t \le \infty\}.$$

The parapeter space $P \subseteq R_+^{27}$,

$$P = \{p = (p_j) : p_j = \eta_x, \eta_h, \delta_x, \delta_h, \alpha_{yx}, \alpha_{sh}, \alpha_{yh}, \mu, \alpha_{sa}, \alpha_{mh}, \kappa_{yx}, \kappa_{sh}, \kappa_{yh}, \kappa_{sa}, \kappa_{mh}, m_y, m_a, m_b, m_s, m_m, r_y, r_s, \gamma_a\}$$

η_x is the rate of aminal offspring and restocking while η_h is the rate of human offspring and susceptible migration. α_{yx} and α_{yh} are *ZV* transmission rates from infectious animals to susceptible aminal and humans, respectively. κ_{yx} and κ_{yh} are the number of effective contacts between infectious animals per-unit time with susceptible animals and humans, respectively. α_{sh} and α_{sa} are *HV* transmission rates from infectious human with *HV* to susceptible human and asymptomatic humans, respectively. κ_{sh} and κ_{sa} are the number of effective contacts between infectious humans with *HV* per-unit time with susceptible humans and asymptomatic humans, respectively. δ_x and δ_h are the death rates from natural incidences for animals and humans, respectively. γ_a is the probability of an infected human having no symptoms. m_y, m_a and m_b are *ZV* virulences among infectious animals, asymptomatic humans, and symptomatic humans, respectively. r_y and r_s are the recovery rates of infectious animals and humans, respectively. μ is the probability of a recombination of *ZV* and *HV* resulting in a mutan-avian flu virus. α_{mh} is the transmission rate of mutant-avian flu. It was proved in [24] that the dynamic system (1) (2) is a well behaved, i.e. having non-negative real solutions. It has six disease equilibria, they will be listed in Subsection 2.5.

2.3. Reproduction numbers

A reproduction number is the expected number of secondary infections produced in a completely susceptible population by a typical infected individual during entire period of infection [25], [26]. Let the state variable be reordered so that the first five elements of the new state variable correspond to infected subpopulations,

$$\bar{Z} = (Y, S, A, B, M, X, H).$$

In term of ordered variable $\bar{Z}$, (1) (2) can be written as

$$\dot{\bar{Z}} = f_1 + f_2 \tag{3}$$

where

$$f_1 = \begin{bmatrix} \alpha_{yx}\kappa_{yx}YX \\ \alpha_{sh}\kappa_{sh}SH \\ \gamma_a\alpha_{yh}\kappa_{yh}YH - \mu\alpha_{sa}\kappa_{sa}SA \\ (1-\gamma_a)\alpha_{yh}\kappa_{yh}YH \\ \mu\alpha_{sa}\kappa_{sa}SA + \alpha_{mh}\kappa_{mh}MH \\ -\alpha_{yx}\kappa_{yx}YX \\ -\alpha_{yh}\kappa_{yh}YH - \alpha_{sh}\kappa_{sh}SH - \alpha_{mh}\kappa_{mh}MH \end{bmatrix}, f_2 = \begin{bmatrix} -(\delta_x + m_y + r_y)Y \\ -(\delta_h + m_s + r_s)S \\ -(\delta_h + m_a)A \\ -(\delta_h + m_b)B \\ -(\delta_h + m_m)M \\ \eta_x - \delta_x X + r_y Y \\ \eta_h - \delta_h H + r_s S \end{bmatrix}.$$

$$\bar{Z}(0) = \bar{Z}_0, \bar{Z}_0 \in \Omega, \tag{4}$$

where $\Omega \subseteq R_+^7$ is the set of all disease states,

$$\Omega = \{\bar{Z}(t) | 0 \le t \le \infty\}.$$

The first part is the rates of new infection vector, while the second part is the rates of transfer vector due to births, deaths, disease mortalities and recoveries.

The dynamic system (3), (4) has a disease-free equilibrium

$$\bar{Z}_1^* = (0,0,0,0,0,\frac{\eta_x}{\delta_x},\frac{\eta_h}{\delta_h}).$$

At $\bar{Z}_1^*$ the deriatives $\Delta f_1(\bar{Z}_1^*)$ and $\Delta f_2(\bar{Z}_1^*)$ can be partitioned as

$$\Delta f_1(\bar{Z}_1^*) = \begin{bmatrix} F_1 & O \\ O & O \end{bmatrix}, \Delta(\bar{Z}_1^*) = \begin{bmatrix} F_2 & O \\ J_3 & J_4 \end{bmatrix}$$

where

$$F_1 = \begin{bmatrix} \frac{\eta_x \alpha_{yx} \kappa_{yx}}{\delta_x} & 0 & 0 & 0 & 0 \\ 0 & \frac{\eta_h \alpha_{sh} \kappa_{sh}}{\delta_h} & 0 & 0 & 0 \\ \frac{\gamma_a \alpha_{yh} \kappa_{yh} \eta_h}{\delta_h} & 0 & 0 & 0 & 0 \\ \frac{(1-\gamma_a) \alpha_{yh} \kappa_{yh} \eta_h}{\delta_h} & 0 & 0 & 0 & 0 \\ 0 & 0 & 0 & 0 & \frac{\eta_h \alpha_{mh} \kappa_{mh}}{\delta_h} \end{bmatrix}$$

and

$$F_2 = \begin{bmatrix} -\delta_x - m_y - r_y & 0 & 0 & 0 & 0 \\ 0 & -\delta_h - m_s - r_s & 0 & 0 & 0 \\ 0 & 0 & -\delta_h - m_a & 0 & 0 \\ 0 & 0 & 0 & -\delta_h - m_b & 0 \\ 0 & 0 & 0 & 0 & -\delta_h - m_m \end{bmatrix}.$$

$F_1 \cdot F_2^{-1}$ is referred to as the *next generation matrix*. Here we compute the reproduction number is as the spectral radius of the next generation matrix [25], [26]. The characteristic polynomials of the next generation matrix is given by

$$X = w^2\left(w - \frac{\eta_x \alpha_{yx} \kappa_{yx}}{\delta_x(\delta_x + m_y + r_y)}\right)\left(w - \frac{\eta_h \alpha_{sh} \kappa_{sh}}{\delta_h(\delta_h + m_s + r_s)}\right)\left(w - \frac{\eta_h \alpha_{mh} \kappa_{mh}}{\delta_h(\delta_h + m_m)}\right).$$

Therefore, the reproduction number for the *ZV* transmission among animals is given by

$$R_{yx} = \frac{\eta_x \alpha_{yx} \kappa_{yx}}{\delta_x(\delta_x + m_y + r_y)}. \tag{5}$$

The reproduction number for *HV* among humans is

$$R_{sh} = \begin{cases} \frac{\eta_h \alpha_{sh} \kappa_{sh}}{\delta_h(\delta_h + m_s + r_s)} & \text{if } R_{yx} \le 1 \\ \left(1 - \frac{1}{(1 - \frac{1}{R_{yx}})R_{yh}}\right) \frac{\eta_h \alpha_{sh} \kappa_{sh}}{\delta_h(\delta_h + m_s + r_s)} & \text{if } R_{yx} > 1, R_{yh} > 1, \end{cases} \tag{6}$$

where

$$R_{yh} = \frac{\eta_h \alpha_{yh} \kappa_{yh}}{\delta_h(\delta_x + m_y + r_y)}. \tag{7}$$

The reproduction number for *ZV* transmission to the infected human who is asymptomatic is

$$R_{sa} = \frac{\eta_h \mu \alpha_{sa} \kappa_{sa}}{(\delta_h + m_s)(\delta_h + m_a)}(1 - \frac{1}{R_{sh}}) if R_{yx} \le 1, R_{sh} > 1. \quad (8)$$

The reproduction number for MV among humans is

$$R_{mh} = \begin{cases} \dfrac{\eta_h \alpha_{mh} \kappa_{mh}}{\delta_h(\delta_h + m_m)} & if\ R_{yx} \le 1 \\ \left(1 - \dfrac{1}{(1 - \frac{1}{R_{yx}})R_{yh}}\right) \dfrac{\eta_h \alpha_{mh} \kappa_{mh}}{\delta_h(\delta_h + m_m)} & if\ R_{yx} > 1, R_{yh} > 1, \end{cases} \quad (9)$$

2.4. Sensitivity analysis

The reproduction numbers are functions estimated from epidemic parameters for which their actual values are not known precisely. Hence, the effects of parameter uncertainties on the predictions of reproduction numbers should be addressed. This can be accomplished by sensitivity and uncertainty analysis using the proposed ranges of the reproduction parameters. Such analysis can be used for various purposes, such as ranking the parameters in order of their relative importance to the results and for assessing changes in the results due to variability in the parameter ranges. This is particularly important in studying the reproduction numbers whose behavior may only be understood by numerical analysis. The analysis of reproduction numbers is carried out by a simple method proposed in [27],[28]. Consider first the reproduction number for ZV transmission among animals.

$$A_p = \frac{p}{R_{yx}} \frac{\partial R_{yx}}{\partial p} \begin{cases} = 1 & if\ p = \eta_x, \alpha_{yx}, \kappa_{yx} \\ < 1 & if\ p = \delta_x, m_y, r_y. \end{cases}$$

Since η_x is fixed, R_{yx} is most sensitive to the change of α_{yx}, κ_{yx}. Similarly, R_{yh} is most sensitive to the change of α_{yh}, κ_{yh}. R_{sh} is most sensitive to the change of α_{sh}, κ_{sh}. R_{sa} is most sensitive to the change of α_{sa}, κ_{sa}. R_{mh} is most sensitive to the change of α_{mh}, κ_{mh}.

2.5. Disease state equilibria

Other then the disease-free equilibrium

$$Z_1^* = (\frac{\eta_x}{\delta_x}, 0, \frac{\eta_h}{\delta_h}, 0,0,0,0),$$

the disease dynamic (1) (2) has five more disease state equilibria. The second is HV epidemic equilibrium

$$Z_2^* = (X_1^*, 0, H_2^*, S_2^*, 0,0,0),$$

where $H_2^* = \frac{\eta_h}{\delta_h} \frac{1}{R_{sh}}$ and $S_2^* = \frac{\eta_h}{\delta_h + m_s}(1 - \frac{1}{R_{sh}})$. The third disease state equilibrium is the state in which there are no animals or humans infected by the ZV and no humans infected by HV but there are humans infected by MV,

$$Z_3^* = (X_1^*, 0, H_3^*, 0,0,0, M_3^*),$$

where $H_3^* = \frac{\eta_h}{\delta_h} \frac{1}{R_{mh}}$ and $M_3^* = \frac{\eta_h}{\delta_h + m_m}(1 - \frac{1}{R_{mh}})$. The fourth, the human epidemic equilibrium state, is

$$Z_4^* = (X_1^*, 0, H_4^*, S_4^*, A_4^*, 0, M_4^*),$$

where

$$S_4^* = \frac{\eta_h}{(\delta_h + m_s)} \frac{1}{R_{yh}},$$

$$A_4^* = \frac{\eta_h}{\delta_h + m_a}\left(\frac{1}{R_{yh}} + 1 + \frac{\eta_h m_s m_a}{(\delta_h + m_s)(\delta_h + m_a)} \frac{1}{R_{yh}}\left(1 - \frac{R_{mh}}{R_{sh}} - \frac{1}{R_{mh}}\right) - \frac{1}{R_{mh}}\right),$$

$$M_4^* = \frac{1}{\delta_h + m_m}\left(\frac{R_{sh}}{R_{yh}} - \eta_h + \left(\eta_h + \frac{m_a m_s}{(\delta_h + m_s)(\delta_h + m_a)R_{yh}}\right)R_{sh}\right)\frac{1}{R_{mh}}.$$

The fifth equilibrium state corresponds to *ZV* epidemic among both animals and humans,

$$Z_5^* = (X_\oplus^*, Y_\oplus^*, H_5^*, 0, A_5^*, B_5^*, 0),$$

where $X_\oplus^* = \frac{\eta_x}{\delta_x} \frac{1}{R_{yx}}$, $Y_\oplus^* = \frac{\eta_x}{\delta_x + m_y}(1 - \frac{1}{R_{yx}})$, $H_5^* = \frac{\eta_h}{\delta_h}\left(\frac{R_{yx}}{R_{yx} + (R_{yx} - 1)R_{sa}}\right)$,

$A_5^* = \frac{1}{(\delta_h + m_a)}\left(\frac{1}{A_{5,1}} - \frac{1}{A_{5,2}}\right)$, with $A_{5,1} = \frac{1}{\eta_h \gamma_a}\left(\frac{1}{R_{sa}} + (1 - \frac{1}{R_{yx}})\right)$, $A_{5,2} = \frac{1}{\eta_h \gamma_a}\left(\frac{1}{R_{sa}} + 1\right)R_{yx}$, and

$B_5^* = \frac{1}{(\delta_h + m_b)}\left(\frac{1}{B_{5,1}} - \frac{1}{B_{5,2}}\right)$ with $B_{5,1} = \frac{1}{\eta_h(1 - \gamma_a)}\left(\frac{1}{R_{sa}} + (1 - \frac{1}{R_{yx}})\right)$,

$B_{5,2} = \frac{1}{\eta_h(1 - \gamma_a)}\left(\frac{1}{R_{sa}} + 1\right)R_{yx}$. The sixth, *ZV* is epidemic among animals and humans and mutant avian flu is epidemic among humans,

$$Z_6^* = (X_\oplus^*, Y_\oplus^*, H_2^*, 0, A_6^*, B_6^*, M_6^*),$$

$H_6^* = \frac{\eta_h}{\delta_h} \frac{1}{R_{mh}}$, $A_6^* = \frac{\gamma_a}{(\delta_h + m_a)} \frac{\eta_h}{\eta_x}\left(\eta_x - \frac{\eta_x}{\delta_x} \frac{1}{R_{yx}}\right)\frac{R_{sa}}{R_{mh}} 4$, $B_6^* = \frac{(1 - \gamma_a)}{(\delta_h + m_b)} \frac{\eta_h}{\eta_x}\left(\eta_x - \frac{\eta_x}{\delta_x} \frac{1}{R_{yx}}\right)\frac{R_{sa}}{R_{mh}}$,

$$M_6^* = \frac{\eta_h}{(\delta_h + m_m)}\left(\frac{1}{\eta_x}\left(\frac{\eta_x}{\delta_x} \frac{1}{R_{yx}} - \eta_x + \delta_h(\delta_x + m_y)\right)\frac{R_{sa}}{R_{mh}} + 1\right).$$

2.6 Stability analysis

The following theorems provide statements regarding the local asymptotically stable (LAS) and global asymptotically stable (GAS) of the six disease equilibria. Outlines of the proofs are given in the appendices, and full proofs are available elsewhere [24].

Theorem 1. If $R_{yx} \le 1, R_{sh} \le 1, R_{mh} \le 1, R_{yh} \le 1$ *then* $Z_1^* = (X_1^*, 0, H_1^*, 0, 0, 0, 0)$ is GAS on Ω_1, where $\Omega_1 = \{Z = (X, Y, H, S, A, B, M) : X \ge 0, Y \ge 0, H \ge 0, S \ge 0, A \ge 0, B \ge 0, M \ge 0\}$.

Proof. See Appendix A.

Theorem 2. If $R_{yx} \le 1, R_{sh} > 1, R_{mh} \le 1, R_{sa} \le 1$, *then* Z_2^* *is GAS on* Ω_2, *where* $\Omega_2 = \{Z = (X, Y, H, S, A, B, M) : X > 0, Y \ge 0, H > 0, S > 0, A \ge 0, B \ge 0, M \ge 0\}$.

Proof. See Appendix B.

Theorem 3. If $R_{yx} \le 1, R_{sh} \le 1, R_{mh} > 1, R_{sa} \le 1$, *then* Z_3^* *is GAS on* Ω_3, *where*

$\Omega_3 = \{Z = (X,Y,H,S,A,B,M): X > 0, Y \geq 0, H > 0, S > 0, A \geq 0, B \geq 0, M > 0\}$.
Proof. See Appendix C.

Theorem 4. If $R_{yx} \leq 1, R_{sh} \leq 1, R_{mh} > 1, R_{sa} \leq 1$ *then* Z_4^* *is GAS on* Ω_4, *where*
$\Omega_4 = \{Z = (X,Y,H,S,A,B,M): X > 0, Y \geq 0, H > 0, S > 0, A > 0, B \geq 0, M > 0\}$.
Proof. See Appendix D.

Theorem 5. If $R_{yx} > 1, R_{sh} \leq 1, R_{mh} \leq 1, R_{sa} \leq 1$, *then* Z_5^* *is LAS.*
Proof. See Appendix E.

Theorem 6. If $R_{yx} > 1, R_{sh} \leq 1, R_{mh} > 1, R_{sa} > 1$ *then* Z_6^* is LAS.
Proof. See Appendix F.

3. Discussion and conclusion

Theorem 1 shows that all diseases will be eliminated if only if the following conditions are satisfied

$$\kappa_{yx} \leq \frac{\delta_x(\delta_x + m_y + r_y)}{\eta_x \alpha_{yx}}, \tag{10}$$

$$\kappa_{yh} \leq \frac{\delta_h}{\eta_x \alpha_{yh}}(\delta_x + m_y), \tag{11}$$

$$\kappa_{sa} \leq \frac{\delta_h}{\eta_h \alpha_{sa}}(\delta_h + m_s + m_a) \tag{12}$$

and

$$\kappa_{mh} \leq \frac{\delta_h}{\eta_h \alpha_{mh}}(\delta_h + m_m). \tag{13}$$

Theorem 2 shows that if conditions (10) (12) and (13) are satisfied but condition (11) is not satisfied then there will be no disease except *MV*. Theorem 3 shows that if conditions (10) (11) and (13) are satisfied but condition (12) is not satisfied then there will be no disease except *HV*. Theorem 4 shows that if the conditions (10) (11) and (13) are satisfied but condition (12) is not satisfied then there are humans who have *HV* and have *ZV* but are asymptomatic for *ZV* and who have *HV*, *ZV* and *MV* but are asymptomatic for *ZV*. Theorems 1 to 4 are satisfied when there is no epidemic among animals while Theorems 5 to 6 are satisfied when there is an epidemic among animals. Theorem 5 and Theorem 6 show that when $R_{yx} > 1$ and there is epidemic among animals, then there will be a *ZV* outbreak among humans. In addition to the condition $R_{yx} > 1$, if $R_{mh} > R_{sh}$ then there will be outbreaks of *ZV* and *MV* among humans. It can be concluded that; (i) to control the spread of disease means to control the effective rate of disease transmissions, (ii) containing ZV in the aminal world does not stop the spread of the implicated diseases in the human world, (iii) the spread of mutant-avian-flu has a bigger magnitude than ZV in terms of the proportion of individuals acquiring the disease, (iv) reducing the contact among animals will reduce the spread of ZV but not HV and mutant-avian-flu, (v) social distancing programs reduce the number human casualties.

Appendices

Appendix A: Proof for Theorem 1

Proof. The characteristic polynomial of the Jacobian matrix of the disease dynamic (1),(2) evaluated at Z_1 is

$$X_1 = w[w+\delta_x][w+(\delta_h+m_a)][w+(\delta_h+m_b)][w+(\delta_h+m_s)(1-R_{sh})] \times[w+(\delta_x+m_y+r_y)(1-R_{yx})][w+(\delta_h+m_m)(1-R_{mh})]=0$$

The first eigenvalue is zero. Since parameter values $\delta_x, \delta_h, r_y m_y, m_s, m_a, m_b$ are all positive numbers, the $2^{nd}, 3^{rd}, 4^{th}$ eigenvalues are negative real numbers. Since $R_{sh} \leq 1$ then 5th eigenvalue $-(\delta_h+2m_s)(1-R_{sh})$ is negative. Since $R_{yx} \leq 1$, the 6th eigenvalue $-(\delta_x+m_y+r_y)(1-R_{yx})$ is negative. Since $R_{mh} \leq 1$ then the 7th eigenvalue $-(\delta_h+m_m)(1-R_{mh})$ is negative. Since all eigenvalues are negative real numbers, the Jacobian matrix is locally stable at DFE Z_1^*. It is easy to show that all minors of the Jacobian matrix are also stable at DFE Z_1^*. Since the Jacobian matrix and its minors are stable at Z_1^*, Z_1^* is LAS.

Consider the third equation of (1). Since the disease transmissions reduce the number of healthy humans $\dot{H} \leq \eta_h - \delta_h H$. Integrating the inequality over $[0,t]$ results in $H(t) \leq H_1^* + |H(0)-H_1^*| e^{-\delta_h t}$. Therefore $\forall \varepsilon > 0, \exists t_1$ such that $|H(0)-H_1^*| e^{-\delta_h t} \leq \varepsilon$ for any $t > t_1$.

Hence $H(t) \leq H_1^* + \varepsilon$ for $t \geq t_1$. Thus for $T_1 \geq t_1, \limsup_{t \geq T_1} H(t) \leq H_1^* + \varepsilon$. Letting $T_1 \to \infty$, results in $H^\infty \leq H_1^* + \varepsilon$, $H^\infty \leq H_1^*$ for $\varepsilon > 0$. Therefore, for any initial disease state $Z(0)$ there will always exist $\omega(Z(0))$, the ω-limit set of orbit through $Z(0)$.

Since Z_1^* is LAS and $\lim_{t\to\infty} X(t) = \dfrac{\eta_x}{\delta_x} = X(0)$ and $\lim_{t\to\infty} Y(t) = 0$ whenever $R_{yx} \leq 1$.

Having $Y=0$ in the equation number 6 of (1) and integrating it over $[0,\infty]$ results in $\lim_{t\to\infty} B(t) = 0$. For any initial disease state $Z(0)$ there will always exist $\omega(Z(0))$, ω-limit set of orbit through $Z(0)$ in G_1, where $G_1 = \{Z=(X,Y,H,S,A,B,M): X>0, Y=0, H>0, S=0, A=0, B=0, M=0\}$. Therefore, Z_1^* is GAS on G_1. Since Z_1^* is LAS on Ω_1 then by the Poincare-Bendixon theorem [12] Z_1^* is GAS on Ω_1.

Appendix B: Proof for Theorem 2

Proof. The characteristic polynomial of the Jacobian matrix of the disease dynamic (1),(2) evaluated at Z_2 is

$$X_2 = w^2[w+\delta_x][w+(\delta_h+m_b)][w+(\delta_x+m_y+r_y)(1-R_{yx})] \times\left[w+(\delta_h+m_m)(1-\frac{R_{mh}}{R_{sh}})\right][w+(\delta_h+m_a)(R_{sa}-1)]$$

The non-zero eigen values are negative real numbers. Therefore, the Jacobian matrix and its all minors are stable at Z_2^*, therefore Z_2^* is LAS. Similar to the previous proof, for any initial disease state $Z(0)$ there will always exist $\omega(Z(0))$, ω-limit set of orbit through $Z(0)$ in G_2 for $G_2 = \{Z = (X,Y,H,S,A,B,M): X > 0, Y = 0, H \geq 0, S > 0, A = 0, B = 0, M = 0\}$. Therefore, Z_2^* is GAS on G_2. Since Z_2^* is LAS on Ω_2 then Z_2^* is is GAS on Ω_2.

Appendix C: Proof for Theorem 3

Proof. At Z_3, the characteristic polynomial of the Jacobian matrix of the disease dynamic (1), (2) is

$$X_3 = w^2[w+\delta_x][w+(\delta_h+m_a)][w+(\delta_h+m_b)]$$
$$\times[w+(\delta_x+m_y+r_y)(1-R_{yx})]\left[w+(\delta_x+m_y+r_y)(1-\frac{R_{sh}}{R_{mh}})\right].$$

The non-zero eigen values are negative real numbers. The Jacobian matrix and its all minors are stable at Z_3^*, therefore Z_3^* is LAS. Z_3^* is GAS on G_3 for $G_3 = \{Z = (X,Y,H,S,A,B,M): X > 0, Y = 0, H \geq 0, S = 0, A = 0, B = 0, M > 0\}$. Since Z_3^* is also LAS on Ω_3 then Z_3^* is is GAS on Ω_3.

Appendix D: Proof for Theorem 4

Proof. The characteristic polynomial of the Jacobian matrix of the disease dynamic (1), (2) evaluated at Z_4 is

$$X_4 = w^2[w+(\delta_h+m_b)][w+2(\delta_h+m_a)][w+(\delta_x+m_y+r_y)(r_0-1)]$$
$$\times[w+\delta_x][w+\frac{-\delta_h R_{mh}-m_m R_{mh}+R_{sh}\delta_h+m_m R_{sh}}{R_{sh}}]=0$$

The non-zero eigen values are negative real numbers. The Jacobian matrix and its all minors are stable at Z_4^*, therefore Z_4^* is LAS. Z_4^* is GAS on G_4 for $G_4 = \{Z = (X,Y,H,S,A,B,M): X > 0, Y = 0, H \geq 0, S > 0, A > 0, B = 0, M > 0\}$. Since Z_4^* is also LAS on Ω_4 then Z_4^* is is GAS on Ω_4.

Appendix E: Proof for Theorem 5

Proof. At Z_5 the eigen values of the characteristic polynomial of the Jacobian matrix of the disease dynamic (1),(2)

$$X_5 = w^2[w+\frac{-\alpha_{mh}\kappa_{mh}\delta_h-\alpha_{mh}\kappa_{mh}m_s-\alpha_{mh}\kappa_{mh}r_s+\delta_h\alpha_{sh}\kappa_{sh}+m_m\alpha_{sh}\kappa_{sh}}{\alpha_{mh}\kappa_{mh}}]$$
$$\times[w+1/2\frac{-\alpha_{yx}\kappa_{yx}\eta_x+r_y\delta_x+\sqrt{Q1}}{\delta_x+m_y}][w+1/2\frac{\alpha_{yx}\kappa_{yx}\eta_x-r_y\delta_x+\sqrt{Q2}}{\delta_x+m_y}]$$
$$\times[w-\delta_h-m_b][w-\delta_h-m_a],$$

where

$$Q1=\alpha_{yx}{}^2\kappa_{yx}{}^2\eta_x{}^2-2\alpha_{yx}\kappa_{yx}\eta_x r_y\delta_x+r_y{}^2\delta_x{}^2+8r_y\delta_x{}^2m_y-8\delta_x\alpha_{yx}\kappa_{yx}\eta_x m_y-4\alpha_{yx}\kappa_{yx}\eta_x\delta_x{}^2+4r_y\delta_x{}^3+12\delta_x{}^3m_y$$
$$+12m_y{}^2\delta_x{}^2+4\delta_x{}^4+4r_y\delta_x m_y{}^2-4\alpha_{yx}\kappa_{yx}\eta_x m_y{}^2+4m_y{}^3\delta_x$$

and

$$Q2=\alpha_{yx}{}^2\kappa_{yx}{}^2\eta_x{}^2-2\alpha_{yx}\kappa_{yx}\eta_x r_y\delta_x+r_y{}^2\delta_x{}^2+8r_y\delta_x{}^2m_y-8\delta_x\alpha_{yx}\kappa_{yx}\eta_x m_y-4\alpha_{yx}\kappa_{yx}\eta_x\delta_x{}^2+4r_y\delta_x{}^3+12\delta_x{}^3m_y$$
$$+12m_y{}^2\delta_x{}^2+4\delta_x{}^4+4r_y\delta_x m_y{}^2-4\alpha_{yx}\kappa_{yx}\eta_x m_y{}^2+4m_y{}^3\delta_x.$$

The non-zero eigen values are negative real numbers. The Jacobian matrix and its all minors are stable at Z_5^*, therefore Z_5^* is LAS.

Appendix F: Proof for Theorem 6

Proof. At Z_5 the characteristic polynomial of the Jacobian matrix of the disease dynamic (1),(2) is

$$X_6=w^2[w+\frac{-\alpha_{mh}\kappa_{mh}\delta_h-\alpha_{mh}\kappa_{mh}m_s-\alpha_{mh}\kappa_{mh}r_s+\delta_h\alpha_{sh}\kappa_{sh}+m_m\alpha_{sh}\kappa_{sh}}{\alpha_{mh}\kappa_{mh}}]$$
$$\times[w+1/2\frac{-\alpha_{yx}\kappa_{yx}\eta_x+r_y\delta_x+\sqrt{Q1}}{\delta_x+m_y}][w+1/2\frac{\alpha_{yx}\kappa_{yx}\eta_x-r_y\delta_x+\sqrt{Q2}}{\delta_x+m_y}]$$
$$\times[w-\delta_h-m_b][w-\delta_h-m_a],$$

where

$$Q1=\alpha_{yx}{}^2\kappa_{yx}{}^2\eta_x{}^2-2\alpha_{yx}\kappa_{yx}\eta_x r_y\delta_x+r_y{}^2\delta_x{}^2+8r_y\delta_x{}^2m_y-8\delta_x\alpha_{yx}\kappa_{yx}\eta_x m_y+4r_y\delta_x{}^3-4\alpha_{yx}\kappa_{yx}\eta_x\delta_x{}^2$$
$$+12\delta_x{}^3m_y+12m_y{}^2\delta_x{}^2+4\delta_x{}^4+4r_y\delta_x m_y{}^2-4\alpha_{yx}\kappa_{yx}\eta_x m_y{}^2+4m_y{}^3\delta_x$$

and

$$Q2=\alpha_{yx}{}^2\kappa_{yx}{}^2\eta_x{}^2-2\alpha_{yx}\kappa_{yx}\eta_x r_y\delta_x+r_y{}^2\delta_x{}^2+8r_y\delta_x{}^2m_y-8\delta_x\alpha_{yx}\kappa_{yx}\eta_x m_y+4r_y\delta_x{}^3+12\delta_x{}^3m_y$$
$$-4\alpha_{yx}\kappa_{yx}\eta_x\delta_x{}^2+12m_y{}^2\delta_x{}^2+4\delta_x{}^4+4r_y\delta_x m_y{}^2-4\alpha_{yx}\kappa_{yx}\eta_x m_y{}^2+4m_y{}^3\delta_x$$

The non-zero eigen values are negative real numbers. The Jacobian matrix and its all minors are stable at Z_6^*, therefore Z_6^* is LAS.

References

[1] Sharp P M and Hahn B H 2011 Origins of HIV and the AIDS Pandemic *Cold Spring Harb Perspect Med.* **1**(1) a006841. doi: 10.1101/cshperspect.a006841

[2] Wenhui Li and Swee-Kee Wong, at.al. 2006 Animal Origins of the Severe Acute Respiratory Syndrome Coronavirus: Insight from ACE2-S-Protein Interactions *Journal of Virology* **80**(9). doi: 10.1128/JVI.80.9.4211-4219.2006

[3] Chan P K 2002 Outbreak of avian influenza A(H5N1) virus infection in Hong Kong in 1997 *Clin. Infect. Dis.* **34**(2)

[4] Louis Y A 2005 *Avian Influenza: Basic Science, Potential for Mutation, Transmission, Illness Symptom, and Vaccines in BIRD FLU - A Rising Pandemic in Asia and Beyond?* (World Scientific Publishing Co. Pte. Ltd) http://www.worldscibooks.com/medsci/6108.html

[5] Yuen KY and Chan PK et al. 1998 Clinical features and rapid viral diagnosis of human disease associated with avian influenza A H5N1 virus. *Lancet* **351** pp 467–471

[6] Kaye D and Pringle C R 2005 Avian influenza viruses and their implication for human health *Clin. Infect Dis.* **40**(1) 108112

[7] Webster R G and Govorkova E A 2006 H5N1 Influenza: continuing evolution and spread *N. Engl. J. Med.* **355** pp 2174–2177

[8] Chowell G and Ammon CE at al. 2006 Transmission dynamics of the great influenza pandemic of 1918 in Geneva, Switzerland: assessing the effects of hypothetical interventions *J. Theor. Biol.* **241** pp 193–204

[9] Sattenspiel L and Herring D 2003 Simulating the effect of quarantine on the spread of the 1918-19 flu in central Canada *Bull. Math. Biol.* **65** pp 1–26

[10] Coburn B J and Wagner B G at al. 2009 Modeling influenza epidemics and pandemics: insights into the future of swine flu (H1N1) *BMC Medicine* **7** p 30
[11] Neumann G and Noda T at al. 2009 Emergence and pandemic potential of swine-origin H1N1 influenza virus *Nature* **459** pp 931–939
[12] Vynnycky E and Edmunds W J 2008 Analyses of the 1957 (Asian) influenza pandemic in the United Kingdom and the impact of school closures *Epidemiol Infect* **136** pp 166–179
[13] Rambaut A and Pybus O G at al. 2008 The genomic and epidemiological dynamics of human influenza A virus *Nature* **453** pp 615–619
[14] Smith G D J and Vijaykrishna D at al. 2009 Origins and evolutionary genomics of the 2009 swine-origin H1N1 influenza A epidemic *Nature* **459** pp1122–1125
[15] Earn DJD and Doushoff J at al. 2002 Ecology and evolution of influenza *Trends in Ecol Evol.* **17**(7) pp 334–340
[16] Treanor J 2004 Influenza Vaccine Out maneuvering Antigenic Shift and Drift *N. Engl. J. Med.* **350**(3) pp 218–220
[17] Pease CM 1987 An evolutionary epidemiological mechanism with applications to type A influenza *Theor. Popul. Biol.* **31** pp 422–452
[18] Iwami S and Takeuchi Y at al. 2007 Avian-human influenza epidemic model *Math. Biosci.* **207** pp 1–25
[19] Martcheva M 2012 An evolutionary model of influenza A with drift and shift *Journal of Biological Dynamics* **6**(2) pp 299–332
[20] Saenz RA and Hethcote HW at al. 2006 Confined animal feeding operations as amplifiers of influenza *Vector Borne Zoonotic Dis.* **6** pp 338–346
[21] Kuiken T and Holmes EC at al. 2006 Host species barriers to influenza virus infections *Science* **312** pp 394–397
[22] Aditama TY and Samaan G at al. 2011 Risk Factors for Cluster Outbreaks of Avian Influenza A H5N1 Infection, Indonesia *CID* **53** pp 1237–1244
[23] Kandun IN and Tresnaningsih E et al. 2008 Factors associated with case fatality of human H5N1 virus infections in Indonesia: a case series *Lancet* **30** e744–9
[24] Sidik W 2012 *Mathematical models for the spraed and control of multi starin influenza-A viruses in Indonesia* (Australia: Phd Thesis, Flinders University)
[25] Diekmann O and Heesterbeek JAP at al. 1990 On the definition and the computation of the reproduction ratio in models of infectious disease in heterogeneous population *J. Math. Biol.* **28** pp 365–382
[26] van-den Driessche P and Watmough J 2002 Reproduction numbers and sub-threshold endemic equilbria for compartmental models of disease transmission *J. Math. Bioscience* **180** pp 29–48
[27] Arriola L and Hyman J 2005 *Lecture notes, forward and adjoint sensitivity analysis: with applications in Dynamical Systems, Linear Algebra and Optimisation.*
[28] Wiggins S 1990 *Introduction to applied nonlinear dynamical systems and chaos* (New York: Springer-Verlag)

Stability and Numerical Simulation of Prey-predator System with Holling Type-II Functional Responses for Adult Prey

Dian Savitri[1]
[1]Mathematics Departement, Faculty of Mathematics and Science, Universitas Negeri Surabaya

E-mail: diansavitri@unesa.ic.id

Abstract. In this paper, we analyze the local stability of the prey-predator model. This model was constructed from two prey involving stage-structure and one predator. The population of prey is divided into two subpopulations, adult prey and immature. The ratio of intrinsic growth of predator population density divided by adult prey conversion factors into a predator. Reduction of predator populations has a reciprocal relationship with the availability of favorite foods determined by the environmental resources. We also consider that Holling type II the functional response for the predator. We have investigated the local stability and the solution of the system. We have studied the existence and feasibility of various equilibrium points. The system has three positive equilibria, namely the original, the extinction of the predator and the interior point. We explore the stabilities of all nonnegative equilibrium point by the real parts of the eigenvalues of the Jacobian matrix at each equilibrium point must be negative. We also use the Routh-Hurwitz criterion to investigate the stability at the interior equilibrium point. This study aims to detect the limit cycle with its phase portrait and its time-series graphs of the prey-predator system. Numerical simulated are represented as phase portraits to draw the stability of the equilibrium point. In our Numerical outcomes confirmed the analytical results and the local behavior prey-predator model which start from their positive initial condition. The parameter value is taken mainly from the literature and assumption. By a computational Python using the fourth-order Runge-Kutta method, we solved the prey-predator system and showed some new phase portraits such as the existence of the stable or unstable equilibrium point under a suitable value of the parameter. The next research, we will explore the global stability analysis behavior of the interior equilibrium point.

1. Introduction

The dynamical behaviour of prey-predator model has investigated the stability of the positive equilibria and the existence of all feasible nonnegative equilibria. In this paper, we discuss the dynamical behaviour of stage-structured prey predator interactions with Holling type II response function for adult prey [1]. Model structure stages in the dynamics of interacting species have long been known [2]. This model has a much simpler way to simulate diversity than existing models and presents new phenomena in the real world.

In a dynamic system, if there is limit cycles, then the interior equilibrium point is center. In these references the limit cycles were all generated by Hopf Bifurcation [3]. Bifurcation is the stability change of a system that occurs due to changes in parameter values [4].

This paper is organized as follows. First, it considers the stability of the equilibria in detail with the conditions for existence. Second, it illustrates the local stability of equilibria and the third, it shows the dynamical behaviour the stability of predator-prey model depending on the parameters by performed numerical simulations. In the last study, several numerical simulations were completed for illustrating the results.

2. Mathematical Model

Huenchucona, Falconi and Vidal [5] proposed interaction of a two prey and one predator with the same predation process using the Beddington De-Angelis functional responses. The response function combines the predator rate of the adult prey population to the young prey populations with density depends on the young prey population.

$$\frac{dx_1}{dt} = rx_1\left(1-\frac{x_1}{K}\right) - \beta x_1 - \frac{\beta_1 x_1 y}{1+m_1{x_1}^2+n_1x_2}$$
$$\frac{dx_2}{dt} = \beta x_1 - \frac{\beta_2 x_2 y}{1+m_2x_1+n_2x_2} - \mu_1 x_2 \quad (1)$$
$$\frac{dy}{dt} = \frac{\alpha_1\beta_1x_1y}{1+m_1{x_1}^2+n_1x_2} + \frac{\alpha_2\beta_2x_2y}{1+m_2x_1+n_2x_2} - \mu_2 y$$

Savitri, Abadi[6] studied the dynamical behavior of prey-predator model with stage-structure for prey. They used different predation rate, Holling type I the functional responses for immature prey and Holling type II functional response for mature prey.

$$\begin{aligned}\frac{dx_1}{dt} &= rx_1\left(1-\frac{x_1}{K}\right) - \beta x_1 - \alpha x_1 y\\ \frac{dx_2}{dt} &= \beta x_1 - \frac{\varepsilon x_2 y}{1+mx_2} - \mu_1 x_2\\ \frac{dy}{dt} &= \frac{\gamma\varepsilon x_2 y}{1+mx_2} - \mu_2 y\end{aligned} \qquad (2)$$

Castellanos and Chan-Lopez [7], analyse a three level trophic chain model. They consider that Holling type I the functional response for the predator in middle and Holling type II for the predator at the top. The presence of the top predator is determinant in the stability of the chain model. The chain model is represented by the following system:

$$\begin{aligned}\frac{dx}{dt} &= \rho x\left(1-\frac{x}{k}\right) - a_1 xy\\ \frac{dy}{dt} &= ca_1xy - dy - \frac{a_2yz}{y+b_2}\\ \frac{dz}{dt} &= \alpha z^2 - \frac{\beta z^2}{y+b_2}\end{aligned} \qquad (3)$$

Before we introduce the mathematical model, let us describe the basic assumptions that we made to formulate it. In this work, we construction from the system (1), (2) and system (3). Motivated by these, in this paper we have presented the dynamical behaviour of three populations. Juvenile-adult stage structure is added to both prey and predator in model.

2.1. *Contruction of Mathematical Model*

The first equation, juvenile prey grows with intrinsic growth rate r cause increasing the juvenile prey population growth rate, as follow

$$\frac{dx}{dt} = rx \qquad (4)$$

Second, predatory predation rate for both of the prey population are different. It is assumed that predator can prey upon either juvenile prey or the adult prey in the structured model. Predatory predation rate used Holling type II for adult prey species, and Holling type I for juvenile prey. The rate of predation is given in the form of Holling type I functional response:

$$f(x,y) = \alpha xz \qquad (5)$$

The rate of predation is given in the original form of Holling type II functional response:

$$f(x,y) = \frac{\eta yz}{y+m} \tag{6}$$

where η and m are the maximum value of the per capita reduction rate of y due to z and the coefficient of environmental protection for the adult prey, respectively. Therefore, the adult prey population growth rate is reduced due to predation, as follow

$$\frac{dx}{dt} = -\frac{\eta yz}{y+m}. \tag{7}$$

The population densities of juvenile prey, adult prey and predator, denoting by $x(t), y(t)$ and $z(t)$, respectively. The mathematical model a two prey one predator interactions are described by the following system of ODEs.

$$\begin{aligned} \frac{dx}{dt} &= rx\left(1-\frac{x}{k}\right) - \beta x - \alpha xz \\ \frac{dy}{dt} &= \beta x - \frac{\eta yz}{y+m} - \mu y \\ \frac{dz}{dt} &= \alpha_1 xz + \rho z^2 - \frac{\eta_1 yz}{y+m} \end{aligned} \tag{8}$$

Changes in population densities of juvenile prey by intrinsic growth rates are reduced by the rate at which juvenile prey is consumed by predators. Juvenile prey populations grow with logistical equations that are limited by carrying capacity (k) to maintain their populations. The predation rate for juvenile prey is proportional to the rate at which predators and juvenile prey find each other or the so-called Lotka-Volterra equation. With parameter $r, \beta, \alpha, \alpha, \rho, m, \mu, \eta, \alpha_1, \eta_1$ and k are positive.

Changes in population density of adult prey, represent the growth of adult prey due to the conversion factor of juvenile prey to adult prey. The η parameter in the predation process causes a decrease in adult prey population which is interpreted as the rate that has an encounter with adult prey per unit density of adult prey. Without predation, the adult prey experiences a natural process of death. The rate of predator predation follows the Holling Type II functional response. The $\frac{m}{\eta}$ ratio is the average time in food processing for predator. The parameter ρ is the rate of population growth of predator population.

The rate of change in predator population is assumed to decrease in predator population following the Leslie-Gower model, with m representing loss of predator residue due to lack of their favorite food. Parameter η_1 is the ratio of intrinsic growth divided by the conversion factor of predation to adult prey. This affects the availability of favorite foods per capita, which is determined by the ability of environmental resources and is proportional to the abundance of their favorite foods [8].

2.2. *Equilibrium point*

In this section, we have investigated the local stability of all possible equilibrium point of the system (8). By the definition of equilibria of a system (8) are satisfied when

$$\frac{dx}{dt} = \frac{dy}{dt} = \frac{dz}{dt} = 0$$

We have studied the existence and feasibility of various equilibrium points. The system has three positive equilibria,

1. The original equilibrium point $E_1 = (0,0,0)$
2. The equilibrium of the predator extinction $E_2 = \left(\frac{k(r-\beta)}{r}, \frac{\beta k(r-\beta)}{\mu r}, 0\right)$
3. The interior equilibrium point $E_3 = (x^*, y^*, z^*)$

2.3. *Local Stability*

The local asymptotic stability of each equilibrium point is analyzed by calculating the Jacobian matrix and finding the eigenvalues assessed at each equilibrium point. The real parts of the eigenvalues of the Jacobian matrix must be negative for the stability of the equilibrium points.

The system (8) is a nonlinear autonomous. The stability of the nonlinear autonomous can be obtained by linearising the system (8) with the Jacobian matrix is given by:

$$J(x^*,y^*,z^*)=\begin{bmatrix} r-\frac{2rx}{k}-\beta-\alpha z & 0 & -\alpha x \\ \beta & -\frac{\eta z}{y+m}+\frac{\eta yz}{(y+m)^2}-\mu & -\frac{\eta y}{y+m} \\ \alpha_1 z & \frac{\eta_1 z^2}{(y+m)^2} & 2pz-\frac{2\eta_1 z}{y+m}+\alpha_1 x \end{bmatrix}$$

The characteristic equation is given by $\det(J(x^*,y^*,z^*)-\lambda I)=0$ where (x^*,y^*,z^*) is the equilibrium point of (8). We explore the stabilities of all nonnegative equilibrium point by the real parts of the eigenvalues of the Jacobian matrix at each equilibrium point.

The characteristic equation of equilibrium point $E_1=(0,0,0)$ is

$$\begin{vmatrix} (-1+r)-\lambda & 0 & 0 \\ \beta & -\mu-\lambda & 0 \\ 0 & 0 & 0-\lambda \end{vmatrix}=0$$

The eigenvalues of the matrix $J(E_1)$ are $((-1+r)-\lambda)(-\mu-\lambda)(0-\lambda)=0$

It has always one negative eigenvalue $\lambda_1=-\mu$, the other eigenvalue negative if $r<1$, and $\lambda_3=0$. Therefore the equilibrium point E_1 is unstable.

The characteristic equation of equilibrium point $E_2=\left(\frac{k(r-\beta)}{r},\frac{\beta k(r-\beta)}{\mu r},0\right)$ is

$$\det\left(J\left(\frac{k(r-\beta)}{r},\frac{\beta k(r-\beta)}{\mu r},0\right)-\lambda I\right)=0$$

$$J(E_2)=\begin{vmatrix} (\beta-r)-\lambda & 0 & \frac{\alpha k(\beta-r)}{r} \\ \beta & -\mu-\lambda & \frac{\eta\beta k(\beta-r)}{r\mu\left(\frac{\beta k(r-\beta)}{\mu r}+m\right)} \\ 0 & 0 & -\frac{\alpha_1 k(\beta-r)}{r} \end{vmatrix}=0$$

The eigenvalue of matrix $J(E_2)$ are $\lambda_1=\beta-r,\ \lambda_2=-\mu$ and $\lambda_3=-\frac{\alpha_1 k(\beta-r)}{r}$. The equilibrium point $E_2=\left(\frac{k(r-\beta)}{r},\frac{\beta k(r-\beta)}{\mu r},0\right)$ is stable under condition if $r>\beta$.

The characteristic polynomial corresponding to the Jacobian matrix, equilibrium point $E_3=(x^*,y^*,z^*)$

$$J(x^*,y^*,z^*)=\begin{bmatrix} r-\frac{2rx^*}{k}-\beta-\alpha z^* & 0 & -\alpha x^* \\ \beta & -\frac{\eta z^*}{y^*+m}+\frac{\eta yz^*}{(y^*+m)^2}-\mu & -\frac{\eta y^*}{y^*+m} \\ \alpha_1 z^* & \frac{\eta_1 z^{*2}}{(y^*+m)^2} & 2pz^*-\frac{2\eta_1 z^*}{y^*+m}+\alpha_1 x^* \end{bmatrix}$$

The characteristic equation of system (8) at the coexistence equilibrium point $E_3=(x^*,y^*,z^*)$ is
$\lambda^2-(\mathrm{trace}J(x^*,y^*,z^*))\,\lambda+\det J(x^*,y^*,z^*)=0$.
it is obvious that $\mathrm{trace}(J(x^*,y^*,z^*))<0$ and $\det(J(x^*,y^*,z^*))>0$.

$$\lambda^3+\gamma_1\lambda^2+\gamma_2\lambda+\gamma_3=0$$

We use the Routh–Hurwitz criterion to investigate the stability at the interior equilibrium point, which has a negative real part if and only if $\gamma_1>0$, $\gamma_3>0$, and $\gamma_1\gamma_3-\gamma_3>0$, the coexistence steady-state E_3 is locally asymptotically stable. The findings local stability condition of all equilibria described in table 1.

Table 1. Stability condition of all equilibrium point

	Type of stability	Local stability condition
$E_1 = (0,0,0)$	unstable	
$E_2 = \left(\frac{k(r-\beta)}{r}, \frac{\beta k(r-\beta)}{\mu r}, 0\right)$	stable	$r > \beta$
$E_3 = (x^*, y^*, z^*)$	Asymptotically stable	

3. Numerical Simulation

We observed the dynamic behavior of the system (8) by confirming the results of analysis and simulation using Python to show the dynamics of the predator model around the positive interior equilibrium point.

We show that the results of interactions in the system can be influenced by the predation process by phase-portrait analysis and numerical simulations.

3.1. Parameter value

Table 2. Parameter value

	r	β	α	α_1	η	η_1	K	ρ	m	μ
S1	0.82	0.87	1.56	1.12	2.41	1.83	12	1.38	0.13	0.11
S2	1.32	0.87	1.16	0.72	1.6	0.41	2.8	0.78	0.23	0.11
S3	1.32	0.87	0.76	0.72	0.6	0.41	2.8	0.78	0.23	0.11
S4	1.32	0.87	1.16	0.72	1.2	0.41	2.8	0.78	0.23	0.11
S5	1.32	0.87	1.16	0.72	0.3095	0.41	2.8	0.78	0.23	0.11

3.1.1. The Dynamic Behaviour the first simulation stable the interior equilibrium point

The first simulation, we set the parameter value $r = 0.82, \beta = 0.87, \alpha = 1.56, \eta = 2.41, \rho = 1.38,$ and $K = 12$. Base on analysis, the model of system (8) only has one equilibrium point is the locally asymptotically stable. The numerical simulation shows that all solutions are going to the interior equilibrium point with different initial value. Initial value 1: [3.01. 5.05. 4.28], initial value 2: [4.6,5.9,3.1] and initial value 3: [12.2,22.1,21.1]. The results of numerical simulation indicate that all populations can survive.

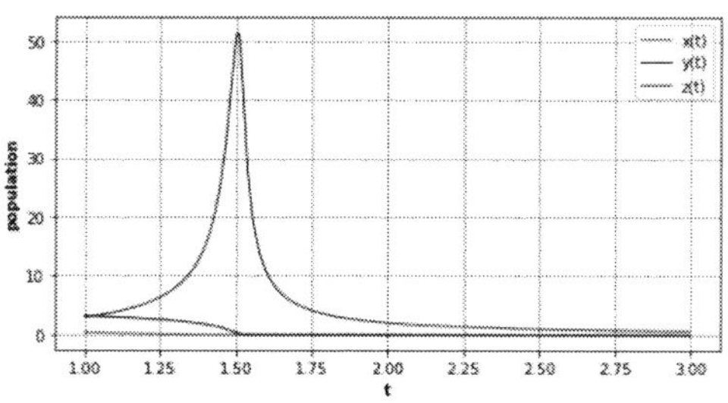

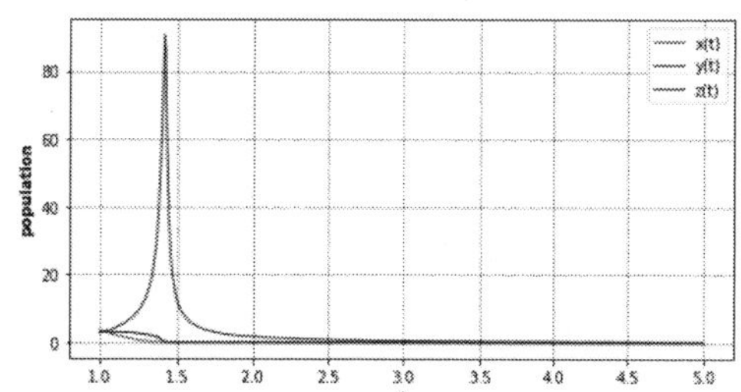

Figure 1 Time series of the solution system in the Interior Equilibrium Point if $K = 12$

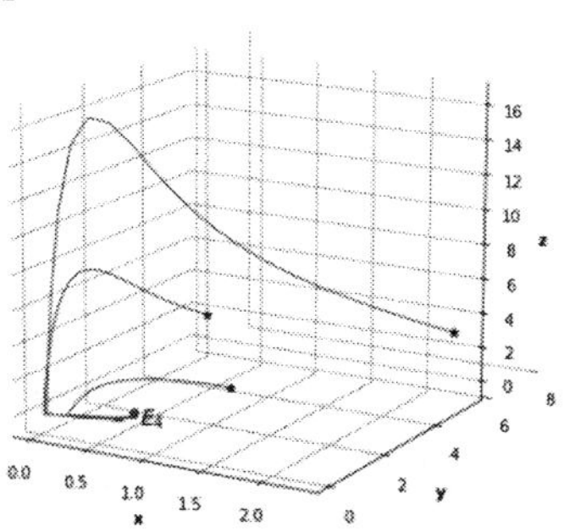

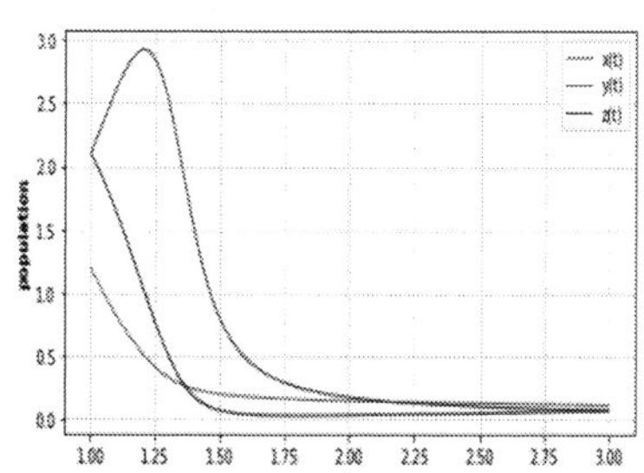

Figure 2 The Dynamic behaviours of the solution system in the Interior Equilibrium Point

Journal of Physics: Conference Series **1417** (2019) 012025 doi:10.1088/1742-6596/1417/1/012025

3.1.2. The Dynamic Behaviour $\eta > \beta$ and $\eta > \alpha$

The second simulation, we set the parameter value $r = 1.32, \beta = 0.87, \alpha = 1.56, \eta = 2.41, \rho = 1.38$, and $K = 2{,}8$. Base on analysis, the model of system (8) only has one equilibrium point is the locally asymptotically stable. The numerical simulation shows that all solutions are going to the interior equilibrium point with different initial value. Initial value 1: [0.3. 2.4. 3.9], initial value 2: [0.6,2.4,4.1] and initial value 3: [2.1,1.2,1.1]

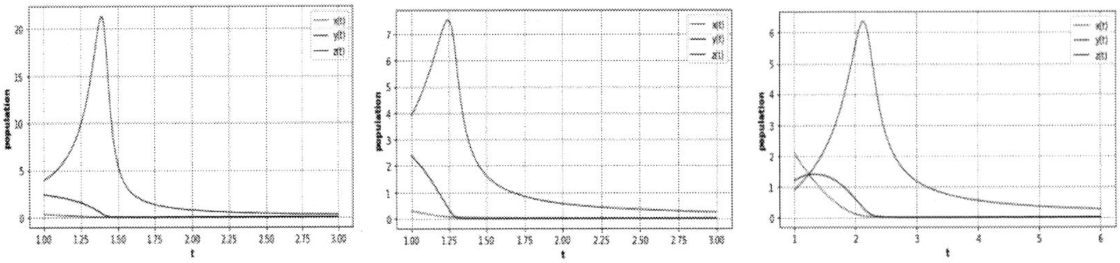

Figure 3 Time series of the solution system in the Interior Equilibrium Point if $K = 2.8$

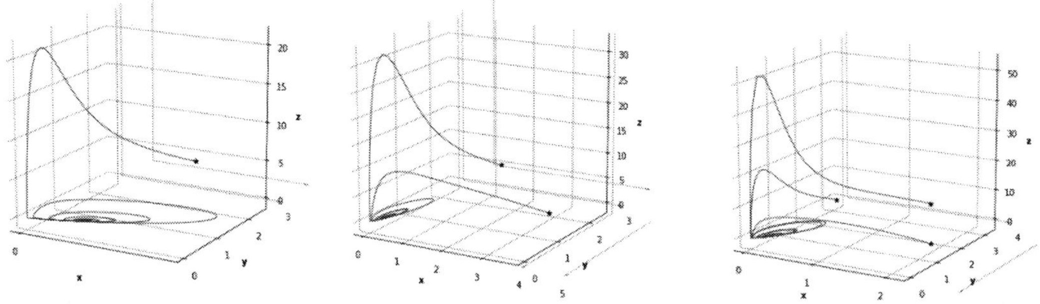

Figure 4 The Dynamic behaviours of the solution system in the Interior Equilibrium Point with $K = 2.8$

3.1.3. The Dynamic Behaviour if alpha < beta

The next simulation $\alpha < \beta$, we set the parameter value $\beta = 0.87, \alpha = 0.76, \eta = 0.6, \rho = 0.78$, and $K = 2{,}8$. The numerical simulation shows that solution are going to the interior equilibrium point with N1: [0.3. 2.4. 3.9]

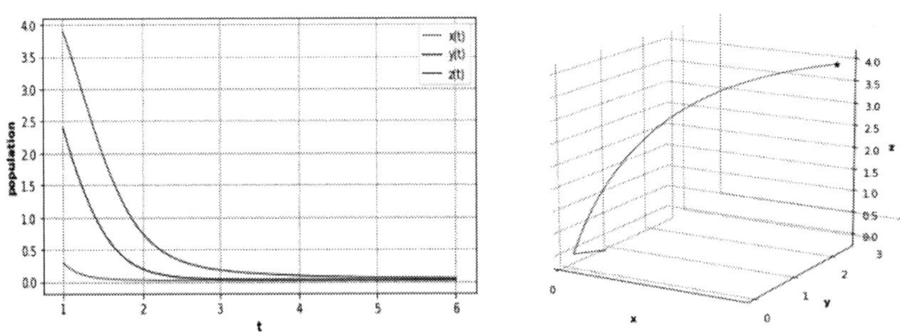

Figure 5 The Dynamic behaviours of the solution system with $\eta = 0.6$ *and* $\alpha < \beta$.

3.1.4. The Dynamics Behaviour $\eta > \beta$ dan $\eta > \alpha$

The first simulation, we set the parameter value $r = 0.82, \beta = 0.87, \alpha = 0.76, \eta = 1.2, \rho = 1.38$, and $K = 2.8$. The numerical simulation shows that all solution are going to the interior equilibrium point with initial value N1: [1.2, 2.1, 2.4]

Journal of Physics: Conference Series **1417** (2019) 012025 doi:10.1088/1742-6596/1417/1/012025

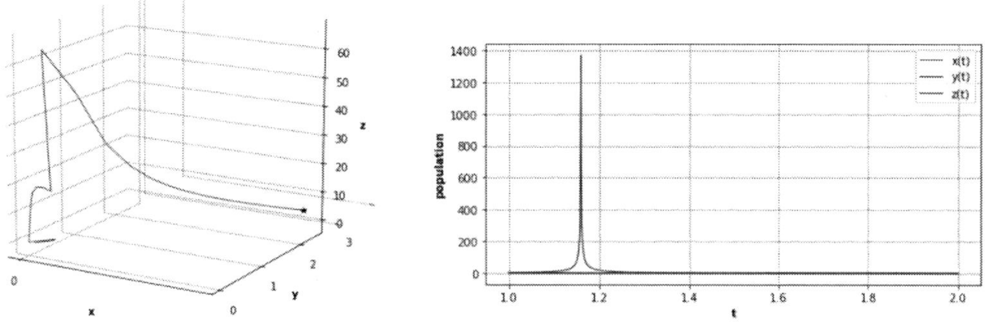

Figure 6 The Dynamic behaviours of the solution system with $\eta = 1.2$

3.1.5. *The Dynamics Behaviour* $\eta = \alpha$

The next simulation, we set the parameter value $r = 0.82, \beta = 1.3, \alpha = 1.2, \eta = 1.2, \rho = 1.38$, and $K = 2.8$. The numerical simulation shows that all solution are going to the interior equilibrium point with N1: [1.2, 2.1. 4.28],

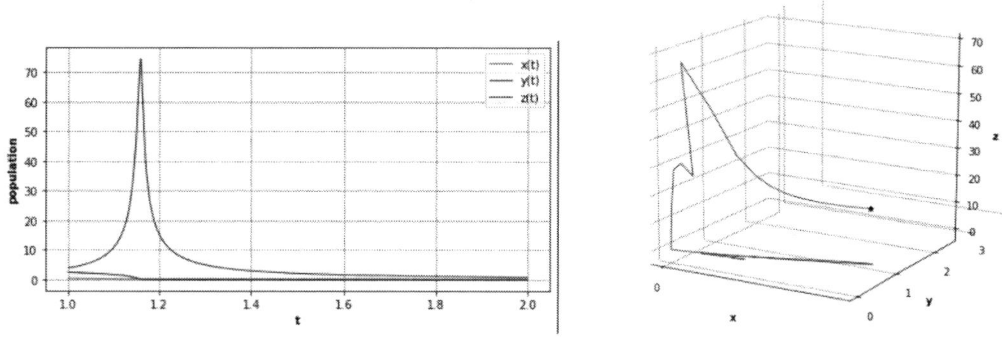

Figure 7 The Dynamic behaviours of the solution system with $\eta = \alpha = 1.2 < \beta = 1.3$.

3.2. *Continuation Numeric*

Then a numerical continuation is carried out on the predation parameter $\eta = 0.6$ which results in one equilibrium point E_3 which is stable. All of parameter are $r = 1.32, \beta = 0.87, \alpha = 1.16, \eta_1 = 0.41, \rho = 0.78, m = 0.23, \mu = 0.11$and $K = 2.8$. Eigenvalues consist of two λ complexes with a real negative portion, and one negative eigenvalue. $\lambda_1 = -0.0519514704367125 + 0.351175046207591\, I$, $\lambda_2 = -0.0519514704367125 - 0.351175046207591\, I$ and $\lambda_3 = -0.445962288036575$

Based on these parameter values, an equilibrium point is obtained

$E_1 = (0; 0; 0)$

$E_2 = (0.9545454547;\ 7.549586778;\ 0)$

$E_3 = (0.1194522936; 0.1667424272; 0.3393851516)$

The results of numerical continuity when the value of $\eta = 0.30951031$ causes a change at the equilibrium point E_3 from the stable spiral to center. If you pass the predation parameter value at $\eta = 0.3$, a change at the equilibrium point E_3 becomes an unstable spiral. Based on these results, it can be seen that the higher the value of the parameter η, the equilibrium point E_3 will experience changes that are stable.
The stability of the equilibrium point changes when Hopf bifurcation occurs when $\eta = 0.30951031$. when the parameter $\eta = 0.30951031$ the stability of the equilibrium point E_3 changes from a stable spiral to a center.

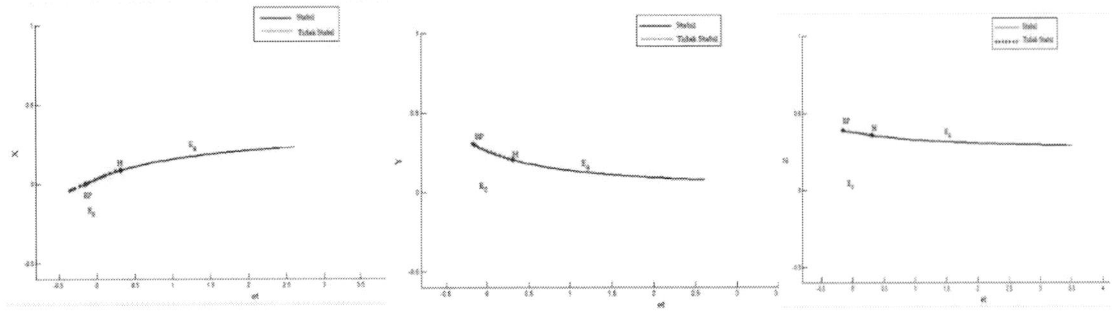

Figure 8. Bifurcation diagram of the juvenile-adult prey and predator population with respect to η

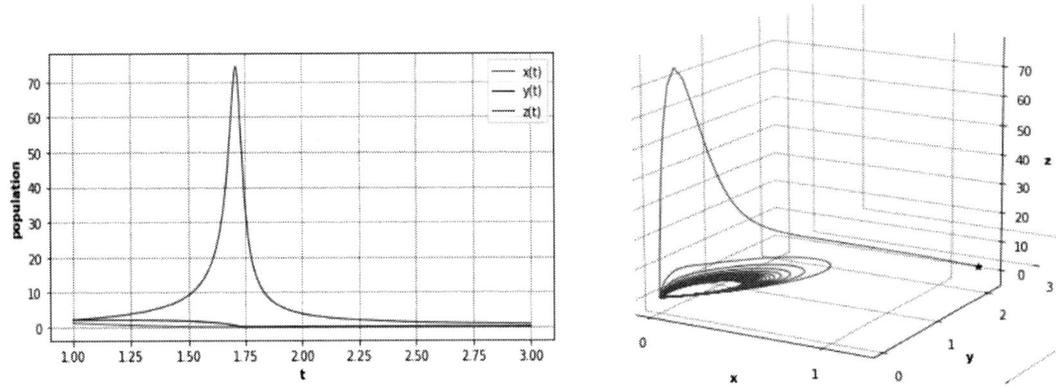

Figure 9 The Dynamic behaviours of the solution system with initial value $N1 = [0.3,2.4,3.9]$

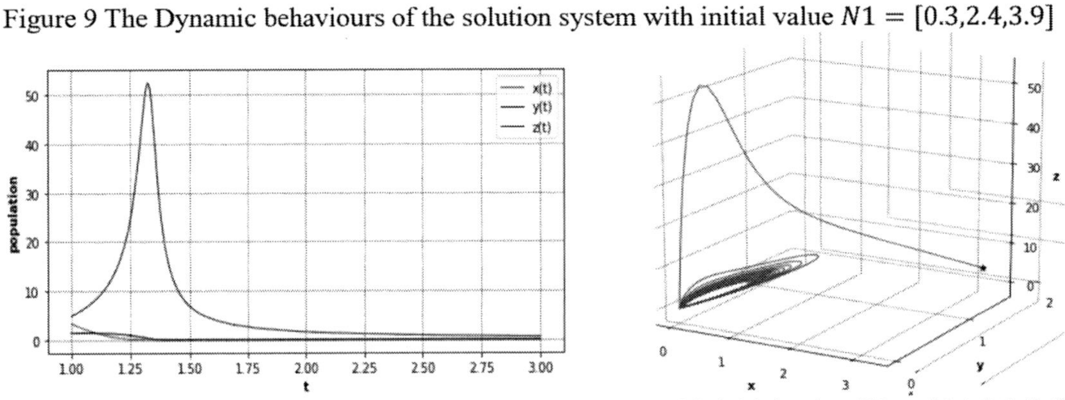

Figure 10 The Dynamic behaviours of the solution system with initial value $N2 = [4.1,2.2,5.1]$

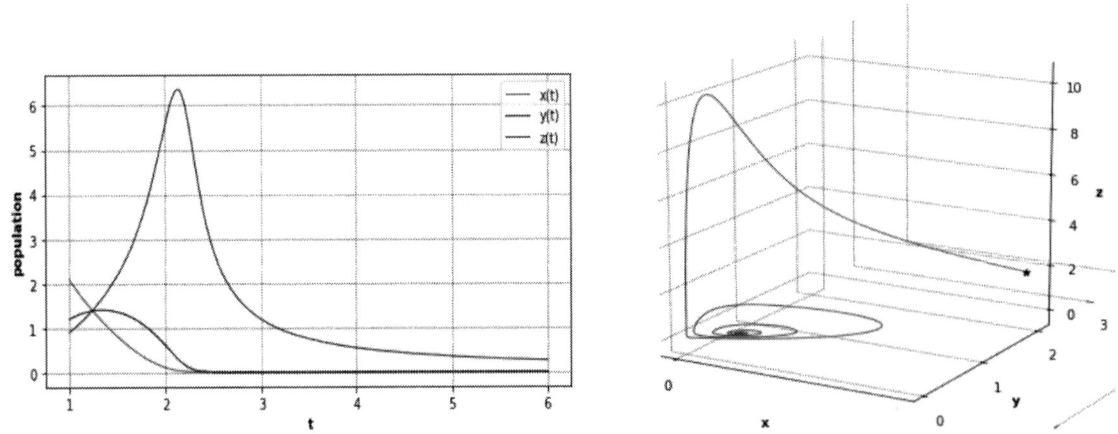

Figure 11 The Dynamic behaviours of the solution system with initial value $N3 = [2.1,1.2,1.1]$

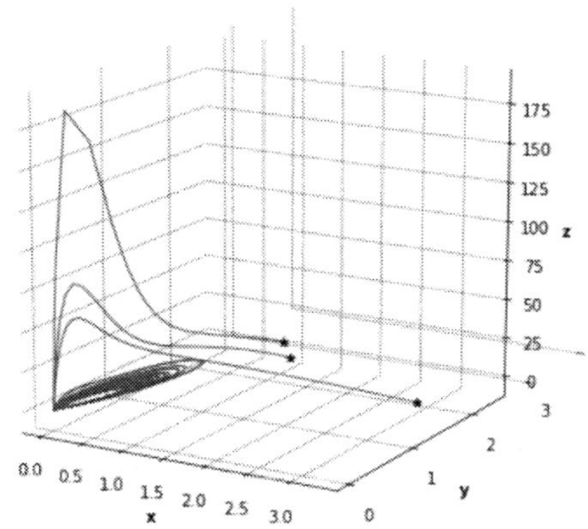

Figure 12 The Dynamic behaviours of the solution system with $\eta = 0.30951031$

It is shown numerically that there exists a stable limit cycle. In this case, the population of juvenile prey, adult prey and the predators will exis and the existence of Hopf bifurcation we will studied further

4. Conclution

In this research, we are investigated with the local stability in predator-prey model with stage-structure on both of the prey population. The model has three e positive equilibria, namely the original, the extinction of the predator and the interior point. The Interior equilibrium point is local asymptotically stable with certain conditions. This study aims to detect the limit cycle with its phase portrait and its time-series graphs of the prey-predator system. Numerical simulated are represented as phase portraits to shows the stability of the equilibrium point. In our Numerical outcomes confirmed the analytical results and the local behaviour juvenile-adult prey and predator model which start from their positive initial condition. The parameter value is taken mainly from the literature and assumption. It is shown numerically that there exists a stable limit cycle. The next research, we will studied further the existence of Hopf bifurcation analysis of the interior equilibrium point.

5. References

[1] Holling C S 1965 The Functional Response of Predators to Prey Density and its Role in Mimicry and Population Regulation *Memoirs of the Entomological Society of Canada* *97*(S45) 5–60. https://doi.org/10.4039/entm9745

[2] Khajanchi S 2014 Dynamic behavior of a Beddington-DeAngelis type stage structure predator-prey model *Applied Mathematics and Computation* **244** 344-360.

[3] Xiao D and Li W 2000 Limit cycles for The Competitive Three Dimensional Lotka Volterra System *Journal of Differential Equations* **164** 1-15

[4] Hofbauer J 1981 On the occurrence of limit cycle in the Lotka-Volterra equation *Nonlinear Analysis* **5**(1981) 1003-1007

[5] Falconi, Huenchucona M, and Vidal C 2015 Stability and global dynamic of stage-structured predator-prey model with group defense mechanism of the prey *Journal Applied Mathematics and Computation* **270** 47-61

[6] Castellanos V and Chan-Lopez R 2017 Existence of Limit cycles in a tree level trophic chain with Lotka-Volterra and Hilling type II functional responses *Chaos, Solitons, and Fractals* **95** 157-167

[7] Savitri D and Abadi 2018 Numerical Simulation in Prey-Predator Model with a structure for prey *Atlantis Highlights in Engineering (AHE)* **1**

[8] Liu S, Chen L, and Liu Z 2002 Extinction and permanence in nonautonomous competitive system with stage structure *J. Math. Anal. Appl.* **274**(2002) 667–684

Parameter Estimation Using INLA for Disease Mapping of Leptospirosis in Bantul Indonesia

R Artiono

Mathematics Department, Universitas Negeri Surabaya, Surabaya, Indonesia

e-mail: rudiantoartiono@unesa.ac.id

Abstract. Agriculture sector is recognized as one important factor for an increased risk of leptospirosis in Bantul, Indonesia. From January to July 2010, there were 70 patients of leptospirosis with 43 patients engaged in agriculture. The aim of this study was to determine the impact of agricultural sector on the disease mapping of leptospirosis in district level. A structured additive regression model was used for statistical analysis associated with agricultural area and agricultural livelihood. In addition, a random effect that captures the spatial heterogeneity in the study region was also included in this study. Model estimation and predictive inference was carried out through the implementation of computer code in the R-INLA package, which makes use of the Integrated Nested Laplace Approximation.

1. Introduction

Leptospirosis is a bacterial disease of worldwide importance. This disease can spread sporadically not only in the human population but also in the domestic animal population. The disease is transmitted by rats, which act as a reservoir for Leptospira bacteria. The bacteria can breed well in the rat body but they do not infect the rat itself. Thus, the favourable environments for rats are associated with the increasing of the risk of leptospirosis.

Generally, incidence is associated with lack of access to clean water supplies [1,2]. A contaminated water provides favourable conditions for the growth of bacteria. It is also well-known that high rainfall can increase the number of human cases of leptospirosis [3-5]. The survival rate of bacteria is known to change seasonally, mortality being lower in rainy season and higher in dry season [6].

The cases can be found not only in tropical areas but also in the sub tropics. The World Health Organisation (WHO) reported that in the period between January 1970 and October 2008, there were 12,033 leptospirosis cases in United Kingdom, France, Italy, and Spain [7]. Another investigation reported a total of 2,694 cases of Leptospirosis in Germany 1962-2003 and a total of 2,553 cases of leptospirosis in Netherland 1925-2008 [8,9]. Moreover, in tropical areas, some countries reported individual years of high incidence of leptospirosis. (e.g. Thailand and Philippines). The Thailand Department of Disease Control (DDC) reported a total of 14,285 cases of leptospirosis in the year 2000 and the National Epidemiology Centre of Phillippines also reported about 2,471 cases of leptospirosis in the year 2012 [10,11].

Indonesia, as a country that lies in the tropical area, also has a significant number of cases. Data obtained from the Health Ministry of Indonesia show that Leptospirosis in humans fluctuated from 2005 to 2010. In the years 2005 to 2007, the number of cases increased dramatically from 115 patients to be 664 patients. However, in 2008 and 2009, it dropped to 426 patients and 335 patients respectively. The number of cases increased again in 2010 to be 409 patients. Although, Indonesia has relatively few cases of leptospirosis compared to Thailand and Philippines, it does not mean that Indonesia has low potential risk to the leptospirosis disease; it is because the disease is well-known that there are undiagnosed cases, misdiagnosed cases, and unreported cases of leptospirosis [12,13].

One area in Indonesia that has a lot of cases of leptospirosis is Bantul, small area in the Yogyakarta province. It is one county with an area of 508.85 km^2. The latitude is -7.88806 and the longitude is 110.32889. Based on the population census done by the Central Bureau of Statistics of Bantul in 2010, Bantul has population as much as 910,572 with the main livelihood is in agriculture. The Health Department of Bantul have reported that in 2009, it was found 9 cases of leptospirosis with 1 person died because of this disease. The number of cases had increased in 2010 became 116 cases with 19 people died and grew up in 2011 became 154 cases with 12 people died. In the year 2010 and 2011, the local government of Bantul stated that Bantul is endemic of leptospirosis. The increasing of the leptospirosis cases occurs more often in the rainy season. It was found that in the period of February-April is the peak season of the spread of leptospirosis disease.

In this study, we used data of 70 patients with leptospirosis in 17 districts from January to July 2010. The data were about the number of cases, the percentage of agricultural land used and the percentage of livelihoods associated with agriculture.

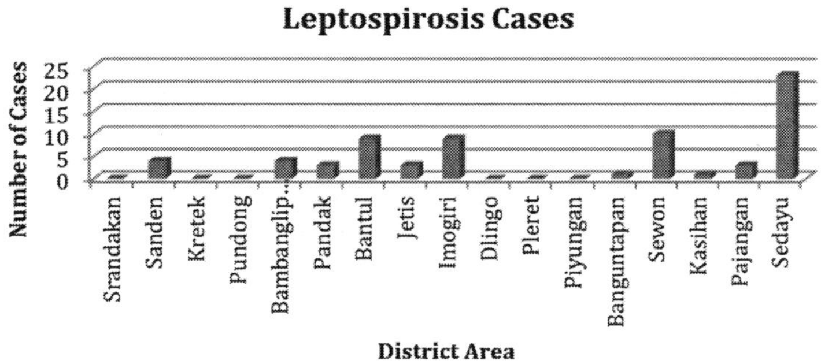

Figure 1. Data of 70 leptospirosis patient in 17 districts

In epidemiology, the most important thing that we need to concern is about how to make inference in order to analyse the spread of disease from some previous data. Unfortunately, it is not easy to get complete data in public health area. There are some aspects which relate to each other. The increasing number of cases is not only depending on the disease itself. It could be some aspects associate with it (i.e. environment, job, life style, etc.). However, it is still possible to construct a stochastic model which can be used as a starting point to inference. Particularly, models attempt to describe the mechanism by which the observed data are generated. Inference then proceeds by attempting to estimate the parameters of the model. One method that can be used to estimate the parameter is Bayesian. This method comes from the extension of the likelihood paradigm. In likelihood paradigm, it is only known that population are random variable. Meanwhile in Bayesian, it is not only population that are random variable but parameters are also random variable. Thus, the parameter also have distribution and we called it as prior distribution. In order to estimate the parameter, we combine likelihood model (data) and prior distribution such that posterior distribution cam be determined. This process can be continuously updated. It means that posterior distribution could be prior distribution for the next estimation.

The primary goal of this study was to determine the impact of agricultural sector on the disease mapping of leptospirosis in district level with Integrated Nested Laplace Approximation (INLA). This approximation was used to find posterior probability and Standard Morbidity Rate. We hypothesized that agricultural area and agricultural livelihood have big influence to the spread of leptospirosis in Bantul.

Calculation and plotting data in this study were conducted entirely by using R. This software also provides INLA package so that in order to find posterior marginal distribution with INLA method could be determined easily. Through shape-file in one area, we could make plot for the posterior probability

Journal of Physics: Conference Series **1417** (2019) 012026 doi:10.1088/1742-6596/1417/1/012026

and the standard morbidity rate. ArcGIS software was used for showing some data related to the Geographic Information System. This software could be used to obtain the shape-file which we need to plotting data in R.

2. Latent Gaussian Model and Hierarchical Structure

Model which used in Bayesian disease mapping is regression models [14-17]. The correlation between dependent variable and independent variable can be determined through this model. Furthermore, regression model that focused on the spatial elements is the spatial models. It is a subclass of structure additive regression models. This model allows us to be able to include some covariates either directly related to spatial or not. This model is more flexible to use. One of the models that exist in the structure additive regression models is Latent Gaussian Models.

Suppose y_i is the number of patients in each district area in Bantul which is modelled as

$$y_i \sim Poisson(\lambda_i);$$
$$\lambda_i = \rho_i . E_i$$

with link function $\eta_i = log\ log\ (\rho_i)$, define latent Gaussian field as $x = (\eta, \alpha, f, \beta)$. Since η, α, f and β are random variable then it should have parameter. Suppose $f(i)$ is parameter for y_i and ψ is parameter for x then we can build a hierarchical structure for this.

Stage 1

Latent Gaussian Model

$$\eta_i = log\ log\ (\rho_i) = \alpha + \upsilon_i + \nu_i + \beta_1 Z_{1i} + \beta_2 Z_{2i}$$

where:

λ_i is the Mean
E_i is the Expected Number of Cases
α is Intercept Term
$\upsilon_i = f_1(i)$ is Spatially Structured Residual
$\nu_i = f_2(i)$ is Spatially Unstructured Residual
$\beta_1 Z_{1i}$ is linear effect of covariate Agricultural Area
$\beta_2 Z_{2i}$ is linear effect of covariate Agricultural Livelihood

Stage 2

Considering spatial structure, it gave two distributions as follows:

1. $\upsilon_i = f_1(i)$ Spatially Structured Residual
   Modeled using an intrinsic conditional autoregressive structure
   $$\upsilon_i | \upsilon_{j \neq i} \sim Normal(m_i, s_i^2)$$
   $$m_i = \frac{\sum_{j \in \mathcal{N}(i)} \upsilon_j}{\#\mathcal{N}(i)} \text{ and } s_i^2 = \frac{\sigma_\upsilon^2}{\#\mathcal{N}(i)}$$
   Where: $\#N(i)$ is number of areas which share boundaries
2. $\nu_i = f_2(i)$ Spatially Unstructured Residual
   Modeled using an exchangeable prior
   $$\nu_i \sim Normal(0, \sigma_\nu^2)$$

Stage 3

$$\psi = \{\sigma_\upsilon^2, \sigma_\nu^2\}$$

Furthermore, informative prior are specified by INLA's default, as follows:

$$\sigma_\upsilon^2 \sim Gamma^{-1}(1{,}0.0005)$$
$$\sigma_\nu^2 \sim Gamma^{-1}(1{,}0.0005)$$

The framework can be seen in the diagram below

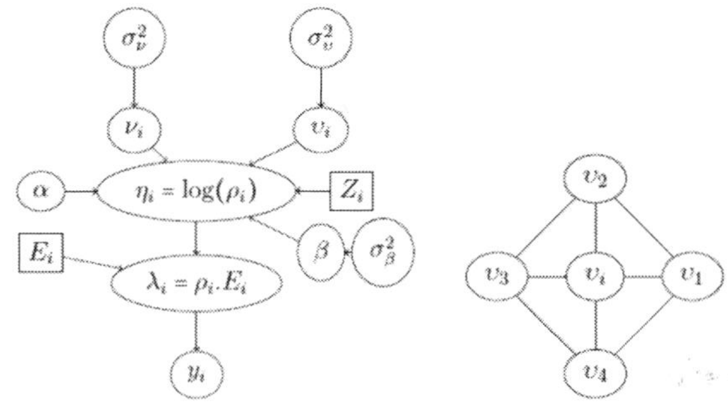

Figure 2. Framework of Hierarchical Structure

It represents spatial structure in each area satisfy Gaussian Markov Random Field (Conditional Independence).

3. Parameter Estimation

The fixed parameters effect (α,β_1, β_2) estimated by INLA are presented in Table 1. If it is exponentiated, they can be interpreted as relative risks: an increase of one unit in the agricultural land and in the agricultural livelihood is associated respectively with an increase of around 3% = exp (0.036) and around 0.3% = exp (0.0035) in the spread of disease.

Table 1. Parameters estimated by INLA

	Mean	Sd	0.025quant	0.5quant	0.975quant
α	-2.421907361	2.18184235	-7.33178841	-2.258992729	1.51788256
β_1	0.035703917	0.04279122	-0.04250809	0.032720051	0.13180982
β_2	0.003532001	0.002851445	-0.05580304	0.004001843	0.0601778

Agricultural land and agricultural livelihood just give a small effect to the spread of the leptospirosis diseases. Both parameters only give effect less than 10% in every increasing of one unit.

4. Posterior Probability

Furthermore, by using INLA, we can produce the posterior probability that represent the probability of finding cases of leptospirosis disease in one area after some factors (i.e. agricultural area and agricultural livelihood) are considered. The proportion of people which probably will be infected of leptospirosis can be obtained. Agricultural area represents the percentage of land area which is used for agricultural purposes, while agricultural livelihood is the percentage of population number which works in agriculture sector. The interpretation of the probability value is divided into three levels:

a. Area with the probability value between 0.6-1, it means that we are sure in those area will have a big probability to the spread of the Leptospirosis disease or in the other word we will find a lot of new cases of leptospirosis disease on those area or we are going to find a new patient of leptospirosis disease easily on those area. Area with this probability value is defined as areas with high risk to the spread of the Leptospirosis disease.
b. Area with the probability value between 0.3-0.6, it means that we are sure those area will have probability to the spread of the Leptospirosis disease but it is not too dangerous or in the other word we will find new cases of leptospirosis disease on those area but it is not as many as area with the probability value 0.6-1. Area with this probability value is defined as areas with medium risk to the spread of the Leptospirosis disease.
c. Area with the probability value between 0-0.3, it means that we sure those area will have a small probability to the spread of the Leptospirosis disease or in the other word we will not find new cases of leptospirosis disease on those area or it is not easy to find a new patient of leptospirosis disease on those area. Area with this probability value is defined as areas with low risk to the spread of the Leptospirosis disease.

The posterior probability, which was obtained from arcGIS and INLA, can be seen in Figure 3, 4, 5, and 6.

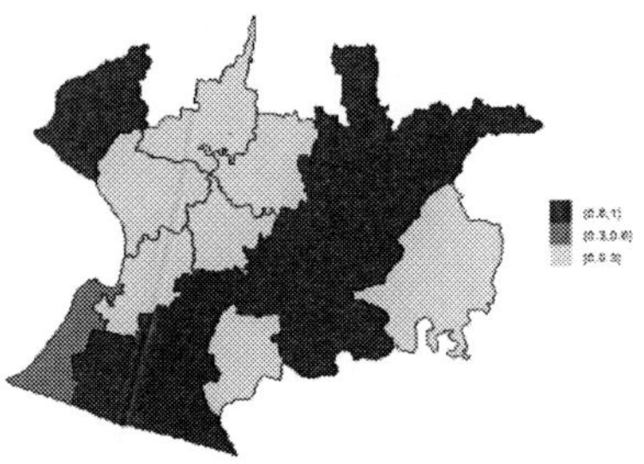

Figure 3. PP with no covariate

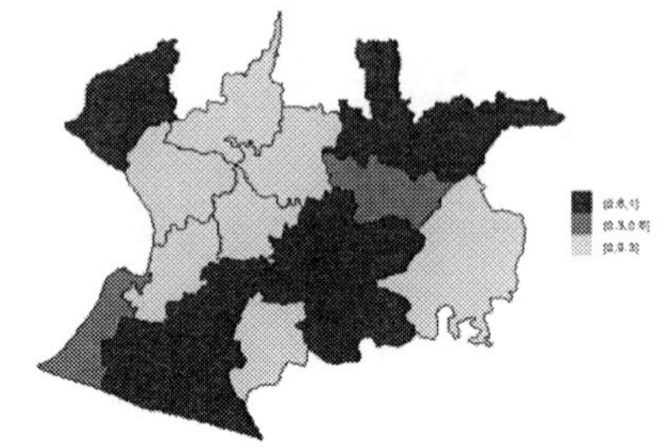

Figure 4. PP with Covariate: Agricultural Area

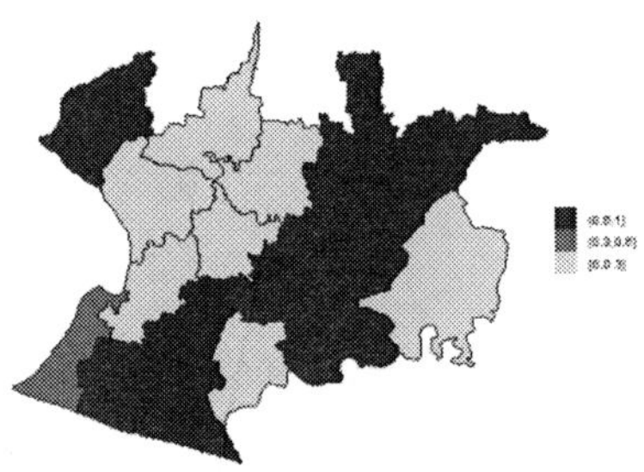

Figure 5. PP with Covariate: Agricultural Livelihood

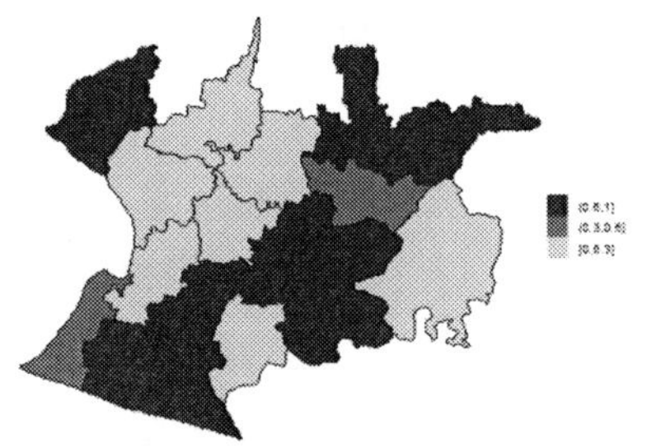

Figure 6. PP with Covariate: Agricultural Area and Livelihood

Figure 3 represents the probability of finding case of leptospirosis disease in each district of Bantul Area without considering of Agricultural Area and Agricultural Livelihood. Figure 4 represents the probability of finding cases of leptospirosis disease in each district of Bantul Area by considering of Agricultural Area. Figure 5 represents the probability of finding cases of leptospirosis disease in each district of Bantul Area by considering of Agricultural Livelihood. Figure 6 represents the probability of finding cases of leptospirosis disease in each district of Bantul Area by considering the considering of Agricultural Area and Agricultural Livelihood. The conclusions which can be derived from figure 3-6 are as follows:

a. Areas such as Sedayu, Banguntapan, Pleret, Imogiri, Jetis, Bambanglipuro, Sandan and Kretek have a high risk for the spread of the leptospirosis disease. Although some areas such as Pleret, Piyungan, and Kretek did not reveal any leptospirosis patients, but this area includes in the area with high risk against the spread of the disease leptospirosis.
b. Some areas with a relatively large number of patients as Sewon and Bantul did not have a high risk to the spread of leptospirosis disease.
c. There is no significant change in the posterior probability when agricultural area is added to the model as covariates. All area has same risk to the spread of leptospirosis disease. It means that agricultural area cannot be a major influence in the spread of the leptospirosis disease.
d. There is also found no significant change in the posterior probability when agricultural livelihood is added to the model as covariates.
e. The significant change appears when both of them are combined. Posterior probability in Bambanglipuro area is decrease from the range 0.6-1 to the range 0.3-06 and Dlinggo area is increase from the range 0-0.3 to the range 0.3-0.6. It means that Bambanglipuro area possible to

decrease the level of risk of the leptospirosis disease while conversely Dlingo area increase the risk of the spread of Leptospirosis disease.

f. Areas such as Kasihan, Sewon, Pajangan, Bantul, Pandak, and Pundong remain secure against the spread of leptospirosis disease because it has a posterior probability in the range 0-0.3. These areas are classified as areas with the low risk for the spread of disease.

5. Standard Morbidity Rate

Standard Morbidity Rate is the proportion of new cases of disease that found in one area in a given time. By run the INLA program, SMR in Bantul area can be seen in Figure 7, 8, 9, and 10.

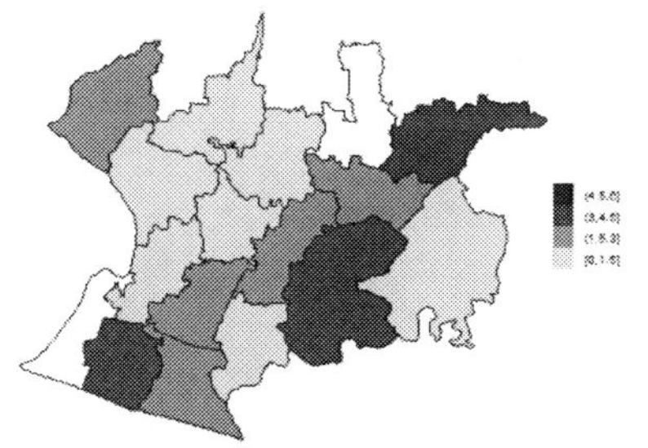

Figure 7. SMR with no covariate

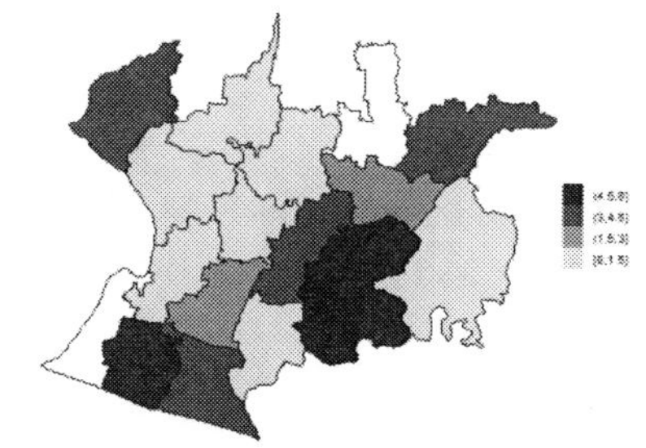

Figure 8. SMR with covariate: Agricultural Area

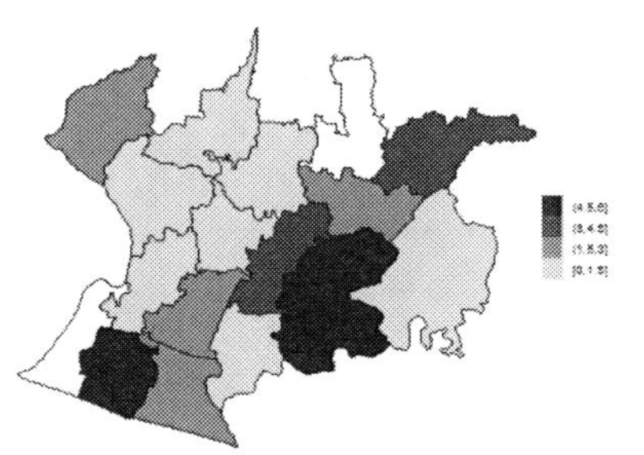

Figure 9. SMR with covariate: Agricultural Livelihood

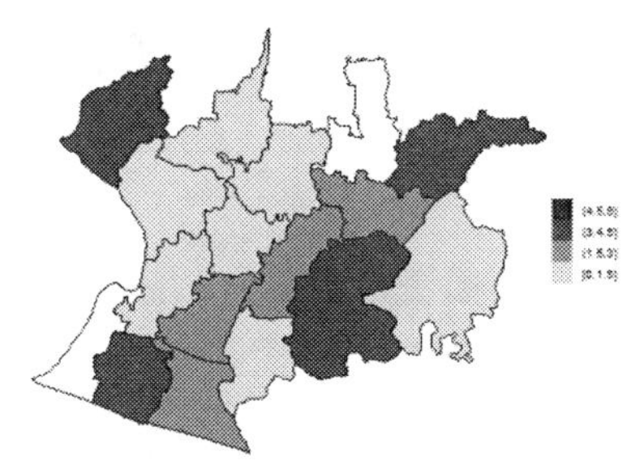

Figure 10. SMR with covariate: Agricultural Area and Livelihood

Figure 7 represents the Standard Morbidity Rate of leptospirosis disease in each district of Bantul Area without the considering of Agricultural Area and Agricultural Livelihood. Figure 8 represents the Standard Morbidity Rate of leptospirosis disease in each district of Bantul Area by considering Agricultural Area. Figure 9 represents the Standard Morbidity Rate of leptospirosis disease in each district of Bantul Area by considering Agricultural Livelihood. Figure 10 represents the Standard Morbidity Rate of leptospirosis disease in each district of Bantul Area by considering Agricultural Area and Agricultural Livelihood. The conclusions which can be derived from figure 6-9 are as follows:

a. If it does not involve the covariate, then Banguntapan has the highest SMR in the interval 9-15; Piyungan, Imogiri, and Sanden have SMR in the interval 6-9; Sedayu, Kasihan, Sewon, Pleret, Dlingo, Jetis, Bantul, Pajangan, Pandak, Bambanglipuro, Pundong, and Kretek have SMR in the interval 3-6, and Srandakan has the lowest SMR in the interval 0-3.

b. There is significant change that occurred in Banguntapan when agricultural area and/or agricultural livelihood was considered as covariate in the model. SMR of this area has decreased to the range 0-3.

c. The number of SMR in some areas such as Sedayu and Kretek increase when agricultural area was included in the models as covariates. While other areas did not change significantly.

d. When the agricultural livelihood is included in the model then area such as Imogiri and Jetis have the decreasing of number from 6-9 to 0-3 and from 3-6 to 0-3.5 respectively.
e. There is no significant change when two covariates are included in the model. All of the area has the same SMR when agricultural livelihood was considered as covariate.

6. Conclusions
By using INLA, the parameter effect of agricultural land and agricultural livelihood can be obtained. The effect of agricultural land is relatively bigger than agricultural livelihood. However, both of them have quite small effect to the spread of disease. They only give an effect to the spread of the disease less than 10%. INLA also can be used to explore posterior probability and standard morbidity rate in Bantul area. Some area has high risk of new cases of leptospirosis.

References
[1] Walker M, Wilcox B A and Wong M 2008 Waterborne zoonoses and changes in hydrologic response due to watershed development, Coastal Watershed Management
[2] Messier V, Levesque B, Proulx, J F, Rochette L, Serhir B, Couillard M, Ward B J, Libman M D, Dewailly E and Dery S 2012 Seroprevalence of Seven Zoonotic Infections in Nunavik,Quebec (Canada) Zoonoses and Public Health
[3] Kupek E, de Sousa Santos Faversani M C and de Souza Philippi J M 2000 The relationship between rainfall and human leptospirosis in Florianopolis, Brazil, 1991-1996, *The Brazilian journal of infectious diseases: an official publication of the Brazilian Society of Infectious Diseases*
[4] Ward M P 2002 Seasonality of canine leptospirosis in the United States and Canada and its association with rainfall, *Preventive Veterinary Medicine*
[5] Davis S, Calvet E and Leirs H 2005 Fluctuating rodent populations and risk to humans from rodent-borne zoonoses, *Vector-borne and Zoonotic Diseases* **5** pp 305-314
[6] Holt J, Davis S and Leirs H 2006 A model of Leptospirosis infection in an African rodent to determine risk to humans: Seasonal uctuations and the impact of rodent control, *Acta Tropica*
[7] World Health Organization 2010 Report of the First Meeting of the Leptospirosis Burden Epidemiology Reference Group
[8] Jansen A, Schneberg I, Frank C, Alpers K, Schneider T and Stark K 2005 Leptospirosis in Germany 19622003 *Emerg Infect Dis*. **11**(7) 10481054.
[9] Goris M, Boer K, Duarte T, Klien S, and Hartskeerl R, 2013 Human Leptospirosis Trends, the Netherlands, 19252008 *Emerging Infectious Disease* **19**(3)
[10] Wuthiekanun V, Chierakul W, Limmathurotsakul D, Smythe L D, Symonds M L, Dohnt M F, Slack A T, Limpaiboon R, Suputtamongkol Y, White N J, Day N P, Peacock S J 2007 Optimization of Culture of Leptospira from Humans with Leptospirosis, *J Clin Microbiol* **45**(4) 13631365
[11] Klutzfree 2012 Leptospirosis disease in the Philippines, http://news.wikinut.com/Leptospirosis-disease-in-the-Philippines/inju6u2f
[12] Cachay E R and Vinetz J M 2005 A global research agenda for leptospirosis *J Postgrad Med* **51** p 174
[13] Levett P N 2001 Leptospirosis *Clin Microbiol Rev* **14**(2) pp 296-326
[14] Blangiardo M, Camaletti M, Baio G and Rue H 2013 Spatial and spatiotemporal models with R-INLA *Spat Spatiotemporal Epidemiol* **4** pp 33-49
[15] Lawson B A 2009 Bayesian Disease mapping: Hierarchical Modelling in Spatial Epidemiology, *Chapman & Hall/CRC Interdisciplinary Statistics Series*
[16] Rue H, Martino S and Chopin N 2009 Approximate Bayesian inference for latent Gaussian models by using integrated nested Laplace approximations. *Journal of the Royal Statistical Society: Series B (Statistical Methodology)* **71** 319392
[17] Martins G T, Simpson D, Lindgren F, and Rue H, 2013 Bayesian computing with INLA: new features, arXiv:1210.0333 [stat.CO]

Journal of Physics: Conference Series **1417** (2019) 012027 doi:10.1088/1742-6596/1417/1/012027

The Global Kernel *k*-means Clustering Algorithm for Cerebral Infarction Classification

Z Rustam[1*], S G Fitri [1], R Selsi[1], J Pandelaki[2]

[1]Department of Mathematics, University of Indonesia, Kampus UI Depok, Depok 16424, Indonesia
[2]Department of Radiology, Cipto Mangunkusumo National General Hospital, DKI Jakarta 10430, Indonesia

e-mail: *rustam@ui.ac.id

Abstract. Cerebral infarction is the death of neurons, glia cells and blood vessel systems caused by a lack of oxygen and nutrients. This situation is often called storke. Common causes of neuron damage are hypoxia, which is caused by impaired blood flow, reduced oxygen pressure in blood circulation, toxins, and hypoglycemia which can result in the same morphological changes as morphological changes in hypoxia. Hypoxia is reduced oxygen pressure in the alveoli, resulting in hypoxemia which can cause hypoxic brain tissue. The initial stage of ischemic neurons is characterized by the formation of micro vacuolization, which is characterized by the size of the cells that are still normal or slightly reduced, the nucleus shrinks slightly, vacuoles occur in the perikaryon region. This micro vacuole can be found in neurons in hippokamus and cortical 5-15 minutes after hypoxia. The final sign of cell damage due to ischemia is characterized by the nucleus becoming pyknotic and fragmented. To classify cerebral infarction, the author uses the global k-means clustering algorithm as a classification method that shows that the method has good accuracy, good memory, and good precision in classifying cerebral infarction. In this proposed method, the global kernel k-means clustering algorithm is an extension of the standard k-means clustering algorithm and has been used to identify or classify clusters that are non-linearly separated in space input. This method adds one cluster at each stage through a global search process consisting of several k-means kernel executions from the appropriate initialization. Therefore, this method can make good classification accuracy. In particular, this achieves classification accuracy of up to 78% for the highest accuracy.

1. Introduction

Stroke is a cerebrovascular disease that is often found in developed countries, such as Indonesia [1]. Stroke is a serious type of disease. About 30% of stroke patients die within 3 months, and 20% of stroke patients are more severe, and 50% of stroke patients can restore their self-care abilities. Where factors that influence recovery depend on the severity of brain damage.

Stroke is caused by blockages (infarction) or rupture of blood vessels that carry oxygen and nutrients to the brain [2] so that the brain does not get the blood and oxygen needed and causes the death of brain cells [3]. Stroke is caused by several factors, that is hypertension, smoking, unhealthy diet, lack of physical activity, high blood pressure, increased blood sugar and increased blood lipid profile [4].

Hypertension is the most important risk factor [5], which can increase the risk of stroke by about two to four times [6]. Increased blood pressure will cause cerebral constriction. If blood pressure rises in a period of months or longer than years, it will be accounted for in the cerebral blood muscle layer so that the blood vessel's lumen diameter will remain. If that happens, it can cause the cerebral arteries to not constrict to overcome fluctuations in systemic blood pressure [6].

Stroke is classified into two types, namely hemorrhagic stroke and non-hemorrhagic (ischemic) stroke [7]. Hemorrhagic stroke is a stroke caused by weak blood exploding and bleeding to the brain and its surroundings which causes damage to brain cells so that it cannot work properly [8]. Whereas ischemic stroke is a stroke caused by thrombotic or thromboembolic blockages in the arteries [5]. This blockage of the arteries to the brain causes disruption of blood flow to the brain so that brain cells cannot make enough energy and will stop working [9].

Ischemic stroke is not only caused by clogged arteries to the brain, but can also be caused by the use of drugs, blood clotting disorders, or traumatic damage to neck veins [9]. The number of stroke patients is estimated to reach 85% of the number of strokes that occur [10]. Ischemic stroke is a major cause of disability in the world. In ischemic stroke, cerebral infarction is a more common condition [7]. Cerebral infarction is the death of neurons, blood cells and blood vessel systems caused by a lack of oxygen and nutrients [11].

Stroke affects each person differently, there are some people who need more time to recover [9]. Recovery from physical, social and emotional changes. Treatment for stroke usually starts with hospital treatment. Stroke rehabilitation requires good coordination between patients, families, doctors, nurses, physical therapists, psychologists, and others. Aspects of certain aspects of stroke rehabilitation have been well established in clinical practice and are a standard of care, for example providing physical therapy for patients with early stroke [9].

Treatment or treatment of stroke can also be done through antihypertensive therapy. Antihypertensive administration is given with consideration not only to the brain but also to other damage, such as the heart and kidneys. In addition, strokes need to be avoided, also need to modify the lifestyle of not smoking, not drinking alcohol and not using cocaine [6]. There are several studies that have addressed this resampling technique, including Z Rustam et al., Which classifies unbalanced datasets to predict cerebral infarction. The modeling technique used is Hybrid Preprocessing Method for Support Vector Machine[7].

In this study, we used the Global Kernel k-means Clustering Algorithm to classify datasets of cerebral infarction in the brain using data obtained from the Radiology Laboratory at Cipto Mangunkusumo Hospital, Indonesia. In this proposed method, the global k-means clustering algorithm is an extension of the standard k-means clustering algorithm and has been used to identify or classify clusters that are non-linearly separated in space input. This method adds one cluster at each stage through a global search process consisting of several k-means kernel executions from the appropriate initialization. By using the k-means kernel, the algorithm works in stages by solving all problems $1, \ldots, M$ clusters. This algorithm combines the global advantages of k-means and the k-means kernel so as to avoid the limitations of both k-means [12].

2. Method

2.1 Clustering

Clustering aims to group similar patterns such as features or data items into certain categories and specifically for classification. Many grouping algorithms have been designed to meet needs, such as efficiency and accuracy, each of which has advantages and disadvantages[13].

2.2 Kernel K-Means Clustering

To group data that is not liner inseparable in the original feature space, the k-means grouping is extended to the kernel version. However, the workings of the k-means kernel depend on the choice of kernel functions, so many kernel learning has been introduced to k-means clustering to get the optimal kernel combination[14].

Some basic structures are taken from the existing k-means grouping approach. The resulting kernel k-means clustering algorithm is[13]:

- Initialization: K is chosen randomly in the dataset as the initial centroid c_p, $1 \leq p \leq K$

- Clustering: Similarity $Sim\ (r_i, c_p)$ between each record and centroid is calculated, and record r_i record is assigned to the c_p centroid which is similar.
- Centroid re-calculation: c_p the new centroid is recalculated for each cluster with the average embedding vector of all r_i records assigned to the cluster, namely:
-

$$c'_p = \frac{1}{nc_p}\sum_{i=1}^{nc_p} r_i \tag{1}$$

Where, nc_p is the number of records in the cluster indicated by the centroid c_p

- Termination conditions: the matrix for all old centroids is denoted by

$\Theta = \{c_1, \dots, c_p, \dots, c_K\}^T, 1 \le p \le K$ and the matrix for all new centroids is denoted by $\Theta' = \{c'_1, \dots, c'_p, \dots, c'_K\}^T$. Next, Euclidean distance $d(\Theta, \Theta')$:

$$d(\Theta, \Theta') = \sqrt[2]{\sum_{p=1}^{k} \left(c_p - c'_p\right)^2} = \sqrt[2]{\sum_{p=1}^{k} \sum_{i=1}^{\lambda} \left(v_{p,i} - v'_{p,i}\right)^2} \tag{2}$$

Where $v_{p,i}$and $v'_{p,i}$ with i are daro centroid elements c_p and c'_p.

2.3 K-Means and Global K-Means

K-means is one of the most popular grouping algorithms.

Suppose we are given $X = \{x_1, x_2, \dots, x_N\}, x_N \in R^d$, The dataset will be divided into separate M, namely $C_1, C_2, \dots, C_M$. The algorithm finds the optimal local solution with respect to grouping errors as the sum of the square of the Euclidean distance between each data. Analytically the grouping errors are stated by[15]:

$$E(m_1, \dots, m_M) = \sum_{i=1}^{N} \sum_{k=1}^{M} I(x_i \in C_k)\|x_i - m_K\|^2 \tag{3}$$

The disadvantages of the k-means algorithm are:

- The final solution depends on the starting position of the cluster center
- Clusters are separated linearly

To overcome this problem, we use the global k-means algorithm[15]. Global k-means is a data grouping approach that dynamically adds one cluster center at once by using a deterministic global from the initial position consisting of N data sizes from the k-means algorithm[16].

2.4 Kernel K-Means

The kernel k-means is a generalization of the k-means standard algorithm, where data points are mapped from space input to higher dimensional feature space through non-linear transformation $\emptyset$. In the k-means kernel, data is expected to separate well because overlapping data can be linear in new dimensional space [15] (see table 1).

Table 1. Examples of kernel functions

Polynomial kernel	$K(x_i, x_j) = (x_i^T x_j + \gamma)^\delta$
Gaussian kernel	$K(x_i, x_j) = exp\left(\frac{-\|x_i - x_j\|^2}{2\sigma^2}\right)$
Sigmoid kernel	$K(x_i, x_j) = tanh(\gamma(x_i^T x_j) + \theta)$

The objective function to be reduced by the kernel k-means is equivalent to the grouping error in the feature, which is expressed by[15]:

$$E(m_1, \ldots, m_M) = \sum_{i=1}^{N} \sum_{k=1}^{M} I(x_i \in C_k)\|\emptyset(x_i) - m_k\|^2 \tag{4}$$

Where, $m_k = \frac{\sum_{i=1}^{N} I(x_j \in C_k)\emptyset(x_i)}{\sum_{i=1}^{N} I(x_j \in C_k)}$

$$\|\emptyset(x_i) - m_k\|^2 = K_{ii} - \frac{2\sum_{j=1}^{N} I(x_i \in C_k)K_{ij}}{\sum_{j=1}^{N} I(x_i \in C_k)} + \frac{\sum_{j=1}^{N} \sum_{l=1}^{N} I(x_j \in C_k)I(x_l \in C_k)K_{jl}}{\sum_{j=1}^{N} \sum_{l=1}^{N} I(x_j \in C_k)I(x_l \in C_k)} \tag{5}$$

In this study, we use the Radial Basis Function Kernel (RBF Kernel). Conventional RBF networks employ a number of kernels such as multi-quadrics, inverse multi-quadrics, and Gaussian, with the following formula [18]:

$$K(x_i, x_j) = exp\left(\frac{-\|x_i - x_j\|^2}{2\sigma^2}\right) \tag{6}$$

2.5 Global Kernel K-Means

The global kernel k-means is a combination of global k-means and kernel k-means.
The global kernel k-means minimizes grouping errors in the feature space. The global kernel k-means map datasets from the input space to the higher dimensional feature space using the kernel matrix. Global kernel k-means produces optimal solutions in clustering problems[17].

3. Experiment

The following data are data from ischemic stroke patients with cerebral infarction in their brains. Data were taken from Cipto Mangunkusumo Hospital, Indonesia can be used to build the model of The Global Kernel K-Means Clustering (See Table 2).

Table 2. Cerebral Infarction Dataset

Area	Min	Max	Average	SD	Sum	Length
0.2	-3	38	16.88	9.3	5166	2
0.1	15	44	30.64	7.37	7722	0
0.2	-5	51	19.19	12.44	4797	1.9
0.1,	18	51	32.29	7.84	8136	1.8
0.1	-14	26	8.67	10.25	824	1.2
0.1	25	61	38.99	7.37	6122	1.5
...	...	...	...	...	...	...
0.1	21	51	34.67	6.37	3432	1.6
0	7	33	16.69	8.4	217	0.7
0.1	25	59	44.73	6.16	3847	1.5

Where there are 156 data with 7 features proportional to 70% as training data and 30% testing data from the original data, with the actual number 103 main data showing data classes with no infarction and 53 minor data indicating infarction. Table 4 shows an explanation of the infarction data features examined.

Table 3. The Feature of Cerebral Infarction Dataset

No	Feature	Definition of feature
1	Area	The size of the area from the infarction point
2	Min	The minimum value of infarction
3	Max	The maximum value of infarction
4	Average	The average value of infarction
5	SD	Standard error value of infarction
6	Sum	The sum value of infarction point
7	Length	Length of infarction point

4. Result and Discussion

By using the Global Kernel K-Means Clustering Algorithm, it can be obtained as follows, where k-fold divides the sample data into k parts. To get the result, we use a Radial Basis Function Kernel (RBF Kernel).

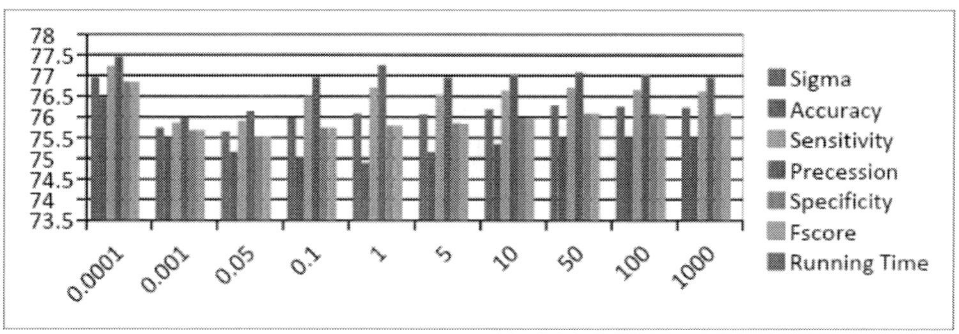

Figure 1. For k-fold= 3

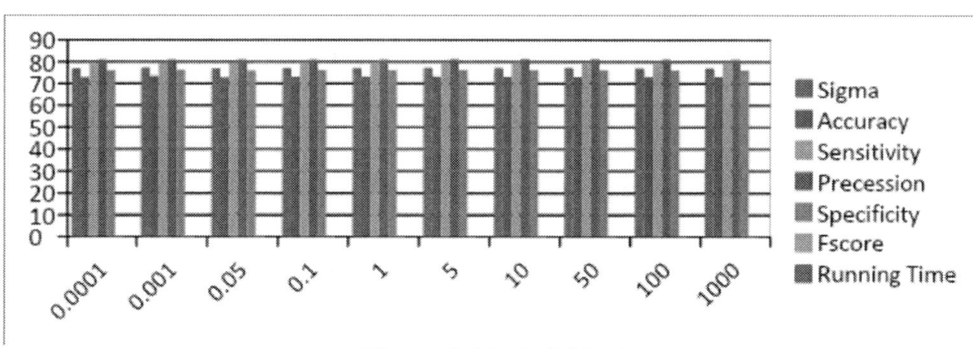

Figure 2. For k-fold= 5

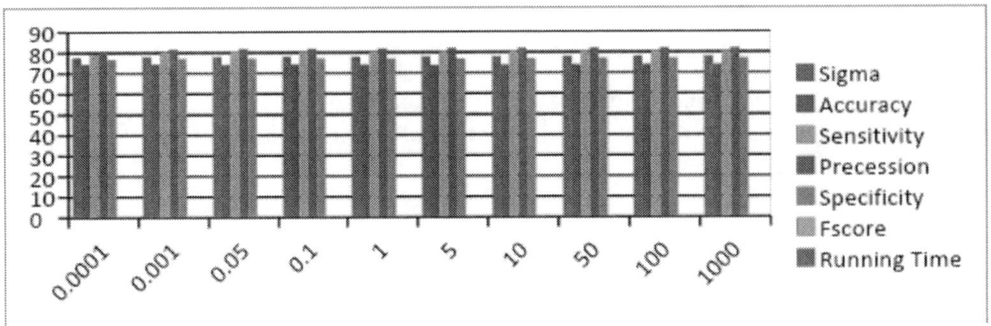

Figure 3. For k-fold= 7

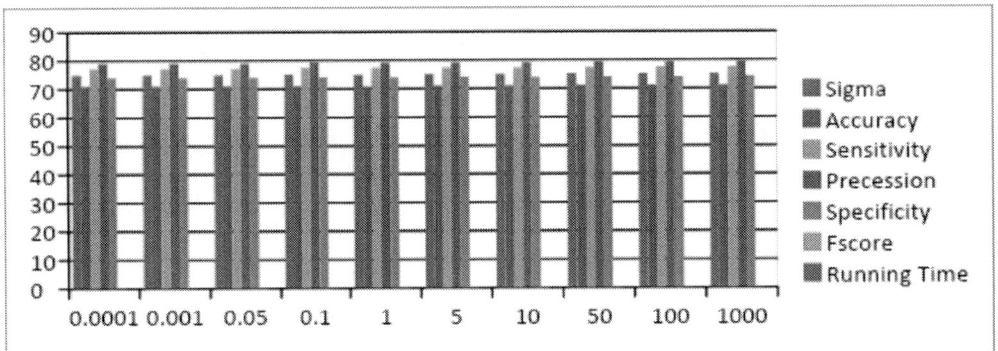

Figure 4. For k-fold= 10

From the chart, we can see that for k-fold=3, the best accuracy is 76.96% while the sigma is 0.0001. For k-fold= 5, the best accuracy is 77.28% while the sigma is 10. For k-fold= 7, the best accuracy is 78.06% while the sigma is 0.001. And last, for k-fold= 10, the best accuracy is 75.25% while the sigma is 1000. So, for all of the k-fold, k-fold=7 with sigma 0.001 gives the best accuracy value, which is 78.06%. While the sensitivity value is 74%, the precession value is 80%, specificity value is 81% and FScore 77%.

5. Conclusion

From the experiments conducted in section 4, we can see that the accuracy achieved by the K-Means Clustering Algorithm Global Kernel method is 78.06%. This method provides high accuracy to classify whether there is an infarction in a person's brain or not. The disadvantage of this method is that it does not produce high accuracy values in terms of classification so that further research is needed by using new methods or with other kernel functions so that it will produce better accuracy.

Acknowledgment

This research is supported financially by the University of Indonesia, with DRPM PIT-9 2019 research grant scheme, ID number NKB-0039/UN2.R3.1/HKP.05.00/2018

References

[1] Darotin R, Nurdiana, and Nasution TH. 2017 Analysis of Predictive Factors of Mortality in Hemorrhagic Stroke Patients at Soebadi Hospital Jember *NurseLine Journal* **2** (2)

[2] Bay V, Kjolby B F, and Iversen N K 2018 Stroke Infarct Volume Estimation in Fixed Tissue: Comparison of Diffusion Kurtosis Imaging to Diffusion-Weighted Imaging and Histology in a Rodent MCAO Model *PLoS ONE* **13** (4) e0196161

[3] Wang G, Jing J, and Pan Y 2019 Does All Single Infarction have Lower Risk of Stroke Recurrence Than Multiple Infarctions in Minor Stroke? *BMC Neurology* **19** (7)

[4] Darotin R, Nurdiana, and TH Nasution 2017 Analysis of Predictive Factors of Mortality in Hemorrhagic Stroke Patients at Soebadi Hospital Jember *NurseLine Journal* **2** (2)
[5] Hanum P, Lubis R, and Rasmalian - *Hubungan Karakteristik dan Dukungan Keluarga Lansia dengan Kejadian Stroke pada Lansia Hipertensi di Rumah Sakit Umum Pusat Haji Adam Malik Medan* (Medan: RSU Pusat Haji Adam Malik)
[6] Qurbany Z T and Wibowo A - *Stroke Hemoragik e.c Hipertensi Grade* II Indonesia
[7] Rustam Z, Utami D A, Pandelaki J, and Nugroho WA - *Hybrid Preprocessing Method for Support Vector Machine for Classification of Imbalance Cerebral Infarction Dataset* Indonesia
[8] Wu B 2017 *What's to Know About Hemorrhagic Stroke?* Available at https://www.medicalnewstoday.com/articles/317111.php
[9] The Internet Stroke Center 2019 *An Independent Web Resource for Information About Stroke Care and Research* Available at http://www.strokecenter.org/patients/about-stroke/ischemic-stroke/
[10] Hadayani D and Dominica D 2018 Gambaran Drug Related Problems (DRP'S) pada Penatalaksanaan Pasien Stroke Hemoragik dan Stroke Non-Hemoragik di RSUD DR M Yunus Bengkulu. *Jurnal Farmasi dan Ilmu Kefarmasian Indonesia* **5** (1)
[11] Japardi I 2002 *Neuropatologi Infark Serebi* Indonesia
[12] Tzortzis G F and Aristidis C 2019 The Global Kerel K-means Algorithm for Clustering in Feature Space *IEEE Transactions on Neural Networks* **20** (7)
[13] Zhang Y, Lu J, Liu F, Liu Q, Porter A, Chen H, Zhang G 2018 Does Deep Learning Help Topic Extraction? A Kernel K-Means Clustering Method With Word Embedding *Journal of Informetrics*
[14] Yao Y, Chen H 2018 *Multiple Kernel K-Means Clustering bt Selecting Representative Kernels* Available at https://arxiv.org/abs/1811.00264
[15] Tzortzis G and Likas A 2008 *The Global Kernel K-Means Clustering Algorithm* Greece
[16] Agrawal A and Gupta H 2013 Global K-Means (GKM) Clustering Algorithm: A Survey *International Journal of Computer Applications* **79** (2)
[17] Tzortzis G F and Likas A C 2009 The Global Kernel K-Means Algorithm for Clustering *Feature Space* **20** (7)
[18] Khan S, Naseem I, Togneri R, and Bennamoun M 2019 A Novel Adaptive Kernel for the RBF *Neural Networks*

Testing of Image Resolution Enhancement Techniques Using Bi-cubic Spatial Domain Interpolation

M A Bustomi

Physics Departments, Institut Teknologi Sepuluh Nopember (ITS), Surabaya 60111, Indonesia

Email: ariefbustomi@gmail.com

Abstract. The technique for increasing digital image resolution from low-resolution image to high-resolution image based on digital image processing is called the super-resolution technique. In this paper, a super-resolution technique is presented using a two-dimensional bi-cubic interpolation method in the spatial domain. The order of the super resolution method applied is as follows: (1) selecting ten images as samples, (2) decrease the sample image resolution to one-fourth of the original resolution by deleting three quarters of the pixel number, (3) increasing the image resolution of a quarter of the part becomes like the initial resolution using bi-cubic interpolation for three quarters of the additional new pixels, (4) testing this bi-cubic interpolated image with the same pixel-sized initial image, (5) using parameters: average value, minimum value, maximum value and standard deviation value as a comparison parameter between bi-cubic interpolated images and the same pixel-sized initial image. The results obtained from the super-resolution technique using spatial bi-cubic interpolation are: (1) The average error value of the bi-cubic interpolation method in image objects in this study is between 4% to 10% or still quite low, (2) Bi-cubic interpolation methods can work well on square pixel-sized images (m = n) compared to non-square pixel-sized images, (3) Bi-cubic interpolation turns out to produce an array of image pixel values that mirror symmetry against the main diagonal lines of the image before being interpolated.

1. Introduction

The application of digital image processing has been very broad and covers various fields of human life. Ranging from the fields of photography, the field of security, the medical field to the field of space [1 - 3]. One of the things that are needed for example in the field of security and space is how to get high-resolution digital images from low-resolution images. Low-resolution images are obtained due to limited capacity and camera facilities. Meanwhile for analysis needs, for example by police and space scientists, high-resolution images are needed. The technique for obtaining high-resolution images from low-resolution images is called the super-resolution technique. Super-resolution technique is a technique to get image resolution by processing it on the image itself. Super-resolution techniques can be a restoration, reconstruction, or continuous image recovery process from samples of images [4].

High-resolution images are needed for various image-based fields of analysis. High-resolution images contain more information contained in them. In the process of increasing image resolution, many methods can be used. Some methods that can be used include the nearest neighbour, bi-cubic interpolation, cubic spline interpolation, maximum loneliness, maximum posteriori. The super-resolution method developed by the researchers mostly uses quite complex methods, such as the use of wavelets [5], simultaneous registration and reconstruction [6], frame fusion techniques [7], non-linear mapping [8], pixel weighting [9] and others.

In this study, the super-resolution method used is quite simple, because it is only based on the results of interpolating pixel values for new pixels resulting from the addition of pixels to increase image resolution. With this fairly simple method, it is hoped that super-resolution images will be obtained with quality that is close enough to the quality of super-resolution images with these complex methods. In this paper, we use the method of increasing image resolution in the spatial domain using bi-cubic interpolation. The bi-cubic interpolation used is two-dimensional bi-cubic interpolation for the entire pixel image whose resolution will be increased.

2. Method

2.1. Image Object

In this study, ten color image objects (RGB images) with various sizes of pixels are used. The ten color images are shown in Figure 1 below:

Figure 1. Research object image: a. Yellow.jpg; b. Red.jpg; c. Laut.jpg; d. Hijau.jpg; e. Pink.jpg; f. Orange.jpg; g. Baby.jpg; h. Imp.jpg; i. Daun.jgp; j. Kapal.jpg

The size of the ten-pixel numbers of the RGB images is presented in Table 1. In Table 1 shows that of the ten images, there are a number of pixel-sized image of the 5 squares (m = n), ie images: Hijau.jpg, Orange.jpg, Baby.jpg, Imp.jpg, Kapal.jpg and 5 other images that are not square in size. This non-square image is divided into an image with the size of m (number of rows of pixels) smaller than n (number of pixel columns), ie images: Yellow.jpg, Red.jpg, Pink.jpg, Daun.jgp and image with size m is greater than n, ie image: Laut.jpg.

Table 1. Size of the pixel number of the research image

Image Name	Size
Yellow.jpg	225, 255, 3
Red,jpg	255, 300, 3
Laut.jpg	300, 255, 3
Hijau.jpg	300, 300, 3
Pink.jpg	150, 200, 3
Orange.jpg	525, 525, 3
Baby.jpg	200, 200, 3
Imp.jpg	255, 255, 3
Daun.jpg	525, 700, 3
Kapal.jpg	400, 400, 3

2.2. Research Flow Chart

The order of work steps in this study is presented in the form of a flow diagram in Figure 1. Broadly speaking, the research flow diagram consists of 4 main steps, ie: forming an original matrix of image A, forming a matrix reduction B, forming a matrix resulting from interpolation of bi-cubic C, and form an accuracy matrix D.

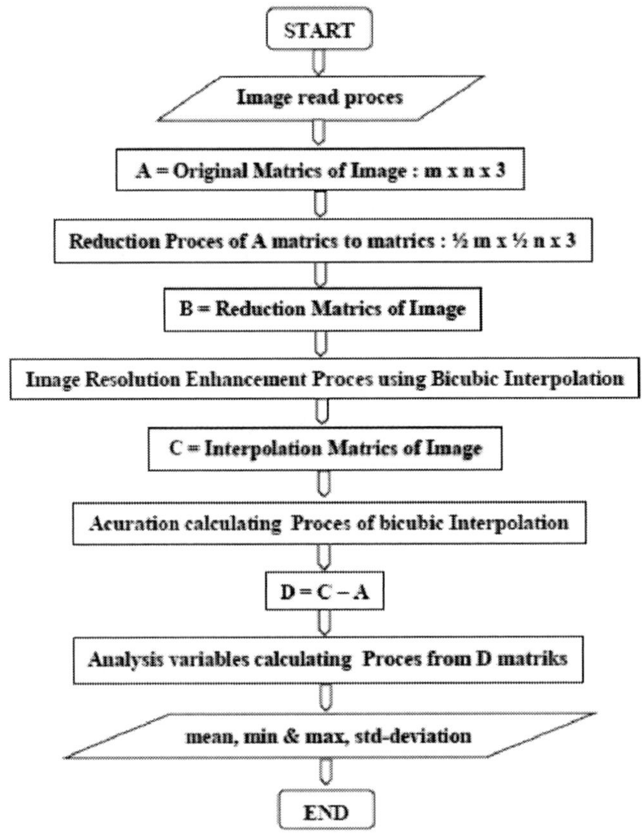

Figure 2. Research flow chart

The original matrix of image A is a matrix obtained from the direct reading of image pixels. Reduction matrix B is matrix A matrix reduction which measures the number of pixels m x n x 3 into a B matrix measuring ½m x ½n x 3. The B matrix formation process is done by removing even-numbered rows and even-numbered columns on matrix A as shown in Figure 3. The interpolation matrix C is a pixel-sized matrix of m x n x 3 obtained from the results of two-dimensional bi-cubic interpolation on matrix B. The process of forming C matrix is done by adding a series of pixels placed as even numbers and columns new rows the new even-numbered column in matrix B as shown in Figure 3.

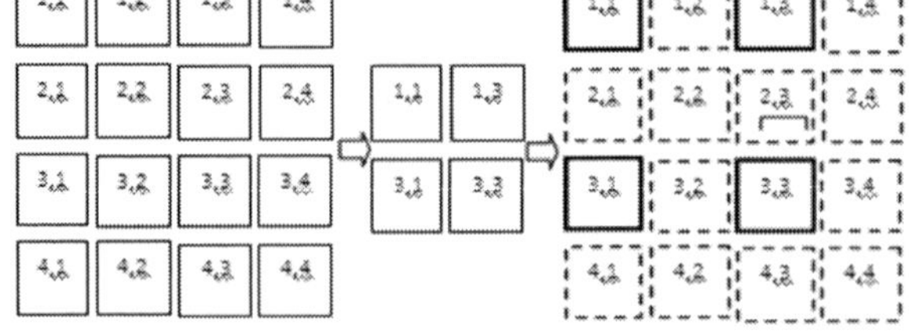

Figure 3. Schema of the reduction and interpolation process

D accuracy matrix is a matrix obtained from the difference between C interpolation matrices to the original matrix A. The D matrix states the difference in the values of each pixel from bi-cubic interpolation to the original matrix for all pixels in the image which is m x n x 3. From the values on the element, This D matrix element can then be calculated statistically the average value, minimum value, maximum value, and standard deviation value as a comparison parameter between the bi-cubic interpolated image and the original image.

2.3. Bi-cubic Interpolation Method

Interpolation (resampling) is a method to get the value of the data you want to search for and not contained in the observation data from several data that have been obtained from observations. There are various interpolation methods, such as linear interpolation (first degree), squared interpolation (second degree), cubic interpolation (three degrees), and others. Bi-cubic interpolation is a cubic interpolation (three degrees) to interpolate data in two dimensions.

In digital image processing, interpolation can be applied to obtain images with a higher resolution in the original image. Bi-cubic interpolation results in finer image enlargement at the edges rather than biliary interpolation (two-dimensional linear interpolation). Bi-cubic interpolation uses 4x4 neighbouring pixels as a data source forming interpolation. The bi-cubic interpolation function P (x, y) can be stated as follows [7]:

$$P(x,y)=\frac{1}{16}\sum_{i=0}^{3}\sum_{j=0}^{3}a_{ij}u(s_x)u(s_y) \tag{1}$$

Where u(s) is a one-dimensional cubic interpolation function with the following equation:

$$u(s)=\begin{cases}\frac{3}{2}|s|^3-\frac{5}{2}|s|^2+1 & for\,0\le|s|\le1\\ -\frac{1}{2}|s|^3+\frac{5}{2}|s|^2-4|s|+2 & for\,1\le|s|<2\\ 0 & for\;\;2<|s|\end{cases} \tag{2}$$

Wheres is the distance between pixels that are interpolated with neighbouring pixels.

3. Results and Discussion

3.1. Research Results

The fourth value of the analysis variable, namely the average value (μ), the minimum value (Min), the maximum value (Max) and the standard deviation value (σ) for each image are shown in Table 2.

Table 2. The fourth value of the analysis variables for each image

Image Name	μ	Min	Max	σ
Yellow.jpg	10,9773	0	221	24,3051
Red,jpg	11,3689	0	195	26,4784
Laut.jpg	13,0656	0	114	26,8827
Hijau.jpg	16,6118	0	234	31,0509
Pink.jpg	15,7511	0	232	35,0783
Orange.jpg	17,7367	0	205	35,0820
Baby.jpg	18,7511	0	217	36,1780
Imp.jpg	16,1993	0	227	41,9467
Daun.jpg	19,0165	0	251	43,1915
Kapal.jpg	25,2071	0	231	45,6477

3.2 Analysis of Variable Accuracy

3.2.1 Average value

In this study, the mean is the average value of the difference in the value of the results of the interpolation of the original value (D matrix element) for the entire pixel that is raised in the process of increasing image resolution.

$$\mu = \frac{\sum_{i=even}^{m} \sum_{j=even}^{n} x_{ij}}{\frac{3}{4} m.n} \tag{3}$$

By:

μ = mean or average value for each RGB of the matrix D
xij = matrix D element i-row j-column for each RGB matrix
= difference value of the interpolation value to the original value for the pixel in i-row and j-column
m = number of pixel lines from the image
n = number of pixel columns of the image

The calculation of the average value of the difference between the results of the interpolation of the original is also done for each image D matrix, namely the D matrix for red (μ_R), green (μ_G) and blue (μ_B). So that the average value for all images is:

$$\mu = \frac{\mu_R + \mu_G + \mu_B}{3} \tag{4}$$

The average value variable (μ) states the average of the difference in the value of the results interpolated with the original value. The small μ value indicates that the interpolation results are good enough. Conversely, the large μ value states that the results of the interpolation are not good enough. Based on the data in Table 2, the average value of the ten image objects ranges from 10 to 25. If the average value of this error is compared with a maximum pixel value of 255, then the value of this error is at a percentage of 4% to 10%. This means that the error value of the bi-cubic interpolation method is still quite low. Average error value is indeed quite low, but this variable is not the only variable for assessing image quality. The quality of the interpolated images can be evaluated directly visually.

3.2.2 Minimum and Maximum Value

In this study, the minimum value is the smallest value of the difference in the value of the results of interpolation to the original value for the entire pixel that is raised in the process of increasing image resolution. Conversely, the maximum value is the greatest value of the difference in the value of the interpolation of the original value for the entire pixel that is raised in the process of increasing image resolution.

The minimum value (Min) and maximum (Max) is the smallest value and the greatest of the difference in the value of the results of the interpolation of the original value. These two variables only provide information about two extreme values, the difference between the results of the interpolation of the original value. Ideally, both of these variables must be zero, meaning that bi-cubic interpolation gives perfect results. Minimum and maximum values that are not ideal do not mean that the results of bi-cubic interpolation are not good. This is because the minimum and maximum variables are only auxiliary variables, not the main variables. The data in Table 2 shows that the minimum value reaches the ideal value (i.e. zero) for all images. Meanwhile, the maximum value varies from the greatest value close to the peak value (ie 255) in Leaf images, jpg to the smallest approaching the half peak value (127) in the Sea image, jpg. This means that an error in bi-cubic interpolation can change a pixel that should be black (pixel value 0) to white (pixel value 255).

3.2.3 Standard Deviation

Standard deviation or standard deviation is one of the statistical techniques used to explain the homogeneity of data groups. Standard deviation is also defined as the variation in the data distribution. The smaller the distribution value means the data is getting the same. Conversely, the greater the distribution value, the more varied the data. The standard deviation is calculated as the root of the average square of the difference in the data to the mean.

$$\sigma = \left(\frac{4}{3m.n} \sum_{i=even}^{m} \sum_{j=even}^{n} (x_{ij} - \mu)^2 \right)^{1/2} \quad (5)$$

By:

σ = standard deviation for each RGB matrix
m = number of pixel lines from the image
n = number of pixel columns of the image
xij = matrix D element i-row j-column for each RGB matrix
μ = mean or average value for each RGB matrix

The overall standard image deviation in this study is expressed as the average value of the standard deviation for red (σR), green (σG) dasn blue (σB). So that the average value for all images is:

$$\sigma = \frac{\sigma_R + \sigma_G + \sigma_B}{3} \quad (6)$$

Standard deviation or standard deviation states the homogeneity of the data group. The smaller the standard deviation value, the more homogeneous the data group is. The data in Table 2 shows that the standard deviation value varies from 24 to 36. The results were pretty good interpolation will generate a standard deviation value which is quite small (homogeneous). Just like the minimum and maximum value variables, the standard deviation value variable is also not the main variable. This variable is an auxiliary variable for the main variable average value which describes how much the difference in the accuracy of the interpolated value of the original value of each pixel in the image.

3.3. Discussion

In contrast to the super-resolution method that has been developed by a number of researchers which tend to be quite complex [5-9], the super-resolution method used in this study is quite simple. This method is only based on the results of interpolating pixel values for new pixels resulting from the addition of pixels to increase image resolution.

The quality of interpolated images can be evaluated directly visually. An example is in Figure 4, two sample images are shown for different cases. The first image is the Yellow.jpg image which is a non-square image and the second image is the image of Daun.jpg which is square. In Figure 4, you can see the results of bi-cubic interpolation in Yellow.jpg images that are not square images resulting in dark areas (black) on the right side of the image. Whereas the results of bi-cubic interpolation in the Daun.jpg image which is square have no dark areas (black) in the interpolated image. This means that the bi-cubic interpolation method can work well on square pixel-sized images, compared to not square pixel-sized images.

(a) (b)

Figure 4. Image reduction (top) and bi-cubic (bottom) interpolation at (a) non-square image and (b) square image

In Figure 4 it also shows that the bi-cubic interpolated images, namely Yellow.jpg and Daun.jpg imagery turned out to be mirror symmetry images of the main diagonal lines of the image before being interpolated (i.e. reduction images). This kind of thing also occurs in all the images that are the object of this research, both square and non-square.

An important result obtained from the application of the bicubic interpolation method in increasing image resolution in this study is the relationship between the quality of the interpolation results and the pixel size of the image before interpolation. Most likely the cause of the bi-cubic interpolation method that only gives good results on square pixel-sized images is in the form of bi-cubic function formulas that are equivalent between the x and y variables. In this regard, there needs to be further research to improve and modify the shape of the bi-cubic function formula that no longer should be equal between the variables x and y variables. Modification of bi-cubic function formula form which is not equal between variable x and variable y is expected to be able to work well on square and non-square images.

4. Conclusion

The average error value of the bi-cubic interpolation method in the image objects in this study is between 4% to 10% or still quite low. Although the average error value is quite low, this variable is not the only variable to assess image quality. The quality of interpolated images can be evaluated directly visually and the results are obtained that the bi-cubic interpolation method can work well on square pixel-sized images ($m = n$) compared to non-square pixel-sized images. Bi-cubic interpolation turns out to produce an array of image pixel values that mirror symmetry against the main diagonal lines of the image before being interpolated.

References

[1] Gonzales R C and Woods R E 2002 *Digital Images Processing* Second edition (New Jersey: Prentice-Hall)

[2] Gonzales R C, Wood R E and Eddins S L 2009 *Digital Image Processing using Matlab* Second Edition (New Jersey: Gates Mark Publishing)

[3] Arioz U 2014 *Introduction to Digital Image Processing* (Turkish: Department of Computer Engineering - Gazi Universitesi)

[4] Thaiyalnayaki K, Karim S A and Parmar V *2010 Finger Print Recognition using Discrete Wavelet Transform* (Sriperumbudur: Electronics and Communication Engineering - Sri Venkateswara College of Engineering)

[5] Cheng H and Liu J 2008 Super-Resolution Image Reconstruction Based on MWSVR Estimation *Proceedings of the 7th World Congress on Intelligent Control and Automation June 25 - 27, 2008*, Chongqing, China

[6] Zhi Y, Peimin Y, Sheng L, Juxia S, Yuhui H 2005 Super Resolution Based on Simultaneous

Registration and Reconstruction *10th Intl. Conf. on Control, Automation, Robotics and Vision 17–20 December 2008*, Hanoi, Vietnam

[7] Bhushan D B, Sowmya V, Soman K P 2010 Super Resolution Blind Reconstruction of Low-Resolution Images Using Framelets Based Fusion *2010 International Conference on Recent Trends in Information, Telecommunication and Computing*

[8] Zeng X and Huang H 2012 *IEEE Signal Process. Lett.* **19** 195

[9] Hamada K, Nakashima R, Takahashi K and Naemura T 2013 Super-resolution with Adaptive Pixel Weighting Scheme and Its Application to Super-resolved Free-viewpoint Image Synthesis *2013 International Conference on Signal-Image Technology & Internet-Based Systems*

Revision Notes :

1. Write more about research gap, state of the art, and the comparison of another related research

Has been revised :

- Research gap : existing super-resolution methods are complex enough so that it needs to develop simpler methods.
- State of the art : the super-resolution method with the bi-cubic interpolation method is quite simple
- The comparison of other related research : the super-resolution method used in this study is a fairly simple process when compared with the methods used by other researchers.

2. The resolution is too low, so it is unreadable flow chart

Has been revised : Flowcharts are made with a higher resolution so that they can be read

3. Typo : bikubik

Journal of Physics: Conference Series **1417** (2019) 012028 doi:10.1088/1742-6596/1417/1/012028

Has been revised : Bikubik becomes bi-cubic

4. Typo : neighboring
Has been revised : neighboring becomes neighbouring

5. What evaluation measure that you used in this study, how you counting the accuracy of the image?
Has been revised : A description of the accuracy variable was added before the discussion

6. Explaining the differences of image resize with this result? The result is very bad, please compare with another methods, and make a synthesis from your result
Has been revised : It has been added to the paper that the probable cause is the bi-cubic function used in the study is equivalent between the x variable and the y variable. There is no research on this and we will study it in the next research.

7. Write without bullet and numbering
Has been revised : It has been written without bullet and numbering

8. How you gte thins number of error?, Your said that it is quite low, but did you know that the result of image isn't good enough
Has been revised : It has been added to the paper that the average error value is indeed quite low, but this variable is not the only variable for assessing image quality. The quality of interpolated images can be evaluated directly visually and has been written in the second and third points of the conclusion

9. The references are too few for international publication, please adding more up to date primary references
Has been revised : References from the latest international papers have been added

Estimation of Super Pairwise Alignment (SPA) parameters on zika virus mutation using Artificial Bee Colony

D Rahmalia[1], A Rohmatullah[2], M S Pradana[3]

[1,2,3]Department of Mathematics, Universitas Islam Darul Ulum Lamongan

Email : [1]dinitarahmalia@gmail.com

Abstract. There are many diseases caused by viruses or bacteria. The structure of virus and bacteria always mutate to create new structures. Sequence alignment is important so that it can be used to research genetic diseases and epidemics. In this reseach, we take case study of zika virus. To see the similarity between original virus and the mutation virus, it is required the alignment process of two virus sequences. The method used for aligning two virus sequences is Super Pairwise Alignment (SPA). Due to the value of objective function depends on SPA parameters, in this research we will apply heuristic method, such as Artificial Bee Colony (ABC) algorithm to optimize SPA parameters minimizing penalty value or maximizing similarity value as objective function. ABC is the optimization method which is inspired by the behaviour of bee colony in which the advantages are there are three types of bees that will update optimal solution in approaching. From the ABC simulations, we can obtain optimal SPA parameters resulting minimum penalty value or maximum similarity value between two aligned zika virus protein sequences in approaching.

1. Introduction

In this time, there are many diseases caused by viruses or bacteria. Based on many researches, the existing of virus and bacteria are from virus and bacteria in the long time ago. They change their structures through mutation process. Mutations can lead to the growth and death of cells, and may lead to disease. Sequence alignment is important so that it can be used to research genetic diseases and epidemics. For example, it is possible to determine the origin, variance, and development of epidemics, and then finding the viruses and bacteria.

In this research, we take case study of zika virus. Zika virus can cause zika disease through *Aedes* mosquito bites with the symptons like dengue fever [1]. Based on previous research, the origin of zika virus is from South Africa and the virus mutate (change genetic sequence) and spread to other countries including Indonesia [2]. To see the similarity between original virus and the mutation virus, it is required the alignment process of two virus sequences. The method used for aligning two virus sequences is Super Pairwise Alignment (SPA) [3].

In sequence alignment using SPA, SPA parameters selection is applied by trial and error [4]. Due to the value of objective function depends on SPA parameters, in this research we will apply heuristic method, such as Artificial Bee Colony (ABC) algorithm to estimate and optimize SPA parameters minimizing penalty value or maximizing similarity value as objective function. In previous research, parameter estimations using heuristic method have been applied in optimization problem [4] and control problem [5],[6].

The Artificial Bee Colony (ABC) was introduced by Karaboga and Akay in 2009. ABC is the optimization method which is inspired by the behavior of bee colony. In the ABC algorithm, there are three types of bees, namely employed bees, onlooker bees, and scout bees in which the advantages are there are three types of bees that will update optimal solution in approaching [7],[8].

Simulations have been applied and we can obtain optimal SPA parameters resulting minimum penalty value or maximum similarity value between two aligned zika virus protein sequences in different countries in approaching.

2. Super Pairwise Algorithm (SPA)

Super Pairwise Alignment (SPA) is the method used for aligning two sequences and computing their penalty or similarity as objective function affected by parameters of SPA (n,θ,ϕ)

Let (A,B) are two sequences with sequence B is the mutation from sequence A. The parameters of SPA are (n,θ,ϕ) with $n>0$, $0.3\le\theta\le 0.4$, and $\theta<\phi\le 1$ [9].

1. Estimate the first mutation position $\hat{i}_1$ in T

   Initialize $i=j=0$ and calculate penalty value $w(A,B,i,j,n)$.

   a. If $(w(A,B,i,j,n)\ge\phi)$ then $\hat{i}_1=0$

   b. Otherwise, $i=j=k_1(n-\tau)$ with k_1 is the integer. Calculate $w(A,B,i,j,n)$. If $(w(A,B,i,j,n)<\phi)$, repeat the step 1(b) by updating $k\leftarrow k+1$ until $(w(A,B,i,j,n)\ge\phi)$, and keep the value of i and j as $\hat{i}_1$

2. Estimate $\hat{l}_1$ based on the estimation $\hat{i}_1$ of the first mutation position in T.

   $w(A,B,\hat{i}_1+l,\hat{i}_1,n)$, $w(A,B,\hat{i}_1,\hat{i}_1+l,n)$, $l=1,2,3,\ldots,l_{max}$

   a. If $(w(A,B,\hat{i}_1+l,\hat{i}_1,n)\le\theta)$, then $\hat{l}_1=-l$ and insert l virtual symbols (gaps) into sequence B following the position $\hat{i}_1$

   b. If $(w(A,B,\hat{i}_1,\hat{i}_1+l,n)\le\theta)$, then $\hat{l}_1=l$ and insert l virtual symbols (gaps) into sequence A following the position $\hat{i}_1$

   Based on these two steps, we can estimate the local mutation mode $T_1=\{(i_1,l_1)\}$, and its corresponding locally uniform alignment (C_1,D_1). It is decomposed as follows:

   $C_1=(C_{1,1},A_{2,1})$ $D_1=(D_{1,1},B_{2,1})$

3. Repeat the above process. The process will terminate at some k_0 such that $C_{k_0}=(C_{1,k_0},A_{2,k_0})$ and $D_{k_0}=(D_{1,k_0},B_{2,k_0})$ have shifting mutations occurring in (A_{2,k_0},B_{2,k_0}).

4. Calculate objective function :

   Penalty value between aligned sequence C_{k_0} and sequence D_{k_0}, $w(C_{k_0},D_{k_0})=\frac{1}{n}\sum_{j=1}^{n}w(c_j,d_j)$

$$w(c_j,d_j)=\begin{cases}0, & \text{if } c_j=d_j\\ 1, & \text{otherwise}\end{cases} \tag{1}$$

   Similarity value between aligned sequence C_{k_0} and sequence D_{k_0}, $s(C_{k_0},D_{k_0})=\frac{1}{n}\sum_{j=1}^{n}s(c_j,d_j)$

$$s(c_j,d_j)=\begin{cases}1, & \text{if } c_j=d_j\\ 0, & \text{otherwise}\end{cases} \tag{2}$$

In this research, penalty value or similarity value optimization simulations are applied separately.

3. Artificial Bee Colony (ABC)

The Artificial Bee Colony (ABC) was introduced by Karaboga and Akay in 2009. ABC is the optimization method which is inspired by the behavior of bee colony. There are three groups in the ABC: employed bees, onlooker bees, and scouts. For every food source, there is only an employed bee. The position of food source represents possible solution of the optimization problem and the amount of nectar represents the fitness of solution [7],[8].

Given objective function $f : X \subseteq R^D \to R$ where D is the dimension of the search space. In the initialization step, generate initial solutions $x_{ij}, i = 1..SN, j = 1..D$ randomly and evaluate the fitness $f(x_{ij}), i = 1..SN, j = 1..D$ [7]. In ABC applied to SPA optimization model, suppose $X = (n, \theta, \phi)$ as decision variable solution with objective function in equation (1) for minimizing penalty value or in equation (2) for maximizing similarity value.

Repeat

1. Employed Bees Step

   In the employed bees step, each employed bee generates solution as follows:

$$v_{ij} = \begin{cases} x_{ij} + \varphi_{ij}(x_{ij} - x_{kj}), i \neq k & \text{if } r_{ij} < MR \\ x_{ij} & \text{otherwise} \end{cases} \quad (3)$$

   with φ_{ij} is random number between $(-1,1)$, r_{ij} is random number between $(0,1)$, and MR is the modification rate.

   Evaluate $f(v_{ij}), i = 1..SN, j = 1..D$. If new solution v_{ij} is better than x_{ij}, replace x_{ij} with v_{ij}.

2. Onlooker Bees Step

   Onlooker bees select one of solutions based on probability $p_i = \dfrac{fitness_i}{\sum_{n=1}^{SN} fitness_n}$.

3. Scout Bees Step

   Determine the abandoned solution for the scout and replace with:

$$x_{ij} = x_j^{\min} + \tau_{ij}(x_j^{\max} - x_j^{\min}) \quad (4)$$

   with $\tau_{ij} \sim U(0,1)$ then eveluate $f(x_{ij}), i = 1..SN, j = 1..D$

4. Keep the best solution

4. Simulation Results

Based on previous research [2], the origin of zika virus is from South Africa and the virus mutate (change genetic sequence) and spread to other countries including Indonesia. Indonesia also derives zika virus from Cambodia. Data used in this research are zika virus protein data in Indonesia, Cambodia as its neighbour country, and South Africa as origin of zika virus. Data are obtained from National Center for Biotechnology Information (NCBI) accessed on May 2, 2019. The characteristics of zika virus protein data can be seen in Table 1.

Table 1. Zika virus protein data.

Access Code	Country	Length (bp)
ABI54480	South Africa	3429
AFD30972	Cambodia	3423
AMK49492	Indonesia	3423

There are two types of ABC simulations used : ABC simulation of penalty value minimization and ABC simulation of similarity value maximization. ABC simulation of penalty value minimization is to find SPA parameters (n,θ,ϕ) minimizing penalty value (including gaps) between two aligned zika virus protein sequences as objective function while ABC simulation of similarity value maximization is to find SPA parameters (n,θ,ϕ) maximizing similarity value (excluding gaps) between two aligned zika virus protein sequences as objective function. ABC parameters used in the simulations are : the number of bees : 10, modification rate : 0.8, maximum iteration : 50.

4.1. ABC Simulations of Penalty Value Minimization

Figure 1(a) shows ABC optimization process of penalty value minimization of aligned zika virus protein sequence in Cambodia and Indonesia. At the early iteration, the bees choose the position of food source as SPA parameters randomly. At the optimization process, we update nectar information as fitness function so that bees choose the SPA parameters resulting minimum objective function. Based on simulation, the optimal SPA parameters are $(n,\theta,\phi)=(246;0.3135;0.5395)$ with minimium penalty value is 0.005259 or 0.5259% and computation time is 36.57 seconds.
Figure 1(b) shows the positions i_k and lengths l_k of optimal SPA parameters. Optimal SPA parameters result iteration $k=2$. From the graphs, positions i_k and lengths l_k can be seen. If $l_k>0$, it occurs insertion, if $l_k<0$, it occurs deletion, and if $l_k=0$, there is no mutation shifting.

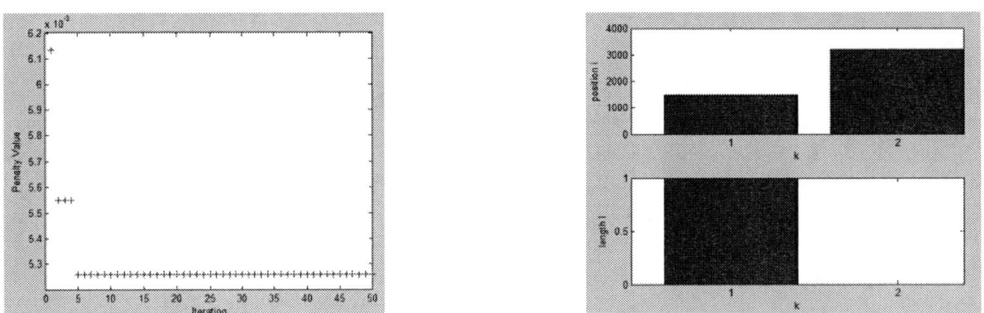

(a) (b)

Figure 1. Penalty value minimization of aligned zika virus protein sequences (Cambodia and Indonesia) (a) ABC optimization process (b) Positions and lengths of optimal SPA parameters

The similar procedure is applied to aligned zika virus protein sequence in South Africa and Indonesia. Based on figure 2(a), the optimal SPA parameters are $(n,\theta,\phi)=(51;0.3719;0.6819)$ with minimium penalty value is 0.262897 or 26.2897% and computation time is 45.02 seconds. Figure 2(b) shows the positions i_k and lengths l_k of optimal SPA parameters.

From the comparison, penalty value of aligned zika virus protein sequences in Cambodia is smaller than penalty value of aligned zika virus protein sequences in South Africa.

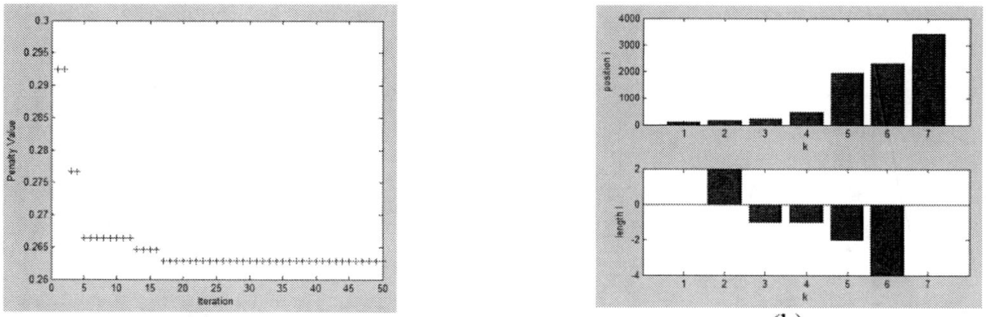

(a) (b)

Figure 2. Penalty value minimization of aligned zika virus protein sequences (South Africa and Indonesia) (a) ABC optimization process (b) Positions and lengths of optimal SPA parameters

4.2. ABC Simulations of Similarity Value Maximization

Figure 3(a) shows ABC optimization process of similarity value maximization of aligned zika virus protein sequence in Cambodia and Indonesia. At the early iteration, the bees choose the position of food source as SPA parameters randomly. At the optimization process, we update nectar information as fitness function so that bees choose the SPA parameters resulting maximum objective function. Based on simulation, the optimal SPA parameters are $(n,\theta,\phi)=(492;0.3682;0.6453)$ with maximum similarity value is 0.995032 or 99.5032% and computation time is 35.53 seconds.

Figure 3(b) shows the positions i_k and lengths l_k of optimal SPA parameters. Optimal SPA parameters result iteration $k=2$. From the graphs, positions i_k and lengths l_k can be seen. If $l_k>0$, it occurs insertion, if $l_k<0$, it occurs deletion, and if $l_k=0$, there is no mutation shifting.

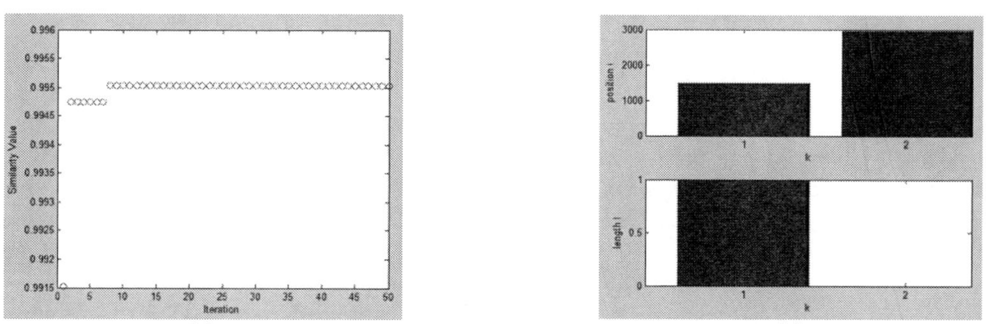

(a) (b)

Figure 3. Similarity value maximization of aligned zika virus protein sequences (Cambodia and Indonesia) (a) ABC optimization process (b) Positions and lengths of optimal SPA parameters

The similar procedure is applied to aligned zika virus protein sequence in South Africa and Indonesia. Based on figure 4(a), the optimal SPA parameters are $(n,\theta,\phi)=(67;0.3959;0.5850)$ with maximum similarity value is 0.737657 or 73.7657% and computation time is 43.90 seconds. Figure 4(b) shows the positions i_k and lengths l_k of optimal SPA parameters.

From the comparison, similarity value of aligned zika virus protein sequences in Cambodia is larger than similarity value of aligned zika virus protein sequences in South Africa.

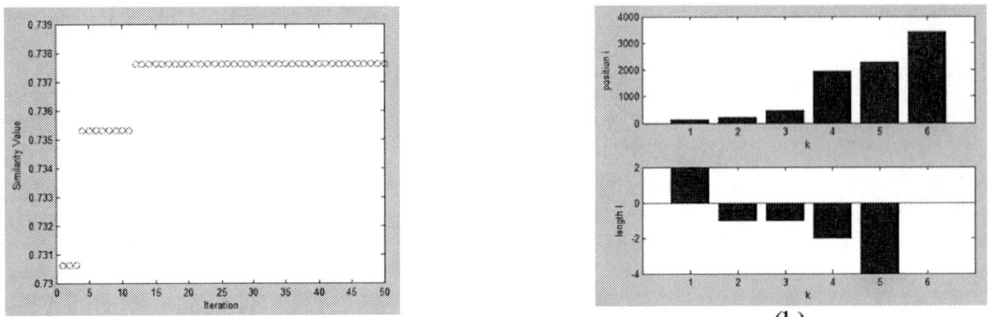

(a) (b)

Figure 4. Similarity value maximization of aligned zika virus protein sequences (South Africa and Indonesia) (a) ABC optimization process (b) Positions and lengths of optimal SPA parameters

5. Conclusion

To see if the similarity between two sequences, it is required the alignment process of two sequences. The method used to align them is Super Pairwise Alignment (SPA). In SPA, the penalty value and the similarity value depend on SPA parameters. The SPA parameters can be optimized by heuristic method, such as Artificial Bee Colony (ABC) inspired from the behavior of bees. From the simulation with case study of zika virus protein, ABC can find optimal SPA parameters resulting minimum penalty value or maximum similarity value in approaching. The development of this research is for optimizing the similarity of multiple sequences.

Acknowlogdement

High appreciation to the Ministry of Research and High Education (Kemenristekdikti), Republic of Indonesia for the Penelitian Dosen Pemula fund for the research conducted in the year of 2019.

References

[1] Pradana M S and Amiroch S 2019 Protein Sequence Analysis of the Zika Virus and the Dengue Virus Using Smith Waterman Algorithm *AIP Conference Proceeding* **2084**

[2] Pradana M S and Amiroch S 2018 Zika Virus Mutation and the Spreading to Indonesia *International Journal of Computing Science and Applied Mathematics* **4** pp 15-18

[3] Isaev A 2004 *Introduction to Mathematical Methods in Bioinformatics* (Berlin : Springer)

[4] Shahab M L and Irawan M I 2017 Sequence Alignment Using Nature-Inspired Metaheuristic Algorithm *International Journal of Computing Science and Applied Mathematics* **3**

[5] Rahmalia D and Herlambang T 2018 Weight Optimization of Optimal Control Influenza Model Using Artificial Bee Colony *International Journal of Computing Science and Applied Mathematics* **4** pp 27-31

[6] Herlambang T, Rahmalia D, Yulianto T 2019 Particle Swarm Optimization (PSO) and Ant Colony Optimization (ACO) for Optimizing PID Parameters on Autonomous Underwater Vehicle (AUV) Control System *Journal of Physics : Conference Series* **1211**

[7] Karaboga D and Basburk B A 2007 Powerful and Efficient Algorithm for Numerical Function Optimization : Artificial Bee Colony (ABC) Algorithm *Journal of Global Optimization* **39** pp 459-471

[8] Liao T, Aydin D, Stutzle T *Artificial Bee Colony for Continuous Optimization : Experimental Analysis and Improvements* (New York : Springer Science+Business Media)

[9] Shen S N and Tuszynski J A 2008 *Theory and Mathematical Model for Bioinformatics* (Berlin : Springer)

Journal of Physics: Conference Series **1417** (2019) 012030 doi:10.1088/1742-6596/1417/1/012030

Magnetic property approach for identification of dominant crystal structure in AISI420 treated by RF-DC Plasma N_2-H_2

H Herdianto

Physics Program, Gajayana University, Malang 65144, Indonesia

e-mail: hengky.herdianto@unigamalang.ac.id

Abstract. The nitridation of RF-DC plasma N_2-H_2 into AISI420 produces several phases namely α' (bcc), ε (hcp), and γ (fcc) which are randomly distributed in a number that is difficult to identify. So the most dominant justification of crystal phase was needed. Based on the concept of isothermal transformation of AISI420 transformed from the α' phase (ferromagnetic), to ε phase (paramagnetic), and to γ phase (nonmagnetic). It can be used as the underlying theory that magnetic property approach can potentially identify the dominant crystal phase. This study measured the magnetic properties of AISI420 treated by RF-DC plasma N_2-N_2 for identification of dominant crystal structure. Based on the hysteresis curve (M-H loop) of VSM method, magnetic characteristics can be produced such as susceptibility. Furthermore, the atomistic simulations were performed with the LAMMPS. Modelling of schematic picture of the structure using VESTA v.3.3.2 64 bit. Susceptibility value of AISI420 treated with RF-DC Plasma N_2-H_2 is range in the notation of 10^{-5} to 10^{-3}, so it meets criteria of $10^{-5} \leq 10^{-3}$ which can be categorized as paramagnetic and super-paramagnetic. It has the dominant ε-Fe_3N phase with trigonal-hcp microstructure based on magnetic property approach and simulations of LAMMPS and VESTA.

1. Introduction

Based on XRD data and microstructure simulations showing the results of nitriding AISI420 with optimization parameters of RF-DC plasma N_2-H_2 in the previous study [1-3] resulted in a heterogeneous random non-homogeneous microstructural phase. Nitridation of RF-DC plasma N_2-H_2 into AISI420 produces several phases namely α' (bcc), ε (hcp), and γ (fcc) [4][5] which are randomly distributed in a number that is difficult to identify. The phase transformation from α' to γ localized at the highly strained slip lines [6]. So the most dominant justification phase which is randomly distributed in a number that is difficult to identify, so the most dominant justification of crystal phase was needed.

Table 1. Summarizes the experimental process parameters for RF-DC plasma N_2-H_2 in the previous study [1-3]

Process	Parameters
Pre-sputtering	DC (-450 V), RF (0 V), pressure (70 Pa), temperature (room standard), duration (400 s), and carrier gas (N_2 only)
Nitriding	DC (-300 V), RF (250 V), pressure (70 Pa), temperature (673 K), duration (5400 s), carrier gas (N_2 + H_2) and with partial pressure (N_2 = 100 ml/min; H_2=20 ml/min)

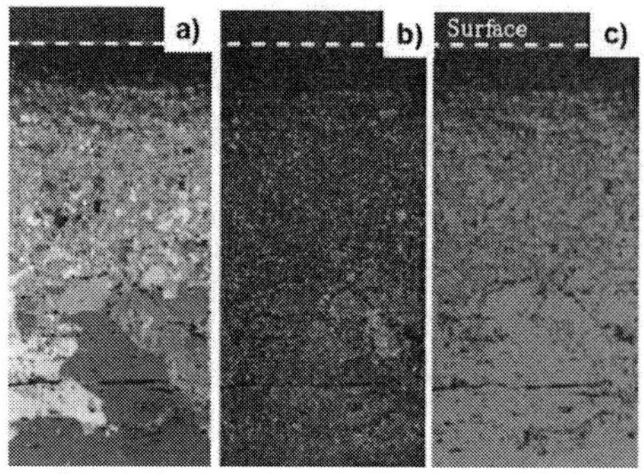

Figure 1. EBSD analysis: a) *IPF-mapping*, b) *KAM* distribution, dan c) *Phase-mapping* [6-9].

This result was confirmed by EBSD analysis using IPF-mapping, KAM, and phase-mapping by Prof. T Aizawa showed that RF-DC Plasma N_2-H_2 nitriding in AISI420 produced several phases namely α' (bcc), ε (hcp), and γ (fcc) which were scattered randomly in quantities that were difficult to identify [6][7][8][9]. In the high nitrogen content layer, the grain size was greatly reduced by this plastic localization to be less than the spatial resolution by EBSD [6][7]. This results in difficulty of justifying the dominant phase of the material.

In the previous research of V Leskovšek, M Godec, and P Kogej [10] resulting that transformation of the martensitic structure $\alpha' \rightarrow \gamma$ formed by deformation can be identified based on magnetic properties and mechanical strength. Increases in the hardness and percentage of the relative magnetic phase in the α microstructure can be attributed to isothermal transformation during plasma nitriding from the ferromagnetic α' martensitic phase, to paramagnetic ε martensitic, and to nonmagnetic γ martensitic.

The α-martensitic structure is a body of cubic (bcc) thus ferromagnetic and as ε is hexagonal close-packed (hcp) which is paramagnetic. It can be used as underlying theory that magnetic property approach can potentially identify the dominant crystal phase. This study measured the magnetic properties of AISI420 (without nitridation and nitridation with the optimization parameters of RF-DC plasma N_2-N_2) for identification of dominant crystal structure using Vibrating Sample Magnetometer.

2. Method

Rectangular samples with $20\times10\times2$ mm^3 of AISI420 specimens were prepared RF-DC plasma N_2-H_2 behaviour with optimization parameter [2][3][9]. External field value data of H_{ex} in (Oe) and σ inner magnetic moment value (emu/gram) of Vibrating Sample Magnetometer (VSM250, Dexing Magnet, Co. Ltd) are presented in the form of hysteresis curve (M-H loop). The curve was analyzed for remanent magnetic moment fluctuations and its coercivity value as an indicator of the occurrence of crystalline anisotropic changes. Based on this method, magnetic characteristics can be produced such as saturation magnetic moment (M_s), remanent magnetic moment (M_r), intrinsic magnetic coercivity (H_k), maximum product energy (H_{max}), and susceptibility (X_v).

The atomistic simulations presented in this study were performed with the Large-scale Atomic or Molecular Massively Parallel Simulator (LAMMPS), the MD simulator developed by the Sandia National Laboratories [11]. In this study, a three-dimensional system with edge lengths of $L_x = 4.56$ nm, $L_y = 2.85$ nm, and $L_z = 2.58$ nm in the x, y, and z directions, respectively, were simulated. Periodic boundary conditions were applied to all three ($x, y, and\ z$) dimensions to model bulk materials such that the total number of atoms was maintained constant during the simulation. The simulation box was filled with iron atoms at the bcc lattice sites (lattice parameter: 0.285 nm). As the energetically preferred sites for nitrogen atoms in an iron monocrystal are octahedral interstitial sites, nitrogen atoms were added randomly to the simulation box at the octahedral sites of the bcc lattice.

The diffusion of nitrogen in iron was modelled using a bcc supercell of 3200 iron atoms with different numbers of nitrogen atoms randomly distributed at interstitials. The interatomic interactions between iron and nitrogen were described with a MEAM formulation in which the total energy of the system is specified by [12]:

$$E = \sum_i \left[F_i(\rho_i) + \frac{1}{2} \sum_{i \neq j} \phi_{ij}(r_{ij}) \right] \quad (1)$$

where F_i is the embedding energy function for atom i embedded in a background electron density ρ_i, and $\phi_{ij}(r_{ij})$ is the pair potential interaction between atoms i and j separated by a distance r_{ij}. In this research, the potential parameters for the Fe-N system were built according to the 2NN-MEAM interaction potential developed by M I Baskes [12] and B J Lee [13] for pure Fe and N. Modelling of schematic picture of the structure using Visualization for Electronic and STructural Analysis (VESTA) v.3.3.2 64 bit [14].

3. Result and Discussion

The AISI420 without nitridation has a low nitrogen content whereas AISI420 which is crystallized with optimization parameters has a high nitrogen content because it has a higher coercivity value with lower saturation magnetic moments. Low nitrogen content forms alloys in the soft-magnetic phase, while in high nitrogen content the average value of saturation magnetic moments decreases and coercivity increases.

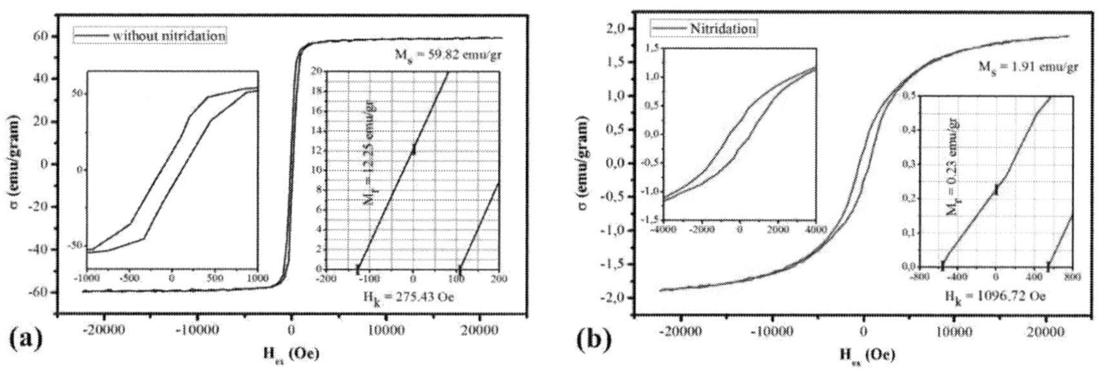

Figure 2. Hysteresis curve of AISI420: (a) without nitridation; (b) nitridation.

The saturation magnetic moment value (M_s) of AISI420 was optimized by the optimization parameter was 1.91 emu/gr.Oe. It can be identified that the microstructure of AISI420 that was nitridation using optimization parameters had the dominant ε-Fe_3N phase. The microstructure of ε-Fe_3N phase is hexagonal close-packed (hcp) type with space P312 (149). The results of this analysis are the same as results of previous studies conducted by S Bhattacharyya [15].

Table 2. The magnetic property of AISI420 without and with nitridation by RF-DC plasma N_2-H_2.

VSM	Without nitridation	Nitridation
H_{max} (Oe)	22334	22364
H_k (Oe)	275.43	1096.72
M_s (emu/gr)	59.82	1.91
M_r (emu/gr)	12.25	0.23
M_r/M_s	0.21	0.12
Min. X_v (emu/gr.Oe)	2.67×10^{-3}	8.48×10^{-5}
Max. X_v (emu/gr.Oe)	0.21	2.52×10^{-3}

Classification of magnetic material properties consists of ferromagnetic, diamagnetic, paramagnetic, and super-paramagnetic. The nature of magnetic material is identified based on the hysteresis curve pattern and susceptibility value (X_v). Susceptibility value of AISI420 treated with optimization parameters of RF-DC Plasma N_2-H_2 is 8.48×10^{-5} emu/gr. Oe up to 2.52×10^{-3} emu/gr.Oe. Susceptibility range in the notation of 10^{-5} to 10^{-3}, so it meets criteria of $10^{-5} \leq 10^{-3}$ according to H D Young and R A Freedman [16] which can be categorized as paramagnetic and super-paramagnetic material.

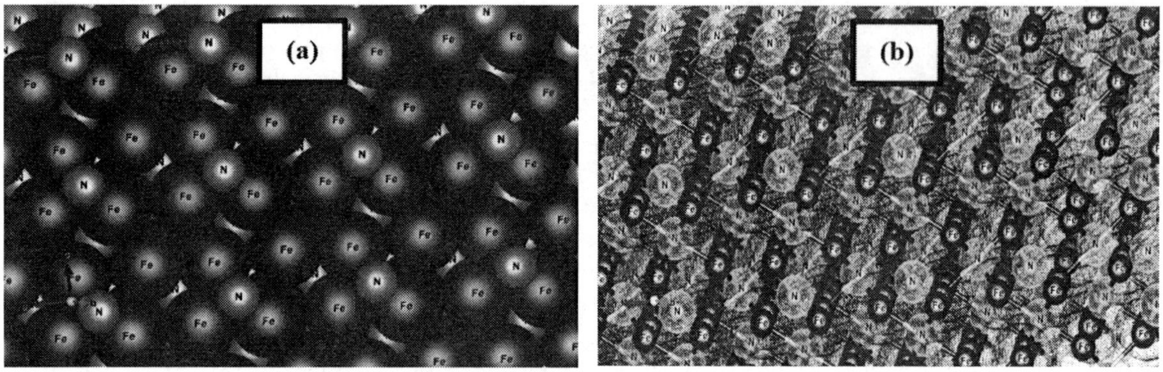

Figure 3. Atomic positions of nitrogen (grey) randomly distributed inside iron (blue), modelling of a schematic picture of the structure using Visualization for Electronic and STructural Analysis (VESTA) v.3.3.2 64 bit: (a) space filling's style; (b) ball-and-stick's style with show dot surface

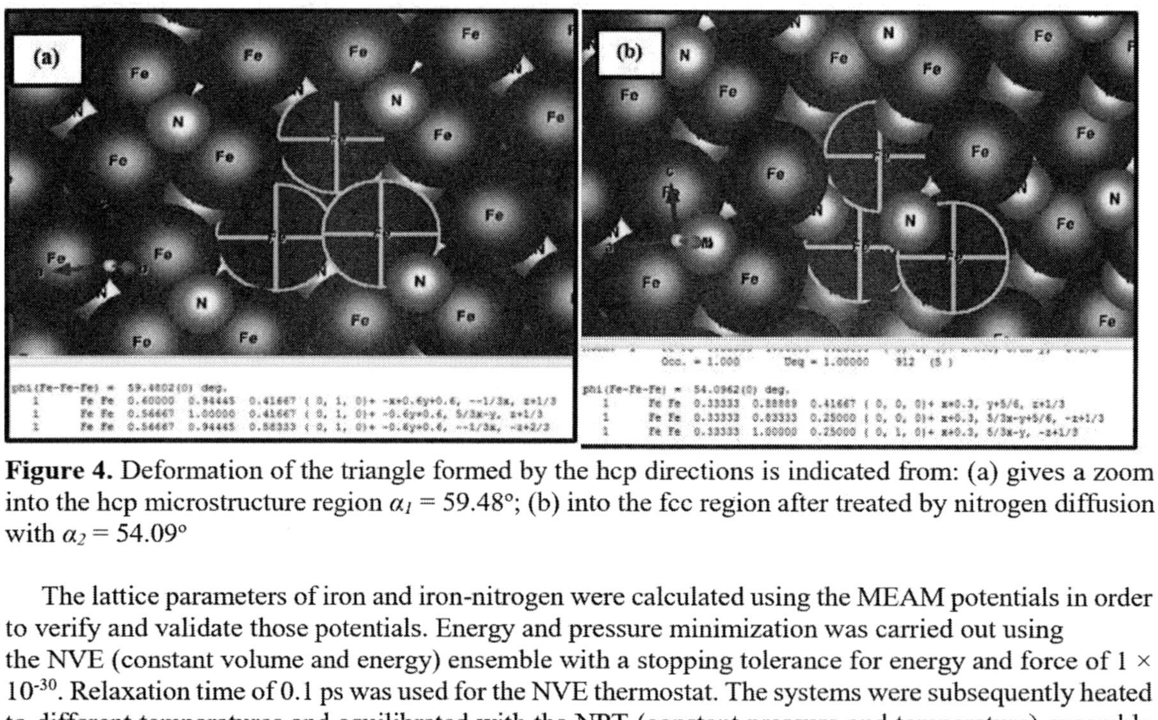

Figure 4. Deformation of the triangle formed by the hcp directions is indicated from: (a) gives a zoom into the hcp microstructure region $\alpha_1 = 59.48°$; (b) into the fcc region after treated by nitrogen diffusion with $\alpha_2 = 54.09°$

The lattice parameters of iron and iron-nitrogen were calculated using the MEAM potentials in order to verify and validate those potentials. Energy and pressure minimization was carried out using the NVE (constant volume and energy) ensemble with a stopping tolerance for energy and force of 1×10^{-30}. Relaxation time of 0.1 ps was used for the NVE thermostat. The systems were subsequently heated to different temperatures and equilibrated with the NPT (constant pressure and temperature) ensemble to achieve the desired temperature and pressure.

The NPT simulations were carried out for 25 ps. In order to calculate the displacement of nitrogen atoms at a constant temperature, the NVT (constant volume and temperature) ensemble was used to keep the volume constant throughout the run. The mean square displacement (MSD) of nitrogen atoms in the simulation box concerning their positions in the equilibrated system was recorded for each time step. The total simulation time was set to 1 ns, with a time step of 0.0001 ps. All simulations were performed under zero pressure.

This atomistic simulation results that deformation of the triangle formed by the hcp directions is indicated from α_1 = 59.48° into α_2 = 54.09° after treated by nitrogen diffusion. The deformations were related to extended martensitic lattice by high nitrogen extraordinary solid solution. The phase transformation from martensitic lattice α'-Fe through expanded phase into ε-Fe_3N lattice-like as in the previous study [1][2].

4. Conclusion

The AISI420 was treated using the optimization of process RF-DC plasma N_2-H_2 has the dominant ε-Fe_3N phase with trigonal hexagonal close-packed (hcp) microstructure in space P312 (149) super-paramagnetic. Classification of magnetic material properties consists of ferromagnetic, diamagnetic, paramagnetic, and super-paramagnetic. The nature of magnetic material is identified based on the hysteresis curve pattern and susceptibility value (X_v). Susceptibility value of AISI420 treated with optimization parameters of RF-DC Plasma N_2-H_2 is range in the notation of 10^{-5} to 10^{-3}, so it meets criteria of $10^{-5} \leq 10^{-3}$ which can be categorized as paramagnetic and super-paramagnetic material. Deformation of the triangle formed by the hcp directions is indicated from α_1 = 59.48° into α_2 = 54.09° after treated by nitrogen diffusion. The phase transformation from martensitic lattice α'-Fe through expanded phase into ε-Fe_3N lattice.

Acknowledgments

The author would like to thank Prof. P Sebayang as a supervisor from Indonesian Institute of Sciences (LIPI) for the research facilities provided as well as guidance for the implementation of the research in the Physics Research Center Laboratory, Puspitek, Serpong, South Tangerang, Indonesia.

References

[1] Herdianto H, Santjojo D J D H, and Masruroh 2017 *AIP Conf. Proc.* **1868** 060004
[2] Herdianto H, Santjojo D J D H, and Masruroh 2017 *J. Phys.: Conf. Ser.* **909** 012027
[3] Herdianto H, Aizawa T, Wasa K, Santjojo D J D H, and Sakti S P 2017 *Proc. of The 11th SEATUC Symposium* 1 (Vietnam-Ho Chi Minh City)
[4] Santofimia M J, Speer J G, Zhao L, and Sietsma J 2011 *J. Sol. Stat. Phen.* **567** 172-174
[5] Katajarinne T and Kivivuori S 2013 *J. K. Eng. Mat.* **1338** 554-557
[6] Aizawa T, Yoshino T, Morikawa K, and Yoshihara S 2019 *Crystals* **9**(2) 60
[7] Farghali A and Aizawa T 2017 *Mat. Tran.* **58**(4) 697
[8] Farghali A and Aizawa T 2018 *ISIJ Int.* **58**(3) 401
[9] Aizawa T 2017 *Proc. of The 7th Annual Basic Science Int. Conf.* **2** 8
[10] Leskovšek V, Godec M, and Kogej P 2016 *J. of Sci. Rep.* **6** 1
[11] Plimpton S 1995 *J. Comp. Phys.* **117**(1) 1
[12] Baskes M I 1993 *Phys. Rev. B* **46** 2727
[13] Lee B J 2006 *Acta Mater.* **54**(3) 701
[14] Momma K and Izumi F 2011 *J. of App. Crys.* **44** 1272
[15] Bhattacharyya S 2015 *J. of Phys. Chem. C* **119**(4) 1601
[16] Young H D and Freedman R A 2015 University physics with modern physics (New York: Addison-Wesley)

The recent advances on potential solid electrolytes for all-solid-state supercapacitors: A short review

Lisda Ainiya

Department of Physisc, Institut Teknologi Kalimantan, Balikpapan, Indonesia

e-mail: lisdaainiya.21@gmail.com

Abstract. The increasing demand to develop renewable energy has increased drastically due to limited resources and environmental problems of fossil fuel. The most widely used energy storage devices are batteries and capacitors. Supercapacitors, on the other hand, can induced higher power density by fast charging/discharging rate which results in higher power density and longer cycle life compared to batteries and fuel cells. To fulfil the demand for safer energy sources, the use of highly flammable organic liquid electrolytes and polymer separators must be replaced by solid electrolytes. Therefore, supercapacitors with solid electrolytes can be alternative renewable energy. Commercially, there are three types of electrolytes of supercapacitors: aqueous electrolytes, organic electrolytes, and ionic liquids. In this paper, there is selective review on solid electrolytes of supercapacitors including chitosan, polyacrylamide, poly(aryl ether ketone), and polyethylene glycol. Overall, this paper aims to provide comprehensive reviews on recent advances in potential solid electrolytes of supercapacitors and the remaining challenges.

1. Introduction

The consumption of fossil fuel to be applied in many aspects of life seems unstoppable. Figure 1 illustrates global fossil fuel consumption ranges from 1800 – 2017. It shows that global fossil fuel consumption reaches its highest plateau in 2017 with more than 120,00 TWh [1].

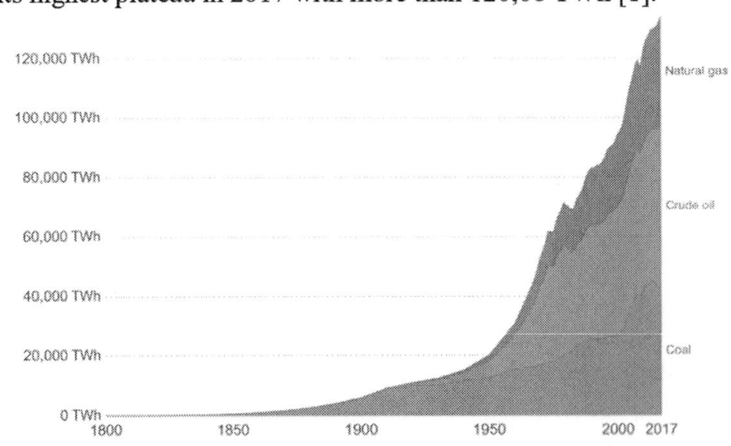

Figure 1. Global fossil fuel consumption measured in terawatt-hours [1].

Recent data for 2018 from IEA of global fossil fuel consumption subsidies by energy source [2]. It shows that the estimated value to be more than $400 billion which is understandable since the rising demand for fossil fuel inversely proportional to the availability of fossil fuel resources. Therefore, the limited resources and environmental problems caused by the use of fossil fuels are the major reasons why renewable energy is being highly developed [3].

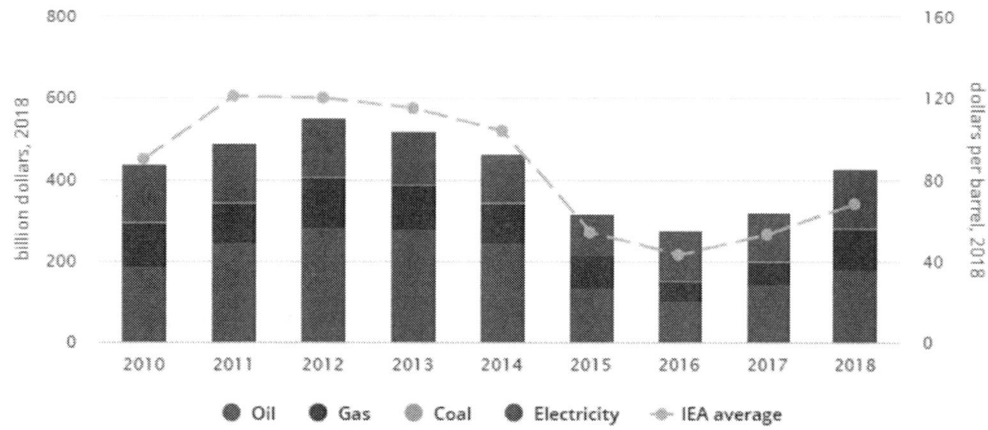

Figure 2. Variation prices of global fossil-fuel consumption subsidies by energy source [2]

However, the focus has been shifted to energy storage devices which contain electrical energy to smooth the intermittency of the energy sources. The most widely used energy storage devices are batteries and capacitors. The latest research of battery has demonstrated better battery with higher energy density and longer life cycle but struggles with power densities. While the limited application of capacitors caused by the small amount of energy thus can be charged and discharged easily and deliver pulses of energy [4].

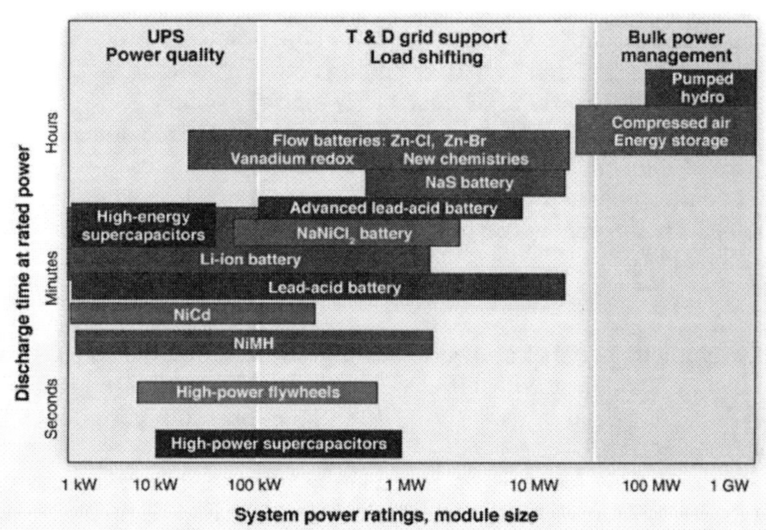

Figure 3. Comparison between energy density and discharge time of energy storages [5]

Supercapacitors, on the other hand, which is one of the words that is used to describe capacitors [6]. Supercapacitors have developed extensively since the first patents of capacitors back in 1957 [7] and first attempts to introduce it to market in 1969 [8]. There are several advantages of supercapacitors making it superior to the rest of energy storage devices. Supercapacitors demonstrate higher power density and longer life cycle (>100,000 cycles) [9]. To fulfil the demand for safer energy sources, the use of highly flammable organic liquid electrolytes and polymer separators must be replaced by solid electrolytes. Therefore, supercapacitors with solid electrolytes can be alternative renewable energy.

Commercially, there are three types of electrolytes of supercapacitors: aqueous electrolytes, organic electrolytes, and ionic liquids. In this paper, there is a selective review on solid electrolytes of supercapacitors including chitosan, polyacrylamide, poly(aryl ether ketone), and polyethylene glycol. Overall, this paper aims to provide comprehensive reviews on recent advances in potential solid electrolytes of supercapacitors and the remaining challenges.

2. A selective review of solid electrolytes of all-solid-state supercapacitors

There are several recent researches on solid electrolytes including chitosan, polyacrylamide, poly(aryl ether ketone), and polyethylene glycol. The advantages and future improvements of these researches are discussed in this section

2.1. Chitosan

Chitosan is a biopolymer composed of glucosamine and N-acetyl glusamine units glucosidic bonds [10][11]. Chitosan has been a hot research topic due to its advantages including safer (non-toxic), water-soluble, and biodegradable which are important to solve the environmental issues caused by the use of fossil fuel [12]. Chitosan is chosen as an electrolyte due to its excellent mechanical properties and high film-forming capability. Material to be applied as an electrolyte must exhibit high ionic conductivity in the order ~10^{-3} S/cm at room temperature. However, chitosan has low conductivity [13]. Therefore, various attempts have been made to increase the conductivity of chitosan [14].

Solid-state electrolytes have three types including solid electrolyte, gel polymer electrolyte, and polyelectrolyte. One of the major drawbacks which limit the application of solid electrolyte is its low conductivity (10^{-8} and 10^{-7} S/cm). Many efforts have been tried to overcome the low conductivity of solid electrolytes. For example, Shujahadeen et al. synthesized chitosan-ammonium triflate (CS-NH4Tf) polymer electrolyte films to see the role of TiO_2 filler [15]. However, the obtained dc conductivity was in the order of~ 10^{-7} S/cm at ambient temperature which is considerably low to be applied as an electrolyte.

Cao et al., on the other side, chose to synthesis a biopolymer-chitosan based supramolecular hydrogel solid-state electrolyte *via* a simple and fast cross-linking between chitosan and Li^+/Ag^+. The electrolyte was flexible and high cyclic stability. SEM characterization shows that MnO2 and KBC particles are evenly distributed among the CNTs which is an important characteristic since it makes the adsorption and desorption of the Li^+ ion easier. EIS characterization shows that chitosan-based hydrogel solid-state electrolyte has high ionic conductivity of 1.6 x 10^{-3} S/cm to 4.6 x 10^{-3} S/cm with temperature ranges from 313 K and 353 K. Their work successfully fulfilled the requirements of a material to be applied as an electrolyte in general [13].

2.2. Polyacrylamide

Polyacrylamide is a synthetic polymer known for its high hydrophilicity formed by polymerization of acrylamide monomer by a catalyst with bifunctional cross-linking agents [16]. Polyacrylamide has considerably high ionic conductivity of 1.7 x 10^{-2} S/cm but still considered to be lower than commercialized PVA/H_2SO_4 [17].

Tabanli et al. Focused their research on rectification of current by synthesizing polyacrylamide hydrogel as an electrolyte. Their work showed that the maximum rectification ratio to be around 850 at 6~ 7 V for Al/PHE/Pt configuration. While recent research from Li et. al. proposed to synthesis polyacrylamide hydrogels as a solid-state electrolyte by cross-linking methacrylated graphene oxide (MGO-PAM). The results showed that polyacrylamide as solid-state electrolyte of supercapacitor could exhibit high specific capacitance, long self-discharge time, and low leakage [18].

2.3. *Poly(aryl ether ketone) and Polyethylene Glycol*

Poly(aryl ether ketone) is one of semicrytalline polymer with a crytalline melting poing around 335°C and a glass transition at around 145°C, possesses excellent mechanical properties, good environmental resistance, thermo-oxidative stability, chemical and radiation resistance [19][20]. While polyethylene glycol is a biocompatible poly(ether) which low cost production and highly soluble in aqueous solutions and organic solvents [21].

Li et al. Synthesized hybrid membranes of branched sulfonated poly (aryl ether ketone) with 1,3,5-tris(4-fluorobenzoyl) benzene as the branching agent to improve the properties of mechanical strength, excellent dimensional stability, and low methanol permeability. The work showed that the highest value of proton conductivity value was 8.2 x 10^5 S/cm^3 [22].

Na et al. Suggested solid-state polymer electrolyte by combining poly(aryl ether ketone)-polyethylene glycol copolymer as a polymer host and $LiClO_4$ as an electrolyte salt aiming to provide alternative to make a flexible solid-state supercapacitor. The results showed that the solid-state supercapacitor has excellent electrochemical performance especially at high temperature, high specific capacitance, and high energy density [23].

3. Conclusion

The recent advances of solid-state electrolytes to produce a solid-state supercapacitors have showed that there are several methods and approaches which are needed in order to come up with better solid-state supercapacitors as energy storage thus yet still needs to be improved and developed by tuning either the electrode mass ratios, increasing the ionic conductivity, energy density, and capacitance.

References

[1] Hanna R and Max R 2019 Fossil Fuels *Published online at OurWorldInData.org* Retrieved from: https://ourworldindata.org/fossil-fuels [Online Resource]

[2] Wataru M and Zaki A 2019 Commentary: Fossil Fuel Consumption Subsidies Bounced Back Strongly in 2018 *Published online at iea.org* Retrieved from: https://www.iea.org/newsroom/news/2019/june/fossil-fuel-consumption-subsidies-bounced-back-strongly-in-2018.html [Online Resource]

[3] Yang Z, Zhang J, Kintner-Meyer M C W, Lu X, Choi D, Lemmon J P and Liu J 2011 Electrochemical Energy Storage for Green Grid *Chem. Rev.* **111**(5), 3577–3613

[4] Temmuz C 2012 *Investigating Solid-State Supercapacitors Constructed with PVA/CNT NanoComposite Electrolytes* (Wichita: Wichita State University)

[5] Dunn B, Kamath H and Tarascon J M 2011 Electrical Energy Storage for the Grid: A Battery of Choices *Science* **334**(6058) 928–935

[6] Kötz, Rüdiger and Carlen M 2000 *Principles and Applications of Electrochemical Capacitors* Electrochim Acta 45 2483-2498 10.1016/S0013-4686(00)00354-6

[7] H E Becker U S Patent 2 800 616 (to General Electric) 1957

[8] D I Boos U S Patent 3 536 963 (to Standard Oil SOHIO) 1970

[9] Zhong, Cheng and Yida, Deng and Hu, Wenbin and Qiao, Jinli and Zhang, Lei and Zhang, and Jiujun 2015 A Review of Electrolyte Materials and Compositions for Electrochemical Supercapacitors *Chem. Soc.* rev. 44. 7431-7920. 10.1039/c5cs00303b

[10] K Soontarapa and U Intra 2006 *Chem. Eng. Commun.* **193** 855–868

[11] J Ma and Y Sahai 2013 *Carbohydr. Polym.* **92** 955–975

[12] S B Aziz, O G Abdullah, M A Rasheed, and H M Ahmed 2017 *Polymers* **9** 187

[13] N A N Aziz, N K Idris, and M I N Isa, 2010 *Int. J. Polym. Anal. Charact.* **15** 319

[14] Cao, Cao and Yang, Mingyang and Wu, Dong and Lyu, Fucong and Sun, Zhifang and Zhong, Xiongwei and Pan, Hui andLiu, Hongtao and Lu, and Zhouguang 2017 Biopolymer-chitosan Based Supramolecular Hydrogels as Solid State Electrolyte for Electrochemical Energy Storage *Chem. Commun.* 53 10.1039/C6CC08658F

[15] Al-Salman Yahya 2018 Conductivity and Electrical Properties of Chitosan - Methylcellulose Blend Biopolymer Electrolyte Incorporated with Lithium Tetrafluoroborate *Int. J. of Electrochemical Sci.* 13. 3185-3199. 10.20964/2018.04.25

[16] Abdullah Omed and B Aziz, Shujahadeen and Al-Zangana, Shakhawan 2019 Solid Polymer Electrolytes Based on Chitosan: NH4Tf Modified by Various Amounts of TiO2 Filler and Its Electrical and Dielectric Characteristics. *Int. J. of Electrochemical Sci.* **14** 1909 – 1925 10.20964/2019.02.31

[17] Banga J P 1998 SDS-Polyacrylamide Gel Electrophoresis (SDS-Page) *Encyclopedia of Immunology* 2143–2144 doi:10.1006/rwei.1999.0541

[18] Jin X, Sun G, Zhang G, Yang H, Xiao Y, Gao J, and Qu L 2019 A Cross-linked Polyacrylamide Electrolyte with High Ionic Conductivity for Compressible Supercapacitors with Wide Temperature Tolerance *Nano Research* doi:10.1007/s12274-019-2382-z

[19] Devesh S, Yuvraj S N, Jugmendra S U and Vijai K 2012 Synthesis and Modification of Poly(ether ether ketone) and Their Properties: A Review *Polymer Reviews* **52**(2) 189-228

[20] Blundell D J and Osborn B N 1983 The Morphology of Poly(aryl-ether-ether-ketone) *Polymer* **24**(8) 953–958 doi:10.1016/0032-3861(83)90144-1

[21] Ivanova E P, Bazaka K, and Crawford R J 2014 Advanced Synthetic Polymer Biomaterials Derived from Organic Sources *New Functional Biomaterials for Medicine and Healthcare* 71–99 doi:10.1533/9781782422662.71

[22] Li Ye & Wang, Xitao & Xie, Min & Liang, Yuan & Liu, Xincai & Wang, Ce 2014 High Proton Conductivity and Low Fuel Crossover Polymer Electrolyte Membranes Derived from Branched Sulfonated Poly(ether ether ketone)s and Silica Sulfonic Acid Nanoparticles *High Performance Polymers* **26** 770-778 10.1177/0954008314528962

[23] Na R., Huo P, Zhang X, Zhang S, Du Y, Zhu K., Wang G 2016 A Flexible Solid-state Supercapacitor Based on A Poly(aryl ether ketone)–poly(ethylene glycol) Copolymer Solid Polymer Electrolyte for High Temperature Applications *RSC Advances* **6**(69) 65186–65195 doi:10.1039/c6ra11202a

Mercury and Cyanide Pollution on the Aquatic Organism in Sekotong People Gold Mining

Suhadi*, Sueb and M Syamsussabri

Department of Biology, Faculty of Mathematics and Natural Sciences, Universitas Negeri Malang, Semarang Street 5, Malang, Indonesia.

* E-mail: suhadi@um.ac.id

Abstract. This study aimed to determine the pollution of mercury and cyanide in gold mining conducted by Sekotong people. Samples of this research were water and the aquatic organism around in people gold mining consisting of *Acetes indicus, Perna viridis, Polymesoda expansa, Plotosus canius, Scylla serrata, Tegillarca granosa, Moolgarda seheli,* and *Crassostrea gigas*. The content of mercury in the samples can be known using Atomic Absorption Spectrophotometry (ASS). In the same way, in order to know the content of cyanide in the samples can be obtained using Ultraviolet-Visible (UV-Vis) Spectrophotometry. The results showed that Sekotong people's gold mining has been contaminated with mercury and cyanide in aquatic and animal environments exceeding the specified threshold value with the highest level of mercury 48.91 ppm and cyanide 90.22 ppm.

1. Introduction

Indonesia is a country with abundant gold resources [1]. There are 90 countries conducting gold mining activities, but of 75% of the world's gold, production only produced by 20 countries. Indonesia is ranked in the top eight with gold production of 73.1 tons, with around 3000 tons of gold reserves [2]. Gold production in Indonesia produced through several sectors including large-scale, medium-scale, and small-scale mining. Small scale mining or called in Indonesia with smallholder gold mining produces about 15-20% gold [3].

Sekotong people's gold mining is one of the small-scale mining points of 900 community gold mining points in Indonesia [4]. A total of 187 mining points has been started since 2001 [5]. Small-scale mining is mining that is carried out by a person, a group, or a family cooperatively with little or no suitable mining mechanism as done in large-scale mining [6]. Geologically, the Sekotong region has the potential to form metal deposits such as gold and base metals, and there are also indications of well-developed mineralisation and relatively even distribution [7] with metal reserves of around 1,686,462 tons with gold potential reaching thousands of tons and silver around three times that amount [8].

The Sekotong region was once explored by large scale mining companies including PT. Newmont Nusa Tenggara from 1986 to 2004 and PT. Indotan from 2004 to 2006 [9]. After the issuance of the West Nusa Tenggara Province, Regional Regulation on Spatial Planning (RTRW) Number 11 of 2006 article 38 paragraph 2 states that the Lombok Island Development Area (SWP) is directed only to mining C class excavated materials [10] which is PT. Indotan stopped his exploration activities. One of the reasons for this mining is done by the community because economically, the income generated is very high. The average income of the community is Rp 4,136,667/month/person [11]. Unlicensed Gold Mining Activity (PETI) which has been going on since 2008 [12,13].

Environmental pollution can be defined as the accumulation and adverse effects of contaminants or pollutants on human health and welfare, and the environment [14]. Pollutants are defined as any substances that are put into the environment that adversely affect the use of resources [15]. Environmental pollution is the entry or inclusion of living things, substances, energy, and/or other components into the environment by human activities so that they exceed the established environmental quality standards [16]. Mining activities carried out by the people are known to use mercury (Hg) and cyanide (CN) which are heavy metals and hazardous compounds if not appropriately managed [17, 18, 19, 20, 21].

Field observations on Sunday, January 21, 2018, showed that the community used simple and inexpensive techniques in gold processing without regard to environmental safety aspects due to pollution of the resulting waste. Gold processing uses amalgamation and cyanidation techniques. The amalgamation technique is a technique used by 10-12 million people's gold mining around the world [22]. This technique is done by mixing rocks with Hg in water media using small gold trommel/spindles to form an amalgam. This tool has a length of 55-60 cm with a diameter of 30 cm, and in it, there are 3-5 iron rods that are useful for crushing stones. Each log is at least able to accommodate 3-5 kg of rock that has been destroyed. This process usually takes between 3-4 hours, and each process is used from 250 to 500 g Hg. The results of the spindle are then filtered and squeezed with a cloth to separate the amalgam from the pulp. Gold separated from the curing process (evaporation of Hg) at very high temperatures, which are around 400°C while the rest of the mining process is discharged directly in the environment. The next technique is the cyanidation technique using a big gold trommel/barrel with a diameter of 1.5 m and a height of 7 meters.

In 2009 the number of spindles found in gold mining in Sekotong District was 1268. In 2011 the number increased by 4630 while the cask was 140 [13]. The cause of Hg pollution is due to the disposal of gold processing tailings into the environment [23]. Further impacts in the form of a decrease in the quality of soil and water will indirectly affect other organisms and end in humans [12]. The mining process is still ongoing until now, and no preventive efforts have been made. Based on these conditions, it is deemed necessary to determine the pollution of mercury and cyanide in Sekotong people gold mining.

2. Method

2.1 Experimental Design

This research is a quantitative descriptive study that aims to find out information related to Hg and HCN pollution in Sekotong people's gold mining. The research data was collected at Sekotong people's gold mining. Observations and analyses were carried out at the Chemical Laboratory of Brawijaya University.

2.2 Population and Samples

2.2.1 Population

The population of this study are as many as six locations, including: Sekotong Tengah village (Telaga Lebur hamlet), Cendi Manik village (Sayong Daye hamlet), Sekotong Barat village (Lendang Re 1 hamlet), Sekotong Barat village (Lendang Re 2 hamlet), Sekotong Barat village (Labuan Petung hamlet), and Pelangan Village (Tembowong hamlet).

Figure 1. Gold processing distribution map

2.2.2 Water Samples

A sampling of water is done by using purposive sampling with consideration of the location of gold processing activities. The number of samples taken at eight points.

Table 1. Coordinates of water sampling points

No	Village	Hamlet	Sample Point	Coordinate	
				Latitude	Longitude
1	Sekotong Tengah	Telaga Lebur	1	S 08°47.101'	E 116°02.423'
			2	S 08°47.081'	E 116°02.423'
			3	S 08°47.073'	E 116°02.416'
			4	S 08°47.071'	E 116°02.400'
2	Pelangan	Tembowong	5	S 08°48.606'	E 115°56.484'
			6	S 08°48.594'	E 115°56.488'
3	Sekotong Barat	Lendang Re I	7	S 08°45.671'	E 116°02.262'
			8	S 08°45.672'	E 116°02.266'
			8		

2.2.3 Animal Samples

A sampling of animals is done by using purposive sampling with consideration of the location of gold processing activities that dispose of waste into the sea. Samples were taken in Sekotong Barat village by considering the distance of taking 100-100 m from the coast. The locations for sampling animals are as follows.

Figure 2. Map of animal sampling

Table 2. Coordinating points for sampling animals

No	Village	Sample Point	Coordinate	
			Latitude	Longitude
1	Sekotong Barat	1	S 08°45.558'	E 116°02.445'
2		2	S 08°45.522'	E 116°02.429'
3		3	S 08°45.476'	E 116°02.407'
4		4	S 08°45.423'	E 116°02.416'
5		5	S 08°45.382'	E 116°02.423'
6		6	S 08°45.296'	E 116°02.389'
7		7	S 08°45.272'	E 116°02.381'
8		8	S 08°45.254'	E 116°02.347'
		8		

Figure 3. Gold processing

2.3 Tools and Materials

Tools and materials used in this study include 100 ml sample bottles (High Density Polyethylene, (HDPE), impermeable), funnels, measuring cups, ice boxes, and Global Positioning System (GPS), tweezers, Atomic Absorption Spectrophotometry (AAS), Ultraviolet-Visible (UV-Vis) Spectrophotometry, 100 ml water/sample, 100 g animals/samples, tissue, cooler, label paper, and stationery.

2.4 Procedures

2.4.1 Sampling

The stages of water sampling are as follows: (1) water samples are taken in reservoirs at a depth of 0.8 d (times depth) provided that the sample points are taken in the centre of the reservoir with the aim of knowing the average level of pollution; (2) Water samples are taken and placed in 100 ml sample bottles; (3) water samples are stored in an icebox at 0°C ≤ 6°C. The stages of animal sampling are as follows: (1) animal samples are taken at predetermined coordinates; (2) animal samples are taken as much as 100 g and put into sample bottles; (3) samples were stored in an icebox at 0°C ≤ 6°C [24, 25, 26].

2.4.2 Sample Observation

Sample observations were carried out with Hg observation guidelines namely SNI 6989.78: 2011 [27], HCN namely SNI 6989.77: 2011 [28], and animal identification sheet [29].

2.5 Assessment
Data analysis was performed by comparing the levels of Hg and HCN in water and animals with the threshold set by WHO, regulation of the ministry of health of the Republic of Indonesia, and regulation of the ministry of environment of the Republic of Indonesia [30, 31, 32, 33].

3. Results and Discussions

3.1 Hg and HCN levels in water
The levels of Hg and HCN in water are carried out at three different locations, namely in the disposal ponds, the sea, and wells.

3.1.1 Drainage Ponds
Analysis of Hg levels in the drainage pond showed that the lowest to highest concentration was 0 to 3.35 ppm. While the analysis of HCN levels in the disposal ponds showed that the lowest to the highest concentration of 5.31 to 13.27 ppm.

Table 3. Hg and HCN levels in drainage ponds

Points	Concentration (ppm)			
	Hg	Information	HCN	Information
1	0.80 ± 0.00	Polluted	7.43 ± 0.00	Polluted
2	0.40 ± 0.00	Polluted	10.08 ± 0.00	Polluted
3	1.45 ± 0.07	Polluted	11.15 ± 0.00	Polluted
4	0.70 ± 0.00	Polluted	6.90 ± 0.00	Polluted
5	1.20 ± 0.00	Polluted	6.16 ± 0.00	Polluted
6	0.00 ± 0.00	-	5.31 ± 0.00	Polluted
7	3.35 ± 0.07	Polluted	13.27 ± 0.00	Polluted
8	0.90 ± 0.00	Polluted	12.21 ± 0.00	Polluted

3.1.2 The Seawater
Analysis of Hg and HCN levels was also carried out in seawater, which has a distance of ± 1 km from the disposal pond.

Table 4. Hg and HCN levels in seawater

Points	Concentration (ppm)			
	Hg	Information	CN	Information
1	0.00 ± 0.00	-	6.37 ± 0.00	Polluted
2	1.57 ± 0.00	Polluted	6.90 ± 0.00	Polluted

3.1.3 Well Water
Analysis of Hg and HCN levels was also carried out in well water within ± 10 m from the holding pond.

Table 5. Hg and HCN levels in well water

Points	Concentration (ppm)			
	Hg	Information	CN	Information
1	0.30 ± 0.00	Polluted	5.31 ± 0.00	Polluted
2	2.18 ± 0.00	Polluted	9.02 ± 0.00	Polluted

3.2 Hg and HCN levels in animals
Analysis of Hg levels in animals showed that the lowest to highest concentration was 0 to 48.91 ppm.

Table 6. Hg Levels in Animals

No	Species	Hg on bone/shell (ppm)	Hg on meat (ppm)	Information
1	*Acetes indicus*	5.28 ± 0.19	0.00 ± 0.00	Polluted
2	*Perna viridis*	6.72 ± 0.17	0.00 ± 0.00	Polluted
3	*Polymesoda expansa*	6.37 ± 0.10	0.00 ± 0.00	Polluted
4	*Plotosus canius*	48.91 ± 0.49	1.36 ± 0.00	Polluted
5	*Scylla serrata*	5.45 ± 0.10	23.43 ± 0.66	Polluted
6	*Tegillarca granosa*	9.99 ± 0.13	20.37 ± 0.36	Polluted
7	*Moolgarda seheli*	43.91 ± 0.49	0.00 ± 0.00	Polluted
8	*Crassostrea gigas*	7.41 ± 0.10	0.00 ± 0.00	Polluted
9	*Menippe mercenaria*	9.73 ± 0.12	17.49 ± 0.25	Polluted
10	*Strombus canarium*	5.45 ± 0.10	11.84 ± 0.29	Polluted

Analysis of HCN levels in animals showed that the lowest to highest concentration was 8.84 to 90.22 ppm.

Table 7. HCN Levels in Animals

No	Species	HCN on bone/shell (ppm)	HCN on meat (ppm)	Information
1	*Acetes indicus*	18.88 ± 0.01	56.51 ± 0.02	Polluted
2	*Perna viridis*	14.15 ± 0.00	50.20 ± 0.02	Polluted
3	*Polymesoda expansa*	8.84 ± 0.00	64.00	Polluted
4	*Plotosus canius*	17.66	25.43 ± 0.00	Polluted
5	*Scylla serrata*	22.99 ± 0.00	18.21 ± 0.01	Polluted
6	*Tegillarca granosa*	17.68 ± 0.00	90.22	Polluted
7	*Moolgarda seheli*	30.21 ± 0.01	51.14 ± 0.02	Polluted
8	*Crassostrea gigas*	20.16 ± 0.00	79.40	Polluted
9	*Menippe mercenaria*	28.30 ± 0.00	22.67 ± 0.01	Polluted
10	*Strombus canarium*	15.91 ± 0.00	34.01 ± 0.00	Polluted

Sekotong people's gold mining is known to have experienced Hg and CN pollutions with levels exceeding the threshold value set by WHO [30, 31], Regulation of the Ministry of Health of the Republic of Indonesia Number 57 of 2016 [32], and Regulation of the Ministry of Environment of the Republic of Indonesia Number 5 of 2014 [33]. Hg and CN pollution in Sekotong people's gold mining is known to have polluted the aquatic environment around the mining site, animals, to humans with very high and dangerous levels. Surely this is a hazardous condition because it is known that heavy metal Hg is a heavy metal that is very toxic and dangerous [17, 18, 19], quickly accumulated [34], can cause digestive disorders, kidney and central nervous system [18] as well as cyanide [20, 21, 35].

Hg is a heavy metal that is very easy to accumulate in the environment, settles in the bottom waters [19, 36], can easily evaporate and pollute the air [37]. This is evidenced by a large number of Hg-contaminated species, especially species that live in the waters, that Hg can easily contaminate aquatic organisms until they accumulate in the human body and enter the blood circulation [38]. Likewise, CN is known to pollute the environment through air, soil, water, and food [30, 39]. CN can settle in waters, contaminate the surrounding environment, and endanger human health [40]. The results showed that the aquatic environment, animals, to humans are known to have experienced Hg and CN pollution, which is very high and certainly very dangerous for the environment.

4. Conclusion

Sekotong people's gold mining has been contaminated with Hg and CN in aquatic and animal environments exceeding the specified threshold value. We also recommend that more in-depth research is needed to find solutions to reduce the impact of Hg and CN pollution due to the Sekotong people's gold mining activities.

Acknowledgements
The authors would like to thank the Ministry of Research, Technology and Higher Education of Republic Indonesia under "*Direktorat Riset dan Pengabdian Masyarakat*" grant in the fiscal year of 2019.

References

[1] Utomo W. H, Suntari R, Arfarita N, Suhartini and Handayanto E 2014 *Am. J. Sustain. Agric.* **8** 34
[2] Lamare A 2013 *Top 10 Gold Producing Countries in the World* (Online) (https://www.celebritynetworth.com/articles/entertainment-articles/top-10-gold-producing-countries-world/)
[3] INCAM 2013 *Indonesian Centre for Artisanal Mining (INCAM) extended concept note* (Internal report)
[4] Ismawati Y 2010 *Mercury Hotspots in Indonesia* (Denpasar: Bali Fokus).
[5] Aspinall C 2001 *Small-scale Mining in Indonesia* (England: International Institute for Environment and Development and the World Business Council for Sustainable Development)
[6] Hentschel T, Hruschka F and Priester M 2002 *Artisanal and Small-scale Mining, Challenges and Opportunities* (London: International Institute for Environment and Development (IIED) and World Bussiness Council of Sustainable Development (WBCSD))
[7] Widi B N and Edya P 2015 *Prosiding Hasil Kegiatan Pusat Sumber Daya Geologi Tahun 2015* (Jakarta: Kementerian Energi dan Sumber Daya Mineral).
[8] Badan Pusat Statistik (BPS) Lombok Barat 2014 *Jumlah Produksi Bahan Galian Golongan C Menurut Jenis Galian.* (Online), (https://lombokbaratkab.bps. go.id/)
[9] Somantri, N. D. Monday, 8 January 2018 (*Personal Communication*)
[10] *Peraturan Daerah Provinsi Nusa Tenggara Barat Nomor 11 Tahun 2006 tentang Rencana Tata Ruang Wilayah Provinsi Nusa Tenggara Barat (Lembaran Daerah Provinsi Nusa Tenggara Barat Tahun 2006 Nomor 11, Tambahan Lembaran Daerah Provinsi Nusa Tenggara Barat Tahun 2006 Nomor 21)*
[11] Suhartini and Abubakar 2017 *J. Degrad. Min. Lands Manag.* **4** 789
[12] Sancayaningsih R P, Sutariningsih E, Hadisusanto S, Purnomo, Mulyanti, Sembiring L, and Sudibyo P 2010 *Studi Ekologi dan Kandungan Merkuri Pada Pertambangan Emas Tradisional di Kecamatan Sekotong, Lombok Barat* (Yogyakarta: Universitas Gadjah Mada)
[13] Krisnayanti B D 2018 *J. Degrad. Min. Lands Manag.* **5** 1077
[14] Pepper I L, Gerba C P, and Brusseau M L 2006 *Environmental & Pollution Science* (California: Elsevier Inc)
[15] Hill M K 2010 *Understanding Environmental Pollution* (Cambridge: Cambridge University Press)
[16] *Undang-undang Republik Indonesia Nomor 32 Tahun 2009 Tentang Perlindungan dan Pengelolaan Lingkungan Hidup (Lembaran Negara Republik Indonesia Tahun 2009 Nomor 140, Tambahan Lembaran Negara Republik Indonesia Nomor 5059)*
[17] Palar H 2012 *Pencemaran dan Toksikologi Logam Berat* (Jakarta: PT Asdi Mahasatya)
[18] Suharto 2011 *Limbah Kimia dalam Pencemaran Air dan Udara* (Yogyakarta: Penerbit Andi)
[19] Widowati W, Sastiono A, and Jusuf R 2008 *Efek Toksik Logam: Pencegahan dan Penanggulangan Pencemaran* (Yogyakarta: Penerbit Andi)
[20] Tiwari D, Kim H U, Choi B J, Lee S M, Kwon O H, Choi K M and Yang J K 2007 *J. Environ. Sci. Heal. Part A* **42** 803
[21] Uppal H, Ttripathy S S, Chawla S, Sharma B, Dalai M K, Singh S P, Singh S and Singh N 2017 *J. Environ. Sci.* **55** 76

[22] Velásquez-López P C, Veiga M M, Klein B, Shandro J A and Hall K 2011 *J. Clean. Prod.* **19** 1125
[23] Herman D Z 2006 *Indones. J. Geosci.* **1** 31
[24] Baird R B, Eaton A D and Rice E W 2017 *Standard Methods for the Examination of Water and Wastewater, 23rd Edition* (Washington, DC: American Public Health Association, American Water Works Association, Water Environment Federation)
[25] Hadi A 2015 *Pengambilan Sampel Lingkungan* (Jakarta: Penerbit Erlangga)
[26] Texas Natural Resource Conservation Commision (TNRCC) 1999 *Surface Water Quality Monitoring Procedure Manual* (Texas: Texas Natural Resource Conservation Commision)
[27] Badan Standardisasi Nasional (BSN) 2011a *Cara Uji Raksa (Hg) Secara Spektrofotometri Serapan Atom (SSA) - Uap Dingin atau Mercury Analyzer* (Jakarta: Badan Standardisasi Nasional)
[28] Badan Standardisasi Nasional (BSN) 2011b *Cara Uji Sianida (CN^-) Secara Spektrofotometri.* (Jakarta: Badan Standardisasi Nasional)
[29] Herbert D G, Jones G J and Atkinson L J 2018 *Phylum Mollusca In: Atkinson LJ and Sink KJ (eds) Field Guide to the of Shore Marine Invertebrates of South Africa* (Pretoria: Malachite Marketing and Media)
[30] World Health Organisation (WHO) 2007a *Cyanide in Drinking-water* (Geneva: WHO Document Production Services)
[31] World Health Organisation (WHO) 2007b *Exposure to Mercury: A Major Public Health Concern* (Geneva: WHO Document Production Services)
[32] *Peraturan Menteri Kesehatan Republik Indonesia Nomor 57 Tahun 2016 Tentang Rencana Aksi Nasional Pengendalian Dampak Kesehatan Akibat Pajanan Merkuri Tahun 2016-2020 (Berita Negara Republik Indonesia Tahun 2016 Nomor 1751)*
[33] *Peraturan Menteri Lingkungan Hidup Republik Indonesia Nomor 5 Tahun 2014 Tentang Baku Mutu Air Limbah (Berita Negara Republik Indonesia Tahun 2016 Nomor 1815)*
[34] Pirrone N, Cinnirella S, Feng X, Finkelman R B, Friedli H R, Leaner J et al 2010 *Atmos. Chem. Phys.* **10** 5951
[35] Dasha R R, Gaur A, Balomajumder C 2009 *J. Hazard. Mater.* **163** 1
[36] Pavlish J H, Sondreal E A, Mann M D, Olson E S, Galbreath K C, Laudal D L 2003 *Fuel Process. Technol.* **82** 89
[37] Wu Q, Wang S, Yang M, Su H, Li G, Tang Y and Hao J 2018 *J. Environ. Sci.* **68** 91
[38] Lee C C, Chang J W, Huang H Y and Chen H L 2012 *Sci. Total Environ.* **424** 316
[39] International Programme on Chemical Safety (IPCS) 2004 *Hydrogen Cyanide and Cyanides: Human Health Aspects* (Geneva: World Health Organization, International Programme on Chemical Safety)
[40] Brüger A, Fafilek G, Restrepo B O J and Rojas-Mendoza L 2018 *Sci. Total Environ.* **627** 1167

The Composition of Water and Ash of Secang Wood's Simplicia and Secang Wood Herbal Drink Powder

Anggarani M A[1*], Ayuningsih A D[1], Yudianto E[2], Prasodi B A[2]
[1]Department of Chemistry, Universitas Negeri Surabaya, Surabaya, Indonesia
[2]Faculty of Mathematics and Science, Universitas Negeri Surabaya, Surabaya, Indonesia

e-mail: mirwaanggarani@unesa.ic.id

Abstract. Empirically, secang wood (Caesalpinia sappan L.) has medicinal properties and is consumed as a health drink. Secang wood contains tannin, gallic acid, brazilin, tannic acid, essential oils, polyphenols and sappanin. Switching people's tastes from consuming the basic form of sappan wood to a more attractive and practical form, which is an herbal drink that contains the properties. The secang drink is measured by water and ash content. Water content is related to the stability index while ash content about minerals is filled. The initial probability of the quality of the secang herbal drink can be compared with that of both aspects. The lower the water content in a product, the better the quality, the higher the stability which can be further improved during storage. While the higher the ash content, the higher the probability of mineral content. Based on the analysis of water content and ash content, it is known that the simplicia of the secang wood is 1.542% while the water content of the herbal drink powder is 0.266%. Meanwhile, based on the analysis of ash content, ash content of simplicia secang bark is 9,449% while the water content of herbal beverage powder is 0,009%.

1. Introduction

Indonesia is a mega diversity country for medicinal plants in the world. Indonesia's tropical forests have high biodiversity. Of the 40,000 species of flora in the world, as many as 30,000 species are found in Indonesia and 940 of them are known to be efficacious as drugs that have been used in traditional medicine for generations. The number of medicinal plants covers about 90% of the number of medicinal plants found in the Asian region [1,2].

The use of traditional medicines in Indonesia is essentially part of the culture. The advantage of using herbs in principle is relatively small side effects compared to modern medicine. Although empirically herbal medicine can cure various diseases, but its efficacy and ability have not been proven scientifically or clinically. Also, it is not yet known what chemical compounds are responsible for the efficacy of these herbs [3].

One of the plants that is widely used in traditional medicine is secang (Caesalpinia sappan L.). The main content is brazilin, which is red-sappan, tanic acid, and gallic acid. Brazilin from sappan wood is scientifically tested to be antioxidant, antibacterial, anti-inflammatory, anti-photoaging, hypoglycemic (reducing fat content), vasorelaxant (relaxing blood vessels), hepatoprotective (protecting the liver) and anti-acne (anti-acne) [4]. Secang wood extract is also suspected to have anti-tumor, anti-virus, immunostimulant and other efficacy [5]. Traditionally, secang wood pieces are commonly used as a mixture of medicinal ingredients.

The factor that most influences a person's health condition is food and beverage intake. This is because food and drinks contain nutrients needed by the body in the form of bioactive compounds [6]. Secang wood in addition to having an attractive color when brewed, also has a distinctive and refreshing taste. Seeing the many benefits of secang, secang drinks were developed in the form of powder so that they could meet the demands of modern society. On the other hand, the demands of a person's need for a healthy life must still be met, therefore also developing food or health drinks, one of which is secang powder drink which is commonly called wedang secang. The main ingredient of wedang wedang secang besides spices is sugar and salt which serves to improve the organoleptic properties, especially taste.

Water in food is not only used to meet human needs, but water also plays a role in the food itself. The presence of water in food is often related to the quality of food, as a measure of dry or solid parts, determinants of stability indexes during storage, and determinants of organoleptic quality, especially taste.

In addition to organic matter and water, food ingredients contain inorganic compounds called minerals or ash. Although the amount is very small, the presence of minerals in food is needed by the human body. In the body minerals function as builders and regulators. Certain minerals are needed as a constituent of bones, teeth, and soft tissues, muscles, blood, and nerve cells, and other metabolisms. Because of the diversity of available mineral sources, ash and mineral analysis is very important to know the nutritional quality of food. Besides being able to know the nutritional quality, ash analysis is very often done as an indicator of food quality. From the ash analysis it can be seen the level of purity of the powder product. Based on the level of importance of knowing the water and ash content in a secang simplicia product and the herbal secang powder, it is necessary to analyze the water content/ash content and the ash content of the two products.

2. Materials and Methods

The moisture content of the material indicates the amount of water content of the weight of the material. In this case there are two methods for determining the moisture content of the material based on dry weight and a wet basis. The moisture content of agricultural products is usually determined based on wet weight. In this calculation the following formula applies: KA = (Wa / Wb) x100% [7]. Meanwhile, ash content is a mixture of inorganic or mineral components found in food. Food consists of 96% inorganic material and water, while the rest is mineral elements. The element is also known as organic matter or ash content. Ash content can indicate the total minerals in a food. Organic materials in the combustion process will burn but the inorganic components do not, because that is called ash content. Determination of total ash content can be used for various purposes, including determining whether or not processing, find out the type of material used, and as a determinant of the nutritional value parameters of a food ingredient.

The sample used was the simplicia wood secang and secang herbal drink powder produced by Griya Sehat SMEs, Trawas, Mojokerto. Both samples were analysed for water content and ash content.

2.1 Organoleptic Test

To find out the characteristics of Simplicia Secang and Secang Herbal Powder, identification was done by organoleptic observation including shape, color, and odor.

2.2 Water Content Analysis (AOAC 2000)

Homogeneous samples were weighed 2 grams and placed in an empty cup which had been weighed, the cup and lid were dried in the oven and cooled in a desiccator. The cup containing the sample is then covered and put in an oven at 100°C for 5 hours or until the weight is constant. The cup is then cooled in a desiccator and after the cold the cup is weighed. Moisture content can be calculated by the formula:

$$Water\ content\ (wet\ basis)(\%) = \frac{W1 - W2}{W1} x100\%$$

Information:
W1 = initial sample weight (g)
W2 = sample weight after drying (g)

2.3 Ash Content Analysis (AOAC 2000)
A sample of 2 grams was put into a grey cup which had been weighed and burned in a furnace and cooled in a desiccator. The cup containing the sample is put in a greying furnace and burned until a greyish ash is obtained. The heating temperature is gradually increased until the temperature reaches 650°C and left for 1 hour. After the furnace temperature drops around 200°C, the ash-filled dish is cooled in a desiccator for 30 minutes and then weighed. This treatment is repeated until it reaches a constant weight. Ash content can be calculated by the formula:

$$Water\ content\ (wet\ basis)(\%) = \frac{Ash\ weight\ (g)}{Sample\ weight\ (g)} x100\%$$

3. Result and Discussion
Simplisia and Secang herbal powder need to be characterized to get simplicia and herbal drink powder which are safe, have good quality, standardized, and tested for stability so that the resulting preparations are guaranteed quality.

Table 1. Organoleptic test results

Parameter	Characteristics
Simplicia secang	
Form	Rough cuts
Color	Dark red
Scent	Fragrant sting
Secang drink powder	
Form	Powder granules
Color	Orange
Scent	Fragrant sting

Water content is one of the physical properties of the material which shows the amount of water contained in the material. Water content is usually expressed as a percentage of water weight to wet material or in grams of water for every 100 grams of material called wet basis water content. The weight of dry or solid material is the weight of the material after having been heated for a certain amount of time so that the weight remains (constant). This is following Suharto (1991), who said that the water content is the percentage of water content of an ingredient that can be expressed on a wet basis or wet basis. Wet weight water content has a theoretical maximum limit of 100%, while water content based on dry weight can be more than 100%.

Ash content is a mixture of inorganic or mineral components found in a portion of food. Food consists of 96% inorganic material and water, while the rest is mineral elements. Determination of total ash content can be used for various purposes, including determining whether the processing is good or not, knowing the type of material used, and determining the nutritional value parameters of a food ingredient. That the ash content is an inorganic substance leftover from the combustion of organic material. The ash content and composition depends on the type of material and how it is ignited.

Simplisia secang wood is a herbal food product (brewed drinks) with little. Fresh secang wood is thinly sliced 0.5 cm thick, then dried in the sun for about 5 days until dry and has a constant weight. Simplisia secang wood produced can be directly consumed by brewing with warm water. While the secang herbal powder is made by smoothing the sappan wood then squeezing the juice of the starch and cooking the juice of the starch to form crystals. Furthermore, the crystals that are formed are crushed and the resulting secang herbal powder can be consumed by brewing using warm water.

Both herbal beverage products produced must have good food product criteria. One of them is from the parameters of water content and ash content. The lower the water content of a solid food product, the better the quality. While the greater ash content indicates the number of minerals in the food product.

Table 1 outlines the quality requirements for powdered drinks based on SNI 01-4320-1996. Simplisia secang wood produced has a water content of 1.542%, while the water content of herbal drink secang powder is 0.266%. Meanwhile based on the analysis of ash content, it is known that the simplicia ash content of the secang wood is 9,449%, while the ash content of the secang herbal drink powder is 0.009%.

Table 2. Quality requirements for powdered drinks are based on SNI 01-4320-1996

No.	Testing criteria	Units	Requirement
1	Colour		Normal
2	Aroma		Normal, typical spices
3	Taste profile		Normal, typical spices
4	Water level, b/b	%	3.0-5.0
5	Ash level, b/b	%	Max. 1.5
6	Sugar (counted as saccharose)	%	Max. 85%
7	Food additives		
8.1	Artificial sweeteners		
	Saccharine		Must not exist
	Cyclamate		Must not exist
8.2	Food coloring additives		As SNI 01-0222-1995
9	Metal contaminant		
9.1	Lead (Pb)	mg/kg	Max. 0.2
9.2	Copper (Cu)	mg/kg	Max. 2.0
9.3	Zinc (Zn)	mg/kg	Max. 50
9.4	Tin (Sn)	mg/kg	Max. 40
10	Mercury (Hg)	Mg/kg	Must not exist
11	Arsenic contaminant	Mg/kg	Max. 0.1
12.1	Microbial contaminant		
12.2	Total plate	Colony/g	3×10^3
12.3	Coliform	APM/g	< 3

If referring to table 1, the water content of the two products is below the required numbers. On the other hand, the ash content of herbal secang powder is 0.009% which means it is in the permissible range. Whereas ash content of simplicia wood products of secang is 9,449% where the number is above the permitted number. Thus between the two products, which are more recommended for consumption is the secang herbal drink powder.

From the analysis of ash content, these results are within the range set by Indonesian Herbal Pharmacopoeia that the ash content does not exceed 2% for simplicia and does not exceed 1.4% for extracts. Therefore, the ash content of simplicia produced in this study was in line with these rules, while for the secang herbal powder was above it. The water content of simplicia and herbal powder drink produced in this study was occurdance with the rules. This water content determines the stability of the extract and the subsequent dosage form.

4. Conclusion

Water content and ash content, it is known that the simplicia of the secang wood produced from this study is 1.542% while the water content of the herbal drink powder is 0.266%. Whereas based on the analysis of ash content, ash content of simplicia secang bark is 9,449% while the water content of herbal beverage powder is 0,009%. Therefore, the ash content of simplicia produced in this study was in accordance with Indonesian Herbal Pharmacopoeia, while for the secang herbal powder was above it. The water content of simplicia and herbal powder drink produced in this study was accordance with the rules.

5. Acknowledgment

We would thank all the students of Universitas Negeri Surabaya participating this research and also thank to Indonesian Government for the support for this research.

6. Reference

[1] Conservation International 1997 Megadiversity New York (USA)
[2] World Wide Fund for Nature 2012 Living Planet Report 2012 Netherlands (NL)
[3] Wijayakusuma H 2002 Tumbuhan Berkhasiat Obat Indonesia Rempah, Rimpang dan Umbi Jakarta: Prestasi Instan
[4] Nirmal N P, Rajput M S, Prasad R G, and Ahmad M 2015 *Asian Pacific Journal of Tropical Medicine* **8**(6) 421-430
[5] Badami S, Moorkoth S, Suresh B 2004 *Caesalpinia sappan a medical and dye yielding plant Natural Product Radiance* 3(2):75-82
[6] Nutrisia A S, Indri K D, Youstiana D R 2018 Jurnal Terpadu Ilmu Kesehatan **7**(1)
[7] Taib G 1988 Operasi Pengeringan pada Pengolahan Hasil Pertanian (PT Mediyatama Sarana Perkasa: Jakarta)

Journal of Physics: Conference Series **1417** (2019) 012034 doi:10.1088/1742-6596/1417/1/012034

Physical Characterization of Injectable Bone Substitute Associated-3D Printed Bone Scaffold for Spinal Tuberculosis

I F Wardhani[1], R M R Samudra[2], Katherine[3], D Hikmawati [4], Aminatun [5]

[1,2,3] Biomedical Engineering Study Program, Universitas Airlangga, Indonesia.
[4,5] Physics Department, Universitas Airlangga, Indonesia.
(usi.ifw@gmail.com, rofimega400@gmail.com, katherinehouw@yahoo.co.id)
Co-author : dyah.hikmawati@yahoo.co.id

Abstract. There are two important factors to overcome spinal tuberculosis, *i.e.* killing and preventing the *Mycobacterium tuberculosis* to spread, also replacing and reconstructing the damaged bone. This study focused on an innovation to overcome the spinal tuberculosis problem by combining the IBS and 3D printed bone scaffold. This innovation could provide a local drug delivery system along with tissue regeneration and bone reconstruction. In this study, 3D printed bone scaffold which have been injected with IBS were tested its physical characterizations. The bone scaffold was fabricated from Polylactide acid (PLA) filament using FDM method through an ordinary commercial 3D printer. The IBS was synthesized from nanohydroxyapatite, gelatin, streptomycin, and HPMC. The physical characterization performed in this study were porosity test functional group test, and degradation test. From those characterizations, it could be concluded that the IBS associated-3D printed bone scaffold is a potential alternative to overcome spinal tuberculosis.

1. Introduction

It was reported that Indonesia still be the 3rd country worldwide with the most tuberculosis (TB) sufferers in 2017. This disease was one of 10 main causes of the death among the people in all over the world [1]. In particular cases of TB, when it got more serious, the bacteria could also infect the other organs, including the musculoskeletal organs. The plenty vascularization to the spinal bone causes the bacteria, which previously had been lodged in the lungs, to reactivate in this spot. This case is named as spinal tuberculosis, which was reported to reach up to 50% of the musculoskeletal tuberculosis case in total [2].

Spinal tuberculosis could cause the disturbance of nerve system and permanent deformation of the spine. The medical practitioners offers a conventional treatment through the drugs. This kind of treatment needs a high level dose of medicine, yet it also needs a long time treatment to be able reach the spot of bacterial infection [2]. While the bacteria grew in the spine, it will cause the bone destruction which has to be corrected by the surgical treatment. The surgical treatment was necessarily done to remove the infected bone tissue, then replace it with the bone graft [3].

Scaffold was one kind of bone graft which offers to help the new bone cell to grow as well as to reconstruct the geometry of the spine. The scaffold must meet some characteristics, i.e. biocompatible, an appropriate pore interconnectivity, also suitable pore size and shape [4]. To provide those characteristics, a new technology, which has been developed in the era of Industrial Revolution 4.0, known as additive manufacturing, offers an efficient and low cost way to fabricate the bone scaffold by using 3D printing technology [5]. The fabrication could simply be done, started by designing the bone scaffold using CAD (Computer Aided Design) application. This application allowed us to design a custom bone scaffold based on the patient's case need.

Injectable bone substitute (IBS) paste which was synthesized by loaded the antituberculosis drug, *i.e.* streptomycin, had been proven to be an alternative to overcome the spinal tuberculosis [6]. This research was aimed to synthesize the 3D printed bone scaffold, then combine it with IBS paste to be a more effective alternative to be able overcoming spinal tuberculosis. This idea was due to the need of bone geometry reconstruction and urge to kill the bacteria so that it could not spread to the wider area.

2. Materials and Method

2.1. Materials

The materials used in this study included Polylactid Acid (PLA) filaments to make bone scaffold. While, the main ingredient for making IBS paste, including nanohidroksiapatit powder, gelatin, Hydroxypropyl Methylcellulose (HPMC), and streptomycin sulfate (powder for injection) packaging 1 gram vial. In addition, materials that are useful as solvents, namely aquades, are also needed.

2.2. Bone Scaffold Manufacturing

The manufacturing of bone scaffold was begun with designing process using Autodesk AutoCAD 2019-Student Version. The scaffold was built by ordering a single fibre to one another. The space between 2 fibres would be the pore size of the scaffold. The fibre was designed using a 2-dimensional line with 14 mm length and a circle with 1.2 mm diameter. The circle was then swept using "sweep" command along the line as shown in the Figure 1.

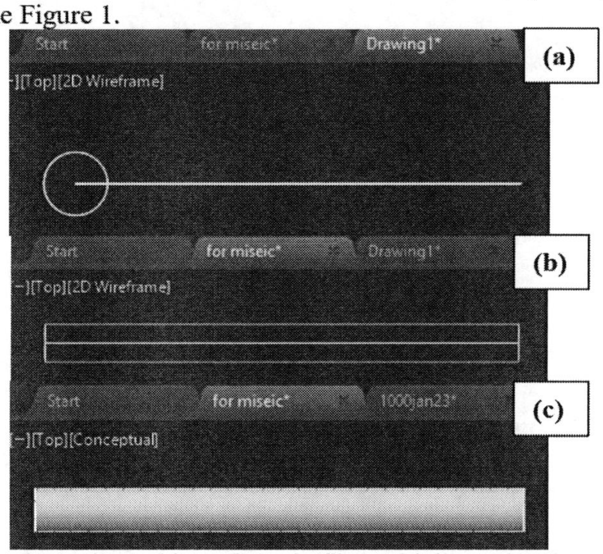

Figure 1. Top view of (a) The making of single fibre which consisted of a line and circle; (b) The circle that had been swept along the line; (c) The fibre in conceptual view

The single fibre was then copied and ordered side by side with the 1.2 mm space between one another to make the 1st layer. The 2nd layer was made from the 90° rotation of the 1st layer, as could be seen in Figure 2. Furthermore, the 3rd and 4th layer were the copying result of the 1st and 2nd layer. The process was then repeated until the 14th layer had been made. The final design which had been made could be seen in Figure 3.

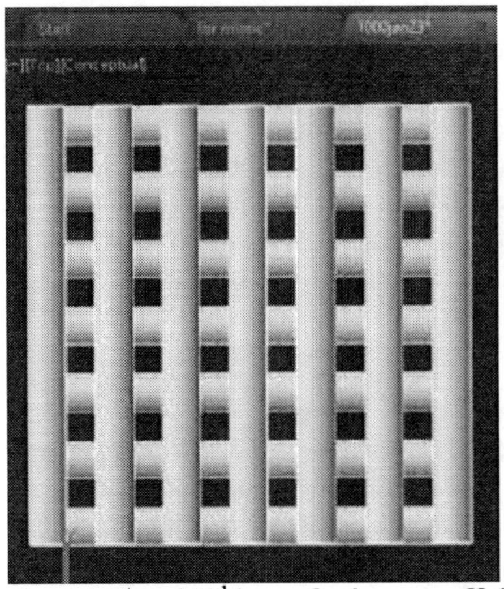

Figure 2. The design of 1st and 2nd layer for bone scaffold manufacturing

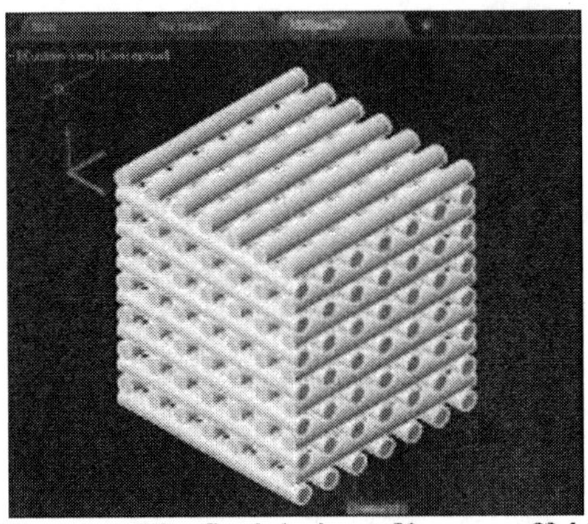

Figure 3. The final design of bone scaffold

Extension from the AutoCAD 2019 application in the form of .dwg was then converted to the .stl form via the Cura from Ultimaker application by looking at the slicing parameters of the design. This stereolithography (.stl) extension could later be read by a 3D printer machine, so the design can be printed. Slicing parameters themselves were a layer-by-layer display that will be printed using a 3D printing machine. This was important to know because from here we could determine whether the design can be printed by a 3D printer machine. Figure 4 would show the slicing parameters of the scaffold design. The filament used is PLA with a diameter of 1.75 mm. To make a good print accuracy, a nozzle is used with hole diameter of 0.2 mm.

Figure 4. 3D design slicing parameter of bone scaffold

2.3. IBS Paste Synthesis

Based on research conducted by Maulida, et al (2015) IBS paste was made with the main ingredients nano hydroxyapatite, gelatin, HPMC, and streptomycin. At first 20% w / v gelatin was dissolved in distilled water at 40°C for 1 hour. In the same time, HPMC 4% w / v was dissolved in distilled water at a temperature of 90°C for 1 hour then left to stand until the temperature of the solution became 40°C. In the gelatin solution that had been made, hydroxyapatite nano powder was added according to the ratio of the composition of HA: gelatin 65:35 (w / w), this mixture was then stirred for 1 hour. After both were homogeneous, the next step was adding 10% streptomycin of the total mass, then stir until homogeneous. Furthermore, the HPMC solution was mixed into HA-gelatin-streptomycin at a mixing temperature of 40°C with stirring for 6 hours.

2.4. Sample Characterization

Porosity test was conducted to know the percentage of pores contained in the scaffold. Scaffold porosity can be calculated using Equation 1 [7]. The density value of the material used, namely PLA, is 1.24 g / cm^3 [8]. Where $v_{scaffold}$ was volume of scaffold and $w_{scaffold}$ was mass of scaffold.

$$porositas\ (\%) = \frac{\left(v_{scaffold} - \frac{w_{scaffold}}{\rho}\right)}{v_{scaffold}} \times 100\% \qquad (1)$$

Then, the IBS associated 3D printed bone scaffold were characterized its chemical identity using FTIR (Thermo Scientific Nicolet iS10). The method used was attenuated total reflectance (ATR). The 3D printed scaffold was tested before and after the IBS injection. Both samples were placed in sample holder, then recorded.

The in vitro degradation test was conducted using *Phosphate Buffer Saline* (PBS) solution which represented the body liquid. The 3D printed scaffold was soaked in PBS solution. The scaffold mass and PBS pH value were then evaluated every 6 days in 18 days.

3. Result

3.1. IBS Associated 3D Printed Bone Scaffold

The 3D printing for bone scaffold fabrication was shown satisfactory result as could be seen in Figure 5. It could be seen that macroscopically, the pore and its interconnectivity were very well printed. Using image J application, the scaffold pore size measured and resulted 1172.614 µm in average. From that measurement, it could be known that the capability of the 3D printing to print the scaffold compared to the design was 97.718%. While the porosity of the bone scaffold was measured and resulted 65.329 ± 0.603 % value of porosity. The pores of the bone scaffold were then filled by the IBS paste which had been able to be made and loaded with antituberculosis drug. The IBS associated 3D printed bone scaffold could be seen in Figure 6.

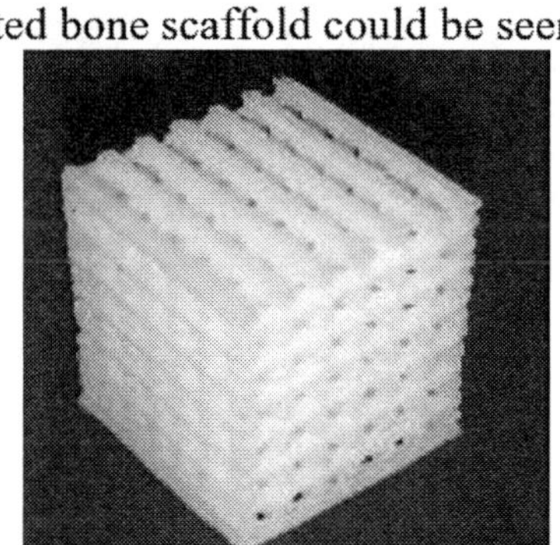

Figure 5. 3D printed bone scaffold before IBS injection

Figure 6. 3D printed bone scaffold after IBS injection

3.2. FTIR Test

The FTIR test was conducted to know the functional group of the bone scaffold before and after the IBS injection. In the Figure 7 provided the FTIR test result of both samples. The chemical identity of scaffold showed typical functional group spectrum of PLA. Stretching C-H spectra were found at wavenumber 2999.93 and 2944.39 cm^{-1}, C=O bonding spectrum was at 1744.57 cm^{-1}, and C-O was found at range 1179.69 until 1041.74 cm^{-1}.
What made different between scaffold before and after the IBS injection were shown by existence of O-H functional group at 3266.24 cm^{-1}, stretching N-H at 1636.77 cm^{-1}, and PO_4^{3-} bonding at 599.33 cm^{-1}. Those functional groups were the specific IBS paste ingredients' spectra.

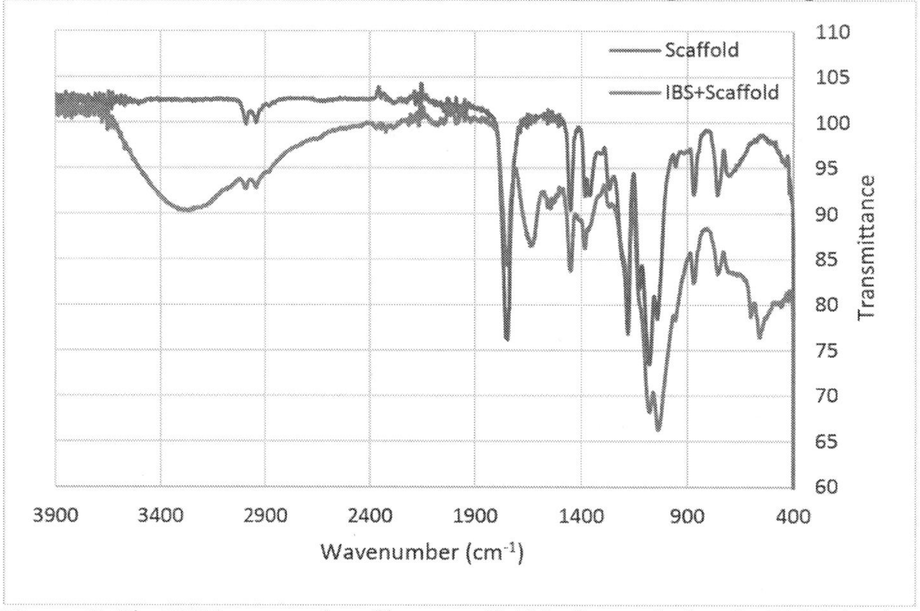

Figure 7. The FTIR test results of bone scaffold before and after the IBS injection

3.3. In Vitro Degradation Test

The degradation test was conducted to ensure that scaffold based on PLA was a biodegradable material. After 18 days, the mass of scaffold had not changed significantly, as could be seen in Figure 8. Otherwise, there was pH value decreasing over time, as shown in Figure 9.

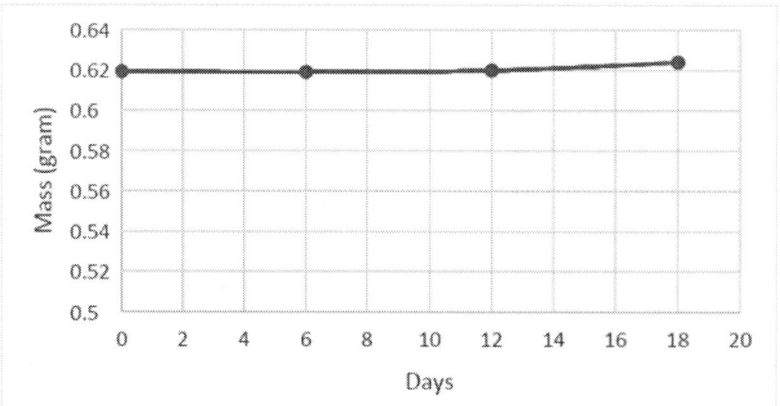

Figure 8. The changing of bone scaffold in 18 days

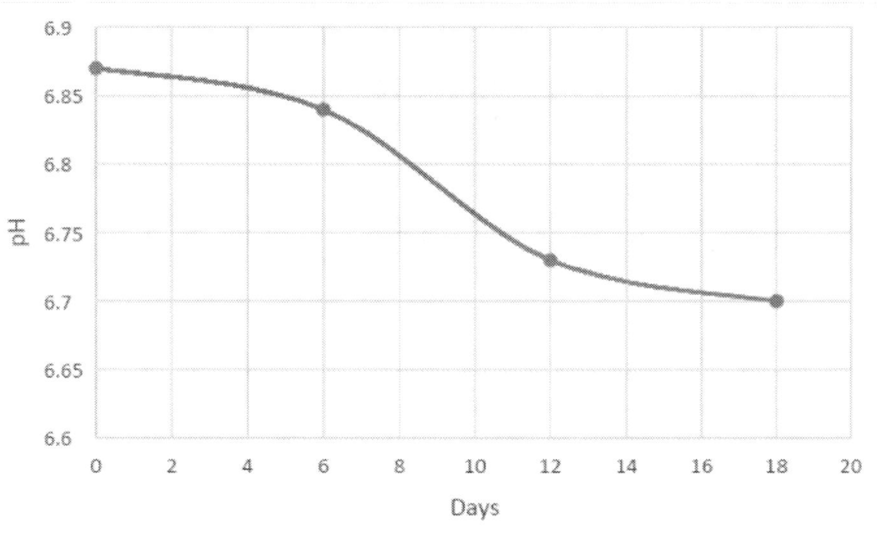

Figure 9. The PBS solution's pH value decreasing in 18 days of scaffold soaking

4. Discussion

It had been proven that 3D printing technology could be used to fabricate bone scaffold with 97.718% accuracy compared to its design. The scaffold, which was fabricated in this study, had porosity above 50% that meant it was like any common scaffold which could be found in the market [5]. The fabrication of IBS associated-scaffold was successfully done, indicated by its chemical identity. From the FTIR test, it was known that the sample had specific functional group spectra from its ingredients.

One of important indicators of scaffold is the biodegradable character. Scaffold was implanted to be an extracellular matrix that could help the new osteoblast growth [4]. When bone has already grown appropriately, the bone scaffold itself has to be degraded by human body fluid. It had been proven in this study that bone scaffold based on PLA was a biodegradable material, which could be seen on the data explained in section 3.3. Although, there had not been mass decreasing in 18 days, the pH value decreasing could be shown as a proof that the scaffold was biodegradable. The decreasing pH value indicated that there was monomerization process during the scaffold soaking. PLA would be degraded by PBS and produced lactic acid.

The degradation profile of IBS paste itself was not studied more deeply in this study. From the results of the study of Maulida, et al (2015) showed that the dried IBS paste could be degraded within 48 hours in a Simulated Body Fluid (SBF) solution. This can be a reference that IBS paste would be degraded first and released the drug to kill bacteria before the PLA scaffold degradation process happened. This ability of killing bacteria had been proven by Maulida, et al (2015) in their research. It had been tested by using *Saccharomyces aureus* (SA) and resulted very sensitive results on the SA inhibition zone [6]. The degradation of IBS paste would also decompose the minerals contained in its constituent ingredients, especially the minerals contained in nano HA. These minerals would help the process of bone regeneration through the process of osteoconductivity [9].

After this IBS paste degradation process, it was hoped that the bacteria that caused spinal tuberculosis had been resolved and new bone cells (osteoblasts) could begin to grow. The next healing mechanism is to make bone scaffold as an extracellular matrix for osteoblasts to attach to and develop into bone tissue as before. Until finally the scaffold will degrade by itself when new bone tissue has formed to replace bone damaged by bacteria.

5. Conclusion

In conclusion, 3D printing technology could be potentially utilized in medical area, especially to overcome spinal tuberculosis. The IBS associated- 3D printed bone scaffold is a potential alternative regarding to this case. For future study, there is still a need to develop this research to know whether the scaffold could really control the drug release by its pore size and design.

6. Acknowledgments
The authors would like to express appreciation for the support to Kemristekdikti in the form of research grant through Program Kreatifitas Mahasiswa (PKM) program.

References
[1] World Health Organization 2018 *Global Tuberculois Report Executive Summary* (Geneva: WHO)
[2] Shaikh N S and Sawakar S P 2017 *iMedPub J.,* **4** 1
[3] Do Brito J S, Tirado A and Pedro F 2014 *The Iowa Orthopaedic J.,* **34** 129
[4] Dorati R, DeTrizio A, Modena T, Conti B, and Benazzo F 2017 *Pharmaceuticals,* **10** 96
[5] Gregor A *et al* 2017 *J. of Biol Eng.,* **11** 11
[6] Maulida H N, Hikmawati D, and Budiatin A S 2015 *J. of Spine,* **4** 266
[7] Tanaka Y *et al* 2010 *Biomaterials,* **31** 4506
[8] Rodrigues N *et al* 2016 *Procedia CIRP,* **49** 33
[9] Kattimani V S, Kondaka S, and Lingamaneni K P 2016 *Bone Regeneration. Bone and Tissue Regeneration Insights,* **7** 9

Synthesis and Characterization of Collagen-Chitosan-Sodium Hyaluronates as Artificial Cornea

P Widiyanti*[1,2], N D W Rini[1], M N Kirana[1] and T Astutik[1]

[1]Biomedical Engineering, Faculty of Science and Technology, Universitas Airlangga, Surabaya, Indonesia
[2]Institute of Tropical Disease, Universitas Airlangga, Surabaya, Indonesia

*e-mail: pwidiyanti@fst.unair.ac.id

Abstract. As reported by Riset Kesehatan Dasar, Indonesia facing a serious problem of blindness with 966.326 cases of 2 million population in 2013. Gold standard such as keratoplasty couldn't completely heal corneal injury due to the recipient's rejection and limited donor. In order to overcome those concerns, artificial corneal tissue needs to be developed to produce an artificial cornea which is biocompatible, supports corneal function, and applicable. This research was conducted to provide an initial description of Collagen – Chitosan - Natrium Hyaluronate (Col-Chi-NaHA) as an artificial cornea through FTIR analysis, hydrophilicity and water absorption measurement. The artificial cornea was synthesized by forming ratio of collagen 20% w/v and chitosan 10% w/v, both solutions were dissolved in 0,1 M and 0,6 M acetic acid separately. Into the mixed solution, NaHA was dripped slowly with variation of 0%, 0,3%, and 0,6%. The homogenous solution was cast and heated for 24 hours at 35°C. The dry membranes were neutralized to obtain a pH around 7. From the analysis known that all of the characteristic bands of Col-Chi-NaHA could be observed in obtained membranes. Water absorption measurement demonstrated percentages over 90% on all membranes. The contact angle of obtained membranes was decreased by the addition of NaHA. The results showed that artificial cornea made from Col-Chi-NaHA successfully fabricated has hydrophilic properties to support cellular interaction and provide hydration to keep eye's clarity and permeability. Therefore, Col-Chi-NaHA could may be a suitable material for the tissue-engineered cornea.

1. Introduction

Corneal blindness still becomes a major problem with high prevalence in Indonesia. According to Riset Kesehatan Dasar by Indonesian Ministry of Health, there are about 966.329 cases of corneal blindness in 2013. This type of blindness is considered as the second most common blindness after cataract [1]. The cornea is a translucent protective layer for the eye. Cornea injury happened because of trachoma and lepra infection, insufficient nutrition to the cornea and trauma [2]. Gold standard such as keratoplasty couldn't completely heal corneal injury due to limited donor. On the other hand, the rejection rate of implant cornea is about 13,3% to 65% four months after keratoplasthy. This could lead to unwanted reactions from the patient because of the non-biocompatibility [3]. Indonesia Eye Bank in 2017 stated that there are only 5-10 percent of blindness patients performed corneal transplantation. To overcome those problems, tissue-engineered cornea which is biocompatible, supports corneal function, applicable and biodegradable needs to be developed.

Chen [3] succeeded in creating tissue-engineered cornea with collagen type 1 which is the main component of cornea. He discovered that collagen 20% + chitosan 10% + 0,5% sodium hyaluronate is better in the terms of biocompatibility than the one without collagen. It was found that collagen 1% + chitosan 2% + 0,5% sodium hyaluronate could transmit 95% of light. This result is in line with the ability of the human cornea to transmit UV light in about 30% to 75% (with wavelength 310 – 400 nm) and visible light in about 75% to 95% (with wavelength 400 – 700 nm). However, it is discovered that during in vivo study, the same concentration of sodium hyaluronate performed different transparency. Therefore, we need to perform other research to see the impact of sodium hyaluronate to corneal transparency.

To overcome those problems, this research is conducted to produce artificial cornea based on collagen with addition chitosan and sodium hyaluronate. Chitosan has a good bio adhesion, permeability, toxicity, and ocular tolerance properties. Addition of chitosan to collagen can improve the stability of collagen and maintain the integrity of the structure. Liang et al in 2011 stated that chitosan can accelerate corneal endothelial cells regeneration in rabbits. Chitosan has many amines and hydroxyl functional groups so it could bond with collagen. On the other hand, sodium hyaluronate can improve light transmittance [3].

2. Material and Method

2.1. *Materials*
Materials used in this research are collagen (Col), chitosan (Chi), sodium hyaluronate (NaHA), chloric acid (HCl), acetic acid, hydroxypropyl methylcellulose (HPMC), phosphate buffer saline (PBS) and distilled water.

2.2. *Sample Preparation*
The artificial cornea was made by dissolving 20% of collagen in 0,1 M acetic acid and 10% chitosan in 0,6 M acetic acid. Collagen solution that has cross-linked with HPMC then added and mixed with chitosan solution for 60 minutes. Then, NaHA with variance concentration of 0%, 0,3%, and 0,6% was added to the col-chi solution and stirred for 30 minutes. Concentration ratio of collagen, chitosan and sodium hyaluronate can be seen in Table 3.1. The homogenous col-chi-NaHA solution was casted on Perspex plate and heated with an incubator for about 24 hours and temperature 35°C until the membrane become dry. Then, the dry membrane was dipped into PBS until the pH becomes normal.

2.3. *Characterization Procedures*

2.3.1. FTIR Test. Dry membranes were cut in 1 cm x 1 cm and evaluated using Perkin Elmer Spectrum One Fourier Transform Infrared (FT-IR) Spectrometer over the range 4000 cm^{-1} – 650 cm^{-1}.

2.3.2. Hydrophilicity Test. One droplet of water was dropped on dry membranes, then images were taken by a digital camera perpendicular to the sample. Contact angle from the images was evaluated using ImageJ software to create a graph of the contact angle vs time. From the results, we can obtain surface hydrophilicity value inversely proportional to contact angle.

2.3.3. Water Absorption Test. Dry membranes were cut in 1 cm x 1 cm, then immersed in distilled water for 24 hours. Water content rate (R_1) was assayed by weighing dry membranes in hydrated form film (G_1) and dry form (G) [3]

$$R_1 = \frac{G_1 - G}{G_1} \, x\, 100\% \quad (1)$$

3. Result and Discussion

3.1. FTIR Test

Col-Chi-NaHA solution went through FTIR test to see the functional groups and any chemical interaction. FTIR results explained in Figure 1. NaHA spectral showed N-H and C=O groups at wavenumber 1650-1500 cm^{-1} and 1700-1725 cm^{-1} respectively. Collagen itself has shown similar functional groups with NaHA indicated by overlapping between spectral. The widening of the absorption band showed that NaHA functional groups are present.

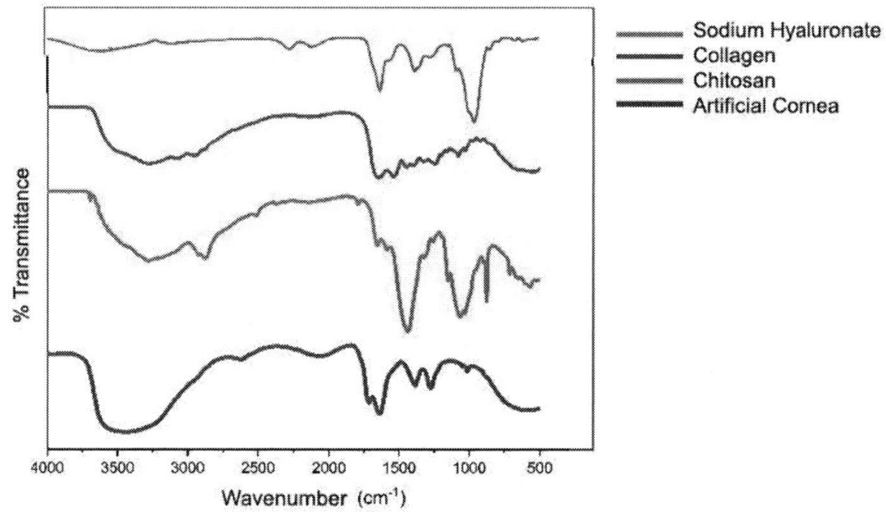

Figure 1. FTIR Spectra

3.2. Hydrophilicity Test

Hydrophilicity test was done by measuring the contact angle of the artificial cornea. Forch (2009) stated that hydrophilic material has a degree of contact angle below 90°, meanwhile contact angle >90° considered as hydrophobic material. Figure 2 showed the degree of contact angle of the artificial cornea is below 90°. Therefore artificial cornea is hydrophilic and supports cellular interaction [4].

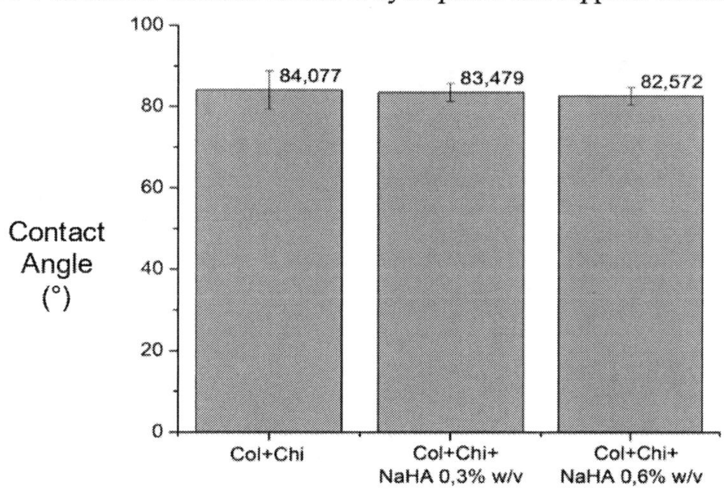

Figure 2. Degree of Contact Angle

The lower the degree of contact angle, the higher the hydrophilicity [4]. In this study it was found that the addition of NaHA was able to reduce the value of the degree of contact angle, in other words, the addition of NaHA was able to increase the hydrophilicity of the artificial cornea.

In previous studies, it was also known that the degree of contact angle of the collagen membrane was 90° [5]. In this study, all three samples of the artificial cornea had lower contact angles than collagen-based sample. It means that the hydrophilicity of artificial cornea is better than samples which used only collagen material.

3.3. Water Absorption Test

Water absorption test was done to determine the ability of artificial cornea provides hydration to the eye to maintain the clarity of the eye. Figure 3 showed that artificial cornea has percentages of water absorption higher than 77%. This result corresponds to the characteristic of the human cornea where the epithelium and endothelium must absorb water at least 77% per weight of the cornea [6].

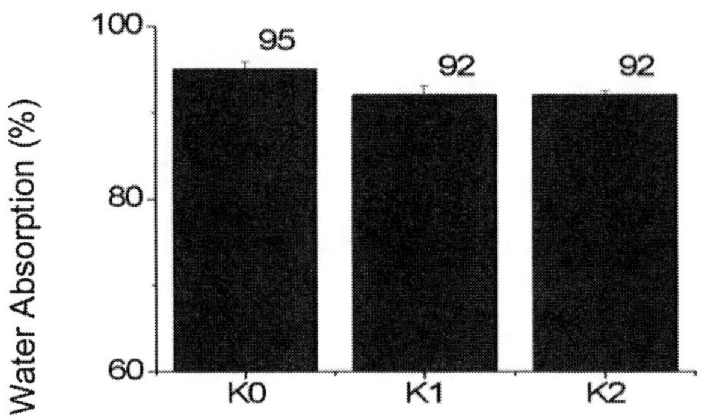

Figure 3. Percentages of Water Absorption

4. Conclusion

Artificial cornea based on Collagen – Chitosan – NaHA showed no changes in chemical interaction which indicates that those three materials were successfully fabricated. Artificial cornea had hydrophilic properties because the degree of contact angle was below 90°. The artificial cornea also has percentages of water absorption at 90%. Those properties support cellular interaction and provide hydration to keep eye's clarity and permeability. Therefore, Col-Chi-NaHA could be a suitable material for the tissue-engineered cornea.

Acknowledgments

The authors would like to thank Indonesia's Ministry of Research for Higher Education and Technology for partially funding the research through Students Creativity Program 2019 and Universitas Airlangga in Surabaya for the knowledge and opportunities given throughout the writer's academic experience in the institution.

References

[1] Whitcher J P, Srinivasan M, Upadhyay M P 2001 Corneal Blindness: A Global Perspective *Bull World Health Organ* **79** (3) 214–221

[2] Sinha R, Sharma N, and Vajpayee R B 2005 Corneal Blindness–Present Status *Cataract Refract. Surg. Today* 59–61

[3] Chen J, *et al.* 2005 Study on Biocompatibility of Complexes of Collagen-Chitosan-Sodium Hyaluronate and Cornea *Artif. Organs* **29** (2) 104–113

[4] Kim E Y, Tripathy N, Cho S A, Joo C K, Lee D, and Khang G 2015 Bioengineered Neo-corneal Endothelium using Collagen Type-I Coated Silk Fibroin Film *Colloids Surfaces B Biointerfaces* **136** 394–401

[5] Ding C, Zhang M, and Li G 2015 Preparation and Characterization of Collagen/Hydroxypropyl Methylcellulose (HPMC) Blend Film *Carbohydr. Polym.* **119** 194–201

[6] Ehlers N and Hjortdal J 2004 Corneal Thickness: Measurement and Implications *Exp. Eye Res.* **78** (3) 543–548

The Effect of Local Micro Organism and Mycorrhizal Fungi on Anatomical and Morphological Responses of Red Chili (*Capsicum annuum* L.) at Different Soil Water Level

T N Naafi[1], Y S Rahayu[2*]

Department of Biology, Universitas Negeri Surabaya, Indonesia

e-mail: yunirahayu@unesa.ac.id

Abstract. Red chilli as a vegetable commodity plays an important role for Indonesian people and its cultivation depends on the level of water availability, which is still a problem in certain area in Indonesia. This study aimed to describe the effect of local microorganism (LMO) concentration and different soil water levels on the anatomical and morphological responses of red chilli by adding the mycorrhizal fungi in all treatments during the cultivation. This experimental study used Randomized Block Design (RBD) with two factors, namely the concentration of LMO (0mL/L, 25 mL/L, 50 mL/L, 75 mL/L) and different soil water levels based on field capacity measurement (100%, 75%, 50%, and 25%) with 3 replications. The parameters obtained included anatomical parameters (stomatal density and trichome density) and morphological parameters (plant height, leaf number, wet biomass, root length, mycorrhizal infection and relative water content of leaves). Data obtained from this study was in the form of morphological response of red chilli and anatomic parameters. Data obtained in this study were analyzed by two-ways ANAVA and continued by the Duncan test. The results showed that there was an effect of LMO concentration and different soil water levels by the addition of mycorrhizal fungi on plant height, wet biomass, relative water content of leaves with the highest effect of LMO concentration at 75 ml/L with water level of 25%. In the stomatal density, the LMO concentration and water levels that significantly effect was 25 ml/L with water level of 25%. Although there was no significant effect on leaf number, root length, mycorrhizal infection and trichome density parameters due to the treatments. These findings recommended that red chilli could be cultivated better in the interaction of 75 mL/L of LMO concentration and soil water level of 25%.

1. Introduction

The use of red chilli in Indonesia based on the calculation of the Food Balance in 2002 until 2013 increased from 654 to 1.72 thousand tons, which shows an increase of 11.14% [1]. Red chilli plant widely cultivated on agricultural land both on a large and small scale and its cultivation depends on the water availability [2]. Water has an important role in plant growth because it plays a role in composing protoplasm, mineral nutrient solvents, medium for all chemical reactions, plays a vital role in photosynthetic reactions, maintains cell turgidity, and cell growth [3]. Mycorrhizae fungi play an important role to alleviate plant growth in drought conditions due to its role in regulating various host aquaporins through transporting water and facilitating membrane diffusion processes for other molecules such as CO_2, silicon, boron, urea, or ammonia [4]. In addition to being affected by water, plant growth was also influenced by the advantages of fertilizers such as local microorganism (LMO) from golden snail because it contains N-available, higher C/N ratio and P nutrient content compared to LMO from banana weed and LMO from rabbit urine and also contains nutrient content such as K, Ca,

Mg, Cu, Zn, Fe, and Mg [5]. Therefore, it was necessary to know more about the application of interaction between the LMO concentration and different soil water levels with the addition of mycorrhizal fungi in all treatments to the anatomic and morphological responses of red chilli (*Capsicum annuum* L.).

There is a connection between LMO, water availability, and mycorrhizae fungi. Water availability has a great impact on microbial community structure [6]. Where water availability have an effect on increasing host-microbe interactions [7]. In this case, mycorrhizal fungi have an ability to improving water availability. Mycorrhizal fungi can stimulate water lifting from roots to shoots via higher water apoplastic flow [8]. Increasing the availability of water will have an impact on increasing interaction between LMO with plants.

2. Methods

2.1. LMO Preparation

Golden snail (*Pomaceacaniculata*) in 1.25 kg has been ground than added rice-washing water as much as 2.5 L, 0.25 kg brown sugar, 2.5 L of coconut water and 2 L of water [5], mix for all of this and incubating for 21 days.

2.2. Water Level Determination

Determination of water level based on field capacity measurement as gravimetrically and simple lysimeter method by taking 5 kg soil from depth on 0-20 cm [9], and was added the water to the soil until saturated and the water stopped to drip out of the polybag. Then the soil was weighted after giving water and 100% field capacity was determined by reducing the final weight of the soil with the weight of the soil at the beginning.

2.3. Planting of Red Chili Seeds

Good quality of red chilli seeds was chosen and then soaked in warm water for about one hour. After that, the seed was stored for 24 hours. Basic fertilizer was added and mixed with 20 grams of mycorrhizae fungi-contain soil. This soil will be as a medium for planting the red chilli seeds.

2.4. Treatment

LMO was applied at the age of 10-15 days after seedling. After the red chilli plants reached the age of 21 days, the level of water supply could be applied and the LMO has applied again at the age of 30-35 days after seedling, 40-50 days after the seedlings are approximately 250 mL/plant.

2.5. Measurement of Stomata and Trichome Density

The abaxial leaves from the third to fifth sample leaves were coated from shoots [10] with colourless nail polish then dry at the air ± 40 minutes, coated polish again uses 1.5 cm tape to be lifted/peeled. The taped sticking parts were affixed to the object-glass and observed using a 400 magnification with diameter of 0.52 mm^2. The calculated by equations:

$$\text{Stomatal density} = \frac{Number\ of\ stomata}{broad\ field\ of\ view(\ mm2)}$$

$$\text{Trichome density} = \frac{Number\ of\ trichome}{broad\ field\ of\ view(\ mm2)}$$

2.6. Measurement of Plant Morphology

Measurements on plant height from the base of the stem to the main stem tip which was calculated at the age of 40, 50, 60 and at 71 days after seedling because at this age the red chilli was in the generative stage starting at 30-115 days after seedling which is obtained in the afternoon. The number of leaves that counted in this study was the number of leaves of red chilli that opens perfectly with clear leaf reinforcement which was calculated at the age of 40 days after seedling, 50 days after seedling, 60 days after seedling and at 71 days after seedling because at this age the plants were in the generative stage. The number of leaves was counted in the afternoon at the same time. The root length was calculated from the base to the end of the stem at 71 days after seedling. Plant biomass was measured at 71 days after seedlings in the form of fresh roots, stems, and leaves. Relative water content of the leaves carried out at 71 days after the seedling. Measurements were modified by weighing fresh leaves to determine fresh weight (FW), then put in a cup containing distilled water until all the leaves were submerged, covered with filter paper and stored at room temperature for 18-24 hours. If the remaining water was left, the remaining water was removed while the sample was drained with tissue, then weighed to obtain the weight of turgid (TW). Then it dried using oven in 60°C for 2x 24 hours to get dry weight (DW) [11]. The relative moisture content of the leaf was obtained from the formula:

$$\text{RWC} = \frac{FW-DW}{TW-DW} x\ 100\%$$

The percentage (%) of mycorrhizal infections was obtained at 71 days after seeding through calculations [12] :

$$\%\ \text{Infection} = \frac{Total\ root\ infection}{Total\ root\ infection + Non\ infection} x\ 100\%$$

2.7. Statistical analysis

Statistical analysis was performed using Two-Ways ANOVA to determine the effect of LMO concentration and different levels of soil water with the addition of mycorrhizal fungi to the anatomical response and morphology of the red chilli plant. If there are significant statistical differences, then was proceed with the Duncan test with a significance of 5%.

3. **Results and Discussions**

The effect of LMO concentration and different soil water levels with the addition of mycorrhizal fungi to morphological responses of red chilli in a) height plant, b.) number of leaves, c.) wet biomass, d.) root length, e.) mycorrhizal infection and f.) relative water content of leaves, were shown in **Figure 1**.

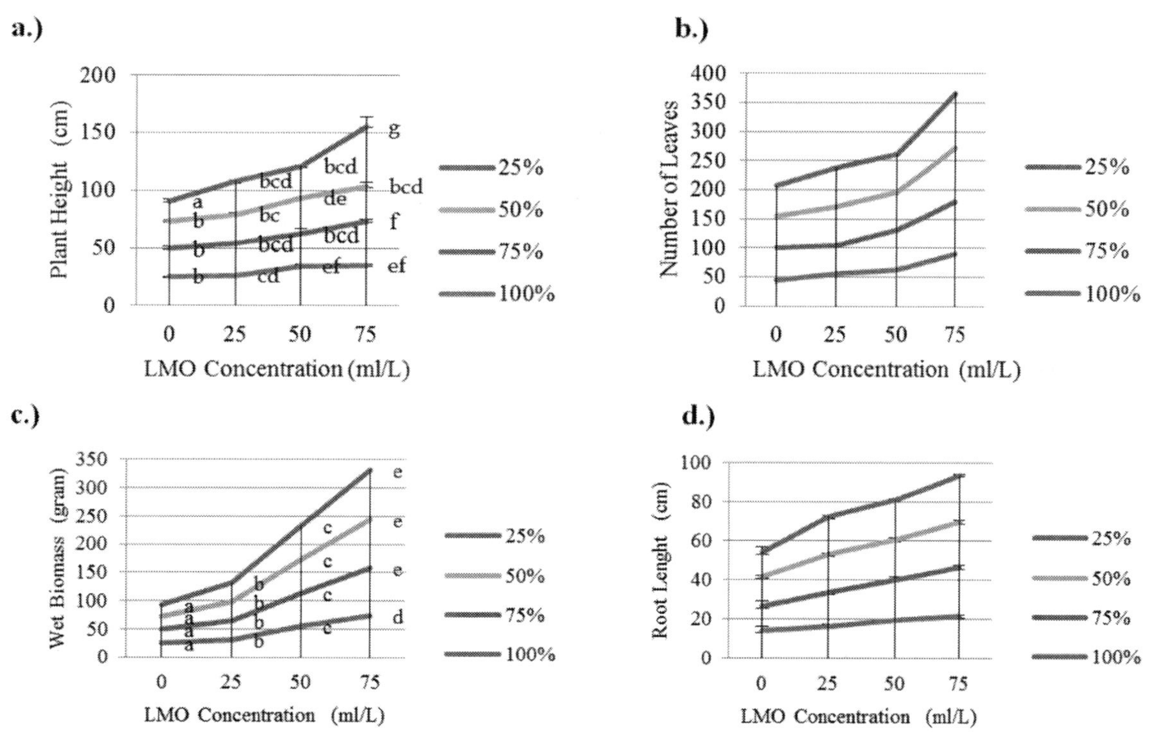

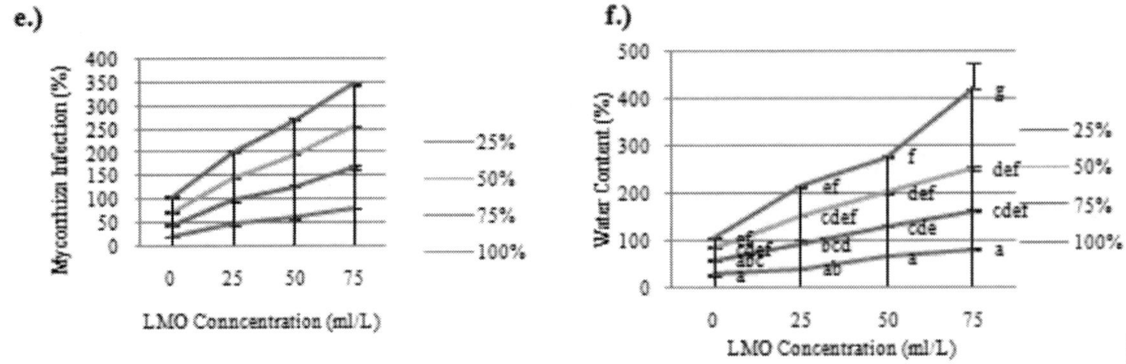

Figure 1. The effect of LMO concentration and different soil water levels with the addition of mycorrhizal fungi to the morphological response of red chilli plant in a) height plant, b.) number of leaves, c.) wet biomass, d.) root length, e.) mycorrhizal infection and f.) relative water content of leaves.

There was a significant effect of the interaction between the LMO concentration and different water soil level with the addition of mycorrhizal fungi to height of red chili plant (Figure 1a), wet biomass (Figure 1c), and relative water content of leaves (Figure 1f), with the highest effect of the LMO concentration of 75 ml/L with a water level of 25%. However, there was no significant effect of the interaction between the LMO concentration and the different water levels to the parameters of leaf number (Figure 1b), root length (Figure 1d), and mycorrhizal infection (1e).

Other data that obtained from this study was the effect of LMO concentration and different water level with the addition of mycorrhizal fungi on anatomic parameters a.) stomatal density and b.) trichome density that is shown in **Figure 2**.

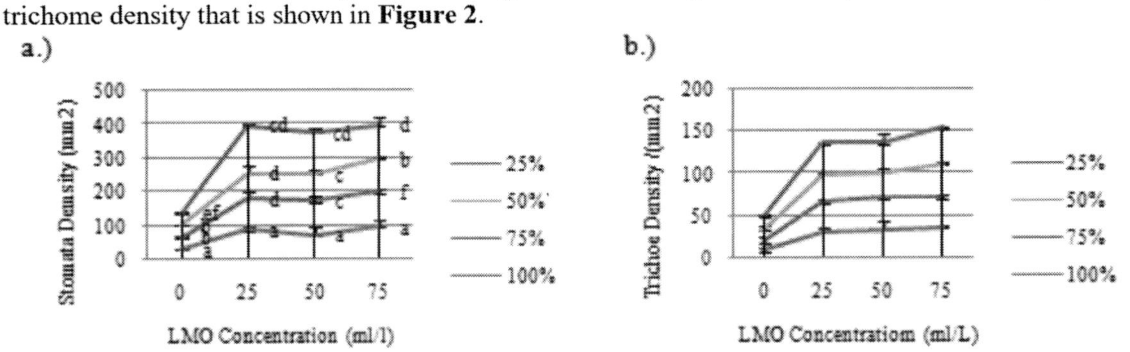

Figure 2. The Effect of LMO concentration and different water level with the addition of mycorrhizal fungi on anatomic parameters a.) stomatal density and b.) trichome density

There was a significant effect of the interaction between LMO concentration and different water level with the addition of mycorrhizal fungi to the stomatal density (Figure 2a) with the highest effect found in the 25 ml/L LMO concentration with a 25% water level. However, there was no significant effect of the interaction between the LMO concentration and different water level on the trichome density (Figure 2b).

Plant growth was defined as a series of events that can be identified through qualitative changes such as germination and flowering or quantitatively such as the number of leaves, number of flowers, etc. in plant structures [13]. Growth occurs because of the increase in the number of protoplasms followed by the increase in the size of the weight and the number of cells that will affect in the meristem tissues to produce new cells that divide, enlarge and differentiate [14]. The division in meristem tissues occur because of mitotic division at the root tip and the end of the stem that depending on the availability of carbohydrates, protein mainly water in the big capacity and the effect will be seen in vegetative growth of plants such as increasing plant height, leaf number, stem diameter and root growth [15]. Plant growth besides being influenced by the level of water availability also influenced by hormones that found in plants such as the auxin, cytokinins, and gibberellins which play an important role in cell expansion [16].

LMO from golden snail used in this study was known to contain nutrients such as N and Fe which play an important role as a constituent of proteins, amino acids, nucleotides, and enzymes [16] for cell division related to protein synthesis. Mycorrhizae fungi play a role in increasing water and nutrient uptakes such as N, P and K [17]. If the availability of N nutrients in plants increases, cell division in plants will also increase. The symbiosis between mycorrhizal fungi and root will increase root growth, architecture and functional hydraulic roots for water and nutrient uptake, absorbs and transportation [18] which has an impact on increasing the relative water content in plant tissues (Figure 1f.). The higher relative moisture content of the leaves indicates that the plant is more tolerant. The increase in the water supply will increase the potential of turgor so that the process of plant cell division will increase too. In addition, mycorrhizal fungi will increase also the diffusion of ions for cell elongation related to the action of auxin hormones in cell formation [16]. This event can be observed with the increasing of plant height (Figure 1a) and the increasing of the stomatal density (Figure 2a.).The density of the stomata was related to the number of stomata that influence the photosynthesis and transpiration. The increase of the density of stomata will increase the transpiration so that there was an increase of CO_2 uptake for dark reactions (Calvin cycle) in photosynthesis by fixation of CO_2 which produces carbohydrates [16].

The results of this study showed that the increase of stomatal density was not followed by an increase in leaf trichome density (Fig. 2b) caused by plant responses further minimized cell division for stomata formation compared to trichomes. Although with high stomatal density, plants will not lose much water through transpiration due to the absorption process which compensates for the rate of water loss through transpiration due to the presence of root symbiosis with mycorrhizal fungi [19].

LMO also contain nutrients Mg, Fe, Zn and Cu which play a role in arranging leaf chlorophyll for photosynthetic materials mainly in bright reactions. If the chlorophyll content of leaves was high, the photosynthesis process will increase so that the photosynthetic compounds produced such as carbohydrates will also increase. These photosynthetic compounds were used for cell division and composting plant biomass (Figure 1c). If the biomass value of a plant was greater, it indicates that the nutrient content absorbed by plants was also getting bigger [20] and more water content in plants because the weight of the plant tissue about 90-95% consisting of water [16,18].

The application of LMO and mycorrhizae fungi with the different water level availability did not appear to have an effect on increasing the number of leaves (Figure 1b). It was because more photosynthetic assimilates are distributed to the roots than to the leaves for the expansion of the root system. Another factor that affects the decrease in the number of leaves is the presence of pest attacks *Aphis gossipy* causes leaf malformations to become curved, perming, shrinking, thickening, chlorosis and declining before adulthood. Then *Thrips* which attack the main leaf on young leaves with symptoms of colour change from light green to brown copper, thickened surface, leaves become wrinkled and when touched will fall/fall out [21, 22]

LMO and mycorrhizae fungi with the different water availability also had no effect on root length (Figure 1d) and mycorrhizal infections (Figure 1e) but these two parameters had other parameters due to their important role in transporting water and nutrients from the soil. Microscopic observations showed that most of the roots were infected by mycorrhizae fungi by finding internal hyphae in the roots. Mycorrhizae fungi are able to aggregate soil grains so that the ability of the soil to store water increases [23] to be absorbed by plants and the effects of various biochemical reactions on plants can be carried out.

4. Conclusion

Based on the data analysis can be concluded that there was an effect between the interaction of the LMO concentration and different water levels by the addition of mycorrhizal fungi on plant height, wet biomass, and relative water content of leaves with the highest effect of LMO concentration at 75 ml /L and water level of 25%. There was no significant effect of the interaction between the LMO concentration and different water level with the addition of mycorrhizal fungi on the response of red chilli (*Capsicum annuum* L.) at parameters leaf number, root length, mycorrhizal infection and leaf trichome density.These findings recommended that red chili can be cultivated better in the interaction of 75 mL/L of LMO concentration and soil water level of 25%.

References

[1] Suwandi, Nuryati L, Noviati I D, and Victor S B 2015 *Outlook cabai komoditas pertanian subsektor hortikultura*
[2] Swastika S, Pratama D, Hidayat T, and Andri K B 2017 *Teknologi budidaya cabai merah.*
[3] Hamim 2008. Fungsi air dan perannya pada tingkat selular dan tumbuhan secara utuh
[4] Li G, Santoni V, and Christophe M 2014 *J. Biochemica et Biophysica Acta* 1574-1582
[5] Suhastyo A A, Anas I, Santosa D A , and Lestari Y 2013 *J Saintek* **10** 29-39
[6] Taketani R G et al 2017 Microb. Ecol. **73** 153–161.
[7] Aung K, Jiang Y and He S Y 2018 *Plant J.* **93** 771–780.
[8] Bárzana G et al 2012 *Ann. Bot.* **109** 1009–1017.
[9] Yamani A 2010 *J Hutan Tropis* **11** 37-46
[10] Yang, Y M, Tang R., Sulpice, and Ban Y 2014 *J Plant mycorrhizal Regulation* **33** 612-625.
[11] Barr H D, Weatherley P E 1962 *Aust. J. Biol. Sci.* **15** 413-428
[12] Phillips J M and D S Hayman 1970 *J. Trans. Brit. Mycol. Soc.* **55** 158–161
[13] Dambevrile A et al *J. Annals Botany* **115** 93-105
[14] Salisbury F and Cleon W R 1992 *Fisiologi Tumbuhan 3th* (Bandung: Institut Teknologi Bandung Press) p 2-3
[15] Masria 2015 *Jurnal Partner* **48** 48-56.
[16] Taiz L and Zeiger E 2002 *Plant Physiology 3th Edition* (Sinauer Associates: Annals of Botany Company) p 424-462
[17] Musfal 2008 *Potensi cendawan mikoriza arbuskula untuk meningkatkan hasil tanaman jagung*
[18] Rapparini F and Peñuelas Josep 2014 *J. Microbes for the Alleviation of Soil Stresses* **1**
[19] Sasli I 2004 Peranan mikoriza vesikula arbuskula (MVA) dalam peningkatan resistensi tanaman terhadap cekaman kekeringan (Internet)
[20] Duaja N, Deviani A and Yan R 2012 *Analisis tumbuh selada (Lactuca sativa L.) pada perbedaan jenis pupuk organik cair J Bioplantae* **1**
[21] Johari A, Herlinda S, Irsan C., and Yulia P 2016 Phenomenon of thrips (Thysanoptera) attack on chili plant (*Capsicum anuum* L.). *Am. J. Agr. Biol. Sci.* **11** 103-109
[22] The Invasive Species Compendium (ISC). *Aphis gossypii* (Cutton Aphid).
[23] Augé R M 2001 Water relations, drought and vesicular-arbuscular mycorrhizal symbiosis. *J. Mycorrhiza* **1** 3-42.

The Effect of Gibberellin Extracted from *Eichhornia crassipes* Root on the Viability and Duration of Hard Seed Germination

K Ummah, Y S Rahayu*

Department of Biology, Universitas Negeri Surabaya, Indonesia

e-mail: yunirahayu@unesa.ac.id

Abstract. The germination process is influenced by external and internal factors including the type of seed. Hard seeds require special treatment in stimulating the germination process such as scarification and stratification (for example applying exogenous hormones to seeds during the germination process). This study aimed to describe the effect of exogenous gibberellin hormone extracted from *Eichhornia crassipes* root on the viability and duration of germination of three different hard seeds. This study used RBD (Randomized Block Design) with two treatment factors, namely root extract concentration from *Eichhornia crassipes* (0 ppm, 200 ppm, 300 ppm, 400 ppm, and 500 ppm) and hard seed types (sapodilla seeds =*Annona muricata*, soursop seeds =*Manilkara kauki*, and longan seeds = *Dimocarpus longan*) with three replications. The observed parameters for seed viability were germination percentage and vigour index, while duration of germination parameters was germination index. Data were analyzed using ANOVA and followed by Duncan Test. The results showed that there was a significant effect of the root-extracted gibberellin concentrations and types of seeds on viability and duration of germination seeds. The optimal concentration of root-extracted from *Eichhornia crassipes* was 500 ppm, with the highest germination percentage, vigour index, and germination index were 80.55%, 387.33%, and 0.254%/etmal respectively. The longan seeds revealed the best viability and shortest duration of germination due to applying the exogenous gibberellin hormone extracted from *Eichhornia crassipes* root. The findings revealed that there was a positive synergism between exogenous and endogenous gibberellin on the effect of germination process in three different hard seeds tested.

1. Introduction

Every plant has the ability to reproduce, to maintain its type and offspring. Angiosperms are able to reproduce sexually or asexually, leading to the formation of seeds that enable them to survive by producing a new generation [1]. Seed germination is important life stages which contribute to plant distribution and adaptation [2]. The germination processes are influenced by internal and external factors. Cardoso *et al.* [3] report that the presence of light and temperature are the external factors that influenced the seed germination. Other external factors are salt stress and pH [4]. Rezvani and Zaefarian [5] stated that germination was restricted by pH lower and higher than 5 and 8, respectively. Another factor that has an effect on seed germination is endogenous hormone. The endogenous hormones were reported to influence plant growth and development such as seed germination [6, 7].

One of the endogenous hormones that influenced on seed germination is gibberellin [7, 8]. Gibberellic acids are a group of plant hormones that act in the elongation and cell division [9]. In addition, Zang *et al.*[10] stated that Gibberellic acid (GA_3) have an important role in many plant growth and development processes such as increasing chlorophyll content, the level of chlorophyll a and b, individual fruit weight, and the number of fertile seeds. The seed germination is a natural process in plant life stage, however, in the coat seeds (hard seeds) have to undergo pretreatment to accelerate the germination [11]. The pretreatment that can be given to coat seeds (hard seeds) are scarification treatments as well as by low-pH solutions to increase the permeability of seed coat [12]. Beside the scarification, another pretreatment that accelerates coat seeds germination is stratification. Stratification is a pretreatment to break down seed dormancy [13].

Based on that description, this study was conducted to determine and to describe the effect of the application of exogenous gibberellin on hard seed germination. In this study, gibberellin was obtained from the extraction of *Eichhornia crassipes* root. Lalitha *et al.* [14] reported that gibberellin was found on *Eichhornia crassipes* root. Although the roots of the *Eichhornia crassipes* plant contain gibberellin, it needs to be studied further to find out and describe the effect of gibberellin from *Eichhornia crassipes* root extract on viability and duration time of hard seed germination.

2. Methods

2.1 *Extract of Eichhornia crassipes Root*

The extraction process started with the sorting of the water hyacinth (*Eichhornia crassipes*) roots and then the roots were washed and dried under sunlight. After that, the water hyacinth roots were dried in oven for 3 days at 60°C. Then the roots were grained to become Simplicia, macerated with a ratio of 1:4 (500g: 2000mL of methanol 60%). After that, the roots were extracted using $NaHCO_3$ and ethyl acetate to separate chlorophyll compounds and other polar compounds.

2.2 *HPLC Test (High-Performance Liquid Chromatography)*

HPLC test was carried out to detect the gibberellins with a wavelength of 230 nm, then dissolved in 10 ml methanol-water with the ratio of 1:1 with result 2995,50 ppm.

2.3 *Dilution of Giberelin Extract of Eichhornia crassipes*

Eichhornia crassipes root extract was diluted to become 500 ppm (0.5 g diluted in 1 L of water); 400 ppm (0.4 g diluted in 1 L of water); 300 ppm (0.3 g diluted in 1 L of water); 200 ppm (0.2 g diluted in 1 L of water).

2.4 *Treatment of Seed Scarification*

Three different hard seeds (sapodilla seeds, soursop seeds, and longan seeds) were rubbed using sandpaper around the eye of the seed (seed embryo) until they were brownish white.

2.5 *Immersion Seeds (Stratification of seeds)*

Three different hard seeds (sapodilla seeds, soursop seeds, and longan seeds) were immersed in solution of gibberellin extracted from water hyacinth root (*Eichhornia crassipes*) for 24 hours; 50 mL each solution.

2.6 *Maintenance of Seeds in Planting Media*

Seeds were planted in media sand and soil with a ratio of 1:1 in a polybag of 3 kg, each polybag contains 4 seeds with a period of observation for 30 days.

2.7 *Data analysis*

The data parameters (germination percentage and vigour index, and germination index) were analyzed by Two-Way Variant Analysis (ANOVA), then followed by Duncan Test.

3. Results

Journal of Physics: Conference Series **1417** (2019) 012037 doi:10.1088/1742-6596/1417/1/012037

The application of gibberellin extract from water hyacinth root (*Eichhornia crassipes*) at various concentrations (0 ppm, 200 ppm, 300 ppm, 400 ppm and 500 ppm) have an effect on germination percentage of three hard seeds (sapodilla seeds, soursop seeds, and longan seeds) (Figure 1).

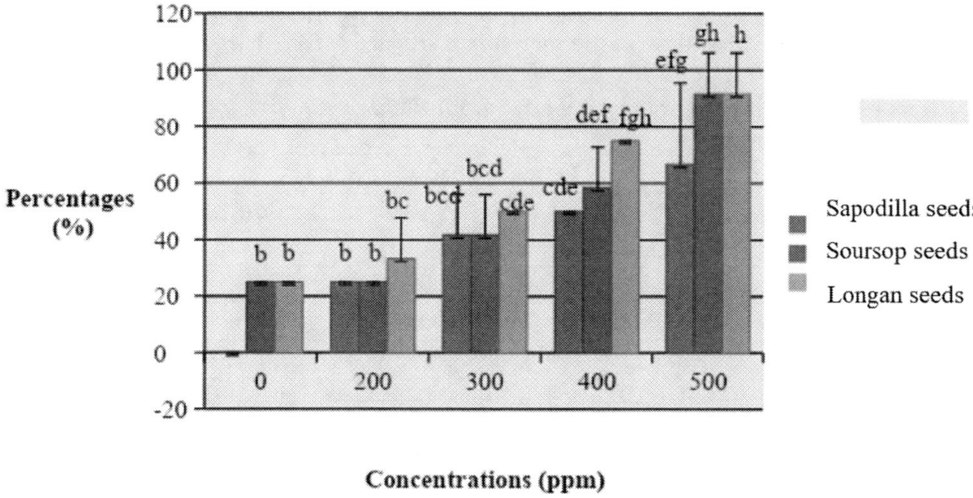

Figure 1. Effect of Gibberellin Hormone Extracted from *Eichhornia crassipes* Root on the Germination Percentage of Three Hard Seeds. Data represent mean values and standard error of the means (n=3)

The highest percentage of seeds germination at a concentration of 500 ppm of 80.55%, while the lowest percentage of seed germination was indicated at concentrations of 0 ppm and 200 ppm, with the percentage of 16.66% and 27.77% respectively (Figure 1). Longan seeds showed a greater value of germination percentage compared to sapodilla seeds and soursop seeds.

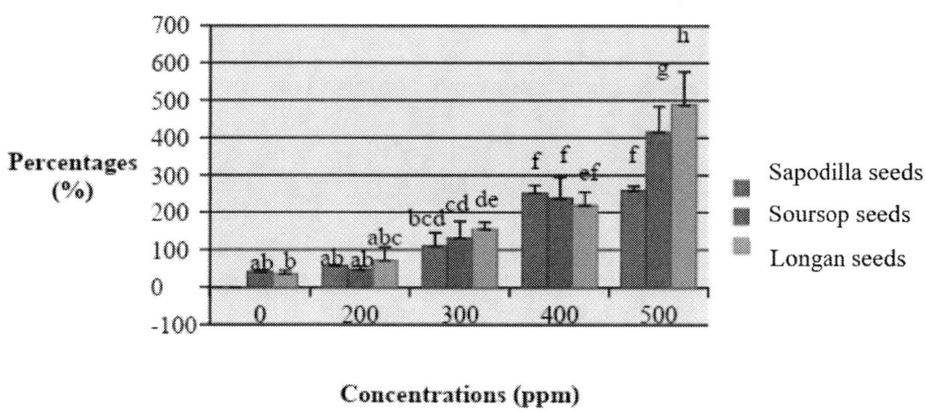

Figure 2. Effect of gibberellin Hormone Extracted from *Eichhornia crassipes* Root on the Vigor Index of Three Hard Seeds. Data represent mean values and standard error of the means (n=3)

Figure 2 reveals that the highest vigour index was the application of 500 ppm concentration of gibberellin hormone extracted from root with a percentage of 387.33% and the low vigour index value was shown by the application of 0 ppm and 200 ppm concentration of gibberellin hormone extracted from root with percentages of 25.55% and 58.66% respectively. Longan seeds showed the highest vigour index of 193.73% compared to sapodilla seeds and soursop seeds with the percentages of 135.60% and 174.53% respectively. Based on ANOVA followed by Duncan Test, it was known that the application of the gibberellin hormone extracted from water hyacinth root with various concentrations (0 ppm, 200 ppm, 300 ppm, 400 ppm, and 500 ppm) have a significant effect on the vigour index in three hard seed tested.

Duration of germination of three hard seeds was obtained by the parameter of germination index by calculating the total germination days in the three types of hard seeds (**Table 1**).

Table 1. Effect of Gibberellin Hormone Extracted from *Eichhornia crassipes* Root on the Germination Index of Three Hard Seeds

Extract Concentration (ppm)	**Types of seed (%)/etmal**			**Average (%)/etm**
	Sapodilla Seeds	**Soursop Seeds**	**Longan Seeds**	
0	0.00±0.00[a]	0.042±0.006[ab]	0.043±0.003[ab]	0.028
200	0.0407±0.004[ab]	0.0403±0.002[ab]	0.058±0.024[ab]	0.046
300	0.0817±0.027[bc]	0.099±0.022[bcd]	0.0993±0.010[bcd]	0.093
400	0.146±0.025[d]	0.139±0.032[cd]	0.1483±0.024[d]	0.144
500	0.1573±0.017[d]	0.258±0.054[e]	0.3467±0.101[f]	0.254
Average (%)/etm	0.0851	0.1156	0.139	0.113

Different notations (abc) indicate a significant difference between treatments, 0.05 significance level.

Table 1 was shown that the highest germination index value was at 500 ppm concentration of gibberellin hormone extracted from root with 0.254%/etmal. The low value was indicated at 0 ppm and 200 ppm concentration of gibberellin hormone extracted from root with percentages of 0.028%/etmal and 0.046%/etmal respectively. Longan seeds showed the highest germination index value of 0.139%/etmal compared to sapodilla seeds and soursop seeds, 0.085%/etmal and 0.115%/etmal respectively. Based on ANOVA followed by Duncan Test it was revealed that the application of the gibberellin hormone from water hyacinth root extract with various concentrations (0 ppm, 200 ppm, 300 ppm, 400 ppm and 500 ppm) have a significant effect on the germination index of three hard seeds.

4. Discussions

Based on the data analyzed, the viability and the duration of the germination of three hard seeds were influenced by the application of the gibberellin hormone extracted from water hyacinth root in various concentrations (0 ppm, 200 ppm, 300 ppm, 400 ppm and 500 ppm) (Fig. 1, 2 and Table 1). These results showed that gibberellin extracted from water hyacinth root has a significant effect on hard seeds germination. Gibberellins are growth hormones that accumulate in the fibre root [15]. Gibberellins are one of the hormones that are needed by plants in their development and growth, especially in the process of germination of seeds before becoming new individual. Gibberellin is an isoprenoid (diterpenoid) compound which is a derivative of the entangling process [16]. The efficiency of exogenous gibberellin via seed treatment may improve seedling initial growth, whereby gibberellin stimulates the break down of endosperm of the seed and stimulates cell divisions and proportion of stem and root part of seedlings [17]. It means that there is a positive synergy between endogenous and exogenous gibberellin to stimulate the germination process in hard seeds tested.

It is known also that gibberellin also triggers the formation of the enzyme α-amylase which will break down the starch so that the sugar levels in the cell will increase and cause the water outside of the cell to enter the cell and the cells will be extended. The results of the breakdown of starch will also be used for the process of respiration by mitochondria so that it will produce ATP which will be used in the process of cell growth and growth and cell expansion [18]. The addition of exogenous gibberellins derived from the extraction of water hyacinth roots containing gibberellin [19] will be able to increase the synergy between exogenous gibberellins and endogenous gibberellins that are in the seeds so that it will be able to influence seed viability and seed germination time.

The seed viability is the ability to grow through the germination process. The seeds viability can be demonstrated through metabolic symptoms or growth symptoms [20]. Parameters for seed viability that used were germination percentage and germination vigour index. The germination of seed normally has a difference in the level of viability of these seeds [21]. These measured parameters were obtained by comparing between one sprout and the other sprouts based on several criteria, namely normal sprouts, abnormal sprouts, and dead sprouts [22]. It can also be influenced by several factors, namely the level of seed maturity, seed size, and seed dormancy period. In this case, these parameters were not recorded yet as supporting data.

To determine seed viability also followed by the duration of seed germination with germination index parameters. Germination index used to determine the speed of seeds in starting and doing the germination process by calculating the duration for seeds to begin to accentuate the radicle and to do the germination process [23]. In this study also showed that longan seeds and soursop seeds were not significantly different based on the germination index value compared to sapodilla seeds. As it was known that germination process was affected by internal factors and external factors [24]. One of internal factors that affect the germination seed was the condition of seed itself. If the seeds have not reached ripe physiologically it will have low viability. Even in certain plants, these seeds will not be able to germinate [25]. Besides that the addition of exogenous gibberellin extracted from the roots where the gibberellin was produced also affect the germination process.

The results of this study indicated that the application of the exogenous gibberellin hormones extracted from water hyacinth root has an effect on the germination process of three types of hard seeds. The greater amount of gibberellin concentration applied on seed will have a greater effect on the germination. These findings revealed that exogenous and endogenous gibberellins can be a synergy to stimulate the physiological processes of seeds before they become new individuals by germination process [16].

5. Conclusion

Based on the results, it can be concluded that the application of the gibberellin hormone extracted from *Eichhornia crassipes* root in various concentrations (0 ppm, 200 ppm, 300 ppm, 400 ppm, and 500 ppm) showed a significant effect on the seed germination percentage, vigour index, and germination index of three hard seeds (sapodilla seeds, soursop seeds, and longan seeds). Longan seeds showed the highest value compared to soursop seeds and sapodilla seeds in germination percentage, vigour index, and germination index. The findings revealed that there was a positive synergism between exogenous and endogenous gibberellin on the effect of germination process in three different hard seeds tested.

References

[1] Pereira, A M and Coimbra S 2019 *J. Exp.Botany* **70** (11) pp. 2933–6
[2] Yuan, X and Wen B 2018 Seed germination response to high temperature and water stress in three invasive Asteraceae weeds from Xishuangbanna, SW China. PLoSONE **13**(1) e0191710.
[3] Cardoso A A, et al 2015 *J. Seed Sci.* (**37-2**) p 111-116, 201
[4] Laghmouchi Y, Belmehdi O, Bouyahya A, Skali Senhaji N and Abrini J 2017 Effect of temperature, salt stress and pH on seed germination of medicinal plant Origanum compactum. *Biocatalysis and Agricultural Biotechnology*, **10** p 156–160
[5] Rezvani M and Zaefarian F 2017 *Acta Botanica Brasilica* - **31(2)**: 220-228
[6] Yuan X K and Yang Z Q 2018 *Hort. Sci. (Prague)*. **45, (3)** 131–138
[7] Vishal B and Kumar P P 2018 *Front. Plant Sci.* **9**:838.doi: 10.3389/fpls.2018.00838

[8] Nasiri A A, Mortazaeinezhad F and Taheri R *Electromagnetic Biology and Medicine*, **37(1)** 50–56
[9] Buchanan B B, Gruissem W and Jones R L 2015 *Biochemistry and molecular biology of plants* **2** 1280 Rockville, American Society of Plant Physiologists
[10] Zang Y X, Chun I J, Zhang L L, Hong S B, Zheng W W and Xu K 2016 *Scientia Horticulturae* **200** 13–18
[11] Mensah S I and Ekeke C 2016 Effects of Different Pretreatments and Seed Coat on Dormancy and Germination of Seeds of Senna obtusifolia (L.) H.S. Irwin & Barneby (Fabaceae). *Int. J. Biol.* **8-2**
[12] Chaves, Izabel de Souza, Silva N C Q and Ribeiro D M 2017 *J. Seed Sci.*, **39-2**, 114-122
[13] Hilooğlu M, Sözen E, Yücel E and Kandemir A 2018 *Not Bot Horti Agrobo,* **46(2)** 376-380
[14] Lalitha P, Sripathi S K and Jayanthi P 2012 *Natural Product Communications* **7**(9) pp 1249 - 1256
[15] Chen K, Liu J, Ji R, Chen T, Zhou X, Yang J, Tong Y, Jiang C, Zhou J, Zhao Y, Jin Y, Yuan Y and Huang L 2019 *Frontiers in Plant Science*, **9**
[16] Taiz L and E Zeiger 2010 *Plant Physiology.* 5th Edition. Sunderland: Sinauer Associates.
[17] Jaques L B A, Carvalho I R, Szareski V J, Pimentel J R, Troyjack C, Dellagostin S M, Mendonça M T, da Rosa T C, Villela F A, de Souza V Q, Aumonde T Z and Pedó T 2019 *J. Agr. Sci.* **11(2)**
[18] Hedden, P and Sponsel V 2015 *J. Plant Growth Regul* **34**: 740–760
[19] Tulika T and Mala A 2015 *Pistia stratiotes* (L.) and *Eichhornia crassipes. Journal of Plant Sciences* **3(1-1)**: 10-18
[20] Carpici E.B, Celik N and Bayram G 2009 *African J. Biotech.* **8(19)** p 4918-4922
[21] Jahangir M M, Amjad M, Afzal I, Iqbal Q and Nawaz A *J. Agr. Sci* **46(1)** p 1-5
[22] Van der Walt K and Witkowsk E T F 2017 *South Africa J. Botany* **109,** p 237 - 245
[23] Talukdar D 2011 *Current Research Journal of Biological Sciences.* **3(2)** p 116-123
[24] Widyawati N, Tohari P, Yudono dan I Soemardi 2009 *Jurnal Agronomi Indonesia* **37(2)** 152- 158
[25] Juhanda Y Nurmiaty dan Ermawati 2013 *Jurnal Agrotek Tropika* **5(1)** 45-49

The effectiveness of nutrient variation to hydroponic *Caisim* (*Brassica juncea L.*) growth

S K Dewi[1], Y S Rahayu[1], dan A Bashri[1]

[1]Department of Biology, Universitas Negeri Surabaya, Indonesia

e-mail: saridewi@unesa.ac.id

Abstract. One type of vegetable that is popular among the Indonesian is mustard greens (*Brassica juncea L.*), which is also known as *Caisim* because it is often processed into various dishes and contains various ingredients and vitamins which are good for health. The last few years *Caisim* production has decreased, one of the causes is the constraints in the cultivation process, especially in the use of fertilizers due to the relatively long-term use of inorganic fertilizers. To improve the cultivation of *Caisim* without looking at the soil conditions, we can use a hydroponic system. However, hydroponics need more cost because the AB MIX nutrition is still relatively expensive, so it needs an alternative nutrient as a substitute for AB MIX which has the same ability. AB Mix is a mixture of fertilizer A and fertilizer B. Fertilizer A contains potassium while fertilizer B contains sulfate and phosphate. An alternative nutrient as a substitute for AB MIX which has the same ability, such as liquid fertilizer based on local microorganisms (LMO) which has properties that are easily absorbed by plants so that the use of organic nutrients can be used as a source of substitute nutrition, which is more economically affordable and has a very good influence on *Caisim* and is expected to increase production in *Caisim* cultivation. This study used a Completely Randomized Design with a single factor experimental design consisting of 4 treatments, namely a combination of AB MIX solution and various kinds of LMO-based organic nutrients as follows: A = LMO A 5 ml + AB MIX 5 mL/L of water; B = LMO B 5 mL + AB MIX 5 mL/L water; C = LMO A 5 mL + MOL B 5 mL/Liter water; D = AB MIX 10 mL/L water. The variables observed were plant height, number of leaves, and biomass of *Caisim*. The results showed that the variation of AB MIX nutrition and liquid organic fertilizer based on LMO effectively increased the growth of *Caisim* plants. The highest average plant height was given by *Caisim* with AB MIX nutrition, which is 69.2 cm. The highest average number of leaves was obtained by *Caisim* which was grown with a combination of LMO A and LMO B nutrients, which was 27.8. The largest average wet weight was given by *Caisim* with AB MIX nutrition, which is 32.3 g. According to the three parameters, the use of ab mix nutrients is 10 mL / L water and a combination of LMO A 5 mL + LMO B 5 mL / L water is the most effective combination of nutrients for hydroponic *Caisim* growth.

1. Introduction

Vegetables are important for a healthy diet. Eating vegetables can prevent various diseases [1]. Along with population growth, Indonesia is coping with food and nutrition security concerning household hunger and malnutrition. The production of vegetable crops in Indonesia has grown both for domestic consumption and in exports to some extents [2]. One type of vegetable that is quite popular among the Indonesian is mustard greens (*Brassica juncea L.)*. The mustard plant (*Brassica juncea* L.) is a type of plant leaf vegetables that has high economic value, including in the *Cruciferae* family, which is also known as *Caisim* because it is often processed into various dishes and contains a variety of ingredients and vitamins that are good for health [3]. Based on data from the Central Statistics Agency in 2015, Mustard Greens production in Indonesia from 2011 to 2013 increased from 580,969 tons to 635,728 tons, but from 2014 to 2015 has decreased from 602,468 to 580.51 tons. This data shows that there were fluctuating production, even the reduction has lasted these three years [4]

The cause of the reduction in *Caisim* production is in its cultivation, one of which is the use of fertilizer. The use of inorganic fertilizer in a relatively long period of time has a bad impact on soil conditions, which in turn will further decrease mustard greens production. Based on these problems, it is necessary to improve and increase the production of *Caisim*. One solution to improve the cultivation of *Caisim* without checking the soil condition is by the hydroponic system either with or without a greenhouse. Cultivation with hydroponics has its own problem, that in order to get AB MIX nutrition, it needs expensive costs, so it needs a substitute for AB MIX which has the same ability. Organic nutrients such as liquid fertilizer based on local microorganisms (LMO) have features that are easily absorbed by plants that use organic nutrients which can be used as a cheaper and efficient source of nutrition for *Caisim*, and will increase production in *Caisim* cultivation. Using liquid fertilizer that is used on local microorganisms (LMO) is an alternative to support the needs of macro and micronutrients. Those Liquid fertilizer have a positive impact on increasing N and P content that are essential for plant growth [5]. Based on the description above, this study was conducted to determine the effectiveness of variations in organic nutrient based on Local Microorganisms (LMO) with inorganic nutrition AB Mixed on the growth of *Caisim* (*Brassica juncea* L.).

2. Methods

This research was an experimental study in which organic nutrient based on local microorganism was given to observe the *Caisim* (*Brassica juncea* L.) growth using NFT model hydroponic installations. While the ingredients used were nutrition AB MIX, LMO of *Tape* (fermented cassava), LMO of the rind, rock wool, water, flannel cloth, label paper, *Caisim* seeds (*Brassica juncea* L.). The *tape* LMO and rind LMO were deposited for 2 weeks before using. The study was arranged in a Completely Randomized Design with a Single Factor experimental design consisting of 4 treatments, namely a combination of AB MIX solution and various kinds of organic nutrients, as follows: A = LMO A 5 mL + AB MIX 5 mL/L of water; B = LMO B 5 mL + AB MIX 5 mL/L water; C = LMO A 5 mL + LMO B 5 mL/L water; D = AB MIX 10 mL/L water.

2.1 Production of liquid organic fertilizer based on local microorganism

Tape LMO is made by weighing 100 g of cassava, then put into a plastic bottle with a volume of 1500 mL. Fill the water in the bottle that has been filled with cassava tape earlier, it does not have to be full. Then put 5 tablespoons of sugar, mix and then stir and shake until the sugar water dissolves. The bottle is left open for 4 or 5 days. Furthermore, after 5 days of incubation, if the smell of alcohol appears, the LMO is ready and can be used as fertilizer.

2.2 Plant seed watering

Before starting planting with a hydroponic system, the first thing to do is to seed. The seeds used are *Caisim* seeds (*Brassica juncea* L.). The seeding using Rockwool (a type of sponge) as a seeding medium. Seeds are in seedlings for ± 14 days until the first 4 leaves appear.

2.3 Seedlings planting into installation and given treatment

The 4-leaf *Caisim* sprouts are transferred to the net pot, then the seeds are ready to be put into the NFT system hydroponic installation that has been made previously. Nutrition is made according to treatment. Substitution of water and nutrition is done every three days, or when the nutrients in the body have shrunk.

2.4 Maintenance

Maintenance is done every 2 days, observing the continuity of the work of the hydroponic installation system, replacing water and nutrients, and keeping the seeds away from pests.

2.5 Observation variables
Observations of the results carried out after the seventh day after the planting (DAP) process. Plant height, measured from the base of the stem to the tip of the longest leaf when the plant is on their 7th DAP, 14th DAP, 21st DAP, 28th DAP. The number of leaves is the leaves that are formed perfectly calculated when the plant is on their 7th DAP, 14th DAP, 21st DAP, 28th DAP. Wet weight, *Caisim* wet weight was measured in the fourth week (28th DAP).

2.6 Observation
Plant height is measured every week by measuring it from the top surface of the media to the tips of leaf buds. The number of leaves is calculated as leaves grow perfectly, done once a week starting from the first week after planting until harvest. The activity was carried out for one month, namely 7th DAP, 14th DAP, 21th DAP, and harvested in the fourth week (28th DAP).

3. Results
The research began with the manufacture of LMO-based organic fertilizer. LMO is made from fruit rind waste and tape (fermented cassava). After 14 days of the incubation period, organic fertilizer is ready to be applied. Mustard greens seedlings which have 4 leaves are placed at the net pot and then put into a hydroponic installation. Each installation consists of 4 pipes (representing 2 treatments), each has 6 holes where the seeds are placed. Seedlings were observed by their height, the number of leaves every week until the fourth week, while the wet weight was measured once at harvest in the 4th week (28th DAP).

Based on observations that have been made for 4 weeks (28 Days) about the effectiveness of nutritional variations on the growth of hydroponic *Caisim* (*Brassica juncea* L.) observation obtained were in the form of plant height, the number of leaves, and wet weight as follows.

Table 1. The Average of the Growth of *Caisim* (*Brassica juncea* L.) Hydroponic

Treatments	The Average of the Growth of *Caisim* (*Brassica juncea* L.) Hydroponic After 4 Weeks		
	Plant Height (cm)	Number of Leaves	Gross Weight (g)
A (LMO A + AB MIX)	54,5	23,2	21,7
B (LMO B + AB MIX)	54,0	23,7	21,5
C (LMO A + LMO B)	67,8	27,8	30,3
D (AB MIX)	69,2	27,2	32,3

**LMO: Local Micro Organism*
AB MIX: the types of hydroponic nutrient

From the observational data, it can be seen that variations in nutrition affect the height of *Caisim* plants, it can be seen from the average height of different *Caisim* plants. The results can be seen that the treatments A (AB mix + LMO A) and B (AB mix + LMO B) give almost the same results, namely 54.5 cm and 54.0 cm. Whereas the provision of LMO nutrition alone in treatment C (LMO A + LMO B) gives an average height of 67.8 cm and the average height of the largest plants obtained in treatment D (AB Mix), which is 69.2 cm. While the parameter results for the number of leaves, the average number of leaves in 4 treatments gave different results. The treatments A (AB mix + LMO A) and B (AB mix + LMO B) gave almost the same results, namely 23.2 and 23.7. While giving only the AB Mix nutrition in treatment D gives an average number of leaves 27.8 and the largest average number of leaves obtained in treatment C (LMO A and LMO B), which is 27.2. The results of measurements of plant wet weight can be seen, that there is an effect of different nutritional benefits on the *Caisim* wet weight. Treatment A and B showed almost the same results, namely 21.7 g (treatment A) and 21.5 g (Treatment B).

Whereas the giving of nutrition in the form of LMO alone (Treatment C) gave a greater wet weight result compared to the previous treatment (A and B), which was 30.3 g. The highest wet weight was produced by *Caisim* who were treated with AB Mix only, which was 32.3 g. The results of the analysis of variance both also gave a significant effect, marked by the value (F <0.05) for each parameter, namely the number of plant leaves and its Wet Weight.

4. Discussions

The results showed that the increase in the concentration of organic fertilizer Tape Local Microorganisms did not show an increase in wet weight, the number of leaves, and plant height. The concentration of LMO fertilizer that has a real effect is the combination of LMO fertilizer with a concentration of 10 ml / L water or treatment C. While the combination of LMO fertilizer with AB MIX (Treatment A and B) does not show a better effect than treatment C and D.

Where it is known that there is a relationship that is directly proportional between the wet weight of plants, the number of leaves, and plant height. The more leaves the wet weight of the plant is also greater as well as the height of the plant, the higher the plant's wet weight is also greater. Local microorganisms (MOL) is a fermented liquid that uses locally accessible resources. MOL contains bacteria of organic matter, plant growth stimulants, pest control agents, and nutrients needed by plants [6]. The local microorganism is the result of own production of natural materials around us (local) which means fertilizer solutions of local microorganisms has the potential to remodel organic matter so that local microorganisms can be used both as decomposers [7].

The combination of organic fertilizer (LMO) and inorganic fertilizer (AB MIX) which is not accompanied by an increase in the wet weight of the plant, allegedly because of this combination of fertilizers results in a large number of deposits resulting from the activity of microorganisms contained in the fertilizer so that the growth of mustard greens becomes inhibited. In another study, it was explained that the presence of microorganism activities that produce residues so that sediment occurs and causes obstruction of water flow and nutrients in the planting media [8].

The low wet weight of mustard plants as a result of the high dose of fertilizer used is also thought to be caused by very high electronic conductivity (EC) making it difficult for plants to absorb nutrients. Electronic conductivity (EC) is a measure of the ability of a solution to conduct electrical current. An electric current in solution is delivered by ions contained in it. The amount of ions in solution is affected by the solids dissolved in it. The greater the number of dissolved solids in the solution, the more likely the number of ions in the solution will also be greater, so that the EC value is also greater [9].

Application of organic fertilizer into the soil can improve soil structure making the soil looser so that the root system can develop better and the absorption process of nutrients runs more optimally [10]. The effect of the use of nutrients in the soil is not only received by plants but will also be received by microbes that have previously been in the soil so that they support each other in the process of plant growth [11]. Microbes disintegrate organic compounds so they can be reabsorbed by plants as nutrients [12]. Cultivation of agricultural land using organic fertilizer appropriately is believed to significantly increase agricultural output, especially vegetables [13].

Different nutrients providing is effective for increasing the growth of *Caisim* (*Brassica juncea* L.). The highest average plant height was given by *Caisim* with AB MIX nutrition, which is 69.2 cm. The highest average number of leaves was obtained by *Caisim* which was grown with a combination of LMO A and LMO B nutrients, which was 27.8. The largest average wet weight was given by *Caisim* with AB MIX nutrition, which is 32.3 g.

5. Conclusion

Based on the data analysis and discussion can be concluded that the variation of AB MIX nutrition and liquid organic fertilizer based on LMO effectively increased the growth of *Caisim* plants. The highest average plant height was given by *Caisim* with AB MIX nutrition, which is 69.2 cm. The highest average number of leaves was obtained by *Caisim* which was grown with a combination of LMO A and LMO B nutrients, which was 27.8. The largest average wet weight was given by *Caisim* with AB MIX nutrition, which is 32.3 g. The use of AB MIX nutrient is 10 mL / L water and a combination of LMO A 5 mL + LMO B 5 mL / L water is the most effective combination of nutrients for hydroponic *Caisim* growth.

References

[1] Sari A D, Basuki B, Mariska I, Budiningsih S 2016 Correlation of physical activity and other factors to fruit and vegetable intake among civilian pilots in Indonesia *Health Science Journal of Indonesia* **7**(2)

[2] Mariyono J, Dewi H A, Daroini P B, Latifah E, Zakariya A Z, Hakim A L, Afari-Sefa V 2018 Farming practices of vegetables: A comparative study in four regions of East Java and Bali Provinces *AGRARIS: Journal of Agribusiness and Rural Development Research*. **4**(2)

[3] Ilhamiyah, Salamiah, Samharinto, Halim H 2019 Agroecosystem Management on Green Mustard Plants in Wetlands during Dry Season. *Journal of Wetlands Environmental Management* 7(1) 41 49

[4] Badan Pusat Statistik (BPS) 2015 *Statistik Indonesia* (Jakarta: Badan Pusat Statistik)

[5] Widjajanto D W, Purbajanti E D, Sumarsono, and Utama C S 2017 *J Applied Chem. Sci*. **4** 325 329

[6] Manullang R R, Rusmini, and Daryono 2019 Combination microorganism as local bio activator compost Kirinyuh *International Journal of Scientific & Technology Research* **7**(6)

[7] Suyanto A and Agnes Tutik Purwani Irianti 2016 Efektivitas *Trichoderma Sp* dan Mikro Organisme Lokal (Mol) sebagai dekomposer dalam meningkatkan kualitas pupuk organik alami dari beberapa limbah tanaman pertanian *Jurnal Agrosains* **12**(2)

[8] Haryanto, Veranica In 2015 Pemanfaatan limbah cair industri tepung aren dan mikroorganisme lokal sebagai larutan nutrisi terhadap pertumbuhan dan hasil tanaman Baby Kailan (*Brassica oleracea*) dengan sistem hidroponik. *EL-Vivo* **3**(2)

[9] Irwan F and Afdal 2016 Analisis hubungan konduktivitas listrik dengan Total Dissolved Solid (TDS) dan temperatur pada beberapa jenis air. *Jurnal Fisika Unand* **5**(1) 85 93

[10] Rahayu S and Tamtomo F 2017 Efektivitas Mikro Organisme Lokal (Mol) dalam meningkatkan kualitas kompos, produksi dan efisiensi pemupukan N, P, K pada Tanaman Ubi Jalar (*Ipomoea batatas* L.) *Jurnal AGROSAINS* **13**(2)

[11] Geisseler D and Kate M Scow 2014 Long-term effects of mineral fertilizers on soil microorganisms–a review *Soil Biology and Biochemistry* **75** 54 63

[12] Idham I S Sudiarso, N Aini, and Y Nuraini 2016 isolation and identification on microorganism decomposers of Palu Local Cow Manure of Central Sulawesi, Indonesia *Journal of Degraded and Mining Lands Management* **3**(4) 625 29

[13] Li S, Li J, Zhang B, Li D, Li G, and Li Y 2017 Effect of different organic fertilizers application on growth and environmental risk of Nitrate under a vegetable field *Scientific reports* **7**(1) 17020

The Potency and Food Safety of Lamp Shells (Brachiopoda: *Lingula* sp.) as Food Resources

R Ambarwati[1,1] D A Rahayu[1,2] and U Faizah[1,3]

[1]Biology Department, Faculty of Mathematics and Natural Sciences, Universitas Negeri Surabaya, Indonesia.

E-mail: reniambarwati@unesa.ac.id

Abstract. Lamp shells is one of primitive brachiopods that can be found in the tropical region. One species of brachiopod has been reported found in Indonesia. This invertebrate is collected by local people for consumption as well as traded in the local market in the area of the southern shore of Madura Island and the northern shore of East Java, Indonesia. This research aimed to analyze the potency of lamp shells as food resources based on the protein and fat content and to evaluate the food safety based on the content of lead (plumbum) and cadmium. The sample of lamp shells was collected from Probolinggo beach. Fresh samples of flesh and pedicle of lamp shell were kept in a cool box for laboratory analysis. The protein content was analyzed using Kjeldahl method, while the content of fat was analyzed using soxhlet method. In addition, the level of plumbum and cadmium in the flesh and pedicle of lamp shell were tested using atomic adsorption spectroscopy. The results revealed that both of the flesh and the pedicle of lamp shells contained protein, namely 14.02±0.62% and 12.06±0.44% respectively. The flesh contained 1.12±0.13% of fat, meanwhile, the pedicle contained 0.84±0.07% of fat. These nutrient content were comparable other marine invertebrates. It seems that the nutrient content outweighs the content of heavy metal that can be accumulated in the sessile invertebrate. The content of plumbum in the flesh and pedicle of lamp shells were 0.10±0.01 ppm and 0.08±0.01 ppm respectively. Meanwhile, the content of cadmium in the flesh and pedicle of lamp shells were 0.09±0.01 ppm and 0.06±0.01 ppm respectively. These numbers are considerably lower that the level permitted. Hence, it can be concluded that lamp shells has potency as food recources.

1. Introduction

The coastal waters of Indonesia are rich in invertebrate animals, one of which belongs to Phylum Brachiopoda, which is a primitive invertebrate that is well-known as living fossils. Previous publications recorded the presence of brachiopod from the middle to Upper Jurasic [1] and from Middle Triasic [2]. Nowadays, living brachiopods consisted of 116 genera and 391 species [3]. Current publication reported the distribution of brachiopod from some locations, for instance Queensland, Australia [4], North Western Philippines [5], South China Sea [6], India [7–10].

Brachiopods are also found in Indonesia. The Siboga Expedition (1899-1900) collected 13 species of brachiopods from the coastal waters of Indonesia, belongs to genus *Lingula, Discinisca, Tugulorbynchia, Basiliola, Terebratulina, Gryphus, Campages., Jolonica, Frenulina* [11]. Sahidin et al. [12] reported the occurrence of brachiopod in Banten, Indonesia. Darmarini et al [13] also reported new findings of *Lingula* sp from Damar Lubuk (Aceh). Previously, described that the lamp shells can be found in Muara Angke, West Java and the local people called them as "kerang kucuk" or "kerang keco", but they did not consume this brachiopod [14]. However, in fact local coastal communities in East Java have been these animals consuming lamp shells and the lamp shells are traded in the local market. In Probilinggo, coastal communities refer to this animal as the name tebalan and in Tuban the community calls it leaf shells, while in Madura it is called tobelen.

So far there are no scientific publications that reported on the use of brachiopod as food resources. In fact, we need the data about their nutrient content as well as the food safety related to the heavy metals contamination as marine organisms potentially accumulate the heavy metals [15][16] [17]. Besides, previous reports are mostly working on the potential of molluscs as food resources, for instances Santhiya et al [18], Salman & Nasar [19], Ghosh et al [20], Baby et al.[21].

The mains aims of this study were to analyze the potency of lamp shell as food resources based on the protein and fat content as well as to evaluate the food safety based on the content of lead (plumbum) and cadmium

2. Methods

The samples of lamp shells were collected from the beach of Bee Jay Bakau Resort Probolinggo, which is located at the south part of Madura Strait (S7°72.842' E113°22.351') (Figure 1). The specimens were collected by digging the holes of lamp shells using a modified cowbar as usually used by local people to collect this invertebrate (Figure 2). Fresh specimens were kept in a cool box for identification and laboratory analysis. The lamp shells were identified based on the morphological character of the shells and pedicles. Morphometric measurement was carried out using a calliper.

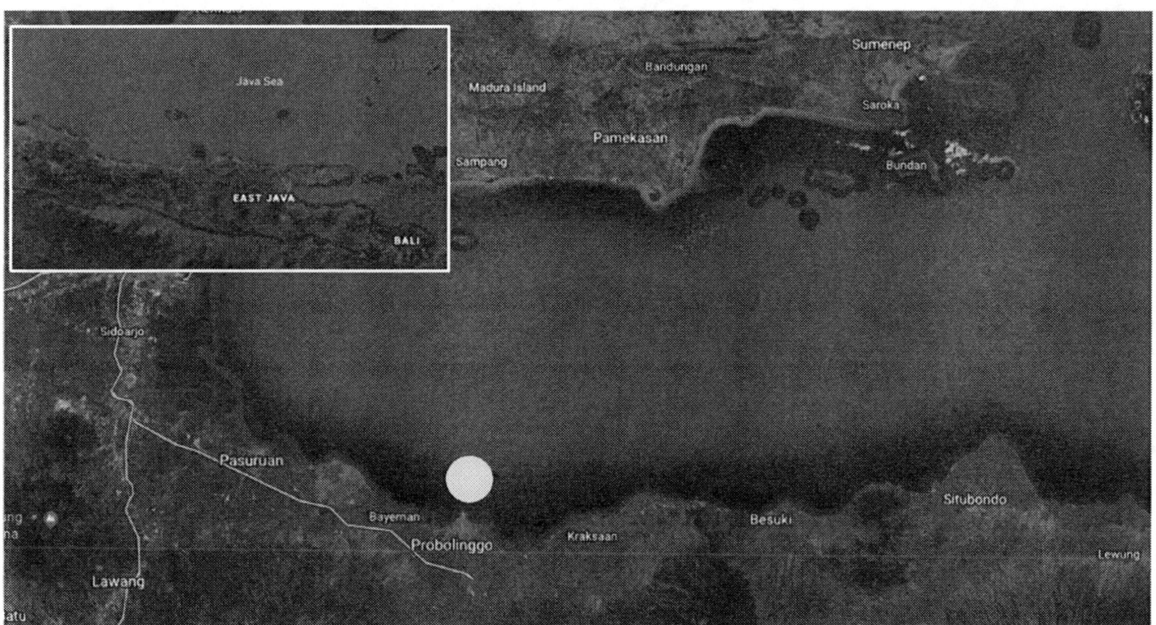

Figure 1. Sampling sites of lamp shells

Figure 2. Local people collected the lamp shells

The flesh and visceral mass were separated from the shells and pedicles. Each sample of flesh (flesh and visceral mass) and pedicle was taken 0.3 grams and prepared for the analysis of lead using Atomic Absorbtion Spectrophotometer (AAS). Meanwhile, each sample of flesh (flesh and visceral mass) and pedicle was taken 0.5 grams and prepared for the analysis of cadmium level using Atomic Absorbtion Spectrophotometer (AAS). The content of protein was analyzed using Kjeldahl method, while the content of fat was analyzed using soxhlet. Three repetitions were applicable for all analyses. Data was analyzed descriptive quantitatively and presented in the histogram.

3. Result and Discussion

Local people in the coastal area of Madura Strait, for example in Probolinggo, collect lamp shells (Brachiopoda) during the lowest tide, for both family consumption and shelling in the local market. The lamp shells occupy muddy sand substratum located in front of mangrove. Local people collects this invertebrate by handpicking using flattened ended cow bar (Figure 2). The habitat of lamp shell are similar to the habitat of brachiopod in Aceh reported by Darmarini et al [13]. However, it is slightly different from the characters of the brachiopod habitat reported from India [7][8][9], which are muddy substratum in the estuary region.

The description of lamp shells from Probolinggo is as follows. Shells shape elongates; lateral margins subparallel; anterior margin slightly convex; smooth external valve surface; distinct growth lines. Colour greenish (varies from translucent green to dark green), some specimens slightly brownish along the lateral and posterior margins. The pedicles are whitish (Figure 3). The length of the shell ranged from 20.2 mm to 49.3 mm (39.45 ± 4.41 mm). The width of the shell was 18.33 ± 2.65 mm. The thickness of the shells ranged from 3.55 mm to 11.4 mm (8.15 ± 1.04 mm). The length of the pedicle ranged from 7 mm to 74 mm (44.5 ± 11.5 mm).

Based on the morphological characters, the lamp shells from Probolinggo belongs to genus *Lingula*, similar to the species from Aceh [13] and India [7][8][9]. Further study are needed to identify the species of this lamp shells, whether the species of this lamp shells is *Lingula* cf. *anatina*

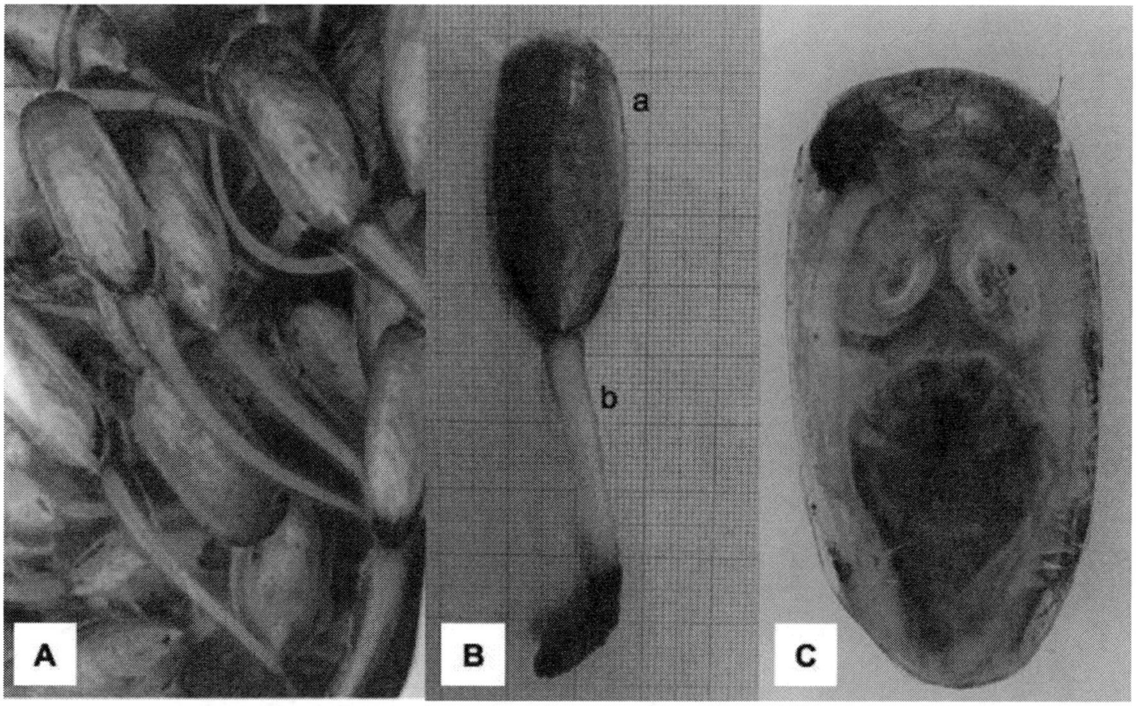

Figure 3. Lamp s from Probolinggo. A: various size of lamp shells; B: Morphology of lamp shells, a: shell; b: pedicle; C. Flesh and visceral mass

The fact that local people are consuming this brachiopod is interesting because there is no scientific publication about the use of this brachiopod as food as well as their nutrient content. The local people described that flesh and the pedicle (they call pedicle as "ekor" or tail) are tasty; the taste and the texture are similar to the taste of bivalves or sea shells. This research revealed that the flesh and the pedicle of lamp shells contained high protein, namely 14.02±0.62%, and 12.06±0.44% respectively. Besides, The flesh contained 1.12±0.13% of fat, meanwhile, the pedicle contained 0.84±0.07% of fat. The flesh (flesh and visceral mass) of lamp shells contained more fat and protein compared to the pedicle. This content was also comparable to the nutrient content of sea shells. Salman & Nasar reported the content of lipid and protein of seashells [19]. Besides, sea shells also contain fatty acid and cholesterols [22].

One parameter of food safety was based on the content of heavy metal in the food [23]. The content of heavy metal lead and cadmium of lamp shells were very low. The content of lead (*plumbum*) in the flesh and pedicle of lamp shells were 0.10±0.01 ppm and 0.08±0.01 ppm respectively. Meanwhile, the content of cadmium in the flesh and pedicle of lamp shells were 0.09±0.01 ppm and 0.06±0.01 ppm respectively. This value was comparatively lower than the maximum limit of heavy metal contamination in food [23]. Besides, the content of heavy metal lead and cadmium of lamp shells were lower than the content of heavy metal lead and cadmium of consumption bivalves fished from Madura Strait [17] [24][25].

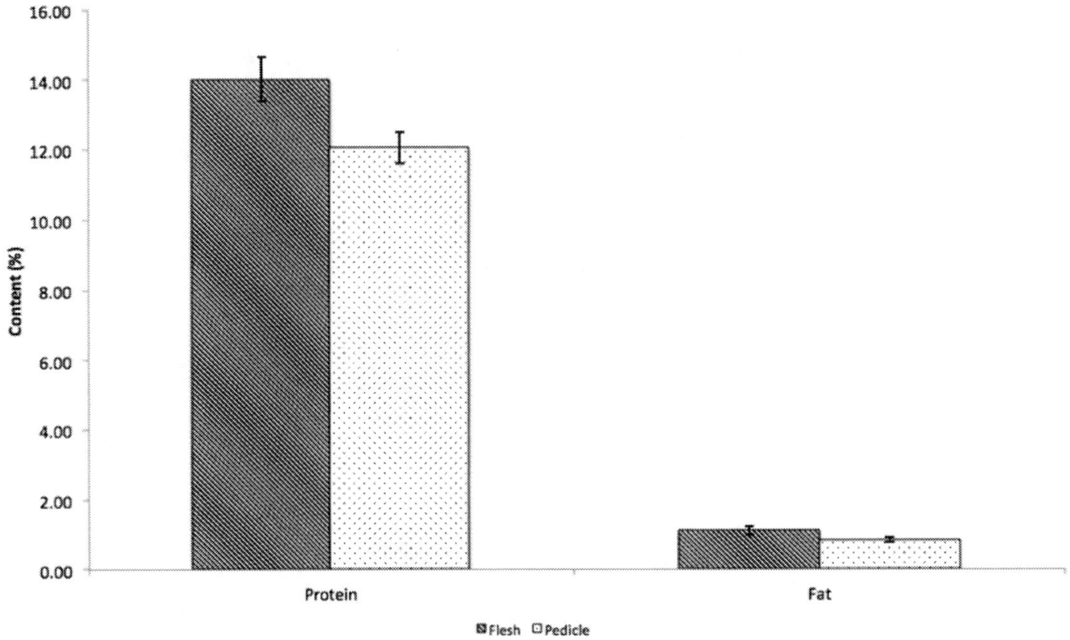

Figure 4. The content of protein and fat of lamp shells from Probolinggo

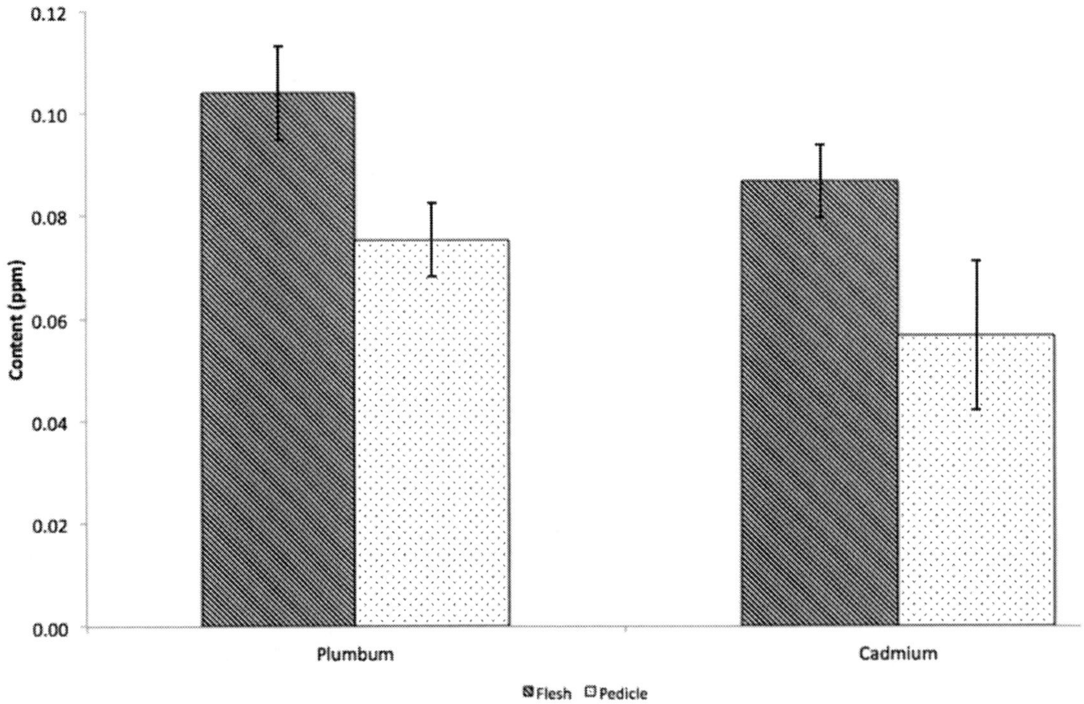

Figure 5. The content of lead (plumbum) and of lamp shells from Probolinggo

4. Conclusions
The flesh (flesh and visceral mass) of lamp shells (Brachiopoda: *Lingula* sp.) from Probolinggo contained more fat and protein compared to the pedicle. The content of heavy metal lead and cadmium in the flesh and pedicle of lamp shells was lower than the maximum limit of heavy metal contamination in food. Based on these parameters, lamp shells that usually consumes by local people are safe as food resources.

5. References

[1] Alberti M, Fürsich F T, Abdelhady A A and Andersen N 2017 Middle to Late Jurassic equatorial seawater temperatures and latitudinal temperature gradients based on stable isotopes of brachiopods and oysters from Gebel Maghara, Egypt *Palaeogeogr. Palaeoclimatol. Palaeoecol.* **468** 301–13

[2] Feldman H R 2017 Tunethyris blodgetti sp. nov. (brachiopoda, terebratulida) from the Middle Triassic of Makhtesh Ramon, Southern Israel *Ann. Soc. Geol. Pol.* **87** 89–99

[3] Emig C C, Bitner M A and Alvarez F 2013 Phylum Brachiopoda *Zootaxa* **3703** 75–8

[4] Kenchington R . and Hammond L . 1978 Population structure , growth and distribution of Lingula anatina (Brachiopoda) in Queensland, Australia *J. Zool. Lond.* **184** 63–81

[5] Leopardas V, Honda K, Go G A, Bolisay K, Pantallano A D, Uy W, Fortes M and Nakaoka M 2016 Variation in macrofaunal communities of sea grass beds along a pollution gradient in Bolinao, northwestern Philippines *Mar. Pollut. Bull.* **105** 310–8

[6] Bitner M A and Romanin M 2018 Recent brachiopods collected during the ZhongSha 2015 expedition to the South China Sea, West Pacific *Mar. Biol. Res.* **14** 551–64

[7] Mitra S and Pattanayak J G 2013 Studies on Lingula anatina (Brachiopoda: Inarticulata) in Subarnarekha Estuary, Odisha with special reference to habitat and population *Rec. Zool. Surv. India* **113** 49–53

[8] Samanta S, Choudhury A and Chakraborty S K 2014 Morpho-anatomical study of Lingula anatina Lamarck, 1801 from West Bengal-Odisha coast, India *J. Mar. Biol. Ass. India* **56** 26–33

[9] Samanta S, Choudhury A and Chakraborty S K 2015 Eco-biology of a Precambrian intertidal benthic brachiopod, Lingula anatina from the confluence of Subarnarekha estuary with Bay of Bengal, India *J. Mar. Biol. Ass. India* **57** 41–6

[10] Nayak A, Charan Kumar B, Lovaraju A, Raut D, Sanghamitra Rout S, Dash B, Patnaik L, Mohanty B and Raman A 2018 Benthic infauna from mudflats of Atharbanki mangrove waterway in Odisha, India *J. Mar. Biol. Assoc. India* **60** 33–9

[11] Jackson J and Stiasny G 1937 *The Brachiopoda of the Siboga Expedition. "Siboga Expeditie" XXVII* (Leiden)

[12] Sahidin A, Zahidah Z, Herawati H, Wardiatno Y, Setyobudiandi I and Partasasmita R 2018 Macrozoobenthos as bioindicator of ecological status in Tanjung Pasir Coastal, Tangerang District, Banten Province, Indonesia *Biodiversitas J. Biol. Divers.* **19** 1123–9

[13] Darmarini A S, Wardiatno Y, Prartono T and Soewardi K 2017 Short communication: New record of primitive brachiopod, Lingula sp. in mangrove ecosystem of Lubuk Damar, Aceh

Tamiang, Indonesia *Biodiversitas* **18** 1438–44

[14] Mudjiono and Suparman M 1992 Sekilas tentang kerang lentera Filum Brakhiopoda *Oseana* **XVII** 159–66

[15] Eugene Ng Y J, Yap C K, Zakaria M P and Tan S G 2013 Assessment of heavy metal pollution in the Straits of Johore by using transplanted caged mussel, Perna viridis *Pertanika J. Sci. Technol.* **21** 75–96

[16] Bosch A C, O'Neill B, Sigge G O, Kerwath S E and Hoffman L C 2016 Heavy metals in marine fish meat and consumer health: A review *J. Sci. Food Agric.* **96** 32–48

[17] Ahyar, Bengen D G and Wardiatno Y 2017 Sebaran dan bioakumulasi logam berat Pb dan Cd pada bivalvia Anadara nodifera, Meretrix lyrata, dan Solen lamarckii *J. Ilmu dan Teknol. Kelaut. Trop.* **9** 631–43

[18] Santhiya N, Sanjeevi S B, Gayathri M and Dhanalakshmi M 2013 Economic importance of marine molluscs *Res. Environ. Life Sci.* **6** 129–32

[19] Salman J M and Nasar A J 2013 *Total Lipids and Total Protein in two Mollusca Species as Environmental Biomarker of Pollution in Euphrates River, Iraq* vol 2

[20] Ghosh S, Jung C and Meyer-Rochow V B 2016 Snail Farming: An Indian Perspective of a Potential Tool for Food Security *Ann Aquac Res* **3** 1024

[21] Baby R L, Hasan I, Kabir K A and Naser M N 2010 Short Communication Nutrient Analysis of Some Commercially Important Molluscs of Bangladesh *J. Sci. Res* **2** 390–6

[22] Purwaningsih S S, Salamah S H and Dewi J R 2015 *Effect of boiling and steaming on the profile fatty acids and cholesterol in muscle tissue of molluscs* **22**

[23] Standar Nasional Indonesia (SNI) 2009 *Standar nasional Indonesia nomor 7387.2009 tentang batas maksimum cemaran logam berat dalam pangan* (Jakarta: Badan Standar Nasional)

[24] Riswanda T, Rachmadiarti F and Kuntjoro S 2014 Pemanfaatan Kitosan Udang Putih (Lithopannaeus vannamei) sebagai Bioabsorben Logam berat Timbal (Pb) pada Daging kerang Tahu di Muara Sungai Gunung Anyar Utilization of Chitosan White Shrimp (Lithopannaeus vannamei) as Bioabsorben of Heavy metals Le *LenteraBio* **3** 266–71

[25] Indasah I, Arbai A, Sugijanto S and Agus S 2011 Citric acid reduces the content of Pb and Cd of kupang beras (Corbula faba) *Folia Medica Indones.* **47** 46–51

The Content Validity of Digital Test Items for Evaluation Courses Based on *Superitem-Wondershare* Using *Aiken*'s Calculation

D G H Divayana[1*], A Adiarta[2], I G Sudirtha[3]

[1]Department of Informatics Education, Universitas Pendidikan Ganesha
[2]Department of Electrical Education, Universitas Pendidikan Ganesha
[3]Department of Vocational Education, Universitas Pendidikan Ganesha

e-mail : hendra.divayana@undiksha.ac.id

Abstract. The digital test item for evaluation courses that is designed with tiered arrangements use super item concept and is showed digitally use the *Wondershare* application makes it easy for students to practice the quality of their critical thinking in working on test questions whenever and wherever they are. But to get the good quality test items, then it is still necessary to test the item validity properly and accurately. The method that can be done to determine the valid test items is through calculating the validity of the items uses the *Aiken* model. Based on those matters, the purpose of this study was to show the calculation process of the test items validity uses the *Aiken* model. Subjects that were involved in the calculated process the validation of each test item were four experts, including two experts in the field of evaluation and two experts in the field of informatics engineering. The tool was used to assess the test items was a questionnaire consisting of 42 question items. The method used to analyze the test items was by comparing the results of the test item validation used the *Aiken* model with the validation classification according to *Guilford*. The study results were showed that from 42 test items tested; there were two invalid items and 40 valid items.

1. Introduction

The cognitive abilities of students in following a learning process in evaluation courses in universities can be measured through tests. A good test to use is a valid test and can accurately measure the cognitive abilities of students from easy to challenging levels. One form of test that can be used to meet the intended test criteria is a *Superitem*-based digital format test.

At present, *superitem*-based digital formatted test items are essential and necessary in making it easier for lecturers to make test questions that arranged in stages from easy to challenging levels with accurate value calculations and easily accessible. One application that can be used to create digital-form test items with a tiered arrangement from easy to difficult is *Wondershare*. Although the application of *Wondershare*, in general, is very suitable to be used as a facility to make digital-form test items in several universities, however, test items that were imported into the *Wondershare* application that was implemented in the Informatics Engineering Departement, Universitas Pendidikan Ganesha wasn't known certainly it content validity yet. For that, so it is necessary to test the content validity of the test items by using an appropriate calculation model. One of the formula used to test the content validity of an instrument is the *Aiken* formula.

Based on that situation, so the problem statement of this research was how the level of content validity of digital test items based on *Superitem-Wondershare* that was used in the Evaluation course at Informatics Education Department, *Universitas Pendidikan Ganesha* with the *Aiken* formula calculation process? The purpose of this research showed the process of calculating the *Aiken* formula used in measuring the level of content validity of digital test items for evaluation courses.

Some of the studies that underlie this research, including research were conducted in 2016 by Balan, and Djuniadi [1] showed the use of the *Wondershare* quiz creator application to make digital tests in English subjects for vocational students. Other research was conducted in 2017 by Gaol, Khumaedi, and Masrukan [2], showed the results of calculating the content validity of the test items using the *Aiken* formula that was conducted by three experts.

The research that was conducted in 2018 by Ningdyah, Greenwood, and Kidd [3], showed the use of *Aiken* formula in determining the coefficient of content validity for measuring instruments characteristic of educational models in Indonesia. The research that was conducted in 2018 by Ikhsanudin, and Subali [4] showed that two quantitative approaches could be used to measure the content validity test, namely *Aiken*'s V and *Lawshe*'s CVR. The research that was conducted in 2018 by Apawu, Owusu-Ansah, and Akayuure [5] showed the use of *superitem* tests to analyze and monitor students' level of thinking in learning algebra. Research that was conducted in 2019 by Bahri and Supahar [6] showed a quantitative approach using the *Aiken* formula in calculating the content validity of religious test items integrated with science. Research that was conducted in 2019 by Rahmawati, Priatna, and Juandi [7] showed the use of *superitem* test models in exploring students' critical thinking abilities by expanding abstract level thinking into thinking with algebraic characters.

2. Method

The approach that was used in this research was the development of instruments. The stages of instrument development consist of 6 stages [8], including 1) determination of variables, 2) determination of indicators, 3) determination of instrument items, 4) instrument trials, 5) instrument analysis, and 6) determination of final instruments. At the stage of determination of variables carried out the determination of the chapters that were used as the main component of test questions. At the stage of determination of indicators carried out by doing the arranged of chapters into sub-chapters. At the stage of determination of instrument, items were carried out making items that were sourced from the sub-chapter. At the trial stage, the instrument was tested for content validity and instrument reliability. However, specifically in this study, the instrument trials focused on content validity testing only. Subjects were involved in conducting the content validity testing were four experts consisting of two experts in the field of informatics education and two experts in the field of evaluation. The tool was used by experts to assess instrument items was a questionnaire consisting of 42 question items. At the instrument analysis stage, was carried out an analysis of the test results of the instrument's content validity. Analysis of the results of the instrument content validation was done by interpreting the results of the content validation (which had been obtained using the *Aiken* formula) based on the *Guilford* content validation reference. The *Aiken* formula [9] can be shown through equation (1), while the categorization of the *Guilford*'s content validation can be shown in Table 1[10].

$$V = \frac{\sum s}{n(c-1)} \quad (1)$$

Notes:

V : Expert agreement index regarding item validity
r_n : score assessment that given by expert-n on item-n
l_o : lowest score in the scoring category
n : number of experts
S : $r_n - l_o$
c : number of choice scores

Table 1. The Categorization of Content Validation Refers to *Guilford*

Validity Score Range	Category
$0{,}80 < r_{xy} \leq 1{,}00$	: Very high validity
$0{,}60 < r_{xy} \leq 0{,}80$	: High validity
$0{,}40 < r_{xy} \leq 0{,}60$	: Medium validity
$0{,}20 < r_{xy} \leq 0{,}40$	: Low validity
$0{,}00 < r_{xy} \leq 0{,}20$	: Very low validity
$r_{xy} \leq 0{,}00$	: Invalid

3. Results and Discussion

Some of the results were obtained in this research was based on the completion of the instrument development stages. The results of this research can be explained in full as follows.

a) Variable Determination Results

The variables in the digital test based on *Superitem-Wondershare* for evaluation course were sourced from the chapters listed in the book entitled "Program Evaluation" authored by Dewa Gede Hendra Divayana. The variables can be shown in full in Table 2.

Table 2. Variables in Digital Test Based on *Superitem-Wondershare* for Evaluation Courses

Chapter	Variable
I	The basic concept of program evaluation
II	Types of evaluation models
III	Strengths and weaknesses of program evaluation models
IV	Research methodology in program evaluation
V	Instruments in program evaluation
VI	Decision-making steps in program evaluation

b) Indicator Determination Results

The indicators in a digital test based on *Superitem-Wondershare* for evaluation courses are sourced from sub-chapters of each variable previously was described in Table 2. Those indicators that intended can be shown in Table 3.

Table 3. Indicators in Digital Test Based on *Superitem-Wondershare* for Evaluation Courses

Chapter	Indicator
I	(1.1) Definition of Evaluation; (1.2) Definition of Program Evaluation; (1.3) Purpose of Program Evaluation; (1.4) Function of Program Evaluation; (1.5) Program Evaluator; (1.6) Requirements of Program Evaluator; (1.7) Elements of Program Evaluation; (1.8) Scope of Program Evaluation
II	(2.1) *Goal Oriented Evaluation Model*; (2.2) *Goal Free Evaluation Model*; (2.3) *Formative Sumatif Evaluation Model*; (2.4) *Countenance Evaluation Model*; (2.5) *Responsive Evaluation Model*; (2.6) *CSE-UCLA* Evaluation Model; (2.7) *Context, Input, Process, Product Evaluation Model*; (2.8) *Discrepancy Model*
III	(3.1) Strengths and Weaknesses of the *Goal Oriented Evaluation Model*; (3.2) Strengths and Weaknesses of the *Free Goal Evaluation Model;* (3.3) Strengths and Weaknesses of the *Formative Sumatif Evaluation Model*; (3.4) Strengths and Weaknesses of the *Countenance Evaluation Model*; (3.5) Strengths and Weaknesses of the *Responsive Evaluation Model*; (3.6) Strengths and Weaknesses of the *CSE-UCLA Evaluation Model*; (3.7) Strengths and Weaknesses of the *CIPP Evaluation Model*; (3.8) Strengths and Weaknesses of the *Discrepancy Evaluation Model*
IV	(4.1) Research Approach; (4.2) Research Methods; (4.3) Research Design; (4.4) Research Instruments; (4.5) Research Subjects; (4.6) Data Collection Techniques and Procedures; (4.7) Data Analysis Techniques
V	(5.1) Closed Instruments; (5.2) Open Instruments
VI	(6.1) Setting of the Evaluation Standards; (6.2) Calculation of Average Scores on Each Evaluation Component and Performing a Conversion to a Scale of Five; (6.3) Changing the Score that Obtained into the Standard Score (Z-Score); (6.4) Changing Z-Score into T-Score; (6.5) Interpretation of T-Score into the Glickman Quadrant; (6.6) Comparing Evaluation Results with Evaluation Standards

c) Results of Determination of Instrument Items

Instrument items in digital tests based on super item-*Wondershare* for evaluation courses were sourced from questions on each of the indicators were described earlier in Table 3. Those instrument items intended can be shown in Table 4.

Table 4. Instrumen Items in Digital Test Based on *Superitem-Wondershare* for Evaluation Courses

Chapter	Indicator
I	1. Explain the evaluation means!
	2. Explain what the program evaluation means is!
	3. Explain the purpose of program evaluation!
	4. Explain the function of program evaluation!
	5. Explain the program evaluator means!
	6. Explain the requirements to be a program evaluator!
	7. Explain the elements of program evaluation!
	8. Explain the scope of program evaluation!
II	9. Explain about the *Goal-Oriented Evaluation Model*!
	10. Explain about the *Goal Free Evaluation Model*!
	11. Explain about the *Formative Sumatif Evaluation Model*!
	12. Explain about the *Countenance Evaluation Model*!
	13. Explain about the *Responsive Evaluation Model*!
	14. Explain about the *CSE-UCLA Evaluation Model*!
	15. Explain about the *Context, Input, Process, Product Evaluation Model*!
	16. Explain about the *Discrepancy Model*!
III	17. Explain the strengths and weaknesses of the *Goal-Oriented Evaluation Model*!
	18. Explain the strengths and weaknesses of the *Free Goal Evaluation Model*!
	19. Explain the strengths and weaknesses of the *Formative Sumatif Evaluation Model*!
	20. Explain the strengths and weaknesses of the *Countenance Evaluation Model*!
	21. Explain the strengths and weaknesses of the *Responsive Evaluation Model*!
	22. Explain the strengths and weaknesses of the *CSE-UCLA Evaluation Model*!
	23. Explain the strengths and weaknesses of the *CIPP Evaluation Model*!
	24. Explain the strengths and weaknesses of the *Discrepancy Evaluation Model*!
IV	25. Explain the approach is used in evaluation research!
	26. Explain the method is used in evaluation research!
	27. Explain the design is used in evaluation research!
	28. Explain closed and open instruments in evaluation research!
	29. Explain the forms of instruments is used in evaluation research!
	30. Explain the steps in determining the research subject in evaluation research!
	31. Explain the techniques and procedures for collecting data in evaluation research!
	32. Explain data analysis techniques in evaluation research!
V	33. Explain the form of closed instruments (questionnaire forms) is used in evaluation research!
	34. Explain the form of the open instruments (forms of interview guidelines, observation guidelines, and documentation) is used in evaluation research!
VI	35. Explain the stages in setting evaluation standards!
	36. Explain the calculation of average scores on each evaluation component and convert to a scale of five!
	37. Explain the calculation of average scores in each evaluation component!
	38. Explain how to convert the calculation results of the average score to a scale of five!
	39. Explain how to change the score that obtained into the standard score (Z-Score)!
	40. Explain how to change Z-Score into T-Score!
	41. Explain how to interpret the T-Score into the Glickman quadrant!
	42. Explain how to compare evaluation results with evaluation standards!

d) Instrument Trial Results

The instrument trials were carried out by four experts. The results of the trial instrument can be seen in full in Table 5.

Table 5. Trial Results of the Digital Test Based on *Superitem-Wondershare* for Evaluation Courses

Items-	Expert				S_1	S_2	S_3	S_4	ΣS	V	Category of Validity
	I	II	III	IV							
1.	4	5	5	4	3	4	4	3	14	0.875	Very High
2.	4	5	5	4	3	4	4	3	14	0.875	Very High
3.	5	5	4	4	4	4	3	3	14	0.875	Very High
4.	4	5	5	4	3	4	4	3	14	0.875	Very High
5.	5	4	4	5	4	3	3	4	14	0.875	Very High
6.	4	4	4	5	3	3	3	4	13	0.813	Very High
7.	4	4	5	5	3	3	4	4	14	0.875	Very High
8.	4	5	4	5	3	4	3	4	14	0.875	Very High
9.	4	4	5	5	3	3	4	4	14	0.875	Very High

Items-	Expert				S_1	S_2	S_3	S_4	ΣS	V	Category of Validity
	I	II	III	IV							
10.	4	5	4	4	3	4	3	3	13	0.813	Very High
11.	4	4	5	4	3	3	4	3	13	0.813	Very High
12.	4	4	5	4	3	3	4	3	13	0.813	Very High
13.	4	4	5	5	3	3	4	4	14	0.875	Very High
14.	4	5	4	4	3	4	3	3	13	0.813	Very High
15.	4	5	4	4	3	4	3	3	13	0.813	Very High
16.	4	4	5	5	3	3	4	4	14	0.875	Very High
17.	4	5	4	4	3	4	3	3	13	0.813	Very High
18.	4	5	4	5	3	4	3	4	14	0.875	Very High
19.	5	4	4	4	4	3	3	3	13	0.813	Very High
20.	5	4	4	5	4	3	3	3	13	0.813	Very High
21.	4	5	4	4	3	4	3	3	13	0.813	Very High
22.	5	4	5	4	4	3	4	3	14	0.875	Very High
23.	4	4	5	4	3	3	4	3	13	0.813	Very High
24.	5	5	4	4	4	4	3	3	14	0.875	Very High
25.	5	4	4	4	4	3	3	3	13	0.813	Very High
26.	4	5	5	4	3	4	4	3	14	0.875	Very High
27.	5	4	5	4	4	3	4	3	14	0.875	Very High
28.	2	3	2	2	1	2	1	1	5	0.313	Low
29.	5	4	5	4	4	2	4	3	13	0.813	Very High
30.	4	4	4	5	3	3	3	4	13	0.813	Very High
31.	4	4	4	5	3	3	3	4	13	0.813	Very High
32.	5	4	5	3	4	3	4	2	13	0.813	Very High
33.	4	5	4	5	3	4	3	4	14	0.875	Very High
34.	5	4	4	4	4	3	3	3	13	0.813	Very High
35.	4	5	4	4	3	4	3	3	13	0.813	Very High
36.	3	2	2	2	2	1	1	1	5	0.313	Low
37.	4	4	5	5	3	3	4	4	14	0.875	Very High
38.	5	4	4	4	4	3	3	3	13	0.813	Very High
39.	4	4	5	4	3	3	4	3	13	0.813	Very High
40.	5	4	4	4	4	3	3	3	13	0.813	Very High
41.	4	5	4	5	3	4	3	4	14	0.875	Very High
42.	4	4	5	4	3	3	4	3	13	0.813	Very High
								Average of V		**0.815**	**Very High**

e) Instrument Analysis

Based on the results of the instrument trials that were shown in Table 5, the interpretation results can be determined with refers to the categorization of the *Guilford*'s content validation. The interpretation results stated that the content validity of the test instruments was included in the very high category because the average of V value was 0.815, in the range of validity scores 0.80 <rxy <1.00.

f) Final Instruments

Based on the results of instrument analysis, there were 40 valid instrument items and two invalid instrument items. Invalid items were item-28 and item-36, while valid items were all other items besides those two items. The valid items were entered into the *Wondershare* application. The display of the final instruments of digital test based on *Superitem-Wondershare* can be shown in Figure 1.

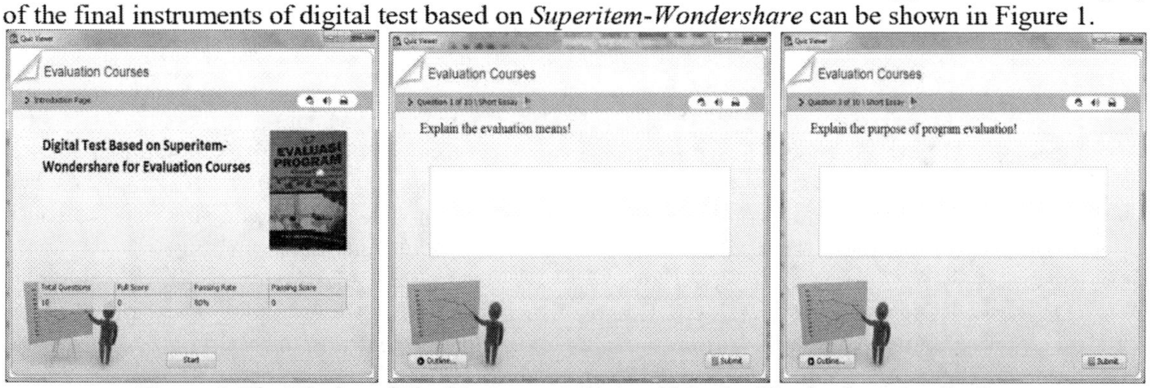

Figure 1. The Display of Final Instruments Entered into the *Wondershare* Application

The results of this study have answered the limitations of the research of Gaol, Khumaedi, and Masrukan which only involved three experts in conducting the validity test of the instrument, whereas in this study it had involved four experts. The results of this study have also answered the constraints of Apawu, Owusu-Ansah, and Akayuure research which only showed the use of *superitem*-based tests for learning algebra, but through this study also demonstrated the use of *superitem*-based tests for evaluation courses. The obstacle that is still found in this study is the source of the test questions, which is still limited sourced from just one book.

4. Conclusions

The content validity of the digital test items based on *Superitem-Wondershare* was included in the very high category based on the categorization of the content validation of *Guilford*, so making it suitable for measuring the student's knowledge in understanding the evaluation courses. The solution as problem-solving to the obstacle in this research can be done by creating more varied test questions sourced from e-books, journals, and other documents related to evaluation.

5. Acknowledgment

The authors would like to express their highest thanks to the Head of the Research and Community Service Institute of *Universitas Pendidikan Ganesha* that had provided the opportunities and had given the funding in this research through research grant contract number 365/UN48.16/LT/2019.

References

[1] Balan Y A and Djuniadi 2016 Implementing media of computer-based test for english language proficiency of vocational high school students at the borderline between indonesia and Timor Leste *SENIT 2016* pp 79–88

[2] Gaol P L, Khumaedi M, and Masrukan 2017 Development of instrument for self-confidence character assessment in mathematics subjects at middle school *J. Edu. Research & Eval.* **6**(1) pp 63–70

[3] Ningdyah A E M, Greenwood K M, and Kidd G 2018 A Training-model scale's validity and reliability coefficients: expert evaluation in Indonesian professional psychology programs *Makara Hum. Behav. Stud. in Asia* **22**(1) pp 56–66

[4] Ikhsanudin, and Subali B 2018 Algebraic thinking characteristics of eighth grade junior high school students based on Superitem Test of SOLO model *The 5th International Conference on Research, Implementation, & Education of Mathematics and Sciences, IOP Conf. Ser.: J. Physics: Conf. Ser.* **1097** pp 1–9

[5] Apawu J, Owusu-Ansah N A, and Akayuure P 2018 A study on the algebraic working processes of senior high school students in Ghana *Euro. J. Scie. & Math. Educ.* **6**(2) pp 62–68

[6] Bahri M F, and Supahar 2019 Content validity and reliability analysis of integrated islamic-science test instrument to measure the student's critical thinking ability *J. Islam. & Soc. Stud.* **5**(1) pp 39–48

[7] Rahmawati D I, Priatna N, and Juandi D 2019 Algebraic thinking characteristics of eighth grade junior high school students based on Superitem Test of SOLO model *International Conference on Mathematics and Science Education (ICMScE 2018), IOP Conf. Ser.: J. Physics: Conf. Ser.* **1157** pp. 1–7

[8] Sugiharni G A D, Setiasih N W, Mahendra I W E, Ardana I M and Divayana D G H 2018 Development of alkin model instruments as evaluation tools of blended learning implementation in discrete mathematics course on stikom Bali *J. Theor. Appl. Inf. Technol.* **96**(17) pp 5803–5818

[9] Retnawati H 2016 Proving content validity of self-regulated learning scale (the comparison of aiken index and expanded gregory index) *Research & Eval.in Educ.* **2**(2) pp 155–164

[10] Suyasa P W A, Kurniawan P S, Ariawan I P W, Sugandini W, Adnyawati NDMS, Budhyani IDAM, and Divayana D G H 2018 Empowerment of CSE-UCLA model based on glickman quadrant aided by visual application to evaluate the blended learning program on SMA Negeri 1 Ubud *J. Theor. Appl. Inf. Technol.* **96**(18) pp 6203–6219

Statistical reasoning or statistical method: Students' preferences for learning Statistics

Kusumarasdyati

English Department, Universitas Negeri Surabaya, Indonesia

e-mail: kusumarasdyati@unesa.ac.id

Abstract. Statistical reasoning should precede the statistical method in the teaching of statistics because the students should master the concepts before they perform computations. To find out whether the students were aware of this principle, the present study explored their preference and the underlying reasons. Twenty undergraduates were taught central tendency and variability by using statistical method (with machine computation) and statistical reasoning (with manual computation) respectively in one session. Most of them turned out to prefer to reason and compute numeric data manually because they understood the statistical concepts and the process of the computation. On the other hand, the rest of the undergraduates preferred the teaching of statistics using the statistical method with machine computation for three reasons: accuracy, efficiency, and convenience. The implications of these findings for the teaching of statistics were presented.

1. Introduction

The teaching of statistics can be approached in two ways: statistical reasoning and statistical methods. Statistical reasoning means "the way people reason with statistical ideas and make sense of statistical information" [1]. Approached in this way, teaching statistics encourages the students to understand the concepts, link one concept to another, and make correct interpretations about the results of the computation. On the other hand, statistical methods refer to the techniques of crunching a large number of data to make them meaningful. Teachers who focus on statistical methods in the classroom only develop the students' skills in computing numbers and probably create graphs based on the results of the computation.

It is important to note, however, that there is more to statistics than computation. Statistics is "the *science* of collecting, organizing and interpreting data" [2] to solve problems in many aspects of life, and the computation is needed in organizing the data. To make the results of the computation meaningful, they must be interpreted carefully. The students definitely should understand the relevant concepts to make accurate interpretations and arrive at sensible conclusions. Unfortunately, some teachers lack knowledge about this. They erroneously equate statistics with computation only, teaching the procedures and the computation without training the students to reason [3]. Moreover, in this age of computerization, some teachers tend to teach the students to operate statistical software rather than facilitating them to understand statistical concepts and reason [4, 5].

Bradstreet [6] argued that for better learning outcomes, statistical reasoning should be taught first, and statistical methods—usually utilize the software for more convenient computation—follow after the students have mastered the statistical concepts. This principle triggers important questions: Do the students know that they should acquire the concepts well before learning to compute numeric data? Do they think they learn statistics better by reasoning the concepts, or are they content with using software to perform computations? The answers to these questions are vital to ensure effective learning of statistics. If the students are aware of the importance of statistical reasoning, the learning objectives are met once they can reason. On the other hand, if they think that learning statistics is completed upon having the ability to run statistical software, the teacher should take necessary actions to make them realize that computations using the software means to an end rather than an end itself.

The present study aimed to explore how statistics should be learned from the students' perspective. The research questions could be stated as follows: (1) Which did the students prefer in learning statistics: statistical reasoning or statistical method? (2) What was the underlying reason for their preference? The procedures for conducting the study to answer these questions will be explained in the next section. Then the results of the study will be reported and discussed. The last part of this paper will present the conclusions drawn from the study and present the implications for the teaching of statistics.

2. Methodology

This study was descriptive research which was approached qualitatively. The participants consisted of twenty undergraduates majoring in the teaching of English as a foreign language (EFL). They took the course "Educational Research Methodology" at the English Department in 2018. In the following semester, they would take the course "Quantitative and Qualitative Analysis" where they were going to learn statistics, so this study attempted to find out whether they knew that there was more to statistics than mere computation.

The data were collected from the students in one session lasting for 100 minutes. This session was divided into two parts: the statistical method (30 minutes) and statistical reasoning (70 minutes). In the first part, the students were taught how to compute central tendency (mean) and variability (standard deviation) using excel. Each student ran this software on his or her laptop and followed my instructions about the procedures of computing the mean and the standard deviation. Afterwards, I taught them how to interpret the results of machine computation. After the first part of the lesson ended, the second part immediately followed. In the second part, the topics remained the same but the way that the materials were delivered entirely differed. The lesson began with explanations about what central tendency and variability meant, what their functions were, and why the students needed to know about them. After that, I showed them how to compute the mean and the standard deviations on the whiteboard step by step, explaining the concepts that came up during the manual computation. When the results of the computation were obtained, I demonstrated how to interpret them by drawing a normal curve with the symbols of mean and standard deviation in the appropriate places and giving detailed accounts of what they meant.

As soon as the lesson ended, the students completed an unstructured questionnaire, in which they had to state whether they preferred the first method of learning statistics (machine computation) or the second one (manual computation) when they learned the two topics *for the first time*. They also had to elaborate on all of the reasons for their preference. The data were analyzed by following these procedures: familiarizing & categorizing, coding & reducing, interpreting & representing [7].

3. Results and Discussion

The results showed that most of the students preferred manual computation when they learned concepts for the first time, indicating they acknowledged the importance of statistical reasoning. Only a few of them supported statistical method, in which the computation became the emphasis of the instruction rather than understanding the concepts. The reasons underlying their preference varied and would be reported below. Because they were still learning English as a foreign language, their answers contained some grammatical errors. These errors, however, did not seriously affect the meaning of the whole sentences.

The reasons for preferring statistical method using software were threefold: accuracy, efficiency, and convenience. The students claimed that the results of the machine computation were more accurate, eliminating the chance of errors. One of the students, Fred, wrote that, "...the result is more accurate than the manual method" and Lisa detailed it further by writing "I think it is easier to compute data using software because it avoided us from human error like when we forget the formula, wrong in counting the scores and typo in computing the scores." It was obvious that machine computation ensured the accuracy of the computation results. The students did not need to worry about making mistakes due to the failure in memorizing complex formulas, miscalculating the numeric data and mistyping the numbers. The second reason was efficiency, as Lisa put it, "... the use of software can help us to compute the data and get the result faster than compute the data manually." Similarly, George stated, "... it is not time-consuming ..." and Chloe wrote, "... it can save time to compute the scores." For these students, it was better to use the software when learning to compute the mean and the standard deviation for the first time because it was more efficient concerning the time spent on the computation. After entering the scores in the computer program, they could obtain the results within a few seconds by clicking certain buttons. The last reason was closely related to efficiency, i.e. convenience. George, for instance, described how convenient it was to learn statistics using the software as follows: "I think using software for calculating the average and standard deviation is much easier. Because we only need to compute/type the formula, and then we can get the result." Likewise, Fred wrote that "In my opinion for the first computing standard deviation it easier use the computer with the Microsoft Excel formula." Finally, Chloe thought that "We do not need a quite complicated formula. In the software, the formula already exists." According to them, using software to learn statistical computation not only saved a lot of time but also provided convenience. Machine computation was easier for them because they did not need to write the formulas and go through the complicated process of calculating the numbers.

On the other hand, the students who preferred to learn the central tendency and variability by manual computation gave two reasons: concepts and process. It was mentioned earlier in this paper that understanding concepts were central in learning statistics and apparently, most students who chose the manual computation over the machine one were aware of this, as shown in the reasons they gave below.

Lucy : For the first time I learn about how to conduct a standard deviation for qualitative research, it is better to learn manually because I will know the concept in depth.

Kent : I think it is easier to study manually for the first time both for means and standard deviation. Because I need to know the concept first and the formula.

Jane : In my opinion, the manual learning material is quite better than using a computer. In manual statistic, we can learn about the background of what is exactly "mean" and "standard deviation" through the formula given by the teacher.

Lucy, Kent and some other students stated explicitly that knowing the concepts was essential when learning statistics. Jane even explained that she understood the concepts of mean and standard deviation by learning the use of the formulas in manual computation. By using the formulas, they got a full grasp of not only the concepts but also the computation process. They revealed this as follows.

Mary : For me, to understand the mean and standard deviation, it is easier to compute the formulas manually. Because it gives me the detail of formulas step by step, while if we use the machine, we only get the result without knowing the detail about how we can get the number. So, it's better to use the formulas manually for a beginner.

Rick : For the first computing the standard deviation and mean, I think I prefer to use manual way because even though that's more complicated than using the software I can know clearly where the number from. I can understand what are the components that I need to get the standard deviation and the mean.

Prue : Because of already learning how to compute the mean of data in high school, I think it is easiest to compute mean by using the software of a computer. However, for learning about standard deviation that unfamiliar, I choose to compute manually. It is because I am going to more comprehend the concepts of standard deviation, the steps to compute it, what it is about, what the function of it. If I do not know the concepts of it and only depend on software, I will find many difficulties in conducting my future research.

Lena : Easier to understand when learning: manual computation. Because I can know the original formula and I can know the step by step such as where those numbers come from comparing to by using software computation. Software computation is easier to use, but I feel like I am "blind" when I use it. It is simple but I don't know where are those numbers come from. I just automatically know the result.

Obviously, for the students, it was compulsory to know how they could obtain the numbers showing the mean or the standard deviation of a distribution of scores, and the only way to know it was by performing manual computations. The process of the computation assisted them to find out where the numbers came from. For example, by computing the mean they learned the terms *the sum of scores* and *the number of scores*, how to use them in the formula, what they obtained as a result of the computation, and how to interpret the result. This process helped them to accumulate pieces of information gradually to construct knowledge about the mean, achieving a good understanding of this concept in the end. The ease of use that the software offered did not deter Rick and Lena from preferring the manual computation. Lena even considered it puzzling (or 'blind', as she called it) to get the result instantly from Excel™ without knowing the process of computation.

It was interesting that the students who preferred machine computation and those who liked manual computation better gave very different reasons. The choice of the former students was motivated by practicality, i.e. the convenience in use and the instant results. It means they viewed learning statistics as gaining computation skills only, indicating that they supported the statistical method. The latter group of students, in contrast, concerned themselves more with conceptual understanding. For them, it was vital to reason and gain knowledge about the statistical concepts, and manual computation became a useful tool to promote their understanding of the concepts. It could be inferred that they put more emphasis on statistical reasoning.

Traditionally, statistics was taught using a statistical method, which enabled the students to compute numeric data. In 1990, however, the teaching of statistics began to undergo reformation and the prominence shifted from computation to conceptual understanding [8]. Therefore, the students who expressed their preference for manual computation to know the concepts and the process of calculating numbers were on the right track in learning statistics. The manual computation was indeed an effective tool to achieve conceptual understanding because it developed their statistical intuitions about the best technique to use, their self-reliance in solving the problems and their quantitative skills [5]. It explained why Mary stated that "if we use the machine, we only get the result without knowing the detail about how we can get the number" and Lena argued, "Software computation is easier to use, but I feel like I am 'blind' when I use it. It is simple but I don't know where are those numbers come from. I just automatically know the result." They felt they were deprived of the opportunity to practice their quantitative skills and sharpen their statistical intuitions. For them, manual computation made learning statistics more meaningful and effective.

It does not necessarily mean that machine computation has no place in the teaching and learning of statistics. In fact, the students should be able to operate Excel™ or other similar computer programs to get accurate results within a very short time when they conduct research in the future, but they may do it only after they have mastered the statistical concepts by learning how to compute manually. It is important to underline that statistical reasoning precedes statistical method [6], so manual computation must be learned before machine computation to ensure conceptual understanding.

4. Conclusion

This study attempted to explore the undergraduates' preference in learning statistics, i.e. statistical reasoning by utilizing manual computation or statistical method by using machine computation. Most of them preferred to compute manually because they could learn the concepts and going through the process of the computation could enhance the conceptual understanding. The rest, however, preferred to compute the numeric data by utilizing software because the results of the computation were more accurate, could be obtained more rapidly, and were easier to get without any complicated process of calculating numbers.

These findings have practical implications for the teaching of statistics. Teachers need to make the students aware of the two approaches—statistical reasoning and statistical methods—in the introductory session of the statistics class. Further, teachers should explicitly emphasize the importance of statistical reasoning when students learn new concepts in statistics. After communicating these verbally to the students, the teachers could begin the instructions by applying statistical reasoning in facilitating the students to get a full grasp of the concepts, using only manual computation to introduce the concepts. When the mastery of concepts and manual computation has been achieved, the teachers may introduce various types of statistical software to ease computation.

References

[1] Garfield J and Ben-Zvi D 2007 How students learn Statistics revisited: A current review of research on teaching and learning Statistics *International Statistical Review* **75**(3) 372–396

[2] Bennett J, Briggs W L, and Triola M F 2014 *Statistical reasoning for everyday life* (Harlow: Pearson)

[3] Ben-Zvi D and Garfield J 2004 *Statistical literacy, reasoning, and thinking: Goals, definitions, and challenges The Challenge of Developing Statistical Literacy, Reasoning and Thinking* (Dordrecht: Kluwer Academic Publishing)

[4] Pimenta R 2006 Assessing Statistical reasoning through project work *Int. Conf. on Teaching Statistics (Salvador)*

[5] Khamis H J 1991 Manual Computations—A tool for reinforcing concepts and techniques *The American Statistician* **45**(4) 294-299

[6] Bradstreet T E 1993 Teaching Introductory Statistics Courses so that nonstatisticians experience statistical reasoning *The American Statistician* **50**(1) 69-78

[7] Ary D, Jacobs L C, and Sorensen C K 2010 *Introduction to Research in Education* (Belmont: Wadsworth, Cengage Learning)

[8] Tu W and Snyder M M 2016 Developing conceptual understanding in a Statistics Course: Merrill's first principles and real data at work *Educational Technology Research and Development* **65**(3) 579-595

Journal of Physics: Conference Series **1417** (2019) 012042 doi:10.1088/1742-6596/1417/1/012042

Ancient China history-based task to support students' geometrical reasoning and mathematical literacy in learning Pythagoras

A D Fachrudin[1], R Ekawati[2], A W Kohar[2], S Widadah[1], I B Kusumawati[1], R Setianingsih[2]

[1]Mathematics Education Department, STKIP PGRI Sidoarjo, Indonesia
[2]Faculty of Mathematics and Natural science, Universitas Negeri Surabaya, Indonesia

e-mail: dhany@stkippgri-sidoarjo.ac.id

Abstract. This study aims to identify the potential learning opportunities provided by ancient china's geometric diagram introduced by Liu Hui into student task in learning Pythagorean theorem. We investigate the role of the history-based task, from *Nine Chapter*, in supporting students' mathematics literacy and geometrical reasoning. Design research was chosen as a method to reach the purpose. Our discussion is limited only from a part of one phase, namely pilot experiment, from three main phases of design research. This research was conducted at three different Junior High schools in Indonesia. The instructional activities were designed based on history-based problem-solving. The analysis was focused on students' learning process, and specifically, the results of their visual reasoning and mathematical literacy. The result indicated that the use of figures or diagrams, with written information act as a support for performing algebraic calculations for students who are unfamiliar using symbol or algebraic operation. In this way, by using geometric manipulation, a new figure is arrived at on which the solution can now be read easier. Furthermore, by using history-based context, students got more opportunities to improve their performance in mathematics literacy by formulating various contextual mathematics problem, sourced from history, into mathematical form.

1. Introduction

History of Mathematics (HoM) provides situations for students gaining a wealthy experience and understanding of the development of mathematical concepts and their connections. Therefore, the integration of HoM in classrooms become an interesting topic for learning designers or researchers to develop an instructional task inspired the HoM. As an example, Fiangga [1] using a historical perspective to design learning activities on the second-degree integral topic. Radford and Guérette [2] developed a set of instructional tasks in learning second-degree equation that inspired by Babylonian Approach. On the other hand, Fachrudin, Putri, Kohar and Widadah [3] also using another perspective of the Babylonian approach to developing a design of learning of quadratic equation. The results of all the studies mentioned, in line with the statement from Man-Keung [4], indicated that the use of HoM in the classroom does not cause students acquire high scores in certain topic overnight, but it encourages learning mathematics to be meaningful, consequently learning process will come easier and encourages. History of Mathematics (HOM) provides conditions for gaining developing rich experience and understanding of the development of mathematical concepts.

One of the HoM topic recommended by Katz [5] is the ninth chapter (*Gou Gu*) of Jiuzhang suanshu or known as "Nine Chapter" for learning Pythagorean theorem which twenty-four problems concerning the properties of right triangle are presented. Liu Hui, in the last chapter of "Nine Chapter", gave commentary concerning geometric manipulation method in solving the various problem related to right triangle in Gou Gu. Guevara-Casanova & Burgués-Flamarich [6] encourage to introduce the geometric manipulation in supporting students' visual reasoning rather than calculating with algebraic symbols. On the other hand, Siu [7] argued that the use of Chinese problem solving from Nine Chapters can support student visual reasoning. Therefore, using Lui Hui's diagram procedure and problems presented in Gou Gu student can associate their symbolic algebraic thinking with visual thinking about area of shape to build their understanding related to the Pythagorean Theorem.

The use of diagrams will be a bridge for students in developing their algebraic knowledge from geometric knowledge. The study of Guevara-Casanova & Burgués-Flamarich [6] indicated that the way of introducing algebra with geometrical interpretations may be profitably applied in the classroom. Meanwhile, Radford and Gerette [2] argued that the use of the geometrical approach from the history of mathematics provides a useful context to help the students developing the meaning of mathematical symbols.

One of the mathematical content categories in the Program International Student Assessment (PISA) for measuring students' mathematics literacy is space and shape. Based on the study of students' mathematical literacy skills conducted by the OECD, the average Indonesian student is at level 1 or below out of 6 levels, lag far behind from some Southeast Asian countries (i.e. Malaysia, Thailand, and Singapore). From the 2012 PISA study, where mathematics is the main domain of assessment, in the context of geometry, known as space and shape in PISA, around 70% of Indonesian students are on level 1 and below. This means that their ability was at the stage of recognizing and solving simple problems in a familiar context using pictures or drawings of familiar geometric objects and applying basic spatial skills.

Through various problems and geometric diagrams of ancient China, we tried to design a series of instructional tasks to study the Pythagorean theorem. Concerning the geometrical manipulation of ancient China, we hope students can associate their symbolic algebraic thoughts with visual thinking about geometric shapes to build their understanding related to the Pythagorean Theorem. On the other hand, using various contexts from the history of mathematics, students have the opportunity to develop their mathematical literacy skills. Specifically, the purpose of this study is to describe how a series of instructional tasks based on ancient china history can support Students' Geometrical Reasoning and Mathematical Literacy in Pythagoras Learning.

2. Ancient China History-based Task

The *Jiuzhang suanshu* or known as Nine Chapter on Mathematical Art is the ancient Chinese that consists of nine different part. In the last chapter, or also known as *Gou Gu* chapter, consists of twenty-four problem solving concerning about right triangle's properties [8]. We will discuss Liu Hui's argument and commentaries of the problem presented in *Gou Gu* which describes the solution to the right triangle problem using diagram and becomes evidence that ancient China recognized the Pythagorean theorem. The geometric interpretation of Liu's argument (from [5]) which is proof of the Pythagorean theorem we present below.

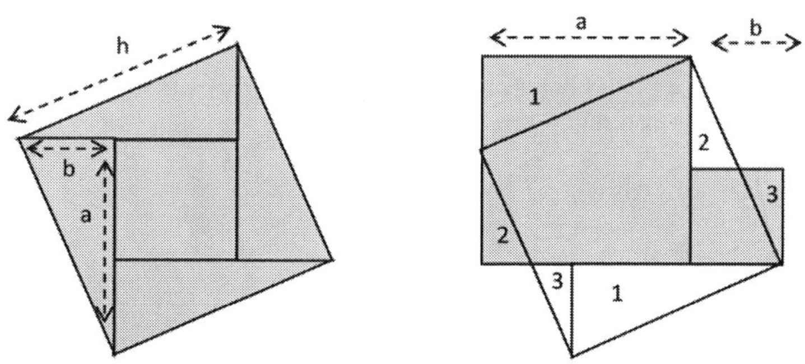

Figure 1. Diagram Representation of Liu's Argument.

Based on figure 1, Liu wants to show that $h^2 = a^2 + b^2$ or $h^2 = (a - b)^2 + 2ab$. By concerning the geometric representation of Liu's argument, we can conclude that mathematician from ancient China has known the basic concepts in the Pythagorean theorem and even one of the simple algebraic identities. In the *Gou Gu* chapter, the sort side adjacent to the right angle known as *Gou*. The longer side adjacent to the right angle is *Gu*. The side opposite the right angle is *Xien*. *Xien* is longer than *Gu*, and *Gu* is longer than *Gou* [5].

The basic of *Gou Gu* Rule: **add the square of *Gou* and *Gu*. The square root of the sum is equal to *Xien*.**

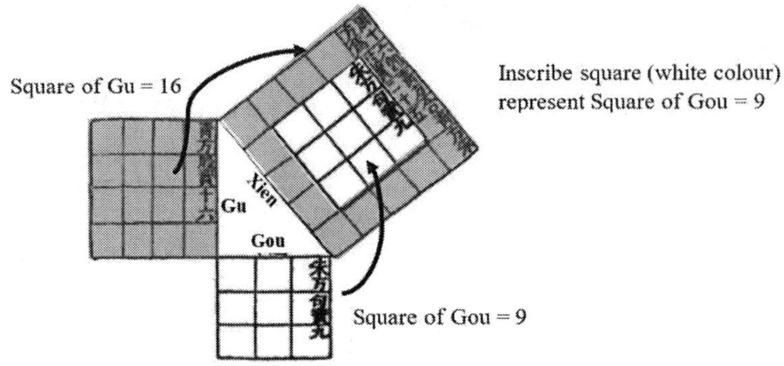

Figure 2. Relation among *Xien, Gu,* and *Gou (image source*: [8]).

Base on the geometric interpretation above, it is known that the *Gou Gu* rule has similarity with Pythagorean Theorem. The pairs of right triangles side, 3, 4, and 5 are used as a case of examples to show the proof and have been widely known by many ancient societies.

Additionally, we will show an example of Liu Hui's explanation of the solution to the sixth problem of Gou Gu in Nine Chapter.

in the center of a square pond whose side is 10 *Chi* (feet), grows a reed whose top reaches 1 *Chi* above the water level. if we pull the reed toward the bank, its top is even with the water's surface. what is the depth of the pond and the length of the plant?

1 Chi = 1 Feet

Figure 3. Problem 6 of *Gou Gu.*

To solve the type of problem using algebraic symbols, indeed students need prior knowledge about product formula of algebraic sum or difference. However, rather than using calculation of algebraic symbols, we may imagine the right triangle devised by Liu Hui as that described by Swetz and Kao [8], Dauben [9], Chemla and Gou [10], Guevara-Casanova & Burgués-Flamarich [6], although Liu Hui did not explain any figures in his commentary.

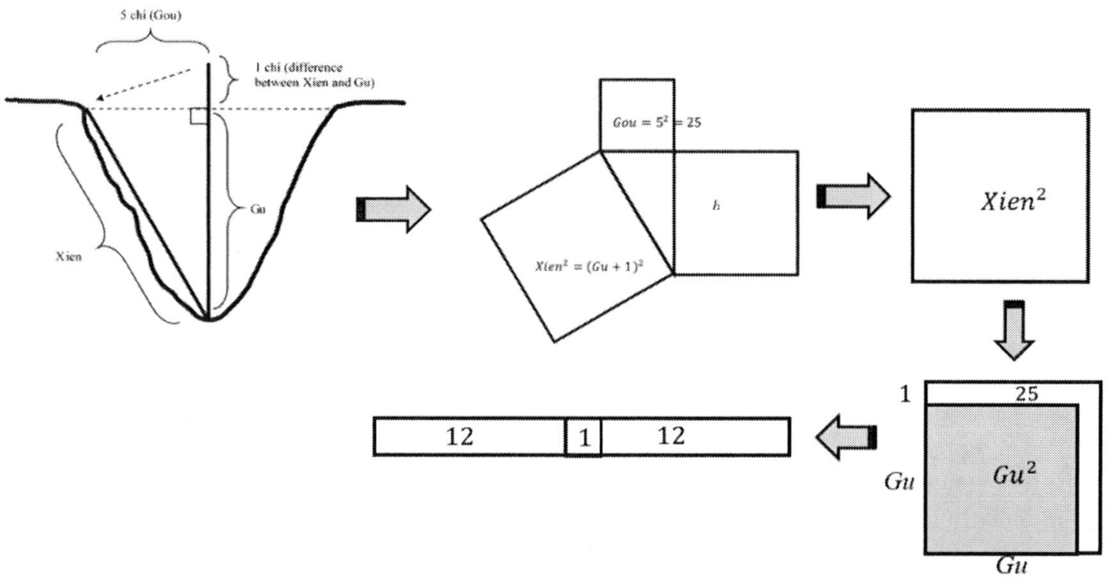

Figure 4. Geometric illustration and method to find *Gou* and *Xien* the problem 6.

We assume that the length of the reed is *Xien* and the depth pond is Gu. Since the sum of the areas of *Gou* square and *Gu* square of the triangle should equal to *Xien* square, by placing the *Gu* in *Xien*, we know that the rest of the area (L shape area or gnomon, see Fig.4) must be 25 chi-square. Then by reshaping the gnomon into a rectangle, with length = 25 ft and width = 1 ft), it will be easily obtained that *Gu* must be 12 ft (see figure 4). Therefore, it will be easy to determine the value of *xien* is 13 Chi. We believe the geometric manipulation based on Liu's argument can support students' geometric reasoning in solving the problems related to the right triangle.

3. Mathematics Literacy and Problem-Solving from the History of Mathematics

OECD [11] defined mathematics literacy as an individual's capacity to formulate, employ, and interpret mathematics in a variety of contexts. If we discuss deeply about context, Van Den Heuvel-Panhuizen [12] stated that context can be used as a first step to build mathematical concepts. On the other hand, CORD [13] argued that knowledge will be meaningful to students if the learning process is carried out by involving a situation or context. Problems in the History of Mathematics can be used as contexts because they are included in the category of educational situations. By involving historical problems in learning, it will also provide several benefits, including alternative solutions to problems that may arise in learning can be identified because the obstacles or problems often faced by students are almost the same as people in the past [14,15]. In other words, integrating literacy into mathematics educations can be done by presenting problem-solving from the history of mathematics as a context. Therefore, it will promote a constructivist vision of learning.

4. Method

The method used in this research was design research. The main aim of this type of research is developing a Local Instruction Theory (LIT) [15] by testing and comparing the Hypothetical Learning Trajectory (HLT) that has been developed with the actual learning through teaching experiments. However, our discussion in this paper is restricted only from a part of the teaching experiment, namely pilot experiment, from three main phases of design research.

This research was conducted at three different Junior High schools in Indonesia and involving 17 students. The data collected was field note, students' work, and video recordings. After testing the HLT, we conducted retrospective analysis to compare the actual learning in every task with the HLT. We use the results of the analysis to revise the HLT. In this paper, we not only conducted qualitative analysis but also quantitative impression of how well the conjecture and actual learning matched based on Bakker & Van Eerde [16]. Moreover, we did qualitative analysis to describe how the instructional tasks based on ancient china history can support Students' Geometrical Reasoning and Mathematical Literacy in Pythagoras Learning.

5. **Result and Discussion**

In this section, we will explain briefly the HLT and student thinking conjectures that we compiled to support students' understanding of the Pythagorean theorem. Furthermore, we will discuss deeply about how a series of tasks we developed can support students' visual reasoning and mathematics literacy skills.

5.1. Learning Trajectory

The following is the explanation of the series of the instructional task we developed and analysis of the comparison between HLT and Actual Learning Trajectories (ALT).

a. Activity I

In the first activity, we ask the students to prove the *Gou Gu* rule using the concept of area. The purpose of this activity is to emphasize that there are various models of geometric interpretation in proving that the area of *Xien* is equal to the sum of the area of *Gou* and *Gu*. The conjectures of student thinking in this activity are following.

C.1.1. Students prove that the sum of the area of Gou and Gu is equal to the area of Xien.

C.1.3. Students explain that the Gou Gu rule applies to right triangles only by showing that the rule does not apply to the acute triangle and obtuse triangle.

b. Activity II

In this activity, students are asked to solve a contextual problem (the vine problem) of right triangle using *Gou Gu* rule. This activity aims to help students achieving a better understanding in mathematics literacy by solving contextual problems. the conjectures of student thinking were:

C.2.1. Students solve problems no.1 in activity II by using the *Gou Gu* rule.

C.2.1. Students solve history-based contextual problems (no.2) in activity II by using the Gou Gu rule.

c. Activity III

In the third part, students were asked to solve several contextual right triangle problems from *Gou Gu*. The case of the tasks is more difficult than the previous ones since the information given involve the difference between two sides of the right triangle. Guevara-Casanova & Burgués-Flamarich (2018) argued that the difficulty can be overcome by introducing geometric approach to support the students to argue visually and follow the procedure for finding the solution to the problem by manipulating the area of shapes.

C.3.1. Students performing geometric manipulation in solving the history-based contextual problems (no.1) in activity III.

C.3.2. Students performing geometric manipulation in solving the history-based contextual problems (no.2) in activity III.

d. Activity IV

The last activity aims to introduce a variety of triple Pythagoras by combining three squares to form a right triangle.

C.4. Students are able to compose a square pair so that they get a Pythagorean triple number and find more than 3 pairs of triple Pythagoras (which are obtained not from multiplying that are known).

Based on the pilot experiment, the following is the quantitative data analysis of comparison between HLT and ALT.

Table 1. ALT result compared with HLT conjectures for the tasks involving a particular type of reasoning

Note: an x means how well the conjecture matched to the observed learning (− refers to confirmation for up to 1/3 of the total students, and + to at least 2/3 of the total students)

Based on the table above, we concluded that in general the actual learning trajectory has been running in accordance with the conjectures in HLT, except for the tasks in activity IV. This is because, during the pilot experiment, we made a mistake in making a square model (as a prop). The difference of the length between the model that we designed in the word file and the printout result caused the triple Pythagoras compiled by students do not match with what we have planned. Therefore, we revise the HLT related to the media we use in activity 4.

5.2. Solving History-based Problems of Right Triangle with Diagram: The Example of Students' Work

We will explain how students solve problem-solving using geometric manipulation or diagram of two problems taken from the test in activity 3 (see reed in the pond problem). During the test, the students worked individually and some of them in pair. The first task in this activity is taken from problem 6 of Gou Gu. The following are examples of students' answers in solving these problems.

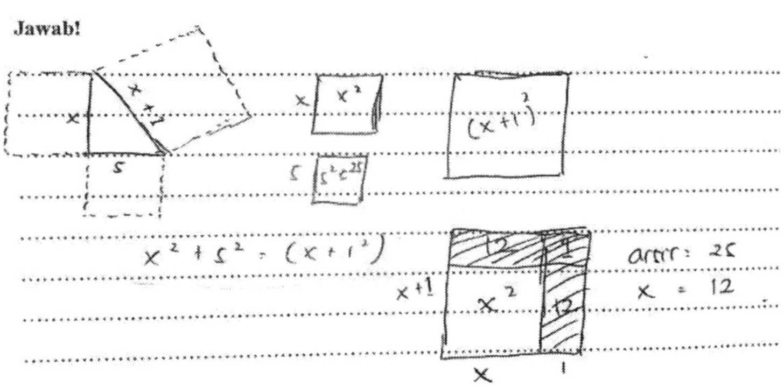

Figure 5. Student's visual reasoning in solving the first problem in activity III.

Based on the picture above, the first step taken by students was formulating the contextual problems into a geometric problem (mathematical form) of a right triangle. The right triangle with one side is known (5 chi) and two sides are unknown. Student determines the deep of the pond as x feet, and the height of the reed as $x+1$ chi. The problem is determining the x such that the deep of the pond and the height of the reed can be determined. When referring to the procedures of solving problem with diagrams of Guevara-Casanova & Burgués-Flamarich [6] we found the similarities. The picture (figure.5 and 6) show that in solving the problem in early-stage student formulate the contextual problem into mathematical form, the right triangle problem. because the stage requires the ability to transform contextual forms into mathematical problems, we call it a *translation* procedure. Secondly, students use the relationship between the area of "square *gou, gu,* and *xien,* and perform a geometric manipulation. The step resulted that the gnomon form (L shape area) has the same area as the *gou* square. We call that step as a *transformation* procedure. In the end, students analyzed the relation between the area of the gnomon and the side is given, they found that $x = \frac{55}{2} \times \frac{1}{3}$ or they wrote $x = 9{,}16$ (actually the most appropriate answer is $x \approx 9{,}167$) (see figure 5).

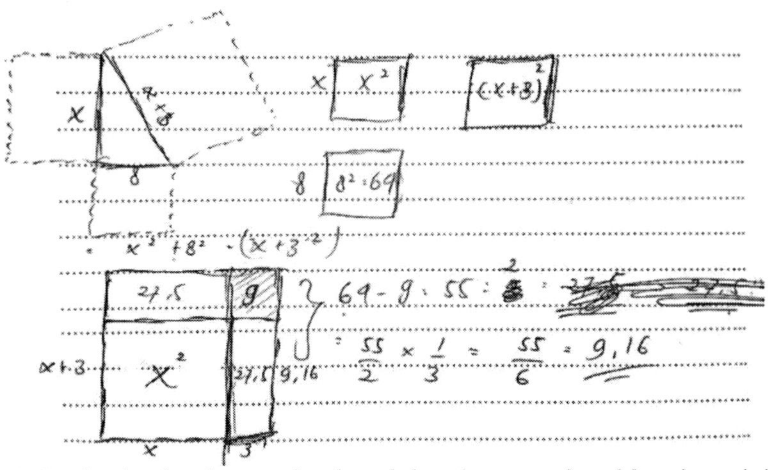

Figure 6. Student's visual reasoning in solving the second problem in activity III.

5.3. *Mathematics Literacy and problem-solving in history-based task*

The following are examples of historical problems that we profile based on content, context and processes in mathematics literacy.

Problem 1.

> A tree of height 20 feet has a circumference of 3 feet. There is a vine which winds seven times around the tree and reaches the top. What is the length of the vine?

Item Profile:

Description	: Using Pythagorean theorem in solving problem related to a context-based task
Context	: Scientific
Content	: Space and Shape
Process	: Employ
Level of context use	: first of context use

This task is categorized to have a context of science since it is related to a discipline of biological science, which show that a vine plant can grow up with watery environment. Furthermore, it is included in the content of space and shape because it demands the solver to use their conceptual understanding of the measurement of triangle. Although this task can be formally solved using the Pythagorean theorem, a solver could also use an informal way to solve the problem. In this task, a solver needs more effort in using mathematical computation in obtaining mathematical results, thus it is categorized in the group of 'employ' task. Regarding the level of context use, this task is indicated to have the first level of context use because the context of satiation describing the geometrical information about the bamboo is needed for solving the problem and judging the answer.

Problem 2.

> In the centre of a square pond whose side is 10 *Chi* (feet), grows a reed whose top reaches 1 *Chi* above the water level. if we pull the reed toward the bank, its top is even with the water's surface. What are the depth of the pond and the length of the plant?

Item Profile:

Description	: Using Pythagorean theorem in solving problem related to a context-based task
Context	: Scientific
Content	: Space and Shape
Process	: Employ
Level of context use	: first of context use

This task is categorized to have a context of science since it is related to a discipline of biological science, which show that a bamboo plant can grow up with watery environment. Furthermore, it is included in the content of space and shape because it demands the solver to use their conceptual understanding of the measurement of triangle. Although this task can be formally solved using the Pythagorean theorem, a solver could also use an informal way to solve the problem. In this task, a solver needs more effort in using mathematical computation in obtaining mathematical results, thus it is categorized in the group of 'employ' task. Regarding the level of context use, this task is indicated to have the first level of context use because the context of satiation describing the geometrical information about the bamboo is needed for solving the problem and judging the answer.

In conclusion, the students 'ability to solve contextual problems from the history of mathematics is in line with the students' mathematics literacy. In this case, it is limited to the specific context, content and process, based on 1 and 2 level of context.

6. Conclusion and Implications

From the result regarding the implementation in pilot experiment of instructional task we developed, we conclude that problem-solving from the history of mathematics can support students' understanding of the concept of Pythagoras. The visual argument by using concept of area in solving the right triangle problem help student improve their visual reasoning through three main procedures namely *translation, transformation,* and *diagrammatic reasoning*. On the other hand, the history-based problem as context students get a chance to learn Pythagorean theorem meaningfully. This is aligning with Freudenthal [17] statements that learning will occur when meaningful for students. On the other hand, HoM plays important role in improving students' mathematical literacy skills through contextual problems in HoM that are aligned with real-world challenges in PISA. Additionally, problem-solving from HoM also support students' mathematics literacy through process categories (i.e. formulate, employ, and interpret mathematics in a variety of contexts) of mathematics literacy.

In conclusion, this study revealed that Ancient China Pythagoras has a plentiful didactical source to improve the quality of mathematics pedagogy, especially for designing the instructional task for learning Pythagorean theorem. Hence, we recommend teachers to integrate the history of mathematics in the classroom and researcher or learning designer to conduct other similar research to enrich teachers' references in using history of mathematics in classroom.

Acknowledgements

The authors would like to thank the Ministry of Research, Technology and Higher Education of the Republic of Indonesia (Kemenristekdikti) for supporting this research funding.

References

[1] Fiangga S 2018 Using historical perspective in designing discovery learning on Integral for undergraduate students In *IOP Conference Series: Materials Science and Engineering* (IOP Publishing) **296** (1) 012042

[2] Radford L and Guerette G 2000 Second Degree Equations in the Classroom: A Babylonian Approach. In V. Katz (ed.) *Using history to teach mathematics. An international perspective* (Washington: The Mathematical Association of America) 69-75

Journal of Physics: Conference Series **1417** (2019) 012042 doi:10.1088/1742-6596/1417/1/012042

[3] Fachrudin A D Putri R I I Kohar A W and Widadah S 2018 Developing a Local Instruction Theory for Learning the Concept of Solving Quadratic Equation Using Babylonian Approach In *Journal of Physics: Conference Series* (IOP Publishing) **1108 (1)** 012069

[4] Man-Keung S 2000 The ABCD of using history of mathematics in the (undergraduate) classroom. *PALEONTOLOGICAL SOCIETY PAPERS* ***6*** 3-10.

[5] Katz V J 2008 *The History of Mathematics: An Introduction* Pearson

[6] Guevara-Casanova I and Burgués-Flamarich C 2018 Geometry and Visual Reasoning. In *Mathematics, Education and History* (Springer, Cham) 165-192

[7] Siu M K 2000 An excursion in ancient Chinese mathematics. In V. J. Katz (Ed.), *Using history to teach mathematics. An international perspective* (Washington, DC: The Mathematical Association of America) 159–166

[8] Swetz F and Kao T I 1977 *Was Pythagoras Chinese?: an examination of right triangle theory in ancient China* (Penn State Press)

[9] Dauben J W 2007 Chinese mathematics. In V. J. Katz (Ed.), *The mathematics of Egypt, Mesopotamia, China, India and Islam. A sourcebook* (Princeton, NJ: Princeton University Press) 187–384

[10] Chemla K and Guo S (Eds.) 2005 Les Neuf Chapitres. Le classique mathématique de la Chine ancienne et ses commentaires (París: Dunod)

[11] OECD 2013 *PISA 2012 Results: Excellence through Equity (Volume II): Giving Every Student the Chance to Succeed*, (Paris: OECD Publishing) https://doi.org/10.1787/9789264201132-en.

[12] Van Den Heuvel-Panhuizen M 2003 The didactical use of models in realistic mathematics education: An example from a longitudinal trajectory on percentage *Educational Studies in Mathematics* ***54***(1) 9-35.

[13] CORD 1999 *Teaching Mathematics Contextually* (Waco:CORD)

[14] Grugnetti L 2000 *The History of Mathematics and its Influence on Pedagogical Problems* In Katz V. (Ed) *Using history to teach mathematics: an international perspective* (USA: The Mathematical Association of America) 29-35

[15] Gravemeijer K 2004 Local Instruction Theories as Means of Support for Teachers in Reform Mathematics Education *Math. Think. Learn* **6**(2) 105–128

[16] Bakker A and Van Eerde D 2015 An introduction to design-based research with an example from statistics education In *Approaches to qualitative research in mathematics education* (Springer Dordrecht) 429-466

[17] Freudenthal 1991 *Revisiting Mathematics Education: China Lectures* (Dordrecht, the Netherlands: Kluwer Academic Publishers)

Students' procedural and conceptual Fraction knowledge based on cognitive style

T D Novianggraeni[1], Masriyah[2] and A Lukito[2]
[1] Postgraduate Mathematics Department, State University of Surabaya, Indonesia
[2] Mathematics Department, State University of Surabaya, Indonesia

e-mail: tressyana.17070785053@mhs.unesa.ac.id

Abstract. Recent research on children's procedural conceptual knowledge has suggested that there are individual differences. Some children have more conceptual knowledge, some children have more procedural knowledge, and some children have an equal level of both. In this study we compared and investigated individual differences knowledge of procedural and conceptual fraction among cognitive style of seventh (n=40) graders in Surabaya, East Java. Data collection in three sessions; all participants answered the same tasks. In the first session, students completed the Group Embedded Figures Test (GEFT) in 12 minutes to investigate their cognitive style (Field Dependence cognitive (FD) or Field Independence cognitive (FI) styles). In the second session, students completed three written measures (procedural and conceptual fraction knowledge test in 45 minutes) individually in a classroom. In the third session, students were individually interviewed for approximately 30 minutes. They answered questions on how they would solve fraction tasks. Students of the FI group performed significantly better than the FD group, on a measuring fraction and estimating number line. Overall, these results provide a more detailed picture of individual differences in procedural and conceptual knowledge in cognitive style.

1. Introduction

Knowledge concept is critically important to learn mathematics. Indeed, students must possess chances and sufficient support of knowledge to study mathematics in-depth [3]. One of the concepts in mathematics is fractions. Fractions are one of the most challenging cognitive lessons which mostly leads to problems of teaching [7]. Not only are fractions essential for learning, but they also benefit to solve problems in daily lives, such as solving financial issues and doing home repairs [4]. Therefore, students need to have knowledge of fractions correctly.

The abstraction of the concept of fragmentation about the whole relationship [10]. Related to this, Charalombus proposes that a general relationship is a condition in which a constant quantity or set of discrete objects is divided into equal parts [2]. Insufficient knowledge of the concepts of fractions leads to difficulties in doing calculations with fractions, decimals and concepts, the use of fractions in measurements, and the concepts of ratios and proportions [10]. Students who have insufficient knowledge in learning fractions will encounter difficulties in other mathematics concepts such as algebra, measurements, ratios and proportions [8].

An important aspect that distinguishes various types of numerical knowledge lies in procedural and conceptual knowledge [6]. Knowledge of fractional procedures consists of four fractional arithmetic operations such as summing, subtraction, multiplication, and division. Conceptual knowledge in this area involves knowledge of the properties of factions, including size (e.g., 4/5 greater than 1/2), principles related to factions (e.g., an infinite number of factions can be found between two factions etc.), and notations to express fractions (e.g., 3/4 = 6/8 = .75) [1].

Individual knowledge as a cognitive activity is highly dependent on different cognitive styles. One of the most widely studied cognitive styles is the cognitive style that is psychologically divided into two fields: Field of Cognitive Dependence (FD) and Field of Cognitive Independence (FI). Individuals with FD-induced cognitive tendencies may have difficulty separating information that is in their context and may be more influenced by external cues and thus less selective in the absorption of information. While FI individuals have little difficulty in separating the most important information from their context, they are more likely to be influenced by internal cues than external cues, and they are selective in entering information [11].

This study aimed to describe the knowledge of first-year high school students' cognitive skills with Field Independence (FI) and Field Dependence (FD). This research is qualitative research with a qualitative approach to data collection techniques through interviews. The results of this study are expected to contribute to the development of learning in procedural knowledge.

2. Method

2.1. *Participants*

A qualitative approach applied in this study to analyse students' procedural and conceptual fraction knowledge with different cognitive style. The sample consisted of 30 on 7th grade students of a school in Surabaya. Only students want to as a volunteer were included in the study. All students are given Group Embedded Figures Test (GEFT) for 20 minutes to investigate their cognitive style (field independence or field dependence) and Procedural-Conceptual Fraction Knowledge Test (PCFUT) to know their procedural and conceptual knowledge in fraction. The result shows, there were 7 field independence students and 23 field dependence students. A volunteer, one student chosen as a participant from each group by considering the criteria of equivalent mathematical ability to analyze deeply about their performance on procedural and conceptual test. In this study, field independence students we labelled as FI and field dependence students we labelled as FD.

Time triangulation is used to determine the validity of the data. In such an area, it is a way of testing the validity of data by comparing data collected at two different times. For instance, data on PCFUT-1 results and interviews 1 compared with data from PCFUT-2 and interview 2. The results of this triangulation indicate that the data is consistent, and therefore obtaining valid data, and PCFUT results and interviews are used as substantial aspects to analyze data to answer research questions.

2.2. *Measures*

Cognitive Style Task

Participants completed the Group Embedded Figures Test (GEFT) for 12 minutes to investigate their cognitive style. There are eighteen complex numbers in GEFT. Based on the correct answer number given by the student, the score on GEFT may be received from 0 (most FD) to 18 (lowest FI). In addition, a preliminary book that contains strict directions together with examples for discussion of subjects, this exam contains three parts: The first section aims to make students able to use the exam, and the other two parts are GEFT. The first section that has a two-minute time limit includes 7 easy problems to practice, and the items in this section are not included in the total score. The real task starts in the second and third sets, where asking for the exam requires making reproach them in every 9 items specified in having a period of five minutes for each one. Those who scored more than 12 out of 18 were labeled as field freedom (FI) and those who had a score of 11 and less than 11 were labeled as field dependency (FD).

Fraction procedural knowledge task

Participants completed eight fraction arithmetic problems (addition, subtraction, multiplication, and division) with equal denominators and unequal denominators. The items were 3/5 + 1/2, 3/5 + 2/5, 3/5 - 1/2, 3/5 - 2/5, 3/5 * 1/2, 3/5 * 2/5, 3/5 : 1/2, and 3/5 : 2/5. The correct answered were scored as correct even if they were not in simplest form.

Fraction conceptual knowledge task

0-1 Number line estimation
Participants completed 10 number lines. Each line's left endpoint was labelled "0", and the right endpoint was labelled "1". The fraction to be estimated were 1/19, 1/7, 1/4, 3/8, 1/2, 4/7, 2/3, 7/9, 5/6, and 12/13. Performance was measured from the correct location of the fraction on the number line.

0-5 Number line estimation
This task and the measure were identical with 0-1 number line estimation task except that the right endpoint of the number line was labelled ''5'' rather than ''1'' and the items to be estimated were 1/19, 4/7, 7/5, 13/9, 8/3, 11/4, 10/3, 7/2, 17/4, and 9/2.

Fraction magnitude comparison
Participants compared eight fractions magnitudes (3/8, 5/8, 2/9, 4/5, 4/7, 5/9, 8/9, and 2/3) with the fraction 3/5 which one is greater than or less than.

2.3. Procedure
Data collection in three sessions; all participants answered the same tasks. In the first session, students completed GEFT in 12 min. In the second session, students completed three written measures (procedural and conceptual fraction knowledge test in 45 min) individually in a classroom. In the third session, students were individually interviewed for approximately 30 min. They answered questions on how they would solve fraction tasks. These interviews were always completed after first and second session.

3. Result and Discussion

3.1. Cognitive style differences in knowledge of fraction procedures
Based on their result of fraction procedural knowledge are shown in Figure 1 and 2.

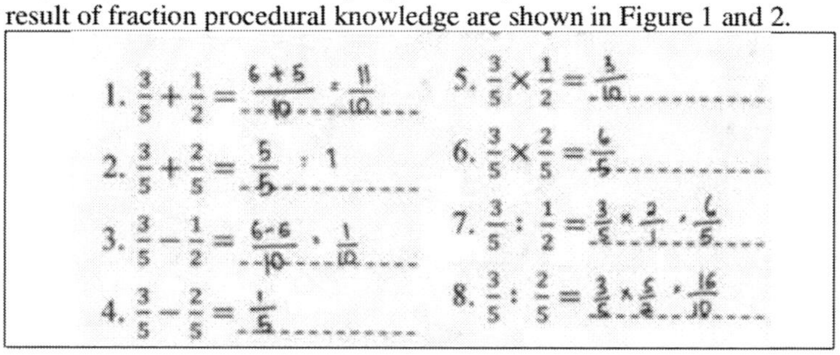

Figure 1. The answer of FI students in fraction procedural knowledge

From the resulting test, he completed all the problems and correct. From the result of the interview, FI students understand the procedure for operating fraction problems. So, we can conclude that he understand how to operating fraction.

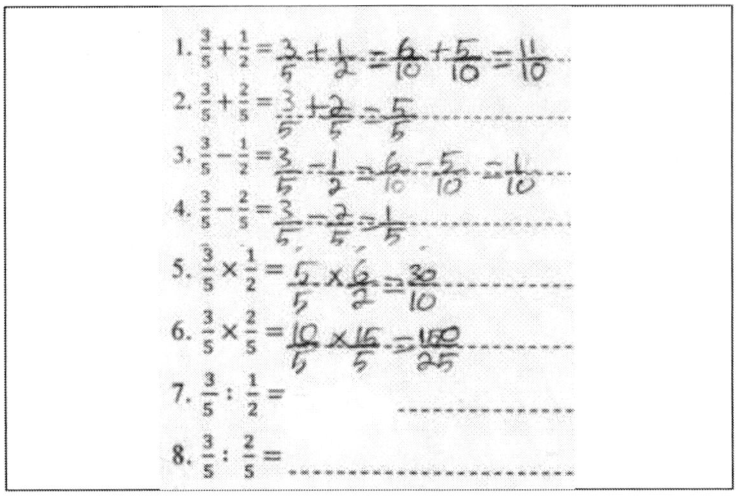

Figure 2. The answer of FD students in fraction procedural knowledge

From the resulting test, we know that FD students completed addition, subtraction, and multiplication problems and correct. But for division problems are uncompleted. From the result of the interview, FD students understand the procedure for operating fraction problems (addition, subtraction, and multiplication). In division problems, he could not solve it.

Based on the resulting test and interview of FI students and FD students we can compare their fraction procedural knowledge. The results show that FI students were represented a complete fractional concept and FD students almost complete fractional concept excepts addition.

3.2. Cognitive style differences in knowledge of fraction concepts

Fraction conceptual knowledge task measured by three different tasks (0-1 number line estimating, 0-5 number line estimating, and fraction magnitude comparison). Based on their result of fraction procedural knowledge are shown in Figure 3.

Performance of fraction estimating was measured by the percentage accuracy of each estimate from the correct location of the fraction on the number line (0-1 number line estimating task and 0-5 number line estimating task) and fraction magnitude comparison.

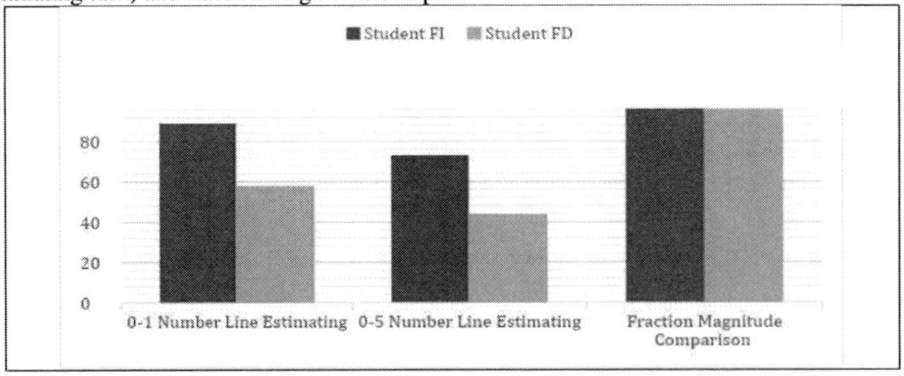

Figure 3. The percentage accuracy of FI students and FD in estimating number line and fraction magnitude comparison

In the model estimating 0-1 number line, there was an effect of different cognitive style, indicating that FI students were more accurate than FD students. From the results of the interview, FI students estimating number line using concept of fraction units. He converting number "1" and number "5" into a fraction that has the same denominator with the fraction in question. Therefore, he can estimate the position of this fraction in the number line. The above answer was justified with the result of the interview as follows.

Q : Do you know what the instruction to answer the problem?
FI : Yes, it is to estimate the location of fraction in the number line
Q : How did you do?
FI : I make the maximum number line to be a fraction with the same denominator with the fraction in the question
Q : Could you explain that?
FI : For instance, in this number, we must estimate fraction 1/7 in 0-1 number line. I make number 1 into a fraction that has the same denominator with fraction 1/7, number 1 is equal to 7/7. After that I just compare the numerator of 1/7 and 7/7, it is 1 and 7. Therefore, I just estimate where is the position of 1 in number line 0 until 7.
Q : Why you estimate like that?
FI : I learn that if we want to compare fractions we must make the fraction into has the same denominator.

From the result of the interview, FD students estimating number line-using concept of part-whole. He imagined a whole number and divided it into 4 parts, one part has become 1/4. He understand about concept is used to estimate fraction, but he had difficulty with large numerator and denominator. The FD students have a dependence on external cues and structures, easily disturbed and easily confused as to lack the ability to solve problems [3].

In the fraction magnitude comparison accuracy, FI students and FD students have no differences of accuracy. The results show that FI students using fraction equivalents and FD student using visualization of concrete objects to estimating a fraction in the number line.

The groups of FI students and FD students differentiated not only seem to match their learning profiles. The differences also relate to problem-solving performance for fractions and intensive quantities [5]. Their understanding of the underlying concepts can we know when students practice solving problems.

4. Conclusion

Based on the results of the study, it is shown that Cognitive Field Independence students measuring fraction very well and express how the procedure with their own words. FI students estimating number line using concept of fraction units. He converting number "1" and number "5" into a fraction that has the same denominator with the fraction in question. Therefore, he can estimate the position of this fraction in the number line. While the cognitive-oriented students of Field Dependence could measuring fraction, but difficulty in addition. FD students estimating number line-using concept of part-whole. He imagined a whole number and divided it into 4 parts, one part has become $\frac{1}{4}$. He understand about concept is used to estimate fraction, but he had difficulty with large numerator and denominator. In the fraction magnitude comparison accuracy, FI students and FD students have no differences of accuracy. The results show that FI students using fraction equivalents and FD student using visualization of concrete objects to estimating a fraction in the number line. These findings suggest that, cognitive style differences are one of the decisive factors that can influence students' knowledge of the concept of fractions.

5. Acknowledgements
The authors would like to express appreciation for the support of parents and lecturers at State University of Surabaya

References

[1] Bailey D H, Zhou X, Zhang Y, Cui J, Fuchs L S, Jordan N C, Gersten R and Siegler R S 2015 *Development of fraction concepts and procedures in U.S. and Chinese children, Journal of Experimental Child Psychology,* vol. **129** 68-83.

[2] Charalambous C Y 2007, *Developing And Testing A Scale For Measuring Students' Knowledge of Fractions,* in Proccedings of the 31st conference of the international group for the psychology of mathematics education, Seoul

[3] NCTM 2000 *Principles and Standards for School Mathematics,* Reston VA: The National Council of Teachers of Mathematics, Inc.

[4] NMAP 2008 *Foundations for success: The final report of the National Mathematics Advisory Panel,* Washington DC: National Mathematics Advisory Panel.

[5] Hallett D, Nunes T, & Bryant P *Individual differences in conceptual and procedural knowledge when learning fractions, Journal of Educational Psychology* **102**395-406

[6] Hiebert J, & LeFevre P 1986 *Conceptual and procedural knowledge in mathematics: An introductory analysis,* In J. Hiebert (Ed.), *Conceptual and procedural knowledge: The case of mathematics,* vol. **47** 1–27, Hillsdale, NJ: Lawrence Erlbaum.

[7] Lamon S J 2012 *Teaching Fractions and Ratios for Knowledge: Essential Content Knowledge and Instructional Strategies for Teachers,* New York: Routledge Taylor & Francis Group.

[8] Razak F A, Noordin N, Alias R and Dollah 2012 *How Do 13-year Olds in Malaysia Compare Fractions, Procedia - Social and Behavioral Sciences,* pp. 100-105.

[9] Rittle-Johnson B & Michael S 2015 *Developing conceptual and procedural knowledge of mathematics* , In R. Cohen Kadosh & A. Dowker (Ed.), *Oxford handbook of numerical cognition,* pp. 1102-1118

[10] Walle J A V 2007 *Elementary and Middle School Mathematics,* London: Pearson Education.

[11] Witkin H A, Moore C A, Goodenough D R and Co P W 1977 *Field-Dependent and Field-Independent Cognitive Styles and Their Educational Implications, Review of Educational Research,* **47** 1-64.

Students' reflective abstraction of middle school in reconstructing quadratic equation concept based on high mathematical ability

L Cahyani [1], Masriyah[1], and E Budi Rahaju[1]

[1]Department of Mathematics, Universitas Negeri Surabaya, Surabaya, East Java, Indonesia

e-mail: lailatul.17070785012@mhs.unesa.ac.id

Abstract. A quadratic equation is one of the mathematics concepts that must be learned in secondary school. Students have many difficulties in solving quadratic equations. One of them is experiencing a process skill error. Students need new abstractions and processes to shape experiences to distinguish whether experience falls into student abstraction or beyond student abstraction. Reflective abstraction is students' understanding of mathematics through problems that reflect and reconstruct mental structures, then the scheme is used to solve problems. The reflective abstraction mechanism that can be applied to metal structures is the APOS theory, namely: Actions, Processes, Objects, and Schemes to construct new mental structures. The process of mental reconstruction of each student is different, so for the formation of mathematical concepts, it needs mathematical abilities. In other words, if someone has good mental reconstruction abilities, hopefully, his concept formation is also good. This research conducted using a qualitative approach. The subject of this research was selected from grade 9 junior high school students in Surabaya. Students must answer two tests, namely tests of mathematical abilities and reflective abstraction tests. Analysis and categorization of scores were carried out on both tests. Then two students who has high mathematical ability was selected as subject. Semi-structured interviews is conducted individually for each subject. The purpose of this study is to find out the reflective abstraction of students that have reconstructing concept of quadratic equations in terms of high mathematical ability. The results of the study indicate that the subject can solve the problem according to the stages of the APOS.

1. Introduction

The material of quadratic equations is one of the mathematical concepts that must be studied by all students during the learning process in secondary education. Studying mathematical material, one of is of quadratic equations requires understanding a good concept, it will make students have more experience in combining and connecting their abilities. There are still difficulties that students make in solving quadratic equations, namely errors of transformation and process skills [1]. Because there are still difficulties in resolving quadratic equations, students need a mathematical concept.

Concepts are abstract ideas that can be used to classify the similarities of ideas, events, or objects that are related or have similarities [2]. The concept is formed from the process of abstraction and classification, where abstraction is a conscious activity of thinking the similarity of ideas of students in classifying an object based on the similarity of the nature of new experiences formed and classification is gathering experiences that have similarities from the results of abstraction [3].

The abstract is a consolidated construction that can be used to create a new construction [4]. There are three types of abstractions, one of them is reflective abstraction [5]. Reflective abstraction is a construction used to reach existing structures to build new structures. Reflective abstraction is an action that reflects or feedbacks students' knowledge and understands various structured problem-solving strategies [6]. Based on above, reflective abstractions can form new concepts of students by identifying problems, then looking for solutions using various appropriate rules in finding solutions from lower stages to higher stages which refer to students' mental structures.

The abstraction mechanism that can be applied to the mental structure is should write the abbreviation of APOS for the first time [7]. Action is the cognitive structure of students in transforming objects learned as an external part in completing operations. The process is the cognitive structure of students who carry out the same operations as the actions but fully students carry out these operations as part of internal construction. The object is the cognitive structure of students who realize the stages of action and process as a totality (one unit) transformation, and actually, build transformations explicitly so it can be said that students have reconstructed the process into objects. Schema is a collection of actions, processes, and objects to form a skeletal framework of students in solving problems related to the concepts learned. APOS theory is a theory of how mathematical concepts and mental constructs can be learned. Students who solve problems by finding new concepts can be said to do construction. Construction is a reference for the emergence of new structures through constructive actions [4]. Reconstruction of different concepts in students depends on the level of reflective abstraction of students that produces new experiences or concepts. The reconstruction process of each student is different, in reconstruction or building mathematical concepts there is a need for mathematical abilities.

The mathematical ability has a significant influence on the thought process in reconstructing mathematical models to solve problems [8]. Having significant mathematical ability on student performance in understanding and finding problem-solving, is the same as constructing mathematical models [9]. In other words, if someone has a good concept of reconstruction ability, it is expected that his concept formation is also good.

Based on the description above, to find out the reflective abstraction that students have about reconstructing the concept of quadratic equations it is necessary to do research. This study can provide an empirical description of the mental structure of students in the quadratic equation material.

2. Method

The research was qualitative research with a type of case study research. This is in line with what was conveyed [10], that is, case studies are suitable when researchers want to answer descriptive questions or explanatory questions. The research subjects were two students in junior high school Surabaya. The two students are those who have high mathematical abilities and can answer the questions given by the researcher. Both subjects were asked to solve problems about mathematical models that must be changed in the form of individual quadratic equations.

The problems in question are as follows.

Doubled the square of a number plus twelve times that number minus thirty-two is equal to zero. Determine the variable, determine the mathematical model in the form of a quadratic equation, then specify the set of solutions!

Fifteen times the square of a number plus sixty times numbers minus four hundred eighty equals zero. Determine the variable, determine the mathematical model in the form of a quadratic equation, then specify the set of solutions!

Besides, to represent his thoughts the subject also provides written answers. During the work process, researchers conduct unstructured interviews to dig deeper into the subject's thinking. So, the data obtained in the form of written answers as well as recording the results of interviews obtained later than analyzed qualitatively through the process of transcription, reduction, classification, interpretation, and presentation of data.

Assessment will be carried out according to the indicators in Table 1.

Table 1. Indicator of reflective abstraction.

Structure	Indicator
Action	a. Determine variables.
Process	a. Determine mathematical models in the form of quadratic equations.
Object	a. Determine the set of resolution from the mathematical model that has been made.
Looking back	a. Linking actions, processes, and objects, to form or build or make a schema b. Using schemes that have been formed through the stages of action, processes, and object that have been in the minds of students to solve more complex problems.

3. Result and Discussion

3.1. Result

To obtain an overview of the mental structure possessed, the presentation of results will be carried out for each research subject on each point of the problem to be given. Then the first subject and the second subject will be referred to as T1 and T2. And for researchers themselves symbolized by P.

Analysis of subject 1

The solution to problem 1 and problems 2 in figure 1 and figure 2 respectively.

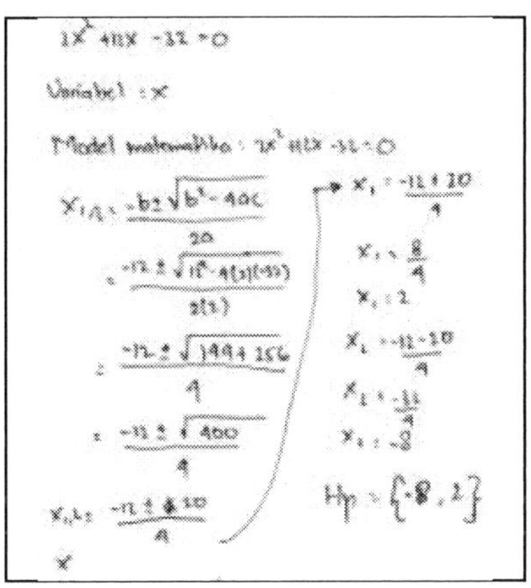

Figure 1. The solution to problem 1.

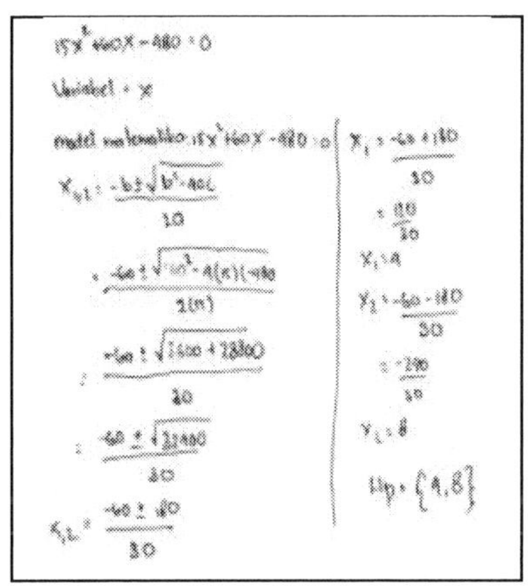

Figure 2. The solution to problem 2.

Subject T1 shows the results on problem 1, where the subject can directly determine the quadratic equation then determine the variable.

P: Which ones are the variables from this mathematical model?

T1: That number. Then I suppose with x

P: How do you change this problem in quadratic equations?

T1: I understand the problem first, right there is "a number", I say a number x

Journal of Physics: Conference Series **1417** (2019) 012044 doi:10.1088/1742-6596/1417/1/012044

P: How many solutions are there in quadratic equations? Can all the equations be done that way? Why!
T1: There are three. Maybe. Because if you can do it with the ABC formula, it can be sure with all three methods that you can
P: Why did you finish using the ABC formula?
T1: I use the ABC formula because I just enter a, b, and c into the formula.

The T1 subject fulfils the stages of the reflective abstraction indicator. Subject forms directly into the equation, then to the action stage. The reason the subject forms directly into quadratic equations is so that it is easy to determine variables. Subject 1 prefers to solve quadratic equation using ABC formula should be consistent, because the subject more easily determines the values of A, B and C and directly enters into the ABC formula. At the action stage, the subject can determine the variable with x. At the process, the stage he can determine in quadratic equations. In the object stage, students determine the set of solutions using the ABC formula.

Problem 2 is, given during interviews which aim to show a more complex scheme. From the results of problem-solving 2, the subject can be said to have a more complex scheme because he can solve the same problem as problem 1.

Analysis of subject 2
The solution to problem 1 and problem 2 in figure 3 and figure 4.

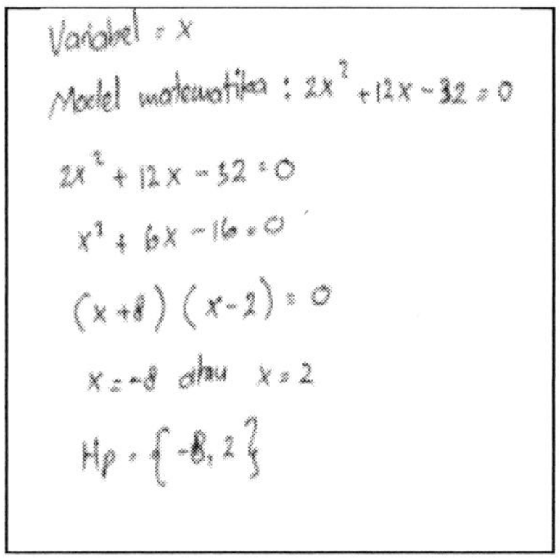

Figure 3. The solution to Problem 1.

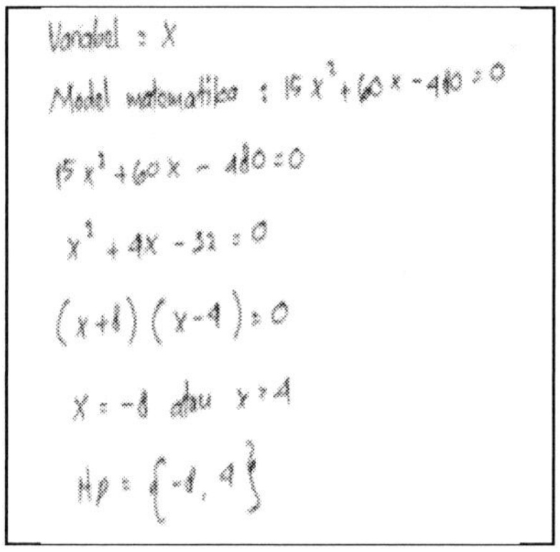

Figure 4. The solution to Problem 2.

Subject T2 shows the results on problem 1, where the subject can directly determine the quadratic equation then determine the variable.
P: Which ones are the variables from this mathematical model?
T2: That number. Then I suppose with x
P: How do you change this problem in quadratic equations?
T2: I understand the problem first, right there there is "a number", I say a number x
P: How many solutions are there in quadratic equations? Can all the equations be done that way? Why!
T2: There are three. Maybe. Because if you can do it with the ABC formula, it can be sure with all three methods that you can
P: Why did you finish using the factoring formula?
T2: I use the factoring formula, in my opinion, is faster factoring, but my first step is to divide so that the equation is easier when factored.

The T2 subject fulfils the stages of the reflective abstraction indicator. At the action stage, the subject can determine the variable with x. At the process stage, he can be determined in the squared equation. At the object stage, the student determines the set of solutions by using factoring.

Problem 2 is given during interviews which aim to show a more complex scheme. From the results of problem-solving 2, the subject can be said to have a more complex scheme because he can solve the same problem with problem 1.

4. Conclusion

The use of sections to divide the text of the paper is optional and left as a decision for the author. Where the author wishes to divide the paper into sections the formatting shown in table 2 should be used. The results of the study show that the reflective abstraction of students in reconstructing the quadratic equation can solve the problem in different ways. In solving the problem of quadratic equations student T1 at the object stage can solve the equation using the ABC formula, students choose the ABC formula because the formula is easier and feels quite fast in solving quadratic equations. While in solving the problem of quadratic equations students T2 at the object stage can solve the equation using the factoring formula, selecting the factoring formula because students feel the formula is easy but still need steps to simplify the equation to be factored. Although in different ways both subjects have met reflective abstraction indicators. Next, to bring up the scheme stage, students are given advanced questions with different numbers but the operation is the same. Students (T1 and T2) can solve advanced problem problems with the same answers, thus students (T1 and T2) can be said to fulfil reflective abstractions even with more complex questions.

Because research is only limited to reflective abstraction in reconstructing the concept of quadratic equations, further research is expected to obtain a more complete picture by adding other concepts. Similar research can also be done for other mathematical concepts.

Acknowledgements

This research reported herein was supported by SMPN 57 Surabaya, especially for 9th class as samples, and Wahyu Wijayaningrum as the teacher in this class who helped the research.

References

[1] O'Connor B R and Norton S 2016 Investigating Students' Mathematical Difficulties with Quadratic Equations *Mathematics Education Research Group of Australasia* 552-559

[2] Morrison R, Ross, Steven M, Kalman, Howard K and Kemp Jerrold E 2011 *Designing Effective Instruction* (USA: John Wiley FC Sons Inc.)

[3] Skemp and Richard R 1987 *Psychology of Learning Mathematics* (Hillsdale: Lawrence Erlbaum Associates Inc.)

[4] Monaghan J and Ozmantar M E 2006 Abstraction and Consolidation *Educational Studies in Mathematics* **62** 233-258

[5] Kërënxhi, Svjetllana, and Gjoci P 2017 Involvement of Algebraic-Geometrical Duality in Shaping Fraction's Meaning and Calculation Strategies with Fractions *J. of Edu. and Social Research* **7**(1) 151-157

[6] Scheiner T 2015 New light on The Old Horizon: Constructing Mathematical Concepts, Underlying Abstraction Processes, and Sense-making Strategies *Educ Stud Mat* **9** 165–183

[7] Cetin I 2015 Students' Understanding of Loops and Nested Loops in Computer Programming: An APOS Theory Perspective *Canadian J. of Science, Mathematics and Education* **15**(2) 155-170

[8] Sybian M 2010 Menumbuh Kembangkan Daya Matematis Siswa *EDUCARE: Jurnal Pendidikan dan Budaya*

[9] Gay L, Mills G and Airasian P 2012 *Educational Research: Competencies for Analysis and Application 10th edition* (USA: Pearson Education Inc.)

Generalization Pattern's Strategy of Junior High School students based on Gender

A M Firdaus[1*], D Juniati[2], P Wijayanti[2]
[1] Universitas Muhammadiyah Makassar, Jalan Sultan Alauddin 259, Makassar 90222, Indonesia
[1]Doctoral Program Student of Mathematics Education, Indonesia
[2] Universitas Negeri Surabaya

e-mail: *andi.mulawakkan@unismuh.ac.id

Abstract. This study aims to describe the Generalisation Patterns Strategy of junior high school students based on gender. The subject of this study is two junior high school students in Makassar City, one male, and one female. Data collection techniques are written tests and interviews which used to obtain data about the strategy of the generalization patterns of students. The validity used time triangulation. The analysis data consists of three stages: reducing data, presenting data, and daring conclusions. The results of this study show that the male subject used the strategy of Counting, Whole-object-No Adjustment, and Difference-Rate Adjustment. Meanwhile, the female subject used the strategy of Countin and Explicit in the generalization of patterns on numbers and pictures.

1. Introduction

Mathematics is one of the subjects taught in the school and have an important role in developing the thinking ability of the students. Besides, mathematics is also mean to think deductive in finding and developing science and technology with Firdaus [1]. A generalization is an important aspect of the learning process. It can take anywhere since as a baby who begins to explore and get acquainted with the world where they live Mason [2]. For example, they show that they naturally tend to induce patterns from objects they encounter in everyday life. They know that the candy was sweet after they taste it first. This process of learning is generalization through the experience. The generalization also occurs in the context of other learning, for example in learning addition. When the children noticed that 1 + 2 and 2 + 1 have the same number, and so are 3 + 4 and 4 + 3 and also 4 + 5 and 5 + 4, they might be able to convey the idea that the order in which two numbers are added is not important. They are said to have the mathematics generalization since they are able to articulate what they had seen only in a few examples for all cases.

On generalization patterns topic, students make inferences depend on the understanding of the concept that students' have and how they used them to generalize. The process of generalization used in enriching the cognitive structures in students' minds, since there is an activity to connect something. Inline, Barbosa & Vale [3], states that the Generalization plays a crucial role in the activity of any mathematician, being considered an inherent ability to mathematical thinking in general (p.58). Warren & Cooper [4], also asserts that the power of mathematics lies in relations and transformation which give rise to patterns and generalizations. Abstracting patterns are the basis of structural knowledge, the goal of mathematics learning in the research literature (p. 305). It means that the power of mathematics lies in relations and transformations in the form of a pattern and generalization.

The generalization pattern is often considered as a difficult topic for students to learn. Generalizing patterns is an important aspect at all levels in mathematics and exists in every material that is often highlighted in teaching Dindyal [5]. The exploration of patterns, especially in students who are in junior high school is a pattern developing. The Pattern developing closely associated with the rows of mathematical objects in the form of numbers and pictures. Mulligan & Mitchelmore [6], said that the pattern in mathematics is represented as a regularity in the form of numerical relationships, spatial, or logical. In line with that, Tikekar [7], in his research shows that when students are given a pattern of numbers, then each student has a different way of generalizing patterns. Different with research conducted by Firdaus, Juniati, & Pradnyo [8], on the image pattern for junior high school students, where male students and female in resolving the problem of generalization patterns is still the stage of action, in which both of the students are still determining n-terms from previous knowledge. While the research conducted by Muzaini, Juniati, & Siswono [9] that junior high school students find patterns that are recursive patterns and get a solution where the subject takes repetitive actions to get a general equation and can reveal new situations that are similar based on certain situations.

Students still experience difficulties when determining the pattern in greater terms Barbosa, Vale, & Palhares [10]. For students, the formation of patterns from greater terms seem more complicated in comparison with determining the result of the nearest terms of the patterns that already exist. It means that students are able to search for nearest terms of the pattern with the use of the strategy of counting (make a model in describing the situation and calculate a desired object) or recursive (describe the sequence using a same-different), rather than look for a general rule or generalization of a pattern of a greater term. Lannin [11] revealed the students found difficulties in determining the number of stickers that must be installed from a series of cubes that are given stickers on every surface. Students just saw a geometry object in front of him without considering the context of the problem when doing the calculation in determining the next pattern. It seems that the students did not pay attention to the context of the problem.

In a problem of generalization of pattern, the student will be asked to define certain terms on a pattern to write down the rules to determine any terms. Stacey [12], show that students generalized patterns using strategies of counting, whole object, and linear methods. Rivera [13], adds another strategy used by the students, they identified the patterns and then simplify or transform into a similar pattern with the same characteristics then make generalizations of the pattern. Lannin, Barker & Townsend [14], conducted research on students in secondary schools on a task about the generalization of patterns. They groups strategies students' used into six strategies: the Counting Recursion, Whole-Object, Contextual, Guess and Check, and Rate-adjust. In his research, there are students who use more than one strategy in solving problems. Furthermore, the research by Lannin [11] and Lannin, Barker & Townsend [15] then developed Barbosa, Vale, & Palhares [10], classified the six strategies and expand the strategy of the whole-object into three parts and the difference-rate into two parts.

According to Lannin, Barker & Townsend [14], the use of strategies in the generalization of patterns students' used influenced by the students' ability to visualize the problems presented. Investigation of the process of generalization patterns performed by the students, especially the strategy used in solving the generalization problems is important to know which strategy will bring the students to the correct result, and which strategy will bring the students to answers that are less precise. In this study, the generalization of patterns of the students refers to the strategy used to make general conclusions from statements The purpose of this study is to describe the strategy of the generalization Patterns of junior high school students in Makassar city based on gender.

2. Method

This is qualitative research using descriptive data. Qualitative is research that describes a phenomenon, event, facts, and events to produce a clear picture and detailed results of the activity the research subject Moleong [15]. The events described in this research is the strategy of the generalization patterns of junior high school students based on gender. Subjects in this study were two junior high school students in Makassar City. The selection of the subject assisted by the mathematics teacher in the junior high school to get accurate information about the mathematical ability of the two subjects. It is also to get information related to the communication skills of the subject in expressing their opinion or the ways in solving problems orally and in writing. Consideration of the researchers chose the male and female subjects with high ability in mathematics and in communication because the two subjects expected to complete the Task of Generalizing mathematical Patterns efficiently and can communicate orally and in writing so the researcher can describe the strategies used in the generalize pattern by the two subjects. The instrument in this research is the task of generalizing the pattern and interview guidelines. Data collection methods used were test, observations, and interviews Lestari and Juniati [16]. Data collection techniques were test of the generalization of the pattern and the interview guide used to obtain data on the strategy of generalizing patterns of student, guidelines for interview-based tasks as semi-structured interview as a guideline for the researcher to get the main objective of the research which is to know the strategy of the generalization patterns of junior high school students. Test validity used triangulation time. The task of generalization of the pattern is designed in the essay test and developed by the researchers and consulted to experts.

The analysis stage data includes reduction, data presentation, and conclusion. Data reduction performed with sharpens, classify and organize the data obtained from tests and interviews that related to the strategy of the generalization patterns of junior high school students based on gender. It means that at this stage, the process of selecting and simplifying the data during the research process performed by removing the data that is not needed and focus on the important information. Data presentation is organizing the data about the information that has been collected for drawing conclusions. At this stage, the researcher presents the data as a result of data reduction. The data is identified and presented descriptively with the chart. The data that has been presented facilitated researchers in understanding the data to draw a conclusion. The conclusion in this research aims to look at the strategy of the generalization patterns of junior high school students based on gender.

3. Result and Discussion

Generalizing pattern's task is presented in the form of a picture, the subject of male (LK) involves the several stages in the process of generalizing a pattern and use different strategies depending on the type of the test that was given and the finding terms. While working on the test in the form of pictures, to determine attributes of the desired value which is relatively small, the first steps of the subject LK were drawing of the model based on the picture in the problem. The model is adjusted with the input term. Furthermore, the subject LK calculated the attributes that are desired directly from the model created. The strategy used is the strategy of "counting". For the value which is large enough (input value = 65) the subject LK using the attribute that is the value of input = 10 as the multiplier factor. Subject LK described the 65 as $6 \times 10 + 0.5 \times 10$. Next, the "10" in the expansion is replaced with the attribute value to the input value = 10 that has been searched before. The strategy used is the strategy of "Whole object-No adjustment".

To determine the general rule, the subject LK realized that he must see and understand the problems presented. On the first test, the subject LK saw that if the base is incremented by one block then a lot of blocks required will be increased by 4 bocks. Meanwhile, in the second test, the subject LK saw that when the rung is incremented by 1, then juliennes needed is 3. Indirectly, the subject LK performed a sequence of numbers of the problems presented and looking for a different constant in the sequence. The subject LK then tried entering some numbers into the calculations and comparing the results of these calculations with the results if he enumerates directly the desired attribute on the model of the pictures previously created. Subject LK found that there is a difference constant between the calculation results and the results of counting on the image. The subject LK though that if the general rule or the n-terms is a different constant time n minus or plus the difference obtained from the results of the calculations and the results enumerate. The strategy used by the subject LK in determining the general rule is the strategy of "Difference-Rate adjustment".

Other information obtained from the results of the work the subject LK was that he tends to use the strategy of different generalization patterns to determine the terms with the value of the different inputs. From the results of the subject LK in solving the test which was in the form of images and numbers, it shows that when asked to determine the attribute with the value of the input is small, the subject LK did the modeling directly and enumerates the desired attributes of the model. This is in accordance with the results from Lannin, Barker, & Townsend [14], about the factors that affect the use of the strategy of the generalization of patterns. The study states that the value of a given input affects the strategy used by the subject and the current value of the input is small enough, students tend to use the strategy of counting.

In the task of generalizing the pattern in the form of a sequence of numbers, the subject LK recognized patterns by finding the difference of two consecutive terms in the sequence. The subject LK found that the difference is a different constant, but the subject LK extending the sequence to determine certain terms, this strategy is the strategy of recursive.

In determining the general rules or n-term, the subject LK using a different constant which is obtained as the multiplier factor. The subject LK realized that he might not extend the sequence until n term because n itself is variable. Then, the subject LK thought that to determine the particular term, the requested terms multiplied by the difference obtained. Subject LK checked the correctness of the answer on the sequence of numbers obtained, and find that there is a difference constant between the results of the multiplication with the terms in the sequence.

In the test of sequence, the subject LK used a similar strategy to determine certain terms but different strategies to determine the n-terms. Of the four terms presented in the sequence, the subject LK already recognized the pattern on the sequence. However, the subject LK has not been able to use the pattern to find a general rule for any terms without extending the sequence of numbers.

In the same case with the subject LK, in solving the task of generalization patterns, the female subject (PR) performed several stages in the process of generalizing a pattern and used different strategies depending on the type of the task and the finding terms.

In working the test in the form in pictures and numbers, the initial steps of the subject PR in determining the attributes for the input value = 10was drawing model corresponding to the picture given. The Model is adjusted with the input value requested. Furthermore, the subject PR did the calculations systematically in accordance with the picture and the context of the problems presented from the calculation. The subject PR saw the calculation pattern to find terms with different inputs.

In solving the first test, the subjects PR found that many of the blocks on the top is one less than a number of blocks on the base part and the number of blocks in inclined sections is twice the number of blocks on the base. In finding the number of blocks, it summed the numbers blocs earlier. On the second test, the subject PR saw that many of the matchstick on the right and left is one more than the number of rungs. In finding the number of matchsticks, she just summed up the numbers of matchsticks before. From the systematical calculation, the subject PR found the other terms without having to do the modeling directly, the subject PR was only able to enter the terms based on the calculations.

In finding the general rules, it is performed using the calculation to determine some of the previous terms because it already checks the answers by entering a few numbers. The subject PR understood that the pattern of these calculations can be used for any numbers. So at the first test, to determine the number of blocks needed to create the skeleton of the bridge with the base n, the subject PR changed the number of bocks on the basis of n (in the previous problems, the base is as a specific number). The subject PR did the same steps on the second test. The calculation systematically carried out beforehand. The subject PR changed rungs which were in the form n. This strategy performed by subject PR in determining certain terms and the general rule is called the explicit strategy.

From the results of the work the subject PR in solving the first test in the form of numbers and pictures, the subject PR tends to look for patterns in finding determining the attribute with the small input value. However, the pattern of the new attribute used when the subject PR determines the attribute with the larger value of the input and the general rule. According to Lannin, Barker, & Townsend [14], the students with a strong visual ability will be able to visualize the problems presented to determine the pattern.

In the task of generalizing the pattern in the form of a sequence of numbers, the subject PR recognized patterns with finding the difference of two consecutive terms in the sequence. The subject PR found that the difference is constant. With the difference, the subject PR extends the sequences until the 15th term and finding 15th term by directly looking at the sequence. After extending the sequence until the 15th term, the subject PR found that the constant difference used as a multiplying factor. Thus, in finding a particular term, the subject PR multiplied the difference with the requested term. The subject PR tried some of the numbers to be multiplied with a difference. The subject PR then compared the result of the multiplication with the terms on the sequence. She realized that there existed a constant difference between the results of the calculation with the terms on the sequence. Therefore, the subject PR dawn a conclusion to determine the particular term by multiplied the requested term by difference and add or subtract the difference obtained from the results of multiplication. The subject PR used the pattern to determine the other terms and the n-term. In determining the n-term, the subject PR just replaced the numbers that should be entered into the calculation with n. Strategy this strategy is the difference-Rate adjustment.

4. Conclusion

From the results of this research, two students from junior high school based on gender, the strategy of generalization patterns that are used by subject LK on the determination of the certain attributes for a small term only through the directly modeling stage and finding the terms. The subject LK did not perform the identification of patterns in the test pattern, the strategy used is the strategy of counting. When the value of a given input is large, the subject LK no longer does the modeling directly. Stages performed were the stages of pattern identification and the determination of the particular terms. The strategy used involves the use of a certain term as one unit, giving the multiplier factor is not done with the final adjustments. This strategy is the strategy of the whole object-no adjustment. To determine the attribute with the value of the input = n, the subject LK performed the stages of the identification pattern, the test pattern and the determination of general rules. The strategy used involves the use of patterns as a multiplication factor and proceed with the final adjustments. Strategy this strategy is the difference - rate adjustment.

The strategy of the generalization of a pattern that involves patterns in numbers and pictures that made the Subject PR in determining the small-term begins with the modeling. The subject PR did not do the pattern identification strategy. She used a strategy of counting. In finding the big terms, the Subject PR was not to the modeling directly. The stage performed is the stage of the identification pattern, the test pattern and the determination of the particular terms. The strategy used involves the discovery of a calculation rule based on the context of the problems presented which allows determining the desired terms with the random value, this Strategy is the explicit strategy. To determine the term with the value = n, the subject PR performed the stages of the identification pattern, the test pattern and the determination of general rules. The strategy used involves the discovery of a calculation rule based on the context of the problems presented and allows us to determine the desired terms. This strategy is an explicit strategy.

Acknowledgments

This work was supported by Lembaga Pengelola Dana Pendidikan (LPDP) Indonesian under grant no. 20161141080709.

References

[1] Firdaus A M 2019 Application Of Cooperative Learning Model Type Thinks Pair Share (TPS) On Mathematical Communication Ability *Daya Matematis: Jurnal Inovasi Pendidikan Matematika* 7 (1) 59-68.

[2] Mason J 2008 Making use of children's powers to produce algebraic thinking In Kaput J J, Carraher D W, and Blanton M L (Eds.) *Algebra in the early grades* (pp. 57 - 94) (New York: Lawrence Erlbaum Associates)

[3] Barbosa A and Vale I 2015 Visualization in pattern generalization: Potential and Challenges *Journal of the European Teacher Education Network* **10** 57-70

[4] Warren E and Cooper T J 2008 Patterns That Support Early Algebraic Thinking in The Elementary School In Greenes, Carole E, and Rubenstein R (Eds.) *Algebra and Algebraic Thinking in School Mathematics* (Reston VA: National Council of Teachers of Mathematics)

[5] Dindyal J 2007 *High School Students' Use of Patterns and Generalizations* Proceedings of the Fifth Congress of the European Society for Research in Mathematics Education 844-851

[6] Mulligan J and Mitchelmore M 2009 Awareness of Pattern and Structure in Early Mathematical Development *Mathematics Education Research Journal* **21** (2) 33-49

[7] Tikekar V G 2009 Deceptive Patterns in Mathematics *Int. J. Math. Sci. Educ.* **2** (1) 13-21

[8] Firdaus A M, Juniati D, and Pradnyo P 2019 The characteristics of junior high school students in pattern generalization *J. Phys. Conf. Ser.* **1157** (4) 042080

[9] Muzaini M, Juniati D, and Siswono T Y E 2019 Profiles quantitative reasoning and students' generalization ability on the topic of direct proportion *J. Phys. Conf. Ser.* **1188** (1) 012034

[10] Barbosa A, Vale I, and Palhares P 2012 Pattern tasks: thinking processes used by 6th-grade students *Revista Latinoamericana de investigación en matemática educative* **15** (3) 273-293

[11] Lannin J K 2003 *Developing Algebraic Reasoning Through Generalization* (The National Council of Teachers of mathematics, Inc. www.nctm.org)

[12] Stacey K 1989 Finding and Using Patterns in Linear Generalising Problems *Educational Studies in Mathematics* **20** 147-164

[13] Rivera F 2013 *Teaching and Learning Patterns in School Mathematics: Psychological and Pedagogical Considerations* (New York London: Springer Science Business Media Dordrecht)

[14] Lannin J K, Barker D, and Townsend B 2006 Algebraic Generalisation Strategies: Factors Influencing Student Strategy Selection *Math. Educ. Res. J.* **18** (3) 3-28

[15] Moleong L J 2012 *Metode Penelitian Kualitatif (Edisi Revisi)* (Bandung: PT. Remaja Rosdakarya)

[16] Lestari N D S and Juniati D 2018 Exploring the Knowledge of Content and Teaching (KCT) of a prospective math teacher in planning mathematical literacy teaching *J. Phys. Conf. Series* **1097** (1) 012150

Students' Problem Solving Behavior On Multiplicative Word Problem

S R Auliatullah, Y Fuad, and R Ekawati
Program Pascasarjana, Universitas Negeri Surabaya, Indonesia.
(sita.17070785007@mhs.unesa.ac.id)

Abstract. Students in junior high school frequently erroneously solving multiplicative word problems. Students in this stage should be able to understand the known and what is asked, what the strategies properly, and how to execute it become the right one. It has close related with student problem solving behavior, it is given provided description of the students' behaviour. The purpose is to investigate students' problem solving behavior on solving multiplicative word problem. This study conducted on a class of 7Ath class in SMPN 39 Surabaya, consist of 27 girls and 15 boys, and had chosen purposively from ninth classes. Students had to answer multiplicative word problem, based on result show that there are three category students when solve the task. Three volunteer students were selected as research subject which have different ability that had interpreted as context this study, subjects needed for deep interview. The problem solving behavior of subject who solve the tasks multiplicative properly tend to Meaning Based Approach-justification (MBA-j), the problem solving behavior of subjects who solve exactly one task multiplicative additively tend to Meaning Based Approach-full context (MBA-fc), and the problem solving behavior of subjects who solve at least two task multiplicative additively tend to Direct Translation Approach-not proficient (DTA-np).

1. Introduction

Multiplicative reasoning is omnipresent in daily life situations: converting currencies from one thing to other thing. Multiplicative reasoning is the capstone of children's primary school arithmetic and the cornerstone of all that is to follow. Learning to reason multiplicatively is a major goal in upper primary mathematics education [2]. In fact, many children read mathematics word problem and directly translate them to arithmetic operations [5]. In the other word, students just going with calculation. There is no reason and there is no justification about the taken step. In case, students' upper primary school frequently erroneously solve additive word problems multiplicatively, while younger children frequently erroneously solve multiplicative word problems additively [2]. The multiplicative word problems can be categorised as various models for multiplication. When students solve word problem, they might construct a model of the problem and demonstrated the way of students reasoning, which typically is an illustration of the specific situation in the problem [6][7]. The way students construct a model and demonstrated can be supported by, the ability to reason multiplicatively. The ability to reason multiplicatively is considered as a significant conceptual leap of students' understanding [8][9]. In order to distinguish when multiplicative its means engaging in a task for which the solution method is not known in advance. In order to find a solution, students must draw on their knowledge, and through this process, they will often develop new mathematical understanding.

Solving problem is not only a goal of learning mathematics but also major means of doing so [3]. Problem solving behaviour play an important role for any student because problem solving behaviour can given provided description of the students' behaviour during the process of solving the mathematics problem [1]. When students solve the task given, the process that appear show how the behaviour students when solve the task. The problem solving behaviour divided into two parts that is direct translation approach and meaning based approach [4]. The two original behaviour above separate again become five category that are 1) Direct Translation Approach-Proficient (DTA-p), this students automatically and efficiently translate the problem elements to a mathematical computation without reading or referring to the problem following the initial reading. They provide no evidence of transformative behaviour in the form of recording given information, use the context during calculations, although the answer may be stated within the context of the problem, or provision of explanations; 2) Direct Translation Approach-Not Proficient (DTA-np), this students show a lack of competence in and have difficulty reading the problem, coming to an understanding of the problem, deciding upon a solution path, and/or performing calculations. They reread the problem frequently, which may be their only behaviour. No evidence is provided that the problems' context supports solution. Their verbal protocols provide no evidence of transformation behaviour. Calculations performed may lack meaning in relation to the problem and serve to complete task; 3) Direct Translation Approach-Limited Context (DTA-lt), this students directly translate the elements of the problem to arithmetic computation. They do so, however, with some, but limited, use of the problem context. Context is limited to single words or rereading followed directly by computation. Limited explanations of arithmetic calculations may be provided; 4) Meaning Based Approach-full context (MBA-fc), this students read or reread each sentence of the problem and record information from the statement with an appropriate context, which is used to support calculation. Final answer are often provided as a complete sentence revealing an understanding of the answer within the context of the problem. Explanations of the mathematical steps, often based on statement of context but not justifications are provided with calculations, and 5) Meaning Based Approach-justification (MBA-j), these students provide evidence of all of the behaviours of the previous category. In addition, they provide justifications for each solution step [5].

2. Methods

2.1. *Participant*

The purpose of this study is to investigate the students' problem solving behavior when solving the multiplicative word problems. This study applied a descriptive-qualitative approach. This study is conducted on a class of the 7th grade of SMPN 39 Surabaya. It consists of 27 girls and 15 boys, and had chosen purposively from 9 available classes. All students had to answer multiplicative word problem test consist of three item test.

Table 1 Grouping Guidelines

Classed	Score Interval	Interpreted in This Study
High	$80 \leq students'\ score$	All item multiplicative properly and given answer right
Avarage	$60\ \leq students' score < 80$	Exactly one item task multiplicative additively and given answer for two other items right
Low	$students' score\ < 60$	At least two item multiplicative additively and given answer for one other item right

Students ability classified [10] interpreted into context of this study, students score above obtained from $\left(\frac{n}{3} \times 100\right)$, where n is the number of items that right such that we had a new categorized. As shown in Table 1, grouping guidelines [10] had interpreted according context of this study. So that students with high ability namely subjects who solve all item multiplicative properly, subjects with medium ability namely subjects who solve exactly one task multiplicative additively, and subjects with low ability namely subjects who solve at least two item multiplicative additively. Three volunteer students were selected as research subject which have different ability that had interpreted as context this study (see Table 1). Interviews were conducted to three subjects to explore more deeply about problem solving behaviour.

2.2. *Analysis*

The data in this study obtained from multiplicative word problem test and semi structured interview based on multiplicative problem solving test. Students had answer the three item of multiplicative word problem test. The different of each indicator of problem solving behaviour briefly separate on the following table.

Table 2 Indicator for problem solving behaviour [1]

Aspects	**DTA-p**	**DTA-np**	**DTA-lc**	**MBA-fc**	**MBA-j**
Fluency of Declaring the questions that have been read using the own sentence	Unfluently	Unfluently	Unfluently	Fluently	Fluently
Relevance of mentioning what is known and asked in the problem	Relevant	Unrelevant	Relevant	Relevant	Relevant
Logical basic used in linking information	Illogical	Illogical	Illogical	Part Logical	Logical
The effectiveness of the used strategies	Ineffective	Ineffective	Effective	Effective	Effective
Write down the final answer as a solution	Not in context	Not in context	In context	In context	In context
Given reasoning and justification	Unrelevant	Unrelevant	Unrelevant	Unreleva nt	Relevant

3. **Results and Discussion**

Multiplication problem solving behaviour test have been analysed. Subject divided into three categories based on grouping guidelines [10] that already interpreted to this study namely, subject who solve all item multiplicative properly, subjects who solve exactly one item multiplicative additively, and subjects who solve at least two item multiplicative additively (See Table 1). Based on analysis there were seven students solve all item multiplicative properly, nineteen students solve exactly one item multiplicative additively, and last solve at least two item multiplicative additively. This result based on multiplicative word problem can be shown briefly in Table 3

Table 3 Result of Multiplicative Word Problem.

Category	**The Number of Students**
All item multiplicative properly	7
Exactly one multiplicative additively	19
At least two multiplicative additively	16

Chosen three subject that represent each category to be as subjects namely, first subject (multiplicative properly), second subject (exactly one multiplicative additively), and third subject (at least two multiplicative additively). The result of multiplicative problem solving test for each subject shown on the following table.

Table 4 Result of problem solving behavior.

	First Subject			**Second Subject**			**Third Subject**		
Item Test	**1**	**2**	**3**	**1**	**2**	**3**	**1**	**2**	**3**
Fluency of Declaring the questions that have been read using the own sentence	fluently	fluently	fluently	fluently	fluently	fluently	Fluently	Fluently	Fluently
Relevance of mentioning what is known and asked in the problem	Relevant	Relevant	Relevant	Relevant	Relevant	Relevant	relevant	un relevant	un relevant
Logical basic used in linking information	logical	logical	logical	logical	Part logical	logical	logical	Illogical	Illogical
The effectiveness of the used strategies	effective	effective	effective	effective	ineffective	effective	effective	Ineffective	Ineffective
Write down the final answer as a solution	in context	in context	in context	in context	in context	in context	In context	not in context	not in context
Given reasoning and justification	relevant	relevant	relevant	relevant	irrelevant	irrelevant	relevant	irrelevant	irrelevant
Category	**MBA-j**	**MBA-j**	**MBA-j**	**MBA-j**	**MBA-fc**	**MBA-fc**	**MBA-j**	**DTA-np**	**DTA-np**

Table 4 show that the problem solving behavior of first subject on all item test include the Meaning Based Approach-justification (MBA-j). The problem solving behavior of second subject on item test 2 and 3 include Meaning Based Approach-full context (MBA-fc), whereas on item test 1 categorized as Based Approach-justification (MBA-j). The problem solving behavior of third subjects on item test 2 and 3 include Direct Translation Approach-not proficient (DTA-np) whereas on item test 1 categorized as Meaning Based Approach-justification (MBA-j).

3.1. Multiplicative Problem Solving of student solve all multiplicative properly

Subject solve all item multiplicative word problem test systematically. This subject solve all item using multiplicative concept and the answer given were right. Here, a part of the result from this subject can be shown in Figure 1 and Figure 2.

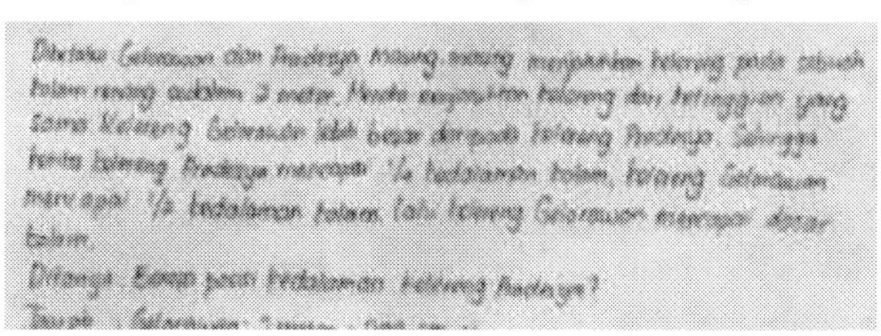

Figure 1 Answer item test 3

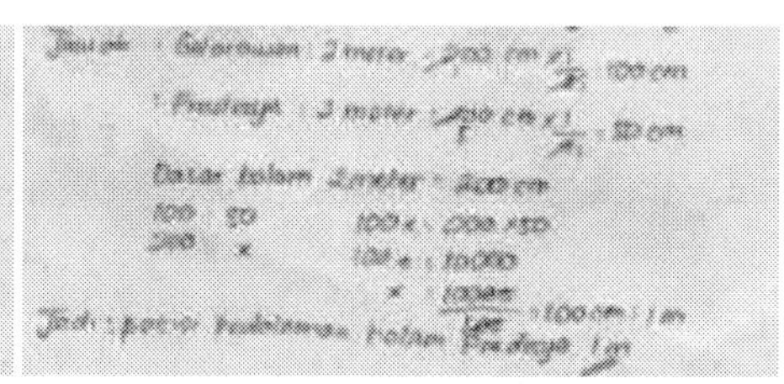

Figure 2 Answer item test 3

From the figure above, subject begin solving the problem with write down the known, write down what asked in problem, choose an appropriate strategy, make calculation based on chosen strategy, write down the result as a context problem. Here some conversations with subject:

R : What information that you got from problem number three?

S1st : There are two marble, but one is bigger than other. Then marbles dropped into pool as deep as 2 meters. And there are a situation that when one marble as deep as one fourth pool, the other one half pool. So I should find the position of marble when the other already reach the bottom of pool.

R : What strategy that you use for solving problem number three?

S1st : First, I find the deep of each marbles by changing the unit become cm. Then, I make a cross multiplication for find the answer.

R : Why you change the unit become cm?

S1st : It would be easier for me than I still using meters, because it will be decimal or fraction if I don't change the unit.

R : Then, why you think to solve problem number three with multiplication?

S1st : Because when there are two marble with different size dropped, the bigger one will be faster than the other.

Based interviews, the problem solving behaviour tend to read the problem, write the known, what asked in problem, make a relevant information, choose an effective strategy logically, write down the answer as context problem, and give the justification of steps that already taken. This behaviour include MBA-j category.

Both result of multiplicative word problem and interview show that students who solve all item multiplicative properly include MBA-j category. This found have the same result that subject with high ability include the MBA-j [1], which is subject with high ability tend to students who solve all item multiplicative properly (see Table 1) include MBA category.

3.2. *Multiplicative Problem Solving of student exactly one multiplicative additively*

Subject solve exactly one item multiplicative additively. This subject solve item 1 and 3 using multiplication concept and the given answer were right. But for item 2 using addition concept and the answer given were wrong. Here, a part of the result from this subject can be shown in Figure 3 and Figure 4.

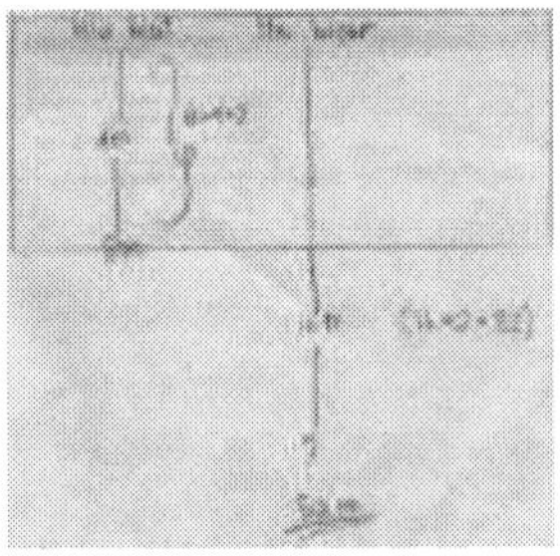

Figure 3 Answer item test 1

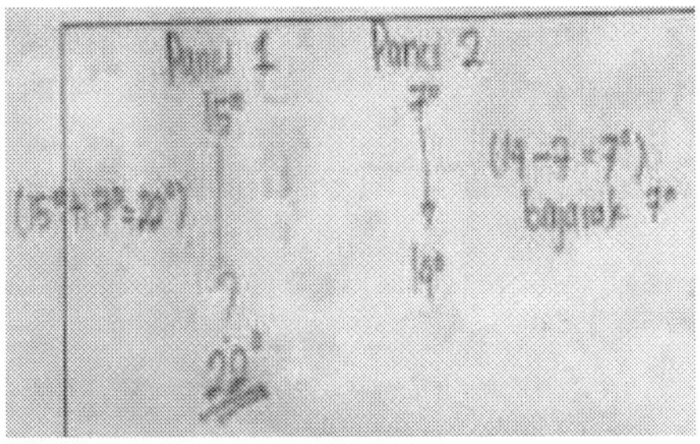

Figure 4 Answer item test 2

From the figure above, subject make a representation using the information in problem. Find the relevancy of these information, make a strategy and got the answer. Here some conversations with subject for item test 2:

R : What information that you got from problem number two?

S 2nd : Aulia will cook water with the same pan and volume but with different stove' flame. And when the pan was 15°C, the other pan 7°C. And what temperature the pan when the other 14°C.

R : Could you linking the information with asked in problem?

S 2nd : Sure, I find the difference between pan one and pan two. Then the value that I got, added to temperature pan.

R : please repeat it, and make it clear?

S 2nd : like I said before, mmm I don't know.

R : That's why you got 22 as an answer

S 2nd : yes, miss.

R : and then how about the initial temperature?

Journal of Physics: Conference Series **1417** (2019) 012046 doi:10.1088/1742-6596/1417/1/012046

S 2nd : means -4°C? mmm, I don't think it usefull

Based interviews, the problem solving behaviour tend to read the problem, write the own known and asked by itself, make a relevant information, the basic logic used in linking information part logical, the chosen strategy were ineffective, write down the answer not in context problem, and there is no reason steps that already taken. This behaviour include MBA-fc category.

Both result of multiplicative word problem and interview show that students who solve exactly one multiplicative additively include MBA-fc category. This found have different result that subject with medium ability include the DTA-p category [1], which is subject with medium ability tend to students who solve exactly one item multiplicative additively (see Table 1) include to MBA category not DTA category like previous research.

3.3. Multiplicative Problem Solving of student at least two multiplicative additively

Subject solve at least two item multiplicative additively. This subject solve item 1 using multiplication concept and the answer given were right. But item 2 and 3 using addition concept and the answer given were wrong. Here a part of the result from this subject can be shown in Figure 5 and Figure 6

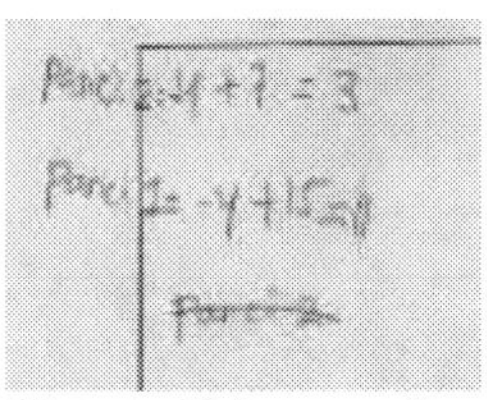

Figure 5 Answer item test 2

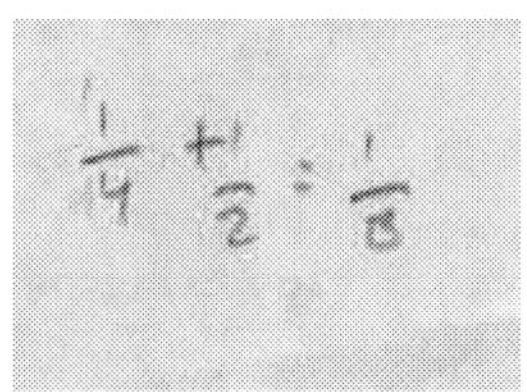

Figure 6 Answer item test 3

From the figure above show that subject directly calculate which is the strategy that used isn't right. It should be using multiplication concept not addition concept. Here some conversations with subject:

R : What information that you got from problem number two?
S3rd : Will cooking water with the initial temperature was -4°C, temperature of second pan is 7°C, temperature of first pan 15°C.
R : What is asked in problem number two?
S3rd : For knowing the temperature
R : What does it mean by temperature?
S3rd : Temperature by pan.
R : then, what is strategy?
S3rd : I just calculated the initial temperature with each temperature pan.
R : what the answer you get?
S3rd : 3°C and 11°C

Based interviews, the problem solving behaviour tend to read the problem, can't make a relevant information, the basic logic used in linking information illogical, not write down the answer, and there is no a reason of steps that already taken. This behaviour include DTA-np category.

Both result of multiplicative word problem and interview show that students who solve all item multiplicative properly include DTA-np category. This found have the a little bit similarity with the result that subject with low ability include the DTA-p category [1], which is subject with low ability tend to students who solve at least two multiplicative additively (see Table 1) include DTA-np category. The found in this study and previous research is students with low ability include in DTA whether in DTA-np category or DTA-p category.

4. Conclusion

There are three category students when solving the multiplicative word problem. Students solve all item multiplicative properly, students solve exactly one item multiplicative additively, and students solve at least two item multiplicative additively.

First, students tend to read problem, write down the known, write down the asked, linking well the information, strategy that used are effective, write down the answer in context, and give justification for each step that already taken. Behaviours of this type subjects include the MBA-j category. Second, students tend to read problem, write down the known and the asked using their own representation, the basic logic used in linking information part logical, strategy that used are ineffective, write down the answer in context, and there is no justification for each step that already taken. Behaviours of this type subjects include the MBA-fc category. Third, students tend to have difficulty to read problem, can't make relation between known and asked, the basic logic is illogical, and directly calculation. Behaviours of this type subjects include the DTA-np category.

5. Acknowledgement

The authors thank to the cooperating teacher and students in SMPN 39 Surabaya especially for students' 7-Ath grade for helping and supporting this study.

References

[1] Azis S N A, Fuad Y and Ekawati R 2018 Problem Solving Behaviors of Grade Seven Students Focusing on Negative Integers. Advance in Intelligent Systems Research. *Atlantis Press* 157

[2] Degrande T, Verschaffel L and Dooren W V 2019 *To add or to multiply? An Investigation of The Role of Preference in Children's Solution of Word Problems.* Learning and Instruction 61 (Newyork: Elsevier)

[3] National Council of Teachers of Mathematics 2000. *Principles and Standarts for School Mathematics.* NCTM.

[4] Pape S J 2003 *Compare word problem: consistency hypothesis revisited.* Contemporary Educational Psychology

[5] Pape S J 2004 *J. for Res. in Math Educ.* **35** 187

[6] Larsson, Kerstin 2016 *Students' Understanding of Multiplication.* (Stockholm: Stockholm University)

[7] Febriyanti F, Fuad Y and Masriyah 2018 *J. of Phy.* Conf. Ser. 1108 012033

[8] Simon M and Blume G 1994 *J. for Res. in Math Educ.* **25** 472

[9] Tzur R, *et al* 2013 *Distinguishing Schemes and Tasks in Childrens Development of Multiplicative Reasoning. PNA* **7** 85

[10] Permendikbud 2016 *Peraturan No 24 Tentang Kompetensi Inti dan Kompetensi Dasar Pendidikan Dasar dan Menengah*

Mathematical Connection Middle-School Students 8th in Realistic Mathematics Education

M Hasbi, A Lukito and R Sulaiman

Department of Mathematics, Universitas Negeri Surabaya, Indonesia.
(muhammad.17070785026@mhs.unesa.ac.id)

Abstract. This study was conducted to notice the effect realistic mathematics education in improving 8th grade Middle-School students' mathematical connection by employing measuring tools of a mathematical connection test and observation sheet, which were developed, was distributed to 37 experimental group students and 37 control group students, which makes 74 learners in total. This study was accompanied experimental by pre-test – post-test control group design. This participant was studying at 8.5th grade as the experimental group and 8.6th grade as the control group from a Middle-School 2 Candi Sidoarjo. Realistic Mathematics Education was implemented in the experimental group while the Control group was implemented to the current Curriculum. As a result of the conducted experimental group, it was being discovered that mathematical connection tests and observation sheets of experimental group students are better than control group students. It was be discovered too, based on the results of the mathematical connection test, 83.78% of students obtained N-Gain more than or equal to 0.7. Therefore, it is possible to say that realistic mathematics education is more efficient than learning in the control group.

1. Introduction

Mathematical connections are one of the abilities that must be possessed to solve problems. This is supported by the statement NCTM, "Mathematics is not a collection of separate strands or standards, even though it is partitioned and presented in this manner" [1]. The ability to recognize, use, understand the interrelationship between mathematical ideas, and form a connection between one idea to another to obtain an overall connection, and to recognize and apply mathematics to situations outside mathematics is called the ability mathematical connection. The ability of mathematical connection as an aspect of mathematical skills that students need to develop. This is contained in the 2013 curriculum mathematics learning objectives, "The purpose of learning mathematics due to students understand mathematical concepts, explaining the interrelationships between concepts and applying concepts or algorithms flexibly, accurately, efficiently, and precisely in problem-solving" [2].

The ability of mathematical connections is very important to be developed in the process of learning mathematics. This is supported by the statement NCTM "To help students build a disposition to use a connection in solving mathematical problems, rather than see mathematics as a set of disconnected, isolated concepts and skills" [1]. The ability of mathematical connections is needed by students, especially in solving problems that require a relationship between mathematical concepts to other concepts in mathematics and other scientific disciplines or in real life [3]. The student's ability in mathematical connections in various schools in Indonesia is still relatively low and moderate. This is based on several research results that say that students' mathematical connection skills are low and students still have difficulty in connecting mathematical concepts [4, 5, 6, 7]. This is also supported by the results of the PISA survey which shows that mathematics achievement in Indonesia at the Middle-School level is always fixed on low numbers and Indonesia is ranked 64th out of 72 countries with a score of 386 [8].

TIMSS research results on students in Indonesia, especially mathematics, are ranked 45th out of 50 countries [9]. Data showed that students were still weak in all aspects of content and cognitive, one of which has not been able to associate three-dimensional concepts and two-dimensional concepts that result in low mathematical connection skills of students. From this problem, a low mathematical connection indicates that students cannot make their mathematical connections. This is consistent with the opinion of Sawyer who said that: "Making connections is the basis for mathematics education which is influenced by their teaching practices in several important ways. Who demonstrates the ability to make connections between mathematical knowledge and other forms of disciplinary knowledge, and between mathematical and real-life knowledge" [10].

Based on the above description of the learning approach that can be applied to improve students' mathematical connection skills namely Realistic Mathematics Education (RME). Realistic Mathematics Education is appropriate for students because it uses contextual problems as the starting point of learning. According to Freudenthal mathematics should not be conveyed to students as a tool (tool) that is ready to be used, but a form of activity constructing mathematical concepts [11]. Under this statement, mathematics learning in the classroom emphasizes a link between mathematical concepts of the student experience in their daily lives and it is very necessary to re-apply the knowledge of students to mathematical concepts in real life. The approach to learning mathematics that directs the mathematical knowledge of everyday students and applies mathematics to their daily lives is Realistic Mathematics Education. So, it can be said that mathematics learning trains students 'logical reasoning by increasing students' ability to mathematical connections. Therefore, the Realistic Mathematics Education approach is appropriate for improving students' mathematical connection skills.

1.1 Realistic Mathematics Education

Realistic Mathematics Education (RME) is a mathematical teaching theory, developed by the Freudenthal Institute in the Netherlands as a reform movement that challenged traditional mathematics education with a mechanistic approach, this approach is based on Freudenthal's assumption "mathematics is a human activity" [12, 13, 14, 15, 16]. This approach characterizes the mathematical activity as an activity of solving problems, finding problems and organizing the subject matter. The main activity is organizing or mathematical. This approach emphasizes student activity and not as passivity. The mathematical activity in question is the activity of rediscovering mathematical ideas and concepts by exploring the real world under the guidance of the teacher as a facilitator. The basic principle of RME refers to guided discovery, didactic phenomenology, and mediating the principle of the model. All these characteristics are inspired by Freudenthal, the basic principle of this RME is 'mathematics as a human activity' [15, 17, 18, 20]. This idea places a heavy emphasis on students who build their knowledge with the guidance of teachers in the learning process of mathematics in the classroom.

Two types of mathematical formulations by Treffers, namely horizontal and vertical mathematics. Examples of horizontal mathematicians are identifying, formulating, and visualizing problems in different ways, and transforming real-world problems into mathematical problems. Examples of vertical mathematics are a representation of relationships in formulas, enhancement, and adjustment of mathematical models, the use of different models, and generalizations. Both types of mathematicians get balanced attention because these two mathematicians have the same value [16]. The real world here is used as a starting point for learning mathematics, such as other subjects, the environment, and everyday life. This approach emphasizes processes more important than results. PMR uses mathematical terms which are interpreted as "Mathematically" consists of two processes namely horizontal mathematical and vertical mathematical. Both processes are interpreted as a guided reinventing activity [19]. In line with Freudenthal, "horizontal mathematical means moving from the real world into the world of symbols to produce mathematical concepts, principles, or models of everyday contextual problems, while vertical mathematical means moving in the world of symbols themselves to produce concepts, principles, or mathematical models of mathematics " [15].

The following are realistic mathematics education principles which are guided reinvention and progressive mathematization, didactic phenomenology, and self-developed models [20]. The following are the characteristics of realistic mathematics education approaches that are using context, using models, using student contributions, interactivity, and interrelationships between topics [16, 20, 21]. Based on the literature review, we make PMR-based mathematics learning steps that are under the principles and characteristics of RME, namely, first step: Understand contextual problems, second Step: Resolve contextual problems, third step: Compare and discuss, fourth Step: Conclude.

1.2 Mathematical Connections

One of the abilities that must be mastered by students in mathematics is mathematical connection skills, as recommended by NCTM, the Standard Process of Problem Solving, Reasoning and Evidence, Communication, Connection, and Representation, highlighting how to acquire and use content knowledge [1]. By statement NCTM," mathematical connections are the relationship between mathematical topics, the relationship between mathematics and other disciplines, and the relevance of mathematics to the real world or in everyday life" [22]. Mathematical connections are connections of mathematics to mathematics itself, mathematical connections to other subjects, mathematical connections to their applications, and mathematical connections to real problems around students through mathematical modeling [23]. In connection with understanding this mathematical connection, we need to note that in mathematics between one concept and the other there is a close relationship, not only in terms of sides but also in terms of the formulas used. One material may be a prerequisite for another material, or a certain concept is needed to explain another concept. This is by Bruner's association proposition which states that in mathematics each concept is related to one another [24].

Mathematical connections are the ability of students to (1) recognize the same representation with the same topic, (2) associate procedures in one representation with procedures in the same representation, and (3) use and value relationships between mathematics and other scientific disciplines" [25]. When students and teachers continue to" think connections "mathematics will grow and become dominant" [26]. In general, suggests that mathematical connection capabilities include "Link conceptual and procedural knowledge, Use mathematics in other curriculum fields, Use math in daily life activities, Look at mathematics as an integrated whole, Applying mathematical thinking and modeling to solve problems that arise in other disciplines, such as art, music, psychology, science, and business, Use and value connections between mathematical topics, Recognise equal representation from the same concept [26].

The purpose of this study was to notice the effect of realistic mathematics education in improving 8th-grade Middle-School students' mathematical connection.

2. Methods

This study used experimental research with the design of the two-groups of pretest-posttest design. The study population was all 8th-grade students of Middle School 2 Candi Sidoarjo. The Sample of this study consists of 8.5^{th} grade and 8.6^{th} grade from the population sampling 8th grade with a total of 74 students. The experimental grade taught with problem-based learning is 8.5^{th} grade which consists of 20 female students and 17 male students, while the control grade taught with conventional learning is 8.6^{th} grade consists of 24 female students and 13 male students. Data collection tools developed in this study uses a mathematical connection ability test sheet (TKMM) and an observation sheet. The TKMM test instrument consisted of 4 questions on the topic of a prism. The observation sheet consists of observing teacher activities and student activities to see student activities during the lesson. Before conducting the research, the TKMM test instrument and observation sheet were validated by experts consisting of 3 lecturers and 1 mathematics teacher. After being declared valid with a validity coefficient of 0.81 and a reliability coefficient of 0.78.

Furthermore, the TKMM description test items were tested to students to get the level of validity and the level of reliability of the instruments to be used, the TKMM items were tested outside the research subject. To measure the validity of an item, you can use the product-moment correlation formula below [27].

$$r_{xy} = \frac{n\sum_{i=1}^{n} x_{ij}y_i - (\sum_{i=1}^{n} x_{ij}\sum_{i=1}^{n} y_i)}{\sqrt{\{n\sum_{i=1}^{n} x_{ij}^2 - (\sum_{i=1}^{n} x_{ij})^2\}\{n\sum_{i=1}^{n} y_i^2 - (\sum_{i=1}^{n} y_i)^2\}}},$$

Then, to calculate the reliability coefficient the Alpha-Cronbach test item is used as follows [28].

$$r_{11}\ r_{11} = \left(\frac{n}{n-1}\right)\left(1 - \frac{\sum_{i=1}^{n} \sigma_i^2}{\sigma_t^2}\right)$$

Furthermore, to determining the level of improvement in the pretest (TKMM) and posttest (TKMM) results, the calculation results with formula below are further categorized and matched with an interpretation based on the criteria in Table 1 as follows [29].

$$<g> \equiv \frac{\% < G >}{\% < G >_{Max}} = \frac{(\% < sf > - \% < si >)}{(100 - \% < si >)}$$

where $<S_f>$ and $<S_i>$ are the final (post) and initial (pre) class averages on the "mathematical connection ability test, "a well-known test of mathematical connection ability".

Table 1. N-Gain Value Criteria

N-Gain	Criteria
$(<g>) \geq 0.70$	"High-g"
$0.30 \leq (<g>) < 0.70$	"Medium-g"
$(<g>) \leq 0.30$	"Low-g"

3. Results

Data Analysis pre-test results of TKMM were compared with the non-parametric Mann-Whitney U test. As a result of the comparison, there was no significant difference between the scores of experimental and control groups in terms of TKMM pre-test scores. Since TKMM post-test scores of experimental and control groups provide parametric testing propositions, they were compared with independent t-test.

Table 2. Mann-Whitney U test results of TKMM pre-test scores

Test	Group	N	Average	Total	U	P
TKMM	Experimental	37	55.53	2776.5	472.05	0.205
	Control	37	51.21	2304.45		

Hypothesis criteria: hypothesis (H_0) accepted if: $T_{count} > T_{table}$ and hypothesis (H_1) rejected. The research hypothesis is as follows:

- H_0: There are no significant differences between students 'pre-test and post-test scores which indicating an effect of teaching on students' mathematical connection.
- H_1: There are significant differences between students 'pre-test and post-test scores which indicating an effect of teaching on students' mathematical connection.

3.1 Results of TKKM

The process of conducting study begins by giving TKMM tests to both classes to see the level of students' initial mathematical connection ability. After applying problem-based learning, a test is given to see the improvement of students' mathematical connection ability. The control grade and the experimental grade are given different learning treatments. Table 3 shows results TKMM to reveal if designed instruction created a significant difference in students' mathematical connection in terms of pre-test and post-test scores.

Table 3. Results of TKMM scores

Test	Class	Number of Subject	Highest Score	Lowest Score
Pre-Test	Experiment	37	70	40
	Control	37	65	35
Post-Test	Experiment	37	100	75
	Control	37	70	50

Based on Table 3 above, it was found that the students' mathematical connection using realistic mathematics education showed better results by looking at the average obtained by students in the experimental grade and the control grade.

Then, the initial analysis is carried out normality and homogeneity test with the purpose to find out data from two groups with normal distribution and homogeneous data obtained. Next, the hypothesis test is performed to show the T_{value} calculated from the pre-test and post-test data. Based on statistical results, T_{count} = 17.656 and T_{table} (N-2) = 1.689. So that $T_{count} > T_{table}$ is obtained, based on this hypothesis (H1) accepted and (H0) rejected, so it can be concluded that the students' mathematical connection using Realistic mathematics education in experiments grade increases and is better than students who get treatment with conventional learning in the control grade. As well as seen from the posttest value of students that fulfills the number that gets a value above the Classical Completion Criteria. Learning completeness achieved by students in classical is more than 75%.

The average N-Gain value obtained by students in the experimental grade is 83.78% more than or equal to 0.7. It can be said that, while in the control grade that is 65% more than or equal to 0.3.

Table 4. N-Gain results of TKMM post-test

N-Gain	**Criteria**
(<g>) ≥ 0,70	83,78%
0,30 ≤ (<g>) < 0,70	16,22%

3.2 Results of Observations

Based on observations during learning on experiment grade, the results indicate that the teacher's activities in learning meet good criteria at each stage in the Realistic mathematics education. Then, student activities that meet active criteria in learning are marked by the ideal time set in the lesson plan to meet the time limit for effectiveness of tolerance and provide a positive response to mathematics learning in class. In mathematics learning takes place, students' responses are positive about mathematical connection skills, because students are given contextual problems to connect the relationship between mathematics and mathematics itself, mathematics with other subjects and mathematics is associated with everyday life.

Based on observations made by observers, the following is a description of the learning process obtained. The stage of treatment or implementation of learning is carried out with several meetings. Each meeting lasts 90 minutes with an estimated learning time from the initial activity to closing. The material discussed in each meeting in this study was prism material. For each meeting, students are given contextual problems related to the prism material, in this case, the surface area and volume of prisms that will be discussed with student group friends. In the initial activity of learning, the teacher arranges classes for learning activities and conveys the purpose of the material to be studied. Next, the teacher forms a group based on gender and the level of ability of students with the aim that students will hold group discussions and solve problems that have been delivered by the teacher. At first, students still have difficulty finding mathematical ideas on the problems presented, but at the next meeting students have begun to recognize mathematical ideas and have been able to represent them in symbolic form, then students make models or procedures, so students can find and draw conclusions from the material that has been taught.

When the material discussion takes place, several students in each group ask the teacher questions about the contextual issues they discussed. In the learning process, the teacher does not directly answer questions from students, but the teacher simply repeats the material and provides reasonable guidance, so students can find their answers. After the group discussion is complete, the teacher asks one of the representatives from each group to present the results of their discussion and write them on the board. After the group presents and presents the results of the discussion, the teacher checks whether the results are by what is expected in the learning objectives. Then the teacher will explain further to confirm the results of the discussion from each group. At the end of the lesson, the teacher and students summarize or conclude the material that has been taught with the guidance of the teacher.

4. Discussion

In general, the results of this study can improve students' mathematical connection skills. This study shows that the average mathematical connection ability of students who use the Realistic Mathematics Education (RME) increases and is better than the class with conventional learning. By using the Realistic Mathematics Education (RME) which is connected with the mathematical concepts themselves, other scientific concepts and other daily life, thus learning will be more meaningful and students' mathematical connection abilities increase and become better.

Learning theory is a guideline for teachers to help students develop their cognitive, social, and spiritual. The development so far with Realistic learning Mathematics Education (RME) has always been based on 4 theories namely Piaget's theory, Vygotsky's theory, Bruner's theory and Ausubel's theory [30]. By the theory of learning, learning and thinking change and develops cognitive structures. In his theory, Piaget suggested that a person's cognitive structure occurs because of the process of adaptation. Adaptation is the process of adjusting schemes in responding to the environment through two processes of assimilation and accommodation. In fact, according to Slavin, "assimilation is an interpretation of new experiences to existing schemes" [30].

The learning process with RME is very closely related to theory, it is based on Piaget's theory, because realistic mathematics learning focuses on how students think, not focusing on the results of student completion. Furthermore, students who are given contextual problems will be allowed to find their ideas to solve problems and represent mathematical symbols. One important theory in the developmental psychology of students is Vygotsky's theory [30]. This theory is following the Realistic Mathematics Education Approach (RME), which is a contextual problem that is presented to students in understanding and completing it designed in the form of groups, and in this group, there are interactions between students and students and teachers in understanding and resolving the contextual problems presented earlier. This approach each student will feel confident and responsible for their knowledge obtained from the results of the discussion.

Learning mathematics is learning about the concepts and structures of mathematics found in the material studied and for discovering the relationship between the concepts and mathematical structures of Bruner [31]. Through these concepts and structures, so that material will be comprehensively understood. Besides, student knowledge is easier to remember and last longer if the material being studied has a structured pattern. So, it can be concluded that Bruner's theory is following RME, the suitability of first guided rediscovery and progressive mathematics, and conformity with the second principle, didactic phenomenology, conformity with the characteristics of the first RME is to use context, third is student contributions and fourth is interactivity.

Learning must be meaningful when the information learned by students is governed by cognitive students [31]. Ausubels' theory, students can associate their new knowledge with cognitive. This way students will have strong and easy memories in transferring their knowledge. If students try to link new information into their knowledge then learning will be meaningful to them. Based on some of the descriptions above, this shows that RME is appropriate and relevant to Ausubels' theory because RME emphasizes understanding more than just memorizing. Besides, the relationship between information to be learned on the cognitive structure of students in the RME approach arises when contextual problems are given that are close to the actual student environment or students can imagine so that they can help students learn meaningfully.

5. Conclusion

Based on the analysis of the results of the above data obtained from the pre-test, post-test, and observation in the experimental class and the control class. The conclusions in this study conducted at Middle-School 2 Candi Sidoarjo are presented below:

- The ability of students' mathematical connection to use realistic mathematics education in the experimental grade increased significantly and was better than the control grade.
- The observations, the activities of the teacher are done well and the students' activities show that they are actively characterized by students who are more enthusiastic in learning mathematics using realistic mathematics education.
- Based on the results of students' mathematical connection tests, 83,78% of students obtained an N-gain more than or equal to 0,7. Therefore, it is possible to say that realistic mathematics education is more efficient than learning in the control grade.

The conclusion obtained above is that learning mathematics using realistic mathematics education can improve mathematical connection students and teachers must master learning models such as realistic mathematics education so that the learning process in the classroom becomes more active, diverse, and learning will be meaningful so that students' knowledge will last a long time and will improve student learning outcomes (Mathematical connection) for the better.

6. Acknowledgments

This research reported here in was supported by SMPN 2 Candi Sidoarjo, especially for 8th grade as samples, and Rante Hanjarwarti as a teacher in this class who helped the research. As well as thanks to Dr. Agung Lukito and Dr. Raden Sulaiman who has provided advice and input in the preparation of instruments and the contents of this manuscript.

References

[1] National Council of Teachers of Mathematics 2000 *Principles And Standards Schools Mathematics* (Reston VA: NCTM)
[2] Depdikbud 2014 *Permendikbud No. 58 Thn. 2014 Tentang Kurikulum 2013 Sekolah Menengah Pertama/Madrasah Tsanawiyah* (Jakarta: Depdikbud)
[3] Rohendi D and Dulpaja J 2013 *J. of Educ and Pract* **4** 17
[4] Hendriana H, Slamet UR, and Sumarmo U 2014 *Int J. of Educ.* **8** 1
[5] Latif S and Akib I 2016 *J. Daya Matematis* **4**
[6] Siregar N and Edy S 2017 *Int. J. of Scie Basic and App Res. (IJSBAR)* **33** 309
[7] Haji S, Abdulla MI, and Maizora S 2017 *J. of Math Educ.* **6** 1
[8] OECD 2018 *PISA 2015 Results in Focus* (New York: Columbia University)
[9] TIMSS 2011 *Kemampuan Matematika Siswa SMP di Indonesia* (Jakarta: Pusat Penilaian Pendidikan Badan Penelitian dan Pengembangan Kementerian Pendidikan dan Kebudayaan)
[10] Sawyer A 2008 Making Connection: Promoting Connectedness in Early Mathematics Education *Annual Conference of the Mathematics Education Research* 429
[11] Wijaya A 2012 *Pendidikan Matematika Realistik: Suatu Alternatif Pendekatan Pembelajaran Matematika* (Yogyakarta: Grahana Ilmu)
[12] De Lange J 1996 *Using and applying mathematics in education,. In A. J. Bishop, et al. (Eds.), International handbook of mathematics education* (Dordrecht: Kluwer)
[13] Freudenthal H 1991 *Revisiting mathematics education: China lectures* (Dordrecht: Kluwer Academic Publishers)
[14] Van den Heuvel-Panhuizen M and Wijers M 2005 *Mathematics Standards and Curricula in the Netherlands* ZDM, **37** 287

[15] Van den Heuvel-Panhuizen M and Drijvers P 2014 *Realistic Mathematics Education In Encyclopedia of Mathematics education* (Dordrecht: Springer Science+Business Media)
[16] Treffers A 1987 *Three dimensions. A model of goal and theory description in mathematics education: The Wiskobas Project* (Dordrecht, the Netherlands: Reidel)
[17] Freudenthal H 1973 *Mathematics As An Educational Task* (Dordrecht, The Netherlands: D. Reidel Publishing Company)
[18] Freudenthal H 1983 *Didactical Phenomenology of Mathematical Structures* (Dordrecht, The Netherlands: D. Reidel Publishing Company)
[19] Hadi S 2005 *Pendidikan Matematika Realistik* (Banjarmasin: Penerbit Tulip)
[20] Gravemeijer K 1994 *Developing Realistic Mathematics Education* (Utrecht: Freudenthal Institute)
[21] Bakker A 2004 *Design Research in Statistics Education. On Symbolizing and Computer Tools*. (Amersfoort: Wilco Press)
[22] National Council Of Teachers Of Mathematics 1989 *Curriculum and Evaluation Standards for School Mathematics* (Reston, VA: Arthur)
[23] Goos M et al 2007 *Teaching Secondary School Mathematics Research and Practice for the 21st Century*. (Australia: Allen & Unwin)
[24] Ruseffendi E T 1991 *Pengantar kepada Membantu Guru Mengembangkan Kompetensinya dalam Pengajaran Matematika untuk Meningkatkan CBSA* (Bandung: Tarsito)
[25] Karnasi and Sinaga M 2014 *J. SAINSAB.* **17** 51
[26] Coxford A F 1995 *The Case for Connections dalam Connecting Mathematics Across the Curriculum*. Ed. House P.A and Coxford A F. Reston (Virginia: NCTM)
[27] Malhotra N K and Birks D F 2007 *Marketing Research: An Research, 3rd European Edition*, Harlow (UK: Pearson Education)
[28] Urbina S 2004 *Essentials of Psychological Testing* (New Jersey: John Wiley & Sons)
[29] Hake R R 1999 *Analyzing Change/Gain Scores. American Educational Research Association Division Measurement and Research Methodology* (USA: Macmillan Publishing)
[30] Slavin R E 2006 *Educational Psychology Theory and Practic (8th Edition)* (USA: Allyn Bacon)
[31] Hudoyo H 1988 *Mengajar Belajar Matematika* (Jakarta: Proyek Pengembangan Lembaga Pendidikan Tenaga Kependidikan Dirjendikti)

Mathematical Representation of Middle School Students in Solving Fractional Problems Based on Sex Difference

N Fadhilah[1], M T Budiarto[2] and E B Rahaju[3]
[1,2,3]Postgraduate Study of Mathematics Education, Postgraduate, Surabaya State University

Email: [1]nurulfadhilah16070785025@mhs.unesa.ac.id,[2]megatbudiarto@unesa.ac.id, [3]endahrahaju@unesa.ac.id

Abstract. Mathematics is one of the subjects taught in schools from elementary education to higher education, in which mathematics cannot be separated from the problem-solving activities. This is because the problem-solving activities is one of the most important skills taught to students early, as a preparation to face problems in the future. Fraction is one material that has a relatively difficult and challenging concept for students. This study aims to describe the students' representation of middle school in solving fractional problems based on sex difference used Polya rules. This research is an exploratory research with a qualitative approach that produces descriptive data. This research was conducted on 7th grade students of SMP Negeri 53 Surabaya. Written data was taken from the results of the written test which produced two research subjects. They are one male student and one female student who have a similar mathematical ability. Data collection method used are the test method by giving the task of mathematics and method of interview. The results of the research showed that the mathematical representation of male junior high school students in solving fractions is by presenting the data provided using symbol manipulation and given verbal information with written words or sentences. While the mathematical representation of female junior high school students in solving fraction questions uses more verbal forms or written sentences as explanatory illustrations. So the representation ability of students can be used as a reference to determine the extent to which a student understands a problem at hand.

1. Introduction

Mathematics learning has one general goal, one of which is mathematical representation. Mathematical representation is an ability that is very important for students and is closely related to communication skills and problem solving. So we need an optimal representation in the form of images, graphs, diagrams, and other forms of representation to be able to communicate something. There are five standard processes that describe the objectives to be achieved in mathematics learning related to understanding mathematics and mathematical competencies include: 1). Problem solving, 2). Communication, 3). Reasoning, 4). Connection, 5). Representation [1]. Representation is referred to as the process of making a concrete model to abstract objects or symbols by describing the relationships between objects, as a result of meaning or operation in the form of illustration or symbolic manipulation and as a result of meaning or operation in the form of illustration or symbolic manipulation [2]. One important material in learning mathematics is fractions. Fraction is one of the important materials because fraction is a prerequisite material in mathematics learning which includes basic concepts to understand other types of numbers such as real numbers and complex numbers. Fractional material is also one of the materials in the domain of number content measured in Trend in International

Mathematics and Science Study (TIMSS). Tunç-Pekkan [3] shows that learning related to fractions is difficult in various countries. So it can be said that learning mathematics in fractions is very important. Sex differences not only result in differences in abilities in solving mathematical problems, but also relates to how to obtain mathematical knowledge. This is consistent with the opinion of Keitel in Ferri and Kaiser [1] who argues that gender, social and culture influence mathematics learning. So, it is also possible for a difference in the ability of representation between women and men in solving problems related to fractions [1]. There are four steps used in problem solving, that is understanding the problem, devising a plan, carrying out the plan, and looking back [4].

2. Method

2.1. Subjects

The subjects in this study were 7th grade students of SMP Negeri 53 Surabaya. This is because fraction material is given to students of class VII at the beginning of the odd semester, so it is hoped that students' memories of fraction material are still strong. The number of subjects chosen to be the subjects in this study were one male subject and one female subject. The subject of research was determined by asking for consideration from the teacher, related to the ability of students to express ideas of thinking verbally and in writing. Subjects that have been selected have equivalent mathematical abilities. Acquisition data related to mathematical abilities obtained from the results of tests of mathematical ability. Two subjects are said to have an equivalent ability if the difference between the values of the two is a maximum of 5 from a total score of 100. It is intended that there are no differences caused by a student's ability.

2.2. Instruments

The main instrument in this study is the researchers themselves and supporting instruments is math ability test, fractional mathematical representation test, and interviews.

2.3. Data and data credibility

The data in this study is qualitative data. The data obtained from the work on fractional mathematical representation tests done by two types of students is male and female. Students work results are assessed on the basis of Polya's problem-solving steps. The data in this study also obtained from the interviews that aims to find out in detail chronological workmanship of students on the second and fourth rare that is devising the plan and looking back. And then to assure the obtained data credibility, the researcher made observations continuously/consistently (to improve perseverance), time triangulation and member check [5].

2.4. Data analysis

Data analysis performed after data collection process and data validation process. Based on the credible data, an analysis with three-flow model of activities were made simultaneously: data reduction, data representation and conclusion drawing [6]. Analysis of the representation test results is divided into two, namely the analysis of the representation results of male students and the analysis of the results of the representation of female students. Analysis of the results of male and female representation tests were carried out on each type of item based on the polya step which included: naming fraction number, making fractions item, reconstructing unit from unit fractions items, reconstructing the unit from proper fractions items, improper fractions items, Reconstructing the unit from improper fractions items.

3. Results and Discussion

3.1. Male students

Table 1. Summary of the work male student

Polya's Steps	Case					
	Naming fraction number	Making fractions item	Reconstructing unit from unit fractions items	Reconstructing the unit from proper fractions items	Improper fractions items	Reconstructing the unit from improper fractions items
Understanding Problem	√	√	√	√	×	×
Devising a plan	√	√	√	√	×	×
Carrying out the plan	√	√	×	×	×	×
Looking back	√	×	×	×	×	×

The result of Male student interview for the second and fourth step can be as follows:

Interview 1:

R : What do you do after understanding the problem?
M : I am trying to find the best plan by symbol manipulation and given verbal information with written words or sentences.
R : After your work is over, will you evaluate your reply process again?
M : Yes, I will do.
R : How did you do that ?
M : I will check each step of the answer, then link the answer with a similar problem.
(Note: R= Research; M= Male; √ = doing well × = not doing well)

3.2. Female students

The result of mathematical representation from male students in solving math problems based on Polya's steps can be seen in the following table:

Table 2. Summary of the work male student

Polya's Steps	Case					
	Naming fraction number	Making fractions item	Reconstructing unit from unit fractions items	Reconstructing the unit from proper fractions items	Improper fractions items	Reconstructing the unit from improper fractions items
Understanding Problem	√	√	√	√	×	×
Devising a plan	√	√	√	√	×	×
Carrying out the plan	√	×	√	×	×	×
Looking back	√	×	×	×	×	×

Interview 2:

R : What do you do after understanding the problem?
F : I am trying to find the best plan by verbal forms or written sentences as explanatory illustrations or sentences.
R : After your work is over, will you evaluate your reply process again?
F : Just for the case of measurement only.
Besides, I'm not sure to do it.

(Note: R= Research; F= Female; √ = doing well × = not doing well)

Based on the data obtained from the mathematical representation test results and interviews it was found that the mathematical representation of male and female students is not much different. Both are not successful in using step polya in solving problem cases related to improper fraction items and reconstructing the unit from improper fractions items

The results of the analysis were supported by previous researchers Halpern [7] who stated that in general women remember better than men, whereas in logical thinking men are better than women. In general, male students are the same as female students, but male students have better abstraction power than female student [3].

4. Conclusion

a) Results Table 1 and interview 1 showed that the mathematical representation of male junior high school students in solving fractions is by presenting the data provided using symbol manipulation and given verbal information with written words or sentences.
b) Results Table 2 and interview 2 showed that the mathematical representation of female junior high school students in solving fraction questions uses more verbal forms or written sentences as explanatory illustrations or sentences.

In general, the mathematical representation of male and female middle school students in solving fractions using Polya's theory is not much different, but male students have better abstraction power than female students.

References

[1]. Ferri R B and Kaiser G 2003 *First Results of a Study of Different Mathematical Thinking Styles of Schoolchildren* In: Burton, L. (Hrsg.) *Which Way? Social Justice in Mathematics Education* (London: Greenwood)
[2]. Salkind G M 2007 *Mathematical Representations: Preparation and Professional Development of Mathematics Teachers, Spring 2007* 1 http://mason.gmu.edu/~gsalkind/portfolio/products/857LitReview.pdf
[3]. Tunç-Pekkan Z 2015 *Educ. Stud. in Math.* **89** 419
[4]. Polya G 1973 *How to Solve It (A New Aspect of Mathematical Method)* (New Jersey: Princeton University Press)
[5]. Moleong l J 2011 *Qualitative Research Methodology* (Revised Edition) (Bandung: Rosdakarya)
[6]. Miles M B and Huberman A M 1992 *Qualitative Data Anarchy: The Source Book About New Methods* Translation by Tjetjep Rohendi Rohedi (Jakarta: UI Press)
[7]. Halpern D F et al 2007 *Psychol. Sci. in the Public Interest* **8** 1

Journal of Physics: Conference Series **1417** (2019) 012049 doi:10.1088/1742-6596/1417/1/012049

Student's Mathematical Representation in Solving Geometry Problems Based on Cognitive Style

Adnan S*[1], Dwi Juniati [2], Raden Sulaiman[2]

[1]Mathematics Education, Postgraduate of Universitas Negeri Surabaya, Indonesia
[2]Mathematics Department of Universitas Negeri Surabaya, Indonesia.

e-mail: adnan.17070785032@mhs.unesa.ac.id

Abstract. This research aims to describe the student's mathematical representation of junior high school who had field independent cognitive style and field dependent in solving geometry problem. Subjects were selected based on the test results of the Group Embedded Figures Test (GEFT), and had equivalent mathematical abilities. The research subjects consisted of two female students who had field independent and field dependent cognitive style. This research uses a qualitative method. The results showed that the subject used mathematical representation in expressing his idea to solve geometry problems by using Polya's problem solving. The mathematical representation of field independent in solving geometry problem is: understanding information and what is asked in verbal, devising a plan in a visual of geometric images, then presenting verbal plans, carrying out the plan in symbolic by doing calculations, and looking back in symbolic by doing recalculation. Where as field dependent mathematical representations to solving geometry problems is: understanding information and asking in visual of geometrical images and verbal, devising a plan in a visual in the form of subsequent images presenting verbal plans, carrying out the plan in symbolic by performing calculations, and looking back in symbolic in the form of recalculation.

1. Introduction

Representation is one of the five standard processes in learning mathematics [1]. Five processes should students know and be able to use as they progress through school: problem solving, reasoning and proof, communication, connections, and representations [1]. We use representational forms to communicate ideas and as tools for reasoning [2]. Problem Solving an important role in mathematics learning, there is a general acceptance of the idea that the primary goal of mathematics instruction should be to have competent students problem solvers [3]. There is a close relationship representation to solve problems such as statements empirical studies suggest that mathematics problem solving competency depend on ones's ability to think in terms of different representational system during problem solving process [4]. The linkage occurs when students begin to instruct the right representation for problems in obtaining the right solution.

The branch of mathematics consists of algebra, geometry, numbers, statistics and opportunities [5]. One of the material that has stronger problems solving at the junior level is a matter of geometry. Geometry provides a rich source of visualization for understanding arithmetic, algebraic, and statistical concepts [6]. Geometry deals with the nature and relationships of points. lines, shapes, and spaces [7]. Geometry has a very important role to learn mathematics, as a link between events in everyday life and mathematical concepts.

Representations which each student is different from the other students, because each student has their own characteristics, such as that complexities of representation as cognitive and social processes and of how it is inextricably linked with the knowledge people have the situation being presented [8]. Various perceptions possessed by someone in exploring possible information for exploration, one of which is cognitive style [9]. That is because the representation as cognitive activities associated with the knowledge of the students, then the style of the students' cognitive affect representations to problem solving. In this study the characteristics of students are seen from their cognitive style. Typical ways of perceiving and remembering information, and typical ways of solving problems and making decisions [10]. Solving a problem is an attempt to determine a way out of a difficulty to get the right solution of the problem [11]. Solving problems in this study using the stages revealed by Polya consisting of understanding the problem, devising a plan, carry out the plan, and looking back [12].

2. Theoretical Framework

Representation plays an important role in efforts to develop and optimize students' mathematical abilities. NCTM includes a representation as the standard process of the fifth after problem solving. reasoning, communication, and connection [1]. According to the context or the use of the representation: a Cartesian graph, for example, can represent a function or the solution on the set of an algebraic equation [13].

Pape & Tchoshanov states that there are four ideas used in understanding the concept of representation. First, representations may be broadly as mental states. These internal representations are mental images of, for example, a set of five objects. Second, representations may more narrowly be thought of as mental representations of a former mental state. Here, the numerals, 5, or the number, five, are examples. Finally, the last two formulations include a structurally equivalent 'pre-sentation' through pictures, symbols or signs, and something 'in place of' something else [14].

The main purpose of mathematical representation as a standard process in school mathematics is to create and use representations to organize, record and communicate mathematical ideas. Translating, choosing, and applying mathematical representations to solve problems. Representation is a basic ability that will help in increasing students' understanding and mathematical abilities.

This study uses a mathematical representation form proposed by Goldin & Shteingold, namely: Symbolic (notations, mathematical models, doing calculations and manipulation of mathematical models), visual (number line, graph, diagram, table or picture), and verbal.

Referring to the explanation of the experts' definitions above, it can be concluded that the mathematical representation in this study is an expression displayed by students as mathematical ideas or concepts that describe, represent, or symbolize something in mathematics in visual, verbal, or symbolic form.

3. Method

3.1. Subject

The subjects of this study were two female students of grade VIII SMP Al-Falah Surabaya, consisting of a student who had field independent cognitive style and field dependent. Students are selected based on the results of the GEFT [16]. Scoring on GEFT is Subject scoring less (0-11) than the national mean for the GEFT (11.4) were categorized as field dependent. Subjects scoring the national mean (11.4) or greater (12-18) were categorized as field independent [17]. The next step students are given a math ability test to choose subjects with equal ability. The subject is said to have an equivalent mathematical ability, if the difference in test results is less than 5 with a maximum score of 100. In order to obtain two female students as subjects with details of field independent students' (FIS) with a score of 18 on the GEFT and a score of 77 on the math ability test, field dependent students' (FDS) with a score of 5 on GEFT and a score of 76 on the math ability test.

Table 1. The Problem Solving Task of Mathematical Representation

For example, ABCD is a isosceles trapezoid. AD and BC are extended so that it intersects at point E. Length $\overline{AB}$ = 30cm, $\overline{CD}$ = 18cm and trapezoid height 8cm. If F and G respectively midpoints of $\overline{AD}$ and $\overline{BC}$. Length $\overline{FG}$ = 24cm, find the area of the triangle EFG!

3.2. Procedure

The design of this study is research with a qualitative approach. This study aims to find out how the student's mathematical representation in solving geometry problems. The description of the phenomenon described in this study is a mathematical representation of junior high school students in solving geometry problems in terms of cognitive style. Moleong explains that qualitative research is a research procedure that produces descriptive data in the form of written or spoken words from people and observable behavior [18].

Data collection procedure begins with giving a GEFT test to students, then students are given a math ability test and then the research subject is determined. Subjects were given the task of solving problems to obtain mathematical representation. After that, researchers conducted interviews based on the results of student work to explore mathematical representations. The data validation process is carried out using triangulation. Data analysis refers to aspects of mathematical representation through the steps of categorizing data, reducing, discussing, and making conclusions about mathematical representation.

4. Result

4.1. Mathematical Representation of Field Independent Student's (FIS)

- Understanding the Problem

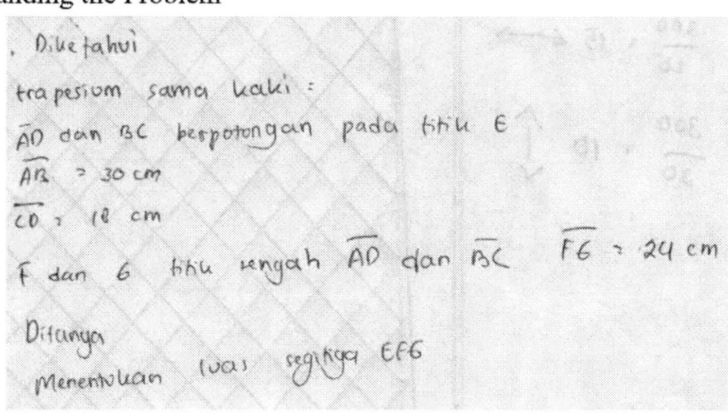

Figure 1. Mathematical Representation of FIS in Understanding the Problem

In fig.1 above, FIS present information that is known in the form of sentences in which there are several symbols . FIS write information on the sides of trapezoid with symbols of line segments such as $\overline{AD}$, $\overline{AB}$, $\overline{CD}$ but not consistently on the side BC are not using symbols segments like the others. FIS write that the sides of the trapezoid $\overline{AD}$ and $\overline{BC}$ intersect at points E, F and G midpoint of $\overline{AD}$ and $\overline{BC}$.

- Devising a Plan

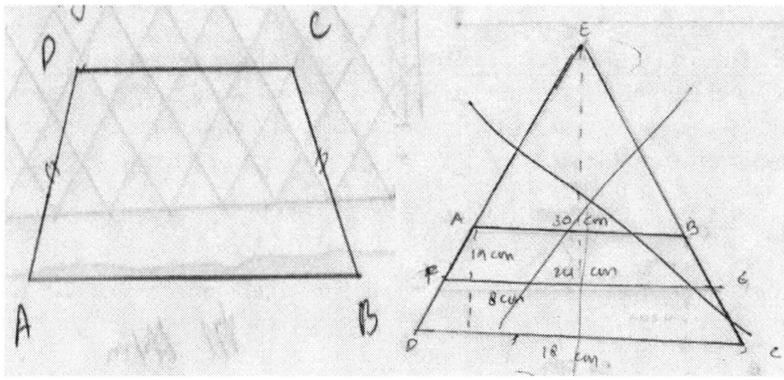

Figure 2. Mathematical Representation of FIS in the Devising a Plan

In fig.2 above, FIS present ideas that help determine the relationship between what is known and asked by using mathematical representations in the form of visual trapezoid images that are complete according to the information obtained at the stage of understanding the problem. There is also a symbol on the trapezoid \\ to represent the same length at the foot of the trapezoid, but there is no symbol that marks the parallel side of the trapezoid.

- Carrying Out the Plan

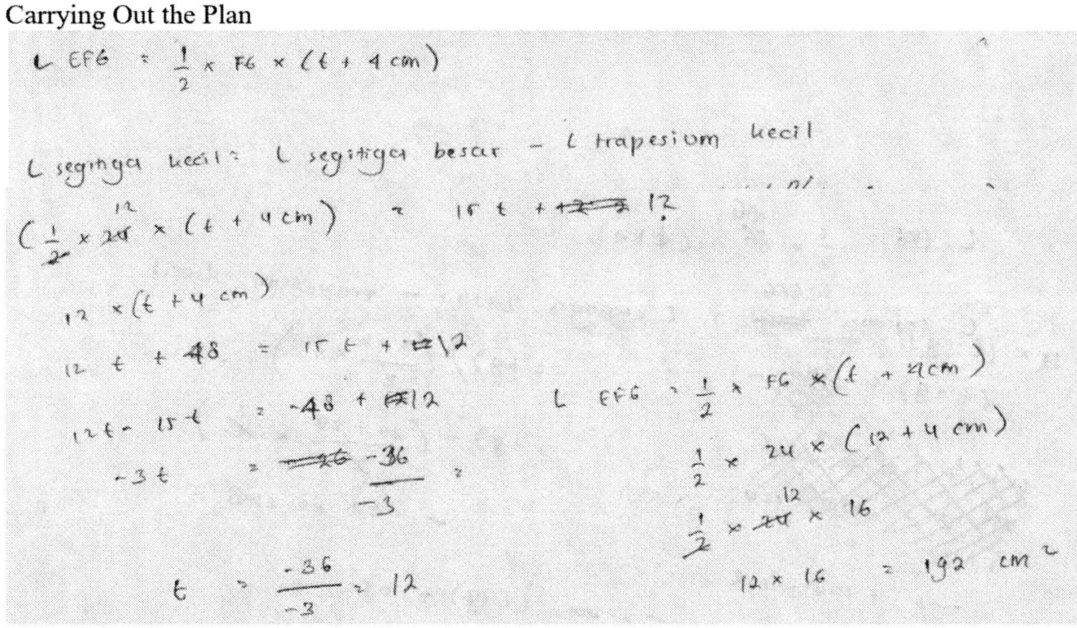

Figure 3. Mathematical Representation of FIS in Carrying Out the Plan

In fig.3 above, FIS calculating based on the mathematical model her has made, namely "*L small triangle = L large triangle - L small trapezoid* ". The purpose of the model is that the area of the small triangle (EFG) is equal to the area of the large triangle (EAB) minus the small trapezoid (FABG). In the calculation process student use the symbol *t* to express the height of a triangle, *t* + 4 for height which is asked the extent. In the final results of calculations FIS write *cm* ^{2} as a unit area of the EFG triangle.

- Looking Back

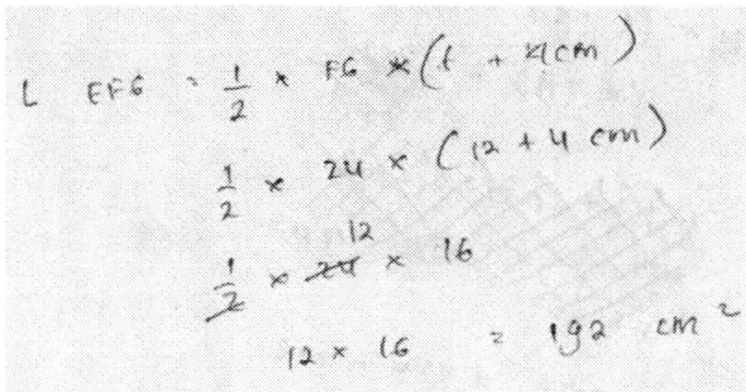

Figure 4. Mathematical Representation of FIS in the Looking Back

In Fig.4 above, FIS re-examine the problem solving, expressing whether or not the problem solving steps in the symbolic form, which is to recalculate the results of the problem solving on the frosted sheet in which there is the *L EFG* symbol as the area of the EFG triangle being asked, the symbol *a* as the base of the EFG triangle and *t + 4* as the height of the EFG triangle.

4.2. Mathematical Representation of Field Dependent Student's (FDS)

- Understanding the Problem

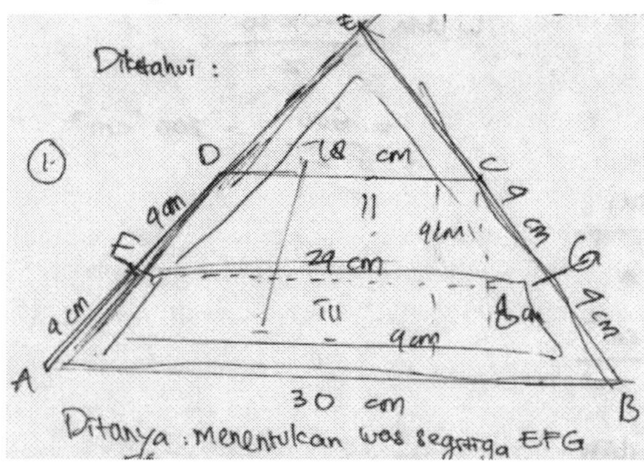

Figure 5. Mathematical Representation of FDS in Understanding the Problem

In fig.5 above, FDS present information that is known using a visual representation of drawing a trapezoid with information that is known in accordance with what is stated in the problem. Student thought that the 4cm in size also in terms of $\overline{AF}$, $\overline{FD}$, $\overline{BG}$, and $\overline{GC}$. FSD wrote were asked to determine the area of the EFG triangle. The triangle is obtained from the extension side $\overline{AD}$ and $\overline{BC}$ the intersecting in E.

- Devising a Plan

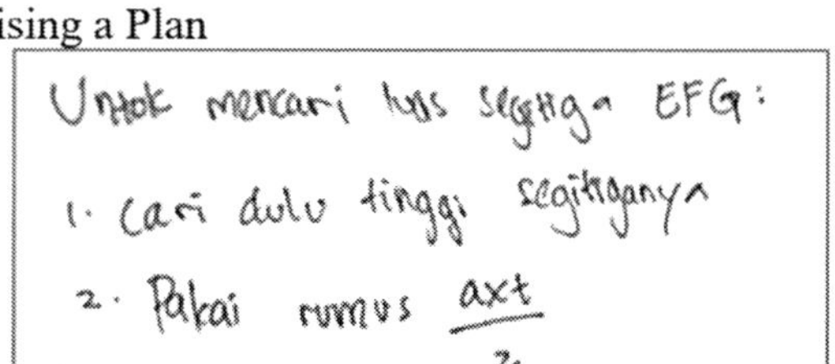

Figure 6. Mathematical Representation of FDS in the Devising a Plan

In fig.6 above, FDS present devising a plans in verbal form in the form of sentences which are sequential steps, FDS write down first determine the height of triangle is asked, then substitute the high value into the triangle area of the formula, in the plan students explain that the symbol a is the base of a triangle is $\overline{FG}$ 24cm.

- Carrying Out the Plan

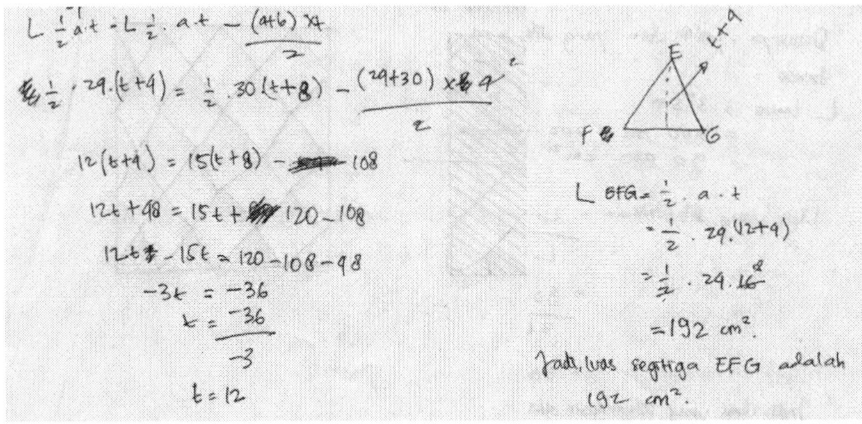

Figure 7. Mathematical Representation of Field Dependent Student's in Carrying Out the Plan

In fig. 7 above, FDS make a mathematical model " $L\ \Delta\ EFG = L\ \Delta\ EAB$ - $L\ ABGF$ " in the model there is a symbol Δ to represent a triangle, but not in the symbol for the trapezoid, FDS only writes $L\ ABFG$, and calculating. In the calculation student also use the symbol t to express the height of the triangle and the trapezoid in the model . After obtaining the value of t, FDS draw an EFG triangle with arrows indicating that the height is t +4. Student write that the area of the EFG triangle is 192cm^2. FDS write in complete unit area, which is cm^2.

- Looking Back

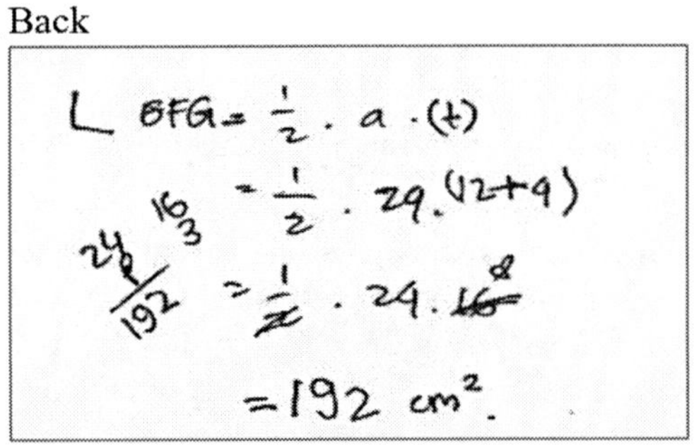

Figure 8. Mathematical Representation of FDS in the Looking Back

In fig.8 above, FDS re-examine the problem solving, revealing whether or not the problem solving step is to do a recalculation of the results of solving the problem in which there is a symbol not as a high value.

5. Discussion

In the stage of understanding the problem, mathematical representation of FIS express information that is known in the problem in verbal. FIS write elements that are known to the problem in accordance with those listed in the problem, in the sentence written by FIS consistently used the symbol of the line segment to express the sides of the trapezoid. Furthermore, FIS revealed what was asked in the verbal form, FIS wrote questions asking to determine the area of the EFG triangle. Mathematical representation of FDS presents information that is known in the form of visual images in accordance with information that is understood in the problem, but in the picture there are no parallel symbols and in the FDS pictures' misplaced a size of side. Furthermore, FDS express information that is asked for problems in the form of verbal. FDS present that in the problem asked to determine the area of the EFG triangle.

At the stage of devising a plan, the mathematical representation of FIS presents ideas that help determine the relationship between what is known and asked in the visual form in the form of a trapezoid image that matches the information that has been obtained at the stage of understanding the problem. Furthermore, FIS presented a devising a plan in the form of verbal words, the FIS said that the area of the triangle in question could be obtained from the process of subtracting a large triangle with a small trapezoid in accordance with the drawing he had made beforehand. Mathematical representation of FDS presents ideas that help determine the relationship between the known and asked in verbal form in the form of written sentences that are steps to solve the problem given, in the sentence there are symbols a as the base and t axle height on the triangle asked.

At the stage of carrying out the plan, the mathematical representation of FIS makes a mathematical model in accordance with the information obtained at the stage of understanding the problem and a plan, then FIS make calculations, and consistent in using the symbol t as the height of the triangle and trapezoid. Mathematical representation of FDS make mathematical models in accordance with information on problems in symbolic form. Furthermore, FDS express interpretations of the results of the solutions obtained in the form of visual images in the form of triangles that are asked and make conclusions on the results of calculations in the form of written sentences. At the stage of looking back, mathematical representation of FIS re-examining problem solving in symbolic form, FIS re-calculate the opaque paper in accordance with the mathematical model they have made. Mathematical representation of FDS re-examining in symbolic forms.

There are some differences in representation between field independent and field dependent students which are in line with the opinion of Witkin which states that differences in cognitive styles affect students in solving problems, namely influencing students' strategies in receiving, thinking problem solving, and learning [16]. Complexities of representation as a cognitive and social process and of how it is inextricably linked with knowledge people have the situation being presented [8]. Representation as cognitive activities associated with the knowledge of the students, then the style of the students' cognitive Affect representations to problem solving.

6. Conclusion

There are several differences in the mathematical representation of students based on the cognitive style they have. Mathematical representation of FIS in solving geometry problems is understanding information and what is asked in verbal form in writing and verbally, while mathematical representation of FDS in expressing their ideas to solve geometry problems by understanding information and being asked in visual form in the form of geometry drawings and verbal and verbal forms. FIS devising a plan in a visual form in the form of a geometry drawing then present the plan in verbal form, while FDS devising a a plan in visual form in the form of a picture then present a plan in verbal form verbally. FIS carrying out plans in symbolic form by doing calculations in accordance with the mathematical model, while FDS carrying out plans in symbolic form by doing calculations in accordance with the mathematical model. FIS re-examine in symbolic form by recalculating, while FDS re-examine in symbolic form in the form of recalculation and writing conclusions.

Acknowledgments
I wish to thank the principal and students of Junior High School Al-Falah Surabaya has participated in this study. Thank you also to the supervisor for giving direction to the researcher.

References

[1] NCTM 2000 *Principles and Standards for School Mathematics* (Reston, VA: National Council of Teachers of Mathematics)
[2] Selling S K 2016 Learning to represent, representing to learn *Journal of Mathematical Behavior.* http://dx.doi.org/10.1016/j.jmathb.2015.10.003
[3] Schoenfeld A H 2011 *How We Think: A Theory of Goal-Oriented Decision Making and its Educational Application* Studies in Mathematical Thinking and Learning (New York & London: Routledge)
[4] Jones A D 2000 *The Fifth Process Standart: An Argument to Include Representation in Standards 2000* (Universuty of Maryland)
[5] Mullis I V S, Martin M O, Ruddock G J, O'Sullivan C Y & Preuschoft C 2009 *TIMSS 2011 Mathematics Framework* (Amsterdam: TIMSS & PIRLS International Study Center Lynch School of Education, Boston College)
[6] Fabiyi T R 2017 Geometry Concepts in Mathematics Perceived Difficult to Learn by Senior Secondary School Students in Ekiti State, Nigeria *TOSR Journal of Research & Method in Education (IOSR-JRME)* **70**(1) pp 83-90 DOI: 10.9790/7388-0701018390
[7] Bergstrom C & Zhang D 2016 Geometry Interventions for K-12 Students with and without Disabilities: A Research Synthesis *International Journal of Educational Research* **80** pp 134-154 http://dx.doi.org/10.1016/j.ijer.2016.04.004
[8] Wu Z 2004 *The Study of Middle School Teacher's Understanding and Use Mathematical Representation In Relation To Teacher's Zone Proximal Developement in Teaching Fractions and Algebraic Functions* Disertation Doctor of Philosophy (Texas A&M University)
[9] Utomo E S, Juniati D, and Siswono T Y E 2017 Mathematical Visualization Process of Junior High School Students in Solving a Contextual Problem Based on Cognitive Style *The International Conference on Research, Implementation, and Education of Mathematics and Science (4th ICRIEMS)*
[10] Zhang L and Sternberg R 2006 *The Nature of Intellectual Styles* (Mahwah, NJ: Erlbaum)
[11] Polya G 1981 *Mathematical Discovery: On Understanding, Learning, and Teaching Problem Solving* (Printed in the United States of America)
[12] Polya G 2004 *How to Solve It: A New Aspect of Mathematical Method* with a New Foreword by John H Conway (United States of America: Princeton University Press)
[13] Godino J D and Font V 2010 The Theory of Representation as Viewed from the Onto-Semiotic Approach to Mathematic Education *Mediterranean Journal for Research in Mathematics Education* **9**(1) pp 189-210
[14] Pape S J and Tschoshanov M 2001 *The Role of Representation(s) in Developing Mathematical Understanding* Theory into Practice **40**(2) Realizing Reform in School Mathematics (Spring, 2001) pp 118-127
[15] Kalathil R R and Sherin M G 2000 Role of Students' Representations in the Mathematics Classroom. *In B. Fishman & S. O'Connor-Divelbiss (Eds), Fourth International Conference of the Learning Sciences* pp *27-28* (Mahwah, NJ: Erlbaum)
[16] Witkin H A, Moore C A, Goodnough D R, and Cox P W 1977 Field dependent and field independent cognitive style and their educational implication *Review of Educational Research* **47**(1) (Winter, 1977) pp 1-64
[17] Davis GA 2006 Learning Style and Personality Type Preferences of Community Development Extension Educators *International Journal of Agricultural Education* **4**(1) pp 90-99
[18] Moleong L J 2017 *Metodologi Penelitian Kualitatif* (Bandung: PT Remaja Rosdakarya)

Student's Multiple Representation in Solving Addition and Subtraction of Fraction Problem

T W Yuliani[1], Y Fuad[2], R Ekawati[3]

[1, 2, 3]Department of Mathematics, Universitas Negeri Surabaya, Indonesia.

E-mail: titis.17070785006@mhs.unesa.ac.id

Abstract. Fraction is one of mathematics' topics that difficult to be understood. Many representation form from fraction. Representation has an important role in mathematic. Representation helps student to create mathematic ideas more concrete. So, multi-representation is needed in fractional problem solving. This study is to describe students' ability of multi-representation in solving fraction problem. The sample of this study is the VII-B grade students of a state junior high school in Munjungan city, East Java. There are eight classes from A to H, VII-B was chosen purposively. There were 10 boys and 12 girls in VII-B. Multi-representation test was given to each student. Three volunteer were chosen as our respondents. The semi-structured interview was utilized to investigate the process and students conceptual of fraction. This study shows that the students still have difficulties to represent fraction. The students were success to solve model representation, but they have difficulty to solve fraction in number line. The error in representation on addition and subtraction of fraction problem shows that student did not understand fraction concept well. Student ability on multi-representations should be conducted through exercise continually.

1. Introduction

Representation was important for student. Representation was one of fifth standard process in mathematics school [1]. Representation could be described the process of modelling the real world to abstract concept or symbol [2]. Representation was needed in problem solving [3]. Representation made mathematics idea more concrete and constructed the understanding; communicated information and reasoning [4]. Representation consisted of visual, verbal and symbolic. Visual representation presented such us picture, number line, and other mathematics picture. Verbal representation presented such us sentence or word problem. Symbolic representation presented such us number or other symbol [2].

Fraction was important in algebraic and other mathematics topic [5]. Fraction was subject material on seventh grade [6]. Therefore, fraction should be mastered by student. There were seven types of representations in fractional concepts: (1) part group; (2) part-whole; (3) part group non-congruent part; (4) part group comparison ; (5) number line ; (6) part whole, comparison ; (7) part whole, non-congruent part [7].

Student of elementary school were difficult to make representation in addition and subtraction of number line [8]. Student of elementary school were difficult to divide the number line into required denominator [8]. The use pictorial models or model representation help student to develop the visualization skills [9]. But, there were student unable to represent symbolic to pictorial or model [10]. The students still have difficulties in the representation of fraction problem especially in pictorial or

model [11]. In model representation, mistakes were found in the denominator and numerator determination of addition operation [8].
Based on the previous research, student failed to represent fraction such us number line and pictorial. Because the representation was important on problem solving, the representation should be mastered by student well. In this study describe about addition and subtraction on number line and model representation in junior high school.

2. Methods

This was qualitative research which aims of this study to describe the students' ability when create each representation in fraction. This study conducted in state junior high school in Munjungan city, East Java. The sample of this study is VII-B, selected from eight classes and chosen purposively. The class consist of 10 boys and 12 girls. Multi-representation test were given to student. The test about addition and subtraction on fraction. This study only discussed about number line and model representation. Multiple representation tests were given to all students. In this research use semi-structured interview. Three students were volunteer to be interviewed to get detail explanation from their answer.

There were 3 questions about about fraction. The criteria of multi representation were measured from the Table 1 (Adapted from [8]).

Table 1. Criteria of number line representation and model representation.

Representation	Number line	Model
	1. Draw the number line	1. Determine the denominator (draw closed figure or picture divided to equal parts)
	2.Determine the denominator (divide equal parts)	2. Determine the numerator (mark requested part)
	3. Show numerator with arrows	3. Perform the operation
	4. Perform the operation	4. Show the result
	5. Show the result	

This was the fraction problem :

1. Use number line to get the answer of $\frac{3}{16}+\frac{6}{16}$
2. Use model to get answer of $\frac{6}{12}+\frac{1}{6}$
3. Use number line to get answer of $1\frac{1}{2}-\frac{3}{4}$
4. Use model to get answer of $\frac{8}{14}-\frac{1}{7}$

3. Results and Discussion

The multiple representation test was analyzed based on criteria table 1. The test only discuss about addition and subtraction on number line and model representation.

3.1. Answer by Student 1

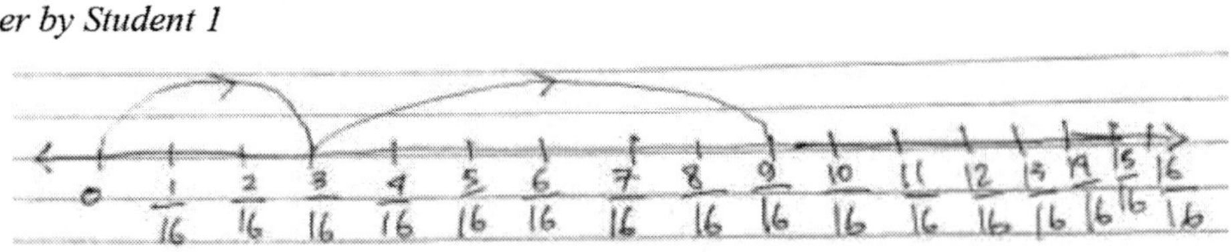

Figure 1. Addition number line

Journal of Physics: Conference Series **1417** (2019) 012050 doi:10.1088/1742-6596/1417/1/012050

Student 1 was able to create the number line correctly. It shown by determine the denominator by divide line with the equal parts. Student showed the arrow to express the numerator. But, Student 1 didn't show the final result by arrowing out the result.

Q : You have answered with the number. Please, do it in number line.
S1 : Yes, mam (students draw the number line and use arrow to show the numerator)
Q : Are you sure ? Show the result. Be careful with the order.
S1 : Yes mam.

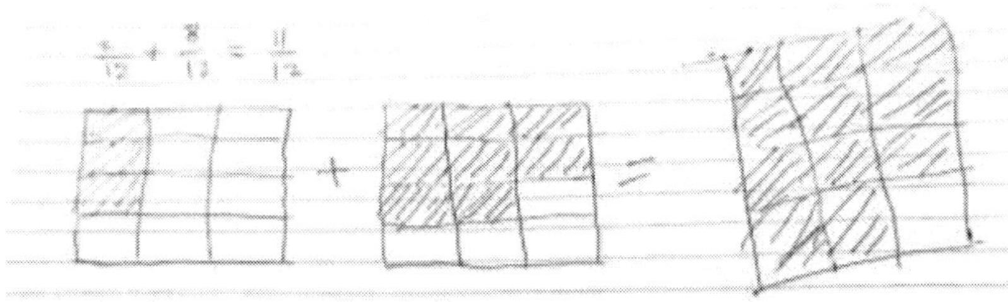

Figure 2. Addition model

Student 1 have drawn the drawing closed figure. Student 1 divided picture with 12 equal parts that show the denominator. Student 1 didn't draw the picture well. From interview Student 1 know that 12 is denominator with same equal parts. Student 1 marked the numerator with concise. Student 1 answer it right and well.

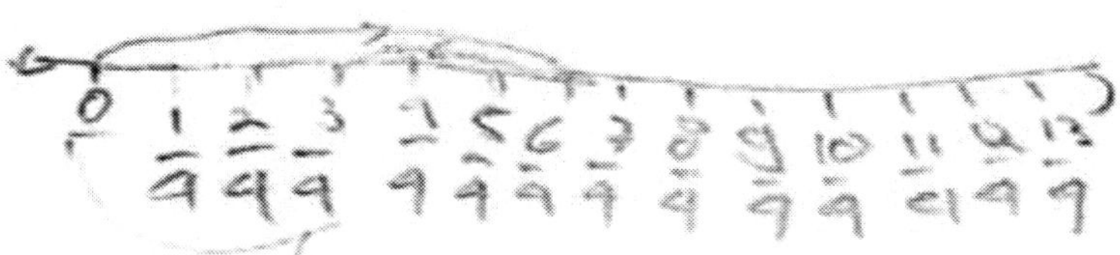

Figure 3. Subtraction number line

Student 1 draw the line well. divide it become 14 part but the denominator were 4 not 14. Student 1 showed the numerator by arrow. Student 1 didn't show the result because there is no arrow out. The result was wrong.

Figure 4. Subtraction on model

Student 1 draw the picture with 14 equal part. Student 1 mark the numerator. Student 1 got right answer.

Q : What did you draw ?
S1 : I draw the picture that represent the fraction.
Q : What did the mark mean on your picture ?
S1 : The mark showed the numerator and the blank show the denominator.

Journal of Physics: Conference Series **1417** (2019) 012050 doi:10.1088/1742-6596/1417/1/012050

3.2 Answer by Student 2

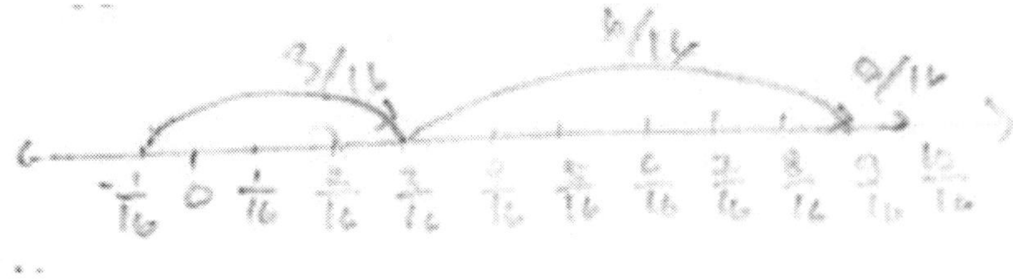

Figure 5. Addition on number line.

Student 2 draw the number line. Student 2 show the numerator by arrow with right direction. Student 2 was not show the result by arrow. There was no final result.

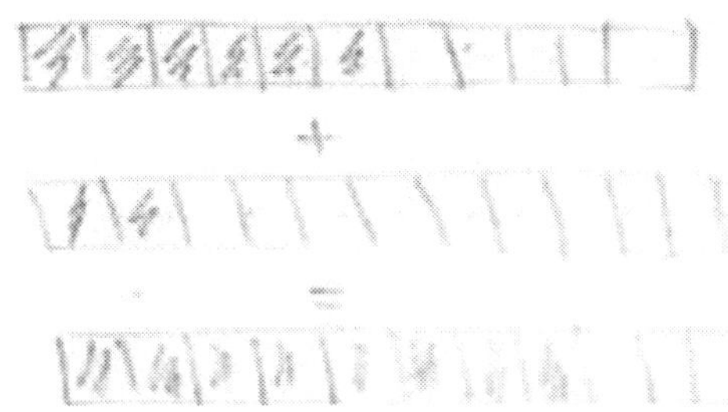

Figure 6. Addition on model.

Student 2 draw closed figure. Student 2 divide the figure become equal part. Student 2 mark the numerator then get right answer.

Student 2 didn't answer the question about subtraction on number line. From interview, Student 2 confused to draw the line and divide it. Confused to show the numerator and result.

Figure 7. Subtraction on model.

Student 2 draw the closed figure. Student 2 showed the denominator by divide it become 14 equal part although not draw it well. Student 2 showed the result of the operation well.

3.3 Answer by Student 3

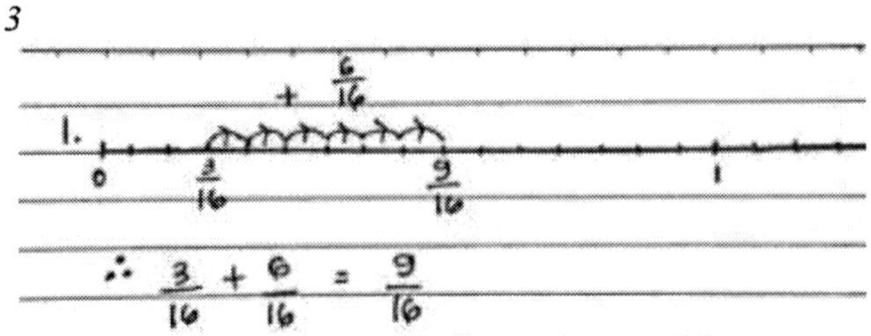

Figure 8. Addition on number line.

Student 3 draw the number line. Student 3 show the denominator by divide the line become 16 part but not placing the integers or number. Student 3 didn't show the numerator well because only show one fraction. Student 3 did not show the result well by arrow out.

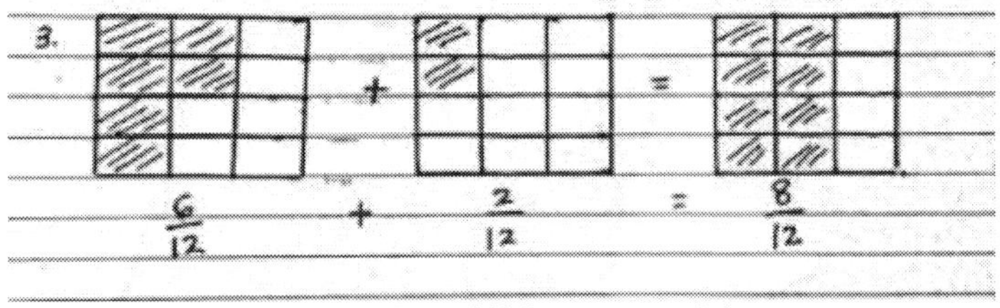

Figure 9. Addition on model.

Student 3 draw the picture well. Student 3 divide it became 12 equal part as denominator. Student 3 mark the numerator by concise. Student 3 got right answer on model representation.

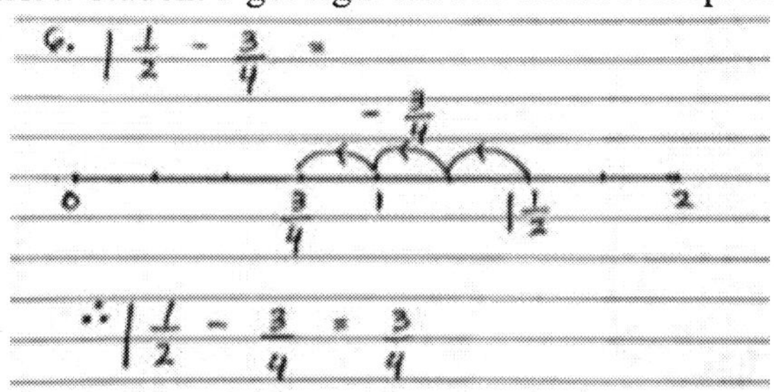

Figure 10. Subtraction on number line.

Student 3 draw the number line well. Student 3 did not place the number completely and well. Student 3 did not show the denominator and numerator well by arrow. Student 3 did not show the result well on number line by arrow.

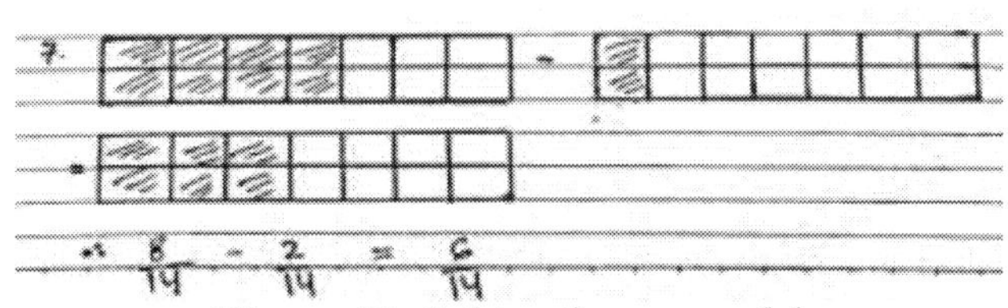

Figure 11. Subtraction on model

Student 3 draw the closed figure. Student 3 divide it become 14 equal parts to show the denominator. Student 3 mark the numerator well. Student 3 got the answer well.
Q : "How did you get the answer ?"
S3 : "First, i draw the figure, then divide the first picture in 14 equal part. Then i did the same on the second picture, i make the same denominator which is 14."
Q : "After that what is your step?"
S3 :"The last step is subtract the first picture by second picture and get the answer is the third picture".
The result from students 3 was contra from the result from research [11] which is the student failed to represent the picture or model.

4. Conclusion

Based on the analysis of the results of the above data obtained from the pre-test, post-test, and Based on the result and discussion, student did well on model representation on addition and subtraction. Student 1 did well on the model representation in addition and subtraction. But student 1 still have difficulty on number line in addition and subtraction. Student 1 confused to show the final result and denominator well. Student 2 did well on model representation both addition and subtraction. Student 2 did not answer the number line representation. It seems student 2 is confused and feel difficult to answer. Student 3 did well on solving model representation both addition and subtraction. Student 3 did not show the final number line well. The conclusion, they draw the closed figure and show the result well. Students still have difficulties on show the result on number line representation. There was student not answer the question. Almost student did not show the result well because not arrow out the final result. So, students still confused about fraction concept. Teacher should explain more about the concept of fraction and gave the exercise on any figure or multiple representations. Because representation will help the student in problem solving.

References

[1] National Council of Teachers of Mathematics 2000 *Principals and standards for School Mathematics* Reston: NCTM.

[2] Huinker, D. 2015 Representational competence: A renewed Focus for Classroom Practice in Mathematics *Wisconsin Teacher of Mathematics* 4-5.

[3] Hwang W,-Y., Chen, N.-S., Dung, J.-J., & Yang, Y.-L 2007 Multiple Representation Skills and Creativity Effects on Mathematical Problem Solving using a Multimedia Whiteboard System *Educational Technology & Society* **10**(2) 191.

[4] Fennel, F. 2001 *Representation: an important process for teaching and learning mathematics.* Teaching children mathematics.

[5] Van De Walle et al 2013 *Elementary and Middle School Mathematics Teaching Developmentally*

[6] Permendikbud no 24 2016.

[7] Novilis C. F 1976 An analysis of the fraction concept into hierarchy of selected subconcepts and testing the hierarchical decencies *Journal for research in Mathematics Education.*

[8] Kara, F., & Incikabi, L 2018 Sixth Grade Students Skills of Using Multiple Representations in addition and Subtraction Operations in Fractions *International Electronic Journal of Elementary Education.*

[9] Packenham, Patricia S.M 2012 Examining Pictorial Models and Virtual Manipulatives for Third-Grade Fraction Instruction. *Journal of Interactive Online Learning* **113**.

[10] Wong, Monica., & Evans, David 2007 Students' conceptual understanding of equivalent fractions. *Proceeding of the 30th annual conference of the mathematics education research group at Australia* **832**.

[11] E Juliangkary & R Johar 2018 Student's representation of fraction through ELPSA framework. *The 6th South East Asia Design Research International Conference.*

Students' Strategies to Solve Reversible Problems of Function: The Part of Reversible Thinking

S Maf'ulah[1], D Juniati[2]
[1]Department of Mathematics Education, STKIP PGRI Jombang, Indonesia
[2]Department of Mathematics Education, Universitas Negeri Surabaya, Indonesia

E-mail: syarifatul.m@gmail.com

Abstract. This study aimed to reveal students' strategies to solve reversible problems, particularly to function course. Individuals' competence to solve reversible problems is the part of reversible thinking. This way of thinking refers to a reversible two-way relationship. Providing reversible problems is a way for teachers to develop their students' reversible thinking. It was qualitative. 105 students who were prospective teachers of mathematics were selected as the subject of this study, and it took a test containing some reversible problems, such as function along with its graphic and between derivative and integral, as the instrument. The students' results were further analyzed to reveal their strategies of solving those reversible problems. The finding showed that among 105 students, 62% of them were capable to draw the graph of an identified function through 4 different strategies, while 28% of them were capable to define the function of an identified graphic through 5 different strategies. The significant difference between those two percentages implied that only few students were capable to both drawing the graphic of an identified function and defining the function of an identified graphic. Overall, it indicated that the students were still less in constructing a reversible relationship between function and its graphic.

1. Introduction

This study was inspired by Jean Piaget's thought on reversible thinking, and motivated by some references from published journals that described the importance of thinking reversibly for students in education field, especially those dealing with mathematics education field [1, 2, 3, 4, 5, 6, 7, 8]. Following Piaget [9], reversible thinking was an individual's mental competence to reverse his/her way of thinking to the initial state of his/her thought. Hackenberg [10] argued that reversible thinking dealt with reversing a scheme. One may not only involve inversion –a reversing action into the initial state, but also involving a compensation to reverse a scheme –taking new action to reverse things into a condition equivalent to the initial state. Wong [11] argued that reversible thinking referred to mental flexibility – individual's competence to anticipate and review in an analytical manner.

Thus, reversible thinking involved thinking for twice in the opposite aspect thus reducing the opportunity for someone to make a mistake in every decision making. Reversible thinking has a role when solving mathematical problems. When a person solves the mathematics problem with involving reversible thinking, he/she would use two solving processes from two opposite directions. In addition, he/she would also do looking back to check the result of problem solving was obtained by him/her. Thus, reversible thinking in mathematics learning should be considered and developed so that students' abilities in problem solving are maximized.

This study aimed to reveal students' strategies in solving reversible problems. The research focused on the subject matter of function. Function is one of mathematics courses that relates to other courses with higher level on their hierarchy, such as limits, continuity, derivative, integral, and the others dealing with function. Hence, students should be able to comprehensively understand the course. Denbel [12] argued that *function* was useful in every branch of mathematics, as it is a unifying concept in all mathematics.

Four reversible problems related to function were given. The first two of them were between function and its graph, and the second ones were between derivative and integral. First, the students were asked to draw the graphic of an identified function. Second, they were asked to define the function of an identified graphic. Third, they were asked to find the derivative of an identified function. Fourth, they were asked to define the function of an identified derivative. The first two problems aimed to reveal the students' competence in constructing a reversible relationship between function and graphic, while the second ones pointed to the reversible relationship between derivative and integral. Such competence was detected through the strategies they used, including the steps, to solving those problems.

2. Method

This study is qualitative. It took 105 students who were prospective mathematics teachers as the subject of the study. Furthermore, it took a test containing some reversible problems, such as function along with its graph and between derivative and integral, as the instrument. The students' results were then analysed to reveal the strategies they used in solving those problems.
This present study applied some procedures consisting of three primary phases as follow. (1) Preparation; in this stage, some theories of thinking reversible were viewed and examined. (2) Implementation; in this stage, the subject of this study was selected. Subsequently, the selected subject might have a test and interview respectively based on their work. (3) Analysis; in this stage, the data was analyzed and reported.

3. Result

The researchers gave a test containing some reversible problems to the students. The test was presented in Figure 1, as follow.

TEST
(Maximum time provided is 15 minutes)

Instruction: Complete the following problems clearly!

1. If $f(x) = 3x - 2$ with $x \in R$, draw the graphic of the function!
2. Look at the following graphic!

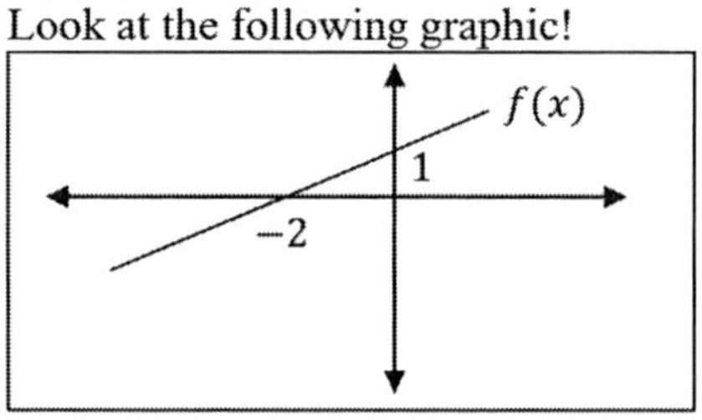

   Define the function $f(x)$ of the graphic!
3. If $f(x) = x^2 - 3x + 1$, define $f'(x)$!
4. If $f'(x) = 2x - 3$, define $f(x)$!

Figure 1. Test

The test as presented in Figure 1 was aimed to reveal the students' strategies in solving the problems. They were usually asked to draw the graphic of an identified function. This test, however, had them define the function of an identified graphic. In addition, they might usually be asked to define the derivative or integral of an identified function. Otherwise, this test had them to find the integral of an identified derivative of a function. This test was designed to see how students construct a reversible two-way relationship as their competence. 105 students who were all prospective teachers participated. The result was presented as follow.

Journal of Physics: Conference Series **1417** (2019) 012051 doi:10.1088/1742-6596/1417/1/012051

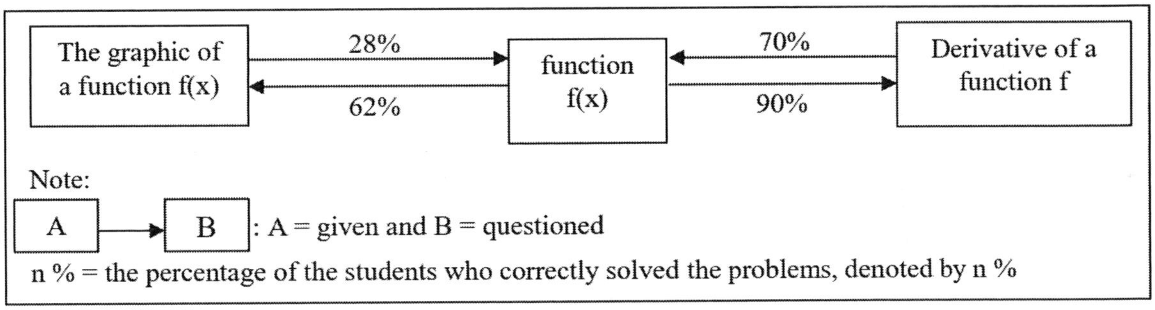

Figure 2. Result of Test

Based on Figure 2, it found the number of students (in percentage) who correctly solved the problems. The test result was then analyzed and classified based on the strategies they used for solving the problems. The following description showed the detail.

1. The students' strategies for drawing the graphic of an identified function.

The students' strategies for drawing the graphic of an identified function were as follow.

1) Two-point strategy without reviewing the final result

The steps of this strategy were as follow.

Step 1 : Defining two points of coordinate through the identified function; the coordinate that cut the axis X and the coordinate that cut the axis Y.

Step 2 : Drawing a graphic based on the coordinates that had been found in step 1 by directly connecting those two points of coordinate using straight line, as it was a linear function.

Step 3 : No review. This information was obtained from interview.

The following figure (i.e., Figure 3) presented an example of the student's work using problem solving through two-point strategy.

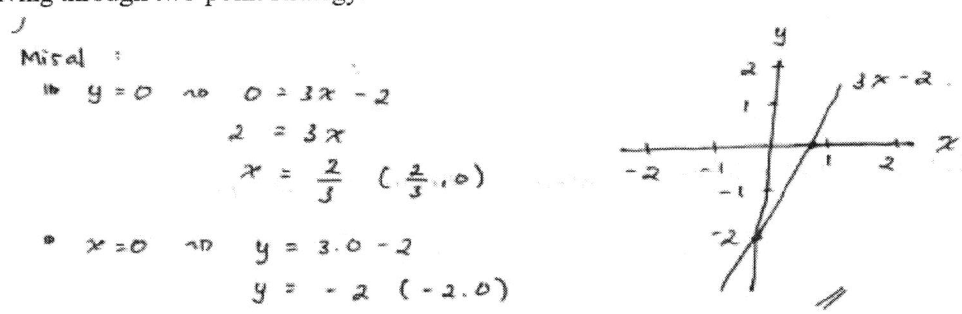

Figure 3. An Example of Problem Solving through Two-Point Strategy.

2) Many-Point Strategy with a Review on the Final Result by Reversing It to the Initial Problem

The steps of this strategy were as follow.

Step 1 : Defining some coordinates (at least 3 coordinates) from a given identified function.

Step 2 : Drawing the graphic based on the identified coordinates by connecting all of them through a straight line, as it was a linear function.

Step 3 : Reversing the result to the initial problem by substituting those identified coordinates (on the graphic) into the function. This information was based on the students' works and interview. A crossed graphic was found in their works. Furthermore, an interview was held to clarify the crossed graphic. The students reversed their work to its initial state by substituting those all points of coordinates on the graphic they have drawn.

The following figure (i.e., Figure 4) presented an example of the student's work using many-point strategy with a review on the final result by reversing the result to the initial problem.

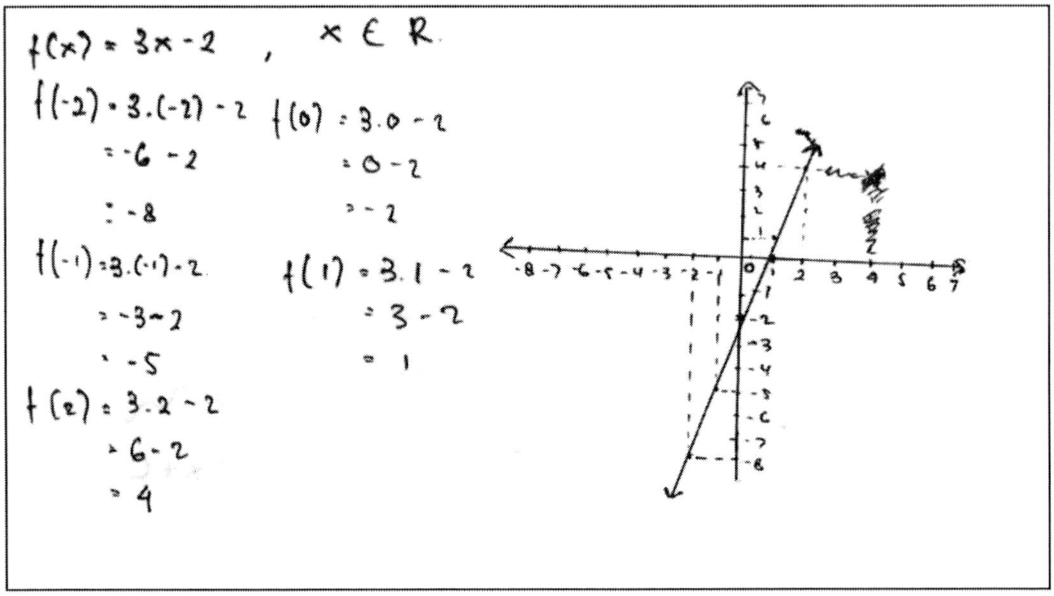

Figure 4: The Example of Many-Point Strategy with a Review on the Final Result by Reversing the Result to the Initial Problem

3) Many-Point Strategy with a Review on the Final Result by Recounting It
The steps of this strategy were as follow.
Step 1 : Defining some coordinates (at least 3 coordinates) from a given identified function.
Step 2 : Drawing a graphic based on the coordinates found in Step 1 by directly connecting those coordinates with a straight line, given that it was a linear function.
Step 3 : Recounting it. This information was obtained by interviewing the students.

The following figure (Figure 5) presented an example of students' work using many-point strategy with a review on the final result by recounting it.

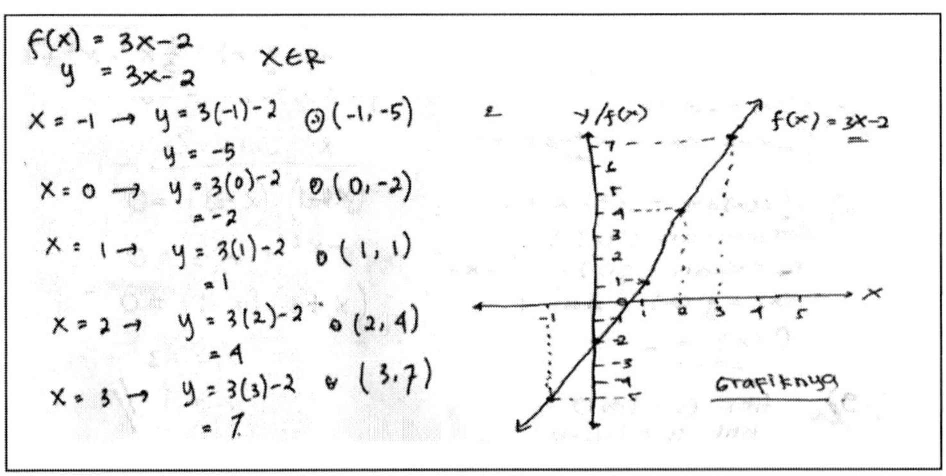

Figure 5. An Example of the Student's Work Using Many-Point Strategy with a Review on the Final Result by Recounting It

4) Many-Point Strategy with no Review
The steps of this strategy were as follow.
Step 1 : Defining some coordinates (at least 3 coordinates) from a given identified function.
Step 2 : Drawing a graphic based on the coordinates found in Step 1 by directly connecting those coordinates with a straight line, given that it was a linear function.
Step 3 : No Review. This information was based on the result of interview.

The following figure (Figure 6) presented an example of the student's work using many-point strategy with no review.

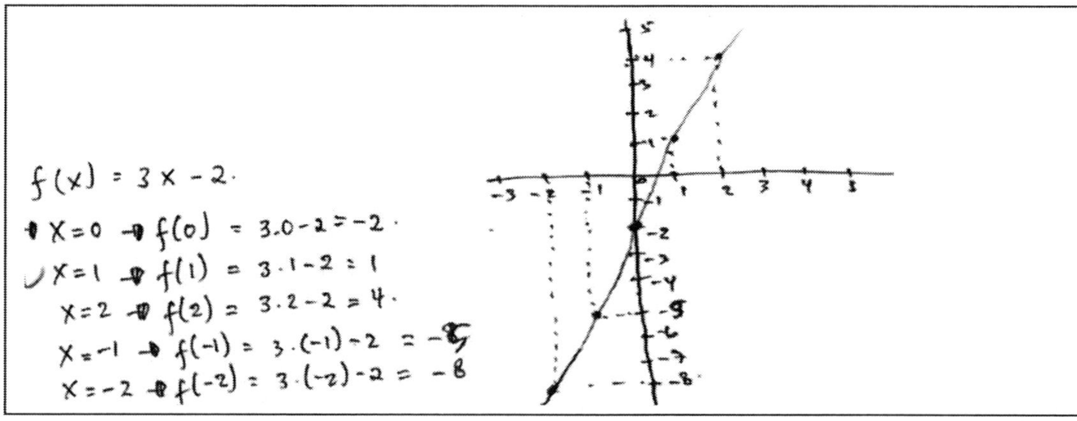

Figure 6. An Example of the Student's Work Using Many-Point Strategy with no Review

2. The students' strategies to define the function of an identified graphic
The students' strategies to define the function of an identified graphic were as follow.
1) Trial-and-Error Strategy with a Review on the Final Result by Reversing It to the Initial Problem
The steps of this strategy were as follow.
a. Defining some coordinates based on the identified graphic.
b. Defining the function of the graphic by having trial and error based on the coordinates identified in step 1. This information was based on the result of interview. The steps of having trial-and-error strategy by students were as follow.
  - Taking f(x)=ax+b as the formula as they got that, in this category, the graphic was in the form of a straight line. Therefore, it must be linear with f(x)=ax+b
  In case that the coordinate cutting the axis Y was (0,1), the f(x)=ax+1, as x=0
  - Defining the coefficient of X in such a way to get a function which graphic was equal to the identified one. Initially, the students wrote f(x)=-2x+1. However, as coordinate (-2,0) substituted to f(x)=-2x+1 resulted in f(-2)=-2(-2)+1=-3≠0, f(x)=-2x+1 was found inappropriate. Then, they wrote f(x)=-2x+1 and the substitution of coordinate (-2,0) to (-2,0) resulted in f(-2)=1/2 (-2)+1=0, they found that f(x)=1/2 x+1 was appropriate as the function of the given graphic.
c. Having a review. It was rechecking the correctness of the function by substituting the coordinates on the graphic into the function found. It was considered correct if the result was equal to the given graphic. this information was based on the result of interview.

The following figure (Figure 7) presented an example of the student's work using trial-and-error strategy with a review on the final result by reversing the result to the initial problem.

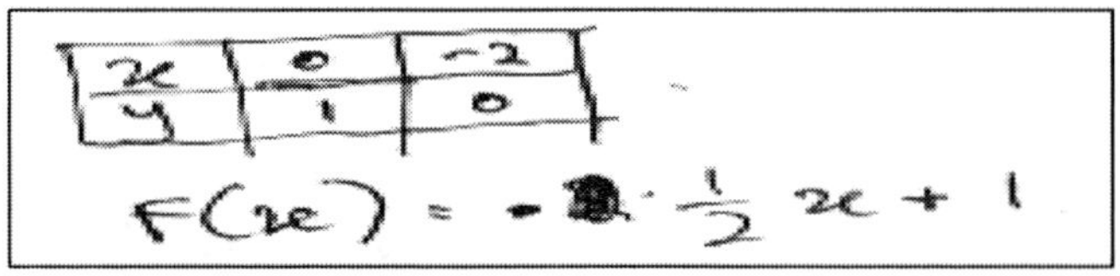

Figure 7. An Example of the Student's Work Using Trial-and-Error Strategy with a Review on the Final Result by Reversing It to the Initial Problem

2) Linear Equation Strategy with a Review on the Final Result by Reversing it to the Initial Problem

The steps of this strategy were as follow.

Step 1 : Defining some coordinates based on the given graphic.

Step 2 : Defining the function of the graphic using two identified coordinates and a formula of linear equation as follow.

$$\frac{y - y_1}{y_2 - y_1} = \frac{x - x_1}{x_2 - x_1}$$

It found $f(x) = \frac{1}{2}x + 1$

Step 3 : Having a review by substituting the coordinates on the graphic into the function. in case that the result was equal to the given graphic, the function was considered correct. This information was based on the result of interview.

The following figure (Figure 8) presented an example of the student's work using linear equation strategy with a review on the final result by reversing the result to the initial problem.

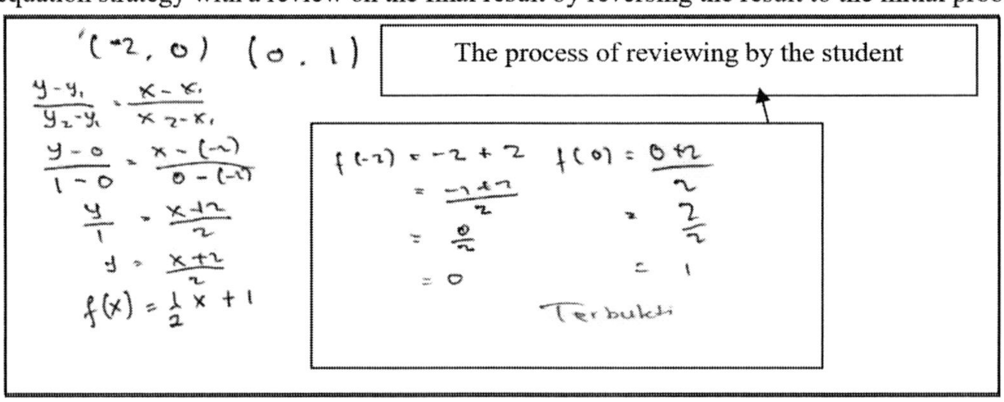

Figure 8. An Example of the Student's Work Using Linear Equation Strategy with a Review on the Final Result by Reversing it to the Initial Problem

3) Line Equation Strategy with no Review

The steps of this strategy were as follow.

Step 1 : Defining the coordinates based on the given graphic.

Step 2 : Defining the function of the given graphic using two coordinates and a formula of line equation as follow.

$$\frac{y - y_1}{y_2 - y_1} = \frac{x - x_1}{x_2 - x_1}$$

It found $f(x) = \frac{1}{2}x + 1$

Step 3 : No review. This information was based on the interview.

The following figure (Figure 9) presented an example of the student's work using line equation strategy with no review.

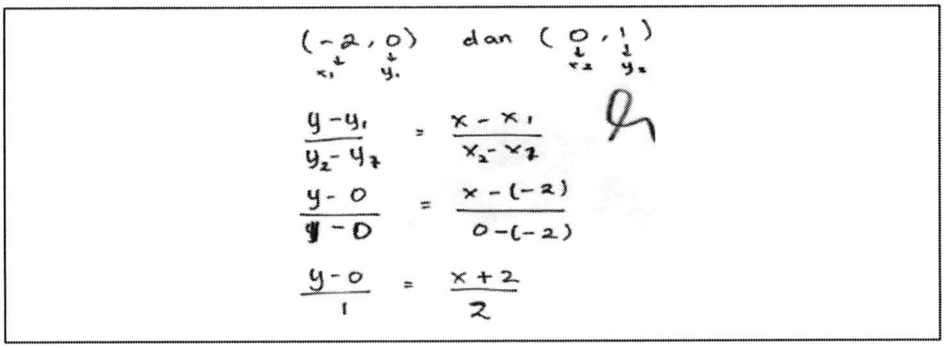

Figure 9. An Example of the Student's Work Using Line Equation Strategy with no Review

4) Linear Equation Strategy with no Review
The steps of this strategy were as follow.

Step 1 : Defining some coordinates based on the given graphic.

Step 2 : Writing a formula $f(x) = ax + b$, given that the identified graphic was in the form of a straight line, which indicated a linear function.

Step 3 : Defining the value of a and b by substituting the coordinates into function $f(x) = ax + b$, and it found $a = \frac{1}{2}$ and $b = 1$

Step 4 : Substituting $a = \frac{1}{2}$ and $b = 1$ into $f(x) = ax + b$, and thus it found $f(x) = \frac{1}{2}x + 1$

Step 5 : No review. This information was based on the interview.

The following figure (Figure 10) presented an example of the student's work using linear equation strategy.

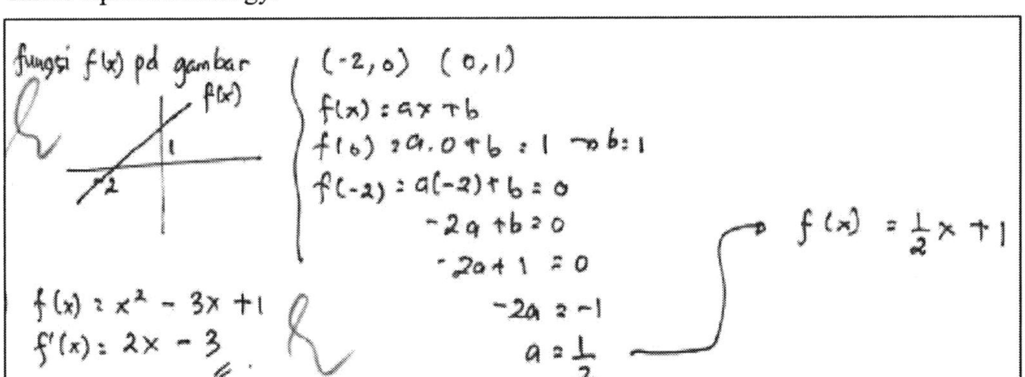

Figure 10. An Example of the Student's Work Using Linear Equation Strategy with no Review

5) Gradient Strategy with a Review on the Final Result by Recounting It
The steps of this strategy were as follow.

Step 1 : Defining some coordinates based on the given graphic.

Step 2 : Defining a gradient (i.e., coded by m) using formula $\frac{\Delta y}{\Delta x} = \frac{y_2 - y_1}{x_2 - x_1}$

Step 3 : Writing $f(x) = mx + c$, given that the identified graphic was in the form of a straight line, and thus, it must be linear.

Step 4 : Defining the value of c through the coordinates found, the gradient m, and $f(x) = mx + c$

Step 5 : Substituting gradient m and value c into $f(x) = mx + c$ in order to find a function of the given graphic.

Step 6 : Having a review by recounting. This information was based on the interview.

The following figure (Figure 11) presented an example of the student's work using gradient strategy with a review on the final result by recounting it.

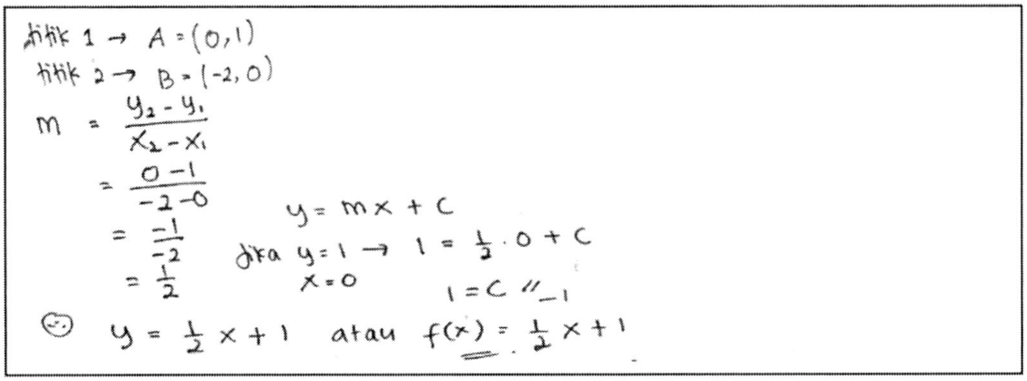

Figure 11. An Example of the Student's Work Using Gradient Strategy with a Review on the Final Result by Recounting It

3. The only strategy the students used to define the derivative of an identified function was through a derivative formula, "*If* $f(x) = ax^n$, *then* $f'(x) = nax^{n-1}$". Using this formula made them able define the derivative of $f(x)$ without having any review on their final result. Therefore, it was called derivative formula strategy. The following figure (Figure 12) presented an example of the student's work using this strategy.

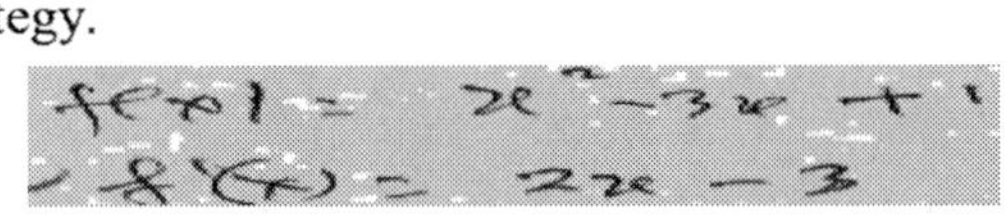

Figure 12. An Example of the Student's Work Using Derivative Formula Strategy

4. The only strategy the students used to define the function of an identified derivative was through an integral formula, "*if* $f'(x) = ax^n$, *then* $f(x) = \int \quad f'(x)dx = \frac{a}{n+1}x^{n+1}$. Using this formula made them able to define function f(x) without having any further review on their final result. Therefore, it was called integral formula strategy. The following figure (Figure 13) was an example of the student's work using this strategy.

Figure 13. An Example of the Student's Work Using Integral Formula Strategy.

In accordance to those all analyses, some information was collected as follow.

1. The students using two-point strategy with no review in drawing the graphic of an identified function were found using line equation formula with no review as well to define the function of the given graphic. However, some others decided to choose linear equation formula without having any review on their final result as their strategy.
2. In drawing the graphic of an identified function, the students who used many-point strategy with a review on their final result by reversing the result to the initial problem were found using line equation formula strategy as well without having any review on their result. In addition, some others decided to use linear equation with no review on their final result.
3. The students who used many-point strategy with a review on their final result by recounting it for drawing the graphic were found defining the function of an identified graphic using gradient strategy with a review on their final result by recounting it. Some others used line equation formula without having any review on their result.
4. The students who used many-point strategy with no review on their final result for drawing the graphic of an identified function were found defining the function of an identified graphic through

line equation formula with a review on their final result by reversing the result to the initial problem. Some others used linear equation formula with no review on their final result.

5. The students who used derivative formula in defining the derivative of an identified function were found defining the function of an identified derivative through integral formula.

Those five points were successively presented in Figure 14 and Figure 15, as follow.

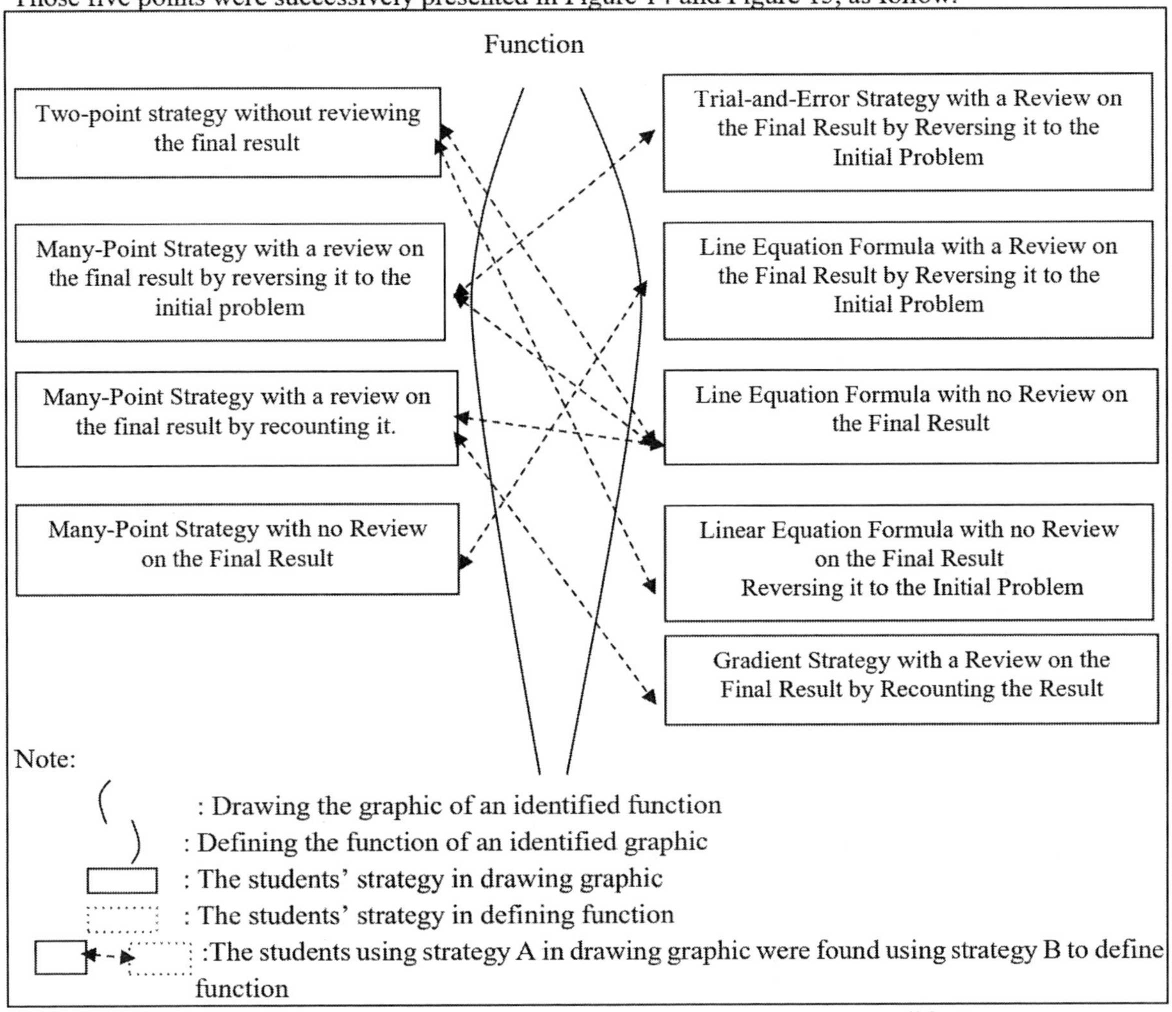

Figure 14. Scheme of the Students' Competence to Construct a Reversible Two-Way Relationship between Function and Its Graphic

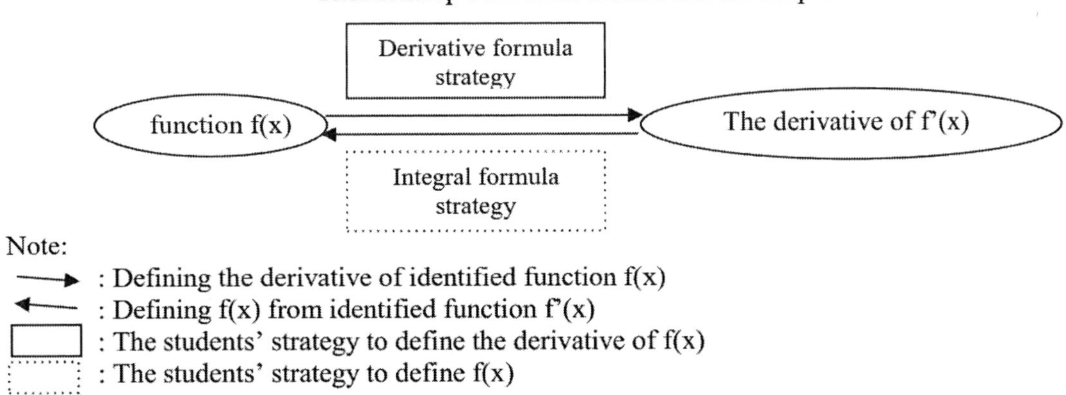

Figure 15. Scheme of the Students' Competence to Construct a Reversible Two-Way Relationship between Function and its Derivative

4. Discussion and Conclusion

This study aimed to reveal students' strategies to solve reversible problems, particularly to function course. This competence was detected through some strategies the students used in solving the given problems. They took some steps to complete their strategies. Finally, the conclusion was made on Table 1 as follow.

Table 1. Students' Strategies to Solve Reversible Problems of Function.

The students' strategy to draw the graphic of an identified function	The students' strategy to define the function of an identified graphic
Two-Point Strategy with no Review on the Final Result	Line equation formula with no review on the final result
	Linear equation formula with no review on the final result
Many-Point Strategy with a Review on the Final Result by Reversing the result to the Initial Problem	Trial-and-error strategy with a review on the final result by reversing the result to the initial problem
	Line equation formula with no review on the final result
Many-Point Strategy with a Review on the Final Result by Recounting the Result	Gradient strategy with a review on the final result by recounting the result.
	Line equation formula with no review on the final result
Many-Point Strategy with no Review on the Final Result	Line equation formula with a review on the final result by reversing the result to the initial problem
	Linear equation formula with no review on the final result

Based on Figure 2 that presented the students' test results, it found that 62% of 105 students were capable to draw a graphic of the given function through four different strategies, while 28% of them were capable to define a function of the given graphic through five different strategies. The significant difference between those two percentages implied that only few students were capable to both drawing the graphic of an identified function and defining the function of an identified graphic. Overall, it showed that students were still lacking in establishing reversible relationships between functions and its graphs. This was due to the fact that their reversible thinking development did not optimal yet on which can be seen from the competence to construct a reversible two-way relationship, as the opposite of one-way relationship which characterized on one single way [8]. Students with less competence in reversible thinking may not be able to define the function of the identified graph even though they are able to draw a graph. However, those who were able to define functions from identified graphs may have the ability to draw graphs from identified functions as well. This finding was consistent to Maf'ulah *et al.* [3] who took high school students as their research subject, especially their reversible thinking in solving function problems. That study found that among 123 high school students, only 5 students were capable to construct a reversible two-way relationship.

Acknowledgements

This article is one of the research results (Post-doctoral grant scheme) funded by DRPM. The researchers thank the Directorate of Research and Public Service, Directorate General of Research Reinforcement and Development, KEMENRISTEKDIKTI for funding this research. Thank also goes to the reviewer who gave feedback and suggestions.

References

[1] Denbel D G 2015 Functions in the secondary school mathematics curriculum *Journal of Education and Practice* **1** 77
[2] Flanders S T 2014 Investigating flexibility, reversibility, and multiple representations in a calculus environment *dissertation* Not Published
[3] Maf'ulah S, Juniati D, and Siswono T Y E 2016 Pupils' error on the concept of reversibility in solving arithmetic problems Educational Research and Reviews 18 1775

[4] Maf'ulah S, Juniati D and Siswono T Y E 2017 The aspects of reversible thinking in solving algebraic problems by an elementary student winning national olympiad medals in science *World Transactions on Engineering and Technology Education* **15** 189
[5] Lamon S J 2007 Rational numbers and proportional reasoning: Towards a theoretical framework for research In F K Lester (Ed.) (Charlotte NC: Information Age Publishing)
[6] Ramful A and Olive J 2008 Reversibility of thought: An instance in multiplicative tasks *The Journal of Mathematical Behavior* **27** 138
[7] Slavin R E 2006 *Educational psychology: theory and practice* (Boston: Allyn &Bacon)
[8] Krutetskii V A 1976 The Psychology of mathematical abilities in schoolchildren (Chicago: The University of Chicago Press)
[9] Inhelder B and Piaget J 1958 The Growth of Logical Thinking from Childhood to Adolescence (New York: Basic Books)
[10] Hackenberg A J 2010 Students' reasoning with reversible multiplicative relationships Cognition and Instruction 28 383
[11] Wong B 1977 The relationship between Piaget's concept of reversibility and arithmetic performance among second graders *ERIC Journal* **136** 962

Identification and Exploration of Elementary School Teacher Misconception in Mathematical Learning

E B Rahaju[1*], D Iriyani[2] and R Artiono[3]

[1,3] Department of Mathematics, Universitas Negeri Surabaya, Surabaya, Indonesia
[2] Department of Biology, Universitas Terbuka, Surabaya, Indonesia

*Email: endahrahaju@unesa.ac.id

Abstract. Mathematics is one of the compulsory subjects that must be taught to students at all levels of education. It creates students not only to have the ability to think logically, critically, analytically, creatively, and meticulously, but also carefully develop the ability to use mathematics in problem solving. Therefore, mathematics teachers are required to convey mathematical concepts correctly. Unfortunately, there are some elementary school teachers whose understanding of the concepts in mathematics learning is incorrect. The aim of this study was to identify the misconception of elementary school teachers in mathematics learning. This type of research was descriptive-qualitative research, where data had been analysed qualitatively. The research data were in the form of interviews and observations of the subjects, namely elementary school teachers in Sidoarjo district. Regarding to the data, there were several misconceptions in elementary school teachers, consisting of 1) Defining the apex point of the cone as the angle formed by the curved side; 2) Defining a cone surface as a side; 3) Defining a cylinder surface as a side, and 4) Defining the cone as a pyramid.

1. Introduction

Mathematics is one of the compulsory subjects that must be taught to students at all levels of education. It prepares students not only to have the ability to think logically, critically, analytically, creatively, accurately, thoroughly but also carefully develop the ability to use mathematics in problem solving [1]. Given the importance of the objectives of this learning, every mathematics teacher is required to be able to convey mathematical concepts correctly. Unfortunately, we still can find some of elementary school teachers who have misconceptions in mathematics. It can be caused by several things, including poor understanding, lack of depth about a concept in mathematics. It can happen not only when the elementary school teachers have undergone their professions as a teacher but also when the teachers take formal education. The understanding of concepts is very crucial for everyone [2]. A person's understanding of a particular concept is one of the indicators of that person's achievement in learning a material. Gagne, Briggs, and Wagner stated that mastery of concepts as well, broadly, and deeply as those possessed by experts in certain fields of study, enables one to apply them in various purposes including for teaching purposes [2].

A person's understanding develops since preschool and changes as a consequence of acquiring new experiences [3]. This allows one to have a different understanding of a concept in the end. For example, suppose someone understanding of the concept of saving money in a bank. Since primary education age, a student has to be taught with the concept of saving money in a bank. However, the difference in one's life experience, there is a shift in understanding of the concept of the bank itself. Similarly, when students learn the concept of area in mathematics. At the beginning, elementary students explore the area by using the number of small squares to calculate the area of a figure drawn in the book. Along with increasing school age that gives new experiences to students, junior high school students no longer need a small square to calculate an area, but enough to use the formulas that apply to each rectangular. While in high school students, the concept of area develops again with the new concept of integrals for an irregular shape.

In the case of introducing concepts in school, Bell states that conceptions in a person need to be clear and identified before they carry out the learning process in school [4]. This applies not only for students but also for teachers. The reasons for the importance of understanding and identifying conceptions are: (a) the conceptions of teachers or students are often not in line with scientific conceptions or conceptions of experts or often called misconceptions. (b) The conception of the teacher or student can influence, help or possibly hinder the understanding of other concepts or can even cause difficulties in learning.

The task of a teacher in the class is not only to teach where the teachers simply transfer information or knowledge to students, but also as a process of changing students' existing conceptions and which may not be in accordance with the opinions of experts [5]. It is believed that the conception owned by a teacher is correct first. If the initial concepts are wrong and those are taught to students; it will contribute to produce new concepts which are irrelevant. It becomes very important then to identify misconceptions among teachers, especially in elementary school teachers. Given their task to teach basic concepts to embed new concepts in higher classes.

The initial conception possessed by elementary school teachers before teaching needs to be known, so that new concepts to be taught by them can be accepted correctly, can be understood, make sense and be useful for students. Teachers who do not understand the basic concepts will not succeed in instilling the concepts in subsequent learning. This will result in reducing the achievement of student learning outcomes in school and in everyday life.

In accordance with the background, the formulation of research question in this study is: "How to identify and explore of elementary school teachers' misconceptions in mathematics learning?".

2. Misconception in Mathematics Learning

It is well-known that in mathematics learning there are many concepts that must not only be understood by students but also be used to learn new concepts. Every mathematical concept learned by students since elementary school will continue to develop until the higher education level. For example, the concept of rectangles; this concept has been introduced to elementary school-age students; and when the students continue their studies at higher institutions with major related to calculation, this rectangular concept will be re-used by these students. Such that, if someone has experienced a misconception in understanding a concept, then it can be ascertained that the person will have difficulty learning new concepts in higher classes.

Misconception itself occurs because of confusion or lack of knowledge about the initial concept which is a prerequisite for learning new concepts. Knowledge is formed by students in contact with the environment, challenges, and material learned [6]. It means, if in constructing knowledge, students experience a mismatch with the concept according to the experts, then the student will experience misconceptions. Misconceptions are deviations or misconceptions that are difficult to change and will take over a long period of time. According to Osborne & Wittrock states that when the conception is considered contrary to the accepted meaning in mathematics, then misconception has occurred [7]. Misconception refers to a concept that is not in accordance with scientific understanding or understanding received by experts in that field [6]. If it relates to learning mathematics, misconception can also be interpreted as a misunderstanding in understanding the material and occurs when students answer math questions.

3. Research Method

This research had been categorized as a descriptive-qualitative research. This study aimed to identify and explore elementary teacher misconceptions in learning mathematics. This is as revealed by Marsigit that descriptive research aims to describe situations and events [8]. Meanwhile, the data obtained were obtained from interviews and observations of the learning process carried out by research subjects. The research subjects were elementary school teachers in Sidoarjo. The elementary school teachers have been grouped according to their employment status and school accreditation. The research instrument used was an elementary mathematical concept mastery test and interview guidelines. This instrument was used to find out whether or not there was a misconception in elementary school teachers, what material in elementary mathematics learning that often occurs misconceptions and what causes these misconceptions. The concept mastery test sheet was used as a justification for misconceptions of elementary school teachers. Interview guidelines were used to determine the causes of misconceptions in learning mathematics. Data collection was carried out directly through tests, interviews and observation sheets of the implementation of learning in class.

4. Results and Discussion

Regarding to the data, several misconceptions were found made by the research subject as follows:

Table 1. Misconception of Elementary Teacher Students

No	Material	Number of teacher	Percentage (%)	Note
1.	Integer Operation	8	15,4	Teachers did not pay attention to the numbers operations, they only see the sequence of operations that lies in the problem.
2.	Application of Integer Operation (story problem)	15	28,8	Teachers were lack of distance concept.
3.	The using of learning media for integer operation	12	23,1	Teachers did something wrong with the procedure.
4.	Determine rational number between two rational numbers	4	7,7	Teachers were lack of information about the problem.
5.	Converting decimal number to rational number	18	34,6	Teachers made a mistake to multiply between some decimal numbers
6.	The application of perimeter and area (story problem)	7	13,5	Teachers made something wrong with the concept of perimeter and area of the plane.
7.	The square root operation	24	46,1	Teachers had problem to convert square root number to decimal number
8.	Cube net	23	44,2	Teachers only knew a cube net 4 out of 11
9.	The element of geometry shape	36	69,2	Teachers made a mistake to decide how many edges, vertex in cube and cylinder?
10.	The relationship between surface area and volume in Polyhedron	15	28,8	Teachers made a mistake to find the cube root
11.	The relationship between surface area and volume in 3D curved space	30	57,7	Teachers made a mistake to find an area of cone surface
12.	The relationship between surface area and volume in polyhedron (story problems)	25	48,1	Teachers made a mistake to find an area of polyhedron which have trapezium base

From the table above, it can be identified that there are still many elementary school teachers in Sidoarjo who have misconception of the material in mathematics learning. The most misconceptions are determining how many edges, vertex in cone and cylinder. In addition, the teacher also made many mistakes in determining the surface area of the cone. The teacher made a mistake in determining the area of a polyhedron that has a base in the form of a trapezoid.

Here are some examples of the results of the work of elementary school teachers in Sidoarjo.

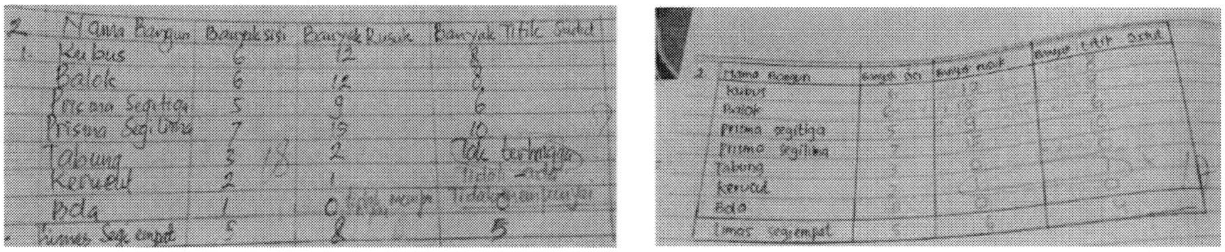

Figure 1. Misconceptions in the Element of Geometry Shape

Regard to the figure 1, it can be seen that teacher could not answer about how many edges and vertex in the element of geometry shape, specifically in cone and cylinder. It was about 36 teacher who made a mistake in this problem.

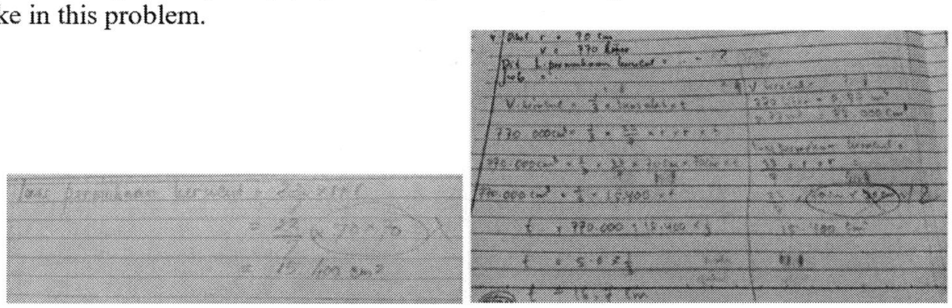

Figure 2. Misconceptions in the area of cone surface

Regard to the figure 2, it can be seen that teacher could not answer about the area of cone surface. Most of the teacher have misconception between the area of cone surface and the area of cone surface without its base. It was about 30 teachers whom made a mistake in this problem.

5. Conclusion

From the data obtained and the results of data analysis, it could be concluded that misconceptions of mathematical materials are still found in elementary school teachers in Sidoarjo district. This material is spread not only on algebraic material but also on geometry. In particular, the teachers had difficulty in determining the elements of geometry shape. Based on the results of this study, further research is needed to address the problems of misconceptions that occur in elementary school teachers in Sidoarjo district. This is considering the role of these teachers in introducing the basic concepts of mathematics to elementary school students.

6. References

[1] Yuanita P, Zulnaidi H, and Zakaria E 2018 *PloS one* **13** 1
[2] Ibrahim M et al 2012 *Seri pembelajaran inovatif konsep, miskonsepsi, dan cara pembelajaran.* (Surabaya: Unesa Press)
[3] Dahar R W 2011 *Teori-teori belajar dan pembelajaran* (Jakarta: Erlangga)
[4] Bell B F 1995 *Children's Science, Constructivism and Learning in Science* (Deakin University: Australia)
[5] Dykstra D I, Boyle C F and Monarch I A 1992 *J. Res. Sci. Teach.* **76** 615
[6] Suparno P 2005 *Miskonsepsi dan perubahan Konsep dalam Pendidikan Fisika* (Jakarta: Grasindo)
[7] Egodawatte G 2011 *Secondary School Student's Misconceptions in Algebra* (Toronto: University of Toronto)
[8] Marsigit 2012 Program Pascasarjana UNY Kajian Penelitian *Rev. J. Int.*

Journal of Physics: Conference Series **1417** (2019) 012053 doi:10.1088/1742-6596/1417/1/012053

Student's Geometric Thinking in Understanding Volume with Three–Dimensional Images of Cubes and Nets

H Setiawati*, D Juniati, S Khabibah

Postgraduate Program of Mathematics Education, Postgraduate, Surabaya State University, East Java, Indonesia

*Email: henysetiawati68@gmail.com

Abstract. This study aimed at describing the thinking skills of junior high school students in solving problems about the volume of cube and cuboid presented in the form of images of nets and solid figure geometry. This research was a descriptive study with a qualitative approach. This research was conducted at eighth grade class in Islamic Junior High School (MTs) Nurul Huda, Sidoarjo, Indonesia. The research subjects were two students, male and female in the same class of seven classes. The instruments used in this study were geometry test and interview. Geometry test was designed to determine student's thinking in understanding volume with three-dimensional images of cubes and nets. Interview was used to make the researcher easier to explore how the students completed the given geometry test on the problem of the volume of cube and cuboid. Student's thinking ability was explained based on three abilities according to Sternberg. The results showed that the female student had analytic and creative abilities in understanding images for the volume of cube and cuboid, but the male student tended to be more thoughtful towards abstract concepts. For practical ability, the male student had practical ability better than the female student.

1. Introduction

This study investigates the relationship between gender and students' thinking abilities in understanding the volume presented with three-dimensional images of cubes and nets of space. Image presentation is a solution to support the goal of problem solving in mathematics, so that it can help shape students' reasoning thinking. Facts that occur during this time students are still experiencing difficulties in mathematics, mathematical problems that are used not as problems from real life, so that the reality and situations of reasoning students are forced, or students do not have the right meta-cognitive skills to solve problems. Liljedahl [1] argued that problem solving not only requires logical and deductive reasoning but also requires extra-logical processes of creativity, insight and illumination to produce solutions. The fact that has happened so far is that students still have difficulties in mathematics specially in problem solving. Students often experience difficulties with word problems used in mathematical statements, mathematical problems that are used not as real-life problems. Such obstacles are often ignored, so the reality and situation are forced, or students do not have the right meta-cognitive skill to solve the problem [2]. In addition, Verschaffel [3] said that the most reported difficulty is that students base their analysis and calculations on a loose association of certain salient quantitative elements of the problem situation with certain mathematical operations. Based on Pattimukay's research results [4], there are also still many prospective teachers who find it difficult to teach problem solving to their students.

Solso [5] states that, "Problems solving is thinking that is directed toward the solving of a specific problem that involves both the formation of responses and the selection among possible responses." Thus, the understanding of mathematical problem solving in this research is the effort of students to solve mathematical problems with respect to logic and problems in calculations.

There is research evidence showing that the representation of the shape of solid figure geometry can have various impacts on the students' reasoning process. Parzysz in Pittalis [6] says that students' reasoning problems can be influenced by representations of the shape image of solid figure geometry and the image of the plane figure will be displayed. According to Mesquita in Pittalis [6] presentation of image-shaped tasks will provide support for geometry intuition, which in some situations can be very strong in helping individuals "to understand the relationship between geometric objects". The use of images to represent situation problems has a better chance of safeguarding students in solving mindset problems than returning to answers to mindset. Images are more easily associated with real life situations, and can be said to feel more authentic for students, and therefore it is possible that the language of images better illustrates problem situations so that it can reduce the difficulty of understanding the text. A good understanding of text is needed by students in building concepts both in mathematics and other subjects. The ability to understand concepts plays a major role in determining student's learning outcomes in mathematics learning. With the abilities possessed by students, this will facilitate learning and solving mathematical problems.

A number of research studies in mathematics education provide evidence from various types of images, which are used in Indonesia in different situations and with different effects on students' mathematical performance. Student's thinking ability to understand, interpret, and build representations of the shape of solid figure geometry will involve stimulation of good performance. However, this does not always mean that students conceptualize the information presented to them. The study of Hanifah[7] showed that In terms of the sex differences, male students imagined the figures yet female students needed the media to solve the problem about spatial.

Thinking can be defined as the process of producing new mental representations through information transformation involving complex interactions among mental attributes such as abstraction assessment, imagination, and problem solving. Solso [5] reveals that, " thought processes occur new mental representation is formed through the transformation of information by a complex interaction of mental attributes of judging, abstracting, reasoning, imagining, and problem solving". In this study, thinking ability is the ability to process the information received and turn it into an idea that produces a decision.

Sternberg [8] in his theory "successful intelligence" states there are three types of abilities, namely analytic, creative, and practical. Student's analytic ability means investigating namely their ability to analyze, compare, evaluate, and assess. Furthermore, student's creative ability means showing namely student's ability to make, find, imagine, and predict. And student's practical ability means using that is the ability of students to apply, use, practice, apply, employ and present practically what they know.

The expected benefits of the results of this study are to find out the extent to which male and female students' thinking abilities based on images of nets and solid figure geometry for the concept of cube and cuboid volume, so that they can be taken into consideration in preparing learning assignments. In addition, it is also as a reference in preparing students' logical thinking regardless of gender differences which serves as the basics in mathematical concepts besides cube and cuboid so that the learning objectives are expected to be achieved.

In this study, the researchers chose to examine the students' thinking ability of class VIII in solid figure geometry, namely cube and cuboid in the form of images of nets and solid figure geometry. So far, the class solves problems in the form of story questions about volume and students are used to using the existing formulas without knowing how to find out the concept of volume by themselves. The purpose of this study is describing the thinking ability between male and female students in solving volume problems in the form of images of nets and solid figure geometry.

2. Research Method

This type of research was descriptive research with a qualitative approach. The research subjects were the students of eight grade class of Islamic Junior High School (MTs) Nurul Huda, Sidoarjo, Indonesia. The subjects in this study were one male student and one female student. The reasons for choosing these two subjects were to represent each student's thinking ability and as a comparison between male student and the other. Information about the selection of subjects based on this commensurate level of ability was got through an interview with the mathematics teacher who taught eight grade class. The mathematics teacher knew more the level of daily ability of students in the learning process which was supported by daily math test scores.

The instruments used in this study were geometry tests and interview. The geometry test was a written test designed to identify student's thinking ability in solving geometry problems especially cube and cuboid. Interview is designed to make the researchers easier to explore information about student's thinking on geometry test given in understanding the concepts of cube and cuboid. The type of interview used in this study was an unstructured interview. Unstructured interview is a free interview where researchers do not use interview guidelines that are arranged systematically and completely for data collection.

The data collection technique used was written test in the form of images of nets and solid figure geometry that was given to the subjects to be worked on, then interviews were conducted to the subjects. The written test used consisted of three items. Data analysis was carried out by comparing the geometry test result and interview data. Thinking ability was described based on three types of abilities according to Sternberg, namely analytic, creative, and practical abilities by considering the results of interviews with the subject.

3. Results and Discussion

3.1 Results

This research was conducted at MTs. Nurul Huda Sidoarjo to eight grade class students by taking a population of eighth-level students is thirty students. The researchers selected two subjects consisting of male and female student from 12 male students and 18 female students in that class. Selection of class and subjects were based on mathematical abilities that are commensurate given by the math teacher who taught eigth grade class.

The results of student work in problem solving based on images of nets and geometry can be seen from the results of the work of the two students as follows.

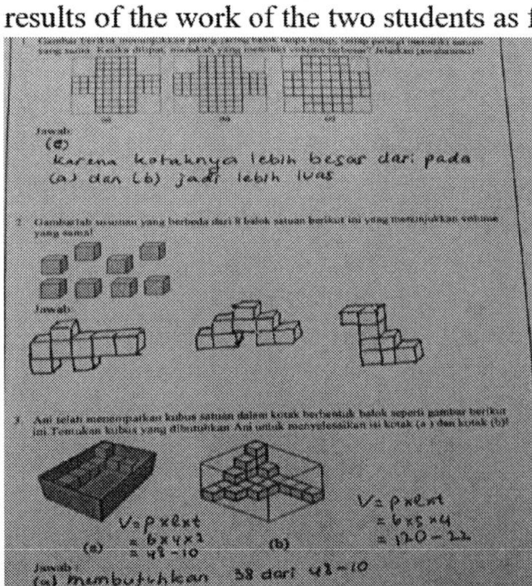

Figure 1. Male student's answer (S1)

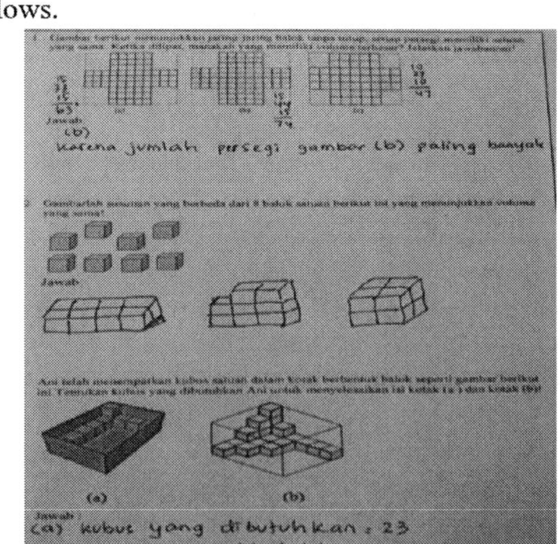

Figure 2. Female student's answer (S2)

3.1.1 Male student's

The male student's answer in determining the volume of the net image (question no.1) was that student's thinking ability was influenced by many considerations in a more detailed direction. It means that it was analytic ability is the ability to compare, assess but without evaluating the problem namely without paying attention to the statement of the problem. For question no. 2 the answer of male student was thinking more abstractly in drawing a set of cube unit. It means that it was creative ability namely making, discovering and imagining. However, in terms of imagining, this student could not imagine whether a cube arrangement could be arranged without a base under the cube unit. While for question no. 3 determining the cubes that were needed to fill the cuboid for practical ability namely using, practicing, applying what they knew with the cube unit of male student's answer showing thinking using the concept of volume obtained from prior knowledge.

To know better the thinking ability of the male student (S1) in solving problems given, below are excerpts of interview with S1

P : After you read the question number 1, what did you do?
S1 : observing the area of a square by looking at the large square of the image, then comparing.
P : Why should you compare?
S1 : To choose the largest area.
P : Then what steps did you take?
S1 : Answering what was asked.
P : Then for question no. 2, what did you imagine?
S1 : Arranging 8 unit cubes
P : What kind of arrangement did you have in mind?
S1 : Interrelated arrangements of the 8 cubes.
P : What do you do if real cubes can be arranged, not fall?
S1 : I don't think the cubes are real.
P : For question no. 3, how did you determine which cubes were needed for the cuboid?
S1 : By counting many cubes for length, width, height and then multiplying continuously minus the many cubes that are already in the beam.

3.1.2 Female student's

Based on the answer of the female student in determining the volume of the net images (question no.1), student's thinking ability by comparing images with others and evaluating what had been answered, it meant that the analytic ability of the female student was the ability to compare and evaluate the answers to the question. For question no. 2, the answer of the female student was simple thinking in drawing a set of unit cubes. It meant that it was the creative ability of making, finding and imagining unit cubes that had to be used so that cubes arrangement could be arranged with a sturdy base. Question no. 3 was determining the cubes needed to fill the cuboid in practical ability of using, practicing, applying what they knew with eight cubes. The answer of the female student used a count of what she saw from the image so that the main factor of accuracy was used.

To find out more about the thinking ability of the female student (S2) in solving problems given, the following is the excerpts of interview with S2

P : After you read the question number 1, what did you do?
S2 : comparing by counting the number of squares of the picture, then rating the number.
P : Why should it be calculated?
S2 : To choose the most.
P : What steps did you take?
S2 : Answer what was asked.
P : Then for question no. 2, what did you imagine?
S2 : Arrangement of unit cubes consisting of eight cubes.
P : What kind of arrangement did you have in mind?
S2 : Neat and not easily broken.

P : Why did you put together neatly?
S2 : So that the cube in the topmost arrangement did not fall down.
P : For question no. 3, how did you determine the cubes needed for the cuboid?
S2 : By observing the image then counting the number of cubes needed.

3.2 Discussion

This study showed the ability of male and female students to solve problems based on images of nets and solid figure geometry. The results showed that based on the answers of male and female students above (Figures 1 and 2), the thinking ability of male and female students was different on the answers generated in terms of analytic ability, creative ability and the ability to use previously known concepts. In question no. 1 the male student, in thinking ability in term of analytic had not evaluated what he had answered and evaluated the statement of the question, while based on the answer of the female student, the thinking ability for analytic started from comparing one image to another, assessing and evaluating the answers to what she wrote with a statement from the question, by calculating the number of unit squares and using the square area concept and the answer already showed the correct one.

In question no. 2 for male and female students, the creative thinking ability showed that in terms of making, arranging and imagining, the male student tended to think abstractly, while the thinking ability in term of the creative one, the female student was more real.

For question no. 3 the answer of the male student showed that the thinking ability was using the concept that had been obtained before so that he could answer easily, while the female student was by counting the number of cubes in the cuboid image without thinking using the concept of volume.

It shows that thinking ability between male and female students is really different. This also corresponds to the results from Pullo's study [9], which showed that visualization process on female student: aspects of generation, inspection, transformation, and use have been carried out by the subject in solving geometry problems, while the male student only does three aspects: aspects of generation, inspection and transformation in solving geometry problems. Fujita et.al [10] also showed that they found different types of thinking when the students solved 3D geometric problems.

4. Conclusion

The results of research on student's thinking ability to solve the problem of volume of cubes and cuboids based on the image of solid figure geometry show that the thinking ability of male student for analytical and creative abilities has been able to compare, arrange but not in the direction referred to in the basic concept of cube and cuboid volume, namely in determining the volume of net images he does not understand the intent of the statement so that he only observes the image and in arranging the unit cube he still leads to the abstract direction without evaluating the statement of the problem. Furthermore, the thinking ability of male student is already capable in practical ability, namely having applied and used the previous knowledge owned.

The female student's thinking ability shows that analytical and creative abilities have been able to solve problems of images of nets and solid figure geometry, but in practical ability she has not used the previous concepts owned. It is different from the answers of male student because the female student uses counts both in determining the volume of the net images and the volume of cuboid with unit cubes. However, in arranging unit cuboid, the female student thinks that the resulted image can be in the form of a solid figure geometry that is formed and which can stand sturdily as it is done when compiling a pile of books or toy cubes in daily reality.

For other researchers who are interested in studying the student's mathematical thinking ability in solving research problems it can be continued with the use of wider problem solving than just geometry tests in the form of images. It can be 'HOTS' questions, story problems, using everyday problems or contextual with another material. Planning of problem-solving tasks helps students understand mathematical concepts. Research can also be done at senior high school level. It is also hoped that further research can accommodate students 'abilities while conducting research using subjects based on students' cognitive abilities.

5. References

[1] Liljedahl P, Santos-Trigo M, Malaspina U and Bruder R 2016 *Problem Solving in Mathematics Education ICME-13 Topical Surveys* (Hamburg: Springer Open)
[2] Hoogland K, Pepin B, Bakker A, de Koning J and Gravemeijer K 2016 *Stud. in Educ. Evaluation* **50** 22
[3] Verschaffel L, Greer B and de Corte E 2000 *Educ. Stud. in Math.* **42** 211
[4] Pattimukay N, Juniati D and Budiarto M T 2018 *J. Phys. Conf. Ser.* **974** 012026
[5] Solso R L 1995 *Cognitive Psychology* (Nevada: University of Nevada)
[6] Pittalis M and Christou C 2013 *The J. of Math. Behav.* **32** 673
[7] Hanifah U, Juniati D, Siswono T Y E 2018 *J. Phys. Conf. Ser.* **947** 012014
[8] Sternberg R J and Grigorenko E L 2004 *Theory into Prac.* **43** 274
[9] Puloo M M L, Juniati D and Wijayanti P 2018 *J. Phys. Conf. Ser.* **1108** 012063
[10] Fujita T, Kondo Y, Kumakura H and Kunimune S 2017 *J. of Math. Behav.* **46** 11

Relationship of Teacher's Content Knowledge on Fraction Topic Toward Student Performance

A H W Kutub*, P Wijayanti and Manuharawati

Faculty of Mathematics and Natural Sciences, Universitas Negeri Surabaya, Indonesia

*Email: akh.17070785042@mhs.unesa.ac.id

Abstract. The aim of this study was to describe the relationship between teacher content knowledge and student performance in Fraction. This study used the mixed method that provided Content Knowledge test for fifteen teachers who taught the seventh class. The test consists of multiple choices to measure facts and procedures, knowledge of concepts and relationships, and knowledge of generalization. The study was divided into two phase, Quantitative phase and Qualitative phase. The Quantitative phase, teachers were mapped in two different levels, low and high based on score of Content Knowledge test. Two samples were classified into low level and high level. Furthermore, two groups/classes of the teacher samples were analyzed using Chi-Square test to investigate the relationship between teachers and student performance. The Qualitative phase of the study were included into three stages: (1) teacher interview, and (2) Student Problem solving, (3) description of the relationship of teacher's content knowledge and student learning performance. The results of the study showed that the Teachers content knowledge in Fraction were affected to student performance, in addition, there were also similarities in the pattern of student performance tests on fraction and the teacher's answers based on the results of the CK test.

1. Introduction

Fractions are one of the richest and most complex topics in mathematics teaching programs [1], and are considered to be one of the most difficult to teach, most cognitive and most important mathematical topics for advanced mathematics [2,3], especially for elementary school students [4]. This is also confirmed by research by numerous studies regarding the low performance of students in algebraic problems involving fractions [5,6]

Teacher knowledge affects to the student performance and achievement [7,8], even becoming one of the most influential factors in student learning [9]. Thus, teachers are expected to be able to teach effectively. The teacher does not only to be a master on the material but also be able to explain using various methods including Get hand On, Use Visuals, Get the Games Out, etc.

Yeo [10] reported that a teacher cannot be expected to be able to explain mathematical concepts if he does not have a complete understanding of the mathematical concepts. A success learning depends on the teacher's content knowledge [11], teachers with a limited content knowledge, have a fewer opportunities to influence student performance than those who conceptually understand. In other words, teacher's mastery of subject matter is very important in the learning system.

However, knowledge of matter alone is not enough, "Saving knowledge of a subject matter is not enough to teach it" [12]. Teachers need knowledge that can be used to transform the knowledge of content (content knowledge) into effective learning, teachers able to present material effectively will help to develop student's knowledge [12].

A number of experts have been developed several works focusing on the teacher knowledge in different domains, such as the definition of numbers [13], Algebra [14], Geometry and Measurement [15], Statistics [16], Ratios and Comparisons [17] and Fractions [18]. However, none of these studies analyzed the relationship between teacher knowledge and student performance.

Other research pathways, such as those carried out by Tchoshanov, [11,19]; Campbell et al., [20] Hill, Rowan & Ball, [21], Olfos R et al., [18] Trobst S et al. [22], provided an alternative that specifically addressed to the relationship between various types of teacher knowledge and student achievement. In general, research on the relationship of teacher and student knowledge is divided into two categories: quantitative research [11,23] and qualitative research [18, 20-22] with indicators of teacher knowledge are CK and or PCK.

These studies indicated that teacher's knowledge has an impact on student achievement or performance, but which teacher's knowledge? Therefore, this study departs from the assumption that teacher's content knowledge, has an impact on the student performance. Therefore, the question of this study is, (1) is there a relationship between teacher's knowledge of certain content on the student performance? And, (2) what is the relationship?

2. Theoretical Background

2.1 Teacher Knowledge

Shulman [12] defines teacher knowledge as a framework of understanding and skills, tools and values, character and performance that shape teaching abilities, which later in 1987, Shulman shared teacher knowledge in seven categories of basic knowledge that must be possessed by a teacher: (1) Content Knowledge, (2) General Pedagogical Knowledge, (3) Curricular Knowledge, (4) Pedagogical Content Knowledge, (5) Knowledge of learners and their Characteristics, (6) Knowledge of Educational Contexts, and (7) Knowledge of purpose , educational purposes and educational values [24].

These seven basic abilities are the basis of research development in the following years. However, from the seven categories of teacher knowledge, the category of content knowledge (CK) and pedagogical content knowledge (PCK) received much attention from a number of experts on the development of teacher knowledge research after 1987, because CK and PCK were the main parts of learning, this also Shulman emphasized that, PCK is a differentiator of teacher professionalism in learning [19,25].

In general, teacher knowledge research is mapped in 3 categories: (1) Teacher knowledge in certain categories (PCK, CK, MCK, etc.), this research was initiated by Shulman and consistently developed by Hill and Ball (Tchoshanov, 2011). In addition, Rowland, Huckstep and Thwaites [26], and Blomeke S et al., [27], In their research, they emphasized the importance of various aspects of teacher knowledge in certain categories, such as knowledge about teaching, teacher knowledge about learning content, knowledge of pedagogical content and others. (2) Teacher's knowledge is related to certain material, such as: Understanding Numbers [13], Algebra [14]; Geometry and Measurement [15], Statistics [16], Ratios and Comparisons [17] and Fractions [18,19]. (3) Teacher's knowledge and relation to student performance. There are at least three names that consistently conduct research and this category, Tchoshanov [11,19] and Campbell [20] and Depaepe, F [22,25].

2.2 Teacher Content Knowledge

Success in learning depends on teacher's content knowledge [11]. Teachers who have limited content knowledge have fewer opportunities to influence student achievement than teachers who conceptually understand material/content. Content Knowledge is defined as "the amount and organization of knowledge per se in the mind of the teacher" or the amount of knowledge contained in a teacher [12]. Such knowledge includes, (1) Subject Matter Knowledge (SMCK) or knowledge of content in the subject matter, (2) Knowledge of pedagogical content, and (3) Knowledge of the curriculum. This involves conceptual knowledge (knowledge of concepts, principles and definitions) and procedural knowledge (knowledge of procedures and problem solving), why this statement is true or false and how knowledge is constructed [11].

Campbell et al., [20] defined Content Knowledge as knowledge related to mastery of material, content or mathematical content; whereas Pedagogical Content Knowledge is knowledge integrated teacher knowledge of certain material and how to integrate it into learning, thus the material is easy to understand and can improve student knowledge. This statement is slightly different from what was stated by Shulman [12] which PCK is part of CK. In contrast to Campbell et al., Olfos et al., [18] provide different research alternatives, to analyze more deeply the specific knowledge of mathematics of teachers and teacher teaching knowledge, Olfos et al., Refers to Vergnaud's conceptual theory which divides CK teachers in two categories, (1) Conceptual content Knowledge (Cck) and (2) Representational Knowledge (RK).

Ball, Hill, and colleagues [21,28] analyzed teacher mathematics knowledge for teaching (Mathematical Knowledge of teaching (MKT)) and its effect on teaching quality and student achievement. In his research, CK was mapped in three categories of knowledge, namely (1) knowledge of general content, (2) knowledge of specific content, and (3) knowledge of content horizon. While for PCK, mapped in 3 categories, knowledge, namely content and student knowledge, content and teaching, and content and curriculum [28].

Tchoshanov [11] used the word cognitive demand as a function of teacher content knowledge (p.145) related to the many opportunities for learning and teaching in the classroom. In the aspect of learning, how much thought is needed in the classroom, or how far the subject matter knowledge of the teacher in the teaching material in the class. while in the teaching aspect, which type of teacher knowledge is needed to support students' thinking in the classroom.

Therefore, Tchoshanov [11,19] used cognitive type as an instrument to measure teacher knowledge content. Cognitive classification used in this study builds the existing research and consists of knowledge of facts and procedures (for example, procedural knowledge), knowledge of concepts and connections (for example, conceptual knowledge), and knowledge of models and generalizations (for example, generalized knowledge).

3. Research Method

3.1 Research Design

This study consisted two phases, quantitative and qualitative. The Quantitative phase is related to the connection performance of each group from the class taught by the selected teachers. The qualitative phase included two stage, (1) collect data from teachers, analyze it and then determine the important point for the interviewed teacher, and (2) Student Problem solving.

3.2 Participants

The participants in this study were all seventh-grade students and all teachers who taught in seventh grade, at the Al-Amien Prenduan Foundation. The details of the participants are as follows, 490 male and 599 females, so that the total number of students is 1089. Meanwhile, the Teachers consist of 6 male and 9 females, the total number of teachers is 15 teachers.

Given to these 15 teachers, the CK-Test, and 1089 students, were given questions about fractions. CK-test scores and student performance scores are the main data for statistical analysis in the quantitative phase. then, based on the CK-test score, teachers are mapped into two groups, low TCK and high TCK. for each group, one was chosen to represent low TCK and high TCK. The low TCK is taken from the teacher with the lowest CK-test score (Ahmad selected), and the high TCK is chosen from the teacher who has the highest score (Muhammad).

The CK-test instrument was designed and developed to measure teacher Content Knowledge based on different cognitive types, according to the Tchoshanov framework [19]. There are three types of Cognitive Teacher Content Knowledge:

1. Type 1: Knowledge of Fact and Procedure.
   This type is called by Procedural Knowledge, the knowledge that emphasizing memorizing facts, definitions, formulas, properties, and rules, performing procedures and calculations; make observations, take measurements, and routine problems solving
2. Type 2: Knowledge of Concept and Connection.
   Called by Conceptual Knowledge, which is emphasizing to the concepts understanding, making relationships, choosing or giving many representations, and resolving non-routine problems.
3. Type 3: Knowledge of Models and Generalization.
   The knowledge that integrates two types of prior knowledge, generalizes mathematical statements, designs and makes mathematical models, tests expectations and proves a theorem.

Type 1 is knowledge focused on recall capabilities, and applies basic mathematical facts, rules and algorithms to routine procedures. For example, students are asked to solve the following fraction questions, $2\frac{1}{3}:\frac{1}{2}=$, if students are able to solve the problem, then they are declared to have procedural knowledge of fraction distribution. If students are able to solve the above questions in more than one way (such as completing in the form of diagrams etc.), then they have type 2. However, type 3 requires integration of the two previous types of knowledge, and is more theoretical [11]. For an example, when is the statement $\frac{a}{b}:\frac{c}{d}=\frac{ac}{bd}$ (for a, b, c and d integers), it is always true? unlike questions in the two previous types, students are asked to have a reason, model and generalize knowledge of the concept of fraction sharing. There are indicated that the difficulties faced by students when performing fraction operations are due to the weak knowledge of the model and the generalization of students to the concept of fractional operations [19].

4. Results and Discussion

4.1 Teacher response to Content Knowledge Question

Two selected teachers is given three questions, focused on the type of teacher's cognitive knowledge based on the results of a test of teacher content knowledge and interviews

4.1.1 Teacher Response to Content Knowledge Question-1

Table 1. Teacher Response to Content Knowledge Question-1

Ahmad	Muhammad
Statements A, C, and D, false and true are B.(The teacher does not give specific reasons, but mentions that in the book it is mentioned $\frac{p}{q}-\frac{r}{s}=\frac{sp-qr}{qs}$)	The answer is B, that $\frac{p}{q}-\frac{r}{s}=\frac{sp-qr}{qs}$ Because in addition operations and subtractions of ordinary fractions, the denominator must be equated with the same number. So that in an ordinary Fraction (here $\frac{p}{q}$) the denominator is multiplied by "s" so that it gets $\frac{ps}{qs}$ (the numerator is also multiplied by s so it doesn't change the value of the Fraction). In the same way $\frac{rq}{qs}$ is obtained. So, the correct statement is $\frac{p}{q}-\frac{r}{s}=\frac{sp-qr}{qs}$

It seems that the two teachers have different approaches in answering questions 1. The response from Ahmad provided a bit of procedural understanding, Ahmad used procedural knowledge in the absence of the development of the concept of procedural knowledge. In contrast, Muhammad gave a better explanation of procedural knowledge than Ahmad. However, in interview session, both of them were able to explain the nature of the reduction of a Fraction, namely by equating the denominator first of a Fraction.

4.1.2 *Teacher response to Content Knowledge Question-2*

In this second question, the teacher was asked to complete $1\frac{1}{4}:\frac{1}{2}$ with two approaches to completion, and then gave an example of a real problem that represents the problem.

Table 2. Teacher response to Content Knowledge Question-2

Ahmad	**Muhammad**
Alternative 1: $1\frac{1}{4}:\frac{1}{2} = \frac{5}{4}:\frac{1}{2} = \frac{5}{4}\times\frac{2}{1} = \frac{10}{4} = 2\frac{2}{4} = 2\frac{1}{2}$ Alternative 2: $1\frac{1}{4}:\frac{1}{2} = 1{,}25:0{,}5 = 125 \times 50 = 2{,}5$ (Ahmad wrote on his paper as above, without giving any information)	Alternative 1: $1\frac{1}{4}:\frac{1}{2} = \frac{5}{4}:\frac{1}{2} = \frac{\frac{5}{4}}{\frac{1}{2}}\times\frac{\frac{2}{1}}{\frac{2}{1}} = \frac{\frac{10}{4}}{1} = \frac{10}{4}$ Alternative 2: 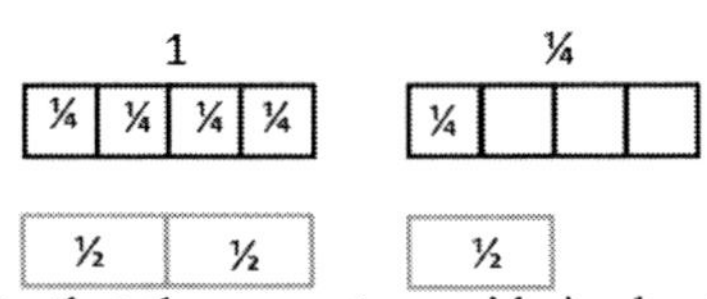 So that there are two mid-sized cubes and ½ of the mid-size cubes. So, the answer is 2½

The response from the two teachers to the second question is correct, but if observed, Ahmad's response shows more on procedural knowledge, using the formula that when a Fraction is divided by another Fraction, the Dividing Fraction is reversed and the "divide" changes into multiplication. Note also on the answer to the second alternative, Ahmad converts Fractions into decimal forms, which means Ahmad means more that Fractions are a value, this is reinforced during the interview session.

> Researcher: Why can the division of Fractions turn into multiplication?
>
> Ahmad: The multiplication rule states that when a fraction is divided by another fraction, then the sign "divide" changes to times and the divider is folded, the denominator becomes the numerator and vice versa.
>
> researcher: In the second alternative, you change Fractions to decimal, Why?
>
> Ahmad: To make it easier for me to work, isn't the answer correct?
>
> Researcher: How do you understand students and illustrate that the statement is true?
>
> Ahmad: yeah ... returned to the original formula.

Muhammad's response, gave a different analysis and approach. Muhammad seemed to be trying to answer like Ahmad did, but was not in a hurry to immediately use "flip". Muhammad, like giving a reason for the "flip" trait, can be used in the division operation in Fractions. Even with the second alternative, Muhammad tried to illustrate Fractions in the form of cubes, 5/4 illustrated with 5 cubes, with each cube being a quarter. Then construct a half-sized cube by grouping two cubes a quarter into one half cube. So that there are two half cubes, and ½ of the half cube. Muhammad's response indicated procedural knowledge supported by adequate conceptual knowledge. This is also reinforced by each teacher's answer to a factual problem that represents $1\frac{1}{4}:\frac{1}{2} =$

Table 3. Teacher response to Content Knowledge Question-2

Ahmad	Muhammad
Suppose Ani has 1 ¼ kg of flour and wants to divide the flour by ½ kg. How many people get the flour?	Pak Umar has a garden planted with coconut trees. One day Pak Umar will give the coconut tree fertilizer. Pak Umar carries 1 ¼ kg of fertilizer. Each coconut tree is given ½ kg of fertilizer. Determine how many coconut trees if the fertilizer that Mr. Umar brings is used up?

Ahmad's conceptual understanding is not as good as Muhammad's, note the illustration of the problem Ahmad gave, Ani has 1 ¼ kg of flour and wants to divide the flour by ½ kg. then the question immediately appeared, how many people got the flour? The problems given are correct but not right, and will be ambiguous to students, especially the answers given, are 2½, how will students illustrate two and a half people, While Muhammad gave problems properly thus easy to understand. This reveals Muhammad's conceptual understanding is better than Ahmad.

4.1.3 Teacher response to Content Knowledge Question-3

The aim of this question is to find out the knowledge of teacher generalization. What is the statement $\frac{p}{q}:\frac{r}{s}=\frac{pr}{qs}$, ever true for All p, q, r, and s are integers, with q and s not zero.

Table 4. Teacher response to Content Knowledge Question-3

Ahmad	Muhammad
The statement is wrong. Look again at the formula for the division of two Fractions. The answer must be: $\frac{p}{q}:\frac{r}{s}=\frac{p}{q}\times\frac{s}{r}=\frac{ps}{qr}$	If an ordinary fraction (in this case $\frac{p}{q}$) in the multiplication or division operation will be equal if multiplied by the identity of a multiplication and division. So, in the $\frac{p}{q}$ Problem: $\frac{p}{q}:\frac{r}{s}=\frac{pr}{qs}$ is a true statement if r = s, because if r = s in the $\frac{r}{s}$ problem will be the multiplication and division identity, namely 1.

Ahmad gave the wrong response to the third question. Ahmad was too focused on procedural knowledge, so he ignored the conceptual analysis of the question. In contrast, Muhammad's answer was correct even though the explanation was not provided a suitable evidence. Muhammad assumed that the statement was true if r = s.

4.2 Student response to Fraction Question

4.2.1 Student response to Fraction Question -1

The first Student questions are identical to the first teacher questions: results for 1 ¾ and ½ are. Almost all students in both groups correctly resolved the fraction distribution problem given: 28 of the 29 students in Ahmad's group, and 25 of 26 students in Muhammad's group. We conducted a linguistic analysis of the terms used by students in both groups while explaining their solutions to the Fraction distribution problem. The most frequently used terms emerge from students' written responses are "flip", "cross multiplication", "divider", "first fraction", and "second fraction". The table below (Table 5) captures the frequency of terms used by students in both groups along with the chi-square value.

Table 5. Frequency of terms used by students while solving the fraction

Term Used by Student	Ahmad Group (n=29)	Muhammad Group (n=28)
Flip	20	18
Cross Multiplication	15	11
Dividend	0	8
Divisor	2	8
First Fraction	11	2
Second Fraction	12	4

Table 5 shows that the two groups are not a much difference in using the word "flip" which are 65% for Ahmad and 61,53% for Muhammad. In contrast, the two groups relatively use the same word of multiplication. Although in quantity there are more Ahmad groups, but proportionally, the two are not much different, but if there is a tendency for the Ahmad group to use the words "first Fraction" and "second Fraction" more than Muhammad's group, many use the word "divisor" and "dividend", which were mathematically accurate.

4.2.2 Student response to Fraction Question -2

Almost all students in both groups answered correctly on the issue of Fraction distribution, 26 of the 29 members of the Ahmad Group answered correctly, and 25 of 26 members of Muhammad's group answered correctly. However, 8 of the 29 members of Ahmad's group were able to present the factual example of the division question correctly, indicating that students had sufficient procedural knowledge, but did not have sufficient cognitive abilities. In contrast to Ahmad's group, 19 of the 26 members of Muhammad's group were able to correctly present the factual example of the issue of Fractions. Some of them use the approach, Fractions as "measurements", and Fractions as part of the whole.

The following are examples, interpretations given by students. “Mother buys 1 3/4 kg of sugar to make a cake. How many cakes can he make if one cake requires 1/2 kg of sugar?”, Students provide fraction interpretations as measurements. While other students gave the example of Fractions as part whole as follows, Fatimah took 1 3/4 bottle of oil. This was only half of what Aishah had. How many bottles of oil were taken by Aishah?

4.2.3 Student response to Fraction Question -3

Based on student work analysis, some responses were obtained, including wrong responses such as "never right" and "impossible", some correct responses such as "true if a = b and c = d" (or "true if a = b = c = d = 1 "), and the response is correct -" true if c = d ". Only one student in Ahmad's group could find a solution that was partly incorrect while in Muhammad's group 8 students offered the correct solution and two students - a partially correct solution. The frequency of student responses along with chi-square values is illustrated in Table 6 below.

Table 6. Frequency of different solution proposed by student in response to question 3

Student response	Ahmad Group (n=29)	Muhammad Group (n=28)
Never true	24	8
Impossible	4	4
True if ps = qr	0	5
True if r=s	1	11
Always true	0	0

According to the results as shown in table 6, there is a statistically significant difference in the response of students in choosing "Never True" in favor of Ahmad's group, and the response of students who choose "right if r = s" in favor of Muhammad's group. But, both groups have the same choice of answers, "impossible" and "always right" answers.

4.3 Connection between Teacher and Student According to Cognitive Knowledge

4.3.1 Connections between teacher and student procedural knowledge

When responding to Question 1, both of teachers emphasized the importance of student skills in applying the rules for the division of Fractions. Based on the above analysis, Ahmad spoke only of the "standard situation", while Muhammad spoke further to "non-routine problem-solving situations" that are very important in developing student reasoning.

Subsequent observations related to mathematical terminology and teacher vocabulary usage. Although Ahmad and Muhammad had the same average score on items that measured teacher knowledge of facts and procedures, the use of mathematical vocabulary and terminology was more accurate in Muhammad's case. He used the terms "dividend" and "divisor" which are mathematically accurate in answering question 2, while Ahmad used inaccurate terms "first fraction" and "second fraction" which are then reflected in student responses: there are statistically significant differences in use the terms "dividend" vs. "first fraction" and "divisor" vs. "second fraction" between groups. More specifically, students in Muhammad's group preferred to use the terms "dividends" and "divisor" which were mathematically accurate compared to the terms "Fractions first" and "Second fractions" which frequently used by students in Ahmad's group. In addition, Muhammad's articulation of the idea of the inverse nature of the operation (see the response to question 2), and reciprocal ideas (see the response to question 3), contributed to his deep understanding of the division of Fractions.

4.3.2 Connections between Teacher and Student Conceptual knowledge

In this second question, it is divided into two approaches, completing the Fraction distribution problem and describing the problem into a factual problem. When completing the Fraction distribution problem, both teachers were able to answer correctly, and almost all students answered correctly. However, the performance shown by Ahmad places more emphasis on procedural knowledge, using the existing formula and nature of the distribution of quotations, without being reinforced by good rationalization. And when giving an example of a factual problem from the question of the operation given, Ahmad was able to answer correctly but not exactly.

Along with Ahmad, the response given by Ahmad's group was not much different. 26 of the 29 Ahmad students were able to answer correctly the Fraction distribution problem, but 3 of 29 were able to provide a correct description of the factual problems involving the Fraction operation.

Muhammad gave a better response from Ahmad, although explicitly Muhammad's answer was not much different from Ahmad's, but Muhammad was able to provide better rationality. This shows that Muhammad's conceptual knowledge is better, and in depth this will also have an impact on students' understanding and performance. It was proved by the response of Muhammad's group on the second question, 25 of 26 were able to answer and 17 of 26 were able to provide a good illustration of factual problems involving fractional operations.

4.3.3 Connections between teacher and student Generalization Knowledge

As discussed in the previous section, Ahmad gave the wrong answer, and Muhammad gave the correct answer but it was not followed by the depth of proof and explanation. Interesting to see is the answer from Ahmad's group, 1 of 29 answered correctly, 24 of 29 students always answered wrongly, and 4 students answered it was impossible. This is also the same as Ahmad's explanation and answers. Meanwhile, 10 out of 26 students from Muhammad's group answered correctly, while the rest were divided into always wrong (7 students), Not possible (4 students), and $ps = rq$ (5 students). And based on the results of student performance, it seems that Muhammad emphasizes good understanding of content in learning, it was seen from the depth of rationality in students' answers.

Overall, we observe differences in the approach used by the teacher to non-routine questions: while Muhammad recognized and accepted the challenge, Ahmad ignored and avoided it. Our observation also shows Ahmad's weak conceptual understanding with questions that explore his conceptual understanding of the division of Fractions. This observation resonates with the difference between the average scores of Ahmad and Muhammad in the Content Knowledge test which measures their conceptual and general knowledge (see the Results section). The most convincing difference in teacher knowledge was related to their students' performance in solving question 3: there was a statistically significant difference between groups at the p-value level <0.01 in choosing "never True" (supporting Ahmad's group) and "True if c = d "(supports Muhammad's group) responses. Lastly, but not least, in-depth observations about statistically significant differences between groups in the number of blank answers (no solution) for question 3: four students in Muhammad's group decided not without any solution (maybe, because of doubts) instead of anything. The solution compared to Ahmad's group: all students in Ahmad's group except one were "sure" to give the wrong solution.

5. Conclusion

The main result of this study is to provide answers and analysis of research questions stated in the introduction. To answer these questions, as we explained in the Methodology section, we deliberately selected two opposite cases based on the average score on the teacher content knowledge test in the Fraction Material to examine carefully the relationship between teacher content knowledge and student performance. While solving a set of three questions related to the division of Fractions. As we expected based on teacher scores on type 1 cognitive items (measuring knowledge of facts and procedures) from Test Teacher Content Knowledge Score (Ahmad score - 80% and Mohammad score - 90%), there were no observed differences between student performance in the two groups at procedural questions 1. With regard to teacher scores on cognitive type 2 items (Ahmad scores - 46% and Mohammad scores - 69%), we observed a statistically significant difference ($\chi 2 = 3.933$, $p<0.05$) in knowledge 2 question measurement students about concepts and connections. The clearest difference between student performance in the two groups observed in question 3 (measuring model knowledge and generalization) which was also statistically significant ((2 = 8.43, p <0.01) and revealed a close relationship with teacher scores on type 3 items cognitive from TCKS (Ahmad score - 30% while Mohammad score - 70%). Student performance is related to their teacher's knowledge: students in Mohammad's group performed better than the students in Ahmad's group, especially in solving questions 2 and 3. Thus, this study contributes to a body of research that claims that teacher content knowledge is very important for student learning with a narrow focus on Fraction topic.

Research on teacher knowledge is very relevant because of its significant relationship with student knowledge. This is consistent with the claims of several experts (Baumert et al., Hill, Shilling, & Ball), that the teacher's content knowledge is very influential on student learning [8,21]. Teachers cannot teach what they do not know [18,19], therefore this research makes a special contribution to the field that shows the relationship between teachers and students' knowledge in certain material, namely Fractions. Teachers need to continue to improve their competence and participate in professional development activities, especially in content knowledge. The findings of this study can add research-based literature and recommendations on developing teacher awareness about a particular knowledge topic and its relationship to students' knowledge and understanding.

References

[1] Aksoy N C and Yazlik D O 2017 *J. of Educ. and Training Stud.* **5** 219
[2] Norton A and Boyce S 2013 *The J. Math. Behav.* **32** 266
[3] Siegler R S, Thompson C A and Schneider M 2011 *Cogn. Psychol.* **62** 273
[4] Trivena V, Ningsih A R and Jupri A 2017 *J. of Phys. Conf. Ser.* **895** 012139
[5] Gabriel F 2016 *Aust. Prim. Math. Classr.* **21** 36
[6] Booth J L, Newton K J and Twiss-Garrity L K 2014 *J. of Exp. Child Psychol.* **118** 110
[7] Goldhaber D and Brewer D J 2000 *Educ. Eval. and Policy Analysis* **22** 129
[8] Baumert J et al 2010 *American Educ. Res. J.* **47** 133
[9] Ma'rufi, Budayasa I K and Juniati D 2017 *J. Phys. Conf. Ser.* **954** 012002

[10] Yeo J 2008 Proceedings of the 31st Annual Conference of the Mathematics Education Research Group of Australasia 621

[11] Tchoshanov M 201 *Educ. Stud. in Math.* **76** 141

[12] Shulman L S 1986 *Educ. Res.* **15** 4

[13] Izsaak A, Jacobson E, Araujo Z D and Orrill C H 2012 *J. for Res. in Math. Educ.* **43** 391

[14] Mccrory R, Floden R, Ferrini-Mundy J, Reckase M and Senk S 2012 *J. for Res. in Math. Educ.* **43** 584

[15] Murphy C 2012 *J. of Math. Teach. Educ.* **15** 187

[16] Groth R E and Bergner J A 2006 *J. Math. Think. and Learn.* **8** 37

[17] Ekawaty R, Lin F L and Yang K L 2015 *Eurasia J. of Math. Sci. and Tech. Educ.* **11** 513

[18] Olfos R, Goldrine T and Estrella S 2014 *Revista Brasileira de Educacao* **19** 913

[19] Tchoshanov M et al 2017 *J. of Math. Behav.* **47** 54

[20] Campbell P F et al 2014 *J. for Res. in Math. Educ.* **45** 419

[21] Hill H C, Rowan B and Ball D L 2005 *Am. Educ. Res. J.* **42** 371

[22] Tröbst S et al 2019 *Unterrichtswissenschaft* **47** 79

[23] Odumosu M O and Olisama O V 2018 *Int. J. of Educ. and Res.* **6** 83

[24] Shulman L S 1987 *Harvard Educ. Rev.* **57** 1

[25] Depaepe, F., et al (2015), Teachers' content and pedagogical content knowledge on rational numbers: A comparison of prospective elementary and lower secondary school teachers. Teaching and teacher education. 47. 82-92

[26] Rowland et al 2005 The knowledge quartet: a tool for developing mathematics teaching. *Paper presented at the European Conference on Educational Research, University College Dublin, 7-10 September 2005*

[27] Blomeke S et all 2016 *Teach. and Teach. Educ.* **56** 35

[28] Ball D L, Thames M H and Phelps G 2008 *J. of Teach. Educ.* **59** 389

Creative Mathematical Reasoning of Prospective Teachers in Solving Problems Reviewed Based on Working Memory Capacity

I Palengka[1*], D Juniati[2] and Abadi[2]

[1] Mathematics Education, Universitas Kristen Indonesia Toraja, Indonesia
[2] Mathematics Education, Universitas Negeri Surabaya, Indonesia

[*] E-mail: inelsipalengka@ukitoraja.ac.id

Abstract. Reasoning requires logical thinking in solving problems, and working memory capacity becomes something important. This study aimed at investigating the creative mathematical reasoning of prospective teachers in solving problems reviewed based on working memory capacity. This research was qualitative. The study conducted at Indonesian Christian University of Toraja to Mathematics Education study program students. The selection of research subjects was done by giving operation span (O-SPAN) assignments to select 1 student with a high category of working memory capacity and 1 student with a low category of working memory capacity. Then they were given a creative mathematical reasoning test and interview. Based on the data analysis, it obtained that the creative mathematical reasoning of prospective teachers in solving problems on the research subject who had low working memory capacity showed *creativity* and *plausibility* criteria but it did not show *anchoring* because he could not provide arguments that supported the selection of strategies based on mathematics. Furthermore, creative mathematical reasoning of prospective teachers in solving problems on the research subject who had high working memory capacity met the three criteria for creative mathematical reasoning, namely *creativity, plausibility and anchoring.*

1. Introduction

Mathematics is an activity of human life, so learning mathematics is learning to solve problems. At present problem solving is a major concern in mathematics learning activities at the elementary, secondary and high education levels. At the higher education level, prospective teachers need opportunities to formulate and solve complex problems that involve a large number of efforts. They must be encouraged to reflect on their thinking during the problem-solving process so they can apply and adapt their strategies in developing them to problems and in other contexts. This is a concern so that reasoning is one of the main objectives of learning mathematics. Also, reasoning ability is critical to understand mathematics and make mathematics meaningful. For this reason, reasoning can be trained from the early age to students by accustoming them to provide arguments or logical reasons in solving mathematics and statistics problems [1] because understanding mathematics is meaningless without emphasising reasoning [2]. Bottou [3] defines that reasoning is manipulating the knowledge gained in algebra to answer new questions. The reasoning defined as the thinking line taken to produce statements and reach conclusions in completing a task [4]. Further reasoning is treated as logical knowledge by Sumpter [5]. To develop student reasoning, the role of problem-solving is very important [6]. This is intended to give them many experiences to develop their reasoning through the problem-solving process [7]. One form of reasonings is mathematical reasoning. One type of mathematical reasoning is creative mathematical reasoning [8], [21]. Creative mathematical reasoning is one of the reasoning processes carried out by a person in solving problems through non-routine procedures [9], [19], [20]. Lithner [4] defines that creative mathematical reasoning meets the criteria:

(a) Creativity: the learner creates a reasoning sequence not experienced previously, or re-creates a forgotten one, (b) Plausibility: there are predictive arguments supporting the strategy choice and arguments for verification, explaining why the strategy implementation and conclusions are true or plausible, (c) Anchoring: the arguments implementation and conclusions are anchored in the intrinsic mathematical properties of the components of the reasoning.

There are many positive impacts experienced by students in solving problems using creative mathematical reasoning. Norqvist [4] explains that students who practice with creative tasks outperform students who practice with algorithmic ones in completing assigned tasks. Norqvist [10] points out that creative mathematical reasoning is more efficient in long-term understanding than algorithmic reasoning. However, the thing that arises is in solving mathematical problems. Prospective teachers still consider problems that have a variety of ways of solving or open questions are problems that are difficult to solve. One reason is, in students' minds, the relevant and irrelevant information to complete a task is very much, so students have difficulty in choosing which information is appropriate in completing tasks that require creative mathematical reasoning. The same thing is stated by Fockert saying that a thing that might happen is that individual with lower working memory capacity has more difficulty in sorting out very much information [11]. For this reason, working memory is needed to sort out the appropriate information to solve the problem by taking into account the working memory capacity as a basis.

Working memory is a limited capacity system that is responsible for the temporary maintenance and manipulation of a series of representations for ongoing cognition [12-14]. Working memory plays a role in processing information and digging back information contained in long-term memory so that working memory capacity becomes indispensable in creative mathematical reasoning. This is consistent with the previous research stating that working memory capacity can reduce the errors in retrieving information that is not needed in problem-solving [22] so Fife's opinion is true that working memory capacity is involved in problem-solving [23]. So, in creative mathematical reasoning, the role of working memory capacity is needed to provide relevant information [4]. Based on the background, this study aimed at investigating the creative mathematical reasoning of prospective teachers in solving problems reviewed from working memory capacity.

2. Method

This research was qualitative. The study conducted at Indonesian Christian University of Toraja to Mathematics Education study program students. Research subjects were determined by using o-span tests to select a subject with high memory capacity (MT) and a subject with low memory capacity (MR). Both subjects were given a creative mathematical reasoning test in the form of an open task consisting of two numbers. Question number 1 raised the problem of how to set paving blocks consisting of four different shapes on a rectangular yard. Question number 2 presented the problem of how-to put-up paintings of four sizes on the wall. After being given a test, interviews were conducted on both subjects. The questions written in the interview guidelines are based on creative mathematical reasoning criteria. In this study, interviews were conducted with the subject about the task that was completed by the subject, namely the creative mathematical reasoning test. The results of mathematical creative reasoning tests and interview results were then analysed to get conclusions.

3. Results and Discussion

The following is the results of creative mathematical reasoning analysis reviewed based on the working memory capacity of the two subjects based on creative reasoning tests and interviews.

3.1. Description of Subject's Creative Mathematical Reasoning Capability Based on Low Memory Capacity (MR)

The creative mathematical reasoning capability of subjects based on low memory capacity (MR) will be described based on three aspects namely *creativity, plausibility and anchoring* as follows:

3.1.1. Creativity

MR showed renewal in new reasoning by assuming the area of the wall with L, the height of the painting from the floor as x, the distance of the painting from the ceiling as y and the distance of one painting to another as z. Suppose the paintings size 100 cm x 100 cm as A, size 50 cm x 50 cm as B, size 50 cm x 100 cm as C, and size 100 cm x 50 cm as D. After that MR described what he thought like the following figure:

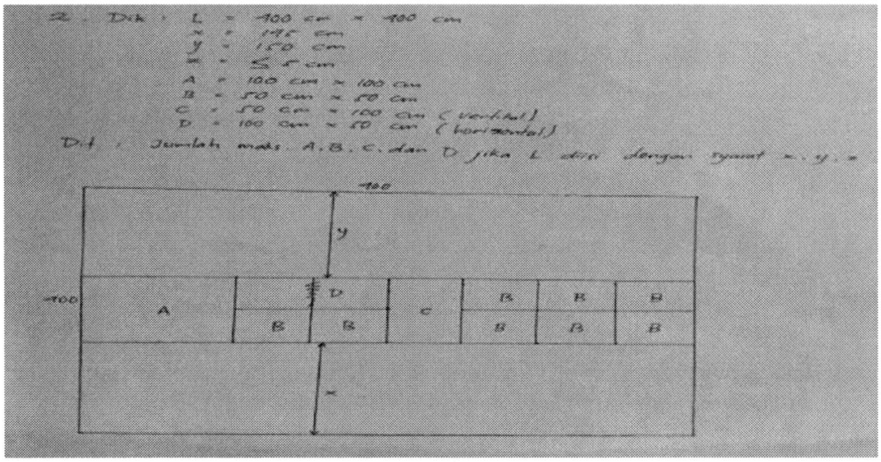

Figure 1. MR's answer to question No. 2

MR showed renewal by giving the correct answer, but MR did not find another answer for that problem. It meant MR's answer did not show fluency. However, MR's answer was quite flexible because MR could provide another solution at the interview.

Q: Do you think there is still a way to put up the paintings other than what you are describing?
A: Yes
Q: How do you do it?
A: by changing the position of the paintings

3.1.2. Plausibility.

Plausibility was shown by placing paintings without giving any distance at all, and it did not conflict with the instructions that the maximum distance of painting was 5 cm. Furthermore, MR stated that "I arrange paintings by considering the maximum number of paintings that must be loaded. In other words, it must contain as many paintings as possible so that I maximise the paintings with the smallest size". It showed that MR could provide reasonable arguments about the strategy he used.

The *plausibility* criterion was only shown in question number 2. MR could not provide a solution in question number 1, so it was difficult to trace the reasons for choosing the solution strategy he used. It also showed that the subject with low memory capacity had difficulty in solving more difficult problems.

3.1.3. Anchoring

The *anchoring* criterion was not shown by MR both in question number one and in question number two. In question number two, MR was able to give the right answer by using a reasonable strategy, but MR could not provide an argument showing that the strategy used was by mathematics.

3.2. Description of Subject's Creative Mathematical Reasoning Capability Based on High Memory Capacity (MT)

The creative mathematical reasoning capability of subjects based on *High* memory capacity (MT) will be described based on three aspects, namely *creativity, plausibility and anchoring* as follows:

3.2.1. Creativity

In question number one, MT was able to understand the problem and showed the creativity criterion with new reasoning. MT stated that he had ever solved almost the same problem, but solving the problem of arranging paving blocks with various types required different thoughts. It was known from the interview as follows:

Q: Do you use the method you used before to answer this question?
A: No Ma'am
Q: why?
A: Because what I have ever done was to arrange rectangles, so it was easy. Now it's very different because in addition to the different shapes, the sizes are also different. That's what makes it difficult.
Q: how do you solve it?
A: I draw the yard first then I draw a hexagon in the middle because it is the most difficult to set and then I set the pentagon and set the rectangle and square around it.

From this interview, it could be seen that in order to solve the problem, MT thought of the form of yard and forms of paving blocks, then he arranged the paving blocks by starting from the form that according to the subject was the most difficult to set with other forms of paving blocks. The MT's answers are seen in the following figures:

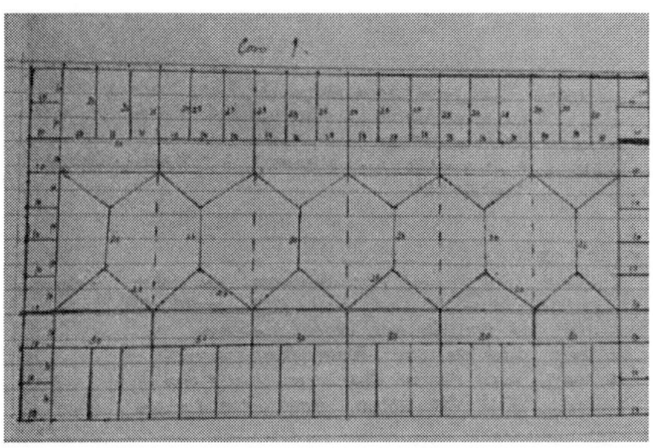

Figure 2. The MT's 1st answer for question No. 1

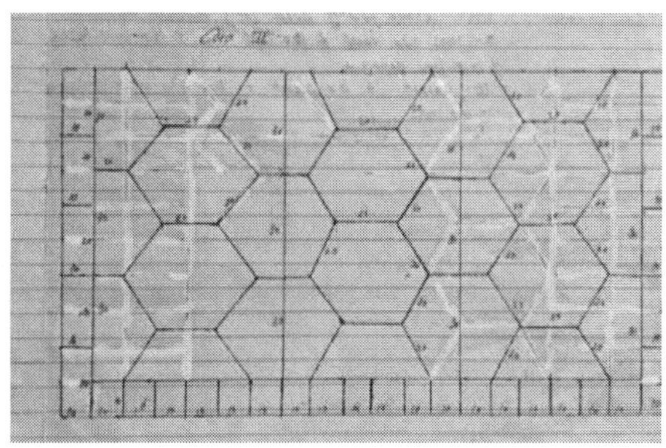

Figure 3. The MT's second answer to question No. 1

From the figures, it can be seen that the MT showed creativity with fluency in giving two different and flexible answers in showing two different ways of setting.

For question number 2, MT stated that it was the first time for him to encounter the question like in question number two so that the reasoning is done by MT could be categorised as new reasoning for the subject. MT also showed creativity with fluency shown by giving two answers and MT was flexibly shown by describing two different ways of putting up paintings on the wall as in the following figure:

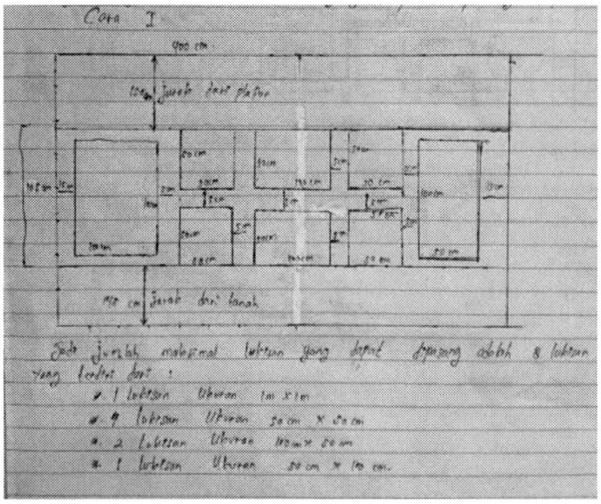

Figure 4. The first solution to question No. 2

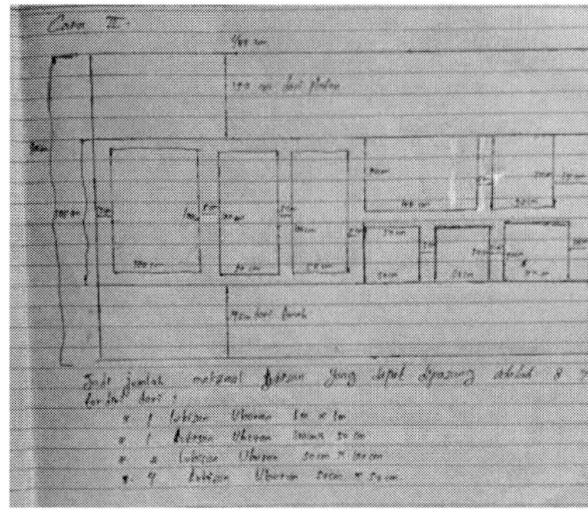

Figure 5. The second solution to question No. 2

3.2.2. Plausibility

The plausibility criterion for question number 1 was shown by MT by writing the yard area equal to the total area of all paving blocks covering the yard. MT determined the area of each type of paving block and multiplied it by the number of paving blocks of each type and summed up the total area to get the yard area as shown below

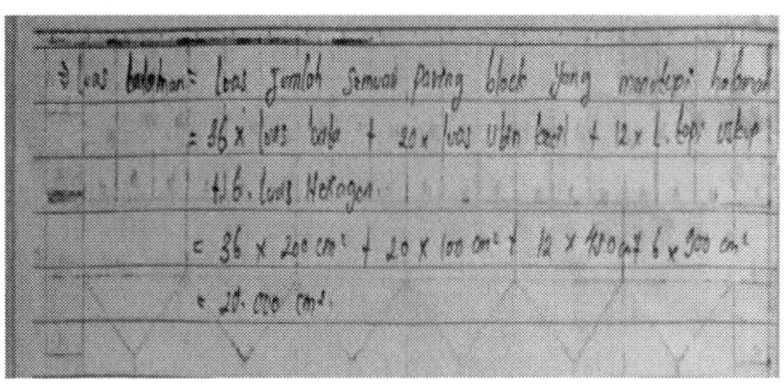

Figure 6. Plausibility criteria of solution 1

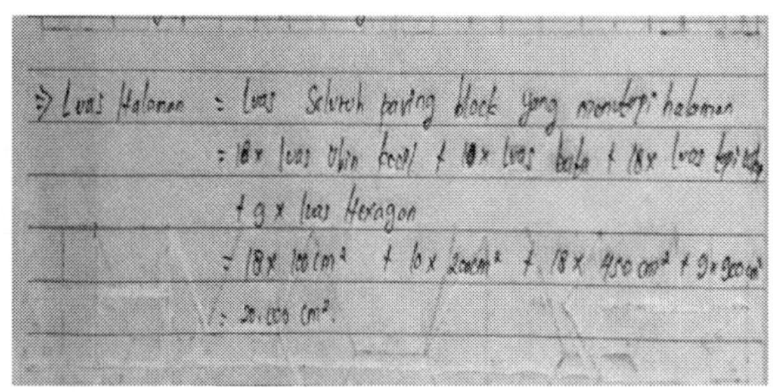

Figure 7. Plausibility criteria of solution 2

For question number 2, the criteria for plausibility were shown by MT by placing paintings whose length did not exceed 400 cm as described in the interview.

Q: Can you explain how you put the paintings as in your answer?
A: The lengths of the painting are summed, not to be more than 400 cm in width
Q: The second one?
A: The height is reduced ma'am, which is from the ceiling and the ground. The remaining is 105 cm.
Q: How do you fill a 105 cm one with paintings like this?
A: The longest size is 1m x 1m Ma'am. There are still 5 cm above, or 2.5 cm below and 2.5 cm above.

3.2.3. Anchoring

The anchoring criterion was shown by MT in question number 1 by using the concept of plane figure area. To calculate the area of a yard and the area of a brick type paving block MT used a rectangular area, to calculate the area of a hexagon he used 2 times the area of a trapezoid, to calculate the area of a bishop's hat he used a rectangular area plus the area of a triangle and to calculate the area of a small tile type paving block he used a square area as in the following figure:

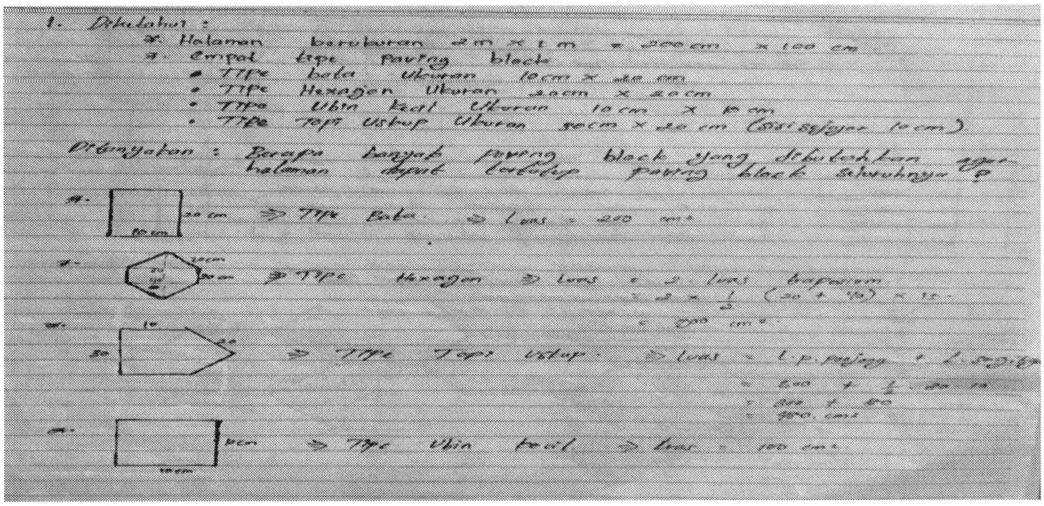

Figure 8. Anchoring criterion for question number 2

Anchoring criteria met if information about matters relating to reasoning can be investigated through argumentation. Lithner [9] states meet anchoring if it meets the relevant mathematical properties of the components in reasoning. For question number 2, anchoring was shown by using addition and subtraction operations. MT reduced the area of the wall by a distance of 150 cm from the ceiling and a distance from the floor by 145 cm and only depicted the paintings to be placed in the reduced area and took the maximum distance and placed the paintings at a distance of 5 cm from contiguous paintings.

4. Conclusion

After analysing the mathematical creative reasoning test and the results of the interview, the following results are obtained: Subject with low working memory capacity demonstrates creative criteria by creating a new set of reasoning in solving problems. Moreover, the subject can provide arguments supporting the logical solution choice, so the subject meets the criteria of *plausibility*. However, the subject cannot provide arguments that support the strategy selection based on mathematics (mathematical basis) that is relevant. It means that the subject does not meet the *anchoring* criteria. And the research subject with high working memory capacity meets the three criteria of creative mathematical reasoning in solving problems, namely the creativity criteria and also meets the criteria of *Plausibility* and *anchoring*.

For further research, it is expected that researchers conduct research on creative mathematical reasoning with anchoring criteria in terms of the components of reasoning, namely objects, transformations and concepts.

References

[1] Rosidah, Budayasa I K and Juniati D 2018 *J. Phys. Conf. Ser.* **1028** 012125.
[2] Mueller M, Yankelewitz D and Maher C 2010. *Math. Enthus.* **7** 307
[3] Bottou L 2014 *Mach. Learn.* **94** 49
[4] Norqvist M 2019 *J. Math. Behav.* **40** 104
[5] Sumpter, L 2013 *Int. J. Sci. Math. Educ.* **11** 35
[6] Muzaini M, Juniati D and Siswono T Y E 2019 *J. Phys.: Conf. Ser.* **1157** 032093
[7] Cai J and Knuth E 2012 *Res. Math. Educ.* **14** 5
[8] Bergqvist T and Lithner J 2012 *J. Math. Behav.* **31** 69
[9] Lithner J 2008 *Educ. Stud. Math.* **67** 76
[10] Norqvist M 2018 *International J. Math. Educ. Sci. Tech.* **49** 15
[11] de Fockert J W, Rees G, Frith C D and Lavie N 2001 *Sci.* **291** 1803

[12] Baddeley A D 2000 *Sciences.* **4** 417
[13] Oberauer K 2009 *Psychol. Learn. Motiv.* **51** 45
[14] Oberauer K, Farrell S, Jarrold C and Lewandowsky S. 2016 *Psychol. Bull.* **142** 99
[15] Daneman M and Merikle P M 1996 *Psychon. Bull. Rev.* **3** 33
[16] Süß H M, Oberauer K, Schulze R, & Wilhelm O 2005 *Psychol. Bull.* **131** 5
[17] Conway A R A, Kane M J and Engle R W 2003 *Trends Cognit. Sci.* **7** 52
[18] Singh K A, Gignac G E, Brydges C R, and Ecker U K H 2018 *J. Mem. Lang.* **101** 18
[19] Jonsson B, Kulaksiz Y C, and Lithner J 2016 *Int. J. Math. Educ. Sci. Tech.* **47** 25
[20] Jonsson B, Norqvist M, Liljekvist Y, and Lithner J 2014 *J. Math. Behav.* **36** 20
[21] Lithner J 2017 *ZDM*, **49** 49
[22] Wiley J and Jarosz A F 2012 *Current Direct. in Psychol. Sci.* **21** 258
[23] Fyfe E R, Eisenband Matz, L Hunt K M and Alibali M W 2019 *J. Commun. Disord.* **77** 17

The Influence of Cognitive Style on Mathematical Communication of Prospective Math Teachers in Solving Problems

I K Budayasa and D Juniati*

Department of Mathematics, Faculty of Mathematics and Natural Sciences, Universitas Negeri Surabaya, Indonesia.

* E-mail: dwijuniati@unesa.ac.id

Abstract. Communication is important to connect with others. It is one of the important aspects that must be mastered by the prospective math teacher as a tool of knowledge transfer. They are expected to have good mathematical communication skills to facilitate them in the mathematics learning process. Problem-solving in learning mathematics is something that cannot be ignored. Through problem-solving, students can have a better understanding of concepts because through mathematical problem solving, and students are required to be more active in using the concepts they have in a variety of new situations encountered. FD/FI cognitive style is the extent to which a person perceives parts of a field discrete from the surroundings rather than embedded in it as a whole. Based on differences from cognitive style, it certainly affects the way of communication, especially mathematical communication. So, the aim of this research was exploration the mathematical communication of prospective math teacher in solving problem viewed from FI/FD cognitive style. This type of research was descriptive research with a qualitative approach. The subjects of this research were 86 prospective math teachers. Data were collected through tests, assignments, and task-based interviews. The results of this study indicated that mathematical communication from FD and FI students were not different when solving routine problems, but were different when solving non-routine problems, which require strategy and decision making. FD students had difficulty communicating their strategies clearly and coherently. Meanwhile, FI students explained the strategies they used in more detail.

1. Introduction

Communication is important to connect with others. If a person cannot communicate his thoughts correctly then it can lead to misunderstanding. Especially for a teacher, communication is very important, it is important to express what is in the mind of the teacher, it is important to explain the material to students so that students are not misconceived, it is important to express what they want clearly so that students do not misunderstand. Pourdavood said that communication helps the teacher to recognize the student's ability, understand their mathematical understanding, and then make the appropriate decisions to help them [1]. Based on some of the opinions above, it is clear that communication is important for teachers. However, communication is also important for students, better communication makes it easy for students to learn the material, convey ideas correctly and fluently following their mind so that teachers can understand their thinking. In other words, communication is important in every learning, including mathematics learning. NCTM also stated the importance of communication in learning mathematics. In Principles and Standards for School Mathematics, the

National Council of Teachers of Mathematics (NCTM) states: "Students who have opportunities, encouragement, and support for speaking, writing, reading, and listening in mathematics classes reap dual benefits: they communicate to learn mathematics, and they learn to communicate mathematically" [2]. Mathematical communication means organizing mathematical thinking coherently and clearly to the others and using the language of math to express mathematical ideas precisely. Mathematical communication includes a variety of cognitive skills because it is an exchange of ideas, which includes both listening, reading (understanding), speaking and writing (expressions) about mathematics, mathematical expressions include the representation of mathematical ideas in a non-linguistic way.
Indicators of mathematical communication, according to NCTM are as follows:
(1) skill to express mathematical ideas and demonstrating them and visualising them visually;
(2) skill to understand, interpret, and evaluate mathematical ideas;
(3) skill to use terms, mathematical notations and structures to present ideas, describe relation, and situational models.

Prospective math teachers are specially prepared to become qualified mathematics teachers. Teachers have an important role played in helping students to learn and affecting the educational outcomes-achievement and attitudes of their students. They are expected to have good mathematical communication skills to facilitate them in the mathematics learning process. So, it is interesting to study the mathematical communication of prospective math teachers.

The research about mathematical communication has been investigated in many studies. Wichelt and Kearney [3] conducted research about communication in mathematics learning using open-ended questions. They claimed that students' mathematical vocabulary comprehension was better when using open-ended questions. Pantaleon et al [4] studied the communication of prospective teacher in proving problems. They found that in the proof of geometry problem, the subject explains what is understood, presents the idea in the form of drawing and symbols, and explains the content/meaning of a representation accurately and clearly, but the subject can't convey the argument systematically and logically. Kabael [5] found out that junior high school math teachers had no communication skills in the mathematical language as expected of a mathematics teacher. Kabael put more highlighted on the skill of mathematical communication than process. In this study, the focus was on the written math communication of prospective math teachers.

Problem-solving in learning mathematics is something that cannot be ignored. Through problem-solving, students can have a better understanding of concepts because through mathematical problem solving, and students are required to be more active in using the concepts they have in a variety of new situations encountered. Problem-solving places the focus on the student making sense of mathematical ideas. When solving problems students are exploring the mathematics within a problem context rather than as an abstract. Liljedahl [6] argued that problem-solving not only requires logical and deductive reasoning but also requires extra-logical processes of creativity, insight and illumination to produce a solution. Lestari [7] studied the differences in the reasoning of prospective mathematics teachers in solving context problems in shipping companies. The result showed female prospective teachers were more capable of logical reasoning, using concepts, facts and procedures and algebraic operations to conclude, make interpretations and evaluations.

Many aspects affect a person's thinking, reasoning, and communication, some of which are their beliefs, gender, cognitive style, and memory capacity. Muhtarom et al. [8] found that prospective teacher's success in solving the problem is in line with their belief in solving the problem. Based on research conducted by Juniati [9], it was found that the strategies used by students in completing technology-based assignments varied and using the software in completing assignments made it easier for students to understand concepts and communicate their results better. Sun [10] studied the effects of thinking style on design strategies, carried out in Taiwan. The results described that the thinking style showed a positive correlation with frequency of substantial change in structure and the goal of strategy, it meant that thinking style affects strategy designed. FD/FI cognitive style is the extent to which a person perceives parts of a field discrete from the surroundings rather than embedded in it as a whole.
A field-dependent learner is one who processes information globally. This kind of learner is less analytical, not attentive to detail, and sees the perceptual field as a whole. Meanwhile, a independent

field person can easily break the field down into its parts. He/she is typically not influenced by the existing structure and can make choices independent of the perceptual field. Based on differences from cognitive style, it certainly affects the way of communication, especially mathematical communication

In this study, the communication aspects that will be explored from prospective math teacher were accuracy, coherence, completeness and clarity of written math in solving mathematical problems.

2. Method

The type of this research was descriptive research with a qualitative approach. A total of 86 prospective math teachers (Mathematics education students of Universitas Negeri Surabaya) participated in this study, 17 were male, and 69 were female students, aged 18-23 years old.

The instruments used in this study were: (1). The Group Embedded Figure Test (GEFT) by Witkin was used to classify subjects based on their cognitive style namely FI (Field Independence) and FD (Field Dependence). The GEFT test consists of 18 items, which are questions about finding a figure in more complex figures. The range of GEFT score was from 1 to 18, the subject was said to have an FD cognitive style if the GEFT score was less than 10 and if the score was 10 or more then the subject was classified as had FI cognitive style; (2). The Mathematical test was used to determine the mathematical abilities of subjects; and (3). The Problem-solving test was used to investigate the mathematical communication of subjects.

The research procedures carried out were: GEFT test was given to subjects and by their score of GEFT test, subjects were classified into FI and FD groups. Mathematics tests were given to subjects to determine the subject's mathematical abilities. Then problem-solving tests were given to the subjects to investigate the subject's mathematical communication in solving problems. The subjects' problem solving were analysed based on aspects of the accuracy in using mathematical symbols, mathematical terms, mathematical expressions and the coherence in explaining the mathematical ideas used, also the completeness and the clarity of the explanations used in solving problems. The results of mathematical communication subjects were grouped based on differences in FI and FD, and then two subjects with the same mathematical abilities were selected, namely one FI subject and one FD subject to be interviewed for further study of mathematical communication. The criterion of each aspect of mathematical communication used in this study were: the written results were said to be accurate if the notations, terms and formulas and the mathematical writing used are correct and mathematically accurate, was said to be coherent if the steps used, the reasons given were logical and interrelated, and said to be complete and clear, if all information needed in the process was given and the explanation given was clear.

3. Results and Discussion

From the results of the GEFT test given to 86 subjects, 11 FD and 75 FI subjects were obtained. The problem solving provided consists of integral problems, geometry problems and maximal strategy problems. Integral problem was the problem of determining the integral of a function if the derivative of the function was given. The geometry problem used is the problem of determining the circumference of the snowflake curve in the tenth iteration if the circumference of the first iteration is equal to 1. The problem of the maximum strategy was the problem of determining the maximum number of apples (in kg) sent to the store on the hill using a cart under certain conditions, subjects were asked to determine the strategy so that the apple sent to the store was maximum. The following general description of written mathematics communication was grouped by FI and FD:

3.1. Integral Problem

The results of the mathematical communication analysis of the work of the FI and FD groups for integral problems showed almost the same results, namely around 64% of the FD group and 71% of the FI group that wrote their answers accurately, the steps that were carried out coherently and logically and the information needed complete. Apart from groups that wrote answers accurately, coherently and completely, many subjects wrote integral forms without "dx", forgot to add constants in the process of integration, changed variables of functions for no apparent reason. Some of them got the function f only

Journal of Physics: Conference Series **1417** (2019) 012056 doi:10.1088/1742-6596/1417/1/012056

by integrating the given function, even though the requested function was integral off when the derivative of f was known. They were not careful in understanding the problem because usually the questions they had encountered so far were determining the integral of a function or determining the derivative of a function.

3.2. Geometry Problem

Students were asked to calculate the circumference of the Koch curve (in Figure 1) in the 10th iteration when the circumference of the first iteration was 1.

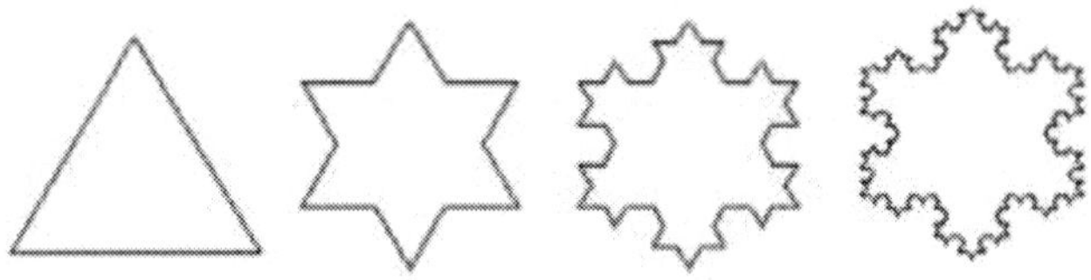

Figure 1. Iteration of the Koch Curve

Unlike the Integral problem which was solved by simply using the formula or integral knowledge of the subjects, the geometry problem required carefulness in analysing the change of the length of the sides of the geometry figure when iteration increased. After the scale changed of the side length was known when iteration increased, this problem could be solved either by using the growth of the circumference with the scale changed in the length of the sides directly or by using a geometric sequence formula.

Only 20% of the FD group and 40% of the FI group answered this geometry problem by writing out the steps that were understandable and logical. They wrote accurate mathematical terms, symbols and formulas and wrote down the information needed to solve the problem completely.

Most FD students had difficulty in determining the pattern of changing sides of a geometry figure as the iteration increases, and they focused more on the number of sides rather than the length. They could not explain or express what they thought, what the steps they took, and some of them just gave the signs in the form of fractions or numbers at the bottom of the figure without any explanation. The works of more than half FD students were incomplete, so it was incoherent and difficult to understand the way they thought. In general, FI students wrote the steps they used to solve this problem more clearly than FD students, although some of them were wrong in determining the ratio of change inside length, the symbols and mathematical formulas that they used were also accurate. However, some of them did not write down the information needed completely. They knew that the problem could be solved by geometric sequences, but the formula used without explanations, why and how they determined the first term and the ratio.

3.3. Maximal strategy problem

The problem is given:

Andre will send 100 kg apples using a cart to a store located on a hill 10km from his warehouse. The cart can carry a maximum of 10 kg, and each time is travelling 1km up the hill, Andre consumes 1 kg of apples to eat. Discover the strategy so that Andre can send as many apples to the store as possible and determine how many kg of apples reach the destination. At each km from the warehouse to the store, there is a stop so that Andre can temporarily leave the apples there if needed.

From the results of students FD work, no one got the optimal answer, 40% of them chose the strategy of sending apples directly so that the apples sent were 0 kg because they were consumed in shipping.

The other 40% used the strategy of trying two different ways, and they got the apples sent were 25 kg, 20% of them had difficulty determining the strategy, they just described the problem in a picture. Almost all FD students did not explain the strategy used coherently and clearly, and they only used the arrows and charts without explanation. 25% of FI students found the optimal results, that was 32 kg apples were sent to the store located in the hills, and they explained the strategies they used either by using words or in the form of schema with little information. Another 50% use a trial and error strategy, but they did not get the optimal results, while 25% could not determine the strategy to be taken.

The students thought a lot when solving problems related to finding strategies, but they could not communicate it by writing their ideas clearly and accurately.

The results of the mathematics test with the range score 0 - 100 showed that of 84 students who took the test, the minimum score of students was 20, and the maximum score was 100. Based on the results of the analysis, the researchers selected one FD subject and one FI subject of 84 students with moderate mathematics ability, that was between 65 to 75.

3.4. *Mathematical communication of FD student*

The following description was the mathematical communication profile in the integral, geometry and maximal strategy problems of one FD student.

3.4.1. *Integral Problem*

Figure 2 below shows the answer of FD subject on integral problem.

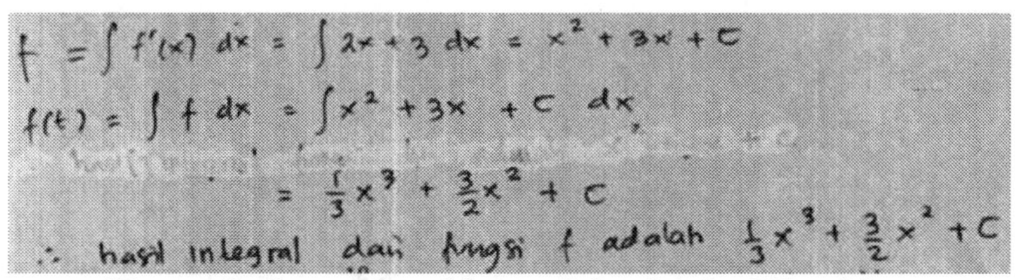

Figure 2. Integral problem solving by the FD subject

In solving the Integral problem, the FD subject initially determined the function *f* by integrating the derivative of the function *f*, then reintegrated the function that had been obtained to get the integral of function *f*, but she did not integrate the constants part of the function. The steps used in solving problems understandable, but some of the information used was not equipped with explanations such as the use of the *f(t)* function. Based on the interview, the subject said that the function *f(t)* was intended as a function that must be found, but the subject could not provide a reason why he changed the variable from *x* to *t*. Therefore, the mathematical communication of FD subjects on integral problem was coherent, less accurate and incomplete.

3.4.2. *Geometry Problem*

Figure 3 below shows the answer of FD subject on the geometry problem.

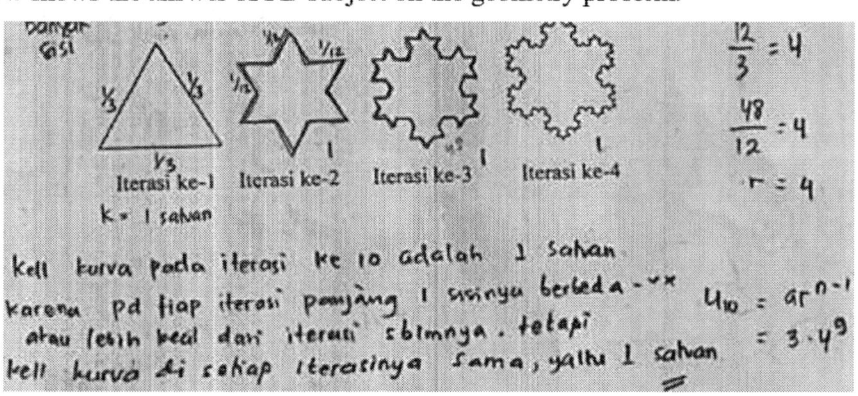

Figure 3. Geometry problem solving by the FD subject

From the answers of FD subjects, it was difficult to understand what she wants to explain. The FD subject only gave a mark on the side of the figure with a fraction and wrote "the circumference of the figure for each iteration is the same that is 1". Then, on the right side she wrote the geometric sequence formula and determined the 10th term with the initial term value was 3 and the ratio value was 4. When interviewed, the subject of FD said that the initial step she took was to determine the scale of the side changes for the next iteration. In the first iteration, because the circumference was 1, the side length was one third. In the second iteration, she counted the number of sides in the figure were 12, so she calculated the length of the sides was one-twelfth. Also, at the third iteration, the side length of one forty-eighth was obtained. Then she concluded that the circumference was always 1. However, she also remembered that this problem could also be solved by using the geometric sequence formula by determining the 10th term. She determined the initial term is the number of sides in the first figure, 3 and the ratio obtained from the number of sides of the second feature divided by the number of sides in the first figure so that it is obtained 4. She realised that she found two different answers but she did not know which one was true, and how to checked it. Therefore, the mathematical communication of FD subjects on geometry problem was incoherent, inaccurate and incomplete.

3.4.3. Maximal strategy problem

Figure 4 below shows the answer of FD subject on maximal strategy problem.

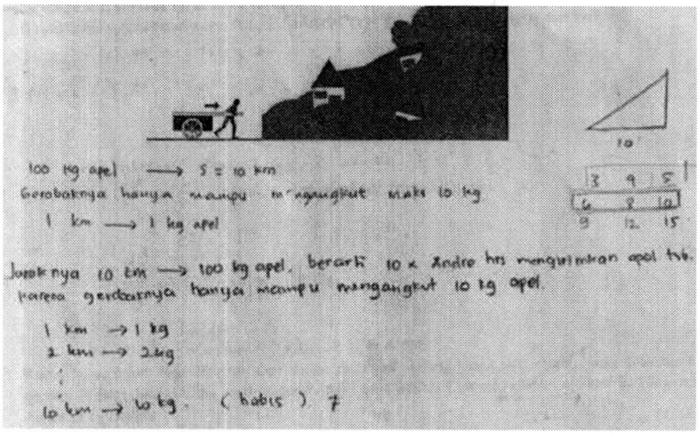

Figure 4. Maximal strategy problem solving by FD

The subject of FD wrote down the important points of the problem in diagrammatic form, she also wrote the three sequential numbers that satisfy the Pythagorean theorem, then made a chart that links the number of apples consumed with the distance travelled to send the apples. Then she concluded the distance between the warehouse and the store was 10 km and the apple consumed was 10 kg equal to the maximum apple that can fit in the cart so that the apple sent was 0 kg. In determining the maximum strategy for sending apples, the FD subject initially thought of using the Pythagorean theorem because it was related to "a store on the hill", but she finally realised there was no need to use it. She made a diagram of the relationship between the apple consumed, and the distance travelled to send the apple. She could not think of another way. Therefore, the mathematical communication of FD subjects on maximal strategy problem was incoherent, inaccurate and incomplete.

3.5. Mathematical communication of FI subject

The following description was the mathematical communication profile in the integral, geometry and maximal strategy problems of FI student.

3.5.1. Integral Problem

Figure 5 below shows the work of FI subject on the integral problem.

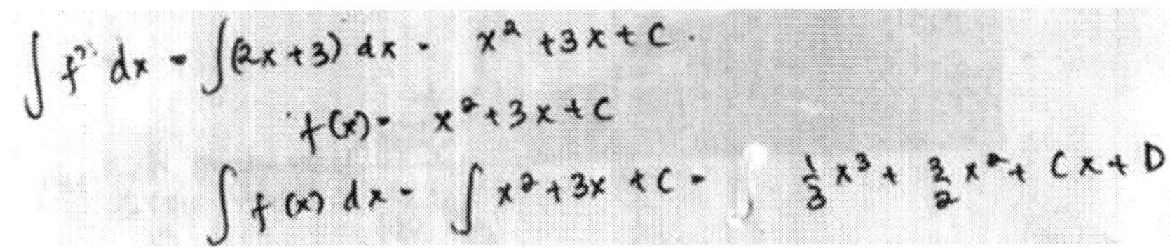

Figure 5. Integral problem solving by the FI subject

The FI subject solved the integral problem by integrating the function known, and she noted this by function *f*. Then, she integrated the new function to find the answer. She wrote the solutions coherently so that the steps could be understood, and it's just that there was a little inaccuracy when writing integrals at the end of the answer where the sign *dx* was missed. Therefore, the mathematical communication of FI subjects on integral problem was coherent, less accurate and complete.

3.5.2. Geometry Problem

Figure 6 below shows the work of FI subject on the geometry problem.

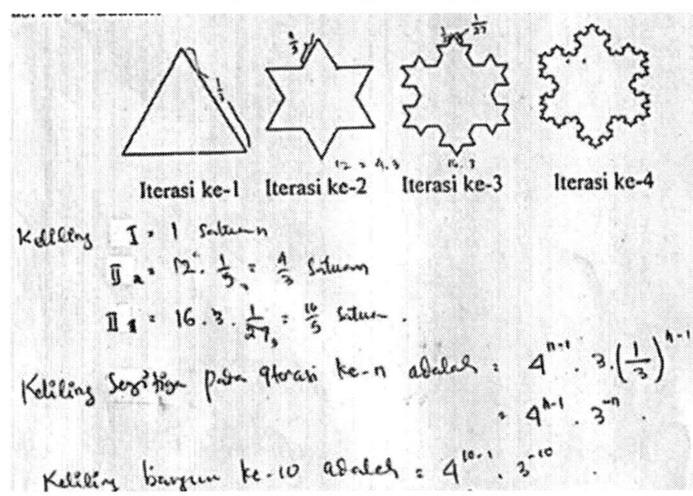

Figure 6. Geometry problem solving by the FI subject

The steps used by FI student: at first, she calculated the length of the side of the triangle, that was $\frac{1}{3}$, and then she determined the length of the side for the second iteration, and the third iteration respectively was $\frac{1}{9}$ and $\frac{1}{27}$. Then she counted the number of sides in the first, the second and the third iteration that she got respectively was 3, 4x3 = 12, 16x3 = 48. She noted the circumference of the object for the first iteration, the second and the third iteration, and from these, she found the pattern of the circumference. She used the formula of the pattern to find the circumference of the tent iteration the curve. She explained her way of thinking by writing down the strategies she used coherently, and used the marked in the side of the curve to show the length of the side but was not careful in entering the calculation results into formulas. When entering the value in the formula, she did not recheck with the data she had obtained so that she did not realise that there was an error in writing the exponent. Therefore, the mathematical communication of FI subjects on geometry problem was coherent, less accurate and incomplete.

3.5.3. Maximal strategy problem

Figure 7 below shows the work of FI subject on the geometry problem.

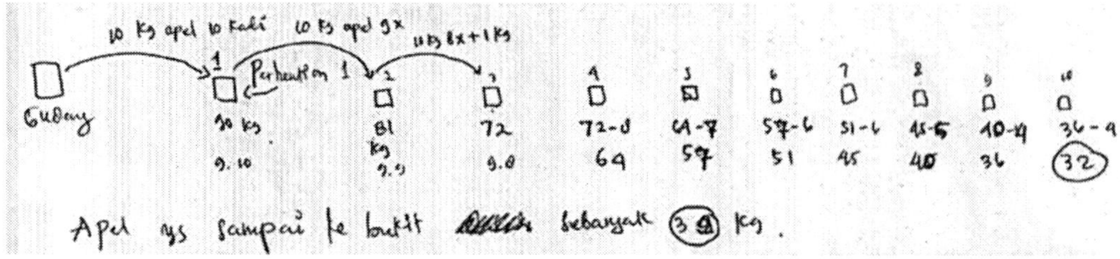

Figure 7. The answer to the maximal strategy problem by FI subject

FI subjects solved the problem by describing the stages of the strategy with a schema by writing down the number of times the apple was sent and the rest of the apple at the next stop. Even though the answer given was correct, it was the optimal result, but there was no written statement that it was an optimal result and the explanation about that. From students' explanations, it was known that after experimenting with various strategy, she wrote the maximum answer, but she could not provide justification logically. Therefore, the mathematical communication of FI subjects on maximal strategy problem was coherent, less accurate and incomplete.

4. Conclusion

Based on the results of this study, it can be concluded that most prospective math teacher in solving problems could not express what was in their minds, could not explain the strategies used clearly, accurately and coherently. Especially if the problem-given was a non-routine problem. Because, to solve routine problems, not too many explanations were needed. This matter could be seen from the percentage of the number of students answering integral problems accurately and coherently was more significant than when answering geometry problems even more with the maximum strategy problem. Related to differences in mathematical communication between FI students and FD students, based on the results of data analysis, it was found that for the integral problems, FI and FD students were relatively similar, they could write their answers understandable, only that there was little inaccurate writing. While for geometry problem and the maximal strategy problem, FI students explained the strategies they did more coherently and clearly than FD students, although some of them used charts, tables to solve the problems. FI students were more detailed in explaining what they did than FD students, and this may be due to the nature of FD who thinks globally. This research was based on written mathematical communication from prospective math teachers, it is suggested that the oral mathematical communication of prospective math teacher also needs to be investigated considering one of the teacher's tasks is to explain the subject matter to students.

Acknowledgements

This article is one of the research results (Fundamental Research scheme) funded by DRPM. Therefore, the researchers would like to thank DRPM Kemenristekdikti for funding this research and Universitas Negeri Surabaya that have assisted, as well as to all parties so that the research results in the form of this article can be completed.

References

[1] Pourdavood R G and Wachira P 2015 *Glob. J. Sci. Front. Res. F Math. Decis. Sci.* **15**
[2] NCTM 2000 Principles and Standards for School Mathematics (Virginia: Reston)
[3] Wichelt L and Kearney N E 2009 *Action Res. Proj.* **18**
[4] Pantaleon K P, Juniati D, Lukito A and Mandur K 2018 *J. Phys.: Conf. Ser.* **947** 012070
[5] Kabael T 2012 *Procedia-Soc. Behav. Sci.* **55** 809

[6] Liljedahl P, Santos-Trigo M, Malaspina U and Bruder R 2016 *Problem Solving in Mathematics Education* (Cham: Springer International Publishing)
[7] Lestari N D S, Juniati D and Suwarsonoe St. 2018 *J. Phys.: Conf. Ser.* **1008** 012074
[8] Muhtarom, Juniati D and Siswono T Y E *Glob. J. Eng. Educ.* **19** 130
[9] Juniati D and Budayasa K 2017 *World Trans. Eng. Technol. Educ.* **15**
[10] Sun C T, Wang D Y and Chang Y Y 2013 *Educ. Tech. Soc.* **16** 309

Analysis of Mathematical Literacy on Students' Metacognition in Conic Section Material

E Salsabila, W Rahayu, S A Kharis* and A Putri

[1] Department of Mathematics, Faculty of Mathematics and Science, Universitas Negeri Jakarta, Jl. Rawamangun Muka, Jakarta, 13220, Indonesia

* E-mail: sellyanas@yahoo.com

Abstract. Mathematical literacy is one of the components that use mathematical concepts and applies them to an everyday situation. Mathematical literacy helps people to identify and understand the role that mathematics plays in the world, and to make the well-founded judgments and decisions required in life by constructive, engaged, and reflective citizens. By contrast, not everybody acquires sufficiently. This study aims to describe the ability of mathematical literacy especially geometry literacy in terms of student metacognition on conic section material. The problems in this study were: 1) How is the ability of mathematical literacy in terms of student metacognition on conic section material 2) What types of mistakes do students make in solving the problem of cone section and 3) What factors cause students to make mistakes in solving conic section question. This research is a qualitative descriptive study. The subjects in this study were 36 mathematics education students who took courses in Analytical Geometry. The data collected using observation, written test, and interview. Before the researcher conducted the analysis, the researcher examined the validity of the data using validity and reliability tests in order to obtain valid data. Furthermore, the valid data were analysed descriptively through identification, grouping, clarification, explanation and conclusion.

1. Introduction

Mathematical literacy is important for students to know and understand the role of mathematics in real life [1,2,3]. Mathematical literacy is an individual's capacity to formulate, employ, and interpret mathematics in a variety of contexts [4]. It includes reasoning mathematically and using mathematical concepts, procedures, facts, and tools to describe, explain, and predict phenomena. It assists individuals to recognise the role that mathematics plays in the world and to make well-founded judgements and decisions needed by constructive, engaged and reflective citizens [5]. By contrast, not everybody acquires sufficiently. According to Programme for International Student Assessment (PISA) organized by the Organization for Economic Cooperation and Development (OECD), Indonesia student's average scores for mathematical literacy in 2000, 2003, 2006, 2009, 2012, and 2015 were 367, 360, 391, 371, 375, 386 points respectively [6,7]. These scores were below the average scores of OECD of 500 points for all the years. From 72 members of world countries, Indonesia is on the 65th place with the percentage of 42.3%, level 1-2 of mathematical literacy, meanwhile, only 0.8% who were on the level 5 – 6 in 2015. Besides that, the survey result of TIMSS' study, which is done in every four years started in 2011, Indonesia placed on the 36th of 40 countries in 2015. It showed that the students in Indonesia still had not shown any satisfying prestige [8].

According to the aforementioned importance and problems, it was necessary to intensively analyse students' mathematical literacy. Students' mistake in solving the problem can be a hint to find out how far they acknowledge the mathematical literacy. Students' mistake can also be useful to find out the students' way of thinking about the concepts which they had learned. The mistake done can be reviewed from students' metacognition. Metacognition is learner's ability to think consciously of their cognition and has control over their cognitive processes [9]. Metacognition puts the designing, monitoring and controlling consciousness about the known information needed to solve a problem and focus on the learning activity. When students realise that they have misunderstood or made a mistake, they can correct it. Students' mistake can be review with metacognition because metacognition investigates and measure the degree of relationship amongst mathematical literacy and student ability.

Based on the results of the observation and interview with lecturers and students in the Mathematics Department, Universitas Negeri Jakarta, one of the materials that students make many mistakes is conic section. Students find difficulty in working with the equations, the complexity of the many new terms and realising the graphs of the equation and the reflection properties of the standard conic section. Students usually think that they fail to integrate the concepts and apply geometry concepts in a real-life situation. They do not understand the question and make mistakes in the test. Some research has established too that conic section is one of the topics that students have many mistake and misinterpretations. [10,11]. The mentioned mistakes and the factors that caused them need to be determined, especially in students' metacognition, to find a way to help to prevent any similar problems. In this paper, we will present an analysis of mathematical literation base on students' metacognition in conic section material.

2. Materials and Method

2.1. *Materials*

The data collected from 36 students studying in Mathematics Education, Faculty of Mathematics and Sciences, Universitas Negeri Jakarta. All students were enrolling Analytical Geometry.

2.2. *Method*

This research-based on descriptive research with a qualitative approach. Descriptive research is a research used to describe, explain, and answers the questions of the phenomena as what they are and also the correlation of various variables in a phenomenon [12]. A qualitative is research which aims to understand the phenomena happened to a subject, such as the behaviours, perceptions, motivations, actions, and so on holistically, and by describing through words and languages, on a specific context which is scientific and taking advantages of various scientific methods [13].

Before being used for this research, the validity and reliability test was applied. The validity test in this research were content validity, construct validity, and empiric validity. Content validity measure represents all facets of a given construct, both of each question and overall questions. Construct validity is about how far the test items can measure what needs to be measured by the specific concept which was set [14]. The content validity and construct were tested by the lecturers. The empirical validity test used the Pearson Product-Moment Correlation shown in equation (1).

$$r_{xy} = \frac{N\sum_{i=1}^{N} x_i y_i - (\sum_{i=1}^{N} x_i)(\sum_{i=1}^{N} y_i)}{\sqrt{\{N\sum_{i=1}^{N} x_i^2 - (\sum_{i=1}^{N} x_i)^2\}\{N\sum_{i=1}^{N} y_i^2 - (\sum_{i=1}^{N} y_i)^2\}}} \quad (1)$$

Where r_{xy} is Pearson's correlation coefficient, N is the number of paired scores, $\sum_{i=1}^{N} x_i$ is the score of the first variable, $\sum_{i=1}^{N} y_i$ is the score of the second variable, $\sum_{i=1}^{N} x_i y_i$ is the product of the two paired scores. Correlation coefficient value which obtained for each item of question compared with the correlation coefficient value which was on the r table with $\alpha = 0{,}05$. If $r_{xy} > r_{table}$, then the correlation coefficient value is significant and considered as empirical valid [15]. The reliability test measured with *Alpha Cronbach* formula shown in equation (2). Where r_{11} is alpha Cronbach coefficient, n is number of items, s_i^2 is the variance of the total score formed by summing all items, s_t^2 is total variance. If the value of alpha is > 0.9 = Excellent, > 0.8 = Good, > 0.7 = Acceptable, > 0.6 = Questionable, > 0.5 = Poor, and < 0.5 = Unacceptable [16].

$$r_{11} = \left[\frac{n}{n-1}\right]\left[1 - \frac{\sum_{i=1}^{n} s_i^2}{s_t^2}\right] \quad (2)$$

After the instrument had checked as valid and reliable, then the collecting of data was done. The methods used to collect data in this research are observation, written test, and interview—the observation conducted by the researchers during the learning process in class. The test given to students consisted of questions of conic section, which were validated by experts. The test aimed to analyse the ability of students' mathematical literacy in geometry base on students' metacognition. The ability of mathematical literacy process in this research analysed based on the PISA indicator. They are communication, mathematically, representation, reasoning and argument, designing the strategy to solve the problem, the usage of symbols, formal language and technical, the usage of operation, and the usage of mathematics tools [17]—the analysing data of subjects' metacognition process based on awareness, evaluation, and regulation. Metacognitive awareness is an expression of how to think mathematically. Metacognitive regulation is a student's position in solving the problem. Metacognitive evaluation is an assessment which was created by the ways of thinking mathematically [18]. The written test takes 120 minutes for six essay questions. After the written test, the interview done by the researchers to lecturers and students.

3. Results and discussion

3.1. Instrument Analysis

The test instrument used in the research consists of six essay questions. The research instrument was tested to 30 students outside of the research samples. The result obtained from empirical validity can be seen in Table 1. From Table 1, it can be seen that the validity test of six items of essay questions fulfilled the validity criteria so that all of them could go on to the next test: the reliability test. Based on the reliability measurement instrument, about 0.879 was considered as Good, so that the instrument can be the measurement equipment because it's valid and reliable.

Table 1. Empirical validity result

	Item Question					
	1	2	3	4	5	6
r_{xy}	0.649	0.715	0.822	0.723	0.890	0.636
r table	0.361	0.361	0.361	0.361	0.361	0.361
Criteria	valid	valid	valid	valid	valid	valid

3.2. Observation

The study implemented during the second semester of the Academic Year 2019/2020 in a regular class environment. Before starting the course of analytical geometry, students gave a set of paper and pen to create their concept map in the geometry topics. Based on the created concept map, students know the basic formula of conic section and its characteristics. They can also mention the shapes of the conic section by the given pictures. They can give examples of a conic section in daily lives. They can write the general form and characteristics of circle and parabola correctly. However, they have difficulties in writing the formula of ellipse, hyperbola, and sphere. The interview was done to know the cause of their difficulties. Some of them confessed that they had not learned the conic section material at all except circle and parabola.

The beginning of observation was continued by asking students to create the writing of "UNJ" by using *desmos* application (www.desmos.com) in groups. With *desmos*, the beginning of students' ability in mathematical literacy of mathematics tool can be seen. The researchers informed the students that they are not obliged to use all the concepts in their concept mapping, and they could also integrate any other concept that they believed was relevant to the problem. We leave the student free in his choice of formula and language. In the making of the "UNJ" writing on *desmos*, the students had difficulties in representing the algebra equation into the geometric form. Some groups tried to find a certain function to create the U. Some groups tried to create the U by uniting semicircle, and the equation x = c, with c is a constant number. At the beginning of this research, it can be seen that mathematical literacy of planning the strategy to solve problems. Some students created the vertical line of N, and then they determined the equation line, which was suitable to connect both vertical lines. Students started to think of what gradient they would use, whether the value was positive or negative. Then, they measured the position of the line. After that, they were able to create the border of the line by the interval of N. From the making of those words, the communication between them happened (which was also considered as one of the mathematical literacy). Students created the letter J by using semicircle and vertical line. The process of making the letter J was not as hard as before, and the students did not have any difficulties. The students then presented ways of making the UNJ letters. The communication happened again by having two ways of communicating; the exchange of questions and answers between one group and the other ones.

After two weeks, a student enrolled in analytical geometry. The researchers did the observation in the class during the learning session and gave examples of questions with answers. Lecturers observed all of the students, reminded and approached the students who were not paying attention to the explanations, then helped them. The analytical geometry lesson was still having various problems, such as some students still could not concentrate during the learning process so they could not understand well about the material, students made mistakes in solving the problems, students still memorise the algorithm and given formulas. The learning process took 14 sessions with 150 minutes for each session. After following the learning session, six questions were applied to 36 students. The test took 120 minutes. With these questions, students' determination of the mistakes was evaluated in line with their mathematical literacy and metacognition.

3.3. Findings and Interpretation

From various students' answers, they got a different kind of mistakes. The classification of mistakes done based on Ashlock criteria [19] shown in Table 2.

Table 2. Type of students' mistakes in solving the questions

Item of Question	Conceptual Mistake	Operating Mistake	Algorithm Mistake	Symbolism Mistakes	Arbitrary Answer	Incomplete Answers	Mean
1	19.44%	0.00%	8.33%	2.78%	2.78%	5.56%	6.48%
2	11.11%	8.33%	8.33%	8.33%	2.78%	16.67%	9.26%
3	25.00%	11.11%	8.33%	8.33%	8.33%	25.00%	14.35%
4	30.56%	2.78%	5.56%	13.89%	13.89%	19.44%	14.35%
5	19.44%	2.78%	16.67%	5.56%	8.33%	19.44%	12.04%
6	25.00%	11.11%	19.44%	16.67%	11.11%	13.89%	16.20%
Mean	21.76%	6.02%	11.11%	9.26%	7.87%	16.67%	

Based on Table 2, the common mistake is a conceptual mistake with an average of 21.76%. To know the details of the mistakes' cause effects done by students in solving given questions of Analytical Geometry and the relationship between the students' metacognition, the author interviewed students and lecturers as respondents. The samples were the students who made mistakes in solving the given problems. The interview's result was combined with the metacognition indicator and then be analysed. Here is the analysis' result of mistakes done by the students.

Question 1: Prove the ellipse polar line equation of $a^2x^2+b^2y^2=a^2b^2$ drawn from the point $T(x_1,y_1)$ is $a^2x_1x+b^2y_1y=a^2b^2$

Question 1 consists of reasoning components and arguments. Mathematical literacy involved the ability of reasoning and give reasons. This ability is based on thinking logically and doing analysis of the information and resulted reasoning conclusion. The mistakes are done in this question and students'

Journal of Physics: Conference Series **1417** (2019) 012057 doi:10.1088/1742-6596/1417/1/012057

view about the reason shown in Figure 1. On Figure 1, it could be seen that the student could not understand the given questions well. During the interview, the student confessed that they did not know about the polar ellipse equations and they made conceptual mistakes. They only wrote the given information they knew from the lesson, without understanding the question. From the interview, the student realises that they made a mistake but still writes it down. It can be said that the student has done a metacognitive evaluation because they know the difficulties and can make an assessment of their abilities and understanding.

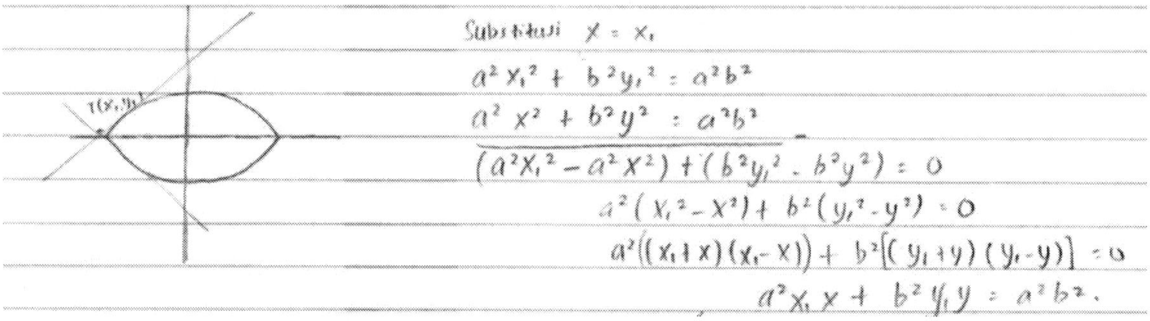

Figure 1. Student's answer for question 1

Question 2: Determine the sphere equation through the circle of intersection of the sphere $S_1 \equiv x^2+y^2+z^2=1$ and $S_1 \equiv x^2+y^2+z^2-4x-5=0$ through the point O(0,0,0).

Question 2 involved the reasoning and argument components, planning strategy to solve the problems, the usage of symbols, formal language and technical, and the usage of mathematics operation. This ability involved understanding, interpreting, and the ability to manipulate a mathematics context which was used to solve problems. The mistakes are done in this question shown in Figure 2. In Figure 2, students made a mistake by stating that the coefficient was equal to 0. They thought that the right side was equal to 0, so the others were the same and could be 0 = 0. Students did not use the information from given questions which were to pass the point O (0,0,0), so there was a mistake. This shows that the metacognitive awareness of students was still not optimally used. They wrote the points of the sphere but did not use the information correctly.

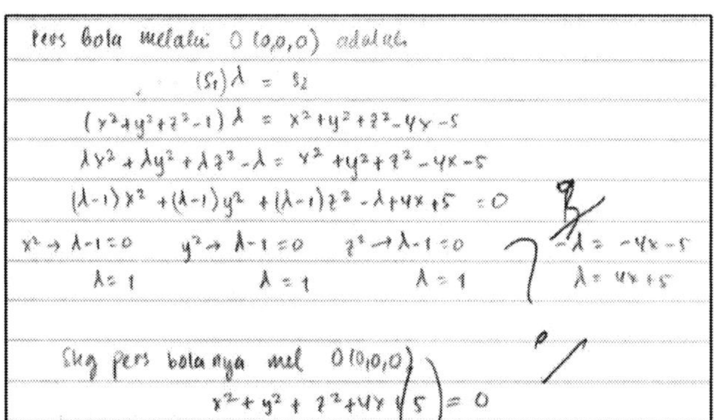

Figure 2. Student's answer for question 2

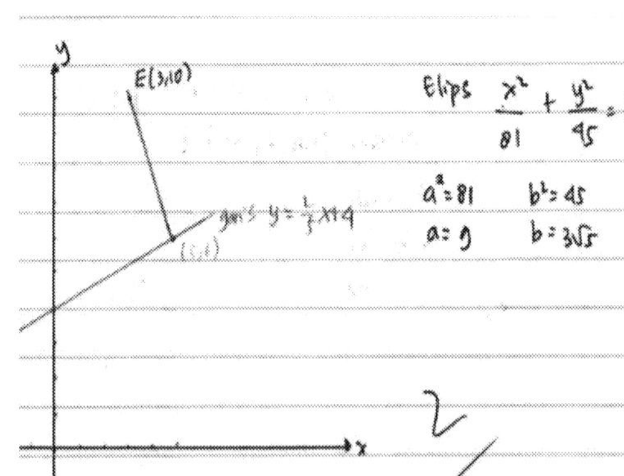

Figure 3. Student's answer for question 3

Question 3: There is a line crossing the point E(3,10) and perpendicular to $g \equiv y = 2/5\,x+4$ so it cut the ellipse $x^2/81 + x^2/45 = 1$ on point E and F. Find the length of EF.

Question 3 involved the representation, reasoning and argument, planning a strategy to solve problems, the usage of symbols, formal language and technical, and the usage of mathematical operations. One of the student's answers shown in Figure 3. Based on the interview, the student could not understand the meaning of the question. The student also would not know what to do with the given information in question 3. So, the student could not reach the level of metacognition because they even did not know how to use the information to solve the problem.

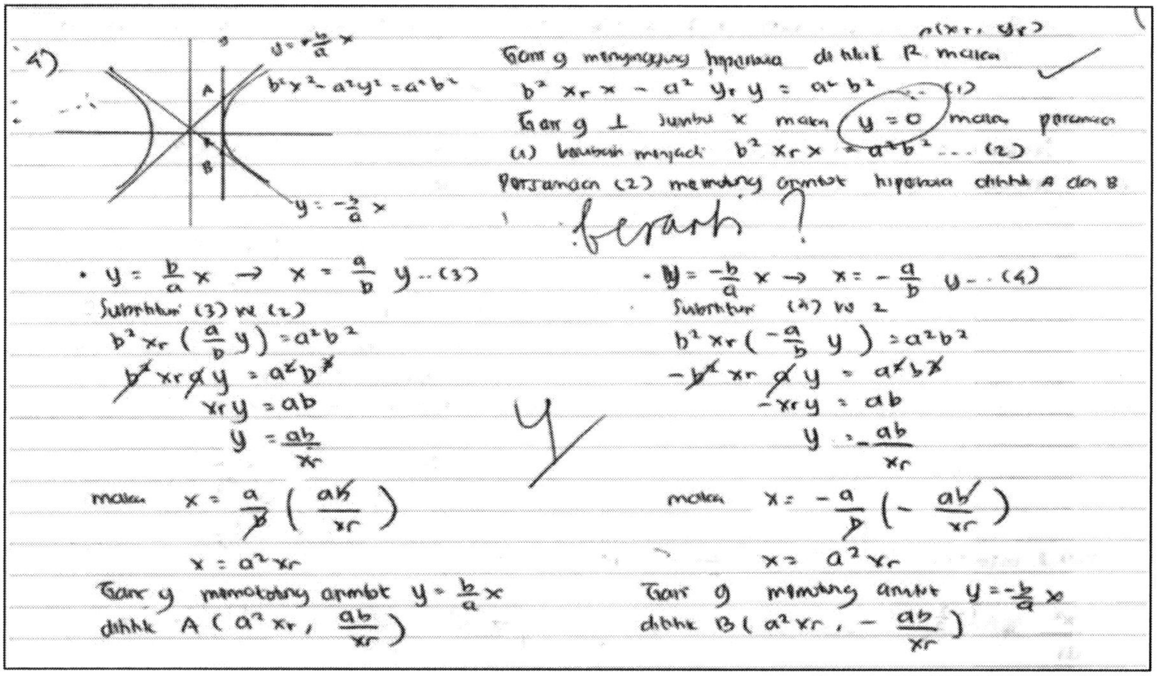

Figure 4. Student's answer for question 4

Question 4: Line g is the line that's crossing the point R on the hyperbola $b^2x^2\text{-}a^2y^2 = a^2b^2$ and perpendicular with the x-axis. Line g is cutting the hyperbola asymptote on point A and B. Determine the distance multiplication from R to A and R to B.

Question 4 involved the representation, reasoning and argument, planning a strategy to solve problems, the usage of symbols, formal and technical language, and the usage of mathematical operations. One of the student's answers shown in Figure 4. Based on the interview, the student could understand the meaning of the questions. They also could determine the first step of solving it, which was to find the intersection point of asymptote hyperbola. On this level, the metacognition regulation had been going on well. Nevertheless, on the next step, the student makes an algorithm mistake. This happens because the student was careless that means the metacognition evaluation not optimal yet: student did not recheck the steps of solving the problem.

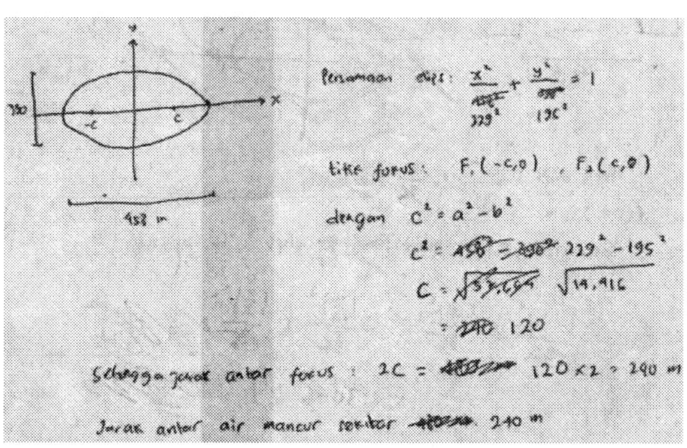

Figure 5. Student' answer for question 5

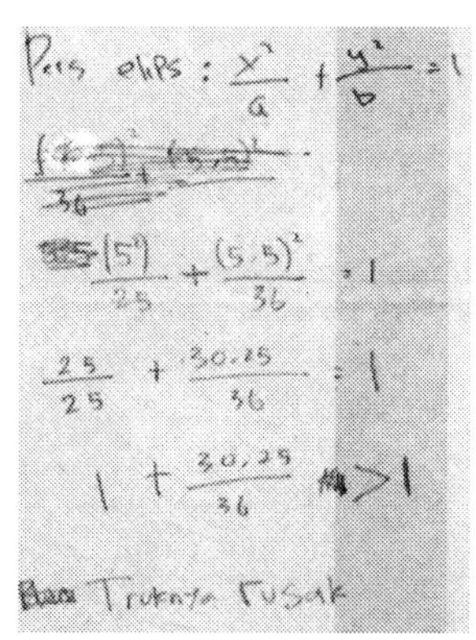

Figure 6. Student' answer for question 6

Question 5: In Washington, D.C, there was a park named "The Ellipses". This park was located between the White House and Washington monument. The park was surrounded by a street which was ellipse-shaped with the length of minor and major 458 and 390 meters respectively. If the manager of the park wants to build a water fountain on each focus point, determine the distance between the water fountains and explain your answer.

Question 5 involved the representation, reasoning and argument, planning a strategy to solve problems, the usage of symbols, formal language and technical, and the usage of mathematical operations. One of the students' answers shown in Figure 5. On this question, the student already fulfils the planning phase. This could be seen in how the student can explain the information from question 5 and presented the problem in Cartesians plane. The student could also explain the first step, which was to look for the ellipse focus. This first step of ability is so important because, in a good understanding of comprehension, the student can also know his/her thought of solving the problem. So the awareness and regulation metacognition take apart well in the process of their learning. This could also happen because the student rechecked the answer, student could do better and realised the mistake and could fix it.

Question 6: An ellipse bridge was built above the public street. The length and height of it were 10 and 6 meters respectively. If a truck with the width 5 meters and height 5.5 meters would pass the bridge, could it pass without having any damages?

Question 6 involved the representation, reasoning and argument, planning a strategy to solve problems, the usage of symbols, formal language and technical, and the usage of mathematical operation. One of the students' answers shown in Figure 6. In Figure 6, the student did not write the full version of the given information on the question. The algorithm in solving problem was not in the correct order, and the student could not understand the question well. The student only writes the answer arbitrary, so the awareness and regulation metacognition did not apply well.

According to the answers, despite the students did not know the conceptions of conic section well. Based on the interviews done, the factors of students' mistakes were classified into two: internal and external factors. The internal factor is the physical and psychological factor. Physical factor: full schedules of college life and side job make the students lack time to rest, so their study activities were also lacking and not maximal. Psychological factor: the students did not understand the materials well and could not solve the given questions if they were kind of different than the examples. The students were not coordinative with the lecturers, so when they had any difficulties in learning Analytical Geometry, they would not let the lecturers know and preferred to do it alone. These were accumulated, and the students would still be unable to solve the problems. During the test, there were careless students and got in a hurry when they were answering the questions. They also lacked doing exercises of Analytical Geometry problems. The external factors are that some students did not have a support system of studying at home, and they were not comfortable when they studied at home.

4. Conclusion

The mistakes done by the students in solving the geometry questions were 21.76% of conceptual mistakes, 6.02% of measurement mistakes, 11.11% of algorithm mistakes, 9.26% of symbolism mistakes, 7.87% of arbitrary answers, 16.67% of incomplete answers. The factors of students' mistakes in Analytical Geometry are classified into two: internal and external factors. Internal factors are full schedules of college life and side job, the students did not understand the materials well and could not solve the given questions if they were kind of different than the examples, the students were not coordinative with the lecturers. The external factors are some students did not have a support system of studying at home, and the crowd in their homes made them feel uncomfortable.

Acknowledgements

This research is fully funded and supported by *Lembaga Penelitian dan Pengabdian Masyarakat* (Institute of Research and Community Services), Universitas Negeri Jakarta.

References

[1] Sumirattana S, Makanong A, Thipkong 2017 *Kasetsart J. Soc. Sci.* **38** 307
[2] Devlin K 2000 *Learning Mathematics for a New Century* ed M J Burke and F R Curcio (Reston, VA: National Council of Teachers of Mathematics) pp 16
[3] Watson A 2002 *Aspect of Teaching Secondary Mathematics: Perspective on practice* ed L Haggarty (London: Routledge Falmer) pp 153-162
[4] Eva J 2014 *Encyclopedia of Mathematics Education* ed Lerman S, Sriraman B, Jablonka E, Shimizu Y, Artigue M, Even R, Jorgensen R and Graven M (Dordrecht: Springer) pp 391-396
[5] De Lange J 2003 *Quantitative Literacy: Why Numeracy Matters for Schools and Colleges* ed Madison B L and Steen L (Princeton, NJ: National Council on Education and the Disciplines) pp 75-89
[6] Argina A W, Mitra D, Ijabah N and Setiawan R 2017 *The 1st Education and Language International Conference Proceedings* 69-79
[7] Julie H, Sanjaya F and Anggoro A Y 2017 *J.Phys.: Conf. Ser.* **890** 012089
[8] Mullis I V S, Martin M O and Foy P 2016 *TIMSS Advanced 2015 International Results in Advanced Mathematics and Physics* (Chesnut Hill, MA: TIMSS & PIRLS International Study Center, Boston College)
[9] Williamson G 2015 *J. Initial Teach. Inq.* **1** 25
[10] Chen W -H 2013 *Int. J. Educ. Inf. Technol.* **7** 73
[11] Salinas P and Pulido R 2017 *EURASIA: J. Math. Sci. Tech. Educ.* **2** 341
[12] Haryanto A 2018 *J. Lang. Teach. Res.* **9** 197
[13] Mohajan H K 2018 *J. Econ. Dev. Environ. People* **7** 23
[14] Taherdoost H 2016 *Int. J. Acad. Res. Manag.* **5** 28
[15] Chee J 2015 *Pearson's Product-Moment Correlation: Sample Analysis*
[16] Namdeo S K and Rout S D 2016 *Int. J. Community Med. Public Heal.* **3** 1371
[17] Programme for International Student Assessment 2015 *PISA Results in Focus* 1
[18] Huda N Sutawidjaja A, Subanji and Rahardjo S 2018 *J.Phys.: Conf. Ser.* **1008** 1
[19] Ashlock R B 2006 *Error Patterns in Computation 9th ed.* (Upper Saddle River, NJ: Merril)

Journal of Physics: Conference Series **1417** (2019) 012058 doi:10.1088/1742-6596/1417/1/012058

Profile of Students' Strategy in Senior High School with Cognitive Reflective and Impulsive Style in Solving the Combinatorial Questions

N Aini*[12], D Juniati[3], T Y E Siswono[3]
[1]Doctoral Program of Mathematic Education Departement, Universitas Negeri Surabaya,
[2]Mathematic Education Departement, Sekolah Tinggi Keguruan dan Ilmu Pendidikan Jombang
[3]Mathematic Departement, Universitas Negeri Surabaya,

E-mail: nurulaini10@mhs.unesa.ac.id

Abstract. The goal of this research is to describe the strategies used by the Senior High students with the reflective and impulsive cognitive style in solving combinatorial questions. The strategy is seen from three aspects, namely: formulating, representing, and solving aspect. The subject of the research is the eleventh grade students based on the test of MFFT (*The Matching Familiar Figures Test),* to determine the reflective and impulsive cognitive style and also the communicative of research subject. The data is collected by giving the combinatorial questions and identifying the subject's strategy in solving the questions through interview. The validation used is time triangulation. The data analysis used is data reduction, data representation, and conclusion. The result shows that the students with reflective type are different from the students with impulsive type. When it is in the formulation stage, student with reflective cognitive style (RCS) reads the question slowly with strong intonation and gives the information by using his own words. In the representation, RCS draws four parallel boxes as many as the representation of the digits. In solving the question, RCS chooses the strategy based on the question rule, by memorizing and using strategy, RCS ways checks each step. While, in the formulation, ICS reads the question first, then makes parts of the questions by not using his own words. In the representation, student with impulsive cognitive style (ICS) uses the mathematics symbol. In solving the questions, ICS chooses a strategy fast without seeing the question rules overall and uses the strategy spontaneously without rechecking so that there are many mistakes found.

1. Introduction

Solving the problem in the form of question is a part of mathematical process. Standard process of mathematics at school includes problem solving, reasoning and proof, connections, communication, and representation [1]. Problem solving is the way of thinking, analyzing, and reasoning by using the experience and knowledge related with the problem [2]. When the students answer the questions, it will be influenced by the students' cognitive style. The impulsive cognitive style and reflective cognitive style are important variables which influence students' response in the class, behaviour in doing the test, and the result of the test.The students with impulsive cognitive style respond fast and dare to take the risk, while the reflective one tends to more slowly in responding, more slowly, and more carefully [3].

The question in this research is related with the combinatorial lesson. The combinatorial lesson is the branch of mathematics discrete focuses on countable discrete structures [4]. There are many advantages of learning combinatorial, such as: 1) it can train the students in enumeration concept, making assumption, optimization, and systematic thinking; 2) it can be applied in the branch of other field such as physics, chemistry, probability, and many others; 3) it can make the students more creative; [5]. However, there are many researches state that the students find difficulties in solving combinatorial questions because there are many dangerous traps when answering the questions so that the students do not understand the questions [6,7,8].

Therefore, there must be a way to solve combinatorial questions by using appropriate strategy [9]. Strategy is important point in solving the problem because it can ease to answer the questions precisely. There are three kinds of strategies in solving the combinatorial questions by listing, combinatorial operating and counting principle [9]. Combinatorial operation includes pattern/combinatorial formula or permutation. Counting principle is the principle in multiplication or addition [9]. Listing has four ways, namely: 1) it can be trial and error; 2) start strategy, where the students use certain pattern but it is not used overall; 3) cyclical pattern, where the students use all opposite patterns; 4) mistaken odometter where the students determine one constant variable but they fail in doing enumeration overall and it is too many mistakes; 5) perfect odometer, where the students determine one constant variable and find enumeration overall [10].To use the strategy above, the students must have competence in using a strategy. Strategic competence is mental activity in using the strategy of formulating, representing, and solving [11]. Fomulating sub aspect seen is the strategy used in understanding the problem to formulate the information known and strategy in formulating the information used to question. Representation of sub aspect seen is strategy used in representing problem situation. Solving the sub aspect is the strategy used to solve it [11]. While the representation meant is the model used by the students as the effort in looking for the solution and interpret the problem [12]. The form of representation is the real object representation, mathematics symbol representation, oral or verbal representation and picture or graphic representation [13,14].

The strategy in this research is development of Sukriyani's sub aspects, where the first aspect in formulating is two sub aspects seen, namely formulating in what kind of strategy used in understanding, formulating the information known and formulating the question and how to use the strategy. The second aspect is representing where there are two sub aspects seen, namely what kind of strategy used to represent the problem and how the process in using the strategy. The third aspect is solving, sub aspects seen, namely the strategy used and the reason why they use the strategy and how to use the strategy. Therefore, the researcher is interested in describing the strategy used by the Senior High School students with reflective cognitive style and impulsive cognitive style in solving the combinatorial questions.

2. Methods

This study is qualitative research. There are 140 students as the respondents. The researcher uses the instrument of *Matching Familiar Figure Test* to get the information between the students who use reflective cognitive style and impulsive one. There are 35% students with reflective cognitive style, 32% students with impulsive cognitive style, 10% students with fast and accurate cognitive style and 23% with slow and inaccurate cognitive style. The combinatorial questions are given to the student who has reflective cognitive style and the student who has impulsive one that has the same mathematical and communicative ability. The combinatorial questions are used after va lidated by the experts.

The combinatorial question is:

Reno and Tomi are students of SMA Negeri 2 Jombang who won the creativity competition in national level. They create a vacuum cleaner robot. To avoid the blame of other people, they designed the robot carefully where the robot would obey its master after he pressed the password button in the robot chess. Reno and Tomi use numbers with the date and the month when the robot is used to ease them to remember. The robot was created in July, 12 2016. The password consists of four digits and does not repeat the same numbers. In your opinion, how many numbers in the password made by Reno and Tomi? Explain how you determine it!

The research subjects with reflective and impulsive cognitive style are given the combinatorial question and they are identified for the strategies used when they solve the combinatorial questions through interview. The validity uses the time triangulation. The analysis technique uses three steps, namely: data reduction, data representation, and conclusion.

3. Results

The researcher will explain the data which inform about the strategy used in solving combinatorial question. The data is got through test and deep interview. The data is presented in two parts namely the strategy of students in Senior High School with reflective cognitive style and students with impulsive cognitive style in solving combinatorial questions.

3.1 *Student's strategy in Senior High School with reflective cognitive style* (RCS) *in solving combinatorial question.*

RCS in the beginning understands the question first by reading the question slowly and stressing intonation by using their own words. RCS gets the information that there are four digits with different numbers, which are arranged of the number 1,2,0,7 from the date when the robot reated. RCSdraw patterns the information questioned that the passwords are searched from the number 1,2,0,7 while pointing the numbers with a pen. The beginning strategy used by RCS to solve the question is drawing. RCS draws the first box, then goes to the next second box, third and fourth box in a parallel line to represent the numbers of the digits in the password as shown in the following Picture 1:

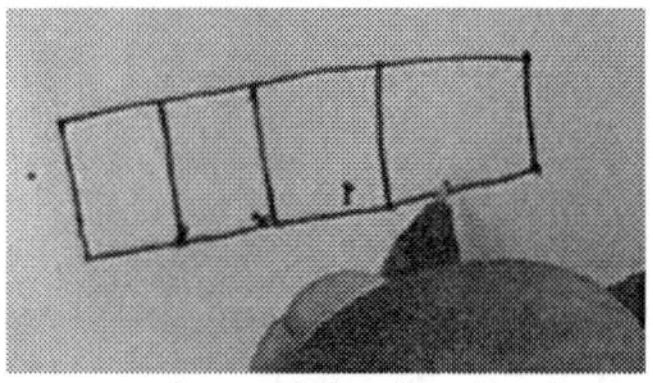

Picture 1. Representation of the beginning strategy of RCS

Then RCS inputs "4" in the first digits. When he writes the next numbers,RCS shakes his pen while memorizes the question rule that states no numbers are the same. RCS inputs in the second digit with "3" as numbers left, then the third digit with two numbers in the rare last digit. Finally, RCS multiplies the numbers based on the multiplication rule so that RCS gets 24 arrangements formed from the number 1, 2, 0, 7. It can be seen in Picture 2 below.

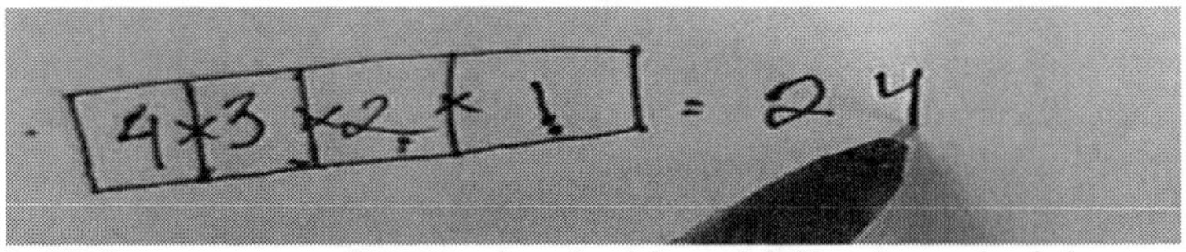

Picture 2. Written answer of RCS related with multiplication principle strategy

RCS looks at the answer, then decides to use another way to make sure his answer. The next strategy used by RCS is using the pattern. In the beginning, RCS determines the pattern will be used by memorizing the question rule by shaking the pen that is no same numbers. RCS determines the pattern of different object permutation that will be used. RCS writes the different object of permutation pattern, namely P(n,r) = n!/(n-r)!. RCS represents the available numbers and the numbers asked are as many as the numbers in the password by using the mathematics symbol, namely n and r. It is shown in Picture 3. And the interview result in Table 1 below.

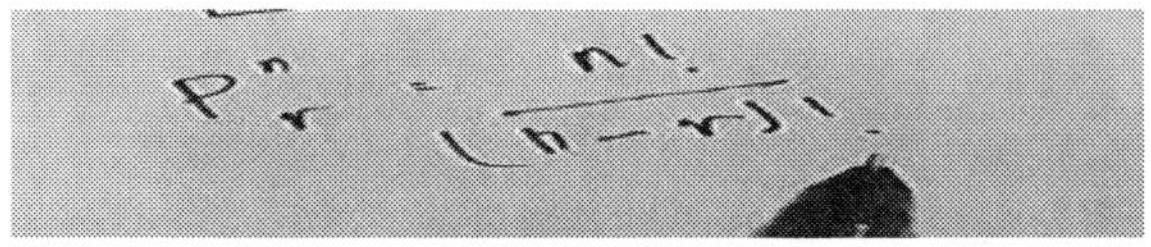

Picture 3.RCS' mathematics symbol representation

Table 1. Interview sample related with mathematics symbol representation by RCS

Researcher	:	What is n?
RCS	:	n is the available numbers
Researcher	:	How about r ?
RCS	:	r is the numbers wanted

RCS understands the exclamation mark in the pattern as the factorial which means the multiplication in order from 1 or decrease. Then, RCS inputs number four as n because there are four numbers, namely 1, 2, 0, 7, then inputs number four as r because the passwords of the object are different with four different numbers. RCS continues it with counting process namely 4!/0!. RCS understands 0! is same with 1 for factorial rule. RCS continues the counting process to become 4!/1= 4!. RCS explains 4! It becomes $4 \times 3 \times 2 \times 1$ so that he gets 24 orders. It can be seen in Picture 4 below.

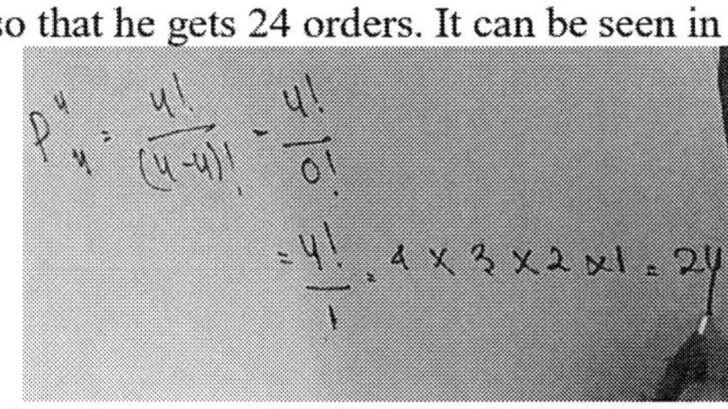

Picture 4.The written answer of RCS related with the pattern of the strategy

After that, RCS makes sure that the result is correct by having multiplication principle and the pattern by listing the number of the password one by one. RCS writes the numbers arranged in the beginning namely 1, 2, 0 and 7. Then, RCS determines the first constant variable namely 1. RCS lists all the possibilities of the passwords such as: 1270, 1072, 1027, 1702, 1720 and 1207. After listing all of them, then RCS determines the second constant variable namely 2. RCS lists all the possibilities of the passwords such as: 2170, 2107, 2701, 2710, 2071 and 2017. Then RCS determines the third constant variable namely 0. RCS lists all the possibilities of the password such as: 0712, 0721, 0217, 0271, 0172 and 0127. At last, RCS determines the fourth constant variable namely 7. RCS lists all the possibilities of the password such as: 7102, 7120, 7210, 7201, 7012 and 7021. It can be shown in Picture 5 below.

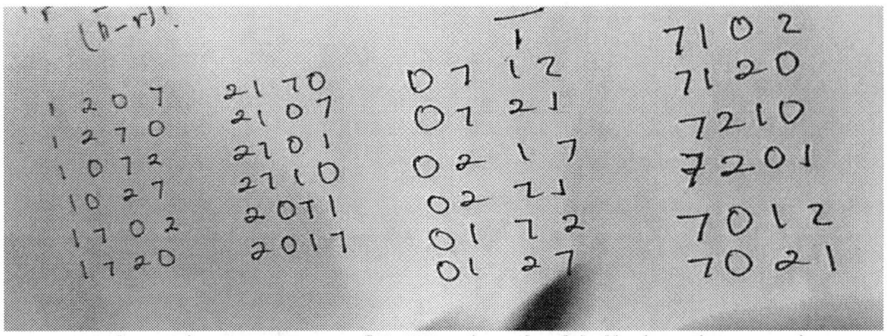

Picture 5.The written answer of RCS related with listing the members

3.2 Student's strategy in Senior High School with impulsive cognitive style (ICS) in solving combinatorial question.

In the beginning, ICS understands the question by reading the question rule first then he takes parts of question by not using his own words. While looking at the question, ICS gets information that Reno and Tomi makes the robot with four digits from the date and the month the robot made that the four digits may not be the same. ICS formulates the information questioned by reading the question command namely: in your opinion, how many passwords that can be made by Reno and Tomi and explain how he can find out. The beginning strategy used by ICS in solving the pattern is that ICS mentions the password in order spontaneously with non random, so that he decides the permutation pattern used. ICS keeps silent because he forgets the name of permutation that is chosen but he writes it correctly, namely $P_{n,r} = \frac{n!}{(n-r)!}$. ICS represents the password and the digits that must be fulfilled by using mathematics symbol, namely n and r. It can be shown in Picture 6 and the interview result below.

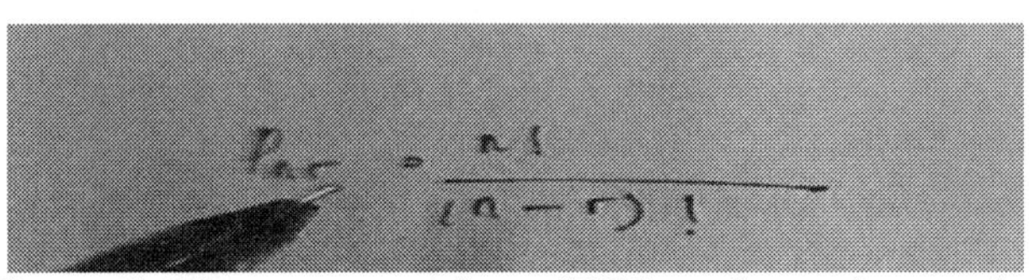

Picture 6. The representation of ICS mathematics symbol

Table 2.The interview sample related with the representation of ICS mathematics symbol

Researcher	:	What is n?
ICS	:	n is the number of the password
Researcher	:	ow
ICS	:	Eh, do you mean they are the numbers for the password?
Researcher	:	How about r?
ICS	:	r is the digit that must be fulfilled

ICS thinks that the exclamation mark is the factorial. ICS thinks that factorial is the multiplication in order from 1. After that, ICS decides the n value is same with 4 because the password is made of the date and the month the robot made and ICS determines the r value is same with 4 because there are four digits. After that, ICS counts it $\frac{4!}{(4-4)!}$. ICS decreases$(4-4)!$, to become $\frac{4!}{1}$. ICS understands that is same with 1. He answers quickly based on the permutation rule. Then, ICS explains 4! to become $\frac{4\times3\times2\times1}{1}$, the he crosses the same number namely 1. Then ICSmultiplies 4 with 3 to become 12 and finally multiplies12 with 2 to become 24. ICS states that 24 is the number of possibilities in making the password. It can be seen in the Picture 7 below.

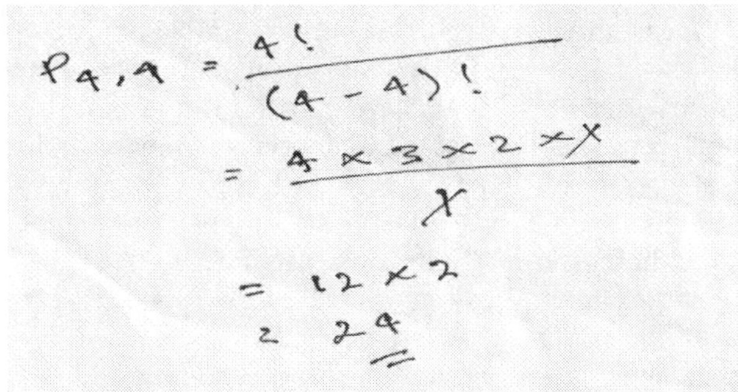

Picture 7.The written answer by ICS related with the pattern strategy

Then ICS makes sure his answer by using the strategy of listing its members. ICS writes first the lottery numbers made. After that ICS lists one by one such as the first 2170, the second 1720, the third 7210, in the fourth step ICS changes 2170 to be 2701, then ICS continues to the fifth 7201, the sixth 2710, the seventh 7021, and the eighth 2710. When ICS writes the ninth and the tenth, he makes a mistake because of carelessness such as the ninth is 7201 that repeats the fifth and also same for the tenth 2710 that repeats the sixth. Without realizing his mistakes, ICS continues listing the members that the eleventh 2017, twelfth 1207, thirteenth 7102, fourteenth0172, fifteenth 0721, sixteenth 0271, seventeenth 2701. ICS repeats the fourth, eighteenth 2107, nineteenth 7021 repeats the seventh, twentieth 2017 repeats the eleventh, twenty first repeats the fifth and the ninth, twenty second 0217, twenty third 1027, twenty fourth 2710 repeats the sixth and the tenth. ICS thinks that he is right by seeing the result of permutation strategy without rechecking. It can be shown in Picture 7 below.

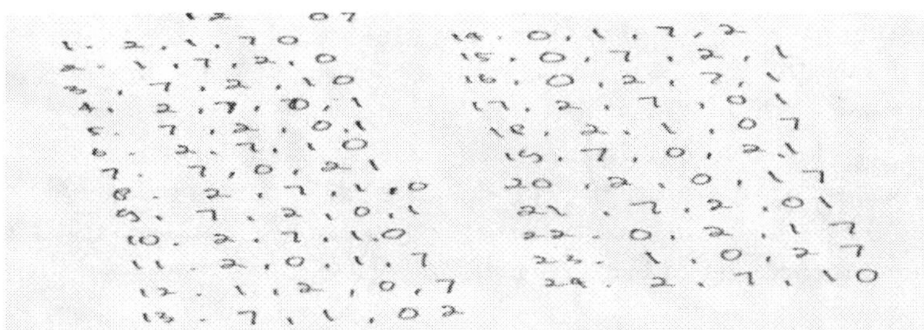

Picture 8.The written answer of ICS related with listing member strategy

4. Discussion

This research finds that there is relation between the strategy of RCS and ICS in solving the combinatorial question. Students of RCS and ICS in solving the question are same in answering the question by understanding the question, identifying the information known, formulating the question, doing the representation, and answering the question but they have different strategy and process. It shows that RCS and ICS does the strategy competence based on [11] that competence strategy can be seen from three aspects namely formulating, representing and solving.

In formulating, RCS tends to read slowly with stressing intonation then retell it what he understands by using his own words for the information known and questioned, while ICS reads by taking parts the question without intonation based on the question language, consistent with [3] who states that student with impulsive cognitive style responds faster more carefully. In the representation, RCS represents the beginning situation by using four boxes picture because there are four digits where RCS draws the first box, the second box, the third and the fourth in parallel way and RCS uses the mathematics symbol representation namely P, n and r. Pn, r is the representation form amount of password orders, n represents the number of the password and r represents the digits. It equals with [12,13] that representing is the interpretation of the question to find out the solution.

In solving stage, RCS uses three strategies namely multiplication rule strategy, pattern strategy, and listing member strategy while ICS uses two strategies namely pattern strategy and listing member strategy. The strategy can be used to solve the combinatorial question refers to [9] who states that solving combinatorial problem needs listing, combinatorial operation, and counting principle. In the counting principle, RCS starts by drawing four parallel boxes as the representation of the password numbers then inputs the numbers in the box base on the rule that the numbers must be different. The numbers are then multiplied based on multiplication principle while ICS does not use this strategy. RCS mentions the strategy as the multiplication principle as [9] states that the multiplication principle is the counting principle. In the pattern strategy, RCS and ICS are similar in using permutation and counting process but they have different reason in determining the pattern. RCS chooses the permutation with different object by seeing the condition that there must be different object, while ICS mentions the order password spontaneously. This refers to [9] that permutation pattern is combinatorial operation. In listing member strategy, RCS and ICS have different listing strategy where RCS tends to be more careful in listing one by one. RCS determines the constant variable first then lists one by one overall. It refers to [10] who states that Odometer is a perfect strategy where students determine variable and enumeration. It refers to [3] that the reflective cognitive style tends to be slower in responding and more careful while ICS tends to be careless so that there are many mistakes in listing by using trial and error strategy. It refers to [10] who states that they can use trial and error way and ICS carelessness is consistent with [3] that impulsive cognitive style responds fast and dare to take a risk.

5. Conclusion

The finding shows that the subject of RCS is more careful in formulating, representing and solving. In formulating, RCS reads the question slowly with stressing intonation and delivers the information by using his own words. In the representation, RCS draws four parallel boxes which represent the number of the digits. In solving, RCS chooses the strategy based on the question rule by memorizing and rechecking one by one. While ICS tend to be spontaneous and careless in doing formulation, representation, and solving. In the formulating step, ICS reads the question first then takes parts the question by not using his own words. In representation, ICS uses mathematics symbol. In solving stage, ICS chooses the strategy fast without seeing the question rule overall and use the strategy spontaneously without rechecking so there are many mistakes found.

References

[1] National Council of Teachers of Mathematics 2000 *Using NCTM's Content and Process Standards* http://www.math.vt.edu/people/lloyd/curriculum/teaching_activity2.pdf. [Accessedon 4 Oktober 2013]

[2] LeshR 1981 Applied mathematical problem solving *Educational Studies in Mathematics* **2**(12) pp 235-265

[3] Acharya Ms Chandrama 2002 Students' Learning Style and Their Implications for Teacher *Center for Development of Teaching and Learning (CDTL)* **5**(6) pp 1-3

[4] Hanali A 2013 *Students' Ways of Thinking about Combinatorics Solution Sets* (America:Arizona State University) p 316

[5] Kapur J N D 1970 Combinatorial Analysis And School Mathematic *Reidel publishing company* **3** pp 111-127

[6] Batanero C, Navarro-Pelayo V, and Godino J 1997 Effect Of The Implicit Combinatorial Model On Combinatorial Reasoning In Secondary School Pupils *Journal Educational Studies In Mathematics* **32** pp 181-199

[7] Syahputra E 2017 Combinatorial Thinking (Analysis Of Students Difficulties And Alternative Solution) *Proc Int Scien and Scien. Edu 7th – 8th October* pp 1-12
[8] Hadar1981 The Road To Solving A Combinatorial Problem Is Strewn With Pitfalls *Educational Studies in Mathematics* **12** pp 435-445
[9] Höveler K 2016 Children´S Combinatorial Counting Strategies and Their Relationship to Mathematical Counting Principles *Jurnal International Congress on Mathematical Education* **1** pp1-8
[10] English L D 1993 Children's strategies in solving two- and three-dimensional combinatorial problems *Journal for Research in Mathematics Education* **3** (24) pp 255-273.
[11] Sukriyani A, Juniati D, and Siswono TYE 2017 Strategic competence of senior secondary school students in solving mathematics problem based on cognitive style *AIP Conference Proceedings* **050009** pp1-6
[12] Ainswort S 2006 Deft: A Conceptual Framework For Considering Learning With Multiple Representations *Elsevier* **16** pp 183-198
[13] Hwang W Y, Chen N S, Dung J J, and Yang Y L 2007 Multiple Representation Skills And Creativity Effects On Mathematical Problem Solving Using A Multimedia Whiteboard System *Educational Technology & Society* **10**(2) pp191-212
[14] Johnson S 1998 What's In A Representation, Why Do We Care, And What Does It Mean? Examining Evidence from Psychology *Automation in construction* **8** pp 5-15

The Investigation of Blind Students' Misconception in Constructing Quadrilateral Analytic Definition Using Geometry's Puzzle

Andriyani[1], D Juniati[2]

[1]Universitas Ahmad Dahlan, Yogyakarta
[2]Universitas Negeri Surabaya, Surabaya

e-mail: andriyani@mpmat.uad.ac.id

Abstract. Many students still possessed misconceptions and not yet know the relation between each quadrilateral, no exception blind students that have abnormal physiological structures. Lack of student' ability in relational thinking makes it difficult to construct analytic definitions that contain the closest type (genus proximum) and special differentiator (differentia specifica) on the objects. The aim of this study is to investigate the misconceptions of blind students in constructing the analytic definition of a quadrilateral using geometry's puzzles. Misconception investigations focus on misconceptions of students' concept or differences of student' scientific understanding with the expert conceptions, especially misconception in constructing analytic definitions. This research is an exploratory study with a qualitative approach and the subjects of this study are one seventh grade student who experienced totally blind and one student who low vision from MTs Yaketunis Yogyakarta. Data collection through task-based interviews and observations. Before being analyzed, the obtained data then transcribed, classified, reduced and validated using the time triangulation method. Based on the results of data analysis, it is known that both of blind students have misconceptions namely classificational, relational and theoretical misconceptions with the level of misconception in low vision student lower than totally blind.

1. Introduction

Changed the educational paradigm to a learning society requires the existence of universal learning principles that underlie education to deal with these changes [1]. To cope with the world changed into the Twenty-First Century, students need to know how to use their knowledge and skills by thinking critically, applying knowledge to new situations, analyzing [2].

The results of the PISA (Program for International Student Assessment) showed that Indonesian students still had difficulty in applying mathematical concepts in contextual issues which are one of the demands of Twenty-First Century learning [3]. This is caused by learning which emphasizes procedural and monotonous knowledge, so mathematics is considered a difficult subject by most students [4]. One branch of mathematics that is subject to study in school mathematics subjects and has the potential for difficulties for students in geometry. Geometry is more familiar and has a great opportunity to understand for students because of the introduction of basic concepts and geometry models they already get through objects that often saw in their informal education environment. Even so, there are still many students who make mistakes in learning geometry especially in a two-dimensional shape [5]. One of the mistakes is caused by the misconception of the concept meaning that is characteristic of learning itself. Conception is generally built on common sense or intuitively to make sense of something.

The misconception is very difficult to change because a person constructs his own knowledge which is influenced by differences in individual thinking, methods of delivering material by the teacher or learning resources used by students. The misconception that comes from the student is caused by the incompatibility of the initial concept student have with scientific concept. While misconception that comes from the teacher or learning resource is caused by a lack of understanding concepts taught and lack in the resources learning preparation. These misconceptions can be an obstacle to students understanding concepts well. Several types of misconceptions are viewed from students' visual-spatial intelligence, namely theoretical misconception, classificational misconception and correlational misconception [6].

According to [7], students experience misconception on the quadrilateral definition and assume that the quadrilateral only in a regular shape. If the quadrilateral concept of students experiencing misconception, students can have difficulty understanding more complex concepts because the geometry concept has a hierarchical structure and is spiral. According to [8], generally students were very capable of solving the calculating or measuring problem in the quadrilateral and perimeter, but whenever teacher asked questions that required them to analyze or synthesize the relationship between the measurements, they ran into problems. Their typical response that they had some misconceptions which needed to be addressed. It means they not really understanding about the quadrilateral concept but recall only. Students haven't been able to identify similarities or differences in the characteristics of each quadrilateral and then be classified in the closest type (genus proximum) and special differentiator (differentia specifica), so it can be seen the relation between quadrilateral with each other. The students' inability in relational thinking can make it difficult for students to construct analytic definitions that contain the closest type (genus proximum) and special differentiator (differentia specifica) on the related object. This shows that quadrilateral learning has problems for students to learn especially the student that can't optimize their learning ability like blind students who have a limitation of visualization or visual sensory acceptance [9].

With these physiological structure abnormalities, blind students possess the problems in understanding geometry picture [10]. This constraint can also be difficult for a teacher who teaches concept especially those related to geometry learning strategy instruction, although using physical models [11]. Limitation and obstacle in learning geometry aren't obstacles to continue teaching geometry to blind students because in the Minister of Education and Culture Regulation No. 157 of 2014 has been arranged the content of the special education curriculum for blind students is equated with the content of the regular education curriculum. Thus, the content of geometry material such as quadrilateral must also be taught to blind students.

The reasons about quadrilateral that is part of the geometry in the mathematic subject and there are still problems in learning geometry for blind underlie researcher to conduct next investigation about the conceptual misconception that might occur in quadrilateral learning by blind students. The misconception investigation is focused on misconception related to identifying the closest type (genus proximum) and special differentiator (differentia specifica) in the quadrilateral which is then used to construct the analytic definition constructed by the closest type (genus proximum) and special differentiator (differentia specifica). So this study aims to investigate the misconceptions of blind students in constructing quadrilateral analytic definition using geometry's puzzle. Misconception investigation is defined as an investigation conducted to detect student learning that is estimated to experience misconceptions of concept so that students' conceptions differ scientifically from the conception of experts.

2. Method

This research is an exploratory study with a qualitative approach that aims to investigate the misconceptions of blind students in constructing the analytic definition of a quadrilateral by using geometry's puzzle. Interview questions consist of the subject's construction about quadrilateral analytic definitions namely, parallelogram, rhombus, rectangle, and square. The subjects chosen as the source of qualitative data were one of the seventh-grade who experienced total blindness and one of the low vision blind students from MTs Yaketunis which is a school for exceptional children in Yogyakarta. Data on how blind students' constructing about quadrilateral analytic definitions were collected through task-based interviews and observations that conducted twice using equal problems at two different times. Subjects were assigned to recognize the quadrilateral shape models in the form of a geometry's puzzle for a parallelogram, rectangle, rhombus, and square given by the researcher. Then, they are assigned to define each of these shapes in terms of the elements of the closest genus and its specific differentiator.

Data collected from interviews are then transcribed, classified, reduced and validated using the time triangulation method to produce credible data. Credible data were analyzed using qualitative data analysis research methods including interpretation of data display and conclusion that can be inferred [12].

3. Result and discussion

Based on the results of the data validation of interviews and observations of two blind students, obtained data about various misconceptions experienced by students in constructing quadrilateral analytic definitions. From several indications of misconceptions in terms of visual-spatial intelligence, namely classificational misconception, correlational misconception, and theoretical misconception, each student has a different misconception in constructing quadrilateral analytic definitions that contain the closest type (genus proximum) and special differentiator (differentia specifica). The following presented the results of the misconception analysis of blind students in constructing analytic definitions using a geometry's puzzle consisting of parallelogram, rectangular, rhombus and square models as follows.

Figure 1. Geometry's puzzle

The totally blind student has a classificational misconception which is a form misconception of facts classification into organized charts. This is seen when a student asked to determine which geometry's puzzle belongs to parallelogram. The student made a mistake in determining the geometry's puzzle included into a parallelogram, he didn't insert rectangular, rhombus or square into the parallelogram. The student even put parallelogram into the rectangle. In classificational misconception, the student also makes a mistake when determining geometry's puzzles that are included in the rectangle. He didn't put the square into parallelogram but instead, put the rectangle into the square. Furthermore, the student also makes a mistake in determining the geometry's puzzle included into the rhombus, he doesn't insert a square into the rhombus. The student just the opposite inserts the rhombus into the square.

Besides classificational misconception, totally blind student also possesses a correlational misconception which is a form of the misconception of special events that are interconnected. This can be seen when the student makes a mistake in explaining the relation between the parallelogram and the rectangle like to interview result below.

Researcher: In your opinion, do the rectangles and parallelogram have similarities or differences in the characteristics of the shape?
Subject (A): Yes
Researcher: What characteristics are the same or different in both?
Subject (A): The same side of the opposite is equally long and some are short
Researcher: What do you mean?

Subject (A): There are two opposite sides that are equally long, then there are two opposite sides that are both shorter
Researcher: How about the difference?
Subject (A): The difference is that the parallelogram is a rectangle whose sides are slanted
Researcher: What about rhombus? Does he have the same similarities or differences as parallelogram and rectangle?
Subject (A): There are no similarities because of both shapes different in all their characteristics
Researcher: How about square and rectangle?
Subject (A): The rectangle is a square that has a longer side measure.
Researcher: What about rhombus and square?
Subject (A): Both are the same side and angles the rhombus is a square that positioning slanted

From the interview, it known that student assumes about the rectangle isn't parallelogram, but the parallelogram is a rectangle. In addition, the student also can't explain the relation between parallelogram and rhombus because both of there are different. Another correlational misconception is the student's mistake in explaining the relation between a square and rectangle because he assumes that the rectangle is a square that has a longer side. Others, the student also make a mistake in explaining the relations between rhombus and square because he assumes that rhombus has the same side length and angle type. The student also assumes the rhombus is a square that is positioned slant

The totally blind student has a classificational misconception which is a form misconception of facts classification into organized charts. This is seen when a student asked to determine which geometry's puzzle belongs to parallelogram. The student made a mistake in determining the geometry's puzzle included into a parallelogram, he didn't insert rectangular, rhombus or square into the parallelogram. The student even put parallelogram into the rectangle. In classificational misconception, the student also makes a mistake when determining geometry's puzzles that are included in the rectangle. He didn't put the square into parallelogram but instead, put the rectangle into the square. Furthermore, the student also makes a mistake in determining the geometry's puzzle included into the rhombus, he doesn't insert a square into the rhombus. The student even inserts the rhombus into the square.

Besides classificational misconception, totally blind student also possesses a correlational misconception which is a form of the misconception of special events that are interconnected. This can be seen when the student makes a mistake in explaining the relation between the parallelogram and the rectangle, he assumes that the rectangle isn't parallelogram, but the parallelogram is a rectangle. In addition, the student also can't explain the relation between parallelogram and rhombus because it is different. Another correlational misconception is the student's mistake in explaining the relation between square and rectangle because he assumes that the rectangle is a square that has a longer side. Others, the student also make a mistake in explaining the relations between rhombus and square because he assumes that rhombus has the same side length and angle type. The student also assumes the rhombus is a square that is positioned slant.

Another misconception possessed by totally blind is a theoretical misconception which is a form of misconception about the explanation of facts or events in an organized system and concerns the development process from that known until that unknown by the student. It has seen when the student not exactly in making an analytic definition of parallelogram shape. According to the student, the parallelogram is a rectangular shape that is slanted by its sides and the angles are all pointed or have no obtuse angles. These results are in line with the Ningrum's result [7] related to students' misconceptions that still assume parallelogram is a sloping rectangle. The student also assumes that all angles that have pointed ends are acute angles, while angles that have curved lines are obtuse angles. Students' interpretation of the acute angle and obtuse angle is a form of incompatibility of the concept of angle that student has with the scientific or mathematical concept because the acute angle is an angle whose magnitude less than 90 degrees while the obtuse angle is an angle whose magnitude greater than 90 degrees and less than 180 degrees. Misconceptions about acute and obtuse angles are seen from the similarity of the student show to acute and obtuse angles in the parallelogram puzzle as one type of acute angle as follows.

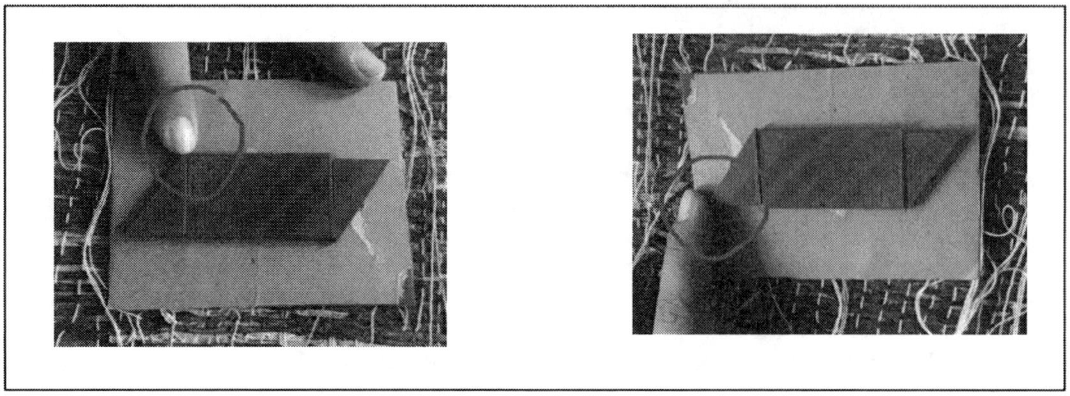

(a) (b)

Figure 2. (a) - (b). Angles that student interpreted as acute angles

Another theoretical misconception that student isn't exactly in makes an analytic definition of rectangular shape because according to him the rectangle isn't a parallelogram otherwise, the parallelogram is a rectangle. Students are also not exactly to makes an analytic definition of rhombus because according to him rhombus isn't a parallelogram. Inaccurate to make analytic definitions also done by the student when he said that the square isn't a rectangle, but the rectangle is a square. He said too that the square is not a rhombus, but the rhombus is a square. All of this interpretation indicates inaccurate of student-constructed quadrilateral analytic definitions because the analytic definitions that should be student construct are as follows:

The rectangle is a parallelogram whose one angles is the right angle
The rhombus is a parallelogram whose four sides are the same length
The square is a rectangle whose four sides are the same length or a parallelogram whose one angle is the right angle and has four the same length sides.
The square can also be defined as a rhombus whose one angle is the right angle or a parallelogram which has four sides that same length and a right angle.

This in line with the analytic definition criteria by [13, 14] about the analytic definition that rests upon must be stated in term of the closest type (genus proximum) and special differentiator (differentia specifica). The analytic definition of a quadrilateral with the closest type (genus proximum) is "parallelogram" while it special differentiator (differentia specifica) is the information contained rearward the word "whose".

Different misconceptions possessed by low vision student. Low vision student also has a classificational misconception which is a form misconception of facts classification into organized charts. This can be seen when the low vision student asked to determine which geometry's puzzle belongs to parallelogram. In determining the geometry's puzzle that is included into the parallelogram, the student makes not the same mistake as a totally blind student because he doesn't put the parallelogram into the rectangle but he distinguishes between the rectangle and the parallelogram. In classificational misconception, the student makes a mistake when determining geometry's puzzles that are included in the rhombus. He didn't put a square into the rhombus but instead, put a rhombus into the square.

Besides classificational misconception, low vision student also has a correlational misconception which is a form of the misconception of special events that are interconnected. This can be seen when the student makes a mistake in explaining the relation between the parallelogram and the rectangle like to interview result below.

Researcher: In your opinion, do the rectangles and parallelogram have similarities or differences in the characteristics of the shape?
Subject (B): Yes, I think
Researcher: What do you mean? Do you think there are have the same or different characteristics?
Subject (B): Both are different. The parallelogram is not a rectangle or otherwise, but both shapes have the opposite sides which the same length
Researcher: What about rhombus? Does he have the same similarities or differences as parallelogram and rectangle?
Subject (B): Yes, there are have similarities because both have the same characteristic related to the four acute angles it has
Researcher: How about square and rectangle?
Subject (B): The square is a rectangle that has a shorter side measure. I think a square is a shorter rectangle
Researcher: What about rhombus and square?
Subject (B): Both are the same length of side and type of angles, so I think the rhombus is a square that positioning slanted

From the interview, it is known that the student not knowing that a rectangle is a parallelogram. He also can't exactly to explain the relation between parallelogram and rhombus because he assumes that both have the same characteristic related to the sum of acute angles that both have. Another correlational misconception is the student's mistake in explaining the relation between square and rectangle because he assumes that the square is a rectangle that has shorter sides. Nevertheless, implicitly student can already be said to be able to classify a square as a rectangle. Others, the student also makes a mistake in explaining the relation between rhombus and square because he assumes that rhombus has the same side length and angle type. The student also assumes the rhombus is a square that is positioned slant.

Another misconception possessed by low vision student is a theoretical misconception which is a form of misconception about the explanation of facts or events in an organized system and concerns the development process from that known until that unknown by the student. It has seen when the student not quite exactly in making an analytic definition of a parallelogram. According to him, the parallelogram is a quadrilateral that sides are slant or have acute angles and other is non-acute angles. Implicitly student can already identify that the parallelogram has two pairs opposite angles that equal magnitude that is a pair of acute angles and a pair of obtuse angles that are represented as a non-acute angle. Non-acute angle interpreted by the student as another type of angle that his unknown.

Another theoretical misconception that student is not quite exactly in making an analytic definition of a rectangle because according to him the rectangle and parallelogram have the same characteristics, namely the opposite sides which the same length but the rectangle isn't a parallelogram. The student also not quite exactly in making an analytic definition of rhombus because according to him rhombus is a parallelogram whose all angles are the acute angle with the same magnitude, not related to the length of the sides measure. Misconceptions about all the same acute angles on the rhombus and the similarity of the acute angles between the parallelogram and the rhombus can be seen from the similarity of the student show the acute angles to parallelogram' puzzle and one of the acute angles that assumed represent fourth other acute angles on the rhombus puzzle as follows.

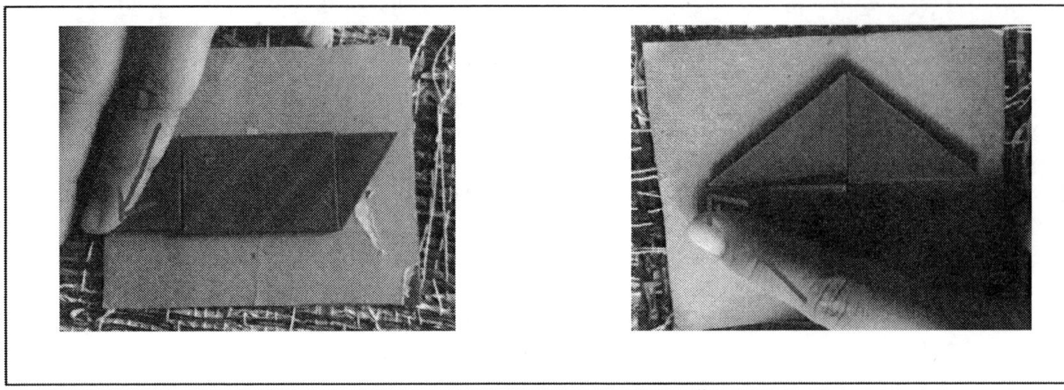

(a) (b)

Figure 3. (a) - (b). The acute angles on the parallelogram and rhombus are assumed to be equal

Inaccurate to make analytic definitions also done by the student when he said that a square is a shorter rectangle. Implicitly it also shows that student has been able to identify the closest type of a square as a parallelogram whose one angle is the right angle or a square is a rectangle. The last theoretical misconception is inaccurate to make analytic definitions of a rhombus that said is a square that is positioned slant.

The misconceptions above originated the characteristics of the quadrilateral shapes where student have a misconception of the characteristics contained in each quadrilateral shapes. This in line with the theory explained by [15] that misconception is defined as the perception that has very different or not appropriated meaning with experts' opinion on a certain scientific concept.

Although the analytic definition constructed by blind students is not entirely correct but blind students can already identify the relation between several quadrilateral shapes based on the characteristics of each shape. This in line with the result of Andriyani' research [16] which shows that blind students not only respond to the syntax of the definition but also try to be able to interpret the object imagined even though the formal definition formed hasn't entirely correct as of the formal definition of mathematics.

In addition, the differences of misconception in both of blind students are caused the construction difference of meaningfulness when students interpret each quadrilateral in their own way according to their language, culture or previous learning experiences. The blind student's way to construct an analytic definition with a geometry's puzzle mostly through manipulated the quadrilateral model using its tactual ability to recognize quadrilateral characteristics. This in line with the results of Argyropoulos' research [in 17] that the uniqueness of blind students' thinking activities related to their cognitive development and understanding of shapes concept that carried out using their tactual perception.

4. Conclusion

From the results of research, it can be concluded that both of blind students have misconceptions in constructing quadrilateral analytic definitions, be it classificational, correlational or theoretical misconceptions. Misconceptions by totally blind student tend to be inaccurate to construct analytic definitions and inability' student to explain the relationship of a certain quadrilateral with other quadrilaterals. While low vision student tends to be less precise to construct analytic definition or explaining the relation between each of quadrilaterals because in implicitly he has been able to identify the characteristics of almost all quadrilateral shapes but in classification, there is a less precise in classifying quadrilateral into quadrilateral which is assumed to be the closest genus. As a result, low vision student has classificational, relational and theoretical misconceptions that lower-level misconception than the totally blind student.

Acknowledgements
The author thanks Graduate Program of Mathematics Education at the Ahmad Dahlan University for the funding of research of publication

References

[1] Komara E 2018 *SE Asian J. for Youth, Sports & Health Edu.* **4**(1) 17-26
[2] Sahin M C 2009 *World Conference on Educational Sciences* Procedia Social and Behavioral Sciences **1** 1464–*1468*
[3] Stacey K 2011 *J.on Math. Edu.* **2**(2) 95-126
[4] Hartini S, Somakim S and Kesumawati N 2016 *Numeracy Journal* **2**(2) 72-90
[5] Ismail Z and Rahman S N A 2017 *Int. J. Emerging Math. Edu.* **1**(2) 121-134
[6] Putri B R K, Nurhilaliati and Kurniawati K R A 2017 *Paedagoria* **8**(2) 24-31
[7] Ningrum R W and Budiarto M T 2016 MATHEdunesa **1**(5) 59-66
[8] Carle S M 1993 *Critical and Creative Thinking Capstones Collection.* Paper 46. http://scholarworks.umb.edu/cct_capstone/46
[9] Andriyani 2018 *5 th ICRIEMS Proceedings* ME-1-6
[10] Vianna C S, Barbosa P M, Rocha D F and Silva B 2006 *International Congress on Mathematical Education*
[11] Pritchard CK & Lamb JH 2012 *J. of Math. Teacher* **106** 22-27
[12] Miles MB & Huberman AM 1994 Qualitative data analysis: an expanded sourcebook 2 nd edition, Sage Publication
[13] Maccia E S 1964 Empirical Theory Construction and Definition (The Ohio University: Bureau of Educational Research and Service)
[14] Hjorth E, Madsen B N, Norling-Christensen O, Jacobsen J R & Ruus H 1987 Descriptive Tools for Electronic Processing of Dictionary Data (Tubingen: Max Niemeyer Verlag)
[15] Hammer D 1996 *American J. of Phys.* **64** 1316-1325
[16] Andriyani, Juniati D & Budayasa K 2018 *Int. J. Eng. Tech.***7** 1-5
[17] Andriyani, Budayasa I K & Juniati D 2018 *J. Phys: Conf.* **947** 1-6

Proving the Pythagorean Theorem to Junior High School students through the Asian method

P Wijayanti[1], A N Hidayanti[1]

[1]Department of Mathematics, Universitas Negeri Surabaya, Ketintang Street, Surabaya 60231, Indonesia

e-mail: pradnyowijayanti@unesa.ac.id

Abstract. The development of right triangle concept has been known since ancient Egypt and is still be used in daily life until now. One of the most popular theorems that relate to the right triangle is Pythagorean Theorem. In its development, the Pythagorean Theorem has been proven by mathematicians with different methods. Based on historical developments there are approximately 200 proofs of the Pythagorean Theorem that has been found. Nevertheless, history still has interesting mysteries to study especially for the development of the Pythagorean Theorem in Asia. In student textbooks, the proof of Pythagorean Theorem introduction often comes from the European continent such as proof by Pythagoras, Euclid, and Leonardo Da Vinci. This caused Pythagorean Theorem' development in Asia as if it were excluded, the proof of Pythagorean Theorem in Asia also developed by mathematicians such as Bhaskara and Zhao Shang with different names. Even so, the discussion of those Pythagorean Theorem proof can be introduced to students as an initial concept in Pythagorean Theorem chapter and presented pleasantly in the form of mathematics historical comic to make it more attractive. The discussion result in this journal is also in line with the 2013 curriculum which emphasizes student activity and fun learning model.

1. Introduction

Since the enactment of government regulation number 24 of 2016 by the Ministry of Education and Culture of Republic Indonesia about curriculum, the curriculum that applies in Indonesia is the Curriculum 2013 [1]. In this curriculum, students are required to be more active in learning and find information through discovery-based learning. To facilitate this, the student textbook has been given various activities that invite students to actively think and find their own concepts. These activities are often inspired by historical developments in order that the concepts taught will be more meaningful.

Correspondingly, Ernst Haeckel [2] suggests the theory that individual development (ontogeny) is a reflection of the history of its evolution (phylogeny). Then if it is associated with mathematics and the history of mathematics, the development of mathematics in a person is in line with the development of the history of mathematics itself so that it will be easier for students to understand mathematical concepts if they are related to the history. One of the mathematical concepts at the junior high school level that is close to the history of mathematics is the concept of the Pythagorean Theorem. The history of the Pythagorean Theorem begins with the development of the history of right triangles which have been known by ancient Egyptian civilization a long time ago.

In its development, the Pythagorean Theorem has been proven by mathematicians with different methods. Based on historical developments there are approximately 200 proofs of the Pythagorean Theorem that has been found. However, the history of the Pythagorean Theorem still has an interesting mystery to study especially the development of the Pythagorean Theorem in Asia. In textbooks, proof of the introduction of the Pythagorean Theorem often comes from the European continent, for example, proof by Pythagoras, Euclid, and Leonardo Da Vinci. This resulted in the development of the Pythagorean Theorem in Asia as if it were excluded, the proof of Pythagorean Theorem in Asia also developed by mathematicians such as Bhaskara and Zhao Shang with different names. Although with different terms, both of the Pythagorean Theorems in Europe and in Asia both examine the same thing, the hypotenuse Theorem in the right triangle. Therefore, the history of mathematics regarding the Pythagorean Theorem in Asia can be used in introducing and proving the Pythagorean Theorem to students at the Junior High school level as one of the activities in mathematics learning.

2. Method

This study uses a literature review method of various historical resources of the Pythagorean Theorem, especially in Asia. Before searching for information about the history of the Pythagorean Theorem, the researchers conducted a study of student textbook on the Pythagorean Theorem taught at the Junior High School level in the Curriculum 2013. Then the researchers focused on looking for information on the history of Pythagorean Theorem in Asia proposed by several mathematicians such as Bhaskara and Zhao Shang. The study was conducted through various books, journals, proceedings, and the internet to find out the historical development of the Pythagorean Theorem. Then the information obtained is re-explained so that they can be understood and used to be one of the choices in introducing and proving the Pythagorean Theorem to grade 8 of Junior High School students.

3. Result

Through the study of student books, it was found that the student book only introduced proof of the Pythagorean Theorem from Europe and America only (Figure 1).

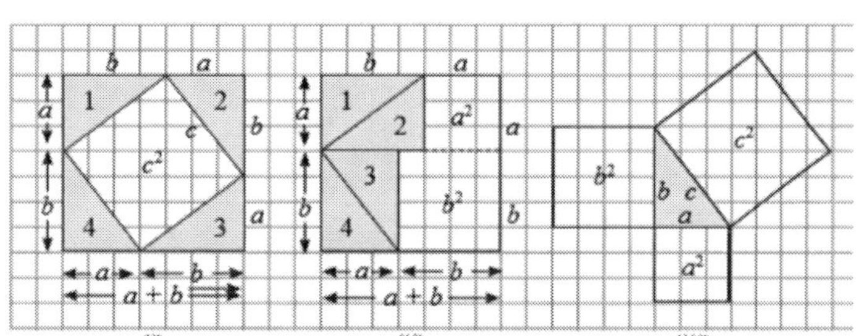

Terdapat lebih dari 200 pembuktian dari teorema Pythagoras. Elisha Scott Loomi mempublikasikannya pada tahun 1927, termasuk di dalamnya pembuktian oleh Pythagoras sendiri, Euclid, Leonardo da Vinci, dan Presiden AS James A.Garfield. Salah satu pembuktian yang terkenal akan kita pelajari di bawah ini.

Berikut ini salah satu pembuktian lengkap untuk memeriksa kebenaran teorema Pythagoras.

(i) (ii) (iii)

Gambar 6.4 Pembuktian teorema Pythagoras

Berdasarkan **Gambar 6.4**, kita bisa menyusun empat segitiga siku-siku pada **Gambar 6.4(i)** ke dalam persegi pada **Gambar 6.4(ii)**. Kita perlu membuktikan bahwa $a^2 + b^2$ sama dengan c^2. Perhatikan bahwa luas persegi yang terbentuk oleh empat segitiga siku-siku pada **Gambar 6.4(i)** adalah $(a + b)^2$, atau

Figure 1. Proof of the Pythagorean Theorem snippet in the student textbook.

In Figure 1, it can be observed that the proof of the Pythagorean Theorem presented in the student book has not shown any of the names of mathematicians from Asia or their proof of the Pythagorean Theorem. This resulted in the assumption that Pythagorean Theorem did not develop in Asia because it did not introduce the proof of Pythagorean Theorem from Asia. However, in fact, the theorem regarding the hypotenuse of a right triangle develops in the Asian Continent even though in different terms. To get to know the brief journey of Pythagorean Theorem in Asia, this study will explain the development of the Pythagorean Theorem in Asia through proof which made by Asian mathematicians.

At first, the Pythagorean Theorem was known as the Theorem regarding the length of the hypotenuse of a right triangle, usually known as Pythagorean Theorem. However, it is not certain who the first originator of the Theorem was, whether Pythagoras discovered the Theorem himself or not. Nevertheless, Pythagoras was the person who first put forward and proved the Theorem formally in ancient times [3]. Based on historical records, the form of the Pythagorean Theorem appeared in ancient Egyptian civilization (2000 BC) which can be found in the use of worker tools. At that time, the workers used a rope 12 tied nodes which formed into a 3-4-5 right triangle. Some sources claim that knowledge of right triangles began to enter Asia through the silk trade route which was later developed and proven by mathematicians in Asia.

In Asia, the knowledge of right triangles spread and developed among mathematicians and astronomers, so that the development of the Pythagorean Theorem developed which can be observed through the following simple timeline [4][5][6][7].

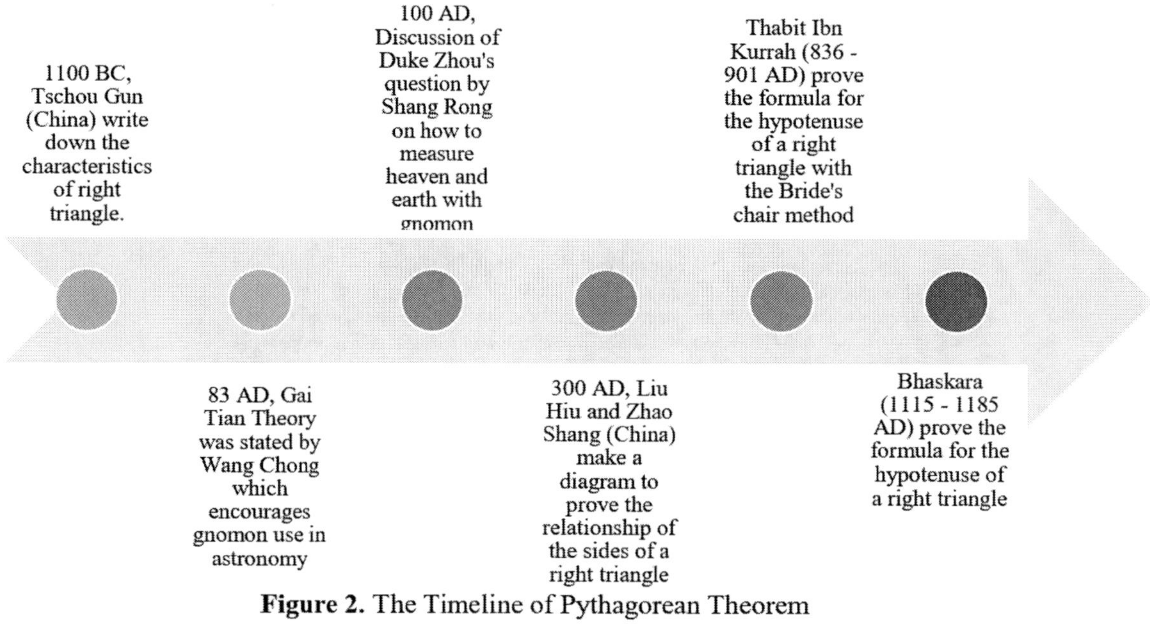

Figure 2. The Timeline of Pythagorean Theorem

The timeline began in 1100 BC after the concept of a right triangle entered Asia, Tscou Gun from China wrote about the characteristics of a right triangle which later developed and utilized in the field of astronomy through a simple tool called gnomon. In the year 100 AD in the Zhou dynasty there was a discussion about the use of gnomon in astronomy which became the beginning of the Pythagorean Theorem' proof in China. At that discussion, Shang Rong answered questions from Duke Zhou by explaining the use of gnomon in estimating the distance of the earth and sky using size 3-4-5 ratio. When explaining this, Rong also mentioned the assumption that the earth was flat, Gai Tian model. Gai Tian model has been trusted by mathematicians and astronomers in China since 83 AD.

Furthermore, the use of gnomon in astronomy was increasingly developed which later inspired Liu Hiu and Zhao Shang in 300 AD to prove the relationship between the sides of a right triangle (Gougu Theorem) which known as the current formula of the Pythagorean Theorem. Besides in China, in other

Journal of Physics: Conference Series **1417** (2019) 012060 doi:10.1088/1742-6596/1417/1/012060

Asian countries, mathematicians began to prove the theorem in their own way. In Turkey, a mathematician named Thabit Ibn Kurrah (836 - 901 AD) proved the theorem with a method names Bride's Chair. Then in India, Bhaskara (1115 - 1185 AD) also tried to prove the Theorem using diagrams. The diagram used by Bhaskara is similar to the diagram used by Zhao Shang, but with a different point of view.

The development of the history of the Pythagorean Theorem in Asia can be used in learning as an activity of students in proving the Pythagorean Theorem. In order to be used and easier to understand, here are some proving steps that can be used in learning that are in accordance with the methods used by the mathematicians.

3.1. *Proof by Zhao Shang*

Zhao Shang's proof of the Gougu Theorem was written in a notebook entitled Zhou Bi Suan Jiang (The Arithmetic Classic of the Gnomon and the Circular Paths of Heaven) or Zhou Bi in the Zhou dynasty in 300 BC [8]. Shang explained his proof through a diagram, following the diagram used by Shang.

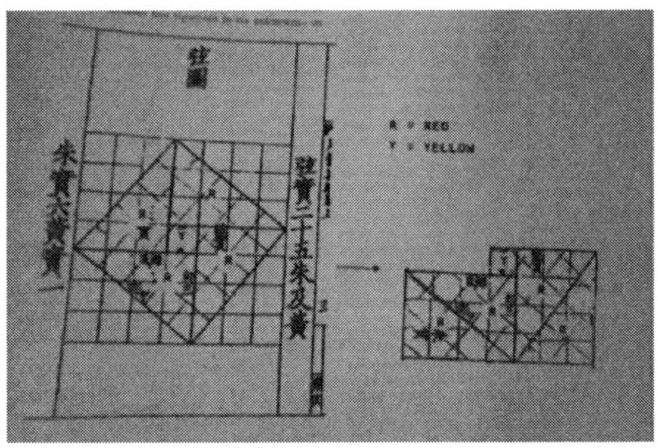

Figure 3. Diagram of Zhao Shang proof.

From the picture above, it can be arranged steps to prove the Pythagorean Theorem as follows:

i. Make four right triangles of the same size, where the length of the triangular legs is made different. The goal is to be easily differentiated.
ii. Arrange the four triangles so that they form a new plane with the side length represents the hypotenuse of the triangle. From this step, the square shape will be formed from the hypotenuse of the right triangle.
iii. Turn the four triangles towards the square that is formed so that the four triangles appear to cover the square as shown in the following figure.

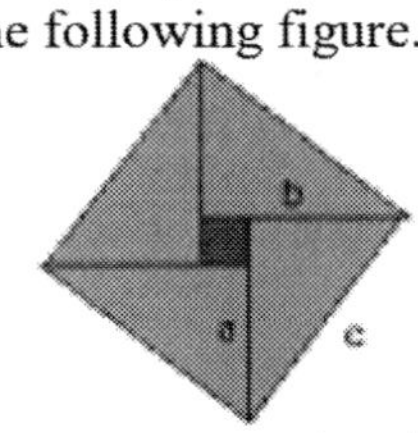

Figure 4. Representation for step iii.

iv. From Figure 4 above, it can be seen that in the middle the square is formed a new square which is represented by a red square. Make a square which resembles the red square to fill the empty space on the red square so that it is obtained that

$$Area\ of\ c - square = 4.\text{Area of triangles} + Area\ of\ small\ square$$

v. Arrange two new square shapes from the four right triangles and the previous red square as Figure 5.

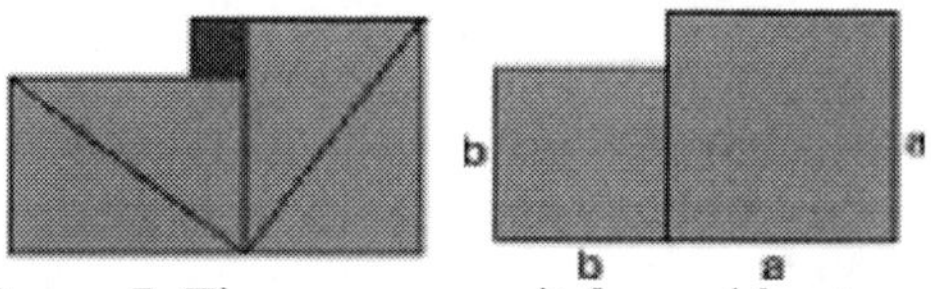

Figure 5. The square result formed in step v.

vi. In Figure 5 above it can be can be concluded that

$$c - side\ square\ area = a - side\ square\ area + b - side\ square\ area$$
$$c^2 = a^2 + b^2$$
$$c = \sqrt{a^2 + b^2}$$

The Pythagorean Theorem that "In any right triangle, the sum of the areas of the squares on the legs equals the area of the square on the hypotenuse" [9] which can be represented algebraically as $c^2 = a^2 + b^2$.

3.2. *Bride's Chair concept of Thabit Ibn Kurrah*

The steps of the proof are:

i. Bride Chair consists of two square pieces that have different sizes. Thus, first make two squares that have different sizes, for example, sizes a and b.

ii. Then, cut the bottom edge of the two square into two right triangles with the two triangular legs representing the size of the two squares (Figure 6).

Figure 6. Representation of steps i and ii.

iii. Turn the two triangles to the top so they form a new large square with the sides being the hypotenuse of a right triangle (Figure 7).

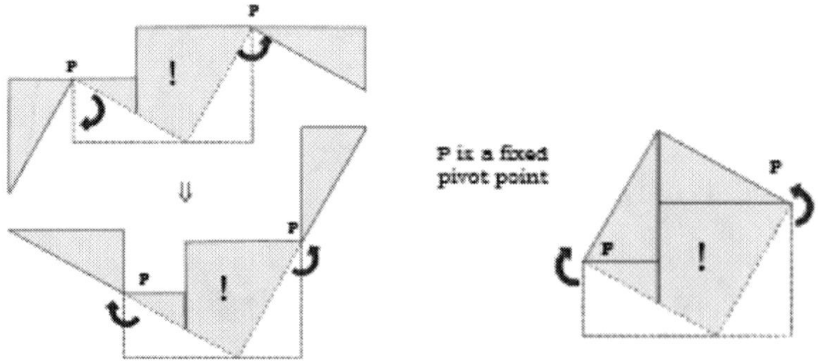

Figure 7. Representation of step iii.

iv. If the hypotenuse of a right triangle is c, it is obtained that $c^2 = a^2 + b^2$.

3.3. *Proof by Bhaskara*
In his proof, Bhaskara used a diagram which is similar to Zhao Shang diagram. The difference is in the algebraic step used, where Zhao Shang only concludes his findings that the area of a large square with sides c equals the number of two squares with sides a and b so that $c^2 = a^2 + b^2$ proved. Whereas in Bhaskara, he uses the quadratic equation where side c equals the difference between sides a and b, then it gets:

$$A_{big\ square} = A_{little\ square} + 4.A_{triangle}$$

$$c^2 = (a-b)^2 + 4\left(\frac{1}{2}ab\right)$$

$$c^2 = a^2 - 2ab + b^2 + 2ab$$

$$c^2 = a^2 + b^2$$

Those methods above can be inserted into learning through worksheet so students can practice and prove the Pythagorean Theorem as the mathematicians did.

4. Discussion
Based on the development of the history of mathematics regarding the Pythagorean Theorem in the results above, it can be seen that the development of the Pythagorean Theorem also developed in Asia even though with a different name. These developments can be introduced to students so that they will not only understand the concept of Pythagorean Theorem but also know that in Asia the theorem also develops. In order to make the presentation more attractive to students, the development of Pythagorean Theorem' proof can be presented in educational comic form.

The use of comic as an information delivering tool to students has recently been highlighted. It is due to the interest in reading for comics is quite high among the public, especially students. Based on the survey, the reading appeal of comic types in Indonesia gets the highest ranking in Asia according to the Line Webtoon version in 2018 with 17 million active readers per month [10]. The use of comics as a media for delivering information has been used for a long time. Thorndike stated that comics have advantages in terms of improving vocabulary and children's reading skills which are comparable to reading textbooks every year [11]. In addition, through images in the comic, students can develop their imagination in imagining the information provided through the fun method. Therefore, it stimulates an idea to combine the history of mathematics and comic into one to convey mathematical concepts to students. By combining comic with the development of the history of mathematics, it is expected that students can better understand the concepts given through the course of the history of mathematics inserted in the story.

5. Conclusion
Based on the results and discussions that have been discussed before, it can be concluded that:

- The development of the Pythagorean Theorem in Asia also developed and spread even with different names. The development of the concept in Asia can be proven by the appearance of mathematicians who tried to prove the theorem such as Bhaskara, Zhao Shang, Thabit Ibn Kurrah, and Liu Hui.
- The history of mathematics can be inserted into learning activities through worksheet so students are able to find their own concepts to be built. Insertion of the history of mathematics can add information which is obtained by students so that the concepts can be more meaningful.
- In order to make the presentation of the history of mathematics into learning more enjoyable, the media for educational comics can become an option. By combining comic with the development of the history of mathematics, it is expected that students can better understand the concepts given through the course of the history of mathematics inserted in the story.

References

[1] Kemendikbud 2016 *Permendikbud Nomor 24 Tahun 2016 Tentang Kurikulum* (Jakarta: Kemendikbud)
[2] Reece and Mitchel 2003 *Biologi Jilid II* (Jakarta: Erlangga)
[3] Sparks J C 2008 *The Pythagorean Theorem: Crown jewel of Mathematics* (Ohio: Author House)
[4] Cullen C 2006 *Astronomy and Mathematics in Ancient China: Zhou Bi Suan Jing* (Cambridge: Cambridge University Press)
[5] Robertson and Connor 2003 *Chinese Mathematics* (online) (http://www-history.mcs.st-andrews.ac.uk/HistTopics/Chinese_overview.html, Accessed on June 06 2018).
[6] Tiles M 1997 Astronomy and Mathematics in Ancient China: The Zhou Bi Suan Jing *China Review International* **4**(2) 367-369
[7] Zitarelli D E 2017 Was Pythagoras Chinese? *Temple Mathematics.* n.d., (Online), (https://math.temple.edu/-zit/Zitarelli/Pythag_Chinese.pdf, obtained on November 14 2017)
[8] Burton, David M 2011 *History of Mathematics An Introduction 7th Edition* (New York: Mc Graw Hill)
[9] Katz J V 2009.*A History of Mathematics: An introduction* (USA: Pearson)
[10] Praditya D 2018 *Tech in Asia: Line creative ekonomi kreatif,* (Online), (https://id.techinasia.com/line-creativate-ekonomi-kreatif/amp/, accessed at 1 Oktober 2018).
[11] Sudjana N 2015 *Media Pengajaran* (Bandung: Sinar Baru Algesindo)

Journal of Physics: Conference Series **1417** (2019) 012061 doi:10.1088/1742-6596/1417/1/012061

Investigating Covariational Reasoning: What Do Students Show when Solving Mathematical Problems?

Y Fuad[1], R Ekawati*[1], A Sofro[1], L D Fitriana[1]

[1] Department of Mathematics, Universitas Negeri Surabaya, Surabaya, Indonesia.

e-mail: rooselynaekawati@unesa.ac.id

Abstract. Covariational reasoning plays a significant role for solving problems. This study examines the covariational reasoning of master program students when solving mathematical problems such as fraction, velocity and acceleration, proportion, and integral. 26 students of mathematics education master program, Universitas Negeri Surabaya, are involved in this study. Generally, mental action of students is more prominent on fraction and proportion than on the other issues. On fraction and proportion problems, most students are able to fulfill all mental actions such as coordinating the value, the direction, and the amount of change of one variable, and also coordinating the average and the instantaneous rate of change of the function. However, on velocity, acceleration, and integral problems most students cannot show their mental actions well. They only fulfill 3 of 5 mental actions of covariational reasoning. Generally, the shape of their graph related to those problems are irrespective with initial point. These findings suggest that learning in mathematics should place increased emphasis on problem involving graph to promote covariational reasoning of the students.

1. Introduction

It has been recognized that one of primary aims of education is to foster student's reasoning which isthe main activity in mathematical activities [1]. Therefore, it becomes crucial task for mathematics teachers to involve reasoning activities in mathematical learning and make mathematics meaningful. Thus, the ability to reason which is an important achievement for mathematics education students, especially prospective mathematics teachers can be fulfilled.

One type of reasoning that is closely related to mathematical problem solving is covariational reasoning. It is defined as cognitive activity which involves coordinating two varying quantities while attending to the ways in which they change in relation to each other [2, 3]. Covariational reasoning provides powerful insight and builds powerful mechanisms to improve reasoning abilities in solving mathematical problems [4].

Based on study result of [3], it found 5 levels of covariational reasoning development. A level will be reached only if underneath level has been fulfilled. For each level, there is indicator as follows:

Table 1. Level and Indicator of Covariational Reasoning

Level	Indicators
Coordination	Coordinating the value of one variable with changes in the other

Direction	Coordinating the direction of change of one variable with changes in the other variable
Quantitive Coordination	Coordinating the amount of changes of one variable with changes in the other
Average Rate	Coordinating the average rate of change of the function with uniform increments of change in the input variable
Instantaneous Rate	Coordinating the instantaneous rate of change of the function with continuous changes in the independent variable for the entire domain of the function

Some previous studies have shown relationship between covariational reasoning and calculus [5, 6, 7, 8]. According to [3], covariational reasoning is foundational for understanding major concepts of calculus. This is because the main of the material studied in calculus is related to the activity of analyzing the speed of a change. The activity is closely related to the activity of coordinating changes in quantity which is a covariational reasoning activity. This study examines student's covariational reasoning in solving calculus problems such as velocity and acceleration, integral, fraction, and proportion.

Moreover, research on covariational reasoning becomes crucial to study, reference to [9] who investigated the covariational reasoning abilities of a group of mathematics teachers and emphasizes that even though covariational reasoning plays an important role but in fact, not only teachers' covariational reasoning abilities were weak and lack depth but also their predictions about students' reasoning abilities bounded by their own thoughts related to the problem. Research on covariational reasoning is expected to provide important input to build covariational reasoning capabilities for both teachers and students.

2. Method

A Qualitative approach was used in this study to investigate student's covariational reasoning in solving mathematical problems which formed velocity and acceleration, integral, fraction, and proportion questions. A purposive sampling technique was applied to select the subjects (26 postgraduate students of mathematics education from Universitas Negeri Surabaya). All of students were given covariational reasoning test in order to promote as much student's covariational reasoning as possible, then were individually interviewed based on their answer. The obtained data was qualitatively analyzed based on covariational reasoning indicator which was done through three steps including data condensation, data presentation, and data interpretation.

3. Result and Discussion

The covariational reasoning tests are described below.

Velocity and Acceleration

1. Look at the picture of Andre's route from City A to City B below.

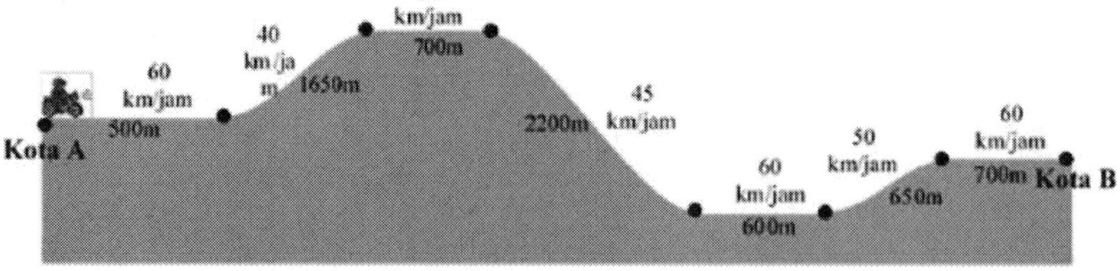

Describe in your words, the speed change of the motorbike driven by Andre from city A to city B! If the acceleration of the motor is expressed by the change of motor speed per unit of time, what is the average acceleration of the motor that Andre drives from city A to city B? How much time does Andre need from city A to reach city B?

2. Look at the picture of Maya's route from City X to City Y below.

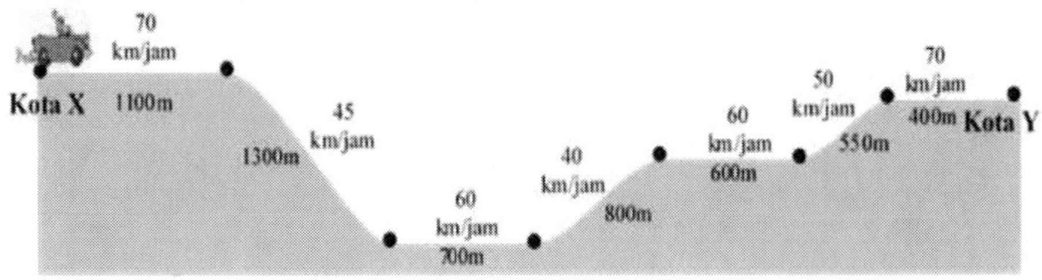

Describe the speed change of the car driven by Maya from city X to city Y! If the acceleration of the car is expressed by the change of car speed per unit of time, what is the average acceleration of the car that Maya drives from city X to city Y? How much time does Maya need from city X to reach city Y?

Fraction

1. Mom divides a tart into 5 equal portions. Dito gets one part, while Dera gets 2 parts.
   a. In fraction form, how many parts of cake do Dito and Dera get?
   b. Then, Dito and Dera divide their tart into 2 equal parts. Each of them give 1 part of their cake to Dini. In fraction form, how many parts of cake does Dini get?

   After being given to Dini, Who gets more cake? (Dito, Dera or Dini)? Please explain it!
2. Amir and Badrul have the same amount of marbles. Each has red and blue marbles. 2/3 of all Badrul marbles are red and the rest are blue. While 2/5 parts of the number of Badrul's red marbles are the same as the number of Amir's red marbles. Apparently, each of them lost one red marble.
   a. Whose red marbles are more after being lost?
   b. How many parts red marbles of Amir and Badrul after each of their marbles disappear?
3. A grocery store holds a promo for purchasing eggs. Each egg buyer will get a bonus as much as 1/5 kg from the store. Ani buys 5/4 kg of eggs and Dian buys 3/2 kg of eggs. The price of eggs is IDR 24,000/kg.
   a. How many kg of eggs did Ani and Dian buy before getting a bonus?
   b. Who gets more eggs after getting a bonus? Please try to explain it!
   c. How much difference of money must Dian and Ani pay?

Proportion

1. Out of 100 students, half of them keep dogs and the rest of them keep cats. Based on your opinion, which animals are the most kept by 100 students? If 2/5 students who keep dogs are male, how many female students keep cats? If the number of male students who maintain dogs are equals to the number of female students who maintain dogs, what is the ratio of the number of male students to the number of female students who have pets?
2. In a stall, Agus buys $\frac{7}{5}$ kg sugar, while Berta buys $\frac{5}{3}$ kg sugar. Each of them plans to give $\frac{1}{2}$ kg sugar to their mom. The price of 1 kg of sugar is Rp 15.000,-.
   a. How many kg of sugar did Agus and Berta buy before each of their sugar was given to his mother?
   b. After giving it to her mother, who has more sugar?
   c. How much money do Agus and Berta have to pay to buy the sugar?

Integral

1. A swimming pool is initially contained of 900 m^3 water. The pool is connected to 2 types of pipes, namely pipes to fill water into the pool and pipes to remove water from the pool. $Q(t)$ states the change of water flow at time t.

   $Q(t) = \{t\,3\ -3t+15\ -t+5\,t-7 \quad if\ 0 \le t \le 3\ if\ 3 < t \le 4\ if\ 4 < t \le 6\ if\ 6 < t \le 7\ if\ 7 < t \le 10$

If $Q(t) = \frac{dV(t)}{dt}$, sketch graph $V(t)$ which states the volume of water at time t in the pool for 10 hours!

2. An object is at its starting point ($t = 0$). Graph $v(t)$ expresses the movement speed of an object (in $\frac{meter}{second}$).

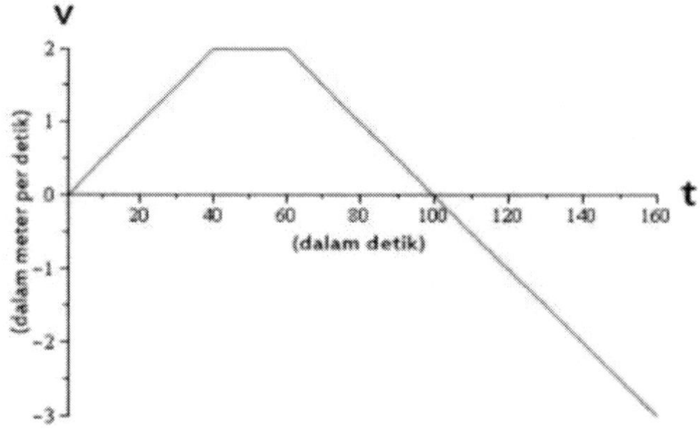

If $v(t) = \frac{ds(t)}{dt}$, sketch graph $s(t)$ which states the distance traveled by object from $t = 0$ to $t = 160$!

3. The swimming pool is initially contained 400 m^3 water. The pool is connected to 2 types of pipes, namely pipes to fill water into the pool and pipes to remove water from the pool. The following table is the result of observing changes in water flow for 7 hours. The observation begins at 10:00 and ends at 17:00.

Observation time	Time (t) in hour	Changes in water flow (counted from initial volume $t = 0$) $X(t)$ in m^3 per hour
10.00	0	0
11.00	1	1
12.00	2	2
13.00	3	3
14.00	4	3
15.00	5	0
16.00	6	-1
17.00	7	0

If $X(t) = \frac{dY(t)}{dt}$, sketch graph $Y(t)$ which expresses volume of water at time t in the pool for 7 hours!

A. *Student's Covariational Reasoning on Velocity and Acceleration Problem*

The covariational reasoning of students were explored to determine how students' fulfillment the level of covariational reasoning in solving velocity and acceleration problem. Figure 1 shows the percentage of students who fulfilled the covariational reasoning level. 8% of students were able to have all the covariational reasoning.

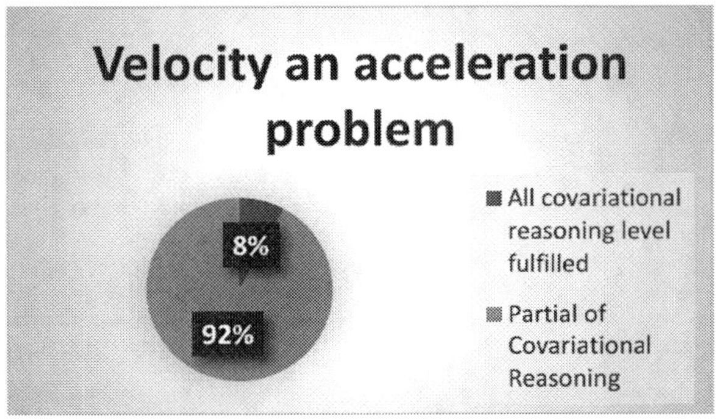

Figure 1. Covariational reasoning level on velocity and acceleration problem

On coordination level, to coordinate the output values (average acceleration and overall travel time) as the values change in input (speed and distance between two points), all students divide the presented data into several intervals. The first interval marks the input that is valid at the first point/beginning of departure until the second point. This interval is used as the basis for determining the first output. The second interval marks the input that applies at the second point to the third point and it is used as the basis for determining the second output, and so on. In this case, input acts as the independent variable while output acts as the dependent variable. This is analogous to the findings of [3] that indicate mental action on coordination level.

On direction level, to coordinate the direction of changes in output as changes in input, 1-2 students identify the presence of output (acceleration) when changes in speed and time changes are both positive or negative, and output (deceleration) appears when changes in speed or changes in time are negative, while other students did not identify this condition. However, in this case all students identify the direction of output (speed/deceleration of travel time) depending on distance and speed. The travel time will be faster if the travel speed is greater than the mileage and slower if the mileage is greater than the travel speed.

On quantitive coordination level, to coordinate the magnitude of changes in output with changes in input, only 1-2 students determine the magnitude of the acceleration/deceleration value by dividing the value of the speed change by the value of the time change for each interval. Nevertheless, all students divide the value of travel distance by the value of speed travel to determine the amount of travel time. To determine the total travel time, students add up the travel time for all intervals. On average rate level, to coordinate the average change in output as changes in input, all students add up the value of the acceleration/deceleration of all intervals and divide by the number of intervals to get the average acceleration/deceleration. For the last level, instantaneous rate, to coordinate instant changes in output (acceleration/deceleration and travel time) when input changes continuously, all students take a look at the output value at the end of each predetermined interval. In this case, all students review temporary travel time. But, only 1-2 students who also review the values of acceleration/deceleration.

B. Student's Covariational Reasoning on Fraction Problem

Fraction problem is considered as more friendly problem than velocity and acceleration problem for students. Figure 2 shows the fulfillment of students in solving fraction problem.

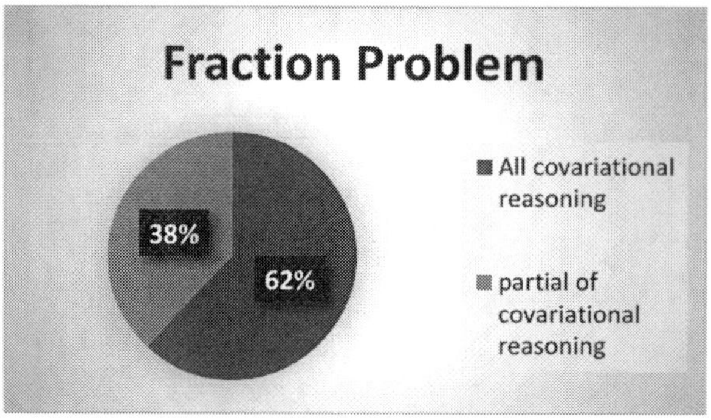

Figure 2. Covariational reasoning on Fraction Problem

On fraction problem, more than 90% students answer the question. Differences in student answers can be categorized into high group and under low group. On coordination level, to coordinate the output values as the values change in input, all students present their parts in fraction form, represent it to images, and divide the condition to be some parts. On direction level, to coordinate the direction of changes in output as changes in input, high categorized students identify that the output will increase when there is addition and will decrease when there is reduction or division. This is analogous to the findings of [3] that "Constructing an increasing straight line" to mark mental actions 2. Meanwhile, under low categorized students did not do it.

On quantitive coordination level, to coordinate the magnitude of changes in output with changes in input, high categorized studens add, subtract, or divide all inputs according to the instructions in the problem. But, under low students didn't do it because of lack of understanding on questions. On average rate level, to coordinate the average rate of change in output as changes in input, more than half of the students determine the rate of output quantity according to the sum value, subtraction value, or divider value. For the instantaneous rate level, to coordinate instant changes in output when input changes continuously, all students take a look at the output value at the end of each predetermined condition.

C. Student's Covariational Reasoning on Proportion Problem

More than 88% of students show their solution on proportion problem. Figure 3 shows the fulfilment of students' covariational reasoning on proportion problem.

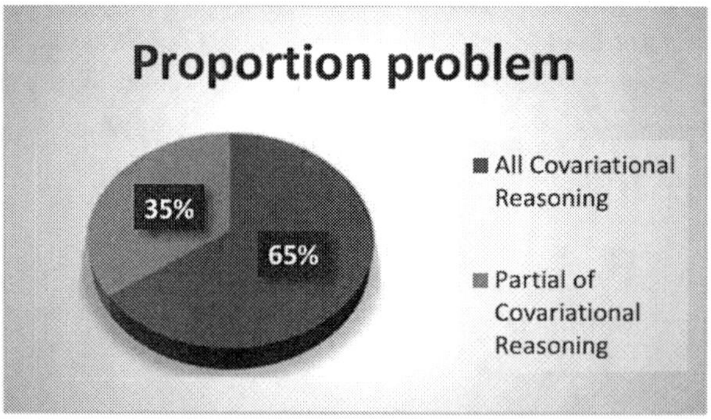

Figure 3. Covariational Reasoning on Proportion Problem

Student's answer can be categorized into 2 groups, namely high and under low. Almost all students can demonstrate the application of the proportion concept well. On coordination level, to coordinate the output values as the values change in input, all students divide the condition to be some parts. On direction level, to coordinate the direction of changes in output as changes in input, high categorized students identify that the output will increase when there is addition and will decrease when there is reduction. They also determine the value of the output in each condition given to the problem, either by adding up or subtracting inputs with the stated values. This condition shows quantitive coordination level. This is analogous to the findings of [3] "Plotting points / constructing secant lines" marks mental actions 3. For the last two level, only a few students who are under low categorized did not show those mentioned behavior.

On average rate level, to coordinate the average rate of change in output as changes in input, more than 80% of the students determine the rate of output quantity according to the condition given such as proportion value. This condition is analogous to the findings of [3], "Constructing contiguous secant lines for the domain" marks mental action 4. For the last level, to coordinate instant changes in output when input changes continuously, students determine output at the end of each predetermined condition.

D. Student's Covariational Reasoning on Integral Problem

Integral problem is considered as the most challenging problem for students. Roundly, just around 15% students fulfill some indicators of covariational reasoning.

Table 2. The Covariational reasoning on Integral Problem

Coordination Level	Direction Level	Quantitative Coordination
Draw axis lines on Cartesian coordinates	Construct the up and down function	Substitute the value of t on the function

To be more specific, on coordination level, to coordinate the output values as the values change in input, students make axis lines on cartesian coordinates. Horizontal lines to mark input values while vertical lines to mark output values. Input acts as an independent variable, while output acts as a dependent variable. On direction level, to coordinate the direction of changes in output as changes in input, students construct the up and down functions (Q(t), V(t), and X(t)) but not construct the asked function. On quantitive coordination level, to coordinate the magnitude of changes in output with changes in input, students substitute the value of t on Q(t), V(t), and X(t) then plot the dots on the cartesian coordinates.

On average rate level, to coordinate the average change in output as changes in input, students connect the points that indicate the value of the function Q(t), V(t), and X(t) which are not the final functions in question. For the last level, instantaneous rate, to coordinate instant changes in output when input changes continuously, students construct a smooth curve that shows the value of Q(t), V(t), and X(t) which are not the final functions in question. The last two conditions complement and appear to support previous findings by researchers such as [3, 10] who find that at the college level, majority of students have difficulty in showing covariational reasoning behavior that is higher than direction level (level 2) or quantitive coordination level (level 3).

4. Conclusion

Overall, students show their covariational reasoning performance well on fraction and proportion problems. For both problems, more than 85% students answer the questions given. The opposite condition appears precisely on two other problems, especially integral. Students only show 3 out of 5 mental actions that correspond to indicators of covariational reasoning. At the third level, students only determine one of two outputs that should be determined. The output determined by the student is the initial output that will be used to determine the final output value. At the fourth and final level, students only focus on the initial output. While the final output becomes the accent at these levels.

References

[1] Ferrini-Mundy J 2000 Principles and standards for school mathematics: A guide for mathematicians *Notices of the American Mathematical Society* **47**(8)

[2] Saldanha L and Thompson P W 1998 Re-thinking covariation from a quantitive perspective: Simultaneous continuous variation, in *Proceedings of the Annual Meeting of the Psychology of Mathematics Education-North America. Raleigh, NC*

[3] Carlson M, Jacobs S, Coe E, Larsen S, and Hsu E 2002 Applying covariational reasoning while modeling dynamic events: A framework and a study *Journal for Research in Mathematics Education* pp 352-378

[4] Strom A D 2006 The Role of Covariational Reasoning in Learning and Understanding Exponential Functions, in *Proceedings of the Twenty-Eighth Annual Meeting of the North American Chapter of the International Group for the Psychology of Mathematics Education* **2** pp 624-630

[5] Thompson P W 1994 Images of rate and operational understanding of the fundamental theorem of calculus *Educational studies in mathematics* **26**(2-3) pp 229-274

[6] Cottrill J, Dubinsky E, Nichols D, Schwingendorf K, Thomas K, and Vidakovic D 1996 Understanding the limit concept: Beginning with a coordinated process scheme *Journal of mathematical behavior* **15**(2) pp 167-192

[7] Zieffler A S and Garfield J B 2009 Modeling the growth of students'covariational reasoning during an introductory statistics course *Statistics Education Research Journal* **8**(1)

[8] Moore K C, Paoletti T, and Musgrave S 2013 Covariational reasoning and invariance among coordinate systems *The Journal of Mathematical Behavior* **32**(3) pp 461-473

[9] Zeytun A S, Cetinkaya B, and Erbas A K 2010 Mathematics Teachers' Covariational Reasoning Levels and Predictions about Students' Covariational Reasoning Abilities. *Educational Sciences: Theory and Practice* **10**(3) pp 1601-1612

[10] Koklu O 2007 *An Investigation of College Students' Covariational Reasonings* (Doctoral dissertation, Florida State University).

Analysis of the Number of Seventh Grade Student Error Type in Solving Two-Dimensional Figure Problems

S F D Iqlima[1] and A Sofro[1]

[1]Mathematics Department, Universitas Negeri Surabaya, Surabaya 60231, East Java, Indonesia

E-mail: ayuninsofro@unesa.ac.id

Abstract. Solving the problem is one of the goals in learning. In reality, there are still many students who make an error in the two-dimensional figure. The style of students in absorbing, processing, and managing information is called learning style. In some studies, the mathematical abilities boys and girls are different. The investigation of the types of student errors based on learning styles and sex has developed massively. Thus, this study will investigate variables that affect the number of types of errors in solving the problem of a two-dimensional figure. The population in this study is a seventh-grade student, and the sample in this study is 32 seventh grade students. Data obtained from questionnaires and tests. There are two variables in this study, the independent variable is learning styles and gender, and the dependent variable is the number of error types. Hence, the appropriate approach is using Poisson regression due to a discrete response variable. The results show that the coefficient of learning style has substantial evidence to reject the null hypothesis with alpha 0.05. It means that the learning style has a significant impact on the number of error types.

1. Introduction

One of the materials taught in mathematics is geometry. The final goal of learning is to solve problems correctly so that the final goal of learning geometry is to solve problems about geometry correctly. However, in reality, there are still many students who still make mistakes in solving problems about geometry. This fact is consistent with previous research results, which show that there are still many students who make mistakes in solving problems about geometry, among the mistakes made are concept errors, principle errors, and operating errors [1]. Mistakes made by students are different [2]. There are four types of errors, namely factual errors, concept errors, principle errors, and procedural errors. Also, based on three years of teaching experience at Al Islamiyah Sidoarjo Middle School, there are still many students who make mistakes in solving questions about a two-dimensional figure where is part of geometry. On the other hand, learning styles have an essential impact during a classroom activity. The learning style characterizes each student on how to learn, to be taught, and to interact in the learning environment. According to research by Joaquin, differences in learning styles make a significant difference in academic performance [3]. Furthermore, research by Rahman and Ahmar shows that learning styles affect mathematics achievement [4]. In addition, the ability to solve geometry problems is related to spatial intelligence [5]. Furthermore, the spatial intelligence of male and female students is different [6].

There are several studies on learning styles, but there is no research on learning styles, sex, and many types of errors so that learning styles are not known to affect the number of errors. There are many mistakes made by students in solving mathematical problems, including a mistake in solving problems of two-dimensional figures. Research by Adhitya shows that many students still make mistakes in solving problems about two-dimensional figures [7].

On the other hand, many qualitative studies have been investigated massively. Meanwhile, there is a big challenge to develop quantitative research due to limited discussion. Therefore this study is a quantitative study that aims to find out the significant factors which affect the student making an error. The data of this study are discrete, and the purpose of this study is to determine the effect of learning styles on many types of errors, so the method used in this study is Poisson regression

2. Literature Review

2.1. *Learning style*

Learning styles are ways that are usually used in learning. It includes how to learn, how to be taught, and how to interact in the learning environment. This fact is consistent with the opinions of several experts, including the following "A learning style is a set of student's characteristics that are reflected in his learning behaviour, how the student learns, how he would be tough, and how the student interacts with the learning environment" [8]. "Learning style refers to an individual's habitual and preferred way of absorbing processing and retaining new information and skills" [9]. "Learning style is the consistent way in which a learner responds to or interacts with stimuli in the learning context" [10]. One type of learning style is learning style according to the sensory modality approach. There are three types of learning styles according to the sensory approach, namely visual, audio, and kinaesthetic (VAK). VAK learning styles provide a simple way to explain and understand the student's learning styles. Besides, the advantages of learning styles with a sensory modality approach that is easily identified, and the use of the senses is the first key in receiving information each learning style has its characteristics. "Visual learners (that include read/write mode also) depend on pictures/words in written form for memorization. They prefer to read than be read. They are good with spellings and are least distracted by noise. They are usually well-groomed orderly and more organized. Most often, their body posture is straight with their heads and eyes up. Auditory learners can repeat conversations easily. They learn by listening and like music. They learn processes sequentially and step by step. The tone of voice affects them. Kinaesthetic learners talk very slowly. They prefer standing close to people and touching them. Their breathing point is the bottom of their lungs. They learn and memorize by doing. They have a physical orientation (move a lot). Common phrases include fell, touch, grasp, hold, and so on." [11]. The instrument used to determine the style of student learning is the VAK learning style of self-assessment. The questionnaire consisted of 30 multiple-choice questions. Each question has three answer choices.

2.2. *Sex*

Sex differences naturally lead to physiological differences and affect psychological differences in learning [12]. In general, the mathematical abilities of boys and girls are the same, but boys having better abstraction strength than girls. Furthermore, girls are generally better at remembering, while boys students are better at logical thinking [13].

Spatial intelligence is a skill to calibrate geometry or in other words, the skill to visualize images [5]. Spatial intelligence consists of skills to change, rotate, and reverse objects provided to be understood in mind [14]. Thus, spatial intelligence relates tightly to geometry. The development of spatial intelligence of boys and girls are different [6]. The spatial intelligence of boys is better than girls [15]. In some studies, boys show better than women on many spatial tasks [16]. Based on some of the opinions above, the ability of boys and girls is different in the two-dimensional figure.

2.3. *Error Type*

Error is a deviation from something real, so error in solving problems about two-dimensional figure are deviations from the correct answer. There are many types of student errors in solving problems. If the error is associated with a mathematical object, then the error is grouped into four types of errors. Error types, according to mathematical objects, are fact errors, concept errors, principle errors, and procedure errors. In eliciting students' errors, they are given a set of problem that consists of four problems about the two-dimensional figure. From the student's answer, we can elicit the error and classified into an error type. The following are indicators used to classify each type of error.

Table 1. Indicator of error type.

No.	Error Type	Indicator
1.	Fact error	Incorrect in giving the name to a two-dimensional figure Incorrect in determining the order of operation
2.	Concept error	Incorrect in determining what is known and what is asked Incorrect in starting the solving step
3.	Principle error	Incorrect in using the right formula Incorrect in using the right rule

2.4. *Poisson regression*

In this study, the analysis technique used is Poisson regression, where response variables are not normally distributed, which is Poisson distribution. The Poisson regression model is a standard model for discrete data and nonlinear regression model [17]. Poisson regression analysis is one of the analytical that used to describe the relationship between the dependent and independent variables. Let y be the number of events that occur in a period with the parameter value of any Poisson λ. The Y variable is a random variable that spreads Poisson with the opportunity mass function is

$$P(y;\lambda) = \frac{e^{-\lambda}\lambda^{y}}{y!}, y = 0, 1, 2, ... \quad (2.1)$$

Assuming the Poisson regression is $E[y] = \mu = Var[y]$. The Poisson regression model is as follows.

$$\lambda_i = exp\ exp\ (x_i^T \beta) \quad (2.2)$$

2.5. *The estimator of Poisson regression parameters*

The MLE (*Maximum Likelihood Estimator*) method is one of the parameter estimating methods used to estimate the parameters of a model known to its distribution. Supposes y_i is a continuous random variable of size n with a function (2.1). The likelihood function of a random sample is a function of the combined density of n random variables which is a function of unknown parameters, so the likelihood function is

$$L(y;\beta) = \prod_{i=1}^{n} f(y_i;\beta) \quad (2.3)$$

Journal of Physics: Conference Series **1417** (2019) 012062 doi:10.1088/1742-6596/1417/1/012062

2.6. Parameter test
Based on the obtained model, a test is conducted to find whether the predictor variables in the model have a significant relationship to the dependent variable. The first test is a simultaneous test, to determine the significance of the β coefficient on the dependent variable simultaneously, with a test like the following.

$H_0: \beta_0 = \beta_1 = \beta_2 = 0$

H_1: at least one coefficient $\beta \neq 0$

G test statistic as follows.

$$G = 2ln\left(\frac{\left(\frac{n_1}{n}\right)^{n_1}\left(\frac{n_0}{n}\right)^{n_0}}{\sum_{i=0}^{N} \hat{\pi_i}^{y_i}(1-\hat{\pi_i})^{1-y_i}}\right) \quad (2.4)$$

G test statistic H_0 rejected if $G > X^2_{(\alpha,p)}$ or p-value $< \alpha$

If H_0 rejected, then the next is the partial test. The partial test hypothesis is as follows.

$H_0: \beta_j = 0$

$H_1: \beta_j \neq 0$

where $j = 0,1,2, ...$

In the partial test, the Wald test is used as follows.

$$W = \frac{\hat{\beta_j}}{\widehat{SE}(\hat{\beta_j})} \quad (2.5)$$

W test statistic H_0 rejected if $|W| > Z_{(\alpha/2)}$ or p-value $< \alpha$

3. Method

3.1. Participant

Based on teaching experience in Al-Islamiyah, there are still many seventh grade students who still make errors in solving a question about the two-dimensional figure. Therefore, participants in this study were 32 students in grade 7. The subject consisted of 11 students with a visual learning style, seven audio learning style students, and 14 kinaesthetic learning style students. From 32 students, there are 11 students with a visual learning style, seven students with an audio learning style, and 14 students with a kinaesthetic learning style. Students with visual learning styles showed fact error, concept error, principle error, and procedure error. These errors also occur in students with audio learning styles and students with kinaesthetic learning styles. The difference is the details of the number of error types in every student. Besides that, they consisted of 16 boys and 16 girls

3.2. Research variable

The dependent variable (y)in this study is the number of error types, while the independent variable (x) is the student's learning style (x_1) and sex (x_2). Student learning styles can be known based on the results of the given learning style questionnaire, while many types of errors are recognized from the results of student work in solving the two-dimensional figure test.

4. Result and discussion

In this study, independent variable is learning style (x_1) and sex (x_2). The dependent variable is the number of error type (y). The result from test is 4 student answer the test correctly, 13 students make 1 error type, 8 students make 2 error type, and 7 students make 3 error type learning style.

The relationship between learning styles (x_1) and sex (x_2) on the number of error types (y) can be seen based on the value of the parameter estimation between learning styles $(x_1)$and sex difference (x_2) on the number of error types (y). The first step is to do simultaneous test, obtained G-value is 42.16069 and X^2 with df 2 and alpha 0.05 is 5.991. Thus G > X^2_{table}, than there is a strong evidence to reject H_0. It means at least one variables that affect the number of error types. After a simultaneous test, the next step is to do partial test, the parameter estimation results are obtained as in table 2 below.

Table 2. Parameter estimation results.

	Estimate	**Std. Error**	**Z Value**	**P (> |z|)**
(Intercept)	1.13502	0.51489	2.204	0.0275
Learning Style	-0.41162	0.16514	-2.493	0.0127
Sex	0.07206	0.28402	0.254	0.7997

From table 2, it can be seen that the P-value for learning styles is 0.0127, and the p-value for sex is 0.7997. In partial testing, H_0 is rejected if P-value < α (0.05), where

H_0 : $\beta_j = 0$ (there are no significant effect on the number of types of errors)
H_1 : $\beta_j \neq 0$ (there are a significant effect on the number of types of errors)
where $j = 0,1,2,\ldots$

Thus it can be seen that for the learning style p-value < α so that H_0 is rejected. As for sex, p-value > α so that H_0 fails to be rejected. So the variables that affect the number of error types are learning styles. While sex does not affect errors. Thus, the model for the number of error types is obtained as follows.

$$\mu = exp\,(1.13502 \quad 0.41162x_1 + 0.07206x_2) \qquad (4.1)$$

Based on the Poisson Regression model above, we know that $\beta_1 < 0$. It means for one unit of increase in learning style, the number of error types will decrease as much as *exp* (0.4078) =1.509261. Learning styles with the most error type made are visual learning styles. So, a student who has a visual learning style needs more attention. In order to reduce the error type of visual learning style students, it is necessary to take action.

5. Conclusion

Based on the result of the analysis, it can be concluded that learning style has a significant impact on the number of error types, while sex has no impact on the number of error types.

References

[1] Mirna M 2018 Error Analysis of Students in Mathematics Department to Learn Plane Geometry *Materials Science and Engineering* **335** D 10.1088/1757-899X/335/1/012116
[2] Sulistyorini Y 2017 Error Analysis in Solving Geometry Problem on Pseudo-Thinking's Students *In: University of Muhammadiyah Malang's 1st International Conference of Mathematics Education (INCOMED 2017)* Atlantis Press pp 103-107
[3] Joaquin C J A 2013 *Student's Academic Performance in Mathematics as Influenced by Teaching and Learning Style (*Roxas City: Filmer Christin University)
[4] Rahman A and Ahmar A S 2017 Relationship between Learning Style and Learning Achievement in Mathematics based on Gender *World Transaction on Engineering and Technology Education* **15** (1) pp 74-77
[5] Harmony J and Theis R 2012 Pengaruh kemampuan spasial terhadap hasil belajar matematika siswa kelas VII smp negeri 9 kota jambi. *Jurnal edumatica* **2** pp 11-19
[6] Sanders J 2011 *Gender Smart*: Memecahkan Teka-teki Komunikasi antara Pria dan Wanita

(Jakarta: Buana Ilmu Populer Kelompok Gramedia)
[7] Adhitya Y 2018 V-A-K Grade 7 Students' Error in Mathematical Problem Solving About Quadrilateral *International Journal of Research in Applied, Natural and Social Sciences (IJRANSS)* **6** (5) pp 157-164
[8] Ocepek U, Bosnic Z, Serbec I N and Rugelj J 2013 Exploring the relation between learning style models and preferred multimedia types *Computers and Education* **69** pp 343-355
[9] Rezaeinejad M, Azizifar A and Gowhary H 2015 The study of learning styles and its relationship with educational achievement among Iranian high school students. *Procedia – social and behavioral sciences* **199** pp 218-224
[10] Manolis C, Burns D, Assudani R and Chinta R 2013 Assessing experiential learning styles: a methodological reconstruction and validation of the Kolb learning style inventory *Learning and individual differences* **23** pp 44-52
[11] Wickramasinghe S and Hettiarachchi S 2017 Relationship among students' learning styles, assessment methods and students' performances *Third International Conference on Education and Distance Learning* **31** pp 6643862
[12] MZ Z A 2013 Perspektif Gender dalam Pembelajaran Matematika *Marwah: Jurnal Perempuan, Agama Dan Jender* **12** (1) pp 14-31
[13] Asmaningtias YT 2012 Undergraduate Student's Academic Achievement, Field Dependent/Independent Cognitive Styles and Attitude toward Computers *Journal of Educational Technology & Society* **9** (1) pp 289-297
[14] Guven B and Kosa T 2008 The Effect of Dynamic Geometry Software on Student Mathematics Teacher's Spatial Visualization Skills *The Turkish Online Journal of Educational Technology* **7** (4) pp 100-107
[15] O'Gorman R 1999 Sex Differences in Spatial Abilities: An Evolution Explanation *Irish Journal of Psychology D* 10.1080/03033910.1999.10558227
[16] Mohler J L 2008 A Review of Spatial Ability Reseach *Enginering Design Graphics* **72**
[17] Famoye F and Singh K P 2006 Zero-Inflated Poisson Regression Model with an Application to Domestic Violence Data *Journal of Data Science* **4** (1) pp 17-130

Student Teachers' Construction of Mathematical Theorem of Set and Cardinality

Evangelista Lus Windyana Palupi[1], Abdul Haris Rosyidi[2], Hani Rizkia Putri[3]
[1,2,3] Mathematics Department, Faculty of Mathematics and Natural Science, Universitas Negeri Surabaya

E-mail:evangelistapalupi@unesa.ac.id

Abstract. Constructivists suggest that teachers should guide and facilitate their students in reinventing and or reconstructing mathematics concepts as in reconstructing a theorem. A theorem in mathematics holds a vital role as a fundamental aspect in building the mathematics itself. Hence, students must learn and acquire this ability. Furthermore, by reconstructing a theorem, students not only learn about the theorem but also learn about problem-solving since the theorem that should be reconstructed can be presented as a problem. In which, this skill is needed and promoted in the 21st century. Hence, the teacher has to be able to reconstruct and prove their construction of theorem. However, many studies were focused on investigating students' mathematical proof ability, not on the ability to reconstruct a theorem. This study is aimed to investigate mathematics undergraduate students' ability to reconstruct and prove a theorem. A problem to make a general statement regarding the numbers of all possibility of the cardinality of A B, with A, B is set, and n(A) is a and n(B) is b, is given to 60 undergraduate students who are majoring mathematics education. The statement then is analyzed and categorized into 0 to 5 based on the developed framework. The result shows that most students' answers (29 answers) are categorized as an incomplete or irrelevant theorem (level 0), while only seven answers can be categorized as level 5 (well structured). As for level 2, 3, and 4, there are 13, 8, 3 answers representatively. Also, there is no student answer, which is categorized as level 1. The mistakes occur because students do not fully understand the mathematics notion. Then, the theorem made is different from the one that is asked, the given condition mentioned by the students is not general enough (does not represent all cases); the conclusion made is false. However, no conclusion is unrelated to the condition.

1. Introduction

Constructivism, the idea that students have to construct their knowledge with the help of their teacher, has still been an ideal view in teaching [1]. In teaching mathematics, based on this view, teachers should guide and facilitate their students in reinventing and or reconstructing mathematics concepts as in reconstructing a theorem [2]. This constructivism means that students have to do the thinking process while the teacher acts as a facilitator and give a scaffolding and guidance in the learning process.

A theorem in mathematics holds an important role as a fundamental aspect in building the mathematics itself. Besides, to construct another theorem, it is a necessity for the construction to be based on or in synch with the other theorem [3]. Hence, students need to have and learn this ability. Furthermore, by reconstructing a theorem, students not only learn about the theorem but also learn about problem-solving since the theorem that should be reconstructed can be presented as a problem [4]. In which, this skill is needed and promoted in the 21st century [5].

The 21st-century learning requires the learner to study in the learning activity actively [5]. The core of this century is skills of high order things, communication, creativity and innovation, problem-solving, and confidence [6]. Mathematical problem solving is a students' process in solving a mathematical problem as a product of creative thinking. Then problem posing is a constructing problem in mathematical terms based on some given information to find a solution [4]. So, problem posing is a counterpart of problem-solving in mathematics problem-posing gives other benefits, such as to increase problem-solving ability, making students active, and enriching fundamental concepts [7]. So, the task of submission a problem should be given in Mathematics Education students to understand the information in axiom or theorem.

Since the teacher should be able to guide and facilitate their students in reconstructing a mathematical theorem, the teacher has to be able to reconstruct and prove their construction of theorem. Siswono [8] finds out that mathematics students teachers find difficulties in constructing a theorem from the given axioms. The results of the study claim that the unsophisticated students' theorem constructions are caused by students' mistakes of a mathematical concept, mistake on the proof, and mistake related to the mathematics sentence [8]. However, the analysis of the theorem structure is missing, This study aims to explore mathematics education undergraduate students' ability to construct a theorem, focus on the structure of the theorem. So that we can find out more data on the mathematics teacher students' theorem construction and its proof.

2. Theorem Construction and Proof Scheme

A theorem in mathematics is defined as a statement that needs to be justified and be proven [9]. As for the formulation, a theorem can be written as a conditional sentence. In other words, it can be presented as *"if ..... then ....."* statement. The *"if"* statement is called premises or condition, and the *"then ..."* statement is called a conclusion of the given premises.

Derived from the definition of theorem and the fact that all theorem can be written as conditional sentences consist of condition and conclusion, a classification level of students' theorem construction is produced. The theorem construction is classified into six levels, start from level 0, which associated with an irrelevant theorem or incomplete theorem to level 5 for sophisticated theorem (Table 1).

Table 1. The descriptor for coding of students' theorem construction

Level	Characteristics	Description
0	Irrelevant or incomplete theorem	Do not know how to construct theorem (leave a blank page) Construct irrelevant theorem with the task
1	Incomplete condition and unrelated conclusion	The condition is incomplete, just using one or more condition, but do not cover all possible condition Relevant to the task Conclude with no relation to the written condition. Unable to relate the conclusion and condition
2	Incomplete condition and wrong conclusion	Relevant to the task The conditions are incomplete The conclusion is related to the condition but wrong
3	Incomplete condition and right conclusion	Relevant to the task The conditions are incomplete The conclusion is related to the condition and right conclusion

4	Complete condition and wrong conclusion	Relevant to the task The conditions are complete, cover all possible cases The conclusion is related to the condition but wrong
5	Well-structured (Complete condition and right conclusion)	Relevant to the task The conditions are complete The conclusion is related to the condition and right

3. Method

This research aims to know how is mathematics education undergraduate students' ability to construct a theorem focus on the structure of the theorem. Sixty mathematics student teachers are given a problem to construct a theorem based on the condition given.

Problem:
Given $n(A) = a$ and $n(B) = b$ with A and B are a set. Construct a general statement regarding the numbers of all possibility of $n(A \cup B)$.

To construct the theorem, students need to define the condition of their general statement since the condition given in the problem is still too general. In other words, students need to think of the condition of the sets, whether they are different or same, dependent or independent. The conjectured condition of the general statement should not to particular and have to cover all the possible cases. The general statement could not only made for a particular case, for instance, only for condition $A = B$. As for the conclusion, students need to understand the difference between the numbers of all possibility of $n(A \cup B)$ and the numbers of all possibility of $(A \cup B)$ thoroughly. Next, the statement obtained needs to be proven. The constructed theorem made by the students then is analyzed and categorized based on the descriptor for coding of students' theorem construction (Table 1).

4. Result and Discussion

In general, the level of students' theorem construction can be divided into four categories, such as failed to understand the problem or construct irrelevant theorem (level 0), construct unrelated conclusion using incomplete condition (level 1), construct wrong theorem with related conclusion using different level in showing the conditions (level 2, 4), and construct the right theorem that have related conclusion using different level in showing the conditions (level 3,5). Table 2 shows the frequency distribution of students' theorem construction on the task.

Table 2. The distribution of students' theorem construction

Level of theorem construction	Frequency	%
0	29	48.33%
1	0	0.00%
2	13	21.67%
3	8	13.33%
4	3	00.05%
5	7	11.67%

Table 3 shows that most students, there are 29 students, are categorized in level 0. This finding means that most of the students failed to construct the relevant theorem. As for level 2 to 4, there are 13, 8, and 3 answers that are categorized in this level, respectively, while no one made an unrelated conclusion. The main cause of the level 0 answers is that the students construct a theorem that is irrelevant to the given task. The factors are that students do not fully understand the mathematics symbol. Hence they mistook the numbers of all possibility of $n(A \cup B)$ as the numbers of all possibility of $(A \cup B)$. A sample of the students' results can be seen in figure 1.

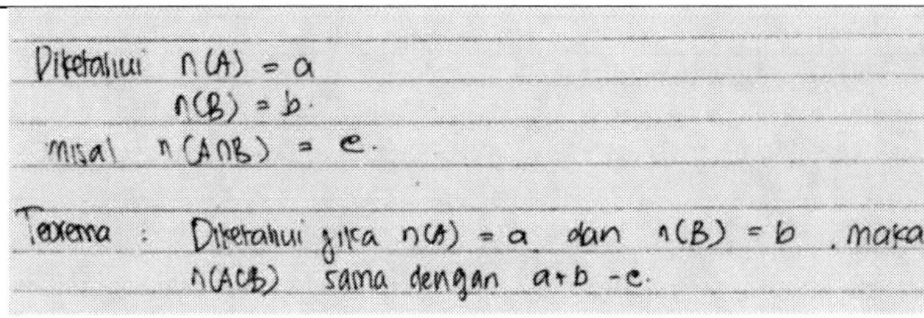

Figure 1. Students' responses which categorized in level 0

From Figure 1, we can see that the students did not understand that the researcher asked the problem. The purpose of the problem is to construct a theorem related to the number of possibilities. However, in this case, the student answered by constructing a theorem related to the irrelevant set. So, the students did not answer the question. Their carelessness or misconceptions cause it in the concept of probability of cardinality. Students conclude that the probability of cardinality has the same meaning as the cardinality of the set chapter.

In other cases, there are no students that the answer was categorized in level 1 that describe there are no related between condition and the conclusion. Furthermore, 13 students answer in level 2. It indicates that students show incorrect theorem because of the incomplete condition, but the conclusion inferred is related to the conditional statement, as in figure 2.

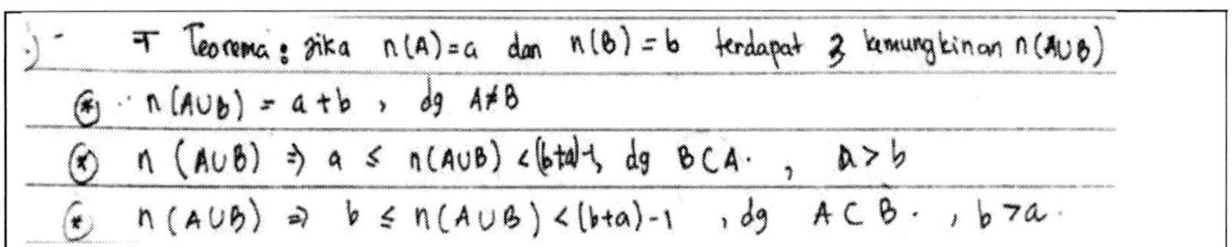

"Theorem: If $n(A) = a$ and $n(B) = b$ there are three possibilities of $n(A \cup B)$

- $n(A \cup B) = a + b; wit \quad A \neq B$
- $n(A \cup B)\ is\ a \leq n(A \cup B) < (b + a) \quad 1; wit \quad B \subset A, a > b$
- $n(A \cup B)\ is\ b \leq n(A \cup B) < (b + a) \quad 1; wit \quad A \subset B, b > a$"

Figure 2. Students' response which categorized in level 2

The example of the incomplete condition of a theorem is shown in Figure 2. The condition should cover all possibility of a relation between $A, B, a,$ and b. However, some students only consider part of it; for instance, is they only wrote the condition for $A \neq B$ with A and B are independent set. Students do not write the condition for $A \neq B$ with A and B are dependent. Also, as for the other condition, they neglect the fact that there is a possibility for $A = B$. Even though students realize that it is possible for $B \subset A\ so\ a > b$ and vice versa, but they do not realize that if $B \subset A\ and\ A \subset B$ then $A = B$ result on $a = b$. This condition is not yet covered in the theorem.

Journal of Physics: Conference Series **1417** (2019) 012063 doi:10.1088/1742-6596/1417/1/012063

Regardless of the given incomplete condition, students failed to make the right conclusion. For the case that $B \subset A$ or $A \subset B$, the cardinality of $(A \cup B)$ should be the same as the number of elements of the more significant set. While the students answer that $n(A \cup B)$ wit $B \subset A$ is $a \leq n(A \cup B) < (b + a)$, so the conclusion is incorrect.

As for level 3, eight students can make a relation to the conclusion and construct the correct theorem, yet the condition is incomplete.

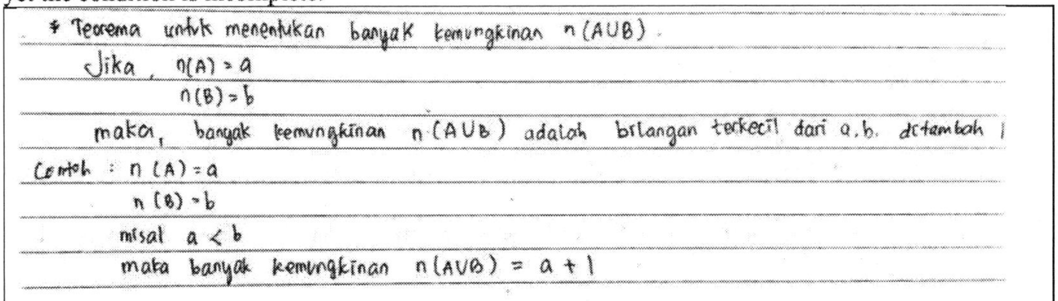
* Teorema untuk menentukan banyak kemungkinan n(AUB).
Jika, n(A) = a
n(B) = b
maka, banyak kemungkinan n(AUB) adalah bilangan terkecil dari a,b. ditambah 1
Contoh : n(A) = a
n(B) = b
misal a < b
maka banyak kemungkinan n(AUB) = a + 1

The theorem of determining the number of possibility of $n(A \cup B)$
If, $n(A) = a; n(B) = b$
Then, the numbers of $n(A \cup B)$ is the smallest of a, b plus 1.
For example: $n(A) = a; n(B) = b$
Assume $a < b$ then the number of possibility of $n(A \cup B) = a + 1$

Figure 3. Students' response which categorized in level 3

From figure 3, we can see that students can construct the correct theorem for the cardinality of $A \cup B$, is $a + 1$. However, the condition is $a < b$. It indicates that the theorem was applicable for $a = b$, so the condition cannot be applied generally or incomplete.

Table 1 also shows that there are only three students can show the complete condition, which is related to the conclusion, but the theorem was incorrect, as shown in the figure below.

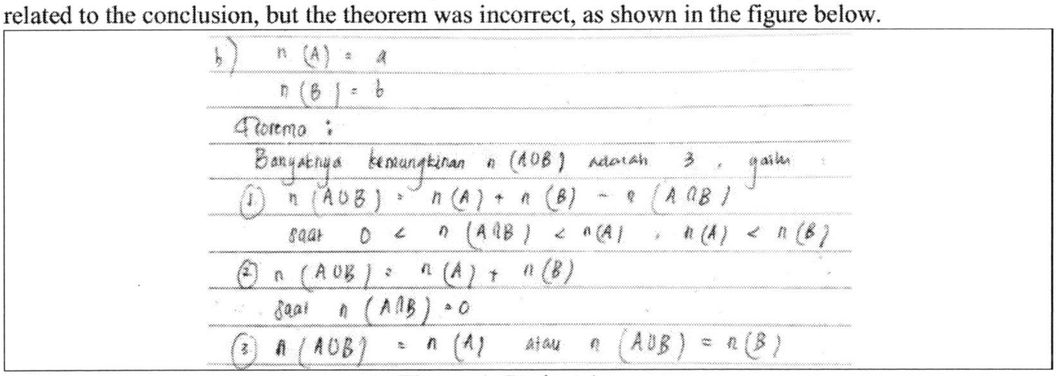
b) n(A) = a
n(B) = b
Teorema :
Banyaknya kemungkinan n(AUB) adalah 3, yaitu
(1) n(AUB) = n(A) + n(B) − n(A∩B)
saat 0 < n(A∩B) < n(A), n(A) < n(B)
(2) n(AUB) = n(A) + n(B)
saat n(A∩B) = 0
(3) n(AUB) = n(A) atau n(AUB) = n(B)

Figure 4. Students' responses

Figure 4 shows that students at level 4 mention all the conditions of the theorem. However, the students' answers, which are categorized as level 4, indicate the wrong conclusion. Students have the right thinking process, yet for a condition in which $0 < n(A \cap B) < n(A)$ with $n(A) < n(B)$ (case 1 figure 4), students do not consider all the possibilities of $n(A \cap B)$ itself. Instead, students see the condition as one whole possibility. This situation leads to the wrong conclusion.

Using the complete condition related to the conclusion, seven students can construct the theorem correctly. The students in this situation can be categorized in level 5 (highest level) of the constructing mathematical theorem. The students' artifact for level 5, shown in figure 5.

Journal of Physics: Conference Series **1417** (2019) 012063 doi:10.1088/1742-6596/1417/1/012063

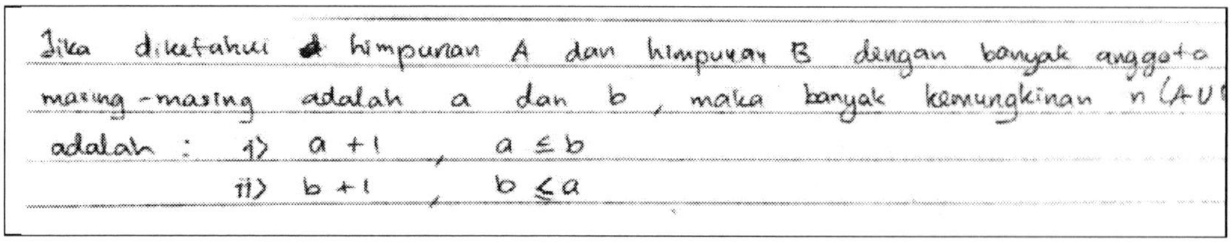

Figure 5. Students' responses which categorized level 5

At the highest level, students can construct the theorem correctly based on a complete condition which related to the conclusion. They mention all the conditions using notation $a \leq b$ and vice versa, which means the theorem will be applied generally. There is no misconception between the concept of cardinality and cardinality of possibility, and the conclusion also obtained mathematically right.

5. Conclusion

The result shows that most students failed to construct the theorem. Even though some of the constructed theorems provide the right conclusion, but if the condition is not cover all cases, the theorem not yet to be called well-structured and vice versa. The same category is also applied for the students who able to construct a well-structured theorem, yet the theorem is irrelevant to the task. There are only a few students (7 out of 60) who can construct a well-structured theorem. This finding suggests that the teacher or lecturer still need to improve student's ability in theorem construction, especially in writing or listing all cases for conditional sentences as well as understanding mathematics symbol. While regarding whether or not the constructed theorem can be proven as true still needs further investigation.

Acknowledgments
Authors wishing to acknowledge assistance or encouragement from colleagues.

Reference

[1] Bada S O and Olusegun S 2015 Constructivism learning theory: A paradigm for teaching and learning *J. Res. Method Educ.* **5** 66–70
[2] Budiarti E and Rosyidi A H 2015 Alur Berpikir Siswa SMP dalam Membuktikan Teorema Pythagoras melalui Tugas Pengajuan Soal Ditinjau dari Perbedaan Jenis Kelamin *MATHEdunesa* **3**
[3] Koichu B 2019 Problem posing in the context of teaching for advanced problem solving *Int. J. Educ. Res.*
[4] SISWONO T Y E 2015 Improving elementary teacher competency to develop the abilities of students' creative thinking through mathematics problem posing and problem solving strategy *7th ICMI-East Asia Regional Conference on Mathematics Education* (Cebu, Philippines)
[5] Jan H 2017 Teacher of 21 st Century: Characteristics and Development *Res. Humanit. Soc. Sci.* **7** 50–4
[6] Warner S and Kaur A 2017 The perceptions of teachers and students on a 21st Century mathematics instructional model *Int. Electron. J. Math. Educ.* **12** 193–215
[7] Siswono T Y E 2005 Student Thinking Strategies in Reconstructing Theorems. *Int. Gr. Psychol. Math. Educ.* **4** 193–200
[8] Siswono T Y E 2004 Problem Posing: Melatih Kemampuan Mahasiswa dalam Membangun Teorema *Proceeding of National Seminar "Penelitian Pendidikan dan Penerapan MIPA", UNY*
[9] Masriyah 2016 *Dasar-dasar Matematika* (Surabaya: Universitas Negeri Surabaya)

The Instructional Design of Blended Learning on Differential Calculus Using Successive Approximation Model

A Wintarti*[1], Abadi[1] and D K Fardah[3]

[1]Mathematics Department, Universitas Negeri Surabaya, Indonesia.

* Email: atikwintarti@unesa.ac.id

Abstract. The combination of traditional learning where teacher and students are in the same classroom and e-learning using technology is known as blended learning. This paper outlines the process of designing blended learning on Differential Calculus. The Successive Approximation Model (SAM) is used as the instructional design upon the course. In the preparation phase, the information gathering is starting to ensure an agreement between the learner's need and learning solution. Afterward, the design and prototype of a blended course was implemented for Mathematics Education students on Mathematics Department Universitas Negeri Surabaya. Along with iterative development phase, the evaluation is conducted in order to make the course move to Alpha (a complete version of the instructional design blended learning) then evolve into Beta (a better version after Alpha revised) and then the Gold as finale version is reached.

1. Introduction

The technological developments in recent years have changed human civilization. Everything that is a product of technology has helped humans in meeting their needs. This happens also in the world of education. One technology product that can be used to help teaching and learning in class is the use of the internet [1]. It does not only help a lecturer to complete the references needed in lecturing but can also use it as a learning medium that bridges the limitations of space and time between lecturers and students on design instructional.

This paper intends to explain the instructional design on Differential Calculus for Mathematics Education students of the Department of Mathematics, Universitas Negeri Surabaya using a blended learning system. This lecture will be held in 30 meetings in 15 weeks with the form of face-to-face and on-line lectures conducted at the Universitas Negeri Surabaya Online Learning System which is integrated with Siakadu (Vi-learn). For this reason, it is necessary to develop teaching materials using information and communication technology. Teaching materials will be delivered face-to-face and through e-learning both off-line and on-line as a blended strategy of learning [2]. The method used in the form of lecturing, independent learning through teaching materials delivered in the form of files, slides, simulations or learning videos and discussions either directly or through information and communication technology. For the assessment of lectures will be evaluated in the form of assignments and quizzes. In addition, students are asked to provide responses in the form of input or comments about lectures conducted with this blended learning system.

2. Background Theories

2.1 Blended Learning

Blended learning is an instructional methodology that using technology to provide a more personalized approach to education and to give students control over the time, place, path and pace of their learning [3]. Blended learning is an innovative concept that embraces the advantage of both traditional teachings in the classroom and ICT supported learning including both offline and online learning [4].

The term "blended learning" was used first by Cooney et al. in 2000 to combine elements of play and work in prekindergarten [5]. Today, there are many articles and practices about blended learning using any kind of technology and variant of media.

2.2 Instructional Design

Instructional design is defined as a systematic procedure in which educational or training programs are developed and composed for a substantial improvement of learning [6]. Instructional design contains both scientific and technological theories on the design of learning environments.

The advancement of technology, institutions are able to reach more students than ever with the help of quality and accessible online courses. ‘e-learning’, ‘distance education’, ‘blended learning’, ‘online campuses,’ and other related programs have grown more prominent in higher education institutions [7].

There are many models of instructional design to facilitate the development of learning system. The abbreviation of Analysis, Design, Development, Implementation, and Evaluation (ADDIE) was a five-phase model that developed in the mid-1970s with the purpose of formalizing the process of developing military inter-service training [8]. There are many other instructional designs such as Dick and Carey, Borg and Hall, Hannafin and Peck, ASSURE, Kemp, Brown, and SAM as a model of initial ADDIE models [9].

2.3 The Successive Approximation Model (SAM)

SAM was introduced by Michael W. Allen in 2012 with the intent of providing increased flexibility with more agile development, responsiveness, and collaborative opportunities than offered by traditional ADDIE [8]. SAM also need collaboration between instructional designer, program manager, facilitator, and students.

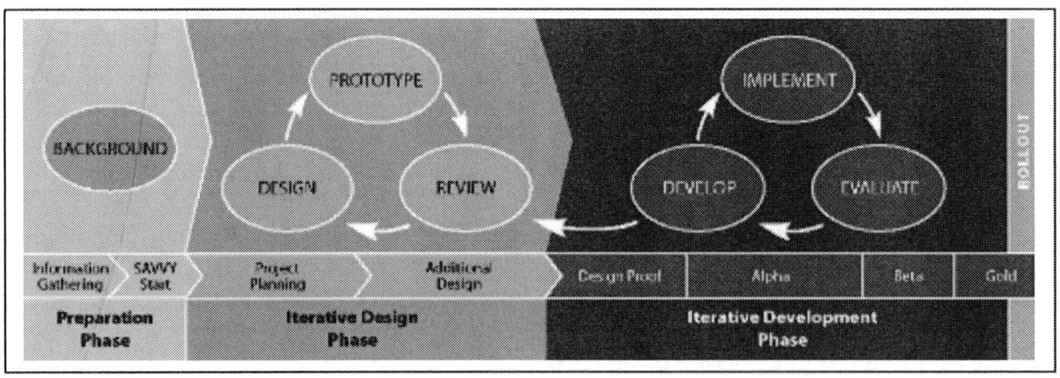

Figure 1. The Phases of SAM (alleninteractions.com/sam-process).

There are three phases of SAM and each could be cycled in order to be the best instructional design that can be represented as Figure 1. SAM starts with the Preparation phase to get information and all background knowledge about objects' needs and ways of solutions. This phase was usually known as a "Savvy Start" where brainstorming and prototyping of the system were initializing [1].

Then in the Iterative Design phase, the initial prototype will be rotating through a cycle of review-design-prototype again as a project goal meeting. This phase is an emphasis on an iterative approach to creating the right product.

The last phase is Iterative Development with the develop-implement-evaluate cycle work together. This phase will start with design proof to implement then evaluate to get the Alpha design as a complete version of the instructional application. Then the design will evolve into the Beta version after revised based on the Alpha cycle. Finally, it will go into the Gold version after feedback of the Beta step.

SAM allows us to analyze and evaluate throughout the process. At any point, if a change needs to occur, it can happen quickly to limit any risk of moving out budget or time [10].

3. Methods

SAM starts with the preparation phase where information and all background knowledge are gathered. The characteristics of the student must be considered. Differential Calculus lectures with blended learning systems are delivered face-to-face and e-learning using both offline and online. The Lesson Plan (Rencana Perkuliahan Semester or RPS) was prepared in advance based on the description of Differential Calculus courses in 30 meetings for 15 weeks. These steps are recalled as a Savvy Start in the SAM model.

The next phase, based on the RPS a storyline is made at each meeting so that the lecture can be determined face-to-face, off-line or on-line as well as the type of media required. This storyline will be validated by media experts to determine the suitability of media used. After the storyline is declared valid the content output will be made then validated by the material expert to find out the validity of the Differential Calculus material content. These works are carried out in the framework of making a prototype then reviewing.

In the Development phase, the reviewed prototype would be implemented for one class of Mathematics Education at Mathematics Department of Universitas Negeri Surabaya use online system (Vi-learn). Data analysis techniques to evaluate will be carried out by:

a. Every process in the research method has been carried out, otherwise, there must be an alternative step taken or there is a relevant reason that does not affect the lecture process.
b. Differential Calculus lecture using blended learning is said to be able to use by students if the average score of Assignments and Quiz students above the passing grade of a course i.e. 56. Student responses after attending Differential Calculus lecture using blended learning is said to be positive if the percentage of positive responses is more than 50%.

If all the above criteria are fulfilled, then it can be considered as Alpha version of Blended Learning Differential Calculus.

4. Results

Calculus Differential is the first part of Calculus lesson that gives to the new enrollment student from fresh high school graduated. Being involved as a university student, they should adapt to learning behavior from teacher-centered to student-centered. Although most students are used to using gadgets, they rarely used them for learning.

Based on these findings and RPS of Differential Calculus, we arrange a storyline to make a prototype of blended learning. Then we implemented blended learning of Differential Calculus on Vi-learn (https://vi-learn.unesa.ac.id/course/view.php?id=3361). One of the online parts of this course can be presented in Figure 2.

Figure 2. One of the online part of Differential Calculus Blended Learning.

The blended learning of Differential Calculus was delivered for 2019A class that taken by 40 students. The first meeting happened on August 19th, 2019, and the next will be held every Monday at 10.20 - 12.00 and Friday at 13.00-14.40.

In every meeting, there are some materials, video, and assignment can be found. Student responses are pretty good because they can access Vi-learn even through smartphones. Students can also chat and upload the results of their work on the given assignment. However, the video uploaded is not loud enough if heard via smartphone. It should be improved for the next video.
SAM model allows us to make a review and revised even change teaching methods and media used. Based on the students' responses as depicted in Figure 3, 36% students feel passive in learning because all the material has been provided

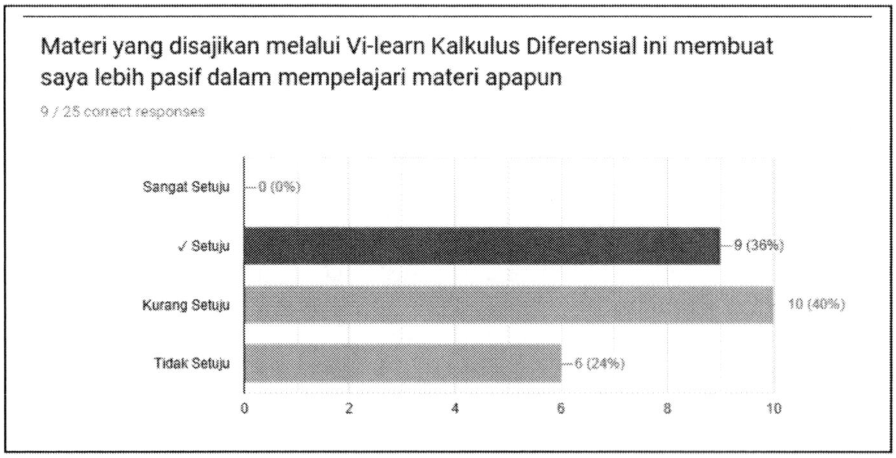

Figure 3. A negative response about blended learning.

In this case, we try to give some assignment or activities not only in face-to-face but also on Vi-learn as show in Figure 4 that we conduct a live chatting to discuss about domain.

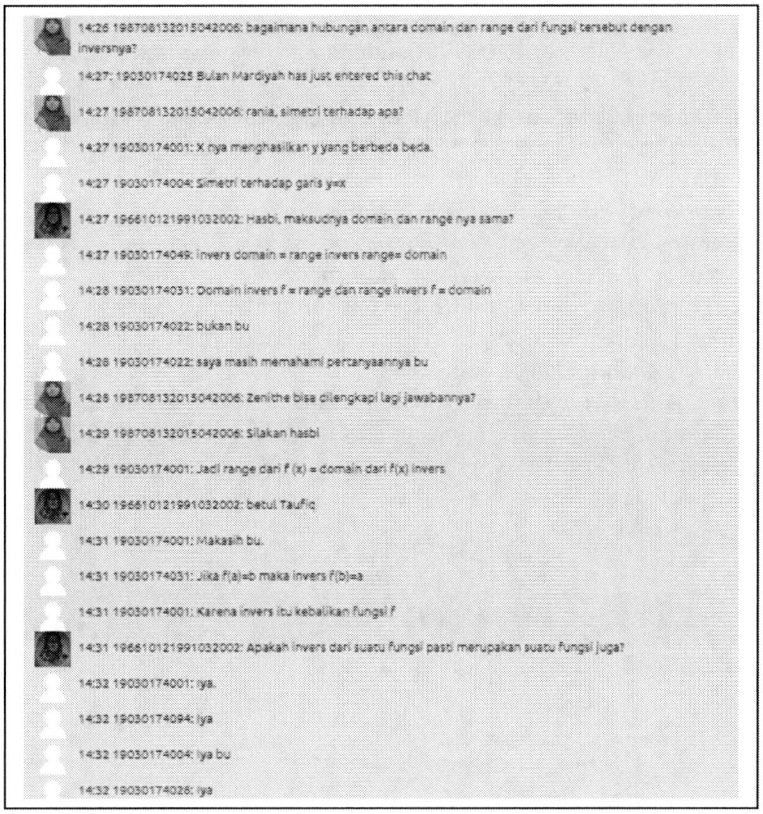

Figure 4. Live chatting to discuss about domain.

5. Conclusion

As a beginning of instructional design on Differential Calculus course using blended learning, it showed that a fairly good development. It is expected with the use of SAM model for instructional design will be produce the blended learning of Differential Calculus that are appropriate to the learning purpose and students' requirements as Alpha version.

For the future works, it could be increased until Beta version and up to Gold version as an e-learning course in SPADA (Sistem Pembelajaran Daring Indonesia), an Online Learning System in Indonesia.

Acknowledgment

The authors would like to thank Universitas Negeri Surabaya upon the foundation of this research via PNBP.

References

[1] Gros B and García-Peñalvo F J 2016 Future Trends in the Design Strategies and Technological Affordances of E-Learning in *Learning, Design, and Technology*, Springer International Publishing pp 1–23
[2] Okaz A A 2015 Integrating Blended Learning in Higher Education *Procedia - Soc. Behav. Sci.*,

186 pp 600–603
[3] Borba M C, Askar P, Engelbrecht J, Gadanidis G, Llinares S and Aguilar M S 2016 Blended learning, e-learning and mobile learning in mathematics education *ZDM - Math. Educ.* **48**(5) pp 589–610
[4] Lalima D and K. Lata Dangwal 2017 "Blended Learning: An Innovative Approach," *Univers. J. Educ. Res.* **5** (1) pp 129–136
[5] Güzer B and Caner H 2014 The Past, Present and Future of Blended Learning: An in Depth Analysis of Literature *Procedia - Soc. Behav. Sci.* **116** pp. 4596–4603
[6] Seel N M, Lehmann T, Blumschein P and Podolskiy O A 2017 *Instructional Design for Learning: Theoretical Foundations* Sense Publishers
[7] Intentional Futures 2016 *Instructional Design in Higher Education: A report on the role, workflow, and experience of instructional designers*
[8] Payne D L 2016 *Mapping SAM to ADDIE* California State University
[9] Siragusa L 2006 Quality eLearning: An instructional design model for online learning in higher education *AARE Annual Conf.*
[10] Jung H, Kim Y R, Lee H and Shin Y 2019 Advanced instructional design for successive E-learning: Based on the successive approximation model (SAM) *Int. J. E-Learning Corp. Gov. Heal. High. Educ.* **18**(2) pp 191–204

The Performance of Prospective Teachers in Interpreting Graphs on context-based tasks

D Hidayat*[,1], TYE Siswono[1], S Hartono[1], NR Prihartiwi[1], AW Kohar[1]
[1]Mathematics Department, Universitas Negeri Surabaya

*E-mail: dayathidayat@unesa.ac.id

Abstract. Graphing ability facilitates students to understand and to connect graphs of data with real-world situations. This ability related to mathematical competencies such as logical thinking and visual representation. One aspect of graphing ability that is important for the student to get meaning from the graph is the interpreting graph. The purpose of this study is to investigate the performance of prospective teachers in interpreting graphs on context-based tasks. This study uses a descriptive-explorative method involving 24 prospective teacher participants at the Universitas Negeri Surabaya. This article presents the performance of prospective teachers' interpreting graph abilities based on five levels of graph interpretation by Aoyama. The prospective teachers' graph interpretation is explained along with the difficulties.

1. Introduction

The ability to understand qualitative information from quantitative information is the goal of statistical literacy in the 21st century [1]. Furthermore, the development of information and technology in the modern era makes the presentation of data more useful and attractive such as the use of tables, graphs, and diagrams. Also, the many different forms of information require an increased ability to evaluate what is presented [2] critically. Therefore, each individual is required to have the ability to interpret data from quantitative information such as graphs to qualitative explanation.

A qualitative explanation of a graph in fullest meaning requires someone to look at the entire graph and capturing meaning about the relationship between variables and get their pattern of covariation [3]. The ability to interpret graphs is closely related to the curriculum in mathematics education. In mathematics education, students must be able to use representations of graphs, tables, or formulas to draw conclusions and make models or solve contextual problems using various representations [4]. Unfortunately, recent studies have indicated that students' understanding of the graph is limited. Some conclude that students still have difficulty in extracting information on the graph [5]. Others conclude that students have difficulties in making connections among graphs, physical concepts, and the real world [6]. The difficulties are due to students often consider graphs to be only pictures of a situation [7], not a graph that adequately illustrates the behavior of the variables involved. To overcome these problems, students must get used to interpreting contextual graphs, especially when studying mathematics in school. The teacher, as a facilitator, must also have excellent graph interpretation skills to be able to accommodate students in developing the ability to interpret graphs.

Based on the discussion above, the purpose of this study is to investigate the performance of prospective teachers in interpreting graphs on context-based tasks. The abilities to interpret the graph of prospective teachers were grouped according to the hierarchy of graph interpretation by Aoyama. The results and discussion of this study can be a reference for teachers or educators to understand better the levels of interpreting graphs of students.

2. Methodology

This descriptive-explorative research aims to investigate prospective teachers' performance in interpreting graphs on context-based tasks. Participants were 24 prospective teachers in a university in Surabaya. Data were collected through a graph task. The task is a context-based task to assess the ability of graph interpretation of students (figure 1). The task consists of two questions. Question (a) is to ask about the interpretations of the horizontal line on the graph. Question (b) is to ask about interpretations of the phenomenon of lines that always go down and form angles on the graph.

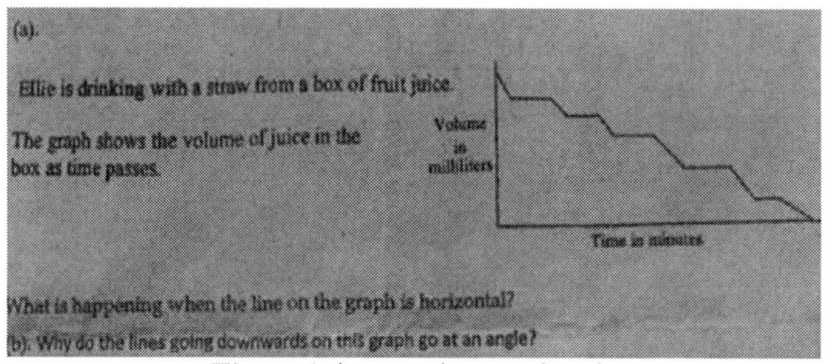

Figure 1. interpreting graph task

Data were analyzed descriptively and grouped according to 5 levels of data interpretation hierarchy [1] (Figure 1). *Idiosyncratic* is the level that students cannot read the values or trends in a graph. *Essential Graph Reading* is the level where students can read values and trends from a graph but are unable to explain the contextual meaning of a trend or feature. *Rational/Literal* is the level that students can read data and trends from a graph and explain contextual meanings, but only limited to the meanings presented. *Critical* is the level where students can read a graph and realize contextual meanings by evaluating the reliability of meanings. *Hypothesizing and modeling* is the level that students are able to read values and trends from a graph and provide accurate contextual explanations. In addition, they can form their explanatory hypotheses or models.

Table 1. Hierarchy of graph interpretation

Level 1	**Idiosyncratic**
Level 2	Basic Graph Reading
Level 3	Rational/Literal
Level 4	Critical
Level 5	Hypothesizing and Modelling

3. Result and Discussion

Table 2 shows the classification of the ability of prospective teacher students to interpret charts based on hierarchy [1]. Based on the graph, overall, it appears that the highest presentation lies in the level of *hypothesizing and modeling*. This presentation means that most prospective teacher students are able to read, evaluate, and explain graphs well. Nevertheless, there are still some prospective teacher students who cannot read values and trends from graphs and are unable to explain contextual meanings.

Table 2. Classification of graph interpretation of prospective teachers

	Idiosyncratic	Basic Graph Reading	Rational/Literal	Critical	Hypothesizing and Modelling
Question (a)	8.33%	12.5%	54.17%	12.%	12.5%
Question (b)	4.17%	16.67%	33.33%	25%	20.83%

The idiosyncratic level is a level where students are not able to read values and trends from a graph. Students belonging to this level are 8.33% for questions (a) and 4.17 for questions (b). In the case of the question (a), they cannot interpret the meaning of horizontal lines correctly on the graph. In the first case, students assume that a horizontal line means the child drinks juice with a constant volume (figure 2a), and the second case is the student considers the horizontal line means the child drinks juice at a constant speed (figure 2b). Students at this level have difficulty in both understanding the meaning of the graph and connecting it with variables on the x-axis and y-axis. Theoretically, students at this level have deficiencies in their ability to think logically because proficiency in understanding graphics is influenced by students' logical thinking structures [8].

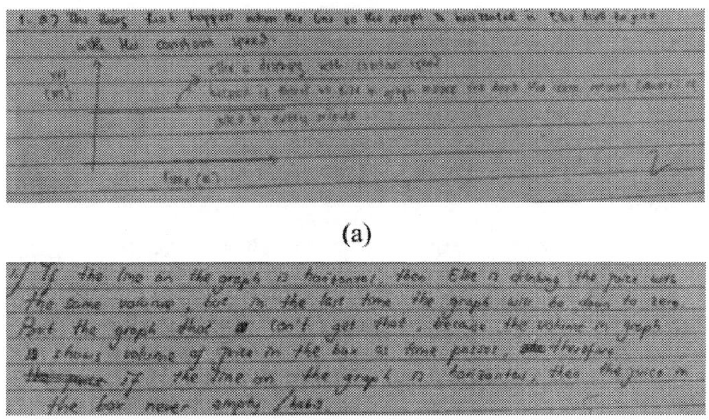

(a)

(b)

Figure 2. Students' responses on level 1

The basic graph reading level is a level where students can read values and trends from graphs but cannot explain contextually. Students belonging to this level are 12.25% for questions (a) and 16.67% for questions (b). In the case of problem (a), they can see the horizontal line as a state where the volume of juice is constant at a particular time but does not contextually illustrate that the child stops drinking juice at that time (figure 3a). In the case of problem (b), they are able to explain the formation of angles due to the meeting between horizontal lines and descending lines, but they cannot contextually explain the contextual meaning of the graph (figure 3b). Students included in this level have problems connecting between graphs and contextual meaning. If students have general notions that a line going downwards means "decreasing" or horizontal line means "steady/no change" and relating to volume of juice, then they can describe the contextual meaning of the graph. This condition occurs when students use reasoning in connecting or comparing facts on the graph with real-life contexts. Therefore, proportional reasoning becomes essential in understanding and interpreting graphs [8].

Rational / Literal level is a level where students can read values and trends from a graph and provide contextual explanations but only limited to the meaning presented. Students belonging to this level are 54.17% for questions (a) and 33.34% for questions (b). In the case of questions (a), they can give contextual explanations the meaning of horizontal lines meaning the child stops drinking juice and connects with the magnitude of the angle formed (figure 4a). In the case of questions (b), they are able to explain to contextually why the line goes down but only limited (figure 4b). Their limited explanation might occur because they only write what they see without evaluating and developing the facts that appear on the graph. Moreover, in interpreting graphs, everyone should be able to fully understand how graphs are related to the actual event [9].

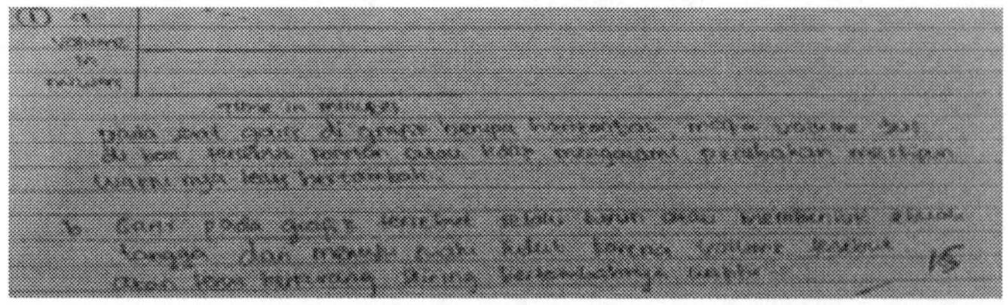

(a)

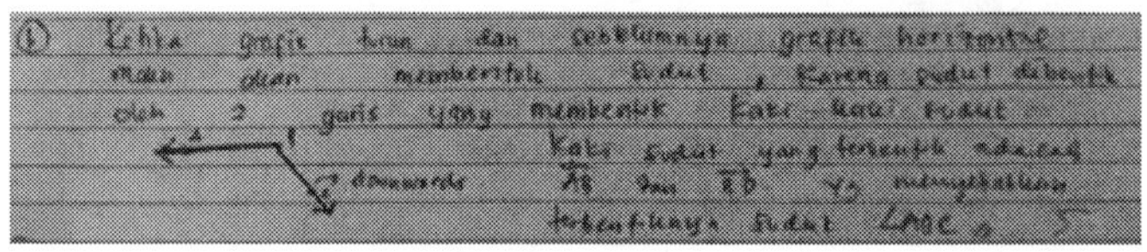

(b)

Figure 3. Students' responses on level 2

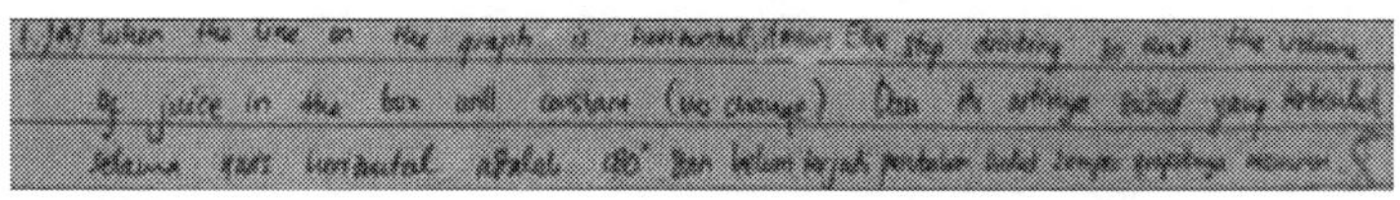

(a)

(b)

Figure 4. Students' responses on level 3

The critical level is a level where students are able to read values and trends from a graph and provide contextual explanations. Furthermore, students can evaluate the reliability of the contextual meanings presented. Students who are at this level are 12.5% for questions (a) and 25% for questions (b). In the case of the question (a), one of the students evaluates the reliability of the child in drinking the juice (figure 5a). In the case of the question (b), a student explains that when a line falls, the child drinks his juice and when the horizontal line, the child stops drinking juice so that, as a result of these two things, an angle is formed.

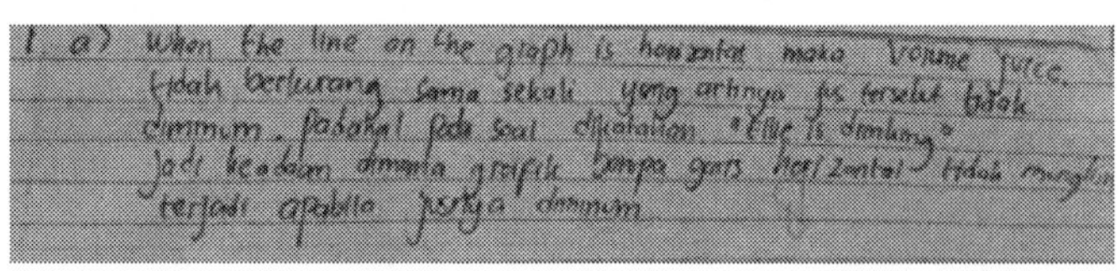

(a)

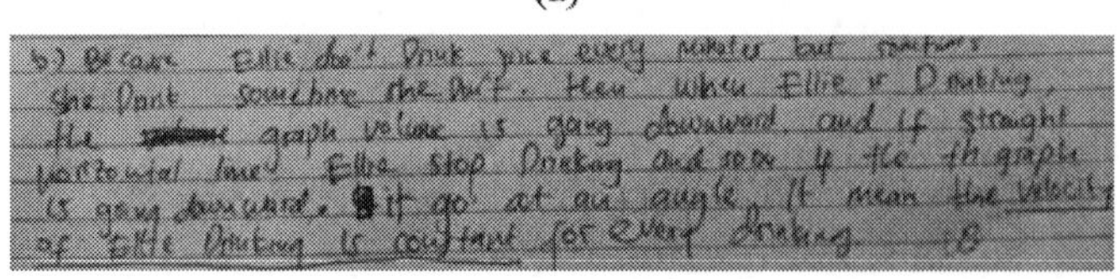

(b)

Figure 5. Students' responses on level 4

Hypothesizing and Modeling level is a level where students are able to read grades and trends from a graph and provide accurate contextual explanations. In addition, they can form their explanatory hypotheses or models. Students belonging to this level are 12.25% for questions (a) and 20.83% for questions (b). In the case of the question (a), one of the students gave an accurate contextual explanation marked by the time information and formed a hypothesis when the graph was a horizontal line (figure 5). In the case of the question (b), a student explains an accurate contextual meaning of lines going downwards and forming an angle. Students who are included in this level are students with high proportional reasoning. They are able to provide a logical description related to the contextual meaning of a graph.

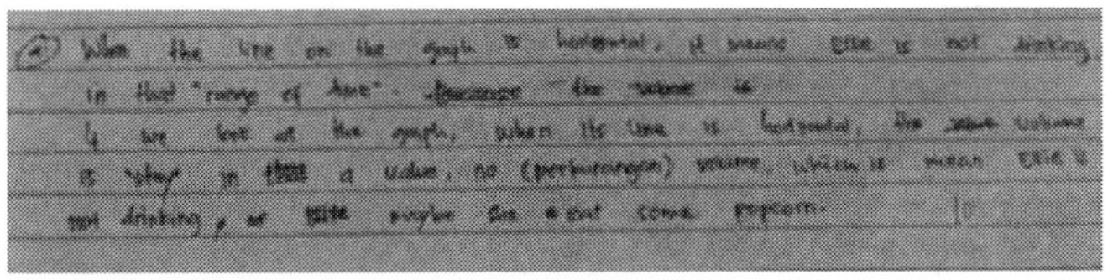

(a)

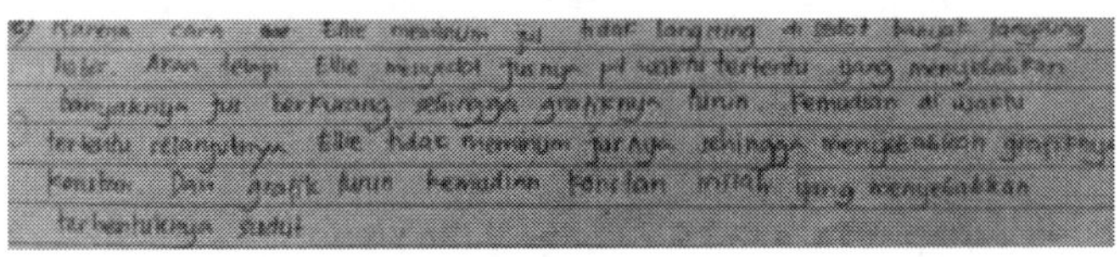

(b)

Figure 6. students' responses on level 5

4. Conclusion

Based on the results and discussion above, the ability to interpret graphs of prospective teachers is mostly at the rational/lateral level. This fact is undoubtedly still not optimal, considering the critical and Hypothesising and Modeling is a level that is the main goal in improving the ability of logical thinking and proportional reasoning. Teachers' expertise in interpreting graphs might be a barrier to the execution of meaningful practice in graphing competence [10]. Therefore we need a sustainable program to be able to develop graphs interpretation skills of teachers and students for the achievement of statistical literacy in the 21st century.

References

[1] Aoyama K 2006 Investigating a hierarchy of students' graph interpretation *Proceedings of the Seventh International Conference on Teaching Statistics. Salvador, Brazil: International Statistical Institute. Recuperado el* vol 2

[2] Aoyama K and Stephens M 2003 Graph interpretation aspects of statistical literacy: A Japanese perspective *Math. Educ. Res. J.* **15** 207–25

[3] Leinhardt G, Zaslavsky O and Stein M K 1990 Functions, graphs, and graphing: Tasks, learning, and teaching *Rev. Educ. Res.* **60** 1–64

[4] National Council Of Teachers Of Mathematics 2000 Principles and Standards for School Mathematics *Sch. Sci. Math.* **47** 868–279

[5] Wainer H 1992 Understanding graphs and tables *Educ. Res.* **21** 14–23

[6] Kwon O N 2002 The effect of calculator-based ranger activities on students' graphing ability *Sch. Sci. Math.* **102** 57–67

[7] Berg C A and Phillips D G 1994 An investigation of the relationship between logical thinking structures and the ability to construct and interpret line graphs *J. Res. Sci. Teach.* **31** 323–44

[8] Barclay W L 1985 Graphing Misconceptions and Possible Remedies Using Microcomputer-

Based Labs.

[9] Planinic M, Milin-Sipus Z, Katic H, Susac A and Ivanjek L 2012 Comparison of student understanding of line graph slope in physics and mathematics *Int. J. Sci. Math. Educ.* **10** 1393–414

[10] Glazer N 2011 Challenges with graph interpretation: A review of the literature *Stud. Sci. Educ.* **47** 183–210

Number sequence as a context for creating a mathematical problem

N R Prihartiwi[1], T Y E Siswono[1], A W Kohar[1], S Hartono and D Hidayat[1]

[1] Mathematics Department, Universitas Negeri Surabaya, Indonesia

E-mail: ninaprihartiwi@unesa.ac.id,

Abstract. Creating a problem is one cognitive process of Bloom's taxonomy. By making a problem and the answer on their own, causes students to understand more complex context problems. Applying number sequence challenges students to apply higher-order thinking skills and emphasize exploration, investigation, conjecture, and generalization. From the previous explanation, the purpose of this study is to explore the ability of students in creating mathematical problems in the context of numbers based on the 'degree of reality' and based on innovation on mathematics problems. This descriptive explorative study is collected from 24 prospective mathematics teachers that consist of 4 males and 20 females in a university. All students have to answer one question about creating word problems related to number sequence and the answer. Those problems then were analyzed and categorized based on the modification of innovation on mathematics problems by Vistro-Yu and based on 'degree of reality' by de Lange. The result of this study, modification is the most emerging technique among all, and the least are addition and replacement. Based on 'degree of reality,' only one-fourth of the number of students successful in creating a camouflage context. This result leads to only forth of the number of students successful in creating word problems. The implication of this study is the choice of students in making questions at least provides an overview of the highest level of difficulty, according to students. In mathematics learning, the lecturer should plan assessments and evaluations that need to be developed to build students' abilities to make mathematics as a whole by connecting subjects in mathematics and to make mathematics more relevant to reality.

1. Introduction

Bloom's taxonomy is the right approach for the teacher in learning to understand the structure of the cognitive process that the teacher wants students to achieve. The highest structure is creating, which is a crucial aspect to be mastered by students. Creating is putting elements together to form a novel, coherent whole, or make an original product [1] that does not yet exist previously based on students' knowledge.

In creating, students apply higher-order thinking skills. This skill corresponds to the use of patterns that challenges students to apply higher-order thinking skills and emphases exploration, investigation, conjecture, and generalization [2]. Another study also suggests that patterns can contribute to the development of mathematical ability and creativity of students [3]. Therefore, using a pattern is a powerful problem-solving strategy.

One aspect of problem-solving that students need to engage in is problem posing. Problem posing can be a powerful strategy to develop problem-solving skills and to have excellent problem solvers; on the other hand, to formulate meaningful mathematical problems, it is necessary to be a good problem solver [3]. Problem posing and problem-solving influence each other, and together they can serve to develop prospective teachers' conceptual understanding [4]. Indeed, problems posed by teachers can contribute to students' mathematical thinking skills and improve students' internalization of mathematical concepts [5].

Silver [6] has explained that there are three kinds of problem-posing based on problems' occurrence: (a) pre-solution posing, generating problems from a given information; (b) within-solution posing, reformulating a problem as it is being solved; and (c) post-solution posing, examining an already solved problem to generate alternative related problems by modifying the goals or conditions. This study applies to pre solution posing. The focus is on creating a word problem related to the number sequence and the answer.

Vistro-Yu [7] suggested innovation techniques in generating problems. The innovation originally is aimed to recycle old problems. Since this study using pre solution posing [6], some modifications in the innovation are needed, as explained in Table 1. Innovation techniques by Vistro-Yu are chosen in this study since the students' choices of innovation techniques are considered can exemplify an overview of the highest level of difficulty according to students.

Table 1. Modification of Innovations on Mathematics Problems Used in This Study

Innovations in Mathematics Problems	Explanation [7]	Modification in This Study
Replacement	Posing the same problem but changing quantities, amounts, units, shapes, so on.	The same problem here is making sequence and students changing the term or difference between consecutive terms.
Addition	Posing the same problem but adding a new given or constraint or adding an obstacle.	Making sequence with more constraint, e.g., changing more complex difference between consecutive terms.
Modification	Takes the same given but modifies the problem.	The sequence is as given, and modification can be finding the nth term or n^{th} sum.
Contextualizing	Make the problem to be more relevant to students.	Applying number sequence related to daily life.
Turning problem around	Taking the same problem but taking the end goal as the given and the given as the end goal.	Since the problem and end goal are needed, this innovation cannot be applied. There is no end goal in the given situation since applying pre solution posing.
Reformulation	Posing the same problem in a different type.	Connecting number sequence with other subjects in mathematics.

Context plays an essential role in fulfilling the requirement of having meaningful and informative assessment problems in assessing students' mathematical understanding [8]. In an assessment situation, 'context' can refer to either the environment in which students are assessed, the tools and formats that are used for it, or the rules that apply to the assessment [8]. Context also differ with respect to their degree of reality. The real world can be used as an initiation to connect mathematical concepts and ideas. De Lange [9] proposed problem classification based on 'degree of reality,' i.e. (a) no context, only involve the translation of textually packaged mathematical problems; (b) camouflage context, the context in this situation is only used to 'camouflage' or 'dress up' the mathematical problem; and (c) relevant and essential context, the contexts make a relevant contribution to the problem. In this study, word problems made by students are classified based on 'degree of reality.' The aim is finding the proximity of mathematics and reality out of students.

2. Methodology

This descriptive explorative study was collected from 24 prospective mathematics teachers that consist of 4 males and 20 females in a university of sixth semester (the academic year 2016). They were chosen as they have been taught nearly all subjects that were related to pedagogical content knowledge, such as basics of education, assessment, learning instructional development, school mathematics, etc. All students had to create a one-word problem related to number sequence based on a given condition and the answer. Those problems then were analyzed and categorized based on the modification of innovation on mathematics problems by Vistro-Yu [7] and based on 'degree of reality' by de Lange [9].

3. Result and Discussion

The data of the number of students applying particular techniques are presented in percent in Table 2 and based on 'degree of reality' in Table 3, then described descriptively. The given condition is as follows.

Let us look at a number sequence: 2, 4, 7, 11, ... Create one word problem related these sequence and solve it!

Figure 1. The given condition

3.1 Categorization Based on Innovation in Mathematics Problems

The percentage of innovation made by teachers is summarized in Table 2 below.

Table 2. The number of techniques of innovation on mathematics problem used by students

	Replacement	Addition	Modification	Contextualizing	Reformulation
The Number of Students	1	1	20	4	2
Percentage	4,17%	4,17%	83,33%	16,67%	8,33%

From Table 2, modification is the most emerging technique, among others. Fardah [10] also found the same result in the study of problem-posing using number sequence. The example is shown in Figure 2 below.

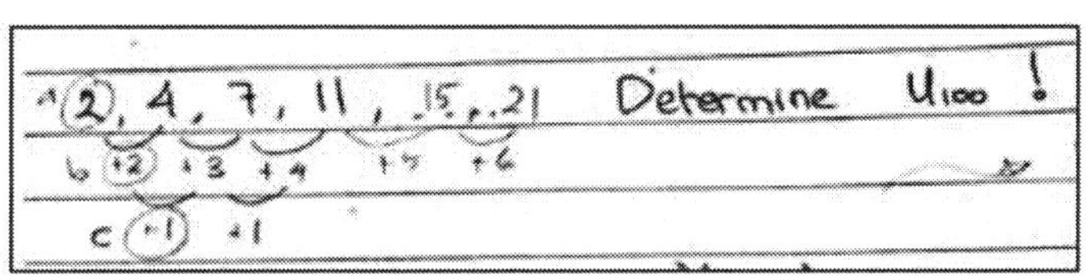

Figure 2. The problem generated using modification

It is reasonable that modification is the most emerging technique since prospective mathematics teachers are familiar with number sequence since in junior high school. They quickly applied the formula for finding the nth term or n^{th} sum.

Four prospective mathematics teachers (16,67%) applied to contextualize. This technique was not stand-alone. They combined modification and contextualizing together. The problems are finding the nth term or n^{th} sum related to daily life. Figure 3 is one of the problems.

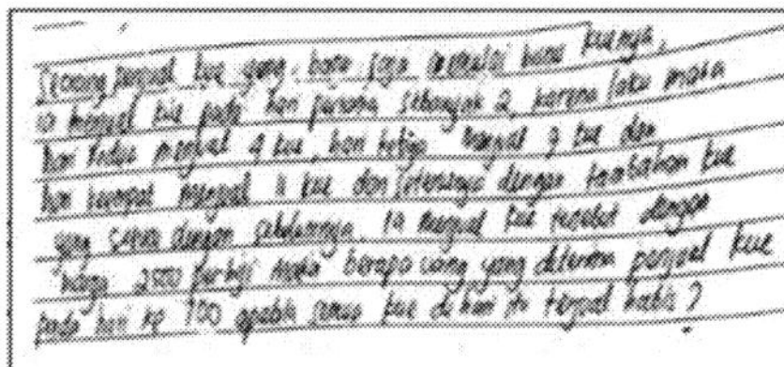

A seller already started her cookies business. She sells 2 cookies on the first day, 4 cookies on the second day, 7 cookies on the third day, dan 11 cookies on the fourth day and so forth with increasing sale corresponds with the pattern before. If the price for every cookie is Rp2500, how much money will she get on 100th day is all of cookies is sold out?

Figure 3. The problem generated using modification and contextualizing

From Figure 3, prospective mathematics teachers applied modification two times. First, modified the given condition by asking 100th term and second, modified amount of money of all cookies if they were sold out on that day by giving information about every cookie's price. They changed the given condition by telling a person engaged in cookies business as contextualizing. All of the problems here were presented in word problems, as requested.

Reformulation technique was also applied by two prospective mathematics teachers (8,33%) in generating the problem. All of them connected the given sequence with a circle. This result shows that prospective mathematics teachers connect number sequences with a relatively new concept — all of them present problems in word problems. The example is shown below.

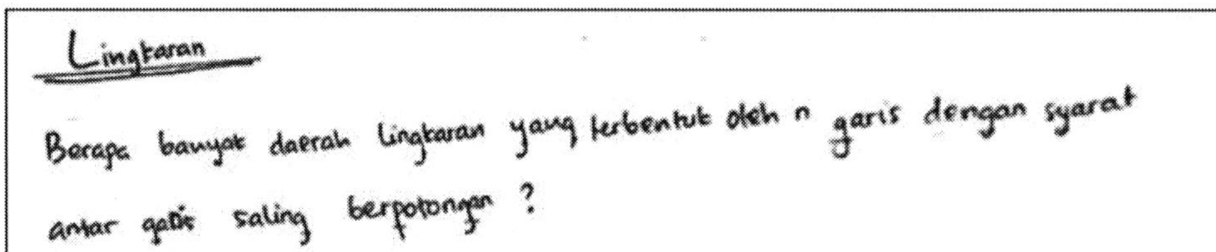

How many areas in a circle are formed by n lines with the condition that the lines intersect?

Figure 4. The problem generated by reformulation (word problem)

The least techniques applied by prospective mathematics teachers are replacement and addition. In replacement, prospective mathematics teachers created a new number sequence, with the difference between consecutive terms was similar to the given condition. Figure 5 shows that the number sequence is not the given, but the difference between consecutive terms is similar to the given. Also, prospective mathematics teachers created a new number sequence with the more complex difference between consecutive terms. This complexity refers to the addition technique. All of them did not present the problem in word problems. The examples are shown as follows.

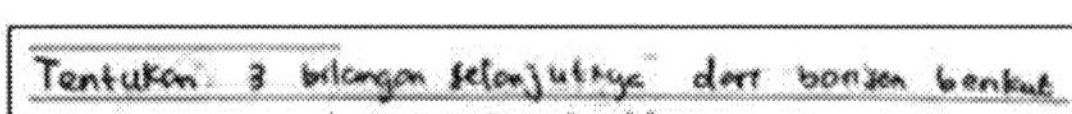

Determine the next three terms of this sequence
-4, -1, 3, 8, 14, ...

Figure 5. The problem generated using replacement

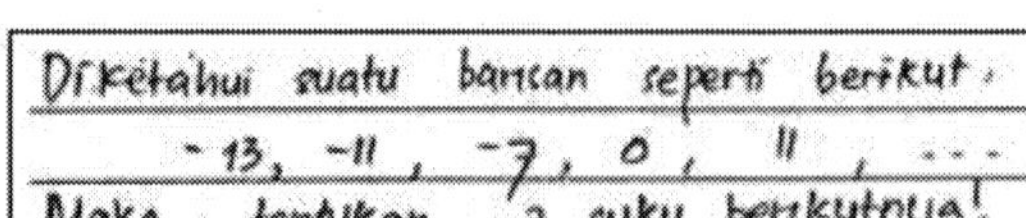

Given a sequence as follows.
-13, -11, -7, 0, 11, ...
Determine the next two terms

Figure 6. The problem generated using addition

3.2 Categorization Based on 'Degree of Reality'

The percentage of 'degree of reality' is summarized in Table 3 below.

Table 3. The number of 'degree of reality' of students' problems

	No Context	**Camouflage Context**
The Number of Students	18	6
Percentage	75%	25%

The 'degree of reality' of problems are associated with contextualizing in techniques of innovation. All problems generated using contextualizing are considered a camouflage context. Problems presented in word problems were also included in these criteria [9], which were generated using contextualizing and reformulation. This low percentage (25%) in the camouflage context shows that prospective mathematics teacher is not familiar with expressing problems in the form of word problems or relating to daily life. The relevant and essential context was not shown in this study due to the limitation of time. This limitation caused prospective mathematics teachers do not have the opportunity to search for exact data about the given condition.

4. Conclusion

From modification techniques used by prospective mathematics teachers, the most emerging technique is a modification. Prospective mathematics teachers because it is taught sequence and series in junior high school. They quickly identified a nth term or n^{th} sum of a sequence. This fact was the reason why when students were asked to create a problem using a given number sequence. They immediately made problems with determining the nth term or n^{th} sum. Modification could be combined with contextualizing in creating the problem. In contextualizing, the modification could be applied twice, for example, determining n^{th} term or n^{th} sum and modifying the amount of money of all cookies if they were sold out on that day by giving information on every cookie's price. Prospective mathematics teachers also not familiar with presenting mathematics problems in the form of a word problem. These results give an overview that prospective mathematics teachers still not associate mathematics with other subjects in mathematics and with daily life. Since the innovation in generating problems originally to recycle old problems, all techniques can not appear, such as turning the problem around. This finding is due to there is no end goal to be turning. When teachers generated new problems, many factors may be influenced by them. The implication of this study is the choice of students in making questions at least provides an overview of the highest level of difficulty, according to students. In mathematics learning, the lecturer should plan assessments and evaluations that need to be developed to build students' abilities to make mathematics as a whole by connecting subjects in mathematics and to make mathematics more relevant to reality.

REFERENCES

[1] Krathwohl D R 2002 A revision of Bloom's taxonomy: An overview *Theory Pract.* **41** 212–8
[2] Vale I and Cabrita I 2008 Learning through patterns: a powerful approach to algebraic thinking *ETEN* 63–9
[3] Vale I, Pimentel T, Cabrita I, Barbosa A and Fonseca L 2012 Pattern problem solving tasks as a mean to foster creativity in mathematics *Proceedings of the 36th Conference of the International Group for the Psychology of Mathematics Education* vol 4 pp 171–8
[4] Xie J and Masingila J O 2017 Examining interactions between problem posing and problem solving with prospective primary teachers: A case of using fractions *Educ. Stud. Math.* **96** 101–18
[5] Konyalioglu A C, Isik A, Kaplan A, Hizarci S and Durkaya M 2011 Vidinli Huseyin Tevfik Pasha's triplets in history of mathematics *Procedia - Soc. Behav. Sci.* **15** 4045–7
[6] Silver E A On mathematical problem posing. For the Learning of Mathematics *Learn. Math.* **14**
[7] Vistro-Yu C P 2009 Using innovation techniques to generate 'new'problems *Mathematical problem Solving: Yearbook 2009, association of mathematics educators* (World Scientific) pp 185–207

[8] Van Den Heuvel-Panhuizen M 2005 The role of contexts in assessment problems in mathematics *Learn. Math.* **25** 2–23
[9] De Lange J 1995 Assessment: No change without problems *Reform Sch. Math. authentic Assess.* 87–172
[10] Fardah D K 2018 PRESERVICE TEACHERS'TECHNIQUES FOR GENERATING NEW PROBLEMS *J. Ilm. Soulmath J. Edukasi Pendidik. Mat.* **6** 37–45

The Model of Visual Thinking of Prospective Math Teacher in Understanding the Formal Definition of Convergent Sequences Based on Gender Differences

Darmadi[1]

[1] Department of Mathematics Education, Universitas PGRI Madiun, INDONESIA

Corresponding e-mail: darmadi.mathedu@unipma.ac.id

Abstract. The model of visual thinking is interesting to learn because the concepts are organized based on the meaning of forming mental imagery in the mind. Thinking by using mental imagery (visual information) is called visual thinking. This paper discusses the model of visual thinking of prospective mathematics teachers in understanding formal definition of convergent sequences based on gender differences as a result of descriptive-explorative research with a qualitative approach. The visual thinking activities of male and female students are found around three aspects: bringing up mental imagery, processing mental imagery (determining mental imagery, improving mental imagery), and utilizing mental imagery. To bring up mental imagery, the mind summons mental imagery or makes mental imagery. To determine mental imagery, the mind collects mental imagery and selects mental imagery. To improve mental imagery, the mind justifies mental imagery or completes mental imagery. To utilize mental imagery, the mind uses mental imagery or represents mental imagery. Processing mental imagery has done according to the needs and purpose of the activity in understanding the formal definition of the convergent sequence.

1. Introduction

Students of mathematics education courses are candidates for mathematics teachers. The characteristics of students entering the program with a high readiness to learn, bringing a lot of experience to the class, less flexible in interacting, want to be treated by maturity and have initiative and independence [1]. The six behaviors of students learning are self-determining the direction and purpose of education, having a set of life experiences, goal-oriented, and relevance, tending to be practical and requiring rewards in learning [2].

One of the materials that must be understood by the students of prospective mathematics teacher is the course of Real Analysis. One of the concepts in the real analysis course is the concept of real numbers sequence. The sequence is the function of the set of natural numbers to a set of real numbers. Each expert can use different notations and symbols in defining. Some notations are $\{s_n\}_{n=1}^{\infty}$ [3], (x_n) [4], $\langle a_n \rangle$ [5], $\{a_n\}$ [6]. In this paper, the sequence of real numbers is denoted by $\{a_n\}_{n\geq 1}$.

One of the real numbers of sequence material that often makes it difficult for students is a convergent sequence concept. Some experts have defined the concept of convergent sequence. If the sequence of real numbers $\{a_n\}_{n\geq 1}$ has the limit L, we say that $\{a_n\}_{n\geq 1}$ is convergent to L [3]. Let $X = \{a_n\}_{n\geq 1}$ be a sequence of real numbers. A real number a is said to be a limit of X if for each neighborhood V of a there is a natural number $K(V)$ such that for all $n \geq K(V)$, the terms a_n belong to V [4]. A sequence $\{a_n\}_{n\geq 1}$ is said to converge to $a \in R$ if given $\varepsilon > 0$, there exist a positive integer m

such that $n \geq m$ implies $|a_n - a| < \varepsilon$ [5]. The sequence $\{a_n\}_{n\geq 1}$ is converged to L and written $\lim_{n\to\infty} a_n = L$, if for each positive number ε there is a corresponding positive number N, Such that $n \geq N \Rightarrow |a_n - L| < \varepsilon$ [6]. In this paper, the sequence of real numbers $\{a_n\}_{n\geq 1}$ is said to be convergent sequence if and only if for every $\varepsilon > 0$ there is $a \in R$ and $n_0(\varepsilon) \in \mathbb{N}$ so for $n \geq n_0(\varepsilon)$ apply $|a_n - a| < \varepsilon$.

To understand definition means to understand the concepts underlying the definitions. Understanding the formal definition of a convergent sequence can be a problem or difficulty for some individuals so it is necessary to look for an immediate solution. One of the goals of mathematics lessons is to understand the concept of mathematics on the content standards behind the concepts [7]. Prolonged difficulty brings a negative impact on brain function. Therefore, it needs creative effort to understand the definition immediately.

One of the creative endeavors to understand is by visualization. To visualize the formal definition of convergent lines, it is necessary to represent the convergent lines and attributes that exist in the formal definition of convergent lines. Visualization requires visual thinking. However, students were found to still lacking in mental imagery [8]. Research results show that vocational secondary students were lacking in visual thinking [9]. Besides, learning is still too focused on formalities.

Visual thinking is information processing in the mind by using mental imagery. Mental imagery is modalities or mental representation (a form of information in the memory) that can be presented in the form of images/graphs. The internal representation of visual codes is mental imagery. Mental imagery is the mental image [10]. Mental imagery is image information or visual knowledge or visual imaging or images or visual imagery [11]. Mental imagery can be called as a picture for perception [12]. Some writers or other researchers call mental imagery as mental images, mental images, shadows or images of the mind, visualization, imagination, imagination, imagery, or visual imagery.

Some research was undertaken by focusing on visual thinking within learning process. Thus, some thoughts regarding this topic are emerging. One such influence is the clarity and stability of their beliefs about the role of visualization [13]. In 2003, the research has examined the increase in limit learning outcomes with visualization [14]. However, the research in 2006 shows that students are still struggling to use visualizations to improve perception [15]. Mental imagery only benefits the learning process' [16]. In 2016, I have researched the visual profile of student thinking with visual [17] and auditory [18] learning style in solving the problem. The stages of students in processing mental shading have not been researching.

In general learning in schools such as SD / MI (Elementary School), SMP / MTS (Junior High School), SMA / SMK (Senior/Vocational High School), even to college. The teachers/educators distinguish students until college students based on gender differences. Gender refers to the concept of male or female based on socio-cultural and psychological dimensions [19]. Gender is distinguished by sex involving the biological dimensions of women or men. Therefore, the university knows for male (vigorous) college students and female college students. Regarding visual thinking scheme, the understanding of the formal definition of convergent sequences based on interesting gender differences are discussed to provide a picture of organizing the concepts of male and female prospective mathematics teachers. By this knowledge, lecturers can give them necessary explanations in understanding the formal definition of convergent sequences.

2. Research methods

The model of visual thinking of prospective math teachers in understanding the formal definition of convergent sequences were presented in complete form with words or sentences according to empirical data. Empirical evidence obtained is natural as they are and focus more on the process than on results. Therefore, a descriptive-explorative research was used with the qualitative approach in this present study.

To obtain the data, a male and a female college students of mathematics education study program of a university in Madiun city, East Java, were selected as the participants of this research. Then, the criteria for selecting the participants are based on some points: 1) having taken the course of real analysis courses, 2) have a GPA above 2.75, 3) The grade/score of the course of calculus value and introduction to basic mathematics minimum B, 4) communicative, honest, and willing to be the participants of research.

The primary instrument in this study is the researchers themselves. Researchers are required: 1) have responsive, adaptive, and holistic nature; 2) aware of the context; and 3) able to immediately process, clarify, summarize, explore, and understand the answers. The research supporting instrument is a task-based interview guide. The task is given by using a duty sheet that contains the formal definition of convergent rows and has validated the expert.

Data were naturally collected by task-based interviews. The task given to students is explaining how they understand the formal definition of the sequence of real numbers converges. The formal definition of the sequence of real numbers converges as follows.

"The sequence of real numbers $\{a_n\}_{n\geq 1}$ is said to be convergent sequence if and only if for every $\varepsilon > 0$ there is $a \in \mathbb{R}$ and $n_0(\varepsilon) \in \mathbb{N}$ so for $n \geq n_0(\varepsilon)$ apply$|a_n - a| < \varepsilon$"

Interviews are conducted in-depth with the semi-structured format. This format is chosen to know the honesty and openness of the participant in conveying information. The visualization of sequence $\left\{-\frac{1}{n}\right\}_{n\geq 1} = \left\{-1, -\frac{1}{2}, -\frac{1}{3}, -\frac{1}{4}, -\frac{1}{5}, -\frac{1}{6}, -\frac{1}{7}, -\frac{1}{8}, \dots\right\}$ is

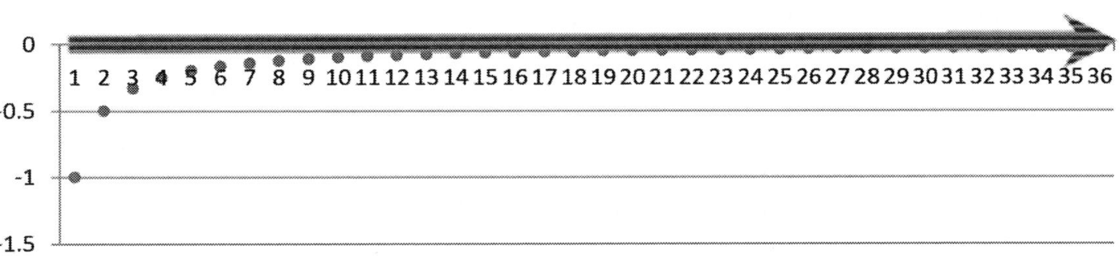

Figure 1. The understanding of the convergent sequence concept of $\left\{-\frac{1}{n}\right\}_{n\geq 1}$

Participants being interviewed are given the freedom to follow their inclination of mind including in determining the direction of the topic of speech to form the focus of the conversation. All activities are recorded with handy cams for the preparation of transcripts.

The result of the interview on the participant is the date of the interview result. The data of the interview result are then triangulated and validated to obtain valid and accurate data. The validity of the data in this study is emphasized more on internal validity test or credibility test derived from extending observation, increasing persistence, triangulation, negative case analysis, and member check. If the data is not valid yet, it is necessary to re-interview, and the existing data are triangulated and validated again.

The steps of data analysis include reduction, categorization, synthesis, and conclusion. Data reduction activity is the activity of identifying the "unit" that is the smallest part found in the data and has meaning when it associated with the focus of the research problem. After it obtained, the next step is to create a "coding" that gives the code on each unit to remain traceable source data. Categorization is the activity of sorting each unit into parts that have similarities. Each classification named after a "label." Synthesis is an exercise of looking for links between one category with another category. The association of one category to another category is given the name/label again. The conclusion was made by formulating a statement that provides the visual thinking scheme of mathematics teacher candidates in understanding the formal definition of the gender-convergent sequence.

3. Research results and discussions

The results of the data analysis show that male and female students understand the formal definition of a convergent sequence by recognizing, imagining, showing (showing the definition picture, showing the attribute definition), and concluding. The male and female participant representation of the formal definition of a convergent sequence is as follows.

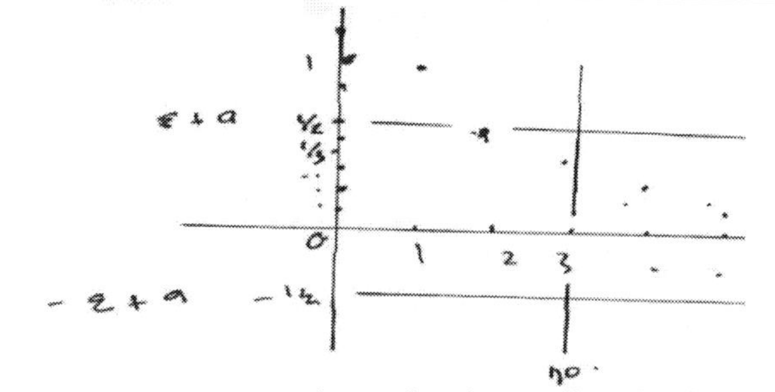

Figure 2. Representations of understanding the formal definition of convergent sequence from the male participant

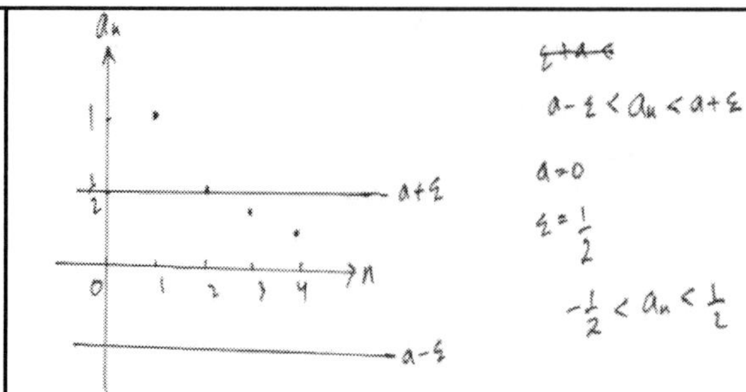

Figure 3. Representations of understanding the formal definition of convergent sequence from the female participant

To recognize, the male participant summarizes the definitions and describes what can describe. To imagine, the male participant makes the general formula, create a help table, and make a general overview. To show a definition picture, the male participant creates a special sequence formula, auxiliary tables, and plotting the members of the row. To show the definition attribute, the male subject specifies a value, determines the value of ε, determines the value of $n_0(\varepsilon)$, draws the $n_0(\varepsilon)$, and draws the line ε.To conclude, the male participant matches and concludes.

To recognize, the female participant encapsulates, determines the keywords, and describes the keywords. To imagine, the female participant remembers about convergent and makes the coordinate axis. To show a definition picture, the female participant specifies a row instance, enrolls a member of a row, and plots a member of the row. To show the definition attribute, the female participant defines the value of ε, determines the value of a, describes the value of ε and a, draws the line of $a \pm \varepsilon$, determines the value of $n_0(\varepsilon)$, and draws the line of $n_0(\varepsilon)$. To conclude, the female participant makes the first conclusion, makes a second conclusion, and makes the third conclusion.

Some male and female activities in understanding the formal definition of the convergent sequence are identified involving the processing of mental imagery for visual thinking. Some male student activities that identified involving the processing of mental imagery are drawing a general picture, creating a unique series formula, plotting, assigning a value, drawing a line $n_0(\varepsilon)$, and drawing a line ε. Some female student activities that identified involving the processing of mental imagery are remembering about convergent, drawing coordinate axis, determining convergence sequence sample formulas, plotting members of the sequence, determining value ε, determining the value of a, drawing the line of $a \pm \varepsilon$, determining the value of $n_0(\varepsilon)$, and drawing the line of $n_0(\varepsilon)$.

To drawing a general picture, the male participant activity is identified involving the processing of mental imagery. Based on data of deep interviews, visual thinking activities to draw a general picture are bring up mental imagery, processing mental imagery (determining mental imagery, improving mental imagery), and utilizing mental imagery. To bring up mental imagery, the male participant remembers the mental imagery of convergent lines and making mental imagery of other converging functions that are still in the form of a conjunctive graph. To processing mental imagery, the male participant collect all mental imagery obtained and justifies it so that it is a plot of points. After that chooses or focuses on mental imagery, and then the male participant utilizes or represents it as a general description.

To create a particular sequence of formulas, the male participant activity is identified involving the processing of mental imagery. Based on data of deep interviews, visual thinking activities to create a particular sequence are bring up mental imagery, processing mental imagery (determining mental imagery, improving mental imagery), and utilizing mental imagery. To bring up mental imagery, the male participant summoning the mental imagery of the general picture and creating another mental imagery. To processing mental imagery, the male participant collect and complete the mental imagery by adding a sequence of rows. Mental imagery by focusing on one mental imagery, then using or using information on selected mental imagery to determine the formula of the unique sequence.

The male participant activity to plot is identified involving the processing of mental imagery. Based on data of deep interviews, visual thinking activities to plot are bring up mental imagery, processing mental imagery (determining mental imagery, improving mental imagery), and utilizing mental imagery. To bring up mental imagery, the male participant pays attention or summoning mental imagery of the coordinate axis and making mental imagery plot sequences, and then utilizing or representing mental imagery in the form of a plot of lines.

To determine the value of *a*, the male participant activity is identified involving the processing of mental imagery. Based on data of deep interviews, visual thinking activities to determine the value of a are bring up mental imagery, processing mental imagery (determining mental imagery, improving mental imagery), and utilizing mental imagery. To bring up mental imagery, the male participant observes or summoning the real mental imagery of the series. Then by creating mental imagery of the plot of the sequence, processing mental imagery collects mental imagery of the emerging plots. Finally, the participant utilizes mental imagery information to determine the value of a.

To draw the line of $n_0(\varepsilon)$, the male subjek activity is identified involving the processing of mental imagery. Based on data of deep interviews, visual thinking activities to draw the line of $n_0(\varepsilon)$ are bring up mental imagery, processing mental imagery (determining mental imagery, improving mental imagery), and utilizing mental imagery. The sequence of activities or visual thinking activities to draw the line of $n_0(\varepsilon)$ is: to bring up mental imagery by paying attention to or summoning mental imagery of the existing image / graph and making mental imagery of line $n_0(\varepsilon)$, based on visual input $n_0(\varepsilon)$ then processing mental imagery by collecting mental imagery of the $n_0(\varepsilon)$, line appearing and selecting one of the mental imagery, and utilizing or representing the mental representation in the form of a line image $n_0(\varepsilon)$.

To draw the line ε, the male participant activity is identified involving the processing of mental imagery. Based on data of deep interviews, visual thinking activities to draw the line ε are bring up mental imagery, processing mental imagery (determining mental imagery, improving mental imagery), and utilizing mental imagery. The sequences of activities or visual thinking activities to draw the line ε are: to generate mental imagery by watching or calling mental imagery of the existing image/graph and making mental imagery of another line of ε. Then processing or determining mental imagery by choosing mental imagery of the line of ε, and then use or represent the selected mental imagery to be expressed in the form of a line drawing ε.

To remember the convergent term, the female participant activity is identified involving the processing of mental imagery. Based on data of deep interviews, visual thinking activities to remember the convergent term are bring up mental imagery, processing mental imagery (determining mental imagery, improving mental imagery), and utilizing mental imagery. The sequence of activities or visual thinking activities to remember the convergent term is: to bring up mental imagery by remembering or summoning mental imagery of convergent lines and creating mental imagery of other convergent lines. Then cultivate the mental imagery by collecting mental imagery that appears and focuses on attention or chooses mental imagery of certain convergent lines. Furthermore, utilizing or using information on mental imagery so that it can give notes to the summary that the convergent is cantered.

To draw the coordinate axis, the female participant activity is identified involving the processing of mental imagery. Based on data of deep interviews, visual thinking activities to draw the coordinate axis are bring up mental imagery, processing mental imagery (determining mental imagery, improving mental imagery), and utilizing mental imagery. A series of activities or visual thinking activities to draw the coordinate axis are: to generate mental imagery by remembering or calling for the mental representation of convergent functions, and then utilizing or representing the axis of mental imagery co-ordinates.

To determine the formulas of convergent sequence example, the female participant activity is identified involving the processing of mental imagery. Based on data of deep interviews, visual thinking activities to determine the formulas of convergent sequence examples are bring up mental imagery, processing mental imagery (determining mental imagery, improving mental imagery), and utilizing mental imagery. A sequence of activities or visual thinking activities to determine the formulas of convergent sequence examples are: bring up mental imagery by watching or calling mental imagery of the coordinate axis and creating mental imagery of the graph and its functions. Then cultivate mental imagery by collecting the emerging mental imagery and selecting the mental imagery corresponding to the shadow of the convergent sequence. Furthermore, utilize or use the information on mental imagery to determine the convergence sequence instance formula.

To plot members of the series, the female participant activity is identified involving the processing of mental imagery. Based on data of deep interviews, visual thinking activities to plot members of the series are bring up mental imagery, processing mental imagery (determining mental imagery, improving mental imagery), and utilizing mental imagery. A sequence of activities or visual thinking activities to plot members of the series are: to generate mental imagery by paying attention to or summoning mental imagery of the coordinate axis and making mental imagery plot sequences, then utilizing or representing mental imagery as a sequence of lines.

To determine the value of ε, the female participant activity is identified involving the processing of mental imagery. Based on data of deep interviews, visual thinking activities to determine the value of ε are bring up mental imagery, processing mental imagery (determining mental imagery, improving mental imagery), and utilizing mental imagery. The sequence of activities or visual thinking activities to determine the value of ε is: to bring up mental imagery by paying attention to or summoning mental imagery of existing images/graphs and making mental imagery of the line ε. Then, participant collect mental imagery of the line ε, choose mental imagery of the line ε, utilize or use mental imagery of the line ε to determine the value of ε.

To determine the value of a, the female participant activity is identified involving the processing of mental imagery. Based on data of deep interviews, visual thinking activities to determine the value of a are bring up mental imagery, processing mental imagery (determining mental imagery, improving mental imagery), and utilizing mental imagery. The sequence of activities or visual thinking activities to determine the value of a: to generate mental imagery by noticing or summoning existing mental imagery of the sequence of lines and making mental imagery of the continuation of the sequence of lines. Then, participant cultivates or collect mental imagery plot sequences that appear, and utilize or use the information on mental imagery to determine the value of a.

To illustrate the line $a \pm \varepsilon$, The female participant activity is identified involving the processing of mental imagery. Based on data of deep interviews, visual thinking activities to illustrate the line $a \pm \varepsilon$ are bring up mental imagery, processing mental imagery (determining mental imagery, improving mental imagery), and utilizing mental imagery. A series of activities or visual thinking activities to illustrate the line $a \pm \varepsilon$, namely: generating mental imagery with summoning mental imagery while paying attention to images/graphs that have no line $a \pm \varepsilon$. Then, the participant creates mental imagery when it describes the line $a \pm \varepsilon$, utilizing or representing mental imagery when describing the lines $a + \varepsilon$ and $a - \varepsilon$.

To determine the value of $n_0(\varepsilon)$, the female subjek activity is identified involving the processing of mental imagery. Based on data of deep interviews, visual thinking activities to determine the value of $n_0(\varepsilon)$ are bring up mental imagery, processing mental imagery (determining mental imagery, improving mental imagery), and utilizing mental imagery. The sequence of activities or visual thinking activities to determine the value of $n_0(\varepsilon)$ is: to generate mental imagery by paying attention to or summoning mental imagery of existing images/graphs and making mental imagery of the line $n_0(\varepsilon)$. Cultivate mental imagery by collecting mental imagery that appears and choosing certain mental imagery. Then use the information on the mental imagery to determine the value of $n_0(\varepsilon)$.

To illustrate the line of $n_0(\varepsilon)$, the female subjek activity is identified involving the processing of mental imagery. Based on data of deep interviews, visual thinking activities to illustrate the line $n_0(\varepsilon)$ are bring up mental imagery, processing mental imagery (determining mental imagery, improving mental imagery), and utilizing mental imagery. To bring up mental imagery, the female subjek pay attention to or summoning mental imagery of existing images/graphs and making mental imagery of the line $n_0(\varepsilon)$. Then use or represent mental imagery of line $n_0(\varepsilon)$.

Based on the analysis of the purpose and usefulness of mental imagery processing, it can be concluded that the visual thinking activities of male and female students are brought up mental imagery, processing mental imagery (determining mental imagery, improving mental imagery), and utilizing mental imagery. To bring up mental imagery, the mind summons mental imagery or makes mental imagery. To determine mental imagery, the mind collects mental imagery and selects mental imagery.

To improve mental imagery, the mind justifies mental imagery or completes mental imagery. To utilize mental imagery, the mind uses mental imagery or represents mental imagery. Processing mental imagery has done according to the needs and purpose of the activity in understanding the formal definition of the convergent sequence. It is in line with the theory of information processing is directed or controlled by a particular purpose [20].

This stage of visual thinking is important because it can assist students in the construction of students in giving meaning to the mathematical concepts of the formal level [21]. Mental imagery is helpful to students in building concepts [22]. Thinking using mental imagery (visual thinking) is very useful in supporting intuition when learning mathematics [23]. Students who are rich with mental shading will be greatly assisted in cognitive development, representing, and proofing [24]. The use of mental imagery (visual thinking) is very important for students during the transition to the mathematical study from concrete to abstract (formal proof) [25].

4. Conclusion

To conclude, visual thinking activities of male and female students are brought up mental imagery, processing mental imagery (determining mental imagery, improving mental imagery), and utilizing mental imagery. To bring up mental imagery, the mind summons mental imagery or makes mental imagery. To determine mental imagery, the mind collects mental imagery and selects mental imagery. To improve mental imagery, the mind justifies mental imagery or completes mental imagery. To utilize mental imagery, the mind uses mental imagery or represents mental imagery. Processing mental imagery has done according to the needs and purpose of the activity in understanding the formal definition of convergent sequence.

References

[1] Kemp 1994 *Teaching Planning Process* (Bandung: ITB Press)
[2] Lieb 1991 *Principles of Adult Learning* (Senior Technical Writer and Planner, Arizona Department of Health Services and part-time Instructor, South Mountain Community College)
[3] Goldberg 1976 *Methods of Real Analysis* (The University of Lowa, United State of America: John Wiley & Sons, Inc)
[4] Bartle & Sherbert 1982 *Introduction to Real Analisis* (University of Illinois: Urbana-Champaign. Illinois. John Wiley & Sons. Inc)
[5] Wasan & Prakash *Real Analysis* (Ramjas College: University of Delhi; Rajdhani College: University of Delhi; New Delhi: Tata McGraw-Hill Publishing Company Limited)
[6] Purcell, Varberg, & Rigdon 2004 *Kalkulus* (Jakarta: Erlangga) p 452
[7] Depdiknas 2006 *Kurikulum Tingkat Satuan Pendidikan (KTSP)* (Jakarta: Departemen Pendidikan Nasional)
[8] Darmadi 2018 Imagery Model of Pre-service Mathematics Teacher on Graphing Rational and Irrational Functions *JRAMathEdu: Journal of Research and Advances in Mathematics Education* **3**(2)
[9] Mohammad Isa Irawan, Imam Mukhlash, Dieky Adzkiya, Darmadi, Sanusi 2018 Development of Trigonometric Visualization Concepts to Increase the Study Motivations of SMK Students *Journal of Physics: Conference Series IOP Publishing* **1218**(1)
[10] Suharnan 2005 *Psikologi Kognitif* (Surabaya: Srikandi)
[11] Sternberg 2008 *Cognitive Psychology* (Yogyakarta: Pustaka Pelajar)
[12] Schunk 2012 *Learning Theories an Educational Perspective* (Yogyakarta: Pustaka Pelajar)
[13] Biza I, Nardi E, and Zachariades T 2009 *Teachers' Views on the Role of Visualisation and Didactical Intentions Regarding Proof* (Lyon France: Proceeding of CERME) INRP 2010 p 261–270
[14] Arcavi A 2003 The Role of Visual Representations in The Learning of Mathematics *Educational Studies in Mathematics* **52** pp 215-241

[15] Presmeg N C 2006 *Research on Visualization in Learning and Teaching Mathematics: Emergence from Psychology* In A.Gutierrez & P. Boero (Eds.) Handbook of research on the psychology of mathematics education (Dordrecht, NL: Sense Publishers) p 205-235
[16] Aspinwall L, Shaw K L, & Presmeg N C 1997 Uncontrollable Mental Imagery: Graphical Connections Between a Function and Its Derivative *Educational Studies in Mathematics* **33** pp 301-317
[17] Darmadi, Handoyo H 2016 Profile of Visual Thinking Student Candidates Mathematics Teacher with Visual Learning Style in Solving Trigonometric Problems *Math Educator Nusantara: Publications of scientific writings in mathematics education* **2**(1)
[18] Darmadi, Handoyo B, Dyah W A P 2016 Profile of Visual Thinking Students of Prospective Mathematics Teachers with the Style of Auditorial Learning In Solving Trigonometric Problems *Jurnal Math Didactic: Mathematics Education Journal* **2**(1) pp 32-40
[19] Santrock 2009 *Educational Psychology* (Jakarta: Salemba) issue 3 book 1
[20] Schunk 2012 *Learning Theories an Educational Perspective* (Yogyakarta: Pustaka Pelajar)
[21] Pinto M M F & Tall D 1999 *Student Constructions of Formal Theory: Giving and Extracting Meaning* (Published in Proceedings of the 23rd Conference of PME, Haifa, Israel) 3 p 281–288
[22] Tall D 1988 *Concept Image and Concept Definition. Senior Secondary Mathematics Education*, (ed. Jan de Lange, Michiel Doorman, OW & OC Utrecht) p 37–41
[23] Tall D 1991 *Intuition And Rigour: The Role of Visualization in The Calculus. Visualization in Mathematics* (ed. Zimmermann & Cunningham) no 19 p 105–119
[24] Tall D 1995 *Cognitive Development, Representations and Proof* (Conference on Justifying and Proving in School. Mathematics, Institute of Education, London) p 27–38
[25] Tall D 2005 *The Transition from Embodied Thought Experiment and Symbolic Manipulation to Formal Proof* (The Delta Conference, on Frazer Island, Australia)

Mathematical Problem Proving and Problem Posing

T Y E Siswono[1*], S Hartono[1], and A W Kohar[1]

[1]Faculty of Mathematics and Natural Science, Universitas Negeri Surabaya, Indonesia

*e-mail: tatagsiswono@unesa.ac.id

Abstract. This article describes students' tasks that concerned to assess students' problem proving and problem posing on triangle tasks. This descriptive-explorative research involves 24 prospective mathematics teachers at a state university in Surabaya, Indonesia. They had to answer three questions. The first question related to problem proving of triangle theorem. The second question related to their perception of the first question, and the last question related to the problem-posing task. A Newman analysis is used to analyse the data. The analysis of the data focused on students' problem proving and students' problem posing of triangle task. The results showed that most of them were correct to answer the question as many as 41%. It can be seen of the first question was performed by the highest number of them. However, some of them still have a perception of the first question is a problem, because it shows as many as 95% of 24 students choose it. They also have some weaknesses in proving problems, mainly misinformation or ambiguity about the problem, it is not a category as a problem proving. They constructed proof by formal proof with a low level of problem difficulty as many as only 38 % of them.

1. Introduction

Today learning mathematics should have oriented to a problem [1][2]. Naturally, a mathematical problem is not just finding a solution but try to convict that proposition is true. There is a problem to finding and a problem to proving. Proving has become a highlight topic in mathematics education. Proving as one of the five standard competencies in mathematics mentioned by NCTM (National Council of Teachers of Mathematics) [3] not only develops individuals' conception of the aspects of mathematics, but also it helps to show the trueness or falseness of statements in their lives. However, it is relatively difficult among students and even college's students. Many students do not know how to prove a mathematical statement. There is a critical question, whether students lack knowledge of proof construction or the proving problem that is given above the ability of students? To answer this question, one way that can be done is giving a freedom to students to pose a problem related to proof. As consequent, students are faced two tasks, namely problem proving and problem posing.

2. Literature Review

2.1. Problem proving

In mathematics, problem proving is a series of logical arguments that explain the truth of a problem. This is stated by Hanna [4] that problem proving is the application of an infinite number of logical steps of what is known (axioms, principles or results that have been proven before) and applying the principles of logic, to create deductive arguments valid for reaching a conclusion of a problem using acceptable inference rules.

Therefore, an explanation of problem proving is a key component in improving mathematics learning as a whole. There are several reasons why teaching problem proving should be given, namely: 1) evidence is an integral part of mathematics, 2) for verification and discovery of facts, 3) for the development of students 'logical and critical thinking skills, and 4) accelerating and improving students' mathematical understanding [5]. In addition, the need for problem proving of mathematics was introduced in several schools recommended by NCTM [3] that problem proving is part of the mathematics curriculum at all levels. The "Reasoning and Proof" section of this NCTM document states that students should be able to: recognize reasoning and proof as aspects of fundamental aspects of mathematics; make a conjecture and check the truth of that conjecture; develop and evaluate mathematical arguments and proofs; choose and use various types of reasoning and methods of proof in solving a problem.

As a recommendation from NCTM [3] it indicates that the problem of proving in mathematics is one aspect that must be considered in learning mathematics in schools. Some students experience in preparing a proof in high school will have an impact on the ability to prove when they attend college at the first level, as stated by Moore [6]; Weber[7]; Sandie[8] that one of the reasons why students encounter difficulties in proof is experience they in constructing evidence is limited to proof of school geometry. Accordingly, based on the results of a study conducted by Sabri [9] on the concept of problem proving in mathematics that prospective teacher students are advised that the high school curriculum should prepare students better in learning problem-based mathematical proof.

2.2. Problem Posing

In learning mathematics, problem posing can occupy strategic positions [10]. Problem posing is said to be the most important core in the discipline of mathematics and in the nature of mathematical reasoning thinking [11]. Graveimeijer [12] mentions the problem posing is a form of creative activity that can operate within tasks involving structured "rich contexts" and uses real-life documents and human interactions [13][14]. Furthermore, Stoyanova and Ellerton [15] said that mathematical problems posing as the process by which, based on mathematical experience, students construct personal interpretations of concrete situations and formulate them as meaningful mathematical problems.

In other words, problem posing is a process that is based on mathematical experience then students construct their own interpretations of concrete situations and formulate them into meaningful mathematical problems. Silver [11] says the problem posing refers to both the generation of new problems and the re-formulation of given problems. Problem posing refers to the formation of new problems and to reformulate given problems. Spillane [16] also defines problem posing as a task designed by the teacher that requires students to make one or more questions. Furthermore, problem posing is seen as a task that asks students to submit or make mathematical problems or problems based on the information provided, as well as solving problems or problems created.

Ulusoy [17] gives the term problem posing applied to three different forms of mathematical cognitive activity, which are as follows: presolution posing is a student makes questions about the situation (information) that is held, within-solution posing is a student reformulates a questions such as those that are being solved, and post solution posing is a student modifies the goals or conditions of the questions that have been solved to create new questions. In addition, Stoyanova and Ellerton [15] also provide a classification of situations in problem posing, namely as follows free problem posing where students are asked to make questions from a given situation or information, situations created, or natural situations, then semi-structured problem posing namely students are given an open situation and invited to explore the structure of the situation and complete it using knowledge, skills, concepts, and relationships from mathematical experiences that have been held, and structured problem posing namely problem posing activities based on a problem special.

3. Method

This is descriptive-explorative research that aims to describe students' task that concerned to assess students' problem proving and problem posing on triangle task.

3.1. *Participants*

The participants of our study were 24 prospective teachers who studied at Department of Mathematics of Universitas Negeri Surabaya in Surabaya city, Indonesia. The research participants were 24 mathematics prospective teacher's students that consist of 10 males and 14 females at Univesitas Negeri Surabaya.

3.2. *Data collection and Analysis*

The data were collected using a simple task of mathematical problem "(1) Prove that sum of measures of triangles are 1800, (2) What do you think it is the mathematical problem? Explain it, and (3) Create other problems that challenges in triangles proving". All students had to answer three questions. The first question related to problem proving of triangle theorem. The second question related to their perception of the first question, and the last question related to the problem posing task. A Newman analysis is used to analyse the data. Analysis focused on students' problem proving and students' problem posing of triangle task. Each prospective teacher has been given the task for 25 minutes. Then the results of data from participants' answers about the correct answer were analysed by classification problem proving and posing framework. The classification problem proving and posing framework, mainly:

Question 1	
No	*Criteria*
1	*Comprehension error*
2	*Transformation/devising strategies error*
3	*Mathematical Processing error*
4	*Interpretation error*
5	*Full understanding*
Question 2	
1	*Students perception that the question is not a problem*
2	*Students perception that the question as problem*
Question 3	
1	*A posed problem is not category as problem proving*
2	*Misinformation/ambiguity (unsolved problem)*
3	*A problem is not related to the initial problem*
4	*Levels of problem difficulty:* *a. Low* *b. Medium* *c. High*

In these cases, the coding was carried out to all the prospective teachers' answers. Because there was more than one possibility of coding given to each answer with different categories, we selected the most significant feature of the response category that emerged from the answer. Thus, each answer only has one code of category.

4. Results and Discussion

In this chapter, the data obtained from the participants were analysed, discussed, and then the results are presented in Table 1.

Table 1. The number of prospective teachers' answers toward task problem

No	Criteria	Quantity	
		Total	Percentage
Question 1			
1	*Comprehension error*	*0*	*0%*
2	*Transformation/devising strategies error*	*3*	*13%*
3	*Mathematical Processing error*	*6*	*25%*
4	*Interpretation error*	*5*	*21%*
5	*Full understanding*	*10*	*41%*
Question 2			
1	*Students perception that the question is not a problem*	*1*	*5%*
2	*Students perception that the question as problem*	*23*	*95%*
Question 3			
1	*A posed problem is not category as problem proving*	*4*	*16%*
2	*Misinformation/ambiguity (unsolved problem)*	*3*	*13%*
3	*A problem is not related to the initial problem*	*2*	*8%*
4	*Levels of problem difficulty:*		
	a. Low	*9*	*38%*
	b. Medium	*6*	*25%*
	c. High	*0*	*0%*

Table 1 shows that at question 1 shows that full understanding is the most found in the prospective teachers' answer, mainly 41% and 0 % for comprehension error. Then, at question 2 shows that student's perception that the question is not a problem obtained as many as 5 % of 24 students and 95 % have a perception is a problem. However, if we look at question 3, there will be shows that level low of problem difficulty is very high than other, mainly 38% of 24 students.

Furthermore, most students able to prove the triangle theorem completely. They constructed proof by formal proof, although they obtained as many as only 10 students. However, the students' proof construction has weakness. In addition, mostly student assumed that the question is classified as a problem. They supposed that the proving task as problem. In fact, this question is not category as problems because the question is commonly used in introducing triangle problem task. Students pose some problems are classified as low level of problem difficulty. Most students proposed questions of similarity problem on a trapezoid. Apart of problems could not be understood and could not be solved.

5. Conclusion

We conclude that most prospective teachers were correct to answer the question as many as 41%. It can be seen at question 1was performed by the highest number of prospective teachers. However, some of the prospective teachers still have a perception question 1 is a problem, because it shows as many as 95% of 24 students choose it. Furthermore, the students have some weaknesses in proving problems, mainly misinformation or ambiguity about problem, it is not category as problem proving. They constructed proof by formal proof with low level of problem difficulty as many as only 38 % of 24 students. This suggestion of this study is that teachers should emphasize learning mathematics through problem proving and problem posing task.

References

[1] Jankvist U T, Clark K M, Mosvold R 2019 *J. Math. Teach. Educ.*, **1** 22
[2] Siswono T Y E, Hartono S, Kohar A W 2018 *Eur. J. of Educ. Res.* **18** 197-212
[3] National Council of Teachers of Mathematics 2000 *Reston* VA NCTM
[4] Hanna G 2000 *Educ. Stud. Math.* **44** 5-23

[5] Dickerson D.S 2008 *Handbook* Syracuse University
[6] Moore R C 1994 *Educ. Stud. Math* **27** 249-266
[7] Weber K 2001 *Educ. Stud. Math.*, **48** 101-119
[8] Sandie S, Purwanto P, Subanji S, Hidayanto E 2019 *Inter. J. Hum. Innov.* **2** 5-30
[9] Sabri 2003 *Thesis* Universitas Curtin, unpublished
[10] Siswono T Y E, Hartono S, Kohar A W 2018 In *J. Phys.: Conf. Ser.* **947** 012008
[11] Silver E A, Mamona-Downs J, Leung S S, Kenney P A 1996 *J. Res. Math. Educ*. 293-309
[12] Gravemeijer K, Terwel J 2000 *J. Cur. Stud.* **32** 777-796
[13] English L D 2009 In Word and World, Brill Sense 351-362
[14] Ekawati R, Kohar A W, Hartono S 2017 *J. Eng. Sci. Technol.* 53-64
[15] Stoyanova E, Ellerton N F 1996 *Techn. Math. Educ.* 518-525
[16] Spillane J P, Coldren A F 2015 *Handbook* Teachers College Press
[17] Ulusoy F, Kepceoğlu İ 2018 *J. Kirs. Educ. Fac.* **19**

Recitation program to improve students' conceptual understanding of Thermodynamics

A S D Adila[1], Sutopo[2], Wartono[2]

[1] Graduate Program of Physics Education, State University of Malang, 5th Semarang Street, Malang, Indonesia
[2] Physics Department, State University of Malang, 5th Semarang Street, Malang, Indonesia

e-mail: agistasintia24@gmail.com

Abstract. The purpose of this research was to analyze the improvement of students' mastery of thermodynamics concepts after they learn three packages of recitation program that cover state equation of ideal gas, The First Law of Thermodynamic, and The Second Law of Thermodynamic. The subjects consisted of 21 undergraduate students enrolling in the fundamental physics course. The results showed that the recitation program improved students' conceptual understanding of thermodynamic with the d-effect size of 2.51 and N-gain of 0.365. The program effectively helps the students in constructing *P-V* diagrams and understanding the relationship between the average particle's kinetic energy and temperature of the gas. However, some contents need to be improved to help the students more fluent in determining total work and heat involved in any cyclic process.

1. Introduction

Mastery of fundamental concepts is the main focus of physics education. An indicator that a student has already mastered a concept is he/she able to apply the concept in various types of problems [1]. However, researches show that most students have difficulty to apply some concept to solve problems [1, 2]. One fundamental concept that is difficult for students is the thermodynamic concept [3].

Several studies have identified student difficulties in mastering thermodynamic concepts. Monteyne et al [4] reported that some students find it difficult to apply the gas state equation if there are three variable changes. Incorrect understanding when identifying pressures in containers that have different temperatures [5], identifying isobaric phenomena [6] and identifying temperatures in the adiabatic compression process are still experienced by students [7]. Students also have not mastered the concepts of work, heat and internal energy presented by PV diagrams [8]. Difficulties in mastering the Second Law of Thermodynamics, especially about the engines, are also often experienced by students. Some students find it difficult to use the Second Law of Thermodynamics to identify engines that can work based on the engines' diagram presented [9]. Besides, students still need to practice to manipulate mathematical equations about heat efficiency and refrigerator engines [10].

One of the ways to improve students' understanding is to provide better learning. The learning must provide some opportunity to construct the concepts broadly and intact because each student has a different speed to understand the concepts. Previous studies have developed various ways to improve students' concepts by using some learning strategies [11, 12, 13] Also, several studies developed some learning that can be done outside of lecture hours [3, 14]. The learning outside of the lecture hours can be done by giving some exercises. So, the previous researchers have developed some exercises about thermodynamic concepts. Erceg et al [7] develop some questions that related to the microscopic gas model or the kinetic gas theory. Sujito et al [15] developed Socratic Dialog Media, including temperature and heat. [3] also developed the question of the First Law of Thermodynamic. Cochran and Heron [9] developed a matter of the Second Law of Thermodynamics about identifying 3 machine diagrams (heat engines, refrigerator engines, and strange devices) that can be realized.

Exercises can improve students' understanding. The exercises must immediately provide feedback so it will be more effective. Feedbacks were used to helps the student knows whether his/her answer is correct or not. It also is used to guide students to correct their concepts. Several studies have provided feedback on their development instrument of kinematic problems [16] and magnetism problems [17]. However, there has been no development of thermodynamic problems with feedback.

University in America always gave some exercises and the feedback outside of the lecture' hour. They provide a deepening of topics in the form of additional lectures which are often called recitation [2]. Their recitation is often accompanied by a teaching assistant. Whereas in Indonesia there were not teaching assistants to give some additional lectures.

Technology can help lecturers in Indonesia to give recitation more efficiently because it can provide immediate feedback. Previous studies also used computer technology so that students immediately get feedback on their answers [18, 16, 15, 17]. Giving recitation by utilizing technology is suitable to be applied to thermodynamic because of the broad and complex scope of the topic. The focus of this study is to checked whether the recitation program can improve students' conceptual understanding or not. There were three sub-topics in the recitation program. There were gas state equations, the First Law of Thermodynamics, and the Second Law of Thermodynamic.

2. Methods

This study uses a quantitative approach to see whether recitation programs can improve students' conceptual understanding. The recitation program in this study was a group of conceptual problems about thermodynamics with different feedback in every choice. The sequence of this study was pre-test, intervention (the usage of recitation program), and post-test. The intervention was conducted for 2 weeks with 3 meetings which take 90 minutes each. Subtopic for every meeting was different. In the first meeting, students learned about gas state equations, thermodynamic processes, and average kinetic energy of particles. In the second meeting, students learned about work, heat and internal energy in thermodynamic processes. In the third meeting, students learned about the heat engines and refrigerator engines' principle. The sum of the problems was 30 problems.

Quantitative data was obtained from pre-test and post-test scores. Then, we tested the normality and paired sample t-tested to finding d-effect size [19]. Besides, we would discuss students' understanding of thermodynamics before and after the intervention. The subject of this study consisted of 21 undergraduate students of Physics Education at the State University of Malang who had just finished discussing thermodynamic concepts in the Second of Fundamental Physics Course.

The instrument was 17 multiple choice problems. This instrument had the same scope of the topic as the coverage of the recitation program topic. The pre-test instrument was the same as a post-test instrument. This instrument has also been tested for reliability with the Cronbach Alpha coefficient of 0.68.

3. Results and Discussion

The results of the normality test show that the pre-test and post-test are normally distributed because the skewness value is between -1 and 1 [20]. The results of the paired sample t-test showed that there were significant differences in the average score between the pre-test and post-test ($t(20) = 9.4$, $p = 0.000$, $d = 2.51$). It meant the recitation program could improve students' conceptual understanding of thermodynamic concepts. In addition, the improvement could be seen from the N-gain. The N-gain value is 3.65. This value indicated the improvement in the lower medium category. Even though it was in the lower medium category but overall there was an improvement of students' conceptual understanding.

Out of the 17 problems were tested, two problems experienced significant increases to answer the correct choices (questions number 2 and 7) and there was one question that none of the students answered correctly (question number 13).

3.1 The Strenght of Recitation Program

Item number 2 shows that students already understand about the isobaric process based on phenomena and constructing *PV* diagrams based on phenomena (Figure 1). The correct answer has increased from 1 to 14. The purpose of this problem was to construct a *PV* diagram based on phenomena. The phenomenon of balloon deflating was due to a decreasing temperature which causes the volume to decrease. While the pressure in the balloon when the balloon deflates was still the same as the atmospheric pressure. So the diagram could be described as an isobaric diagram. With directions from A to B (the decreasing of volume and temperature) (d).

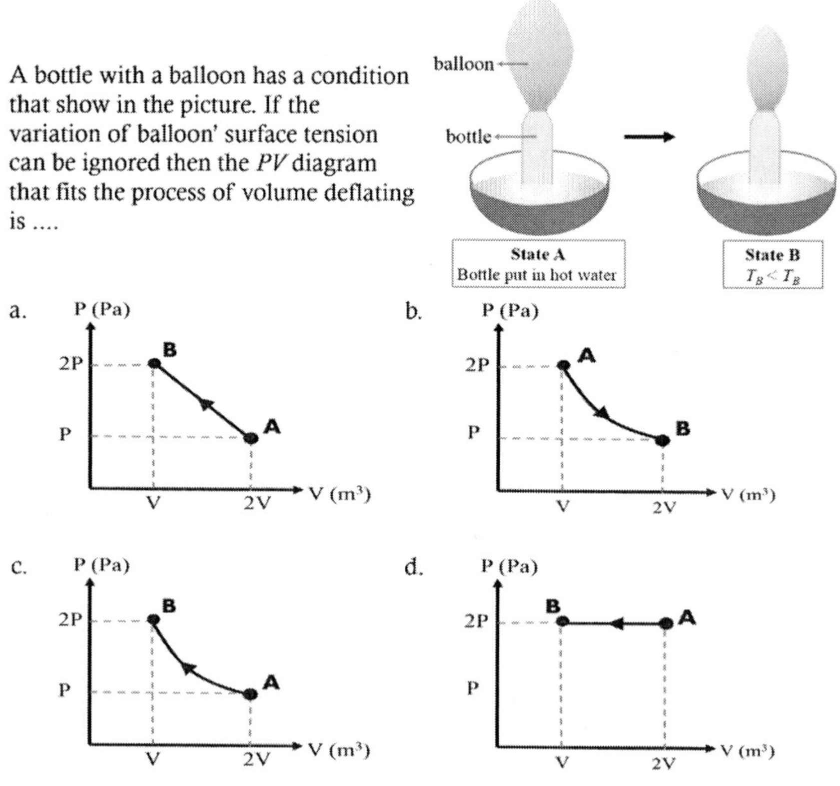

Figure 1. Problem number 2

The pre-test results showed that only one student answered correctly. While the other students did not answer correctly. There were 10 students chose the answer b for various reasons. The first reason was to consider this phenomenon as an isothermal process even though the problem had been implicitly explained the temperature would decrease. The second reason was when in state B the volume decreases so that the pressure would increase $P \approx \frac{1}{V}$. The third reason was to assume that the process occurs as a process of constant volume so that the variables that change are only P and T. From those reasons, we concluded that students have not fully understood the gas state equation. Nine students choose answer c with unclear reason. There was a student who did not answer question number 2.

Table 1. Cross-tabulation of problem number 2.

		Post-test				Total
		a	**b**	**c**	**d[a]**	
Pre-test	b	0	3	0	7	10
	c	1	0	1	7	9
	d[a]	0	0	1	0	1
	-	0	1	0	0	1
	Total	1	4	2	14	21

[a] correct answer

The post-test result showed that 14 students answered correctly for various reasons. Two students understood if the process that occurs is the isobaric process. However, 12 students had not realized that the process was the isobaric process.

R	:	Why did you choose d?
S1	:	The temperature of the system was decreasing. So, the volume would be decreased.
R	:	Okay. And then?
S1	:	I analyze the diagram's choices. The initial and final temperature diagrams a, b and c are equal. While the final temperature diagram d is less than the initial temperature
R	:	From the problem, did you realize if the process is the isobaric process?
S1	:	No. I chose d because it is the only choice with decreasing temperature

Based on the interview, students could analyze the *PV* diagram very well. However, they have not realized that the process was the isobaric process. Whereas, the problem has been stated that we could ignore the balloon surface tension (the air pressure in the balloon was always the same as the air pressure outside the balloon). Nevertheless, the students have been proficient in analyzing diagrams using gas state equations to find temperatures. They knew that the temperature of the initial state was greater than the temperature of the final state (d). There was a student who chose answer (a) with reason $P_1 > P_2$ and $V_1 > V_2$. There were 4 students chose answer b. They thought the pressure in state A greater than state B and the volume in state B greater than state A. From those statements we knew that students did not understand if the volume of the balloon will shrink. Two students chose to answer c. They thought the process was an isothermal process.

There were two reasons for the increase in the correct answer. First, students had understood how to read and analyze *PV* diagrams. Second, they also had already used the ideal gas state equation ($PV = nRT$) to identify the *PV* diagrams. There were 7 problems that they learned in the recitation program. Problem number 1.2 - 1.5 trains students to identify T-V diagrams in isochoric, isothermal, isobaric and unfamiliar processes by using $PV = nRT$. Problem number 1.6 - 1.8 trained students to identify temperatures in four conditions in unfamiliar processes represented by the P-V diagram. Also, all feedbacks were used to train students to use $PV = nRT$ for diagram identification. According to records during the intervention, students faced some difficulty. It was analyzing the diagram, especially the *TV* diagrams and unfamiliar processes. Students were often fooled when answering problems number 1.6 - 1.8. It also happened in previous research [21]. Whereas those problems use the same diagram and ask the same thing, the only difference was the representation. Those problems were designed to make students always careful in analyzing diagrams. As a consequence, they became experts in analyzing diagrams by using $PV = nRT$.

Item number 7 (Figure 2) had a large increase of correct answer. This shows that students understood the concept after the intervention.

Two containers A and B are filled with helium gas. The number of gas particles in container A is double compare to container B and the volume of container B is doubled compare to container A. The temperature of the two containers is the same as the environment temperature. The average kinetic energy of gas particles (*EK*) in containers A and B is...
a. $EK_A = EK_B$ because of the temperature of containers A and B are the same
b. $EK_A = EK_B$ because of the volume of the container is equal
c. $EK_A > EK_B$ because of the number of particles A greater than the number of particles B
d. $EK_A < EK_B$ because it is tighter so that the average distance between particles is shorter

Figure 2. Problem number 7.

Students need to understand the relationship between temperature and *EK*. Based on the equation, the temperature is the macroscopic representation of *EK*. So the number of gas particles and container volume does not affect *EK* (a). At the pre-test, three students chose answers a. But at the pre-test, ten students chose answer c because they did not understand the difference between the average kinetic energy of gas particles ($EK = -k_BT$) and the internal energy of gas (E_{int}=-3/2(Nk_BT)). So they think that the number of gas particles affects *EK*. Besides, three students were mistaken between *EK* and Boltzman's constant (k_B) ($PV = Nk_BT$). Three students who choose the answer d have a reason. If the gas particles are denser then it will affect *EK*. Some of them have miscalculated EK_A and EK_B.

Table 2. Cross-tabulation of problem number 7.

		Post-test			Total
		a[a]	c	d	
Pre-test	a[a]	3	0	0	3
	b	1	0	0	1
	c	9	3	1	13
	d	3	0	0	3
	-	1	0	0	1
	Total	17	3	1	21

[a] correct answer

The post-test results showed that 17 students chose the answer because of *EK* depending on temperature. There is even one student who explains that *EK* influenced by temperature not the number of particles. Three students chose to answer b without writing down any reason. One student chooses the answer d $EK = \frac{1}{2}mv^2$. Based on these reasons, it can be seen that he still does not understand if $EK = \frac{1}{2}mv^2$ it is used in particle objects (solid objects). The increase in the number of students who answered correctly was due to the use of the first package of recitation program. Students learned about the relation of average kinetic energy with temperature in a different context.

The number of correct answers was raised from 3 to 17. It due to the recitation program provides appropriate problems and feedbacks. The problem is problem number 1.12. Students were trained to identify the relationship between temperature and the average kinetic energy of particles. In addition, the recitation program provided feedback to guide students to recognize the differences between *EK* and E_{int}. While students who are still wrong to answer question number 7 still have the same thoughts with previous research [7].

3.2 The Weakness of Recitation Program

Item number 13 (Figure 3) shows the difficulties to analyze cyclic processes. No one answered it correctly either at the pre-test or post-test. This problem aims to determine the total of work and heat in the cyclic process.

The *PV* diagram shows the gas pressure and volume values, starting at state A, continuing to states B, C, and D and returning to state A. The correct statement in one cycle A → B → C → D → A is....

a. the magnitude of the work total is zero because the initial and the final state of the system are same
b. the total heat transferred to the system is zero because the initial and the final state of the system are same
c. as a whole, the work is done on the environment
d. as a whole, the heat release to the environment

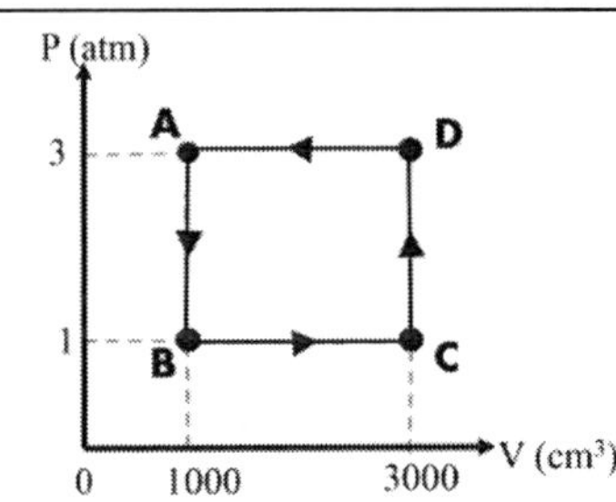

Figure 3. Problem number 13.

To solve problems number 13, we must understand about work and heat concept. Work and heat are process function. Work is a process of energy exchange through mechanical interactions while heat is a process of energy exchange through thermal interactions. The way to find out heat total is to know the work total in the cyclic process. Process A → B → C → D → A shows the work done on the system. When the gas is compressed, the heat in the system will come out. So that overall gas releases the heat to the environment (d).

At the pre-test, there were ten students chose answers a. The reason is the initial and final state is the same so the work total is zero. Meanwhile, five students choose the answer b because Q_{DA} has negative value, Q_{AB} has a positive value, Q_{BC} has a positive value, and Q_{CD} has a positive value. There were three students chose the answer c with unclear reason. And, there were three students who did not answer the problem.

Table 3. Cross-tabulation of problem number 13

		Post-test				Total
		a	b	c	d[a]	
Pre-test	A	4	0	6	0	10
	B	2	1	2	0	5
	C	2	0	1	0	3
	d[a]	0	0	0	0	0
	-	1	1	1	0	3
	Total	9	2	10	0	21

[a] correct answer

The post-test results show that no one has chosen the correct answer (d). There were nine students who chose answers a. Some students still have an understanding that if the work total is zero because it returns to its original state.

R : What is the work total in the cyclic process?
S2 : The work total is zero
R : Why?
S2 : Because the initial and final conditions are the same, so the work total is zero
R : Did you calculate the area under the *DA* and *BC* path?

S2	:	No

There were two students answer b. They thought heat as a state function. So, the total heat which transferred in the system was zero. Ten students chose to answer c. They thought the work done on the system and the heat enters the system. But, they were not writing the work and heat in each process. However, ten students have understood that in the cyclic process the change in internal energy is zero. In addition, students also know that work can be calculated from the area below the path.

R	:	Why the work total of the cyclic process is zero?
S3	:	Because the area under the path *DA* is greater than the area under the path BC so that the work total cannot be zero
R	:	Overall gas does mechanical work into the environment or does gas receive mechanical work from the environment?
S3	:	Maybe gas does mechanical work into the environment
R	:	Are you sure?
S3	:	Yes, because of the answer in recitation program similar to these problems states
R	:	How about the previous recitation problem? Didn't the previous question guide you to understand the work done on the system or work done on the environment?
S3	:	Yes it's already there, but I still confused about it

Almost all students chose answers a and c. Students who choose to answer a thought work as a state function. Even though, the recitation program already trained students to determine the work and heat in cyclic processes (questions number 2.6 and 3.1). Students' thought about work and heat similar to the previous research [22] [23]. The reason for the answers given by students who answer c is more or less the same as the feedback in the recitation program. It indicated that students still not understand about work and heat concepts. To answer this problem, students must understand the concept of work done on system or environment and heat entering the system or heat out of the system first. While answering question number 13 in pre-test and post-test, there were no students who used work and heat concept correctly. So when we gave a problem with the opposite process directly with the exercises, they still consider the process presented in the pre-test-post-test as the same as the problem in recitation program.

4. Conclusion

Three packages of conceptual problems along with the feedbacks (recitation program) could improve students' conceptual understanding in thermodynamics. This is evidenced by the results of d-effect size of 2.51 and N-gain of 0.365. However, students still need to improve their understanding because there were problems that they cannot answer correctly and there were some concepts that they did not fully understand yet. So, we suggest in future research to adding some problems with analyzing phenomena which need thermodynamics process, analyzing work total and heat total in various processes which represented by *PV* diagrams.

References

[1] Lin S Y and Singh C 2011 Using isomorphic problems to learn introductory Physics *Physical Review Special Topics-physics Education Research* **7**(2) 020104

[2] Docktor J L and Mestre J P 2014 Synthesis of Discipline-based Education Research in Physics *Physical Review Special Topics-physics Education Research* **10**(2) 020119

[3] Sriyansyah S P and Suhandi A 2016 Development of a representational conceptual evaluation in The First Law Of Thermodynamics *J. of Physics: Conf. Series* vol. 739 no. 1 IOP Publishing 012125

[4] Monteyne K, Gonzalez B L and Loverude M E 2008 an interdisciplinary study of student ability to connect particulate and macroscopic representations of a gas *AIP Conf. Proc.* vol. 1064 no. 1 AIP pp 163-166

[5] Kautz C H, Heron P R, Loverude M E and McDermott L C 2005 Student understanding of The Ideal Gas Law, Part I: A Macroscopic Perspectivev *American J. of Physics* **73**(11) pp 1055-1063
[6] Talaeb P and Wattanakasiwich P 2010 Development of Thermodynamic conceptual evaluation *AIP Conf. Proc.* **1263**(1) 183-186
[7] Erceg N, Aviani I, Mešić V, Glunčić M and Žauhar G 2016 Development of The Kinetic Molecular Theory of Gases Concept inventory: Preliminary Results on University Students' Misconceptions *Physical Review Physics Education Research* **12**(2) 020139
[8] Loverude M E, Kautz C H and Heron P R 2002 Student understanding of The First Law of Thermodynamics: Relating work to the adiabatic compression of an ideal gas *American J. of Physics* **70**(2) 137-148
[9] Cochran M J and Heron P R 2006 Development and assessment of research-based tutorials on heat engines and The Second Law of Thermodynamics *American J. of Physics 74*(8) 734-741
[10] Smith T I, Christensen W M, Mountcastle D B and Thompson J R 2015 Identifying student difficulties with entropy, heat engines, and the carnot cycle *Physical Review Special Topics-physics Education Research* **11**(2) 020116
[11] Tatar E and Oktay M 2011 The effectiveness of Problem-based Learning on Teaching The First Law of Thermodynamics *Research in Science & Technological Education* **29**(3) 315-332
[12] Iyengar S S and deSouza R T 2013 Teaching Thermodynamics and Kinetics to advanced general Chemistry students and to upper-level undergraduate students using PV Diagrams *J. of Chemical Education* **91**(1) 74-83
[13] Ramabhadran R O 2017 A Mistake based approach probing students' understanding of PV-type work in Thermodynamics *Resonance* **22**(10) 961-970
[14] Koenig K M, Endorf R J and Braun G A 2007 Effectiveness of different tutorial recitation teaching methods and its implications for TA Training *Physcal Review Special Topics-physics Education Research* **3**(1) 010104
[15] Sujito S, Pebriana I N, Pratiwi H Y, Hayyi A, Masjkur K, Asim and Sutopo 2018 Students misconception: The developing of Socratic Dialogue Media on Temperature and Heat. *Jurnal Pena Sains* **5**(2) 87-95
[16] Ryan Q X, Frodermann E, Heller K, Hsu L and Mason A 2016 Computer Problem-solving Coaches for Introductory Physics: Design and usability studies *Physical Review Physics Education Research* **12**(1) 010105
[17] Cahyaningrum R and Hidayat A 2018 Contrasting-cases problems: Learning material to improve students' conceptual understanding on Magnetism *J. of Physics: Conf. Series* **1097**(1) 012028 IOP Publishing
[18] Gladding G, Gutmann B, Schroeder N and Stelzer T 2015 Clinical study of student learning using mastery style versus immediate feedback online activities *Physical Review Special Topics-physics Education Research* **11**(1) 010114
[19] Morgan A G, Leech L N, Gloeckner W G and Barret C K 2004 *SPSS for Introductory Statistic use and interpretation (Second Edition)* (London: Lawrence Erlbaum Associates Inc.)
[20] Leech N L, Barret K C and Morgan G A 2005 *SPSS for Introductory Statistics: Use and Interpretation Second Edition (*New Jersey: Lawrence Erlbaurn Associats Inc.)
[21] Adila A S D, Sutopo and Wartono 2018 Students' reasoning in analyzing temperature from PV Diagram representing unfamiliar Thermodynamics process *J. of Physics: Conf. Series* **1097**(1) 012012 IOP Publishing
[22] Clark J W, Thompson J R and Mountcastle D B 2013 Comparing student conceptual understanding of Thermodynamics in Physics and Engineering *AIP Conf. Proc.* **1513**(1) 102-105
[23] Brookes D T and Etkina E 2015 The importance of language in students' reasoning about Heat in Thermodynamic Processes *Int. J. of Science Education* **37** 759-779

Critical Thinking, Creative Thinking, and Learning Achievement: How They are Related

A Fatmawati[1,2*], S Zubaidah[2], S Mahanal[2], and Sutopo[3]
[1]Department of Biology Education IKIP Mataram, Jl. Pemuda No. 59A Mataram 83125, Indonesia
[2]Department of Biology Education, Universitas Negeri Malang, Jl. Semarang No. 5 Kota Malang Jawa Timur 65145, Indonesia
[3]Department of Physics Education, Universitas Negeri Malang, Jl. Semarang No. 5 Kota Malang Jawa Timur 65145, Indonesia

e-mail: *anyfatmawati@ikipmataram.ac.id

Abstract. This study aimed to investigate the correlation between (1) critical and creative thinking skills; (2) critical thinking skills and learning achievement; (3) creative thinking skills and learning achievement; and (4) critical thinking, creative thinking, and learning achievement. The current study was conducted in April 2019 and employed a correlational research design. The participants of this study consisted of 30 fourth-semester students from the Department of Biology Education of IKIP Mataram, Indonesia, who were currently studying Plant Physiology. Data were collected using a test that contained 19 essay questions on photosynthesis. The critical thinking instrument was composed of five aspects, whereas the creative thinking instrument comprised of eight aspects. Besides, the instrument used to determine learning achievement incorporated six aspects. Each of the elements was represented by one test item. The results of the data analysis indicated correlations between (1) critical and creative thinking skills; (2) critical thinking skills and learning achievement; (3) creative thinking skills and learning achievement; (4) creative thinking, critical thinking, and learning achievement. Since critical and creative thinking skills affect learning achievement, the empowerment of these skills may lead to the enhancement of learning achievement.

1. Introduction

Critical thinking is the most valuable skill that can be passed on by the school to its graduates and becomes a learning goal at all levels of discipline [1]. Critical thinking is one of the high-level skills that is the goal of world education today, but this skill is not well developed in students. This is evidenced by the results of research that shows 75.63% of students in Public Junior High School 1 Turen have critical thinking skills that are still undeveloped or are still low [2]. Critical thinking skills of Public Senior High School students in Batu City in Biology are still not well developed and even need to be improved [3]. Based on the results of observations and interviews with eleventh-grade students in biology class at Public Senior High School 1 Alla in Enrekang Regency, it was obtained information that students' critical thinking skills are qualitatively relatively low [4]. Also, the students' critical thinking skills in science subjects in one of the Public Junior High Schools in Sukabumi City were still low [5]. Moreover, the results of research at Islamic Senior High School in Magetan Regency provide information that critical thinking skills of students are still low [6]. The results of other studies also show a phenomenon that is not much different.

Low critical thinking skills are also experienced by students who are the output of the senior high school. This is seen from the results of the previous researches. Based on the results of the study as a whole in the Biology education program of STKIP Bima, it was found that students' critical thinking skills were in a low category [7]. Besides, the results of research on 139 students from UIN Alauddin Makassar, STKIP PI Makassar, UPRI Makassar, and STKIP Yapim Maros, showed that critical thinking skills were still low and did not develop well [8]. The results of the research at the Biology Education Bachelor degree program at PGRI University Ronggolawe Tuban also showed that students still had difficulty using critical thinking skills [9]. The prospective teachers' critical thinking in Biology Department IKIP Mataram is underdeveloped [10]. The results of research at Biology Department IKIP Mataram showed that prospective science teachers' critical thinking disposition generally neither strength nor weakness category [11].

Another skill that needs to be developed in students is creative thinking. Creative thinking skills are the basis of science, which is very important for students [12]. Optimization of higher-order thinking skills is essential because it is a life skill that needs to be developed. It is required to overcome complex problems along with the times, especially in this 21st-century era [13]. The results of research on students at UNS show their creative thinking skills are insufficient category and still need to be developed [14]. Besides, the effects of research on Chemistry students at Sriwijaya University also showed that their creative thinking skills were at moderate and low levels [15].

Student learning outcomes are also significant to be a concern because one of the benchmarks of learning success is student learning outcomes related to the material being studied. Several research results show that learning outcomes in biology are still low in students and college students. The results of research at Public Senior High School 2 Surakarta show that student learning outcomes in biology subjects are still low and need to be appropriately handled [16]. The results of research at Biology students at the Muslim University of Maros in the Plant Ecology course show that their average learning outcomes are in the moderate category [17]. Therefore, biology learning outcomes need to be improved because they are still in the low and medium categories.

Facts about the low critical thinking skills, creative thinking skills, and learning outcomes also occur in the Biology IKIP education study program Mataram, seen from the results of observations and preliminary studies conducted on April 8-13, 2019, on students and interviews with lecturers of Plant Physiology courses. The course lecturers said that students were not accustomed to using their critical thinking skills and creative thinking well. It can be seen from their answers during midterm that some of which were theoretical because they could not analyze, evaluate, and conclude well. Besides, students still answer precisely like the material in textbooks, so the answers are still monotonous and less varied, especially on complex material, namely material in the Plant Physiology course.

The explanation above shows that critical and creative thinking skills are skills that are very needed in the 21st-century life so that a study of student learning outcomes is needed as a benchmark of learning. Therefore, a research is required in order to see: (1) the correlation between critical thinking skills and creative thinking skills; (2) the relationship of critical thinking skills to learning outcomes; (3) the relationship of creative thinking skills to learning outcomes; (4) the relationship between critical and creative thinking skills on learning outcomes.

2. Method

The design of this study used a correlational research design, which aimed to determine the extent to which one variable is related to other variables. Data collection was carried out in the Biology Education Study Program of IKIP Mataram in the academic year 2018/2019, on 22-27 April 2019. The sampling technique was done by saturation sampling (population sample), where all population members were sampled. Participants in this study were students who took Plant Physiology courses in semester 4, as many as 30 students, four males, and 26 females. The material being tested is Photosynthesis.

Data collection instruments for critical thinking skills using a rubric developed by Greenstein [18]. The assessment of critical thinking skills consisted of 4 scores (1-4) as a reference to examine each item with criteria (4 = perfect, 3 = good, 2 = enough, 1 = less), consisting of apply, evaluate, use data to develop critical insights, analyze, synthesize. The instruments for creative thinking skills are curiosity, fluency, originality, elaboration, flexibility, divergence, messiness/risk-taking, with others [18], the assessment of creative thinking skills consisted of 4 scores (1-4) as a reference to check each item with criteria (4 = very good, 3 = good, 2 = enough, 1 = less). Besides, learning outcomes referred to Benjamin S. Bloom [19], consisting of remembering, understanding, applying, analyzing, evaluating, and creating, while the assessment for cognitive learning outcomes with a range of values from 0-100. All questions were in the form of essay tests taken from photosynthetic material that has been previously validated.

Data collected was in the form of quantitative data and analyzed statistically. Data analysis used multiple regression analysis to reveal the correlation between critical thinking skills, creative thinking skills, and student cognitive learning outcomes. The data analysis process was assisted by SPSS version 18.0 for windows at a significance level of 5%.

3. Results and Discussion

The results of data analysis and discussion were presented sequentially. The results of the correlation between critical thinking skills with creative thinking skills are given as follows.

Table 1. The Correlation Results Between Critical Thinking Skills and Creative Thinking Skills.

Model	Sum of Squares	df	Mean Square	F	Sig 2-tiled
Regression	785.952	1	785.952	7.281	.012
Residual	3022.590	28	107.950		
Total	3808.542	29			

Table 1 shows that p = 0.012, which means that there is a correlation between critical thinking skills and creative thinking skills. The regression coefficient of the relationship between critical thinking skills and creative thinking skills of 0.454 is shown in Table 2.

Table 2. The Regression Coefficient of Critical Thinking Skills and Creative Thinking Skills

Model	Unstandardized Coefficients		Standardized Coefficients	t	Sig.
	B	Std. Error	Beta		
1 (Constant)	29.302	7.513	.454	3.900	.001
Critical Thinking Skills	.360	.133		2.698	.012

a. Dependent Variable: Creative Thinking Skills

The total contribution of critical thinking skills to creative thinking skills is shown in Table 3.

Table 3. The Results Summary of Regression Analysis of Critical Thinking Skills and Creative

R	R Square	Adjusted R Square	Std. Error of the Estimate
.454[a]	.206	.178	10.390

a. Predictors: (Constant), Critical Thinking Skills

The results of the multiple regression analysis show that the regression coefficient value (R) is 0.454, and the determination coefficient is (R2) 0.728. This means that the contribution of critical thinking skills to creative thinking skills is 20.60%, while the remaining as much as 79.40% is influenced by other variables.

Critical thinking skills and creative thinking skills are related to each other. When a person starts thinking, there is a disequilibrium process in his brain so that someone can bring up things that are different from usual. In line with the results of the above study, it has also been found that there is a significant correlation between critical thinking skills and creative thinking skills on cognitive learning outcomes through inquiry learning in environmental science courses [20].

The development of science in the 21st century has given rise to a sense of urgency in students, primarily to obtain specific skills needed in a career [21]. These skills include critical thinking skills, communication skills, teamwork skills, and creativity [22]. Therefore, the empowerment of thinking skills, especially critical and creative thinking, is needed.

Table 4. The Data Analysis Results of Correlation between Critical Thinking Skills and Cognitive Learning Outcomes.

Model	Sum of Squares	df	Mean Square	F	Sig 2-tiled
Regression	3438.328	1	3438.328	84.020	.000
Residual	1145.838	28	40.923		
Total	4584.167	29			

Table 4 shows that p = 0.000, which means that there is a correlation between critical thinking skills with cognitive learning outcomes. The regression coefficient of the relationship between critical thinking skills and learning outcomes of 0.866 is shown in Table 5.

Table 5. The Regression Coefficient of Critical Thinking Skills and Cognitive Learning Outcomes

Model	Unstandardized Coefficients		Standardized Coefficients	t	Sig.
	B	Std. Error	Beta		
1 (Constant)	16.807	4.626		3.633	.001
Critical Thinking Skills	.753	.082	.866	9.166	.000

a. Dependent Variable: Learning Outcomes

The total contribution of critical thinking skills to cognitive learning outcomes is shown in Table 6.

Table 6. The Results Summary of Regression Analysis of Critical Thinking Skills and Cognitive Learning Outcomes.

Model	R	R Square	Adjusted R Square	Std. Error of the Estimate
	.866[a]	.750	.741	6.39709

a. Predictors: (Constant), Critical Thinking Skills

The results of the multiple regression analysis show that the regression coefficient value (R) is 0.866, and the determination coefficient is (R2) 0.750. This means that the contribution of critical thinking skills to cognitive learning outcomes is 75.00%, while the remaining as much as 25.00% is influenced by other variables.

Critical thinking skills are needed in welcoming the 21st century so that students are expected to get meaningful learning and practice these critical thinking skills in learning. In empowering critical thinking skills, there will be learning outcomes that follow as a form of learning outcomes. However, based on the results of the 2018 National Examination, students were still weak in higher-order thinking skills such as reasoning, analyzing, and evaluating [23].

One of the competencies for sustainability is the ability to think critically about norms, opinions, and actions; reflect the values, perceptions, and activities that become his beliefs [15]. Critical thinking will support the process of thinking that is fast, accurate, and without assumptions [24]. The government expects students to achieve various competencies with the application of HOTS or Higher Order Thinking Skills [25]. These competencies are critical thinking, creative and innovative, communication skills, the ability to work together (collaboration), and confidence. In addition to students, college students are also required to be able to analyze, synthesize, and infer the information obtained with the ability to think critically, so students can distinguish between good and bad news, and can make decisions about the information obtained through critical thinking [26].

Table 7. The Correlation Analysis Results Between Creative Thinking Skills and Cognitive Learning Outcomes.

Model	Sum of Squares	df	Mean Square	F	Sig 2-tiled
Regression	676.030	1	676.030	4.843	.036
Residual	3908.137	28	139.576		
Total	4584.167	29			

Table 7 shows p = 0.036, which means that there is a correlation between creative thinking skills and cognitive learning outcomes. The regression coefficient of the relationship between creative thinking skills and cognitive learning outcomes of 0.384 is shown in Table 8.

Table 8. The Regression Coefficient of Creative Thinking Skills and Cognitive Learning Outcomes.

Model	Unstandardized Coefficients		Standardized Coefficients	t	Sig.
	B	Std. Error	Beta		
1 (Constant)	37.224	9.610		3.874	.001
Creative Thinking Skills	.421	.191	.384	2.201	.036

a. Dependent Variable: Learning Outcomes

The total contribution of creative thinking skills to cognitive learning outcomes is shown in Table 9.

Table 9. The Summary of Regression Analysis Results of Creative Thinking Skills and Cognitive Learning Outcomes

R	R Square	Adjusted R Square	Std. Error of the Estimate
.384[a]	.147	.117	11.81424

a. Predictors: (Constant), Creative Thinking Skills

The results of the multiple regression analysis show that the regression coefficient value (R) is 0.384 and the determination coefficient is (R2) 0.147. This means that the contribution of creative thinking skills to cognitive learning outcomes is 14.70%, while the remaining as much as other variables influence 85.30%.

Creative thinking skills that are trained in schools will make students have creativity [16]. Another opinion says that the development of cognitive learning outcomes is considered vital because it is often associated with children's intelligence [27]. Therefore, between creative thinking skills and learning outcomes need to be harmonized in learning, both for students and college students, and this is the task of educators, namely teachers and lecturers.

Table 10. The Correlation Analysis Results of Critical Thinking Skills and Creative Thinking Skills Against Cognitive Learning Outcomes.

Model	Sum of Squares	Df	Mean Square	F	Sig 2-tiled
Regression	3438.839	2	1719.420	40.534	.000
Residual	1145.327	27	42.420		
Total	4584.167	29			

Table 10 shows p = 0,000, which means that there is a significant correlation between critical thinking skills and creative thinking skills on cognitive learning outcomes. The regression coefficient of the relationship between creative thinking skills and learning outcomes is 0.012, while the regression coefficient of the correlation between critical thinking skills and learning outcomes of 0.871 is shown in Table 11.

Table 11. The Regression Coefficient of Critical Thinking Skills and Creative Thinking Skills Against Cognitive Learning Outcomes.

Model	Unstandardized Coefficients		Standardized Coefficients	t	Sig.
	B	Std. Error	Beta		
1 (Constant)	17.188	5.851		2.938	.007
Creative Thinking Skills	-.013	118	-.012	-.110	.913
Critical Thinking Skills	.757	.094	.871	8.070	.000

a. Dependent Variable: Learning Outcomes.

The total contribution of critical thinking skills and creative thinking skills to cognitive learning outcomes is shown in Table 12.

Table 12. The Summary of the Regression Analysis Results of Critical Thinking Skills, Creative and Learning Outcomes

R	R Square	Adjusted R Square	Std. Error of the Estimate
.866[a]	.750	.732	6.51303

a. Predictors: (Constant), Critical Thinking Skills, Creative Thinking Skills

The results of the multiple regression analysis show that the regression coefficient value (R) is 0.866, and the determination coefficient is (R^2) 0.750. This means that the contribution of critical thinking skill and creative thinking skill to cognitive learning outcomes is 75.00%, while the remaining as much as other variables influence 25.00%.

Almost all levels of education have the same problems regarding critical thinking skills, ranging from high school students, college students, and prospective teachers. Research conducted by [28] shows that prospective teachers have a low level of critical thinking disposition in general. Based on the World Economic Forum, to face of the industrial revolution 4.0, the capabilities needed include the ability to solve problems, critical thinking, creativity, management, coordinating with others, emotional intelligence, evaluating and making decisions, service orientation, negotiating, and cognitive flexibility [15]. This capability is needed to adapt to the latest technological advancements. One effort in the field of education that can be done to produce high-quality human resources is by making a culture of critical thinking as a habit for students in the learning process [26]. Therefore, an education expert John Dewey, from the earlier expects that students are taught critical thinking skills [29].

Creative and critical thinking skills are considered essential for students [24]. Crane stated the importance of these two skills when he wrote: "When reason fails, imagination saves you! When intuition fails, the idea keeps you! "The results of meaningful learning will tend to be useful, both in cognitive, practical, and psychomotor aspects [30]. Meanwhile, creative thinking is different from critical thinking that is convergent. Creative thinking tries to create something new, while critical thinking tries to assess value or validity in something that exists. Creative thinking is done by violating accepted principles, while critical thinking is done by applying accepted principles. Creative and critical thinking may be different sides of the same coin as they are not identical. The explanation reveals the importance of critical thinking skills and creative thinking skills to be aligned in learning.

Learning activities that can empower critical and creative thinking skills are innovative learning. One of which is inquiry learning strategies, because it can create meaningful and effective knowledge of critical thinking skills and creative thinking skills [31]. Lecturers can consider inquiry strategies as alternative learning, especially for new students at the university, to empower critical thinking skills and creative thinking skills. It is based on research results that inquiry strategies have a significant contribution to cognitive learning outcomes. Another type of innovative learning that could be an alternative is the Multiple Representation Based Learning (MRL) model [32].

MRL is a learning model of Multiple Representation Based Learning. Based on the results of the study, MRL is the most effective model for improving students' understanding of concepts compared to Problem Based Learning (PBL) and Direct Learning (DL) [32]. Several studies on Multi Representation Based Learning show that MRL has been able to improve students' higher-order thinking skills [33-35]. The MRL model is an effective learning model to optimize students' imagination abilities that can increase the students' ability to think and reason in solving problems [35]. Sunyono suggested applying MRL to material involving macro, submicron, and symbolic phenomena because students can learn with a variety of very important representations, and can stimulate students' interest in seeking information [32]. Those models can encourage teachers to develop critical thinking skills, creative thinking skills, and learning outcomes of students.

4. Conclusion

Correlation analysis results show: (1) there is a significant correlation between critical thinking skills with creative thinking skills of students. The amount of contribution by given critical thinking skills to creative thinking skills was 20.6%, with the equation of the regression line $Y = 0.360\ X + 29.302$. (2) There is a significant correlation between critical thinking skills and learning outcomes. The amount of contribution given by critical thinking skills to learning outcomes was 75.0% with a regression line equation that is $Y = 0.753\ X + 16.807$. (3) There is a correlation between creative thinking skills and learning outcomes. The amount of the contribution given by creative thinking skills to learning outcomes is 14.7% with a regression line equation that is $Y = 0.421\ X + 37.224$. (4) There is a correlation between critical and creative thinking skills in learning outcomes. The total amount of contribution given by critical thinking skills and creative thinking skills to learning outcomes is 75.0% with the regression line equation that is $Y = 0.757\ X + 17.188$ and $Y = 0.013 + 17{,}188$. Overall, it can be concluded that there is a significant correlation between critical thinking skills, creative thinking skills with student cognitive learning outcomes. Thus, it is necessary to empower thinking skills in learning to further enhance critical and creative thinking skills, as well as learning outcomes by using innovative learning. This recommendation is used to improve the quality of learning.

References

[1] Thompson C 2011 Critical Thinking Across the Curriculum: Process over Output *International Journal of Humanities and Social Science* **1** (9) 1-7

[2] Prihartiningsih, Zubaidah S, and Kusairi S 2016 The critical thinking ability of junior high school students on the classification of living things *Prosiding Seminar Nasional Pendidikan IPA Pascasarjana Universitas Negeri Malang*

[3] Kurniawati Z L, Zubaidah S, and Mahanal S 2015 The critical thinking skills of Batu City State High School students in biology *The 2nd National Seminar on Biology/Natural Sciences and Learning*

[4] Falah C M N, Sistiana W, Suhendar 2018 Increasing students' critical thinking skills through ethnics-based search, solve, create, and share (SSCS) learning models *Didaktika Biologi* **2** (1)

[5] Jafar J 2018 The effect of inquiry learning models on biology subjects on the critical thinking skills of students of class XI Science of Public Senior High School 1 Alla Enrekang *Prosiding Seminar Nasional Biologi dan Pembelajarannya*
[6] Susilowati, Sajidan, Ramli M 2017 Analysis of Critical Thinking Skills of State Aliyah Madrasah Students in Magetan Regency *National Seminar on Science Education Universitas Sebelas Maret*
[7] Nurfathurrahmah 2018 The application of the problem-based learning model based on contextual to increase students' critical thinking skills *Oryza Jurnal Pendidikan Biologi* **7** (1) 21-28
[8] Amin A M, Corebima A D, Zubaidah S, Mahanal S 2017 The critical thinking skills profile of preservice biology teachers in animal physiology *3rd International Conference on Education and Training* 179-183
[9] Wulandari T S H, Amin M, Zubaidah S, and Al Mudhar M H 2015 Implementation of critical thinking skills of biology lecturers at universities *The 2nd National Seminar on Biology/ Natural Sciences and Learning*
[10] Fitriani H, Asy'ari M, Zubaidah S, and Mahanal S 2019 Exploring the Prospective Teachers' Critical Thinking and Critical Analysis Skills *Jurnal Pendidikan IPA Indonesia* **8** (3) 379-390
[11] Fitriani H, Asy'ari M, Zubaidah S, and Mahanal S 2018 Critical Thinking Disposition of Prospective Science Teachers at IKIP Mataram, Indonesia *Journal of Physics: Conference Series* **1108** (1) 012091
[12] Baker M and Rudd R 2001 Correlations between critical and creative thinking *Journal of Southern Agricultural Education Research* **51** (1) 173-188
[13] Zubaidah S 2018 The role of education (biology) in the industrial revolution 4.0 and supporting SDGs *Seminar Nasional UMM*
[14] Dewi NR and Masrukan 2018 Improving Creative Thinking Ability of the Master's Program Students *Prisma, Prosiding Seminar Nasional Matematika* **1** 539-546.
[15] Sari D K, Permanasari A, and Supriyanti F M T 2017 Profile of students' creative thinking skills on quantitative project-based protein testing using local materials *Jurnal Pendidikan IPA Indonesia* **6** (1) 71-75.
[16] Pangestika I W, Ramli M, Nurmiyati, and Sapartiwi, 2017 Biology student learning outcomes in grade XI Science through the application of Socrates Dialogue *Proceeding Biology Education Conference* **14** (1)
[17] Murti W 2019 Improving learning outcomes for biology students through the use of flashcard media in plant ecology courses *Jurnal Al-Ahya* **1** (1) 135-143
[18] Greenstein L 2012 Assessing 21st Century Skills *A guide to evaluating mastery and authentic learning* (USA: Corwin A Sage Company)
[19] Krathwohl D R 2002 A Revision of Bloom's Taxonomy: An Overview *Theory Into Practice.* **41** (4) 212-218
[20] Siburian J, Corebima A D, Ibrohim, and Saptasari M 2019 The Correlation Between Critical and Creative Thinking Skills on Cognitive Learning Results. *Eurasian Journal of Educational Research* **81** 99-114
[21] Mahanal S, Tendrita M, Ramadhan F, Ismirawati N, and Zubaidah S 2017 The analysis of students' critical thinking skills on biology students *Anatolian Journal of Instruction* **10** (1) 21-39
[22] Amin A M, Corebima A D, Zubaidah S, and Mahanal S 2016 Analysis of concepts mastery and learning methods of prospective biology teachers learning in Makassar *Prosiding Seminar Nasional ke-3 Biologi, IPA, dan Pembelajarannya*
[23] Setiawati W, Asmira O, Ariyana Y, Bestary R, and Pudjiastuti A 2018 *The assessment book is oriented towards higher order thinking skills* (Jakarta: Kemendikbud)
[24] Caroselli M 2009 50 Activities for developing critical thinking skills *Complimentary Resources from HRD Press*

[25] Ariyana Y, Pudjiastuti A, Bestary R, and Zamroni 2018 *The learning handbook is oriented towards higher-order thinking skills* (Jakarta: Kemendikbud)

[26] Saragih S Z, Sari N F, and Ritonga N 2018 The effect of the learning model on the critical thinking ability of students' environmental knowledge at STKIP Batu Harbor. *Jurnal Pelita Pendidikan* **6** (3) 167–173

[27] Suratno, Komaria N, Yushardi, Dafik, and Wicaksono I 2019 The Effect of Using Synectics Model on Creative Thinking and Metacognition Skills of Junior High School Students *International Journal of Instruction* **12** (3) 133-150

[28] Temel S 2014 The effect of problem-based learning an preservice teachers critical thinking dispositions and perceptions of problem-solving ability *South African Journal of Education* **34** (1) 1-19

[29] Johnson E B 2002 *Contextual teaching and learning* (California: Corwin Press Inc.)

[30] Crane L D 1983 Unlocking the brain's two powerful learning systems *Human Intelligence Newsletter* **4** (4) 7

[31] Corebima A D 2006 *Biology learning which empowers students' thinking skills* Paper presented at training in metacognitive strategies in biology learning for SMA biology teachers in Palangkaraya 23 Agustus

[32] Sunyono and Meristin A 2018 The effect of multiple representation-based Learning (MRL) to increase students' understanding of chemical bonding concepts *Jurnal Pendidikan IPA Indonesia* **7** (4) 399-406

[33] Davidowitz B, Chittleborough G, and Murray E 2010 Student-Generated Submicro Diagrams: A Useful Tool for Teaching and Learning Chemical Equations and Stoichiometry. *Chemistry Education Research and Practice* **11**(3) 154-164

[34] Jaber L Z and BouJaoude S 2012 A Macro–Micro–Symbolic Teaching to Promote Relational Understanding of Chemical Reactions *International Journal of Science Education* **34** (7) 973-998

[35] Yuanita L and Ibrahim M 2015 Supporting Students in Learning with Multiple Representation to Improve Student Mental Models on Atomic Structure Concepts *Science Education International* **26** (2)

Effectiveness of Inquiry-Creative-Process Learning Model to Promote Critical Thinking Ability of Prospective Physics Teachers

Wahyudi[1], N N S P Verawati[1*], S Ayub[1], and S Prayogi[2]
[1] Universitas Mataram, Jl. Majapahit No.62 Mataram 83115, Indonesia.
[2] Institut Keguruan dan Ilmu Pendidikan (IKIP) Mataram, Jl. Pemuda No. 59A, Mataram 83126, Indonesia.

e-mail: *veyra@unram.ac.id

Abstract. Critical thinking has become one of the main competencies of learning at the level of higher education in Indonesia, and teaching interventions that lead to the achieving of these competencies are important to do. One of the learning models developed to promote critical thinking ability is the Inquiry Creative Process (ICP), a learning model. This research aims to describe the effectiveness of the ICP learning model to promote the critical thinking ability of prospective physics teachers. The implementation of the ICP learning model involved prospective physics teachers at four higher education institutions in the province of West Nusa Tenggara - Indonesia, that are the Mataram University (UNRAM), Teacher Training and Education Institut of Mataram (IKIP Mataram), Mataram State Islamic University (UIN Mataram), and the University of Muhammadiyah Mataram (UM Mataram). This study is experimental research, where is for each group performed pretest, implementation of the ICP learning model, and then posttest. The critical thinking ability of prospective physics teachers was measured using a critical thinking ability test instrument. The data of critical thinking ability were analyzed descriptively and statistically, where the homogeneity, normality, t-test, and ANOVA test were conducted. The results showed that the implementation of the Inquiry Creative Process learning model was effective to promote the critical thinking ability of prospective physics teachers. Descriptions of further research results are described in this article.

1. Introduction

Students after graduation must have the ability to think critically, and this has become the general hope of educators, parents, and the general society. Critical thinking is considered important for young people who want to continue further education, compete to get a job, or simply become responsible citizens. For critical thinking to be properly cultivated in schools, we must ensure that teachers have the critical understanding and critical thinking [1], and critical thinking teaching interventions are very appropriate for prospective teachers when they are still educating at the higher education level [2].

Teaching critical thinking to learners especially prospective teachers has attracted attention in recent decades given the role of future teachers as agents of change for the education system in terms of developing learner critical thinking [3]. In higher education institutions that produce prospective teachers, lecturers must teach high-level thinking skills and education before becoming a teacher is the right time for intervention activities that can promote critical thinking [4]. However, teaching to think remains confusing for many teachers [5]. This is due to the lack of clarity of the various methods available to teach critical thinking skills [5, 6]. At the higher education level, Bissell and Lemons [7] show that the average learner does not think critically. Thompson [8] argues that teaching critical thinking requires a holistic approach and must involve a set of appropriate learning models. Therefore, it is necessary to develop a set of specific learning models to improve learner critical thinking skills.

One of the learning models that aimed at how learners to think is the inquiry instruction model [9] and has been developed for specific purposes, for example in this study to improve the critical thinking ability of prospective physics teachers. The inquiry model is integrated with the process of scientific creativity into it, according to Philley [10] critical thinking is a multidimensional cognitive construction, as a result of the creative process. The integration of this study constructed in the learning model is called the Inquiry Creative Process (ICP) learning model. The ICP learning model is a model developed by integrating the attribution of creative processes (scientific creativity) in each syntax of inquiry models. Scientific creativity is creativity in learning science which is attributed to the emphasis on the ability to find the problems (problem finding), creating hypotheses, creatively experiment designing, science creatively problem solving, and creatively product design [11,12,13,14]. The assignment of creativity can expand the reach of creative activities, so the learner can apply, produce, find, compare, connect, imagine, and design creative ideas [15].

Attributions in scientific creativity are then integrated into scientific inquiry activities. The inquiry-based lesson according to Arends [9] consists according to activities; problem identification, formulating hypotheses, planning experiments to test hypotheses, formulating explanations, and reflecting. Inquiry processes need to be integrated and attributed to scientific creativity for the purpose to improve learner critical thinking, as explained by Adams [16] that the creative process or scientific creativity has the potential to exercise critical thinking ability. This integration is also to sharpen the potential to facilitate thinking through inquiry activities.

The ICP model promotes the critical thinking ability of prospective physics teachers. Ennis [17] and Hassard [18] define critical thinking as reasonable and reflective thinking that is focused on deciding what to believe or do. Critical thinking is often called independent thinking, reflective thinking, or evaluative thinking [19]. Critical thinking ability has been seen as a cognitive process, some previous researchers [2, 20, 21, 22] consistently use aspects of analysis, inference, evaluation, and decision making as the main indicators of critical thinking ability. This study aims to implement the ICP learning model and then evaluate its effectiveness in promoting the critical thinking ability of prospective physics teachers.

2. Methods

This study is a type of experimental research that aims to describe the effectiveness of the ICP learning model to promote the critical thinking ability of prospective physics teachers. The implementation of the ICP learning model involved prospective physics teachers (PPT) at four higher education institutions in the province of West Nusa Tenggara-Indonesia, which are the 20 PPT in UNRAM, 9 PPT in IKIP Mataram, 16 PPT in UIN Mataram, and 16 PPT in UM Mataram. The number of PPT involved was 61 people. For each group performed pretest, implementation of the ICP learning model, and then posttest. CT ability of PPT measured using a critical thinking ability test instrument, adapted from the Ennis-Weir Critical Thinking Essay Test with the indicators measured are the ability to analyze, inference, evaluate, and decision making [20]. The data of CT ability were analyzed descriptively and statistically, where the homogeneity, normality, t-test, and ANOVA were conducted. The effectiveness criteria of the ICP learning model if there is an improvement in critical thinking ability after the implementation of the model, and score of CT ability at least "critically" criteria in the posttest. Moreover, the effectiveness criteria according to the hypothesis tested in this study are: a) there is an increase in CT ability of PPT between the pretest score and the posttest score after implementation the ICP learning model in all group, and b) there is not significantly difference in increasing CT ability between classes after the implementation of the ICP learning model.

3. Results and Discussion

The result of the CT ability test of prospective physics teachers who are processed descriptively is shown in Table 1. The result of CT ability involving in 4 higher educational institutions (UNRAM, IKIP, UM, and UIN) shows average CT ability score on pretest of -2.13, with criteria of "not critically" (not critically, if: $X \leq -1.60$), and the posttest score of 16.34 with criteria of "critically" (critically, if: $11.20 < X \leq 17.60$), with n-gain of 0.71 with "high" criteria. From these results, it can be declared that the ICP learning model is effective in promoting the critical thinking ability of prospective physics teachers.

Table 1. The data of CT ability of prospective physics teachers in 4 group tested

Group	*Mean score & criteria*					*N-gain*	*Criteria*
	N	*Pretest*	*Criteria*	*Posttest*	*Criteria*		
UNRAM	20	-2.40	not critically	16.60	critically	0.72	high
UM	9	-1.89	not critically	16.22	critically	0.70	high
UIN	16	-1.38	not critically	15.81	critically	0.67	medium
IKIP	16	-2.88	not critically	16.75	critically	0.76	high
Average		*-2.13*	*not critically*	*16.34*	*critically*	*0.71*	*high*

Statistical analysis was performed to evaluate the effectiveness of the ICP learning model with predetermined hypothesis testing criteria. Homogeneity tests, normality tests, t-tests, and ANOVA tests were conducted. The variance homogeneity test using Levene's test, and the normality test using the Kolmogorov-Smirnov normality test in each of the four groups showed that the data variance was homogeneous and normally distributed with a significance value of all test groups greater than 0.05. The results of the homogeneity test of variance dan normality test provided in Table 2 and Table 3.

Table 2. The results of the homogeneity test of variance (Levene's test)

Data	Group	Levene test statistic	dF	Sig.
Pretest-Posttest	UNRAM	0.209	38	0.650
	UM	0.381	16	0.546
	UIN	0.658	30	0.424
	IKIP	0.668	30	0.287

Table 3. The results of the normality test (Kolmogorov-Smirnov normality test)

Data	Group	Kolmogorov- Smirnov test	N	Sig.
Pretest-Posttest	UNRAM	0.098	20	0.200
	UM	0.170	9	0.200
	UIN	0.155	16	0.200
	IKIP	0.111	16	0.200

The results of testing the critical thinking ability of each group using the t-test show that the significance value of the test (0.000) for all groups smaller than the alpha test (0.05), thus the hypothesis H_1 is accepted, its meaning that there is an increase in critical thinking ability of prospective physics teachers between the pretest score and the posttest score after implementation the ICP learning model in all of these groups. The result of the t-test is provided in Table 4.

Table 4. The result of the t-test (independent samples test)

	Group	*t-test for Equality of Means*		
		t	*df*	*Sig.*
Pretest-Posttest (Equal var. assum.)	UNRAM	-31.996	19	0.000
	UM	-27.651	8	0.000
	UIN	-22.965	15	0.000
	IKIP	-33.358	15	0.000

Analysis of differences in the improvement of critical thinking ability of all classes was tested using ANOVA (F-test). The ANOVA test results show that the significance value (0.473) is greater than the alpha test (0.05), thus that H_0 is accepted (there is no significant difference in the increasing critical thinking ability between all groups after the implementation of the ICP learning model). This means that the ICP learning model has the same effect on improving the critical thinking ability of prospective physics teachers in all groups tested. The ANOVA test results are provided in Table 5.

Tabel 5. The ANOVA test result

	Sum of Squares	*df*	*Mean Square*	*F*	*Sig.*
Between Groups	9.764	7	1.395	0.955	0.473
Within Groups	77.384	53	1.460		
Total	87.148	60			

The results of the statistical analysis show that the ICP learning model has been consistently in promoting the critical thinking ability of prospective physics teachers. The implementation of the ICP learning model to the prospective teachers of physics is in line with the demand that a physics learning has to master CT ability into it to correlate and interrelate between two or more theories and concepts in learning physics. The use of the ICP learning model increased the PPT ability in developing their CT. The result of this current study is parallel with the previous relevant study which found that the intervention of scientific creativity to develop CT ability is significantly effective [23].

The ICP learning model has strongly built critical thinking ability because it is by following the emphasis of the principles of constructivism learning. The constructivist perspective views teaching as not transferring knowledge from a teacher to learner, but an activity that allows the learner to develop their knowledge. Constructivist principles are based on ideas where learning develops knowledge actively, rather than accepting it passively in packages, from teachers (lecturers) or outside sources. The processes of creativity in inquiry activities have underpinned constructivist principles in learning. In addition to emphasizing the aspect of critical thinking, several advantages of the ICP learning model when implemented, which can encourage the curiosity of learners, create collaborative learning, encourage a sense of responsibility, support a positive classroom, and develop skills in learning itself.

The aspects of critical thinking that are promoted through the implementation of the ICP learning model in this study include the ability to analyze, inference, evaluate, and decision making. In the aspect of analysis, prospective physics teachers can identify the actual relationship between statements, questions, concepts, descriptions, or other. In the aspect of inference, prospective physics teachers can identify the elements needed to draw reasonable conclusions, to form guesses and hypotheses, to consider relevant information. In the aspect of evaluation, prospective physics teachers can assess the credibility of statements, representations, descriptions, perceptions, experiences, situations, judgments, beliefs, or opinions. The final aspect of critical thinking is being able to decision making, that is the process of choosing choices or actions between a set of alternatives on the basis of criteria or strategy. Achieving these aspects are caused by learning that emphasizes the exploration of scientific creativity through inquiry activities. Strengthening the scientific inquiry aspects of reasoning has helped prospective physics teachers in their tendency to think critically. According to Bailin [24], the objectives in inquiry activities focus a lot of critical thinking, for example; identifying assumptions, using logical thinking, analyzing direct events and phenomena, critical analysis of secondary sources, analyzing arguments by reviewing current scientific understanding, considering evidence, and examining the logic.

The results of this study, in general, have shown that ICP learning models have been effective in promoting the critical thinking ability of prospective physics teachers. The ICP learning model uses a systematic and well-organized learning activity through some experimental activities which involve scientific creativity and scientific process skills in it. Scientific process skill has a great effect on learning because it helps the learner to improve higher mental skills, such as critical thinking, decision making, and problem-solving [25, 26]. It can be an instrument that improves critical thinking ability. For the sake of a broader teaching and learning results, it is important to teach some steps to reach and conquer the knowledge itself, in which it is definitely needed whey the learners conducting a scientific experiment during the learning process.

4. Conclusion

The ICP learning model has proven its effectiveness in promoting the critical thinking ability of prospective physics teachers. Moreover, this model has also consistently improved the critical thinking ability when implemented in different subjects in several higher education institutions administering the educational programs that produce prospective physics teachers. The findings in this study are also important reasons for using the ICP learning model as an alternative learning model that can be used by educators in general to promote critical thinking ability.

Acknowledgment

This study is a part of a decentralized research grant that organized by the Ministry of Research, Technology, and Higher Education of the Republic of Indonesia, research contract number of 7/E/KPT/2019.

References

[1] Mok F KT & Yuen T W W 2016 A critical evaluation of the understanding of critical thinking by school teachers: The case of Hong Kong *Citiz. Soc. Econ. Educ.* **15** (1) 28-44

[2] Prayogi S, Yuanita L & Wasis 2018 Critical inquiry-based learning: Model of learning to promote critical thinking ability of pre-service teachers *J. Phys. Conf. Ser.* **947** 1-6

[3] Sendag S, Erol O, Sezgin S & Dulkadir N 2015 Pre-service teachers' critical thinking dispositions and web 2.0 competencies *Contemp. Educ. Tech.* **6** (3) 172-187

[4] Warburton E C 2008 Changes in dance teachers' beliefs about critical-thinking activities *J. Educ. Humanit. Dev.* **2** (1) 1-16

[5] Bensley D A & Murtagh M P 2012 Guidelines for a scientific approach to critical thinking assessment *Teac. Psych.* **39** 5-16

[6] Abrami P C, Bernard R M, Borokhovski E, Wade A, Surkes M A, Tamim R & Zhang D 2008 Instructional interventions affecting critical thinking skills and dispositions: A stage 1 meta-analysis *Rev. of Educ. Res.* **78** (4) 1102-1134

[7] Bissell A N & Lemons P P 2006 A new method for assessing critical thinking in the classroom *BioScience* **56** (1) 66-72

[8] Thompson C 2011 Critical thinking across the curriculum: Process over output *Int. J. Humanit. Soc. Sci.* **1** (9) 1-7

[9] Arends R 2012 *Learning to teach (9th Ed)* (New York: McGraw-Hill)

[10] Philley J 2005 Critical thinking concepts *Prof. Safety* **50** 26-32

[11] Türkmen H & Sertkahya M 2015 Creative thinking skills analysis of vocational high school students *J. Educ. Instruc. Stud. in the World* **5** (1) 74-84

[12] Ayas B & Sak U 2014 Objective measure of Scientific creativity: Psychometric validity of the creative Scientific ability test *Elsevier: Thinking Skills and Creativity* **13** 195-205

[13] Hu W & Adey P 2010 A scientific creativity test for secondary school students *Int. J. Sci. Educ.* **24** (4) 389-403

[14] Aktamis H & Ergin O 2008 The effect of scientific process skill education on student's scientific creativity, science attitudes, and academic achievements *Asia-Pacific Forum on Science Learning and Teaching* **9** 1-21
[15] Rotteram K 2014 *Teaching, learning and creativity (TLC) Model for science* SSR **95** 79-84
[16] Adams K 2006 *The sources of innovation and creativity* (USA: National Center on Education and Economy)
[17] Ennis R H 2011 *The nature of critical thinking: An outline of critical thinking dispositions and abilities* Presentation at the Sixth International Conference on Thinking of MIT, Cambridge, MA.
[18] Hassard J 2005 *The art teaching science* (New York: Oxford University Press)
[19] Reid J C 2006 *Mengajar Anak Berpikir Kreatif, Mandiri, Mental dan Analitis* (Jakarta: Prestasi Pustakaraya)
[20] Prayogi S, Yuanita L & Wasis 2018 Critical inquiry-based learning: A model of learning to promote critical thinking among prospective teachers of physics *J. Turkish Sci. Educ.* **15** (1) 43-56
[21] Wahyudi, Verawati N N S P, Ayub S & Prayogi S 2019 The effect of scientific creativity in inquiry learning to promote the critical thinking ability of prospective teachers *Int. J. Emer. Tech. Lear.* **14** (14) 122-131
[22] Prayogi S, Muhali, Yuliyanti S, As'ary M, Azmi I & Verawati N N S P 2019 The effect of presenting anomalous data on improving student's critical thinking ability *Int. J. Emer. Tech. Lear.* **14** (6) 133-137
[23] Wahyudi, Verawati N N S P, Ayub S & Prayogi S 2018 Development of inquiry-creative-process learning model to promote the critical thinking ability of prospective physics teachers *J. Phys. Conf. Ser.* **1108** 1-6
[24] Bailin S 2002 Critical thinking and science education *Sci. Educ.* **11** 361-375
[25] Lee A T, Hairston R V, Thames R., Lawrence T & Herron S S 2002 Using a computer simulation to teach science process skills to college biology and elementary education majors *Comp. Simul. Bioscene* **28** (4) 35-42
[26] Koray Ö, Köksal M S, Özdemir M & Presley A I 2007 The effect of creative and critical thinking based laboratory applications on academic achievement and science process skills *Element. Educ. Online* **6** (3) 377-389

The Effectiveness of the Methods and Approaches of ESD for 2030 Sustainable Development Goals; From Analysis of the Questionnaire Survey to the School Teachers

T Icihnose

Research Institution of Teacher Training and Development, Miyagi University of Education 149 Aobaku Aramaki Azaaoba, Sendai, Japan
E-mail: ichinose@staff.miyakyo-u.ac.jp

Abstract. ESD is a key element of quality education. Its cross-cutting competencies in cognitive, socio-emotional and behavioural dimensions of learning bear relevance to all areas of education. (UNESCO Executive Board 20 February 2019). This study, the result of four years of research in 279 primary and secondary schools in Japan, reveals that in integrated study periods, and with the engagement of local human resources, an increase of activities across different classrooms and grades are a clear result of enhanced ESD and students' collaboration, cooperativeness, cooperative attitude, and ability to communicate are interconnected. From a pluralistic perspective of ESD, teaching students to confront tensions and conflicts and tackling issues with no single answer seems to be difficult. Since the ESD curriculum has been constructed through whole-school educational activities in Japan by linking it with other subjects and areas, utilizing integrated study periods, it is important to examine approaches to understanding the effectiveness of a whole-school approach of ESD.

1. Introduction

"All ESD activities contribute to the achievement of the SDGs. it continues to have particular relevance for the global education agenda enshrined in SDG 4, in which ESD has a distinct place. ESD is a key element of quality education. Its cross-cutting competencies in cognitive, socio-emotional and behavioural dimensions of learning bear relevance to all areas of education" [1]. After the end of DESD, evidence of the improvement of schools addressed ESD and its contribution to the quality of education will be more strongly emphasised. However, there is very little empirical evidence for the extent to which ESD has been implemented at the school level, which could provide a cross-regional discussion.

Sustainable development, the principal concept of ESD, was first widely defined in the Brundtland report entitled 'Our common future', as follows: 'Sustainable development is development that meets the needs of the present, without compromising the ability of future generations to meet their own needs' [2]. This definition implies that there is a necessity of including all three dimensions (environmental, social, economic) of education (McKeown and Hopkins [3]) and focusing on their interrelationship, as well as their interactions over time and space.

In the outline of Education for Sustainable Development as a teaching tradition, Stables and Scott [4] and Öhman [5] identify three essential aspects of holism: connecting the environmental, social, and economic dimensions of Sustainable Development (SD) issues; integrating their past, present, and future implications; and focusing on their local, regional, and global nature.

In addition to holism, the second essential feature of ESD deals with the processes of teaching and learning. ESD focusses on the development of skills and action competence for sustainability; this pedagogy has been labelled pluralism. Thus, with holism and pluralism in mind, this article researches the contributing schools' efforts concerning their implementation of ESD from the viewpoint of schoolteachers. Because there is very little empirical evidence for international comparison, this research digs deeper into the effects of a holistic ESD approach to content and a pluralistic ESD approach to teaching in the context of Japanese formal education. In addition, this research aims to clarify which barriers exist when ESD is addressed in schools.

2. Concepts of Holism and Pluralism

Prior to the promotion of Education for Sustainable Development (ESD), Environmental Education (EE) was often associated with a quest for a holistic worldview. There is a growing consensus that the holistic perspective of sustainable development (SD) should be applied in ESD. When we are looking at ESD, it is necessary to include all three dimensions equally and to focus on the interrelationship between ecological, economic, and social dimensions in a holistic perspective of Sustainable Development [6]-[9]. The UNESCO document 'Framework for the UNDESD International Implementation Scheme' outlines fifteen sub-themes of three dimensions in connection to the education for Sustainable Development: "(1) *Socio-cultural perspectives*-human rights, peace and human security, gender equality, cultural diversity and intercultural understanding, health, HIV/AIDS, and governance; (2) *Environmental perspectives*-natural resources (water, energy, agriculture, biodiversity), climate change, rural development, sustainable urbanisation, and disaster prevention and mitigation; (3) *Economic perspectives*-poverty reduction, corporate responsibility and accountability, and market economy" [1].

Each of these sub-themes are covered in ESD classrooms. However, holism refers not only to the treatment of those topics, but also refers to time and space, because development (SD) issues integrate their past, present, and future implications and focus on their local, regional, and global nature as well [5]. The essential feature of ESD which deals with the processes of teaching and learning is called pluralism. Essentially, ESD aims to facilitate learning in such a way that learners understand the world based on their own observations and develop competences to take action for sustainability.

The pluralistic approach is characterised by a striving to acknowledge and engage different perspectives, views, and values, and by recognizing and accepting our different standpoints through deliberative conversations [11]. This transformative pedagogy calls for learner-centred teaching approaches such as critical thinking, participatory decision making, value-based learning, multi-method approaches, and social learning.

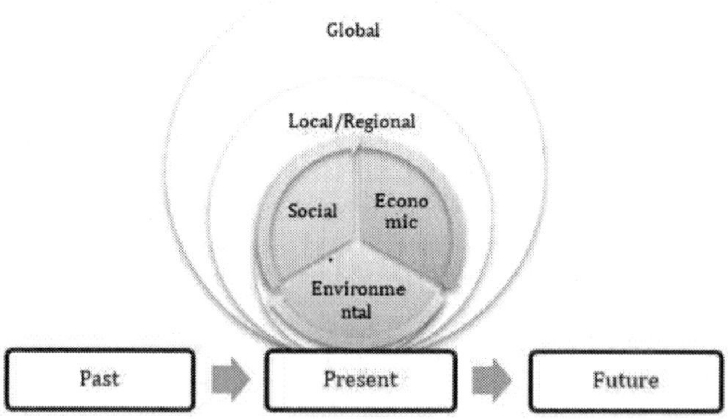

Figure 1. ESD's holism.

In addition, ESD is often viewed as an approach to teaching and learning that aims to build students' abilities to focus on behavioural change. Mogensen and Schnack [12] further emphasise that "a key role of ESD in an action competence approach is to develop students' ability, motivation, and desire to play an active role in finding democratic solutions to SD problems and issues. For this purpose, ESD builds on the strengths of joint, co-created solutions, independent and autonomous thinking, and learning to deal with the reality of change and uncertainty".

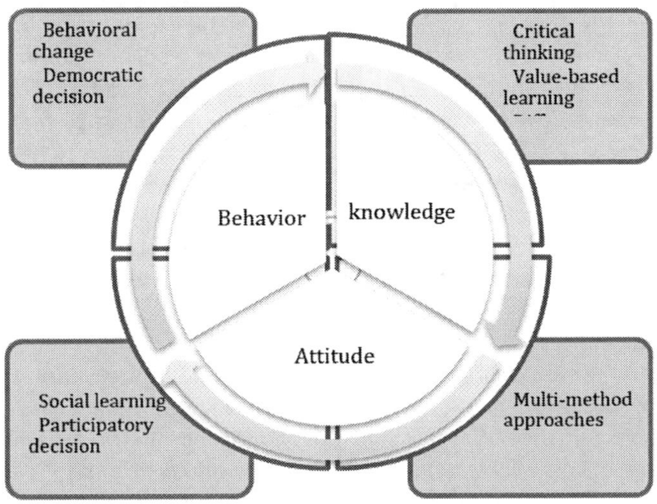

Figure 2. ESD's pluralism.

3. Students' and teachers' recognition of ESD

3.1. Students' conscious recognition of ESD

Internationally, several surveys have focused mainly on the effects of environmental education (EE). Many studies also show that EE mainly influences their students' environmental knowledge; it does not create a transformation of attitude and behaviour. Walshe [13] showed the impact and understanding of sustainable development of 12 to 13-year-old geography students in the UK. There was a wide range in the level of their understanding of the concept of sustainability. Pauw and Petegem [14], based on the research of 1,287 10–12-year-olds from 59 schools (38 eco-schools and 21 control schools), show that eco-schools mainly influence their students' environmental knowledge but do not influence environmental attitude and behaviour. The same authors [15] show that the effect of environmental values on environmental behaviour is different across diverse cultural contexts from a massive research study with a total of 1,833 ten- to thirteen-year-old children from Flanders, Guatemala, and Vietnam. Berglund, Gericke, and Chang [16] indicate that "there are significant differences in sustainable consciousness between students from Swedish schools that teach with an ESD approach compared to students from regular schools. Furthermore, a significant difference between the two groups of students was found in the underlying economic dimension of sustainable consciousness. No significant differences were found in the environmental and social dimensions of sustainable consciousness".

Pauw, Gericke, Olsson, and Berglund [17] gathered a large amount of data from 2,413 students from 51 schools across Sweden to study the effectiveness of ESD. Their results indicate that ESD can actually impact student outcomes in terms of their sustainable consciousness. The results of this study reveal the key role ESD plays in addressing SD, paving the way for a more sustainable future.

3.2. Teachers' understanding of ESD

Studies that have looked at teachers' conceptual understanding of ESD have shown that the ecological perspective is the most commonly recognised, while a relatively small number of teachers have a holistic understanding of ESD. Early in the ESD movement, Summers, Corney,and Childs [18] researched student teachers' conceptions of sustainable development in UK. They showed that a large number of student teachers recognised the centrality of environmental (87%), economic (69%), and social (49%) factors. Öhman [5] found disciplinary differences amongst upper secondary teachers' ways of conducting EE in Sweden. Social science teachers taught environmental issues to a higher extent than science teachers, while only 30% of language teachers taught environmental issues at all. In addition, many teachers did not teach according to an ESD tradition. Summers, Childs, and Corney [19] showed that schools in UK were not yet well developed as sites for the development of student teachers trained to teach ESD. For example, student teachers generally had a greater understanding of sustainable development than their mentors. For both students and mentors, there were significant gaps in the understanding of ESD compared with representations found in the literature. Corney [8] reported that for geography student teachers, there were three main areas of challenge: namely, challenges related to the complexity of subject matter, to teaching and learning approaches and strategies, and to differences between student teachers' beliefs about education for sustainable development and the context for their teaching.

In addition to the research from Sweden and the UK, Spiropoulou et al. [20] studied Greek in-service primary teachers' perceptions about environmental issues and attitudes toward SD. The analysis of data revealed that teachers held misunderstandings or misconceptions of the conceptual meaning of the terms "sustainability' and 'renewable source of energy'. Borg et al. [21] "referred to the different ways that teachers with different backgrounds recognised sustainable development. The greatest uncertainty for teachers is related to the economic dimension of ESD. Science and social science teachers are critical of incorporating economic growth into the concept of sustainable development, while language, vocational, and esthetical–practical teachers are not incorporating it at all". Borg, Gericke, Höglund,and Bergman [22] showed that "Swedish teachers were uncertain with respect to their understanding of all three dimensions". They ranked their personal understanding of SD high, higher than assessments reported in studies in other Western European countries such as the UK. However, most teachers still had an understanding of SD in line with an EE-tradition.

These previous studies focused on three essential aspects of holism: connecting the environmental, social, and economic dimensions of sustainable development issues. For the students' and teachers' recognition, the effect that this type of education has on knowledge, attitude, and behaviour is emphasised. This previous research mostly focused on classroom studies or secondary school teacher perspectives. The results indicate that the ecological dimensions of ESD were most commonly identified among students and teachers. One thing to note is that there may be differences between how secondary school teachers from different disciplines and subjects understand sustainable development.

4. ESD in Japanese Education

During the Decade of Education for Sustainable Development (2005–2014), which was strongly supported by the Japanese government, in 2008, the Japanese Ministry of Education launched 'The basic educational promotional plan of 2008'. This promotional plan positioned UNESCO-Associated Schools as centres for promoting ESD and gave support for the increase of UNESCO-Associated Schools in the global school network of UNESCO-leading ESD. This allowed UNESCO-Associated Schools in Japan to take part in efforts to develop the educational content and methods of ESD. After this initiative, the numbers of UNESCO-Associated Schools in Japan went from 24 in January of 2008 to 1,116 in May of 2019.

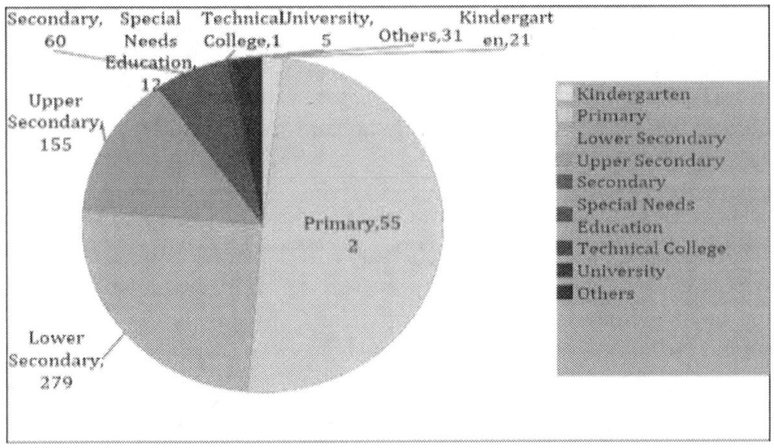

Figure 3. Breakdown of UNESCO associated schools in September 2019.

This Figure 3 shows that of the ESD schools in Japan, public school and compulsory schools are the majority, with 76 percent. Primary schools make up 52 percent of ESD schools. On the other hand, ESD is included as part of the description of all subjects in the curricula. The Revised National Curriculum Guidelines in 2008 contain a significant amount of content connected with ESD. ESD content is most noticeable in geography, civics, natural science, home economics, and physics.

However, most of the ESD in Japan is practiced during the integrated study periods rather than in core subjects. The integrated study periods were established when the Curriculum Guidelines, which are the national standards of the curricula in Japanese schools, were revised in 1998. "These study periods have provided space for teaching and learning approaches that encourage task-based inquiry learning, problem solving, and participatory learning, which are important components of pluralistic approaches of ESD" [1]. The curriculum of ESD has been constructed through whole-school educational activities by linking it with other subjects and areas (such as extracurricular activities); the so-called ESD calendar makes this whole school approach clear. Because of this situation, research on the Japanese viewpoint of ESD is needed; it will certainly require a different viewpoint from the European (e.g. Sweden, UK) concept of ESD, which is mainly focused on secondary education and individual teachers under different principles.

5. Research Questions

Despite a global commitment to ESD as a teaching approach, there is very little empirical evidence for international comparison. One thing that would be useful to compare is the extent to which ESD is implemented in the school and classrooms, and the effects ESD has on the school and students.

In Japan, ESD curriculum through whole school educational activities has been constructed mainly utilizing integrated study periods. With this in mind, this research digs deeper into the effects of ESD in the context of Japanese formal education. The research questions are therefore the effects of a holistic ESD approach to content and a pluralistic ESD approach to teaching in the context of Japanese formal education: What is the effect of ESD for the students and in the school from the viewpoint of schoolteachers?

6. Research Design

The purpose of this paper is to study the effect of ESD from the viewpoint of schoolteachers in Japan. In order to design an evaluative tool for ESD schools, I looked at Transfer 21 [23] , which proposed the formulation of concepts for quality development in German schools. The catalogue of criteria, on the basis of which self-assessments were created, can be divided into nine quality areas; learning culture, learning groups, competencies, school culture, opening of schools to the outside world, school management, school program, resources, and staff development. Another way to measure ESD in schools comes from an evaluation of sustainable schools in the UK known as the 'S3: Sustainable School Self-evaluation' [24]. The S3 allows for the evaluation of ESD schools' performance across four key areas of performance and improvement; Ethos: characteristics of your school, Curriculum: teaching provision and learning, Campus: leadership and management, Community: engaging local people and partner.

Bringing together the key points of Transfer 21 and S3: Sustainable School Self-evaluation, the research employs 8 criteria; Improvement of students' attitude and behaviour, Acquisition of ESD competencies, Recognition of the components for ESD, Improvement of teaching and learning method, Improvement of teaching and learning concepts, Improvement of the school organization, Engagement of the local partner, Teacher training. The following is the analysis of 5 parts of these criteria: Improvement of students' capacity and behaviour, acquisition of ESD competencies, recognition of the components of ESD, improvement of teaching and learning concepts, and improvement of the school organisation with the question of the barriers of teaching ESD for the teachers. In the questionnaire, teachers are asked to grade different aspects of their school practice on a five-point scale.

Questionnaires were distributed in the various prefectures of the north-eastern part of Japan on July in 2014,2015,2016, and 2017. Because there are different teachers each year, the questionnaire respondents were made up of a mixture of teachers who are responsible for the practices of ESD. The target schools in this research–the UNESCO-Associated Project Network (ASPnet) schools–consisted of 3 kindergartens, primary schools, lower secondary schools, and upper secondary schools. Preschools were omitted from the analysis because educational system is completely different from that of other grades.

Table 1. Research target school collection number

	2014 (pilot)	2015	2016	2017
Pre-school	6	6	3	6
Primary school	18	38	50	45
Lower secondary school	13	25	27	28
Upper secondary school	0	5	7	6
Total	31	74	88 (1 invalid)	85

7. Research Analysis

Through the three years of research, the same trend was shown for 3 years as was shown by the pilot research conducted in 2014. This research shows the effectiveness of the Japanese school enhancing ESD from the teachers' perspective.

Table 2. Effectiveness of the Education for Sustainable Development 2014-2017 (relatively high rate, highlighted in grey)

Category	Questionnaire	2014 (pilot)	2015	2016	2017
1 Improvement of students' capacity and behaviour	Improvement of students' reading and writing skills	3.29±0.59	3.26±0.56	3.18±0.59	3.32±0.73
	Improvement of students' calculating skills	3.00±0.45	3.09±0.48	3.08±0.61	3.06±0.72
	Improvement of students' thinking ability	3.87±0.62	3.85±0.63	3.76±0.59	3.90±0.62
	Improvement of students' interest of study	4.13±0.62	3.90±0.60	3.95±0.65	4.01±0.61
	Improvement of students' participation	4.00±0.58	3.93±0.61	3.95±0.67	3.97±0.62
	Improvement of students' leadership	3.65±0.71	3.59±0.65	3.37±0.71	3.72±0.77

Category	Questionnaire	2014 (pilot)	2015	2016	2017
	Improvement of students' collaboration	3.94±0.68	3.99±0.59	3.95±0.68	3.97±0.68
	Improvement of students' active participation in the ECO activities	3.45±0.62	3.63±0.74	3.68±0.71	3.77±0.72
2 Acquisition of ESD competencies	Critical thinking	3.35±0.49	3.32±0.63	3.28±0.65	3.35±0.75
	Ability to forecast the future and plan	3.74±0.68	3.54±0.63	3.64±0.65	3.62±0.72
	Multi-faceted and comprehensive thinking	3.80±0.54	3.65±0.59	3.92±0.60	3.86±0.67
	Ability to communicate	4.13±0.50	4.09±0.60	4.12±0.63	4.10±0.57
	Cooperative attitude	4.29±0.69	4.31±0.55	4.21±0.65	4.25±0.52
	Respect for interconnectedness	4.16±0.64	4.10±0.60	4.07±0.56	4.04±0.69
	Attitude of proactive participation	4.06±0.57	3.97±0.60	4.01±0.62	4.06±0.58
3.Recognition of the components of ESD	Diversity	4.16±0.92	4.04±0.70	4.25±0.54	4.26±0.66
	Interdependence	4.39±0.56	4.27±0.61	4.32±0.58	4.38±0.59
	Finiteness	3.78±0.72	3.66±0.82	3.80±0.74	3.74±0.80
	Fairness	3.97±0.71	3.80±0.61	3.90±0.64	3.96±0.64
	Cooperation	4.42±0.50	4.37±0.60	4.41±0.52	4.38±0.59
	Responsibility	4.23±0.67	4.16±0.59	4.19±0.61	4.17±0.68
4 Improvement of teaching and learning concepts	Adapting a viewpoint of sustainability for the future	N/A	3.99±0.66	4.17±0.60	4.14±0.69
	Utilizing multi-faceted and comprehensive methods	3.91±0.60	3.78±0.59	3.85±0.69	3.90±0.67
	Adapting a global and local perspective	4.06±0.57	4.00±0.77	4.15±0.67	4.15±0.74
	Understanding different types of values which cause conflict	3.48±0.67	3.45±0.74	3.57±0.77	3.36±0.84
	Discussing unanswerable topics on the issues of sustainability	N/A	3.51±0.70	3.73±0.74	3.50±0.75
	Beyond knowledge transfer, challenging students to change values and behaviour	4.06±0.51	3.84±0.61	3.92±0.60	3.86±0.68
5 Improvement of the school organisation	Active use of school buildings and school yard	3.48±0.72	3.51±0.78	3.65±0.76	4.14±0.69
	Respect for energy and water conservation in the school	3.35±0.61	3.53±0.74	3.55±0.74	3.90±0.67
	Making an ESD educational plan utilizing the whole school approach	3.90±0.75	3.82±0.85	3.63±0.78	4.15±0.74
	Settlements with school board, making changes that engage parents of the school	3.26±0.86	3.50±0.85	3.57±0.76	3.36±0.84
	Enhancing open school and collaboration with local organisations	3.54±0.86	3.73±0.81	3.96±0.77	3.50±0.75
	Engagement of local human resources and NPOs	4.26±0.73	4.01±0.86	4.06±0.72	3.86±0.68
	Increase of activities across classrooms and grades	4.03±0.70	3.99±0.78	3.93±0.77	4.14±0.69
	School activities affecting parents and local communities	3.61±0.61	3.40±0.67	3.45±0.72	3.90±0.67

Table 3. Effectiveness of Education for Sustainable Development by school level (Average of 2015-2017)

Category	Questionnaire	Primary	Lower Secondary	Upper Secondary
1 Improvement of students' capacity and behaviour	Improvement of students' reading and writing skills	3.32±0.66	3.31±0.58	3.22±0.81
	Improvement of students' calculating skills	3.11±0.65	3.10±0.54	2.89±0.83
	Improvement of students' thinking ability	3.83±0.65	3.93±0.59	3.89±0.58
	Improvement of students' interest of study	4.00±0.69	3.96±0.62	3.94±0.64
	Improvement of students' participation	3.99±0.57	3.91±0.65	3.94±0.73
	Improvement of students' leadership	3.53±0.76	3.83±0.68	3.78±0.65
	Improvement of students' collaboration	3.91±0.74	4.07±0.57	3.89±0.58
	Improvement of students' active participation in the ECO activities	3.77±0.77	3.61±0.64	3.50±0.62
2 Acquisition of ESD competencies	Critical thinking	3.25±0.68	3.34±0.67	3.72±0.75
	Ability to forecast the future and plan	3.50±0.68	3.70±0.68	3.78±0.65
	Multi-faceted and comprehensive thinking	3.73±0.71	3.87±0.56	4.00±0.59
	Ability to communicate	4.07±0.67	4.10±0.57	4.33±0.69
	Cooperative attitude	4.23±0.68	4.25±0.55	4.33±0.59
	Respect for interconnectedness	4.05±0.69	4.11±0.57	4.17±0.71
	Attitude of proactive participation	4.02±0.76	4.04±0.62	4.00±0.59
3. Recognition of the components of ESD	Diversity	4.10±0.69	4.29±0.59	4.50±0.51
	Interdependence	4.30±0.67	4.39±0.57	4.33±0.77
	Finiteness	3.78±0.83	3.69±0.73	3.78±0.73
	Fairness	3.98±0.69	3.90±0.62	3.61±0.61
	Cooperation	4.34±0.65	4.39±0.57	4.44±0.61
	Responsibility	4.15±0.69	4.24±0.62	4.06±0.73
4 Improvement of teaching and learning concepts	Adapting a viewpoint of sustainability for the future	4.10±0.70	4.11±0.59	4.22±0.73
	Utilizing multi-faceted and comprehensive methods	3.96±0.67	3.92±0.62	3.64±0.45
	Adapting a global and local perspective	3.94±0.81	3.85±0.82	4.22±0.81
	Understanding different types of values which cause conflict	3.45±0.79	3.35±0.83	3.61±0.85
	Discussing unanswerable topics on the issues of sustainability	3.52±0.72	3.55±0.79	3.83±0.92
	Beyond knowledge transfer, challenging students to change values and behaviour	3.91±0.67	3.85±0.65	4.00±0.69
5 Improvement of the school organisation	Active use of school buildings and school yard	3.66±0.83	3.61±0.78	3.33±0.84
	Respect for energy and water conservation in the school	3.65±0.79	3.39±0.73	3.39±0.70
	Making an ESD educational plan utilizing the whole school approach	3.95±0.81	3.59±0.80	3.39±0.98
	Settlements with school board, making changes that engage parents of the school	3.62±0.82	3.34±0.77	3.11±0.90
	Enhancing open school and collaboration with local organisations	3.88±0.79	3.73±0.89	3.89±0.90
	Engagement of local human resources and NPOs	4.11±0.81	4.10±0.83	3.83±0.86
	Increase of activities across classrooms and grades	3.95±0.78	4.03±0.89	3.61±0.78
	School activities affecting parents and local communities	3.54±0.77	3.34±0.74	3.11±0.76

7.1. Improvement of Students' Capacity and Behaviour

In this questionnaire, the first category asked about student capacity and behaviour. We were looking for information about the improvement of students' calculating, reading, and writing skills, and the question asked about the connection of ESD to traditional subject-based learning. The results show that ESD is not a concept which directly connects with the basic skills of traditional subject-based learning. Students' reading and writing skills are not clearly connected with ESD; on the other hand, teachers reported that students' thinking ability and motivation are improved by ESD.

An important aspect of ESD is the concept of a participatory and active citizenship that builds on the strengths of joint, co-created solutions. Category 1 was also concerned with student behaviour, participation, leadership, collaboration, and participation in eco-activities. In answering these questions, teachers reported that participation and collaboration attitudes of students are generally improved.

7.2 Acquisition of ESD Competencies

The learning process of ESD is intended to be transformative, mediated by a critical, open-ended, pluralistic, and democratic approach. In the questions of Category 2, these seven competencies of ESD are those that were identified by the National Institute for Educational Policy Research (NIER) in Japan [25].

Critical thinking is the core competency for striving to acknowledge and engage different perspectives, views, and values, and recognizing and accepting our different standpoints of SD. From the teacher's reports, the rate of critical thinking is low, especially for primary school. On the other hand, this data shows that students 'ability to communicate and cooperative attitude are quite good.

7.3 Recognition of the components of ESD

The questions in category 3 are also defined by the National Institute for Educational Policy Research [25] as the basic component of sustainable development. Recognition of the role of finite environmental resources and fairness to future generations [26] are essential components of ESD; however, concerns about the finiteness and fairness are not highly developed in students, in contrast to interdependence and cooperativeness, which are consistently highly ranked.

7.4 Improvement of Teaching and Learning Concepts

One of the most important subjects of ESD is the understanding of the complex nature of sustainable development. ESD was developed using different sources reflecting different perspectives, and thus includes consideration of social and political issues together with economic and environmental concerns (Holism). ESD in particular relates to differing educational ideologies and has implications for effective student learning. In addition, ESD is often viewed as an approach to teaching and learning that aims to build abilities to focus on behavioural change (Pluralism). These questions were created to ask about the specific ideology of ESD. The questions below are classified as either relating to Holism (HL) or to Pluralism (PL): Adapting a viewpoint of sustainability for the future (HL), Utilising multi-faceted and comprehensive methods (PL), Adapting global and local perspectives (HL), Containing different types of values which cause conflict (PL), Discussing unanswerable topics on sustainability issues (PL), Challenging value and behaviour changes beyond knowledge transfer (PL).

The data shows that the pluralism concepts of ESD, teaching students to confront tensions and conflicts and tackle issues with no single answer, seem to be difficult, especially for primary school students. Global and local perspectives were gained by the students in each type of school. In 2017, this question was asked divided in Two domains: local and global. Thus, ESD seems to be more focused on local rather than on global issues, especially for the primary and lower secondary schools.

Table 4. Adaptation of local and global perspective (Average of 2017).

	Primary	**Lower Secondary**	**Upper Secondary**
Adapting a local perspective。	4.18±0.75	4.25±0.64	3.50±0.84
Adapting a global perspective	3.45±0.76	3.79±0.85	4.50±0.84

7.5 Improvement of School Organisation

Whole-school approaches recognise that sustainability is relevant to all aspects of school life, including in formal curricula and hidden curricula, school leadership and management, and teacher development. Regarding school organisation, the contents that are considered to be improvements are the adaptation of a whole school approach, the opening of a school, and the engagement of local human resources.

The Figure shows that creating an ESD educational plan that utilises a whole school approach is easier for primary schools (3.95) than for upper secondary schools. (3.39); lower secondary schools are in the middle (3.59). This research shows that Engagement of local human resources and an increase of activities across different classrooms and grades are a clear result of enhanced ESD. The effect of the engagement of student parents is not clear. Legault and Pelletier [27] and Volk and Cheak [28] indicate that no significant effects were found in terms of ecological knowledge, competence, importance, motivational subtypes, and behaviours of student parents. We found the same results on the effect of environmental education as did.

7.6 Correlation test results

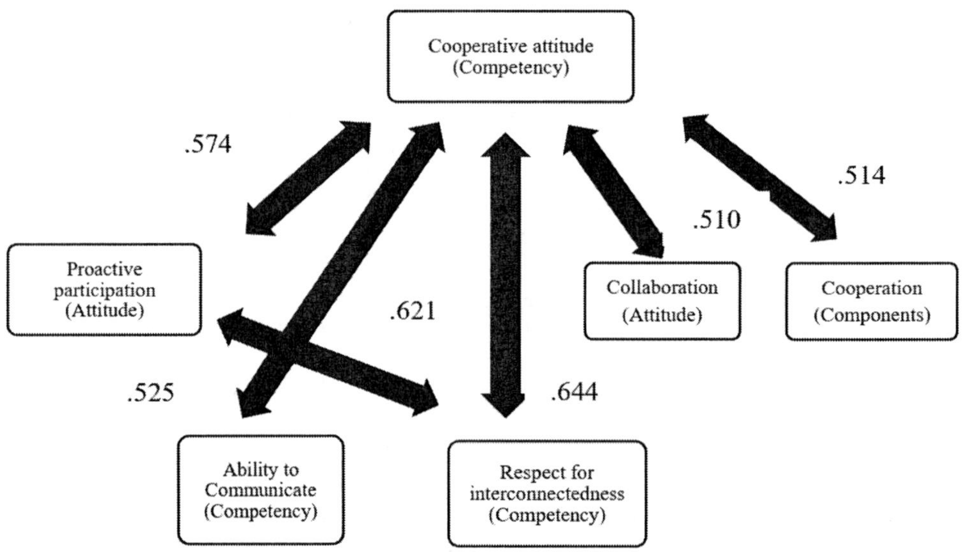

Figure 4. Correlation test for analyses of each of the 5 categories, which were above 0.500 from 2015 to 2017.

For further analysis, the collected data from 2015 to 2017 were downloaded into IBM SPSS Statistics version 20. The results presented in this article are based on Pearson's correlation test for analyses of each of the 5 categories.

Correlation describes the strength of a linear relationship between two variables. The intermediate correlation between Cooperative attitude (Competency) and Respect for interconnectedness (Competency) is (r=.644, p<.001), Ability to Communicate (Competency) is (r=.621, p<.001), Cooperation(Components) is r=.514(p<.001) and Collaboration (Attitude) is (r=.501, p<.001). Respect for interconnectedness (Competency) shows an intermediate correlation with Proactive participation (Attitude) (r=.525, p<.001). These result show that students' collaboration, cooperativeness, cooperative attitude, and ability to communicate are interconnected.

7.7. Barriers for the teachers

The above data show that teachers' recognition about the effectiveness of ESD in Japan is relatively positive and has been across all dates. The next question pertains to the difficulty and barriers which exist for the enhancement of ESD. Borg's study [16] showed a total of 90% of respondents stating that they were passionate advocates or thought that it was a good thing. "Despite this, more than half of the teachers (55%) felt that they had difficulties integrating SD into their teaching. The most common obstacles reported were that they: (a) lacked inspiring examples of how to include SD issues in their teaching (29%); (b) lacked necessary expertise on SD (27%); (c) lacked time to implement necessary changes to their courses (15%); and (d) lacked sufficient support from the school management (7%). Some teachers also felt that SD was not relevant to their subject (14%)"[2].

The following table shows the result of the barriers for the teachers in their implementation of ESD in 2015, 2016 and 2017:

Table 5. Barriers to Education for Sustainable Development

	2015	2016	2017
The concept of ESD is so comprehensive, everything is involved	3.55±1.15	3.63±0.99	3.58±1.05
The aims and purpose of ESD is not clear	3.70±1.07	3.58±1.13	3.58±1.09
School has many research topics, ESD is one of them	3.66±1.14	3.60±1.10	3.66±1.18
Only few numbers of teachers promote ESD	3.00±1.13	3.04±1.13	3.08±1.20
Not enough time for the enhancement of ESD	3.32±1.03	3.29±1.09	3.37±1.25
Not enough funding support for the enhancement of ESD	3.38±0.96	3.28±1.13	3.04±1.10

For effective teaching of ESD, it is necessary to include all three dimensions (ecological, economic, and social) equally and to focus on their interrelationship from a holistic perspective of Sustainable Development. Although holistic and pluralistic perspectives are special characteristics of ESD, a confusion remains of how to integrate ESD into school practice. In addition, ESD is not treated as a top priority in several of the educational topics in school. A comprehensive ESD approach to the issues will address Borg's [16] abovementioned barriers: a lack of inspiring examples of how to include SD issues in their teaching; a lack of necessary expertise on SD; an irrelevancy of SD to a given subject.

For effective teaching of ESD, it is necessary to include all three dimensions (ecological, economic, and social) equally and to focus on their interrelationship from a holistic perspective of Sustainable Development. Although holistic and pluralistic perspectives are special characteristics of ESD, a confusion remains of how to integrate ESD into school practice. In addition, ESD is not treated as a top priority in several of the educational topics in school. A comprehensive ESD approach to the issues will address Borg's [16] abovementioned barriers: a lack of inspiring examples of how to include SD issues in their teaching; a lack of necessary expertise on SD; an irrelevancy of SD to a given subject.

8. Conclusion

Until now, there has been very little empirical evidence for the extent to which ESD is implemented in the school; this paper adds to the current knowledge by presenting results from a regional survey conducted in Japan. Pauw et al. [27] indicate that ESD can actually impact student outcomes in terms of their sustainability consciousness. The questionnaires' analysis shows that teachers' beliefs about the effectiveness of ESD in Japan is relatively high. This research also revealed that in Japan, students' collaboration, cooperativeness, cooperative attitude, and ability to communicate are interconnected. In the integrated study periods, and by engaging local human resources, cooperativeness, cooperative attitude, and ability to communicate are promoted. The goals of ESD are to build on the strengths of cooperation, co-create solutions, and learning to deal with the reality of change. These goals are critical in the context of Japanese public education; In other words, this tendency is recognised as a special characteristic of the promotion of ESD in Japan.

However, from the pluralistic perspective of ESD, teaching students to confront tensions and conflicts and tackling issues with no single answer seems to be difficult. Corney [7] suggested that, as a challenge for teachers, appreciation of differing interpretations of the aims and purpose of ESD (particularly relating to differing educational ideologies and the implications for student learning, which is especially related to the concepts of action competence and behavioural change) should be emphasised. This issue still exists as a challenge for Japanese teachers when they address ESD. For the next step of implementing ESD, the development of skills, methods, and materials which deal with complexity, uncertainty, and the conflict of SD issues, as well as those which deal with the aims and purpose of ESD, are required for teachers.

Previous studies focused on three essential aspects of holism: connecting the environmental, social, and economic dimensions of sustainable development issues. From the aspect of pluralism, the effect of ESD on knowledge, attitude, and behaviour is emphasised. These previous studies mostly focused on secondary school teacher perspectives and note the differences in how secondary school teachers from different disciplines and subjects understand sustainable development. However, as mentioned above, ESD should be interdisciplinary, holistic, and participatory, with 'learning for sustainable development embedded in the whole curriculum, not as a separate subject' [1].

Since the ESD curriculum has been constructed through whole-school educational activities in Japan by linking it with other subjects and areas such as moral education and extracurricular activities, utilising integrated study periods, previous research does not relate well to the Japanese situation. Further research should focus on approaches to determine the effectiveness of a whole-school approach to ESD.

Acknowledgments
This work was supported by JSPS Grant-in-Aid for Scientific Research (C) Grant Number J150100020. I would like to thank Editage (www.editage.jp) for English language editing.

References
[1] UNESCO 2005 *UN decade of education for sustainable development 2005-2014 (International implementation scheme)* (Paris: UNESCO)
[2] WCED 1987 Chapter 2 Report of the World Commission on Environment and Development: Our Common Future http://www.un-documents.net/wced-ocf.htm (accessed 06.06.19)
[3] McKeown R, and Hopkins C 2003 EE / ESD: defusing the worry *Environ. Educ. Res.* **9**117
[4] Stables A, and Scott W 2002 The Quest for Holism in Education for Sustainable Development, *Environ.l Educ. Res.* **8** 53
[5] Öhman, J 2004 Moral perspectives in selective traditions of environmental education: conditions for environmental moral meaning-making and students' constitution as democratic citizens. *Learning to change our world? Swedish research on education and sustainable development* Lund: Studentlitteratur 33
[6] Gough S 2002 Right Answers or Wrong Problems? Towards a Theory of Change for Environmental Learning *The Trumpeter* **18** 1
[7] Scott W, and Gough S 2003 *Sustainable Development and Learning: Framing the Issues* (London: Routledge/Falmer)
[8] Corney G 2006 Education for Sustainable Development: An Empirical Study of the Tensions and Challenges Faced by Geography Student Teachers *Inter. Res. Geogr. Environ. Educ.* **15** 224

[9] Summers M, and Childs A 2007 Student Science Teachers' Conceptions of Sustainable Development: An Empirical Study of Three Postgraduate Training Cohorts. *Res. in Sci. and Tech. Educ.* **25** 307
[10] UNDESD International Implementation Scheme 2006 http://unesdoc.unesco.org/images/0014/001486/148650E.pdf. (Accessed 06.06.19).[11] Rudsberg K, and Öhman J 2010 Pluralism in Practice: Experiences from Swedish evaluation, School Development and Research *Environ. Educ. Res.* **16** 95
[12] Mogensen F and Schnack K 2010 The action competence approach and the 'new' discourses of Education for Sustainable Development, competence and quality criteria *Environ. Educ. Res.* **16** 59
[13] Walshe N 2008 Understanding Students' Conceptions of Sustainability *Environ. Educ. Res.* **14** 537
[14] Boeve-de Pauw J and van Petegem P 2011. The Effect of Flemish Eco-schools on Student Environmental Knowledge, Attitudes and Affect *Int. J. of Sci. Educ.* **33** 1513
[15] Boeve-de P J and Van Petegem P 2013 The Effect of Eco-schools on Children's Environmental Values and Behaviour *J. Biol. Educ.* **47** 96
[16] Berglund T, Gericke N and Chang-Rundgren S-N 2014 The Implementation of Education for Sustainable Development in Sweden: Investigating the Sustainability Consciousness among Upper Secondary Students *J. Res. Sci. Technol. Educ.* 32 318
[17] Boeve-de Pauw J, Gericke N, Olsson D and Berglund T 2015 The Effectiveness of Education for Sustainable Development *Sustain.* 7 15693
[18] Summers M, Corney G, and Childs A 2004 Student Teachers' Conceptions of Sustainable Development: The Starting points of Geographers and Scientists *Educ.l Res.* **46** 163
[19] Summers M, Childs A, and Corney G 2005 Education for Sustainable Development in Initial Teacher Training: Issues for Interdisciplinary Collaboration *Environ. Educ. Res.* **11** 623
[20] Spiropoulou D, Antonakaki T, Kontaxaki S, and Bouras S 2007 Primary Teachers' literacy and attitudes on Education for Sustainable Development *J. of Sci. Educ. and Techn.* **16** 443
[21] Borg C, Gericke N, Höglund H and Bergman E 2012 The Barriers Encountered by Teachers Implementing Education for Sustainable Development: Discipline Bound Differences and Teaching Traditions *Res. Sci. Technol. Educ.* **30** 185
[22] Borg C, Gericke N, Höglund H, and Bergman E 2014 Subject- and Experience-bound Differences in Teachers' Conceptual Understanding of Sustainable Development *Environ. Educ. Res.* **20** 526
[23] The Transfer-21 Programme's 'Quality and Competences' working group 2007 *Developing Quality at "ESD Schools" Quality Areas, Principles and Criteria* Berlin pp38 http://www.transfer-21.de. (Accessed 06.06.19).
[24] "S3: Sustainable School Self-evaluation." 2011 https://www.newcastle.gov.uk/wwwfileroot/legacy/childrensservice/Sustainable_school_self evaulation.pdf. (accessed 06.06.19)
[25] NIER 2012 *The research about the Education for Sustainable Development in the school*, National Institution of Education Policy Research
[26] Weiss B 1992 In Fairness to Future Generations and Sustainable Development *American University Int. Law Review* **8** 1
[27] Legault L, and Pelletier G 2000 Impact of an Environmental Education Program on Students' and Parents' attitudes, Motivation, and Behaviors *Canadian J. Behav. Sci.* **32** 243
[28] Volk T, and Cheak J 2010 The Effects of an Environmental Education Program on Students, Parents, and Community *J. of Environ. Educ.* **34** 12

Understanding the Nature of Science Through a Content Analysis of Dynamic Fluid: Voices of Pre-Service Physics Teachers

N Suprapto*, **H Mubarok and A S Adam**

Department of Physics, Universitas Negeri Surabaya, Indonesia.

*E-mail: nadisuprapto@unesa.ac.id

Abstract. Understanding the Nature of Science (NOS) is useful for pre-service, novice, and senior teachers. The specific purposes of this research were analysing the reasoning of NOS among Pre-service Physics Teachers' (PSPTs) regarding the fact, concept, principle, hypothesize, law, and theory of dynamic fluid, and analysing these NOS variables based on gender, GPA, and score of Philosophy of Science course. This research utilised a mixed-method study, descriptive-interpretive approach with supporting by quantitative analysis of the NOS in relating to PSPTs voices on dynamic fluid. A content analysis was considered in helping the researcher to better comprehend the process of constructing meaning. The study involved of a written questionnaire and followed by a semi-structured interview. A number of thirty one (31) PSPTs in public university in Surabaya, Indonesia were invited to complete the questionnaire form and joined a semi-structured interview. The PSPTs' voices were categorised into four levels: dissatisfied (naïve), intelligible, plausible, and fruitful that represents the levelling of reasoning. The findings indicated the distribution of PSPTs' reasoning in dynamic fluid concept varied from the level of dissatisfied to fruitful. PSPTs female performed better in reasoning of NOS than male of dynamic fluid. Moreover, PSPTs whose GPA is better also performed better in reasoning of NOS. Finally, the score of course had also in line with the level of reasoning of NOS.

1. Introduction

"The phrase 'history and philosophy of science' (HPS) has been used to describe the interplay of disciplines that inform science education (SE)……However, a more encompassing phrase to describe the scientific enterprise for SE is the 'nature of science' (NOS)" [1].

Generally, NOS is an accepting of the characteristics of scientific knowledge that deals with its empirical nature, its creative and imaginative nature, characteristics of its theory, its social-cultural nature, and its tentative nature [1-4]. NOS is also connected to the epistemology and sociology of science, namely science as a way of knowing, or a value and belief that is inherent in science and its development [3]. As a way of knowing, NOS has some implications: (a) knowledge is not a mere picture of the world of reality, but is always a construction of reality through the activities of the subject; (b) subjects form cognitive schemes, categories, concepts, and structures that are necessary for knowledge; and (c) knowledge is formed in one's conception structure [2].

The structure of conception shapes knowledge when that conception applies in dealing with one's experiences. By doing so, NOS includes two aspects, namely body of knowledge which is often referred to as product aspects and methods known as process aspects. So, the nature of science is a bridge for students to uncover and understand the reality of nature. To help students understand the NOS, a good physics teacher will always emphasise the NOS in every teaching. Nevertheless, the teaching about the NOS will be limited to time, but should still be carried out explicitly in the relevant teaching material. This is because students will not have a good understanding of the NOS if they do not have first-hand experiences through empirical scientific methods [4].

Emphasizing the nature of science in learning is an effort to empower students through education, i.e. physics education. There are three important components of science in physics education in schools [5]: products of science, process of science, and attitude of science. Science knowledge that is often called a science product is an accumulation between the results of empirical activity and the analysis of scientists. Science products are produced through scientific investigations involving scientific attitudes and scientific processes. Science products that are built on attitudes and scientific processes will then regenerate new scientific products through scientific processes and investigation of new phenomena in nature.

The product of science included facts, concepts, principles, and theories. Facts are products of empirical activities of science, while concepts, principles, and theories are products of analytical activities [5]. Knowledge, principles, laws and theories are the results of fabrications or man-made in understanding and explaining nature with various phenomena that occur in it. Thus, science as a scientific product includes concepts, laws and theories that are developed as fulfilment of human curiosity. In science learning, product aspects appear in the form of teaching materials that contain the subject matter.

Scientists use a variety of empirical and analytical procedures in their efforts to uncover universal reality. This procedure is better known as the science process. The process aspect refers to a method of gaining knowledge. This method is called the scientific method. The standard scientific method is currently the result of previous developments. The scientific method is a combination of rationalism which believes that knowledge can be obtained through thought and empiricism which believes that knowledge can be obtained through experience. Generally, it has six basic frameworks of procedures, namely being aware of problems and formulating problems, observing and collecting relevant data, compiling or classifying data, formulating hypotheses, deductions and hypotheses, and testing the correctness of hypotheses.

Science process skills are a form of science as a process. In science learning, it is very important to help students learn science process skills or inquiry skills to solve problems. Science process skills are a form of science as a process. The expected intellectual process skills in the nature of science-oriented learning are 1) building principles through induction, 2) explaining and predicting, 3) observing and recording data, 4) identifying and controlling variables, 5) making graphs to find relationships; 6) designing and carrying out scientific investigations, 7) using technology and mathematics during the investigation, 8) describing conclusions from the evidence [3, 4-6]. Scientists study phenomena and events through scientific processes, such as observation, experimentation, and other empirical and analytical activities.

In addition to these two aspects of science, Carin [5] stated that the third aspect of science is scientific attitude. The attitude aspect in question is a variety of beliefs, opinions, and values that must be maintained by a scientist especially when looking for or developing new knowledge, including responsibility, curiosity, discipline, diligence, honesty, and being open to the opinions of others. Physics education that is relevant to the nature of science requires an atmosphere that allows students to be directly involved in the learning process. So that by having a scientific attitude and after going through a series of learning processes, students can arrive at a conclusion that they form themselves.

Based on the authors' experiences in research on the essence of philosophy of science [7] and the physics conception and physics knowledge [8-11,13], then this article manifests the emerging idea the nature of knowledge and the nature of science specifically in fluid dynamic concept. By doing so, this research explores the reasoning of NOS among PSPTs regarding the fact, concept, principle, hypothesize, law, and theory of dynamic fluid and to sightsee the PSPTs' existing knowledge and reasoning of dynamic fluid. In order to make a clear the research direction, the following two research questions (RQ) were proposed:

RQ 1: What reasoning of NOS among PSPTs do regarding the fact, concept, principle, hypothesize, law, and theory of dynamic fluid?
RQ2: What reasoning of NOS among PSPTs do regarding gender, GPA, and Score of Philosophy of Science course?

2. Method

This research used a mixed-method with content analysis was well-organized in serving the researchers to grasp the process of creating meaning [12]. Indeed, previous researchers used four types of content analysis based on some assignments were designed to provide insight into the kinds of understandings each participant of constructing meaning, i.e. reflection journals, a research paper (students' issues and arguments), class assignments (informal interviews, reaction papers, and provocative questions), and a position paper [15]. For the reason of cost and time allocation, this study used a written questionnaire and followed by a semi-structured interview (see Figure 1). Totally, there were thirty one (31) participants contributed in the research. All of them were joined in the course Philosophy of Science with 2 credits. Therefore, they learned about the three foundation of Philosophy of Science: Epistemology, Ontology, and Axiology of knowledge [7]. They have also learned the content of dynamic fluid through the courses Fundamental Physics and Mechanics. The distribution of demographic variables of participants is depicted in Table 1.

Table 1. The demographic factors of participants.

Variable		Frequency	Percentage (%)
Gender	Male	4	12.90
	Female	27	87.10
GPA	$GPA \leq 2.75$	-	0
	$2.75 < GPA \leq 3.00$	2	6.45
	$3.00 < GPA \leq 3.25$	7	22.58
	$3.25 < GPA \leq 3.50$	19	61.29
	$3.50 < GPA \leq 3.75$	3	9.68
	$3.75 < GPA \leq 4.00$	-	0
Score of Philosophy of Science course	A	4	12.90
	A-	21	67.74
	B+	6	19.36

Question:
Lakukan identifikasi dan analisis terkait Fact, Concept, Principle, Hypothesis, Law, and Theory pada topik fluida dinamis, supaya terlihat jelas perbedaannya! **(Translation: Identify and make an analysis in terms of fact, principle, hyphotesis, law, and theory of dynamic fluid topic in order to make a clear meaning of each differences!)**
Answer:
____

(a)

Semi-structured Interview Form:
Choose three among these five terms: fact, principle, hypothesis, law, and theory
Make a brief definition of the terms that you was chosen
What is your conception about dynamic fluid according to your terms was you chosen
Make reasoning according to your term of dynamic fluid topic in order to make a clear each differences

(b)

Figure 1. Sample of written questionnaire (a) and semi-structured interview form (b).

The voices of participants were coded by pseudonyms name. The data gained from the results of the written questionnaire and semi-structured interviews [12,15]. By deducting the data, the PSPTs' reasoning of NOS in terms of fact, concept, principle, hypothesize, law, theory and the existing knowledge and reasoning of dynamic fluid reasoning was categorised into four levels: "dissatisfied (naïve), intelligible, plausible, and fruitful" [7].

3. Result and Discussion

RQ 1: What reasoning of NOS among PSPTs do regarding the fact, concept, principle, hypothesize, law, and theory of dynamic fluid?

Table 2 summarises the level of reasoning performed by PSPTs. There are the different reasoning of PSPTs regarding the fact, concept, principle, hypothesize, law, and theory of dynamic fluid. It was about 12.90% among them have the lowest level or in the dissatisfied (naïve) level.

Table 2. The level of reasoning performed by PSPTs in dynamic fluid concept.

Level of reasoning	Frequency	Percentage (%)
dissatisfied (naïve)	4	12.90
intelligible	9	29.03
plausible	9	29.03
fruitful	9	29.03

The following quotes are the samples of PSPTs reasoning the dissatisfied level.

Fact: *"flowing fluid experiences friction with objects around it" [Inayah]*
"...dynamic fluid in the form of liquid, gas which has a constant speed with time" [Jelita]

Concept: *"Concepts are a series of statements or ideas. Airplane wings are an example of the application of static fluids" [Dian]*

Principle: *"moving fluid is influenced by time and friction" [Inayah]*
"fluid has a constant speed with time, it does not change volume, it doesn't turbulent, and it doesn't experience rotation" [Melly]
"dynamic fluid has limited coverage" [Dian]

Hypothesis:	*"if the fluid doesn't have a constant velocity then the fluid isn't a dynamic fluid" [Melly]*
Law:	*"Bernoulli's law is divided into 2 things: compressed and uncompressed flow" [Dian]*
Theory:	*"Water is liquid that flows" [Jelita]*

All quotes above are categorised into (naïve) level of reasoning because of some reasons:

a. Simple reasoning either wrong or incomplete reasoning, indeed it performs too general sentences. "*Airplane wings are an example of the application of static fluids*", this is an example of general information.
b. There is unclear information, i.e.: "*fluid has a constant speed with time*", so the question is what kind of fluid?
c. Just providing any quantities or variables of physics, however, there are inappropriate quantities, for example, from the expression of: "*moving fluid is influenced by time and friction*",

In the intelligible level, it is about 29% of PSPTs indicating inappropriate statement. However, their performance was better than the previous level.

Fact:	*"airplanes can fly because of the aircrafts' wings" [Intan]* *"water is a type of fluid that easily flows" [Nabilah]* *"fluid has speed when it flows" [Reny]*
Concept:	*"Airplane wings used concept of dynamic fluid" [Nabilah]*
Principle:	*"water is a liquid" [Nabilah]*
Hypothesis:	*"if water flows then there is a pressure of liquid" [Nabilah]*
Law:	*"Bernoulli's law applies constantly" [Ella]*
Theory:	*"Water is the fluid of the liquid that flows is influenced by gravity, density, pressure, and volume" [Nabilah]* *"Fluid is considered to have a constant speed with time, no change in volume, and not thick" [Ella]*

The expressions above are considered into intelligible level of reasoning due to these reasons:

a. Simple correct reasoning, but sometime it performs too general expression; for example: *"Airplane wings used concept of dynamic fluid"; "fluid has speed when it flows"*.
b. Providing quantities or variables of physics, however, there are inappropriate quantities, for example, from the expression: "*Water is the fluid of the liquid that flows is influenced by gravity, density, pressure, and volume*",

In the next level, the PSPTs' reasoning about dynamic fluid reached in the plausible level. They performed better in giving reasoning. The following illustration shows their reasoning.

Fact:	*"Airplanes have wings like those of a bird, which are curved and thicker at the front than at the rear" [Radit]*
Concept:	*"A number of masses of water m at height h have potential energy, Ep = mgh" [Hari]*
Principle:	*"The greatest fluid pressure is on the part with the smallest flow rate, and the smallest pressure is on the part with the greatest flow rate" [Radit]*
Hypothesis:	*"By intuition, we might think that for a horizontal pipe, the fluid pressure is greatest at the point where the flow rate is greatest" [Ruly]*
Law:	*"Bernoulli used the energy-effort theory to derive the energy conservation equation in dynamic fluids" [Hari]*

Theory: *"if the fluid that flows does not change in volume (or density) when pressed, the fluid flow is said to be incompressible" [Ruly]*

If we compare to intelligible reasoning, the plausible level reasoning based on the logical thinking and considering the relationship among physical quantities. Taking for example, *"By intuition, we might think that for a horizontal pipe, the fluid pressure is greatest at the point where the flow rate is greatest"*, this hypothesis considers the relation between variables of pressure and the flow rate (speed) at the point of pipe. Other example is, *"Bernoulli used the energy-effort theory to derive the energy conservation equation in dynamic fluids"*. This law is expressed by participant due to he/she understand how the Bernoulli law was derived.

In the fruitful level, the PSPTs' reasoning about dynamic fluid is getting better than the previous level. The reasoning provided by participants performs more complex and comprehensive. For example, Rina expresses her principle based on scientific principle which proposed by scientist, *"If a fluid flows with steady flow, then the mass of fluid entering one end of the pipe must be the same as the mass of fluid coming out of the other end of the pipe during the same time interval"*.

Fact: *"When we use a water hose, and we put pressure on the end of the hose, the water reaches farther away" [Faradila]*

Concept: *"Discharge is a quantity that states the volume of fluid flowing through a certain cross section in a certain time unit" [Dian]*

Principle: *"If a fluid flows with steady flow, then the mass of fluid entering one end of the pipe must be the same as the mass of fluid coming out of the other end of the pipe during the same time interval" [Rina]*

Hypothesis: *"The greater the cross-sectional area of the pipe, the smaller the velocity of fluid flow" [Ayu]*

Law: *"Implications of Bernoulli's law: In horizontal (horizontal) pipes, the greatest fluid pressure is in the part with the lowest flow velocity, and the least pressure is in the section with the highest flow rate" [Dian]*

Theory: *"If the velocity v at a point is constant with time, the fluid flow is said to be steady. If the fluid that flows does not change in volume (or density) when pressed, the fluid flow is said to be incompressible" [Ayu]*

From the illustration above, the researchers agreed the essence of NOS especially in exploring the PSPTs' understanding of fact, concept, principle, hypothesize, law, and theory of dynamic fluid. All the PSPTs' reasoning above mapped the characteristics of the NOS in science education that were understandable by PSPTs. Indeed, previous researchers analysed recent science education curriculum documents and identified statements about the NOS that are common to most curricula [16-18]: "(a) Science is an attempt to explain natural phenomena; (b) people from all cultures contribute to science.' (c) scientific knowledge, while durable, has a tentative character; (d) scientific knowledge relies heavily, but not entirely, on observation, experimental evidence, rational arguments and scepticism; (e) there is no one way to do science – therefore, there is no universal step-by-step scientific method; (f) new knowledge must be reported clearly and openly; and (g) laws and theories serve different roles in science – therefore, students should note that theories do not become laws even with additional evidence".

RQ2: What reasoning of NOS among PSPTs do regarding gender, GPA, and Score of Philosophy of Science course?

After carefully coding the participants' response regarding their opinion of their NOS in the concept of dynamic fluid, then we classified their answer based on the demographic background: gender, GPA, and score of philosophy of science course, (see Table 3).

Table 3. The level of reasoning performed by PSPTs in dynamic fluid concept.

Demographic variables		Level of reasoning [frequency(%)]			
		dissatisfied (naïve)	**intelligible**	**plausible**	**fruitful**
Gender	Male	1(3.22)	2(6.45)	1(3.22)	0
	Female	3(9.68)	7(22.58)	8(25.81)	9(29.03)
GPA	2.75<GPA≤3.00	1(3.22)	1(3.22)	0	0
	3.00<GPA≤3.25	2(6.45)	5(16.13)	0	0
	3.25<GPA≤3.50	1(3.22)	3(9.68)	9(29.03)	6(19.35)
	3.50<GPA≤3.75	0	0	0	3(9.68)
Score of Philosophy of Science course	B+	4(12.90)	2(6.45)	0	0
	A-	0	7(22.58)	9(29.03)	5(16.13)
	A	0	0	0	4(12.90)

Based on Table 3, we underlined some points:

a. Based on gender, the level of reasoning of dynamic fluid perform female better than male PSPTs, with dominance in plausible and fruitful categories.
b. Regarding GPA, PSPTs whose better GPA performs better level of reasoning.
c. The score of Philosophy Science course have also in line with the level of reasoning. It means, if the score is higher that the level of reasoning is also greater.

4. Conclusion

The conclusion can be prompted from this study includes: (a) The distribution of PSPTs' reasoning in dynamic fluid concept varied from the level of dissatisfied to fruitful; (b) There was no dominant level of reasoning among PSPTs regarding the concept of fluid dynamic; (c) PSPTs female performed better in reasoning of NOS than male of dynamic fluid; (d) PSPTs whose better in GPA performed better level of reasoning; and (e) The score of Philosophy Science course have also in line with the level of reasoning. This result gives the implication that PSPTs need to learn in-depth this concept and overcome the understanding that is still wrong in their mind.

Acknowledgements

This paper is one of the outcomes of the course "Philosophy of Science". Therefore, the researchers would like to thank all PSPTs as participants who took part in the course. Moreover, the authors would also like to thanks the member of Research Group of Philosophy and Physics Education Curriculum, Universitas Negeri Surabaya through its' roadmap this research was manifested.

References

[1] McComas W F, Clough M P and Almazroa H 2002 The role and character of the nature of science in science education. In W F McComas (ed.), *The nature of science in science education: Rationales and Strategies* (pp. 3-39) (The Netherland: Kluwer Academic Publishers)
[2] Schwartz R S, Lederman N G and Crawford B A 2004 *Sci. Educ.* **88** 610
[3] Wenning C J and Rebecca E 2006 *J. Phys. Teach. Educ. Online* **3** 24
[4] Wenning C J 2006 *J. Phys. Teach. Educ. Online* **3** 3
[5] Carin A A 1993 *Teaching science through discovery (seventh ed.)* (New York: Macmillan Publishing Company)
[6] Mubarok H, Lutfiyah A, Kholiq A, Suprapto N and Putri N P 2018 *J. Phys.: Conf. Ser.* **997** 012039
[7] Suprapto N, Deta U A, Lestari N A, Kholiq A, Adam A S, Mubarok H and Jauhariyah M N R 2018 *J. Phys.: Conf. Ser.* **1108** 012097

[8] Jauhariyah M N R, Suprapto N, Suliyanah, Admoko S, Setyarsih W, Harizah Z and Zulfa I 2018 *J. Phys.: Conf. Ser.* **997** 012031
[9] Suprapto N, Syahrul D A, Agustihana S, Pertiwi C A and Ku C-H 2016 *Chemistry* **25** 718
[10] Anggrayni S, Mubarok H, Putri N P, Suprapto N and Kholiq A 2018 *J. Phys.: Conf. Ser.* **997** 012026
[11] Suprapto N, Abidah A, Dwiningsih K, Jauhariyah M N R and Saputra A 2018 *Periodico Tche Quimica* **30** 387
[12] Creswell J W and Plano Clark V 2011 *Designing and conducting mixed methods research (2nd ed.)* (Thousand Oaks, CA: Sage)
[13] Suprapto N, Chang T-S and Ku C-H 2017 *J. Baltic Sci. Educ.* **16** 7
[14] Suprapto N, Suliyanah, Prahani B K, Jauhariyah M N R and Admoko S 2018 *J. Phys.: Conf. Ser.* **997** 012011
[15] Sullenger K and Turner S 2002 Nature of Science: Implications for education (An undergraduate course for prospective teachers). In W F McComas (ed.), *The nature of science in science education: Rationales and Strategies* (pp. 243-254) (The Netherland: Kluwer Academic Publishers)
[16] Abd-El-Khalick F, Bell R L and Lederman N G 1998 *Sci. Educ.* **82** 417
[17] Hodson D 2009 *Teaching and learning about science: Language, theories, methods, history, traditions and values* Rotterdam: Sense Publishers
[18] McComas W F and Olson J K 1998 The nature of science in international science education standards documents. In McComas (Ed.), *The nature of science in science education: Rationales and strategies* (pp. 41–52) (The Netherland: Kluwer Academic Publishers)

University Students' Misconception in Electromagnetism

D S Bestiantono, E H Sa'diyah, R Rachmatya, H Mubarok, A S Adam, N Suprapto*

Physics Department, Universitas Negeri Surabaya, Indonesia

* E-mail: nadisuprapto@unesa.ac.id

Abstract. This aim of the research was to analyse the university students' conception of electromagnetism. This research used a mixed-method design, the combination of a quantitative and qualitative method. For the quantitative part, the research utilised a conception test of electromagnetism, for the qualitative part, the research focused on the descriptive analysis that generated by the university students. The participant in this research was 15 physics students from 9 different classes in the Physics Department of a national university in East Java Indonesia. A conception test included ten different multiple-choice problems and followed by the descriptive analysis of these problems. The participants asked to choose the right answer and write their reasons or arguments that support their answer. In this study, the test was given without giving any treatment for the participant. The analyse method that used in this research was the quantitative and qualitative description. The result showed that the conception of students ranged between 2nd semester (freshman) and 4th Semester (sophomore) in the physics department. Therefore, it is concluded that the students have electromagnetism misconception.

1. Introduction

Electricity and magnetism have been studied separately until 1819 when Oersted found if there is a relation between electricity and magnetism [1]. Nowadays, Electromagnetism becomes one of the most important branches of physics since its use in many ways. Even though it is used in many ways, there are only a few people that understand the concept of electromagnetism well. It indicates that this concept state in abstract concept, need more illustration to explain certain electricity concepts.

In education purpose, electricity and magnetism are available both on a junior and senior high school in the national curriculum. As the pedagogic and pre-service teacher campus, well understanding of the particular concept is inevitable. For that reason, this research aimed to analyse the misconception among students in a public university in Surabaya Indonesia.

The concept is things, phenomenon, situations, or characteristics which have to define nature and represented in cultures by a sign or symbol [2]. The concept explanation by someone called as conceptions. Even though there are clear meanings about the physics concept by physicist, in reality, these meaning commonly different for each student.

The misconception is the student inability to connect knowledge to others. This inability leads the student having contradicted views than what scientist about the concept of physics [3]. There are a lot of causes of misconception such as interaction between students and their environment that lead them to their own analyse which is wrong [4-5]; "dead leaves model" of learning where the teacher teaches only the equations and formulas, not the concept [6]; book; and themselves [7].

Physics is not only about an algorithm, but also the concept and analysing as the concept of 21st-century learning outcome. One of these outcomes supports the analysing part of the environment to the physics learning process. It will realise why some concept needs to learn and applied to their lives. However, this activity has more possibility to get the students understanding into a misconception. Hence there are five common misconceptions in physics. Those are: a) Preconceived notions, b) non-scientific beliefs, c) Conceptual misunderstandings, d) vernacular misconceptions, and e) factual misconceptions [3].

The sample of this research is the pre-service physics teacher. These students have been taught electromagnetism when they were in senior high school or at the beginning of the year. The reason is why students of doing physics department chosen for this research because they are the people who will teach the physics concept. If the teacher themselves has misconceptions, it can be assumed if their students will have misconceptions too. As this research published wishfully, there will be a solution to solve the misconceptions among students in the Physics department.

2. Method

This research used the descriptive quantitative and qualitative method. The main reason was "it wants to get a general description of the misconception that occurred in the physics students about electromagnetism concept" [8-9]. The collecting data was conducted for 3 months, starting from April until June 2019. The research process was organised by preparing proposal, composing instrument in the form of test, judgment instrument by experts and researcher.

2.1. Participants

Initially, the target of the participants was 20 physics students; finally, the participant of the study was 15 physics students from 9 different classes in the Physics Department of a public university in East Java Indonesia. The sampling by the purposive sample characterised of student of level 3 who has received the subject of electromagnetism. For small sample size, we followed the Slovin's formulation with margin of error 0.1(10%) [13,14].

$$n = N/\left(1+Ne^2\right) \quad (1)$$

Where n is the sample size, N is the population size, and e is the margin of error to be decided by the researcher. Therefore, n=20/(1+20.0.01)= 16.67. Thus, the number of 15 physics students was enough for a group with a small sample size according to the Slovin's formulation.

2.2. Instruments

The technique of data collection has been done by using different multiple-choice problems and followed by the descriptive analysis of these problems. The participants asked to choose the right answer and write their reasons or arguments that support their answer. They also asked whether they are sure or unsure about their answer. The test was given without giving any treatment to the participant. Students' answers are categorized into four categories (a) scientific conception; (b) false positive; (c) misconception; (d) false negative; and (e) lack of knowledge. "The criteria for selecting the instrument used in this study are (a) the instrument must be justified by the expert, (b) the instrument must be suitable for use in the science domain adapted to three chemical representations: macroscopic, sub microscopic and symbolic, and (c) uncovering misconceptions accompanied by reason and level of confidence in answering questions" [8].

3. Results and Discussions

The concept is things, phenomenon, situations, or characteristics which have to define nature and represented in cultures by a sign or symbol [10]. Hence, the concept defined as the abstraction of nature that proposed to make communication between people easier and enable people to analyse their environment. The result of those analyse can be different among people, for example, density, friction, or elasticity concept analyses.

The concept explanation by someone called as conceptions. Even though there are clear meanings about the physics concept by physicist, in reality, these meaning commonly different for each student. The most probable reasons about the difference of the meaning by the physicists and students are the complexity of the physicist conception. The physicist conception commonly related to another concept and more sophisticated than the students. If student conception has the same meaning with the simpler version of the physicist's, then it cannot be called as a misconception. However, if the student conceptions are contradictive to the physicist, then it is called misconception [2]. Then misconceptions are strongly held cognitive structures that are different from the accepted understanding in a field and that are presumed to interfere with the acquisition of new knowledge [5].

Electricity and magnetism concept use in instruments follow the basic relating concepts. As the pre-service teacher, the basic concept should be understood well to answer application and practical problem. Commonly, the particular concepts learning method in junior and senior high level focused on a practical solution due to choosing the best answer in the final examination. Furthermore, that question transforms into high order thinking skill level. The teacher with no complete understanding of the basic concept of electricity and magnetism difficult to solve various problems. For gain this purpose, the misconception measurement uses ten questions in the basic concept of this case. The result of the population in this study show in Table 1.

Table 1. Result of recapitulation level of conception based on the problem indicator.

	Indicator of Problem	**Scientific Conception (%)**	**False Positive (%)**	**False Negative (%)**	**Misconception (%)**	**Lack of Knowledge (%)**
1	Identify the Electric System	33.33	0.00	0.00	46.67	20
2	Identify the influence of loop	0	6.67	0.00	53.33	40
3	Identify two parallel wires	20	0.00	0.00	60	20
4	Identify how could the magnetic field exist	26.67	0.00	0.00	46.67	26.67
5	Identify the influence of magnet on the charge	13.33	0.00	6.67	66.67	13.33
6	Identify Lorentz's Law	26.67	0.00	6.67	26.67	40
7	identify the influence of electricity on the magnetic field	26.67	0.00	0.00	33.33	40
8	Identify the influence of material on the magnetic nature	20	13.33	13.33	20	33.33
9	Identify the influence of the material conductivity and its magnetic nature	33.33	0.00	0.00	60	6.67
10	Identify the force of a charge	26.67	0.00	0.00	40	33.33
	Average Percentage	**22.67**	**2.00**	**2.67**	**45.33**	**27.33**

Based on Table 1, most of the sample has a misconception with a range of 20% - 66.67%. It means they have a problem to master electricity and magnetism concept. The significant problem shows in the combination of both concepts, the influence of magnet and to the charge as the first aspect of the electricity. The understanding of this combination makes the more difficult thinking process, need each concept understanding to determine the phenomena from these two parts. Most of the sample missing the fact that the current state from the moving electron, different from the fifth question as to the most significant score on a misconception. The eighth question ignores the current producing and shows the lowest score in misconception. Although those two questions discuss the influence of the two concepts, it shows the different results. In other words, two questions on the most significant and lowest misconception score caused by the incomplete concept understanding of both. Although there is the lowest and biggest score, the other concept mastering shows similar results. All the score classified to predetermined criteria to define the student's conception. Since this research use three-tier questions we use criteria such this:

Table 2. The criteria for scoring [11]

The first stage	The second stage	The third stage	Categories
Correct	Incorrect	Not believed	• Misconception • Lack of knowledge
Incorrect	Correct	Believed	• Negative false or error
Incorrect	Correct	No believed	• Lack of knowledge
Incorrect	Incorrect	Believed	• Misconception
Incorrect	Incorrect	Not believed	• Lack of knowledge
Correct	Incorrect	Believed	• Positive False

The result shows that misconceptions among student reach 45%. Most of the students have a misconception in identifying the influence of magnet to the charge. This question gives 66.67% of students have misconceptions. It means most of the sample believe the inappropriate fact on electricity and magnetism influence. As explained before, that students believe the wrong current concept, these are shown as question and answer below:

Question: *"a particle with a positive charge is placed between two magnets as shown in the figure, will the magnet be giving force on the particle? Giving the explanation!"*

Figure 1. Identify the influence of magnet on the charge

Answer: *"the magnets giving force on the particle"*

Reason: *"the force that is caused is the object of reject or attraction, according to the direction of the poles. Reject it if the poles are same and pull if the poles are different".*

The other question with the simple question but in the third-highest score of misconception is Identify the influence of loop. It was about 53.33% students have misconception to determine the brightest lamp on a closed-loop, connect a new wire to divide three lamps into two sides, one side consists of two series lamp circuit and another consist of one parallel lamp. Most of the sample writes that all the lamp dimmer than before as the wrong answer, even though the parallel lamp circuit will brighter due to the same voltage in every parallel wire. This misconception mostly caused by the direction sign of the current as of the disturbing part of this question, in other reason shows like this:

Question: *"If the circuit as shown in Figure 2 is connected by a wire (k) as shown in the figure, what about the lights L_1, L_2 and L_3?"*

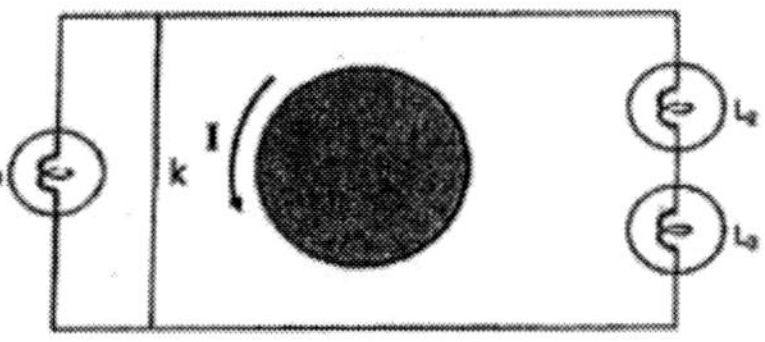

Figure 2. Identify the influence of loop

Answer: *L_1 is dim, while L_2 and L_3 are bright*

Reason: *"Because when the circuit is connected by a wire, the current flowing at L_1 gets smaller (the greater the resistance is, the smaller the current is so that the lights are dim) compared to L_2 and L_3, the light will turn brighter than L_1"*

Identify two parallel wires and identify the influence of the material conductivity and its magnetic nature is the high percentage on misconception score. Two parallel wires identification has been explained in the first-year subject on basic physics laboratory or senior high level. Besides, conductivity on metal is a piece of common knowledge. These two questions have similarity; it has the same result in the experiment setting and should be easy to remember. However, students shouldn't gain this information in meaningful learning. Both of the questions gain 60% of students with misconception.

Question: *"Is this statement true: "All metals are conductor, which means that metals can flow charged particles (electrons) easily. Therefore, all metals will be easily affected by magnets.*
(a) true
(b) wrong"

Answer: *(a) Right*

Reason: *"Because metals are electrical conductors, so it can flow electron charged particles easily."*

Conception and misconception itself commonly formed when students' brain interacts with the environment. Since their childhood, Student has many experiences that related to the concept of force, momentum, speed, velocity, and acceleration. Even though it is not used yet, the conception and misconception formed related to those concepts [10]. The different interpretation between the observable and actual reason for some phenomena is the reason for believing the wrong concept. As a simple example, children will believe that all animals in the ocean are Pisces. They will believe that dolphin is Pisces, even dolphin is exactly a mammal.

This result inclined that there are many misconceptions among physics students. The cause of these students having misconception can come from the wrong learning method until the wrong way to meaning electromagnetic matter. Commonly, they learning a particular physics concept for practical use, answer the equation physics problem in the final examination. It makes physics has an image of the mathematically based subject without any physical phenomena interpretation. Furthermore, analysis skill on natural interpretation will be less trained, and this ability needs a basic understanding of the physics concept. For decrease this misconception, the students should have laboratory activity to synchronise the concept from class to the real phenomena [12].

4. Conclusion

From the study, the result showed that if there are misconceptions in the electromagnetic subject among physics students. The matter that has most misconceptions is "identifying the influence of magnet to the charge". The factor of students' misconceptions is the false meaning that student makes during their experience in the electromagnetic subject and the wrong learning method used by the teacher.

References

[1] Byrne C 2015 *A Brief History of Electromagnetism* (Lowell: University of Massachusetts)
[2] Nurulwati, Arsaythamby V and Ruslan M A 2014 *J. Pendidik. Sains Indones.* **02** 87
[3] Helm H 1980 *Phys. Educ.* **15** 92
[4] Daud N, Syaharudin N, Mohd M, Karim A, Siti W N, Wan H and Nurulhuda A R 2015 *Misconception and Difficulties in Introductory Physics among High School and University Student: An Overview in Mechanics* (Tanjung Malim: Educatum)
[5] Hammer D 1996 *Am. J. Phys.* **64** 1316
[6] Redish E F 1994 *Am. J. Phys.* **62** 796
[7] Bayuni T C, Sopandi W, and Sujana A 2018 *J. Phys.: Conf. Series* **1013** 012086
[8] Suprapto N, Syahrul D A, Agustihana S, Pertiwi C A and Ku C-H 2016 *Chemistry* **25** 718
[9] Mubarok H, Lutfiyah A, Kholiq A, Suprapto N and Putri N P 2018 *J. Phys.: Conf. Series* **997** 012039
[10] Van D B E 1991 *Miskonsepsi Fisika dan Remediasi [Physics Misconception and Remediation]* (UKSW, Salatiga)
[11] Irwansyah, Sukarmin and Harjana 2018 *Jurnal Ilmiah Pendidikan fisika Al-BiRuni* **07** 207
[12] Adam A S, Anggrayni S, Kholiq A, Putri N P and Suprapto N 2018 *J. Phys.: Conf. Ser.* **997** 012020
[13] Israel G D 1992 *Determining sample size* (University of Florida: Florida Cooperative Extension Service)
[14] Yamane T 1967 *Statistics: An introductory analysis, 2nd Ed* (New York: Harper and Row)

An Effort in Teaching Invertebrates and Training Digital Literacy to the Students

I D T Putri* **and R Ambarwati**

[1] Biology Department, Faculty of Mathematics and Natural Sciences, Universitas Negeri Surabaya, Indonesia.

* E-mail: intandeliatp@gmail.com

Abstract. Digital literacy is an ability to understand and use information in multiple formats and sources while presented through computers. The research purposes have analysed the influence of the electronic Invertebrate book on students' digital literacy and learning outcomes, as well as describe student responses upon learning process using electronic Invertebrate book. This research used mixed-method. The research samples selected using purposive sampling. Student's digital literacy and learning outcomes, in the form of pre-test and post-test, were analysed quantitively using normalised gain scores (g). The difference between pre-test and post-test in treatment or control class calculated using Wilcoxon test. Meanwhile, the post-test score and normalised gain score between treatment and control class calculated using Mann-Whitney test. The student's responses through the learning process were analysed descriptive-quantitatively using questionnaire. The result revealed that about 57% of students completed the digital literacy test and 100% of them enhanced in the category of medium to high on digital literacy skills.

1. Introduction

Producing human resources who have 21st-century competencies is one of the education targets. 21st-century competencies consisted of cognitive skills and the skills of technology, media and information; learning and innovation skills, and also life and career skills [1]. Digital technology plays an important role in the development of education, especially the development of the 21st century [1]. Digital literacy is an ability to understand and use information in multiple formats which coming from various sources while presented through computers [2]. Martin [3] stated that digital literacy is the awareness, attitudes, and abilities of individuals to precisely using digital tools and facilities to identify, access, manage, integrate, evaluate, analyze and synthesize digital resources, build new knowledge, create media expressions, and communicate with others, in the context of specific life situations, to enable constructive social action. Thus, the definition of digital literacy is not only the process of students accessing information through digital media, but also the process of students to understand, process, analyse, evaluate, and build digitally available information.

Although currently students are generally considered technology-savvy, many of them find it challenging to do well. The students are labelled as digitally literate when mastering several competencies [4]. Digital literacy should be studied and taught [2]. The teacher needs to train students on how to use the Web correctly and how to be critical, how to assimilate information, evaluate, and then integrate it [2]. Students' digital literacy skills need to be improved because although students are now familiar with digital technology, students generally only know how to access, create and share digital information [5]. Poor understanding of digital literacy can affect the psychology of students who tend to insult, lead to jealousy towards others, moodiness towards negative comments, depression, and get to communicate using politeness languages [6]. The abundance of information spread through the internet made it difficult for audiences to determine the correct information with fake ones. As a result, this wrong information was part of a conflict between groups to defend their opinions [7].

In the previous study [8] an electronic Invertebrate book was developed and stated to be very feasible and practical. Digital literacy has also been successfully trained through the development of digital literacy-based electronic books that integrated with digital technology and information. The material contained in electronic Invertebrates book is according to the latest revised 2013 Curriculum. The selection of material based on Basic Competencies 3.9 namely "Grouping animals into phyla based on body layers, body cavity, body symmetry, and reproduction". The demands of Basic Competencies 3.9 stated that Invertebrate material has a broad scope of material and often makes it difficult for students to describe characteristics and classifies each phylum in a limited meeting, thus influencing students' understanding and interest in Invertebrate material [9]. Based on observations in ten high schools in Gresik showed that 100% of learning still rely on textbooks as the central learning resource and has never used several electronic book references or other learning resources. Based on Putri & Ambarwati's research [10], the content of digital literacy in Biology Class X textbooks on Animalia was still low. Although internet usage has been encouraged in several schools, not all teachers can train students to browse the internet properly, and not all students can find the right and reliable information through the internet. Therefore, Gilster [2] formulated four student competencies which said to be digital literacy used in this study, including internet searching, hypertextual navigation, content evaluation, and knowledge assembly. Based on the problem, it is necessary to follow up the research of electronic Invertebrate book development [8] which have been done by applying the book to find out its role in increasing student digital literacy.

2. Method

This study was mixed-method research [11]. The research was located in the Biology Department, Universitas Negeri Surabaya and SMA Negeri 1 Gresik during January to April 2019. The data collected on February 4 to 18, 2019. The research sample selected using purposive sampling. The samples were class X MIPA 1 and X MIPA 2. Both of them are experiment and control classes which consisted of 30 students. In this study, the components used to measure digital literacy are according to Gilster [2], they are internet searching, hypertextual navigation, content evaluation, and knowledge assembly.

The procedure of this study consisted of preparation, implementation and analysis of data. The research instruments were digital literacy assessment sheets, learning outcome assessment sheets, and student response questionnaire sheets. Data collected from tests and questionnaires. The research results were analysed based on students' mastery of digital literacy, student' mastery of learning outcomes, and student responses. Students' mastery of digital literacy assessed using pre-test and post-test scores of digital literacy aspect based on minimum standard adapted from [12], that was 71. The students were categorized into very poor (<30%), poor (30-50%), sufficient (51-70%), good (71-90%) and excellent (91-100%). Students' mastery of learning outcomes assessed using pre-test and post-test scores of cognitive aspects based on the minimum standard from the school, that was 75. Then, the digital literacy and learning outcomes data calculated using normalised gain scores (g). The difference between pre-test and post-test in treatment or control class calculated using Wilcoxon test. Meanwhile, the post-test score and normalised gain score between treatment and control class calculated using Mann-Whitney test. The student response analysed using Guttman scale assessment criteria, then calculated the percentage and interpreted using student's responses category levels.

3. Results and Discussion

The results showed that students' digital literacy in the treatment class was mastery in the percentage of 57% with 3% student categorised as excellent and 54% of students categorised as good. Students in the percentage of 43% were not mastery with sufficient categories. While in the control class, 100% of students did not mastery digital literacy with 3% of them categorised as sufficient, 7% as poor, and 90% as very poor. Based on the normalised gain score of the treatment class, there was no student who experienced an increase in the low category of digital literacy, 13 students experienced an increase in the medium category and 17 students in the high category. In normalized gain score of the control class, all students did not experience an increase (Figure 1).

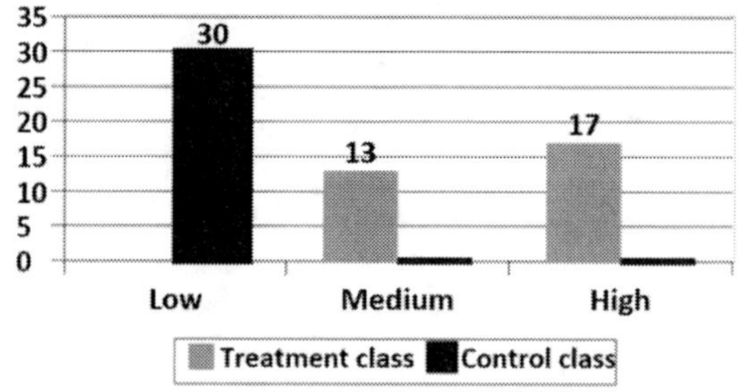

Figure 1. Histogram of student' digital literacy improvement in the treatment and control class based on normalised gain scores (g)

The results of statistical tests showed that there was an influence of learning using the electronic Invertebrate book on students' digital literacy, detailed in Table 1.

Table 1. Statistical test results of student' digital literacy in treatment and control class

No.	Parameters	Explanation
1.	The difference between pre-test and post-test scores of digital literacy in treatment class.	Significantly different, based on the Wilcoxon test with the significance level of 95%
2.	The difference between pre-test and post-test scores of digital literacy in the control class.	Not different, based on the Wilcoxon test with the significance level of 95%
3.	The difference in digital literacy post-test score between treatment class and control class.	Significantly different, based on the Mann-Whitney test with the significance level of 95%
4.	The difference in digital literacy gain score (g) between treatment class and control class.	Significantly different, based on the Mann-Whitney test with the significance level of 95%

Four indicators of digital literacy in the learning process using the electronic Invertebrate book, are internet searching, hypertextual navigation, content evaluation, and knowledge assembly (Table 2).

Table 2. The percentage of digital literacy indicators mastery between treatment and control class

No.	Indicators	Percentage (%)	
		Treatment class	Control class
1.	3.9.7 Formulate the stages of information searching related to Invertebrate material on the Internet using the search engine.	75	4.3
2.	3.9.8 Identify the navigation of hypertext in the web browser related to Invertebrate material.	72.5	27.8
3.	3.9.9 Evaluate the validity of content related to Invertebrate material.	79.5	22.9
4.	3.9.10 Manage information related to Invertebrate material in the form of fact and opinion that has been found online.	56	14.2
Average (%)		**71**	**17**

Based on Table 5, it is known that the average mastery of each indicator in the treatment class was 71%, and the control class was 17%. In the treatment class, the digital literacy indicator that experienced the highest mastery was the 3.9.9 indicator which was 79.5%, and the lowest was at the indicator 3.9.10 which was 56%. The digital literacy indicator in control class, the highest mastery was the 3.9.8 indicator which was 27.8%, and the lowest indicator was 3.9.7, which was 4.3%.

The result also showed that student' learning outcomes in the treatment class were 80% mastery and 20%, not mastery. In the control class is 77% mastery, and 23% is not mastery. Based on the normalized gain score of the treatment class, there was no student who experienced an increase in low category learning outcomes, and six students experienced an increase in the medium category and 24 students in the high category. In normalised gain score results of the control class, there was no student who experienced an increase in low category learning outcomes, and seven students experienced an increase in the medium category, and 23 students experienced a high category increase (Figure 2).

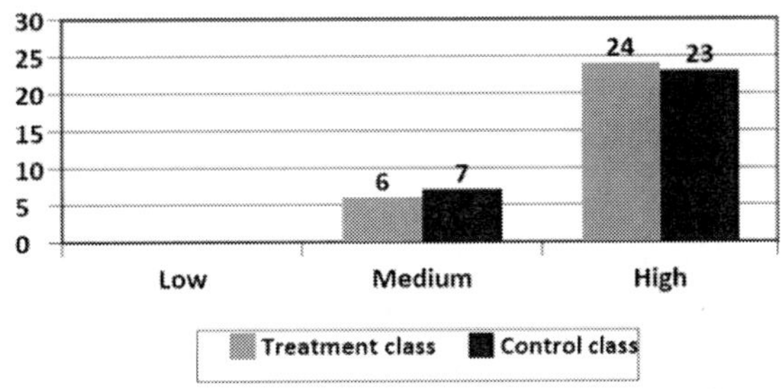

Figure 2. Histogram of student' learning outcomes improvement in the treatment and control class based on normalised gain scores (g)

The results of statistical tests showed that there was an influence of learning using the electronic Invertebrate book on student learning outcomes, detailed in Table 3.

Table 3. Statistical test results of student' learning outcomes in treatment and control class

No.	Parameters	Explanation
1.	The difference between pre-test and post-test scores of learning outcomes in treatment class.	Significantly different, based on t-test with the significance level of 95%
2.	The difference between pre-test and post-test scores of learning outcomes in the control class.	Significantly different, based on the Wilcoxon test with the significance level of 95%
3.	The difference in learning outcomes post-test score between treatment class and control class.	Significantly different, based on the Mann-Whitney test with the significance level of 95%
4.	The difference in learning outcomes gains score (g) between treatment class and control class.	Not different, based on the Mann-Whitney test with the significance level of 95%

They are six indicators of students' outcomes in the learning process using electronic Invertebrate book. The average treatment indicator was 81%. Meanwhile, the control class was 78% (Table 4).

Tabel 4. The percentage of learning outcome indicators mastery between treatment and control class

No	Indicators	Percentage (%)	
		Treatment class	Control class
1.	3.9.1 Analyze the basic classification of phylum levels in animals based on body layers, body cavity body symmetry, and reproduction	92.2	78.9
2.	3.9.2 Explain the general characteristics of Porifera Phylum members.	93.3	82.2
3.	3.9.3 Analyze the differences in several life cycles as a basic classification of class in the Coelenterata Phylum.	54.2	60.4
4.	3.9.4 Describe the life cycle of Platyhelminthes Phylum' members.	79.4	92.2
5.	3.9.5 Analyze the distinguishing features of members of the Phylum Platyhelminthes, Nemathelminthes, and Annelida.	85.6	90
6.	3.9.6 Grouping members of Phylum Arthropod to the class level.	79	63.1
	Average (%)	**81**	**78**

Based on Table 4, the mastery average of each indicator in the treatment class was 81%, and the control class was 78%. In contrast to the control class, the learning outcome indicators that experienced the highest mastery was 3.9.4 indicator in the score of 92.2%, and the lowest was 3.9.3 indicator, in the score of 60.4%. A class was mastery if 75% of students reach a value of ≥ 75. The two classes were mastery because 80% of individual scores in the treatment class were mastery and 77% of the individual scores in the control class were mastery.

Student responses measured using student response questionnaires. Most students (97.4%) gave a positive response to learning using electronic book Invertebrates for enhancing digital literacy skills. Thus, it can be stated that students give a very good response to the learning using the electronic Invertebrate book to improve the digital literacy of high school students grade 10 (Table 5).

Table 5. Students responses to the learning using the electronic Invertebrate book

No	Criteria	Percentage (%)
1.	Learning atmosphere of Invertebrate sub material was entertaining.	96.7
2.	Learning atmosphere of Invertebrate sub material was fun.	93.3
3.	Learning of Invertebrate sub material made students easier to know the characteristics of each phylum of Invertebrate animals.	100
4.	Learning of Invertebrate sub material made students easier to classify Invertebrate animals.	100
5.	Learning of Invertebrate sub material made students easier to identify the phylum of Invertebrate animals.	100
6.	Usage instructions made students easier to operate electronic book Invertebrate.	100
7.	Learning goals and chapter descriptions helped students to use electronic book Invertebrate.	96.7
8.	Learning atmosphere about Invertebrate using electronic book Invertebrate was a new thing for students.	93.3
9.	Learning atmosphere of Invertebrate sub material using electronic book Invertebrate was fun.	96.7
10.	Electronic book Invertebrate facilitated students to study independently outside school hours.	96.7
11.	Electronic book Invertebrate support learning and additional information about Invertebrate.	100
12.	Electronic book Invertebrate helped students to observe parts of Invertebrate animals that were invisible and rarely found.	100
13.	Features in electronic book Invertebrate support the learning process of students about Invertebrate material.	100
14.	Invertebrate material contents in electronic book Invertebrates was appropriate if paired with face-to-face learning in the form of practicum.	96.7
15.	Students were more motivated in learning when using electronic book Invertebrate.	100
16.	Through the learning process using electronic book Invertebrate, students can access material every time and everywhere.	100
17.	Students can look at the stages of information search related to Invertebrate sub material on the internet using search engines through the feature of "*mari kita lihat*" and "*Info +*".	93.3
18.	Students can identify hypertext navigation in web browser related to Invertebrate sub material through the feature of "*Info +*" and "*Mari menjelajah*".	96.7
19.	Students can evaluate the validity of information content related to Invertebrate sub material through the feature of "*Mari menjelajah*".	96.7
20.	Students can managed information related to Invertebrate sub material in the form of new facts and opinions found online through the feature of "*mari menjelajah*".	100
	Total Average (%)	**97.4**

The pre-test results in both control and treatment classes revealed that 100% of students scored under the minimum standard on digital literacy and learning outcomes so that all students declared as not mastery. This was because students have not got the prior knowledge about Invertebrate and digital literacy material. According to Shepherd & Tello [13] a material delivered by the teacher will be fully understood by students if students have gained prior knowledge regarding the material. Therefore, prior knowledge was necessary for students to form material concepts.

After learning using electronic book Invertebrate was applied, digital literacy skills and student learning outcomes had increased. All treatment class students experienced an increase in digital literacy because students learn to use electronic book Invertebrates which contain features that facilitating digital literacy. The difference in students' digital literacy skills is because the level of understanding of each student related to digital literacy taught was different, influenced by various factors. In accordance with Lee [14] that students' environmental factors are related to scores and levels of digital literacy. According to Jan [12], students who are accustomed to processing technology can have high digital literacy. Also, the aspects of learning outcomes between the two classes get a percentage of 100% which has increased. This is because, in learning, improving learning outcomes is caused by several factors, both internal and external. Some of them are mastery the material, efforts of students to learn, ability and motivation, and other factors[15].

In contrast with the control class, 100% of students increased in a low category. This is because, in the control class, electronic book Invertebrate was not applied, so learning took place only in the school. However, there was a student in the control class who has a higher digital literacy pre-test and post-test score than the other students. The students' digital literacy was reflected in the way they answer questions that cannot be answered by other control class students below:

> *"In question number 1, to find relevant information about Porifera, I should enter Porifera key as an antibacterial and choose articles on the website journal.plos.org. I did not choose www.brainly.co.id because it uses a free domain and cannot be trusted. I do not choose the wordpress.com website because it contains answers that can be written by people who are not experts. I do not choose any website that is in the top search because it is only based on the number of webs that are clicked the most." (Student number 4 in the control class)*

The answer written by student number 4 of the control class was different from the other students because the student has prior knowledge about digital literacy at his home. This prior knowledge helped him to elaborate new information with what he had learned before. This is consistent with Yuksel [16] that prior knowledge helped students to understand and draw a framework of the topic. Also, according to Hargittai [17], personal use and experience influenced the level of digital literacy skills. However, other control class students do not have this ability. The average achievement of learning indicators using Invertebrata electronic book in the digital literacy aspects of the treatment class and aspects of learning outcomes are declared as mastery. Whereas in the control class, aspects of digital literacy declared in mastery and aspects of learning outcomes declared as mastery. According to Uno & Koni [18] the achievement of an indicator is said to be mastery if the average percentage of achievement is around 70 to 90%.

The highest achievement of the digital literacy indicator in the treatment class was indicator 3.9.9, which evaluate the validity of information content related to Invertebrate sub material in the score of 79.5%. Previously, the achievement of students' digital literacy indicators on indicator 3.9.9 was 18.8%. Before learning using the electronic book, almost all students cannot answer the questions given correctly. Students do not understand how to evaluate the validity of information content. As in one example of student answers:

> *"The website for accessing information about Coelenterata is https://id.m.wikipedia.org/wiki/cnidaria because Wikipedia often appears at the top of the search and is recommended." (Student number 15 treatment class)*
>
> *"Information from Facebook regarding photos of worm sightings living in sardines, I chose to spread the news to social media. This is because by spreading the news, we can invite people around to be more vigilant." (Student number 24 treatment class)*

The answers written by students number 15 and 24 of the treatment class were a reflection of the most answers to other treatment class students. This indicated that the treatment class students before learning using electronic book Invertebrate, do not have the skills to evaluate the validity of information content. However, after learning using the electronic book, most students were able to choose trusted websites based on educational domains such as .edu, .ac.id, and .ac.uk. As in one example of student answers:

"The website to find out information about leeches is at http://www.ucmp.berkeley.edu/ Hirudinea because on the website there are .edu domains, which is educational websites." (Students number 15 is a treatment class).

Besides, after learning with an electronic book, students can answer correctly when given information that was not undoubtedly true. Students have been able to choose to hold the spreading of the information and find the truth through various trusted sources. As in the number 2 post-test digital literacy item regarding the choice of action against leech information in kale that lives in the human intestine even though it has been cooked. There are students who write the answer:

"It is better to keep the news spread because before spreading the news, we have to look for information related to the truth of the news in order to avoid hoax." (Student number 12 treatment class).

"It is better to keep the news spread because leeches are ectoparasites which can only grow or live outside the body. So that the news is not true or false." (Student number 5 treatment class).

Almost all students in the treatment class answered questions like in number 2 correctly. It was because in electronic Invertebrate book, especially related to leech discussion (Chapter 5 Annelida) there is a menu "*Mari menjelajah*" which facilitate students to find reliable sources of information on the internet through trusted websites such as ww.animaldiversity.org, www.oseanografi.lipi.go.id, www.ucmp. berkeley.edu, www.cdc.gov and scientific articles we can access such as in www.mri.journal.or.id, www.e-journal.biologi.lipi.go.id, and www.citeseer.ist.psu.edu to train the ability to evaluate the validity of information content [8].

The process of finding reliable information can finish well, because there were clear guidelines, systematic paths, and equipped with links in the book. This is consistent with Embong et al. [19] that good electronic book display navigation of hypertext links, multimedia objects, and technology access guides. In addition, the information presented in the book is a contextual phenomenon that commonly happens in the community, so students can easily understand it. Evaluation of information content (Content Evaluation) is one of the critical digital literacy competencies [2]. According to Metzger & Flanangin [20], evaluation of online information content was to determine the information received, whether it is trusty or not. Through the feature of "Mari Menjelajah", students also asked to elaborate their searching results in the form of essay. This menu was found in each chapter of the book so that students were able to evaluate thoroughly.

In addition, the indicator of manage information related to Invertebrates in the form of facts and opinions that have been found online (knowledge assembly), get the lowest percentage, in the score of 56%. Some students answer correctly, but the steps were incomplete, such as:

"In fact, in the human stomach, there is chloride acid which makes the worms that enter the body die. Also, at the time while sardines preserved, it must be through a heating process. Thus, if there is a worm or other animals, it will die. The news presented is not correct." (Student number 15 treatment class).

This is because processing and building information in the form of facts and opinions are indicators that still need to be accustomed to students. There are still many students who provide incomplete answers. Students are less able to elaborate on some data related to the questions presented. Under the statement of Moreno & Park [21] students who have high initial knowledge will be able to integrate various complex information elements with existing schemes, and can process them as one of the memory systems. Also, Ngilawajan [22] revealed that information processing by students is done by linking information received with the knowledge that is possessed. Therefore, strong initial knowledge will help students build and combine information and students need to get used to processing information online so that their abilities increase rapidly.

In the aspect of learning outcomes, the highest indicators mastery in the treatment class was found in indicator 3.9.2, whereas in the control class, the indicator with the highest mastery was indicator 3.9.4. Before the learning process, 14% of students answered correctly on indicator 3.9.4, then increased to 92.2%. The highest indicator in these two was because, explaining and describing was a simple level of cognitive process, which is included in the cognitive level of understanding (C2) [23]. Therefore, many students answer correctly on this indicator. On the other hand, the lowest mastery in the treatment class is found in indicator 3.9.3, with a percentage of 54.2%. Before learning, only 4% of the control class answered correctly on indicator 3.9.3. This is because students are less accustomed to dealing with questions with a higher cognitive level. Analysing in Bloom's Taxonomy is a reasonably high level of cognitive process dimensions (C4) [23]. This is also supported by the research of Surbakti & Napitupulu [24] that the assessment of "analyse" level questions is one of difficult by students in addition to evaluating (C5).

The results of the statistical tests in Table 1 have four parameters tested to prove the hypothesis proposed, whether there is an influence of the electronic Invertebrate book on the digital literacy abilities of high school students grade 10. Based on the non-parametric statistic test, Wilcoxon, with the significance level of 95%, there are differences in the digital literacy of students before and after treatment. This was because 100% of students experienced a significant increase in digital literacy after being given learning using electronic book Invertebrate. Unlike the control class, there was no difference in students' digital literacy before and after treatment. This was because learning in the control class does not use electronic Invertebrate book. According to Vrana's research [25], digital literacy positively influences student academically. Also, digital literacy plays a role in developing students' knowledge of certain subjects by encouraging high creativity and curiosity [26]. Therefore, digital literacy of students can increase after learning using electronic Invertebrate book.

The results of statistical tests in Table 3 have four parameters tested to prove the proposed hypothesis, whether there is an effect of using electronic book Invertebrate on the learning outcomes of high school students grade 10. Based on the t-test in the treatment class with 95% significance level, there was a difference between student learning outcomes before and after treatment. This difference is because the electronic book contains Invertebrate material which visualised into text, images, schemes, videos, and links that connect to other learning resources so that student learning outcomes can increase as a result of learning. According to Pradina & Suyatna [27], the contents of electronic book consist of videos, images, texts, and animations can make it easier for students to visualise abstract concepts. Lai [28] also revealed that electronic book that transformed into three-dimensional forms containing multimedia such as video and animation could stimulate students' enthusiasm and interest in reading and make the learning atmosphere more lively and enjoyable. Research by Herrlinger et al. [29] showed that the presence of images could improve learning outcomes. In addition, with the presence of animation and video, students can quickly learn abstract material and relate it to previous knowledge [27]. Thus, electronic book played a positive role in increasing student learning outcomes. Similar to the control class, there was a difference in student learning outcomes before and after treatment. Both the treatment and control classes experienced a significant increase in learning outcomes, so there were no normalised gain score differences between the learning outcomes of the two classes. This is caused by several factors that can influence student learning outcomes, such as in Soleh et al., [15], learning outcomes can be influenced by internal and external factors of students. In addition, the material content taught in electronic book Invertebrate and school book has the same weight, so the material absorbed by students in both classes was same.

Students respond very well to the learning process and the application of digital literacy skills. Not all students give a positive response to all specified aspects. There were two students who were still having trouble looking at the stages of information searching on internet related to Invertebrate using search engines through the features of "*Mari kita lihat*" and "*Info+*" that existed in the book. Such as the answer to a student:

> *"Electronic book contains many webs which delivered in English, making it difficult." (Student number 2 is a treatment class).*

This is because each web page has a different display and language of instruction, and students have difficulty understanding web pages that mostly use English. Even so, students need to be trained to take advantage of caption options (on videos available on Youtube), Google Translate, and page translation options available in the web browser. This guide is listed in the "Usage Instructions" page.

Knight [30] stated that technology built on reliable content to support students in manipulating material and engaging in learning is a useful educational technology. Thus, technology is useless without valid content. An understanding of information content requires the integration of various aspects, and using technology not only to access information but to identify and evaluate it [31]. Therefore, students need to be accustomed to accessing various types of web browser views so that they can identify and evaluate them properly. Thus, learning using the electronic Invertebrate book that implemented has done very well. Electronic Invertebrate book has proven to be improving digital literacy and student learning outcomes.

4. Conclusions

Students' digital literacy abilities increased. Invertebrate material, especially in the Nemathelminthes, Annelida, and Arthropod chapters, is facilitated using digital literacy learning. Learning using electronic book based on Invertebrate sub material influenced student learning outcomes. The mastery of student learning outcomes individually and classically has increased. Learning using electronic book based on Invertebrate sub material received very good responses from students.

References

[1] Partnership for 21St Century 2008 21st Century Skills, Education & Competitiveness *a Resour. Policy Guide.* 20
[2] Gilster P 1997 *Digital literacy* (Wiley Computer Pub.)
[3] Martin A 2006 *Nord. J. Digit. Lit.* **2** 61
[4] Greene J A, Yu S B and Copeland D Z 2014 *Comput. Educ.* **76** 55
[5] Ting Y L 2015 *Internet High. Educ.* **26** 25
[6] Pratiwi N and Pritanova N 2017 *Semantik* **6** 11
[7] Gumgum G, Justito A and Nunik M 2017 *Pengabdi. Kpd. Masy.* **1** 35
[8] Putri I D T, Ningrum H T R, Diana N P and Ambarwati R 2018 Invertebrata Biologi X untuk Melatihkan Literasi Digital. Patent number: 000140314
[9] Alawiyah H, Muldayanti N D and Setiadi A E 2016 *J. Biol. Educ.* **3** 9
[10] Putri I D T and Ambarwati R 2019 *Proceeding Semin. Nas. Biol. III Unesa* 1
[11] Fraenkel J R and Wallen N E 2009 *How to Design and Evaluate Research in Education* (New York: McGraw-Hill)
[12] Jan S 2018 *Issues Trends Learn. Technol.* **6** 15
[13] Shepherd C and Tello E A 2015 *J. Coll. Teach. Learn.* **12** 95
[14] Lee S J, Kim J M and Lee W G 2015 *Indian J. Sci. Technol.* **8** 81
[15] Soleh A, Pramono and Suratno 2009 *J. Pendidik. Tek. Mesin* **9**
[16] Yuksel I 2012 *World Appl. Sci. J.* **20** 197
[17] Hargittai E and Eshet Y 2005 *Soc. Sci. Comput. Rev.* **23** 371–9
[18] Uno H and Koni S 2012 *Assesment Pembelajaran* (Jakarta: Bumi Aksara)
[19] Embong A M, Noor A M, Hashim H M, Ali R M and Shaari Z H 2012 *Procedia - Soc. Behav. Sci.* **47** 1802–9
[20] Metzger M J and Flanagin A J 2013 *J. Pragmat.* **59** 210–20
[21] Moreno R and Park B 2010 *Cogn. Load Theory* 7
[22] Ngilawajan D A 2013 *Pedagogi. J. Pendidik.* **2** 71
[23] Arends R I 2012 *Learning to Teach* (New York: McGraw-Hill, Inc)
[24] Surbakti J and Napitupulu M A *J. Pelita Pendidik.* **4** 9
[25] Vrana R 2014 *Commun. Comput. Inf. Sci.* **492** 9
[26] Payton S and Hague C 2010 Digital literacy across the curriculum Key to themes
[27] Pradina L P and Suyatna A 2018 *Turkish Online J. Educ. Technol. - TOJET* **17** 17
[28] Lai C 2016 *J. Educ. Sci. Environ. Heal.* **2** 57
[29] Herrlinger S, Höffler T N, Opfermann M and Leutner D 2017 *Res. Sci. Educ.* **47** 685
[30] Knight B A 2015 *Cogent Educ.* **2** 1
[31] Koutropoulos A 2011 *J. Online Teach. Learn.* **7** 38

The Effect of Environmental Pollution Module on Environmental Worldview in Senior High School

M Syamsussabri[*]**, Suhadi and Sueb**

Department of Biology, Faculty of Mathematics and Natural Sciences, Universitas Negeri Malang, Semarang Street 5, Malang, Indonesia.

[*] E-mail: syamsussabri.edu@gmail.com

Abstract. This study aimed to know the effect of environmental pollution module on the environmental worldview in senior high school. This type of research is quasi-experiment. The research instrument was a scale of the New Ecological Paradigm (NEP). This research conducted at SMAN 1 Lembar. The research subjects were 54 students of class XI Mathematics and Natural Science. Samples were taken by cluster sampling with the equality test. The analysis of data used One-way Analysis of Covariance (ANCOVA). The results showed there was the effect of the environmental pollution module on the environmental worldview in senior high school with p-value $< .005$.

1. Introduction

Education has a role in creating quality human resources. Through education, it is expected to develop innovative souls, who can develop their potential and play a role in development in various aspects of life [1]. Education must continue to make better innovations to face future challenges and meet people's needs. One effort that can be done to realise these hopes is by providing facilities that are by the development of science and technology [2]. One effort that can be done to improve the quality of education is by developing teaching materials that are following the curriculum and the conditions of the students being taught [3, 4]. Teaching materials developed according to the needs of teachers and students and adequately used will be one of the essential factors that can improve the quality of learning [5].

Excellent teaching materials can not only improve conceptual understanding, but also develop skills such as analysing, interpreting, summarising, and solving problems with creative solutions [3]. Teaching materials can change the role of teachers and students in the learning process. At first, the teacher was the only source of information in the class and students were only recipients of the data. With the existence of teaching materials, the teacher is not only the only source of learning in the class, but the teacher is more directed to act as a student facilitator in the learning process. Meanwhile, using teaching materials developed according to the needs of students, students are directed to become active learners by studying the material contained in teaching materials first before participating in classroom learning. One teaching material that can be used in learning is the module.

Modules are printed teaching materials that are arranged to be learned by students independently. Modules are also referred to as independent teaching materials because they have been equipped with instructions for self-learning, which means that students can carry out learning activities without direct educators [6]. Modules are teaching materials arranged following the subject matter, design learning, and disseminated to students to be used in learning activities [7]. The use of modules is very important in learning because it is not limited by the time of learning so that it can be used for active and detailed learning from the material being studied [8]. Modules are one of the appropriate teaching materials developed and used by the environmental conditions of students. Environmental pollution module has been developed and is by the analysis of teacher and student needs and environmental conditions around students [9].

Environmental worldview is a collection of assumptions and beliefs about how nature works and how humans relate to the environment, including environmental ethics, or ideas about what is right or wrong with how we treat the environment [10]. Environmental worldview is also interpreted as a human perspective on the situation [11]. This environmental worldview can be measured using the scale of the New Ecological Paradigm (NEP) with five indicators including: (1) the reality of the limits of growth, (2) anti anthropocentrism, (3) damage to the balance of nature, (4) rejecting exceptionalism, and (5) ecological crisis [12]. Based on the statement above, this research needs to be done to explore the effects of the application of environmental pollution modules to NEP support.

2. Method

2.1 Design

The design of the research conducted is a quasi-experimental study, which is a type of research used in education [13]. The design of the research was nonrandomized control pre-test post-test design. The experimental class was conducting learning activities with the environmental pollution module and the control class conducting learning activities with student worksheets. This research was conducted in SMAN 1 Lembar, West Nusa Tenggara Province. Participants in this study consisted of 54 grade-tenth students in academic years 2018/2019. The hypothesis in this study is there is the effect of the environmental pollution module on the environmental worldview in senior high school.

2.2 Procedures

The treatment was a 2- to 3-week environmental pollution module devoted to global environmental problems, with some emphasis on changes in environmental balance, environmental pollution, and management of the environment. Experimental data came from students receiving the environmental pollution module. Control data came from students who did not have the environmental pollution module. The scale of the NEP was distributed at the time before treatment and after the treatment of environmental pollution module.

2.3 Assessment

Measurement environmental worldview used NEP scale. In was designed to identify five possible components of an ecological worldview: (1) limit of growth; (2) anti-anthropocentrism; (3) the fragility of nature's balance; (4) rejection of "exemptionalism"; and (5) the possibility of an eco-crisis [12]. The NEP consist of 15 Likert scale question, three on each component. Possible scores range from a minimum of 15 to a maximum of 75. The odd-numbered NEP statements are coded as follows: Strongly Agree (SA) = 5, Mildly Agree (MA) = 4, Unsure (U) = 3, Mildly Disagree (D) = 2, Strongly Disagree (SD) = 1. Values assigned to even-numbered statements are coded in the exact opposite manner.

In addition, individuals segmented into three groups based on the sum of their NEP scores. The groups are comprised as follows: (1) Anti-environmental: respondents with a NEP score of 50 or less; (2) Moderately: respondents with a NEP score of greater than 50 and less than 59; (3) pro-environmental: respondents with a score of 59 or more [14]. The effect of applying the applied environmental pollution module was analysed using one-way ANCOVA provided that the data is normally distributed and homogeneous [15].

Table 1. NEP scale items

No.	Question
	The reality of limit to growth
1	We are approaching the limit of the number of people the earth can support
6	The earth has plenty of natural resources if we just learn how to develop them
11	The earth is like a spaceship with very limited room and resources

No.	Question
	Anti-anthropocentrism
2	Humans have the right to modify the natural environment to suit their needs
7	Plants and animals have as much right as humans to exist
12	Humans were meant to rule over the rest of nature
	The fragility of nature's balance
3	When humans interfere with nature, it often produces disastrous consequences
8	The balance of nature is strong enough to cope with the impacts of modern industrial nations
13	The balance of nature is very delicate and easily upset
	Rejection of exemptionalism
4	Human ingenuity will ensure that we do NOT make the earth unliveable
9	Despite our special abilities, humans are still subject to the laws of nature
14	Humans will eventually learn enough about how nature works to be able to control it
	Eco-crisis
5	Humans are severely abusing the environment
10	The so-called "ecological crisis" facing humankind has been greatly exaggerated
15	If things continue on their present course, we will soon experience a major ecological catastrophe

Source: Dunlap et al (2000) [12]

2.4 Environmental Pollution Module
The module development model adapts the Analyze, Design, Develop, Implement, and Evaluate (ADDIE) development model [16]. Environmental pollution module has been developed and is by the analysis of teacher and student needs and environmental conditions around students [9]. The environmental pollution module used, shown in Figure 1.

Figure 1. Environmental pollution module

The environmental pollution module developed has a final form in the form of a printed book. The module has fulfilled the eligibility requirements in terms of material, media, practicality, and readability [17]. The main components of this environmental pollution module include titles, learning objectives, material descriptions, learning activities, summaries, formative tests, self-evaluations, and study journals. The component was compiled and adjusted [6, 18, 19]. Other components, modules, identity, introduction, usage instructions, table of contents, list of images, table list, introduction, summative test, self-evaluation, answer key, summary map, reference list, and a glossary about the preparation of teaching materials [19].

3. Results and Discussions
Environmental pollution module is known to have a significant influence to improve environmental worldview students. It is known that the indicator of natural balance damage is an indicator that has the highest increase score with a percentage of 11.8% while the indicator with the lowest increase is rejecting

exceptionalism with a percentage of 5.9%. The comparison between the control class and the experiment showed that the experimental class had an increase in environmental worldview scores by 9%.

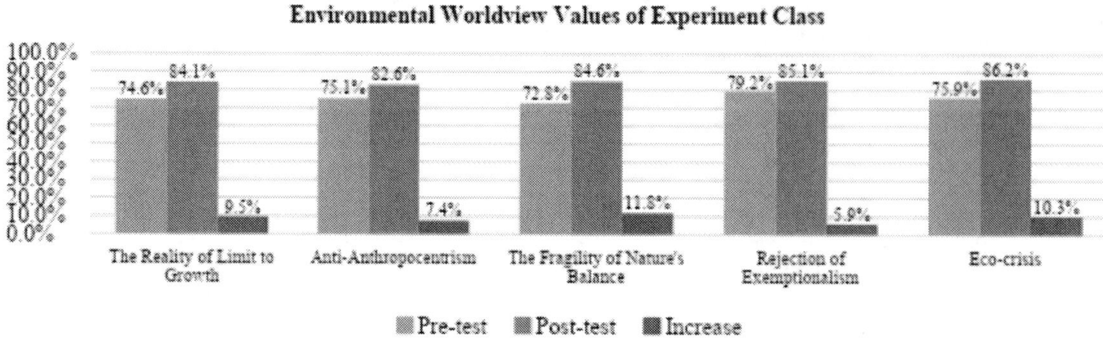

Figure 2. Environmental worldview values of experimental class

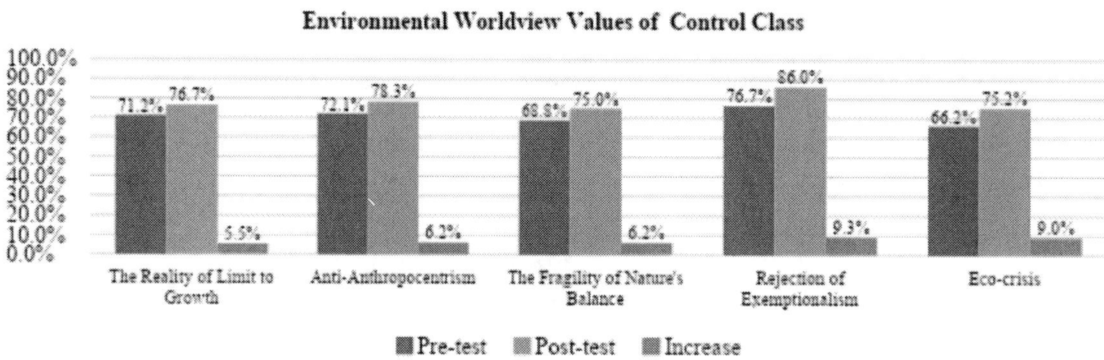

Figure 3. Environmental worldview values of the control class

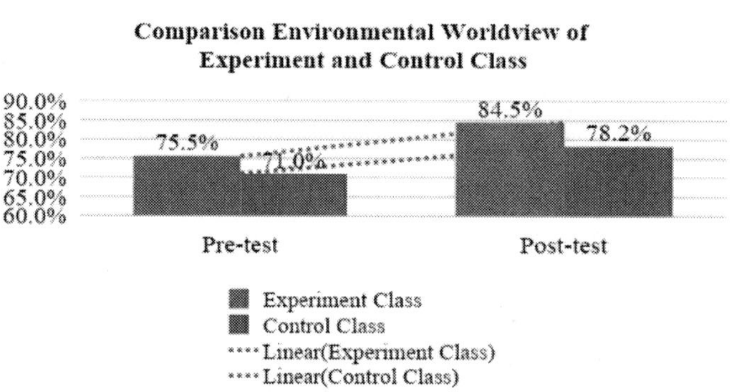

Figure 4. Comparison of environmental worldview in the experimental and control class

Hypothesis test results showed there was the effect of the environmental pollution module on the environmental worldview in senior high school with p-value $< .005$.

Table 2. Tests of between-subjects' effects

Dependent Variable: Environmental Worldview

Source	Type III Sum of Squares	df	Mean Square	F	Sig.
Corrected Model	584.939[a]	2	292.469	16.761	.000
Intercept	748.409	1	748.409	42.891	.000
Pretest	286.366	1	286.366	16.412	.000
Group	130.977	1	130.977	7.506	.008
Error	889.895	51	17.449		
Total	202043.000	54			
Corrected Total	1474.833	53			

a. R Squared = .397 (Adjusted R Squared = .373)

This module is known to have a significant influence on students' environmental worldviews. That modules that examine environmental problems can have a significant influence on increasing the score of NEP [20], which is an indicator that measures environmental worldview. NEP itself is a very good indicator and is suitable for measuring the worldview of one's environment [12, 20, 21, 22]. The NEP scale is already widely used to measure environmental worldview [11, 20, 23, 24, 25, 26]. The results of the analysis showed that the environmental pollution module has a significant influence on environmental worldview, as evidenced by the results of the ANCOVA test with *p-value* < .005.

4. Conclusion

Based on the results of the analysis showed there was the effect of the environmental pollution module on the environmental worldview in senior high school with *p-value* < .005. Based on this result we also recommend other research and teaching materials need to be developed that can increase students' environmental worldview, awareness, and attitude.

Acknowledgement

The authors would like to thank the Ministry of Research, Technology and Higher Education of Republic Indonesia under "*Direktorat Riset dan Pengabdian Masyarakat*" grant in the fiscal year of 2019.

References

[1] Badan Pusat Statistik (BPS) 2017 *Potret Pendidikan Indonesia* (Jakarta: Badan Pusat Statistik)
[2] Padmapriya 2015 *Int. J. Educ. Psychol. Res.* **4** 44
[3] Barron B and Hammond L D 2008 *Teaching for Meaningful Learning: A Review of Research on Inquiry-Based and Cooperative Learning* (San Francisco: The George Lucas Educational Foundation)
[4] Departemen Pendidikan Nasional (Depdiknas) 2006 *Pedoman Memilih dan Menyusun Bahan Ajar* (Jakarta: Depdiknas)
[5] Sadjati I M 2012 *Hakikat Bahan Ajar* (Banten: Universitas Terbuka)
[6] Departemen Pendidikan Nasional (Depdiknas) 2008 *Penulisan Modul* (Jakarta: Departemen Pendidikan Nasional)
[7] Kementerian Riset, Teknologi, dan Pendidikan Tinggi Republik Indonesia (Kemenristekdikti) 2017 *Panduan Penyusunan Perangkat Pembelajaran & Bahan Ajar* (Jakarta: Kementerian Riset, Teknologi, dan Pendidikan Tinggi Republik Indonesia)
[8] Shah S, Cox A G and Zdanowicz M M 2013 *Curr. Pharm. Teach. Learn.* **5** 651
[9] Syamsussabri M, Sueb and Suhadi *2018 International Conference on Mathematics and Science Education (ICoMSE) Proceeding* 1
[10] Miller G T and Spoolman E 2016 *Environmental Science* (USA: Cengage Learning)
[11] Sueb 2015 *Prosiding Seminar Nasional Biologi/IPA dan Pembelajarannya* 335

[12] Dunlap R E, Van Liere K D, Mertig A G and Jones R E 2000 *J. Soc. Issues* **56** 425
[13] Ary D, Jacobs L C, Razavieh A and Sorensen C K 2010 *Introduction to Research in Education Eighth Edition* (Wadsworth: Cengage Learning)
[14] Kotchen M and Reiling S 2000 *Ecol. Econ.* **32** 93
[15] Mertler C A and Reinhart R V 2017 *Advanced and Multivariate Statistical Methods Practical Application and Interpretation* (New York: Routledge).
[16] Branch R M 2009 *Instructional Design: The ADDIE Approach* (New York: Springer)
[17] Syamsussabri M, Sueb and Suhadi 2019 *J. Pendidik. Teor. Penelitian, dan Pengemb.* **4** 175
[18] Daryanto 2013 *Menyusun Modul, Bahan Ajar untuk Persiapan Guru dalam Mengajar* (Yogyakarta: Gava Media)
[19] Prastowo A 2015 *Panduan Kreatif Membuat Bahan Ajar Inovatif* (Yogyakarta: DIVA Press)
[20] Rideout B E 2005 *J. Environ. Educ.* **37** 3
[21] Dunlap R E and Jones R E 2002 *Handbook Environmental Sociology* 482
[22] Mifsud M C 2011 *US-China Educ. Rev. 3* pp 413
[23] Birhanu A 2014 *J. Environ. Earth Sci.* **4** 89
[24] Liefländer A K *Environ. Educ. Res.* **21** 4
[25] Lwo L, Fu J and Chang C 2017 *J. Balt. Sci. Educ.* **16** 706
[26] Sookram R 2013 *J. Values-Based Leadersh.* **6** 1

The Effectiveness of Learning Materials of Inquiry Models to Practice Science Process Skills in Simple Harmonic Motion's Matter

M Sholikhah *[1], T Prastowo[2] and Prabowo[2]

[1] Department of Postgraduate Science Education, Universitas Negeri Surabaya, Indonesia
[2] Department of Physics, Faculty of Mathematics and Natural Science, Universitas Negeri Surabaya, Indonesia

*Email : maratus.17070795021@mhs.unesa.ac.id

Abstract. This research aims to know the effectiveness of learning materials of inquiry learning models to practicing science process skills in simple harmonic motion's matter. This research was conducted on 60 tenth grade students at Muhammadiyah 1 Gresik Senior High School. The effectiveness of this research is seen from science process skills tests, learning achievement, and student response. Science process skills data were obtained from pre-test and post-test. Learning outcomes data were obtained from affective and psychomotor assessment through observation methods and cognitive assessment through the post-test method. While student response data were obtained from the questionnaire method. N-Gain values were obtained from average to high range for each class from the results of pre-test and post-test, this showed the increase in process skills of students. The data analysis of learning achievement indicated that all students have final results that meet the minimum completeness criteria (>75). The data analysis of student responses showed positive results, the highest positive response was the use of student worksheets (LKS) that eases students in learning. Thus, based on the results obtained it can be concluded that this research learning instrument is effective for practice the science process skills in simple harmonic motion's matter.

1. Introduction

Education is an important factor in character building of personal, family, community and nation. The purpose of national education is to develop the potential of a student becoming a human being who believe and fear God, have good moral, knowledgeable, creative, independent, and become a responsible citizen. The cult of technological development requires teachers to continue to plays an active role not only in providing knowledge but also to teach students how to solve problems around them in preparation for the future. Hence special skills in problem solving must be embedded, these skills are nothing but process skills. Science subjects, especially physics aim to develop science and concepts related to daily life, develop student's curiosity, produce positive attitudes, and realize that there is a connection between science and the environment as well as being able to develop student's science process skills.

One of the subject matter of physics that is of concern to researchers is simple harmonic motion. Simple harmonic motion studies the phenomenon of objects that move periodically to their equilibrium position. Mathematical calculations in simple harmonic motion material are one of the factors causing students to avoid this material [1]. The mistakes made by students in working on problems on this topic are allegedly caused by students still having difficulty identifying the quantities used to determine the spring period or determine the swing period, from the test instruments given, students assume that the greater the mass give the greater the swing period [2]. Students also have difficulty in determining the change in the direction of objects that vibrate [3]. Another difficulty occurs when students have to interpret a graph [4]. In addition, this material is also suitable for training science process skills with an inquiry-assisted inquiry model according to Basic Competence 4.11, namely conducting experiments on the simple harmonic motion.

As said before, the cult of technological development requires teacher to continue to plays an active role so that the learning process runs smoothly and shared goals are achieved. But the learning process does not always depend on the existence of the teacher (educator) as a manager of the learning process. This is based on the nature of the learning process, namely the interaction between students and the object being studied. Every learning process requires a learning materials. Learning materials are competencies that refer to actions that are rational and meet certain specifications in the learning process [5]. Learning materials are a product of development research that must meet several conditions, one of which is the effectiveness of these materials. The materials is said to be effective if it gives results in accordance with the goals set by the developer.

Researchers assume that students need to have science process skills to solve problems related to everyday life. In fact, the science process skills in our country are still below average. Learning in the classroom is still centered on teacher (teacher-centered), teacher in the same direction teaches concepts to students without involving students to be active in the learning process [6]. In addition, the results of literature studies on the ranking of natural sciences in Indonesia still show a low level. It is based on PISA (International Student Assistance Program) 2015 which focuses on literacy reading, Mathematics, and Science which ranked Indonesia 69th out of 76 countries. The relatedness of science-focused PISA with skills of science process lied in indicators of skills of science process such as observing, interpreting, etc [7].

The science process skills were an approach to learning, where students get the opportunity to interact in real objects to find their concepts [6]. Science process skills were physical and mental skills related to fundamental abilities that are possessed, mastered, and applied in scientific activities, so scientists can discover something new [8]. The science process skills encouraged students to discover the facts, concepts of knowledge and develop the attitudes and values requested [9]. The science process skills need to be trained on students in science learning, particularly physics. revealed that science process skills are important competencies that must be achieved in the learning process of science. To realize such desire learning in a classroom should actively engage students [10].

The learning process was an activity in which students are given the opportunity to develop, both in terms of knowledge, skills, and character. The learning model is said to be effective if students can be engaged actively in the learning process, the learning model must be adapted to the material and abilities of students and facilities at school [11]. The problem that has occurred so far because students are not able to analyze existing information and tend to receive what information is submitted or written in the books, students tend to be passive in the learning process. The inquiry model encourages students to observe, classify, make guesses, explain, measure, make conclusions, and so on. The inquiry learning model refers to the development of students in the material by collecting data. Whereas the development of process skills needs to be based on working, designing or other things that produce something. The process of collecting data can be obtained in various ways one of them by conducting experiments.

The experimental method is a method that provides opportunities for students either individuals or groups to be trained in performing a process or experiment [12]. By this method students are expected to fully engage in experiments, conduct experiments, find facts, collect data, and solve problems they encounter in real time. Students are motivated when they work together to solve physics problems [1]. By conducting experiments, it eventually makes students going to do science activities in groups and will be trained in process skills and then students can search for knowledge and truth in science concepts that they understand from experiments.

Previous research that revealed the application of inquiry learning was positively responded to understanding students' abilities especially in science learning [1] [13]. Previous studies have also shown that the application of inquiry models can improve science process skills and learning achievement [14] [15] [16]. Based on the reasons that have been presented and see the successful implementation of inquiry learning beforehand, it is important to know the effectiveness of the tools designed based on inquiry models to foster students' science process skills. The materials used in this learning are materials that have gone through the validation process first of 3 validators and have been tested valid. In this study, researchers wanted to see the effectiveness of learning materials of an inquiry model in training process skills on simple harmonic motion's matter. In addition, researchers also want to see the effectiveness of learning materials used based on learning achievement and student response to learning with these materials.

2. Research Methods

This study used one group pre-test – post-test design with 3 replications. This research was conducted on 60 tenth grade students at Muhammadiyah 1 Gresik Senior High School. The effectiveness of learning materials seen from 3 aspects: science process skills, learning achievement and student responses. Retrieval of research data was done through a preliminary test (pre-test) and a final test (post-test) for aspects of knowledge and science process skills, observation of student attitudes during learning and student skills during conducting experiments, as well as questionnaires for analysis of student response during learning.

To find out whether students have trained science process skills could be seen from the N-Gain increase between pre-test and post-test. Student learning results could be seen from the cumulative calculation of 50% aspects of knowledge, 30% aspects of affective and 20% aspects of psychomotor [17]. Student response could be seen from the percentage of student's interest in learning. On this research also conducted parametric statistical tests because 3 replications were conducted. As for statistical tests conducted were normality, homogeneity, ANOVA's test and paired samples T-tests.

3. Result and Discussion

To find out the normality and homogenity of the three replication samples, it is necessary to do a statistical test with the following hypothesis:

H_0: There were no significant differences in the 3 samples

H_1: There are significant differences in 3 samples

Statistical test results showed that the samples used in normal and homogeneous distributed research indicated by significance values of > 0.05 (H_0 accepted) as in table 1 and table 2 below.

Table 1. Results of Normality Test

	Kolmogorov-Smirnov[a]			Shapiro-Wilk		
	Statistics	Df	Sig.	Statistics	Df	Sig.
Pre-test of MIPA 1 Class	0.178	20	0.098	0.928	20	0.144
Pre-test of MIPA 2 Class	0.169	20	0.138	0.934	20	0.184
Pre-test of MIPA 3 Class	0.149	20	0.200*	0.925	20	0.124

a. Lilliefors Significance Correction

Table 2. Results of Homogenity Test

Levene Statistics	df1	df2	Sig.
3.113	2	57	0.052

Data analysis of pre-test and post-test on science process skills showed that MIPA 1 class shows the results of N-Gain in the range 0.42-0.91 (moderate-high), for the MIPA 2 class the results of the N-Gain in the range 0.45-0.85 (moderate-high), while for the MIPA 3 class the results of N-Gain ranged from 0.42-0.84 (moderate-high). The results of the N-Gain proved that using the inquiry learning tools are effective for practicing student's science process skills. This is in line with research [5] [18] which concluded that the application of inquiry models had a positive effect on student's science process skills.

Analysis of learning achievement data is seen from 3 aspects namely knowledge, affective and psychomotor. On those 3 aspects, the results of analysis of learning achievement data for 3 classes showed that after through learning the inquiry model the value of all students meets the minimum completeness criteria (KKM $\geq$ 75) with an average grade of MIPA 1 class is 77.15, the average grade of MIPA 2 class is 77.60, and the average grade of MIPA 3 class is 77.63.

Analysis of student response data was obtained from the results of the student response questionnaire given after all of the learning processes is complete. The questionnaire contained 11 questions regarding the applied learning process. The results of student response analysis classe MIPA 1 are presented in the student response in figure 1 below.

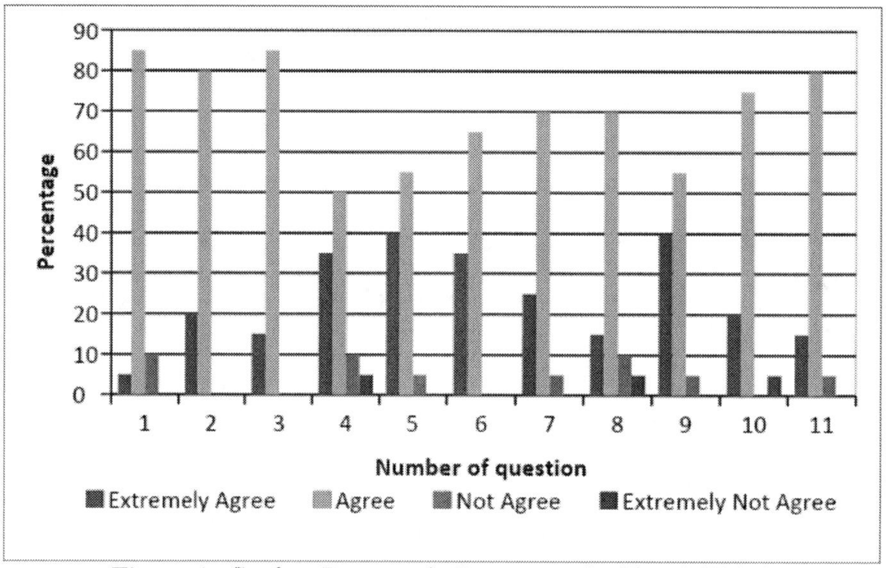

Figure 1. Student Response's Percentage in Class MIPA 1

The results of student response analysis classe MIPA 2 are presented on the student response in figure 2 below.

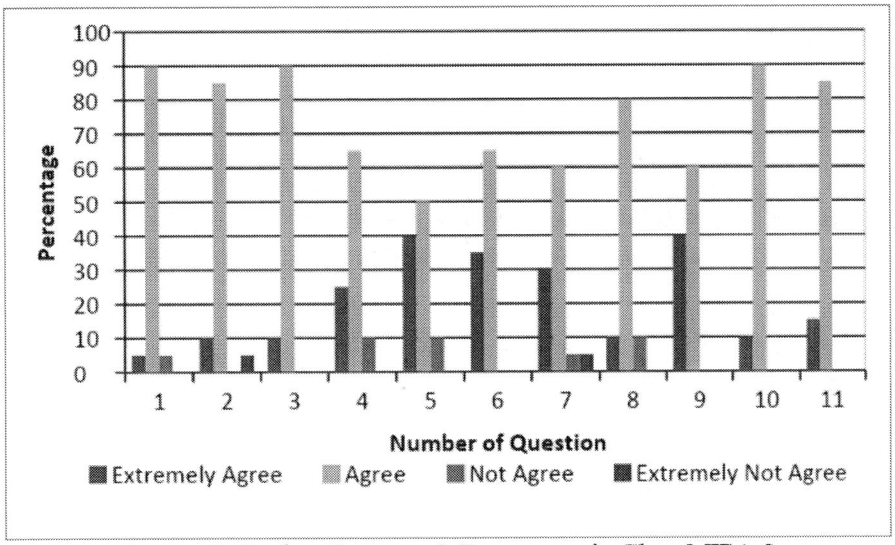

Figure 2. Student Response's Percentage in Class MIPA 2

The results of student response analysis classe MIPA 2 are presented on the student response in figure 3 below.

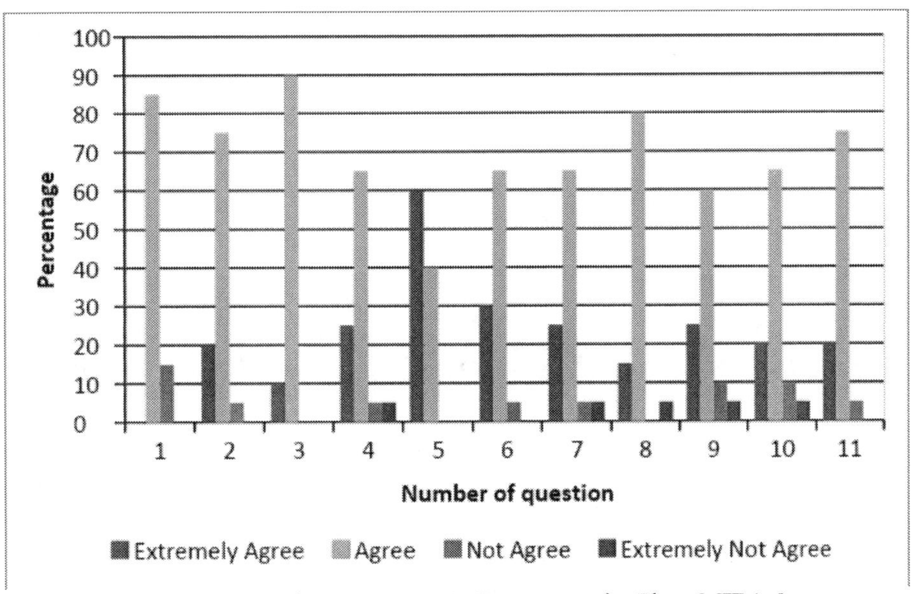

Figure 3. Student Response's Percentage in Class MIPA 3

Based on the figure 1, figure 2 and figure 3, it could be seen that student response to learning are classified as positive response because the majority of student choose 'agree' and 'extremely agree' for the question given. The highest positive response is about the worksheets (LKS) that shared to students as well as many new things students felt in this learning among others formulating problems, identifying problems, formulating variables, formulating hypotheses, designing experimental activities, making table and chart independently, analyzing observational data and concluding experiments based on problems.

ANOVA test results showed a significance value > 0.05 (Ho accepted) indicated a consistency of N-Gain increase in all three classes. For the result of ANOVA test is presented in table 3 below.

Table 3. ANOVA Result

	Sum of Squares	df	Mean Square	F	Sig.
Between Groups	0.050	2	0.025	2.012	0.143
Within Groups	0.713	57	0.013		
Total	0.763	59			

Lastly, to compare two samples in pairs, i.e. samples with the same subject but different treatments so it will be seen that there is an effect due to the treatment given, it is necessary to do a paired samples T test with the following hypothesis:
H_0: there is no effect due to the treatment given
H_1: there is an effect due to the treatment given
Statistical test results showed that there is an effect due to the treatment given indicated by significance values of < 0.05 (H_0 rejected) as in table 4 below.

Table 4. Result of Paired Samples T-Test

				95% Confidence Interval of the Difference				
	Mean	Std. Deviation	Std. Error Mean	Lower	Upper	T	Df	Sig. (2-tailed)
Pre–Post test MIPA 1	-5.17855E1	11.26601	2.51916	-57.05815	-46.51285	-20.557	19	0.000
Pre–Post test MIPA 2	-5.392E1	8.96651	2.00497	-58.12246	-49.72954	-28.896	19	0.000
Pre–Post test MIPA 3	-5.785E1	7.64960	1.71050	-61.43312	-54.27288	-33.822	19	0.000

4. Conclusion

Based on the results and discussion, firstly, seen that N-Gain from pretest and posttest data shown that student has criteria moderate to high value, so it can be concluded that the materials are effective to practice science process skills of student. Secondly, the accumulation of knowledge test data, affective assessment data, and psychomotor assessment data showed that the learning achievement for all students has final results that meet the minimum completeness criteria. Thirdly, almost all of the students give a positive response with the learning process that uses the learning materials. So can be concluded that learning materials of inquiry model used by researchers are effective to practice student's science process skills in simple harmonic motion's matter.

Acknowledgments

The authors would like to thank Muhammadiyah 1 Gresik Senior High School (Gresik, Indonesia) for allowing authors to conduct research, as well as teachers and tenth-grade students who have many roles in this research.

References

[1] Aldophus T, Alamina J and Aderonmu T S B 2013 *Journal of Education and Practice* **4** pp 95-100
[2] Sugara Y D, Sutopo and Latifah E 2016 *Pros. Semnas Pend. IPA Pascasarjana UM* vol 1 (Malang: Pascasarjana Universitas Negeri Malang) pp 506–12
[3] Parnafes O 2010 *The Journal of Science Education and Technology* **19** pp 565-79
[4] Merhar V K, Planinsic G and Cepic M 2009 *Eur J.Phys* **30** pp 163-75.
[5] Rando A R 2016 *Jurnal Pendidikan* **1** pp 1-11
[6] Erikko D, Qurbaniah M and Kurniati T 2018 *Ar-Razi Jurnal Ilmiah* **6**
[7] Gusdiantini L, Aeni A N and Jayadinata A K 2017 *Jurnal Pena Ilmiah* **2** pp 651-60
[8] Fatmi N S 2014 *Jurnal Pendidikan Fisika* **3** pp 47-52
[9] Wulanningsih S, Prayitno B A and Probosar R M 2012 *Pendidikan Biologi* **4**
[10] Hombing R Br, Yennita and Sahal M 2017 *Jurnal Online Mahasiswa FKIP* **4**
[11] Kurniawati L, Akbar R O and Misri M A 2015 *EduMa* **4** pp 62-74
[12] Ningsih C S 2015 *Jurnal Pendidikan Guru Sekolah Dasar* **9**
[13] Wahyuni S, Kosim and Gunawan 2018 *Jurnal Pendidikan Fisika dan Teknologi* **4** pp 240-46
[14] Ainina N 2017 *Makalah Komprehensif* (Surabaya: Pascasarjana Unesa)
[15] Rinarta I N, Yuanita L and Widodo W 2014 *Jurnal Pendidikan Fisika* **2** pp 70-88
[16] Sopacua V, Prabowo and Sudibyo E 2018 *Proc. Int. Conf. on Mathematics and Science Education* vol 3 (Bandung: Universitas Pendidikan Indonesia) pp 124-29
[17] Prabowo 2013 *Prosiding Seminar Nasional 2nd Lontar Physics Forum* (Semarang: IKIP PGRI Press) pp 1-4
[18] Wegasanti N and Maulida A N 2017 *Pensa e-jurnal: Pendidikan Sains* **5** pp 376-80

Learning Management System with Moodle to Enhance Creativity of Candidate Physics Teacher

G Gunawan[1*], H Sahidu[1], S Susilawati[1], A Harjono[1] and L Herayanti[2]

[1]Physics Education Study Program, Universitas Mataram, Jl. Majapahit No. 62, Mataram 83115, Indonesia
[2]Physics Education Study Program, IKIP Mataram, Jl. Pemuda No. 59A, Mataram 83125, Indonesia

*Email: gunawan@unram.ac.id

Abstract. Creativity is one of the main goals of education. Debriefing creativity is essential to win in global competition. Practicing creativity can be done by utilizing the development of information and communication technology, including learning physics for candidate teachers. One innovation that can be done to improve the quality of creativity-oriented learning is to develop a learning management system (LMS) with Moodle. The aim of this research is to produce a learning management system with Moodle that can enhance the creativity of candidate physics teachers. However, the focus of this article is to explain the results of trials of the effectiveness of products that have been developed to increase the creativity of candidate physics teachers. This research is a research development that adopts ADDIE research design (analysis, design, develop, implementation, and evaluate). The research instrument used to measure candidate physics teacher creativity is a creativity test consisting of 80 questions and consists of various types of tests both multiple-choice, short answers, and essays, which have been developed according to creativity indicators. To find out the increase in student creativity, data were analysed using the n-gain test. The results of the study show that the model develops successfully trained the creativity of candidate teachers, including verbal, figural, numerical, and procedural aspects. This model allows lecturers to manage learning and exchange information with students quickly and flexibly.

1. Introduction

The development of information and communication technology is getting faster, and this has an impact on the changing perspective of education. The change in perspective does not only occur in the concept of education, but also in the mechanism of education itself. According to Susanti [1], the need for information technology-based learning mechanisms becomes necessary. Technology in learning is a tool used in formal education practices to disseminate, illustrate, communicate, or maintain students and teachers in activities designed intentionally to sustain learning [2]. Application of technology in learning is expected to increase the effectiveness of learning.

Learning relates to the application of technology in it. Learning can be said to be effective if it utilizes information and communication technology optimally as a tool [3]. One use of information and communication technology in learning is to use computers. Form of computer utilization can be in the form of integration of LMS (Learning Management System) software. LMS provides features that support learning activities such as discussions, practice questions, final exams, etc., online and offline.

Herayanti et al. [4] stated that learning management systems are effective digital learning environments and are designed for learning that continues to develop. An effective digital-based learning environment will help students improve their knowledge through activities that encourage critical thinking, communication, collaboration, and creativity.

Development of LMS requires good planning that is organized so that development is able to convey learning objectives thoroughly. LMS is available in a variety of options, including online tutors, interactive blackboards, online classes, Moodle, and others. Moodle is a free open source software [5]. Moodle is a learning platform designed to be used by educators, administrators, and students as a reliable and secure integrated system so that it becomes a learning environment suitable for each user [6]. Moodle is used by teachers to convey information, give and listen to assignments, deliver electronic journals, and other learning resources.

In addition, student creativity can be enhanced by providing learning that supports the creation of an active learning environment. One of the active learning environments is to apply computer-based or digital-based learning. Computer-based learning requires students to be active in every stage of learning. Moodle-based e-learning also includes digital-based learning. This based learning has proven to be very supportive in developing students' creative abilities [7].

Moodle is electronic-based learning. Electronic-based learning has more roles and benefits compared to traditional learning. The benefits of electronic learning according to Sims [8] and Seok [9] are (1) increasing levels of learning interactions between students and teachers, (2) enabling learning interactions to occur from anywhere and at any time, (3) reaching students in a broad scope, and (4) facilitate the improvement and storage of learning materials. The main advantage of electronic-based learning is the opportunity for students to interact electronically with each other and their teachers during the forum, on the discussion board, via email and in the chat room [5].

Moodle (Modular Object-Oriented Dynamic Learning Environment) which means a dynamic learning place by using an object-oriented model or a dynamic web-based educational environment package. Moodle is open-source software that supports the implementation of learning management systems with an integrated paradigm with various learning support features that can be accommodated in an e-learning portal. According to Sampurno et al. [10], Moodle functions as an effective tool in providing learning facilities because it is equipped with important learning support features such as assignments, quizzes, chat, collaboration, and main features that can upload various formats of learning material and are easier to understand because the information presented is not only in the form of writing but also images.

Moodle can be used to build systems with the concept of e-learning (electronic learning). The benefits of using LMS with Moodle are very important, including overcoming the limited frequency of face-to-face meetings between students and lecturers. Graf [11] stated that this model supports teachers in creating and managing online learning and provides them with various features that can be included in learning such as learning materials, quizzes, discussion forums, assignments, and so on. This model also allows lecturers to manage learning and exchange information with students quickly and flexibly. Lopes [12] stated the benefits of providing online content on learning make students learn anywhere and anytime, faster than other conventional distance education methods. In addition, communication resources can make it more efficient between teachers and students, when compared to other conventional methods. The final goal of this research is to produce a LMS with Moodle model to improve the creativity of candidate physics teachers.

Most of the learning management systems that have been developed are only for increasing mastery of concepts. Though the potential of the LMS is extraordinary. However, for the development of high-level skills (eg creativity), the LMS system must be developed more modern and smarter in supporting the development of those skills. Then was born the product LMS system that researchers developed in such a way that could support the development of creativity, especially in the candidate of physics teacher. The selection of teacher candidates as research samples aims to introduce teacher candidates to the great potential possessed by an LMS system if developed properly and seriously, and of course, the best introduction is to try it out by yourself. Data on increasing creativity will provide a new understanding of the vast potential of LMS.

2. Methods

This research includes development research with ADDIE design (analysis, design, develop, implementation, and evaluate). However, the focus of this article is in explaining the results of product implementation in learning. In this implementation phase, researchers used a pretest posttest design research design to test the effectiveness of the product on improving the creativity of the research sample. An LMS with Moodle model had been developed to increase the creativity of candidate physics teachers. The research data were obtained from the creativity skills test instrument given before and after learning. The research instrument used to measure candidate physics teacher creativity is a creativity test consisting of 80 questions and consists of various types of tests both multiple-choice, short answers, and essays, which have been developed according to creativity indicators. Increased creativity in both classes was based on the verbal, figural, numerical, and procedural aspects. Increased student creativity was determined from the N-gain score. This is done to avoid mistakes in interpreting the acquisition score of increasing student creativity. An N-gain score is obtained from the following formula [13].

$$N - Gain = \frac{Post\ test\ score - Pre\ test\ score}{Maximum\ score - Pre\ test\ score} \text{ x } 100\% \quad (1)$$

N-gain > 70% (high); 30% ≤ N-gain ≤ 70% (medium); dan N-gain < 30% (low).

3. Result and Discussion

The results of the study are in the form of a learning process for candidate physics teachers using LMS with Moodle and increasing student creativity. Student creativity measured in research includes numerical, verbal, figural, and procedural aspects. In general, improvements occur in every aspect. Herayanti et al. [4] concluded that in learning using Moodle-based learning media, students showed an increase in critical thinking skills and provided ease of use and were effective in overcoming learning difficulties that students experienced. Increasing students' critical thinking skills will increase their creativity because in this critical process students must be able to give their logical and creative thinking well.

The average percentage score of N-gain creativity in verbal, figural, numerical, and procedural aspects of physics teacher candidates for 2nd semester, 4th semester, and 6th semester is shown in Figure 1 below.

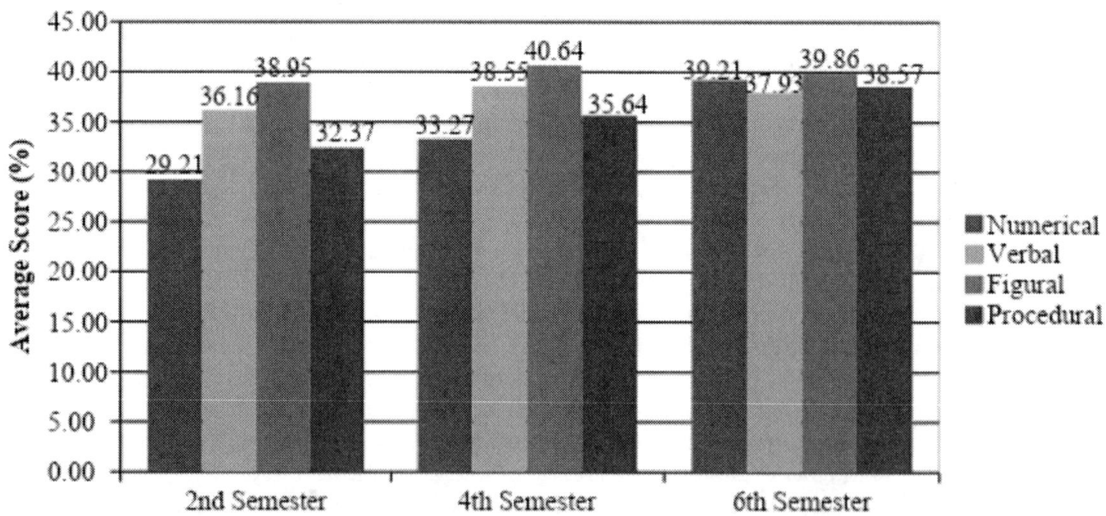

Figure 1. Percentage of N-gain Average Score Creativity of candidate Physics Teachers.

Second-semester students show a variety of increased creativity. In numerical indicators, the group of second-semester students increased by 29.21% in the low increase category. The indicator of verbal creativity has increased by 38.16% with the category of moderate improvement. Figural creativity increased by 38.95%. And the last indicator increased by 32.37%. Second-semester students experienced a fairly low increase in numerical and procedural creativity. This is because the second-semester students are still less reliable in dealing with numerical and conceptual problems. Jiao & Onwuegbuzie [14] stated that junior students are still struggling in adjusting their study habits at university so that sometimes they cannot be faced with problems at a higher level. Moreover, second-semester students are still struggling to build their basic concepts and when faced with higher concepts, they tend to experience difficulties. However, the application of project-based learning has been able to improve their high-level skills a bit, especially in their creativity. Bereiter & Scardamalia [15] stated that project-based learning is very good to help build knowledge in students.

Another result shows that fourth-semester students' creativity percentage is varied. In numerical indicators, the group of second-semester students increased by 32.27% in the category of moderate improvement. Verbal creativity indicator increased by 38.55% with the category of moderate improvement, figural creativity increased by 40.64%, whilst, the last indicator increased by 35.64%. Fourth-semester students also experienced a fairly low increase in numerical and procedural creativity. However, compared to the beginning semester students their ability on numerical and procedural creativity indicators has increased. This is in line with a study by Cavallo et al. [16] stated that both male and female students, over time their understanding of concepts continues to develop. The fourth-semester students struggle to build and apply their basic concepts, and when faced with a higher concept, they still experience difficulties, but their knowledge capital is better than before. However, the application of project-based learning has been able to improve their high-level skills a bit, especially in their creativity.

The sixth-semester students show improvement in creativity is relatively better. In numerical indicators, the group of second-semester students increased by 39.21% in the category of moderate improvement. The indicator of verbal creativity has increased by 37.93% with the category of moderate improvement. Figural creativity increased by 39.86%. And the last indicator increased by 38.57%. sixth semester students show that they have developed a series of knowledge from year to year. Indicators of their creativity are increasing gradually. McNeel [17] stated that the cognitive abilities of students will continue to develop in a supportive environment. Every semester students are faced with higher level subjects, this learning environment will train students and if it is adapted to good learning methods, such as the use of project-based learning, then their chances of developing better will be greater. This was confirmed by the results of this study.

Based on the percentage of increase in creativity it can be seen that the highest average percentage of the second semester is in the aspect of figural 38.95, the fourth semester is in the aspect of figural 40.64, while the sixth semester in aspect of figural is 39.86 in the medium category. Figural aspects are higher than other aspects such as numerical, verbal, and procedural. The highest percentage that occurs in the figural aspect shows the ability of creativity is quite high in bringing up new ideas or ideas through the drawings made. Gunawan et al. [18] said that the implementation of cooperative learning models with virtual media has a positive effect on student creativity on the concept of static fluids and there is an increase in creativity of both groups in verbal and figural aspects after treatment.

The difference in the percentage of N-gain in the numerical aspects of candidate physics teachers has increased at each semester level of physics teacher candidates. This is because candidate physics teachers are accustomed to doing things related to numbers and solving problems with number concepts. For example, in the second semester of basic physics and calculus; the fourth semester of wave material and basic electronics; and the sixth semester of modern physics. This shows the effect of student training activities with the creativity of candidate teachers. Herayanti et al. [19] stated that the use of Moodle as a learning medium in learning could improve students' conceptual understanding of wave concepts.

Students' verbal and figural creativity each semester is not significantly different. This shows the ability to make new combinations based on data, information, or elements found from possible answers to one problem and verbally revealed does not develop in line with increased learning time. This is supported by research by Gunawan [20] which shows that the application of computer-based learning tends to support the development of procedural creativity, but the indicators of verbal creativity and figural creativity do not increase as well as procedural creativity. This is because candidate physics teachers in their learning environment associate more with numbers, mathematical equations and physics concepts, which of course these things rely more on procedural creativity.

This research data shows that the application of LMS can support the development of student creativity. The character of students in each semester can be well supported in accordance with their abilities and needs. Moodle besides being a non-paid application, also provides a variety of interesting learning experiences for students. Students are trained in creativity in every aspect of teaching, both in online and offline learning, and the simulations and animations provided are able to support student development. Hermansyah et al. [21], Gunawan et al. [22], and Herayanti [19] stated that all computer simulations both online and offline have great opportunities to support the improvement of various specific abilities, such as only student creativity.

4. Conclusion

The results of this study indicate that the product that has been developed in the form of a moodle-based learning management system has been able to increase the creativity of candidate physic teachers. The increase that occurs in each semester is different for each indicator. The results also showed that the difference in improvement at each semester level was different. The pattern of improvement shows that indicators of numerical and procedural creativity increase with increasing semester levels. However, verbal and figural creativity indicators show a non-significant increase. This relates to the experiences and learning environments of candidate physics teachers who are mostly faced with learning environments filled with numbers, mathematical formulas, and physics concepts, so of course the ability of creativity on indicators related to these will be greatly affected, especially if supported by the system learning right, like a moodle-based learning management system.

Acknowledgements

The authors would like to thank profusely to the Institute for Research and Community Service, University of Mataram, which has funded this research in 2019. The authors also thank all those involved in the research, data collection, and improvement of this publication.

References

[1] Susanti E and Sholeh M 2008 Rancang Bangun Aplikasi E-Learning *Jurnal Teknologi* **1** (1) pp 53-57

[2] Garrison D R and Anderson Terry 2003 E-Learning in the 21st Century (New York: Taylor & Francis Group)

[3] Hanum N S 2013 Keefetifan E-Learning sebagai Media Pembelajaran (Studi Evaluasi Model Pembelajaran E-Learning SMK Telkom Sandhy Putra Purwokerto) *Jurnal Pendidikan Vokasi* **3**(1)

[4] Herayanti L Fuaddunnazmi M and Habibi H 2017 Pengembangan Perangkat Pembelajaran Fisika Berbasis Moodle *Jurnal Pendidikan Fisika dan Teknologi* **3**(2) 197-206

[5] Al-Ajlan A and Zedan H 2008 Why moodle 12th IEEE International Workshop on Future Trends of Distributed Computing Systems pp 58-64

[6] Hakim A B 2016 Efektifitas Penggunaan E-Learning Moodle Google Classroom Dan Edmodo. I-STATEMENT **2**(1)

[7] Hermansyah H, Gunawan G and Herayanti L 2017 Pengaruh Penggunaan Laboratorium Virtual Terhadap Penguasaan Konsep dan Kemampuan Berpikir Kreatif Mahasiswa Pada Materi Getaran dan Gelombang *Jurnal Pendidikan Fisika dan Teknologi* **1**(2) 97-102

[8] Sims 2008 Rethinking (E) learning: a Manifesto for Connected Generation *International Journal on E-learning*

[9] Seok S 2008 Teaching Aspect on E-learning *International Journal on E-learning*
[10] Sampurno P J, Maulidiyah R and Puspitaningrum H Z 2015 Implementasi Kurikulum 2013: MOODLE (Modular Object Oriented Dynamic Learning Environment) dalam Pembelajaran Fisika melalui Lembar Kerja Siswa pada Materi Optik di SMA *Jurnal Fisika Indonesia*, **19**(56) pp 54-58
[11] Graf S and Liu T C 2008 Identifying Learning Styles in Learning Management Systems by Using Indications from Students' Behaviour *Eighth IEEE International Conference on Advanced Learning Technologies* pp 482-486
[12] Lopes A P 2014 Learning management systems in higher education. In EDULEARN14 Conference *Proceedings of EDULEARN14 Conference-IATED Publications* pp 5360-5365
[13] Cheng K K, Thacker B A, Cardenas R L and Crouch C 2004 Using An Online Homework System Enhances Students' Learning of Physics Concepts in An Introductory Physics Course *American Journal of Physics* **72**(11) pp 1447-1453
[14] Jiao Q G and Onwuegbuzie A J 2001 Library Anxiety and Characteristic Strengths and Weaknesses of Graduate Students' Study Habits *Library Review* **50**(2) pp 73-80
[15] Bereiter C and Scardamalia M 2003 Learning to Work Creatively With Knowledge. *Powerful Learning Environments: Unravelling Basic Components and Dimensions* pp 55-68
[16] Cavallo A M, Potter W H and Rozman M 2004 Gender Differences In Learning Constructs, Shifts in Learning Constructs, and Their Relationship to Course Achievement in A Structured Inquiry, Yearlong College Physics Course For Life Science Majors *School Science and Mathematics* **104**(6) pp 288-300
[17] McNeel S P 1994 College Teaching and Student Moral Development *Moral Development in The Professions: Psychology and Applied Ethics* pp 27-49
[18] Gunawan G, Harjono A, Sahidu H, and Nisrina N 2018 Improving students' creativity using cooperative learning with virtual media on static fluida concept *In Journal of Physics: Conference Series* **1006** (1) 012016
[19] Herayanti L, Gummah S, Sukroyanti B A, Ahzan S and Gunawan G 2018 Developing Moodle in Problem-Based Learning to Improve Student Comprehension on the Concepts of Wave *Advances in Intelligent Systems Research (AISR)* **157**(1) pp 134-137
[20] Gunawan G, Harjono A, Susilawati and Dewi S M 2019 Generative Learning Models Assisted by Virtual Laboratories to Improve Students' Creativity in Physics *Journal of Advance Research in Dynamical & Control Systems* **11**(7) pp 403-411x`
[21] Hermansyah H, Gunawan G, Harjono A and Adawiyah R 2019 Guided inquiry model with virtual labs to improve students' understanding on heat concept *Journal of Physics: Conference Series* **1153** (1) 012116
[22] Gunawan G, Nisrina N, Suranti N M Y, Herayanti L, and Rahmatiah R 2018 Virtual Laboratory to Improve Students' Conceptual Understanding in Physics Learning *Journal of Physics: Conference Series* **1108** (1) 012049

Practicality and effectiveness of SETS based learning materials to trained students' higher-order thinking skills

A F Hasanah*[1], Raharjo[2] and F Rachmadiarti[2]

[1]Department of Postgraduate Science Education, Universitas Negeri Surabaya, Indonesia
[2]Department of Biology, Faculty of Mathematics and Natural Science, Universitas Negeri Surabaya, Indonesia

*Email: auliafirda01@gmail.com

Abstract. At present, there are no many environment-based learning activities with interesting practices that utilize technology and at the same time to train students with higher-order thinking skills. Therefore, a learning instrument is needed to implements these practices. This research aims to describe the practicality and effectiveness of SETS based learning materials for higher-level thinking skills of 1st grader of junior high school in Sciences subject, including (1) describing the implementation of learning, (2) describing student responses toward learning, and (3) describing higher-order thinking skills of students. Research methods used were observation method for learning realization, questionnaire method for student responses, and test method to determine learning achievement (pre-test and post-test). Research instruments used were observation sheets of learning realization, questionnaire sheet for student response, and sheets of the higher-order thinking skills test. Results of the research showed that (1) learning based on the SETS approach was conducted very well; (2) student responses toward learning are highly positive; and (3) higher-order thinking skills of students have increased after SETS-based learning materials have been applied. This is shows that SETS learning materials are stated to be practical and effective to be used in learning to trained higher-order thinking skills.

1. Introduction

SETS approach is a learning approach by relating components of science, environment, technology, and society. Within SETS, the student will be deployed to the community for analyzing occurred problems equipped with knowledge as their capability. By student knowledge, whether it new or already they have, students expected to find the solution of the occurred problem. From this, many ideas expected to be generated to solve the occurred problems by creating a new technology product or by developing existing technology with notice impacts may appear in the environment. Thus, all components of SETS have relations that affect each other [1]. By implements of SETS approach on Sciences material, then various student skills such as thinking skills, problem-solving, activity, and learning achievement also expected to be improved.

Both critical thinking skill and creative thinking to solve the problem are two competencies must behave by student in this 21st century beside skills in collaborating and communicating [2]. Creative and critical thinking skills are two indicators of higher-order thinking skills. Higher-order thinking skill

(HOTS) is an ability of human to thinking at higher cognitive level, where many information coming in can be analyzed critically and creatively to get solution of the problem based on information/knowledge he/she has, which then used to make a decision in various situations [3]. Bloom categorized intellectual behavior become six thinking levels, they are; knowing, understanding, applying, analyzing, synthesizing, and evaluating [4]. Meanwhile, higher-order thinking skill can be trained through activities or giving questions with higher cognitive level, like C4 (analyzing), C5 (evaluating), and C6 (creating) [5].

SETS approach based learning which is based on the surrounding environment that authentic for students expected can train their thinking skill. Learning based on reality is easy to implicated in the concepts of science, environment, technology, and society, because of this the indicators of HOTS can be realized [6]. For the student, learning conducted by using surrounding environment objects will create real experience because conducted through direct observation, creating action patterns and thinking patterns may be reflected in daily life, also can be used as guidance to learn in a group or alone [7]. One of the environmental problems is the waste of batik industry. Batik industry produces chemical waste which in long term will give a very bad impact on both environment and community if no further treatment for such waste [8].

Such a thing needs to become the attention of all parties, without exception for educators to equip their students in order to have a role to prevent environmental damage. One of the ways can be conducted by teachers is by prepared and conducted learning with activities based on waste treatment. SETS approach learning that applying science, environment, technology, and society is expected to appropriate with that goal and to trained student thinking skill at once.

Nevertheless, based on observation and interview with teacher and student at a school in Bangkalan Regency, generally, learning at that school still applying the memorize method. The student never does activities to process chemical waste yet, and the absence of learning instrument based on SETS approach that designed to trained higher-order thinking skill of the student. Therefore, research with activities of waste treatment is necessary to be conducted.

SETS based learning instrument proved valid theoretically according to expert judgment, it also expected to obtain empirical validity to trained higher-order thinking skills. Therefore, this research will describe the practicality and effectiveness of SETS based learning instrument to trained higher-order thinking skill of students.

2. Method

2.1 Research Objects

Objects in this research were 60 graders VII junior high school 1 Arosbaya on Sciences subject in the even semester of the academic year 2018/2019 who selected randomly and divided into three groups. Students in this research were 13-15 years old, where students in this age range are in the development stage of formal operational intellectuals. At this age student can constructs ideas more abstract than at the concrete operational stage and commonly they able to solve the problem [9].

2.2 Research Subjects

SETS based learning instruments in this research limited on lesson plan, student textbook, student worksheet, and test questions of higher-order thinking skill. The lesson plan arranged based on the SETS approach that consists of invitation stage, exploration, solution, application, and conceptualization. Textbook containing materials about Environmental Pollution consists of air pollution, water pollution, and soil pollution, where each material discussed based on causes, impacts, and solutions of the pollution. The student worksheet includes activities of environmental pollution analysis, batik liquid waste treatment, and effectiveness test of waste treatment. Test questions of higher-order thinking skills consist of seven items composed based on cognitive level C4 (analyzing), C5 (evaluating), and C6 (creating).

2.3 Research Design

This research was using one group pretest-posttest design by following scheme O_1□ X □ O_2. In this case, O_1 is the initial stage, where a pretest will be given to student in order to determine the initial student skill of higher-order thinking. The next stage is treatment (X). Within this stage, treatment will be given to student in form of learning by applying learning instrument based on SETS approach. At the last stage (O_2), a posttest will be given to the student in order to determine their final skill in higher-order thinking after the implementation of learning instrument based on SETS approach.

2.4 Data Collecting Method and Research Instrument
This research was using three methods and three instruments in order to obtain data according to the research goals. Observation method used to determine the practicality of SETS based learning. Data collected are about the realization of SETS based learning by using research instrument in form of observation sheet of learning realization. Test method conducted to know the effectiveness of SETS based learning instrument toward student higher-order thinking skill. The research instrument used is a question sheet of higher-order thinking skill. A questionnaire was used to know student responses toward SETS based learning, while research instrument used was student response questionnaire sheet.

2.5 Data Analysis Technique
Data were analyzed in descriptive quantitative by using analysis of learning realizing techniques, analysis of student responses, and analysis of high-level thinking skills tests. Learning is stated well-realized if obtained $\geq$ 60%. Student response is stated positively if obtained $\geq$ 70%. Data of learning effectiveness analyzed through scoring of test question sensitivity, score of individual completeness, completeness of higher-order thinking indicator, and result of the gain score.

Test question stated sensitive toward learning if obtained score 0.3; student stated completed within learning if achieved score 75. This score is adjusted with minimum completeness criteria determined by school policy. The indicator of higher-order thinking skill stated completed if student is obtained score 70. Results of gain score categorized into three groups. The improvement of higher-order thinking skill is categorized lower if obtained score $g \leq 0.3$; categorized medium if obtained score $0.3 < g > 0.7$; and categorized higher if obtained score $g \geq 0.7$.

3. Result and Discussion
The obtained data are including learning realization, student response, and result of higher-order thinking skill test. Data of learning realization and student response are to describe the practicality of SETS based learning instrument, while the result of higher-order thinking skill test describes the effectiveness of SETS based learning instrument. The following is Table 1 that presents the result of learning realization.

Table 1. Realization of SETS approach based learning.

Step of SETS	Group results		
	1 (VII B)	2 (VII C)	3 (VII D)
Invitation	3.75	3.75	3.75
Exploration	3.375	3.5	4
Solution	3.75	3.5	3.75
Application	3.75	4	3.75
Conceptualization	3.55	4	4
Percentage	91.87%	93.75%	96.25%
Average		93.95%	

Journal of Physics: Conference Series **1417** (2019) 012079 doi:10.1088/1742-6596/1417/1/012079

Learning realization shows whether learning based on the SETS approach within five stages including invitation, exploration, solution, application, and conceptualization realized or not. In this research, the subject given was Environmental Pollution with the activity conducted was batik liquid waste treatment. According to data analysis, the average obtained is 93.95%. It shows that learning based on SETS approach with it five stages already realized well.

Within SETS vision learning, in the beginning, the student is directed to find a problem occurred in the community. By this, student pushed their critical thinking skill to perform analysis related to the occurred problem, example, from the aspect of problem origin or problem impact toward other aspects whether community or environment. Such analyzed problems then become references to make the most right decision by regarding to aspects of the environment and community. Thus, student creative thinking skills can be trained in this stage. After those, the student is directed to implements their knowledge to solve the occurred problem, and expected that technology product which regarding environmental aspect is created in order to not giving a negative impact. Further, knowledge and experience of solving the problem already students have then evaluated to strength their concept [1].

Learning based on the surrounding environment can develops the mastery of student environmental concepts, and they can apply their obtained concept in daily life, also growing good attitude and care to their environment [10]. From stages of SETS learning, student higher-order thinking skill also can be trained, among them are critical thinking skills, creative thinking, making decisions, and problem solving [11]. Positive response data from 60 students in three groups is presented in the following Table 2.

Table 2. Student positive response toward SETS based learning instrument.

No.	Assessment Aspects	Group results		
		1 (VII B)	2 (VII C)	3 (VII D)
1	Interesting on learning activity	100	100	100
2	Interesting on student textbook	100	100	100
3	Interesting on student activity sheet	100	100	95
4	Effect of test to the student thinking skill	100	95	95
5	Knowledge acquirement within learning	100	95	100
6	Knowledge implementation in technology	80	80	90
7	Learning based on surrounding problem	100	90	100
8	Advantages for community	100	95	100
9	Advantages for environment	100	100	100
10	Generates an environmental care attitude	100	100	100
11	Activeness to solve the problem	100	90	95
	Percentage	98,18%	95%	97,75%
	Average	96.97% (highly positive)		

Student response can describes perception of student toward learning instrument that based on applied approach. Response questionnaire contains eleven questions that composed of questions about interesting on learning activity and learning material, the effect of higher-order thinking skill test to trained student thinking skill, knowledge acquirement within learning, knowledge implementation in technology, learning based on surrounding problem, advantages for community and environment, also generates environmental care attitude and activeness to solve the problem.

Result shows that student response is highly positive toward learning by obtains percentage 96.97%. SETS based learning is very interesting to conducted in learning, media which used is more varied, problem-solving conducted in learning corresponding to the daily problem around the student, and the interaction between teacher and student within learning is better and harmony [12]. Because of thinking skills on four mentioned components, the student can develop their skill in analyzing, evaluating, and make decisions based on knowledge they already have [13].

According to the results of learning realization and positive student response, proved that learning instruments based on SETS approach are practical to be used in learning to trained higher-order thinking skill of the student.

The effectiveness of learning instrument can be determined based on result of student higher-order thinking skill test. The sensitivity of seven questions of higher-order thinking skills is presented in the following Table 3.

Table 3. The sensitivity of higher-order thinking skill questions.

Group	Result
1 (VII B)	0.39
2 (VII C)	0.40
3 (VII D)	0.40
Average	0.39 (sensitive)

In conducted research, the test of higher-order thinking skill was giving seven questions that composed based on thinking indicators of cognitive domain C4 (analyzing), C5 (evaluating), and C6 (creating). Questions composition was according to indicators of competence achievement on the main material of Environmental Pollution. The questions tested on 60 students obtained results that seven questions categorized sensitive with score 0.30; where the question said sensitive if achieved score ≥ 0.3 [14]. This result shows that SETS based learning instruments applied within learning have an effect on student higher-order thinking skill. Data of student learning achievement at pretest, posttest, and N-gain results are shown in Table 4 and Figure 1. Figure 1 shows the comparison of students' pre-test and post-test results.

Table 4. Learning achievement at pretest, posttest, and N-gain.

	VII B	VII C	VII D	Average
Pretest	48	47	44	46.3
Posttest	86	86	84	85.3
N-gain	0.72	0.72	0.71	0.71

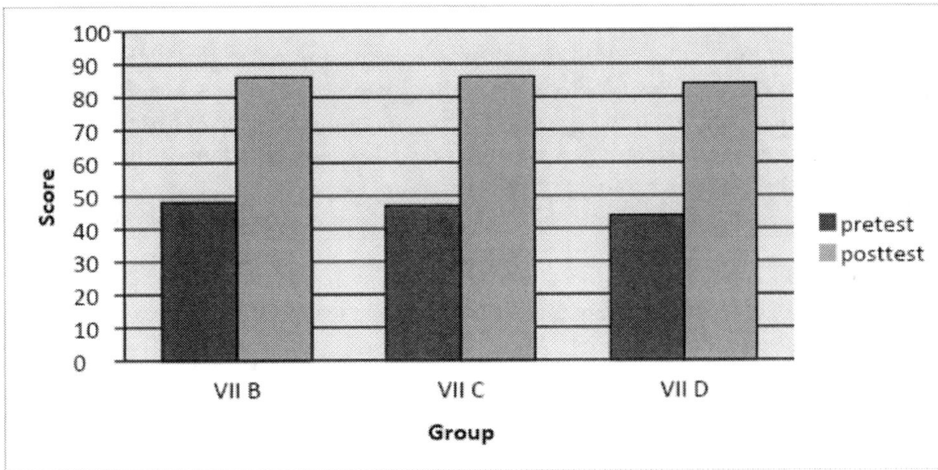

Figure 1. Graphic of learning achievement at pretest and posttest.

Scores of individual tests show the improvement of pretest score to posttest. The individual completeness score based on KKM score of school, it is ≥ 75. Within conducted research, the posttest result of 60 students entirely meets KKM score, while before learning, the student score categorized lower and not meets KKM score. Those result shows that there are improvements in student higher-order thinking skill. This is also proved by N-gain score of three groups ≥ 0.71which categorized high. Completeness of higher-order thinking skill indicator presented in following Table 5.

Table 5. The completeness of higher-order thinking skill indicators.

Cognitive Domain	Results			N-gain average	Category
	1 (VII B)	2 (VII C)	3 (VII D)		
C4	0.75	0.88	0.82	0.81	High
C5	0.76	0.75	0.75	0.75	High
C6	0.74	0.68	0.63	0.68	Medium

Each cognitive domain in three groups shows improvement of cognitive level from higher to lower are C4, C5, and C6. This indicates that each student's cognitive components in higher-order thinking skill already achieved well. From those data also known that average of the gain score at cognitive level C4 is better than C5, and cognitive domain C5 is better than C6. That means, these results indicated that the skill of students in analyzing is better than evaluating, and the skill of students in evaluating is better than creating. This is consistent with Bloom's cognitive theory revised, which stated that in order to evaluate, student must able to analyze and made or creates a product.

SETS components, including science, environment, technology and society, able to push the student to understand science comprehensively and applied it in technology, also studying many impacts caused by it technology product on society and environment [15]. Implements the SETS approach within learning may improve understanding about characteristics of science and enhancement of student attitude on science if it compared with traditional learning [16].

Test results of student higher-order thinking skill being evident that SETS based learning instrument is effective within trained higher-order thinking skill of the student.

4. Conclusion and Suggestion

Learning instruments based on SETS (science, environment, technology, and society) proved practical and effective to trained higher-order thinking skill of the student. This indicated from: a) SETS based learning proved realized well, b) student response toward learning is very positive, and c) there are improvements of student higher-order thinking skill from pretest to posttest with high N-gain result.

Based on conducted research, suggestions could be given are: 1) SETS based learning instrument can be developed with different material or other environmental problem, 2) different method of waste processing can be done for similar type of waste, and 3) learning instrument that focused on one of higher-order thinking skill indicator can be developed.

Acknowledgments
The authors would like to thank Junior High School Arosbaya 1 (Bangkalan Regency, Madura, Indonesia) for allowing authors to conduct research, as well as teachers and seventh-grade students who have many roles in this research.

References
[1] Yulistiana 2015 *Jurnal Formatif* **5** pp 76-82
[2] Hidayah R, Salimi M and Susiani T S 2017 *Jurnal Taman Cendekia* **1**
[3] Heong Y M, Othman W D, Md Yunos J, Kiong T T, Hassan R and Mohamad M M *Journal of Social and Humanity* **1** pp 121-125
[4] Saido G M, Siraj S, Nordin A B B and Al_Amedy O S 2016 *The Malaysian Online Journal of Educational Science* **3**
[5] Kusuma M D, Rosidin U and Abdurrahman A S 2017 *IOSR Journal of Research and Method in Education* **7** pp 26-32
[6] Rahayu S 2016 *Jurnal Penelitian Pendidikan Indonesia* **1**
[7] Juairiah, Yunus Y and Djufri 2014 *Journal of Biology Education* **13 pp** 83-88
[8] Sasongko D P 2010 *Journal of Science and Technology* **27**
[9] Ibda F 2015 *Jurnal Intelektualita* **3**
[10] Aini M H, Rachmadiarti F and Prastiwi M S 2014 *Jurnal Bioedu* **3**
[11] Smitha E T and Aruna P K 2014 *IOSR Journal of Humanities and Social Science* **19** pp 54-58
[12] Sugiyono T, Sulistyorini S and Rusilowati A 2017 *Journal of Primary Education* **6**
[13] Mapeala R and Siew N M 2015 *Springerplus Journal* **4** p 741
[14] Lambertus 2010 *Journal of Mathematic Education* **1** p 154
[15] Usmeldi, Amini R and Trisna S *Journal of Indonesian Science Education* **6** pp 318-325
[16] Akcay 2015 *International Journal of Education in Mathematics, Science and Technology* **3**

The Effectiveness of Guided Inquiry Learning Model to Improve Science Process Skill on Heat Matter

A U Yunianti*[1], Wasis[1] and M Nur[1]

[1]Department of Postgraduate Science Education, Universitas Negeri Surabaya, Surabaya, Indonesia

*Email: adninyunianti@gmail.com

Abstract. This research aimed to describe the effectiveness of guided inquiry learning model to trained science process skill. Research conducted in SMA Kemala Bhayangkari 1 Surabaya at grade XI on heat matter by using one group pre-test and post-test design. Data collecting techniques consist of tests and questionnaires. The effectiveness of the guided inquiry model is obtained based on the science process skill test, learning achievement of knowledge test and student responses. Science process skill test and learning achievement of knowledge test obtained based on the results of the pre-test and post-test. Student responses are obtained based on the results of the questionnaire that includes the student responses to learning. The data analysis using n-gain score to determine the improvement of science process skill and learning achievement of knowledge while descriptive quantitative to describe student's response. Result of this research shows that (1) indicators of science process skill improved with high category (2) learning achievement of knowledge improved with middle category (3) student responses to learning are highly positive. Based on those results indicates that guided inquiry model is effective to improve science process skill on heat matter.

1. Introduction

Learn is the process of effort intentionally conducted to acquiring new behavior alteration entirely, consciously, and that alteration is relatively settled, also bring positive effects and advantages for student in interaction with their environment [1]. Teaching and learning process will be success if it results bringing changes on knowledge, understanding, skill, and norm of attitude within student self. The success of student in learning process is also affecting by teacher role as educator. An educator should able to create conditions that enable students learning activity occurred. Mentioned conditions can be task assignment, motivates student to dare to ask and express opinion, motivate student curiosity on learning material. These conditions may be conducted by teacher through sequence setting of learning activities entirely, start from making learning design, conducting learning activity, and performs learning evaluation including learning process and learning achievement in form of change of learning effect itself [2].

By formed of Curriculum 2013 expected that new spirit and innovation will be created in Indonesian education world. Curriculum 2013 implements scientific approach within learning through activities of observation, questioning, experiencing, reasoning, and communicate. Curriculum should base on reasoning affirmation more, not only on memorization anymore. This change decided by referring to the results of international study on performance of Indonesian student. The reason of Indonesian

Ministry of Education and Culture compiled Curriculum 2013 was education curriculum must be adapted to the demand of times. One reference of mentioned study result is "Trends in International Math and Science" that conducted every four years performed by International Association for the Evaluation of Education Achievement (IEA) that is an international association for achievement evaluation in math and science centered in Lynch School of Education, Boston College, USA. Achievement obtained by Indonesian students in 2003 ranked on 34 of 46 participant countries, in 2007 Indonesia achieved rating 36 of 49 participant countries, in 2011 Indonesia ranked on 36 of 43 participant countries, and in 2015 Indonesia ranked on 45 of 48 participant countries [3]. These results shows average score of science skill of Indonesian student is below the international average score year by year.

To improve the reasoning capability of physic subject, students have to self-learning regularly and learn to find their own understanding. All this time, student accustomed to receive material from teacher explanation, with no searching other learning resources, so their learning is teacher centered and student are less active role in the finding of their knowledge. The obstacle besides learning model applied by teacher in school, there is another obstacle that make student learning was less meaningful, that is about learning instrument used by student within learning activity.

The wise teacher will give chance to student to search and find information by their self or if giving information, teacher only gives basic information as foundation for student in search and find other information by their self. This teaching style will grow student confidence about their done [4]. Besides that, this teaching style may change learning that was passive become active learning, because student will trains their thinking skill and reasoning skill independently in finding learning concepts.

Based on result of performance test of science process skill in SMA Kemala Bhayangkari 1 Surabaya, students were yet to solve questions of science process skill well. Test result obtained using multiple choice questions which loads some indicators of science process skill. Test result shows that 40 % of students can identify variable, 41% of students can state hypothesis, 37% of student can defined operational variable, 42% of students can designed research, and 38% of students can describe and interpret data. This test result shows that science process skill of student is still in lower classification.

Physic teaching and learning process should emphasize more on process skill approach so that student can finding facts, construct concepts, theories, and scientific attitude by their self, which finally can affect positively on the quality of education process or education product. One of process skill can be developed is science process skill. Mukhlis [5] stated that science process skill is skills needed by student to develop their potentials. This potential development can be student skills in problem formulation, proposing question, conducting experiment, processing data, and drawing conclusion [6].

The effort can be conducted to trained science process skill is by using learning instrument based on a learning model allows supporting teaching and learning activities. A learning model that supports within trained science process skill is guided inquiry model. Inquiry strategy is a sequence of learning activities that involving all student skill maximally to find and investigate systematically, critical, logic, analytical, so they able to formulate their finding with confidence. Inquiry learning emphasizes on investigating and finding process [7]. The main plans within inquiry learning are to determining goals and problem identification properly [8].

Rismawati [9] stated that one of learning model can make student active and develop science process skill optimally is by using guided inquiry model. Guided inquiry learning model is an instruction model where the teacher provides guidance for student by giving opening question about concepts will be studied and direct it into discussion topic. This is accordance with research conducted by Mukhlis [5] which stated that guided inquiry regarded as effective way to improve student achievement and science process skill by involving student directly within learning process and resulting meaningful learning more. Along the research, learning process conducted with practice using guided inquiry model, thus student involved in finding process directly and allows student science process skill developments.

This inquiry model is supportive toward Physic learning, because by conducting mentioned activities, student able to find actively their knowledge independently. Thus, in order to improve learning quality, teacher needs to give the chances for student to conducting observation, questioning, proposing hypothesizes, collecting data, and make conclusion independently. Thinking well naturally means able to produce questions to do investigation, developing reasonable hypothesis, designing controlled experiment, collecting and presenting data correctly, using evidences to support the conclusion, and presenting experiment process effectively [10]. Juhji [11] shows that to improving science process skill, student should familiarize to be more self learning in order to find concepts, scientific principles, also develops creativity in science problems solving which in realization is guided by teacher intensively. With supported by some elements, those are qualified learning instrument and proper learning model, expected to support and trained science process skill to the student. Based on mentioned problems above, researcher interested to do research with title "The Effectiveness of Guided Inquiry Learning Model to Improve Student Science Process Skill on Heat Matter".

2. Research Methods

This research conducted in SMA Kemala Bhayangkari 1 Surabaya academic year 2018/2019 at grade XI on subject of Physic on material of heat. Test design was using one group pre-test and post-test design. Pre-test is test conducted before learning, while post-test is test conducted after entire process of learning activities. Based on pre-test and post-test, we can determine how about learning achievement obtained by student before and after conducting learning by using guided inquiry model. Data collecting technique used are test, which used to measure or determine the presence of inquiry learning model contribution in trained science process skill and disseminate response questionnaires which are used to know response of student toward guided inquiry learning model.

Results data of pre-test and post-test of science process skill and learning achievement of knowledge analyzed descriptive quantitative, that is N-gain analysis. Gain indicates the difference of student process skill and psychomotor skill before and after treatment is given. Normalized gain score indicates level of treatment effectiveness than obtained score. *N-Gain* formulated as follows.

$$N - Gain = \frac{Post\ test\ score - Pre\ test\ score}{Maximum\ score - Pre\ test\ score}$$

Category:
high = score $g > 0.70$
middle = score $0.30 < g < 0.70$
low = score $g < 0.30$

Data of student response obtained from questionnaire of student response toward learning activities then analyzed by using descriptive quantitative, that calculates percentage of statements given by student. Equation to calculate data of student response is following.

$$P = \frac{\sum K}{\sum N} x100\%$$

Explanation:
P = percentage of student response
$\sum K$ = Total of student response score
$\sum N$ = Total of all student response score
Data were collected is processed then interpreted with criteria as follows.
PRS ≥ 85% = highly positive
70% ≥ PRS < 85% = Positive
50% ≥ PRS < 70% = less positive
PRS < 50% = negative
PRS = Percentage of Student Responses

3. Result and Discussion

3.1 Science Process Skill

The effectiveness is success action of student to obtain certain goals that bring optimum learning achievement. To know the success of learning process, students are given science process skill questions on pre-test and post-test. Pre-test conducted before learning activity while post-test conducted after student participated in all learning activity. Test was made refer to some indicators of science process skill that trained at learning activities, they are 1) problem formulation, 2) writing hypothesis, 3) variables identification, 4) writing definition of operational variables, 5) making table of observation, 6) analyzing data, 7) drawing conclusion.

In order to measure effect of guided inquiry learning instrument toward the effectiveness of teaching and learning activity, it known based on improvement result of science process skill indicators and the completeness of student learning achievement. The following presented improvement results of each indicator of science process skill before and after participated within learning.

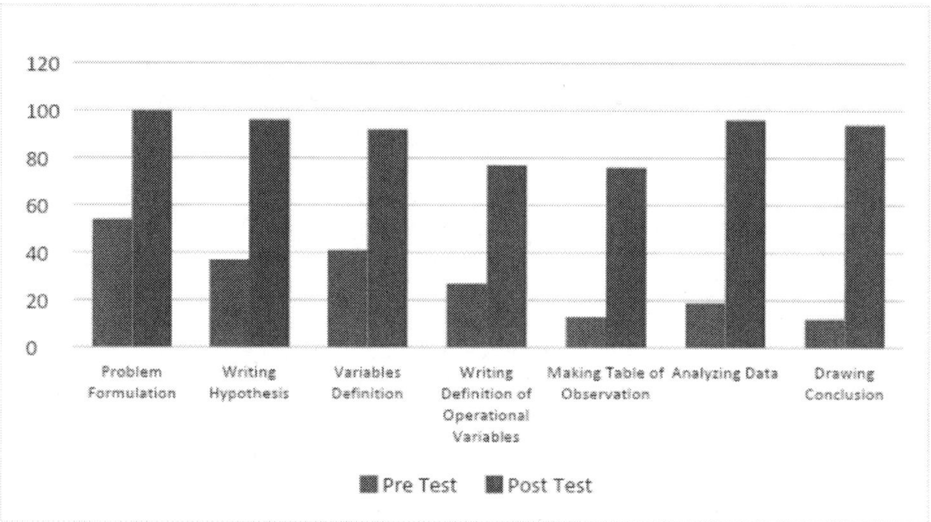

Figure 1. Improvement Result of Science Process Skill Indicators

Based on Figure 1 known that mean of student pre-test score is 29 and mean of post-test score is 90 with score of n-gain is 0.87. Score of this n-gain indicates that high improvement occurred on every indicator of science process skill after students were participated in learning by using guided inquiry learning instrument. Based on research conducted by Rismawati [8] shows that learning process which using guided inquiry model is able to improve science process skill. The improvement of science process skill caused by student participation on steps of guided inquiry learning. Scientific activities may give chances toward student in self-finding of facts, concepts, and principles through direct experience so that learning resulted is more optimal.

The effectiveness of learning instrument also can be obtained based on the learning achievement of knowledge that contains indicators remember (C1), understand (C2), Apply (C3) and Analyze (C4).

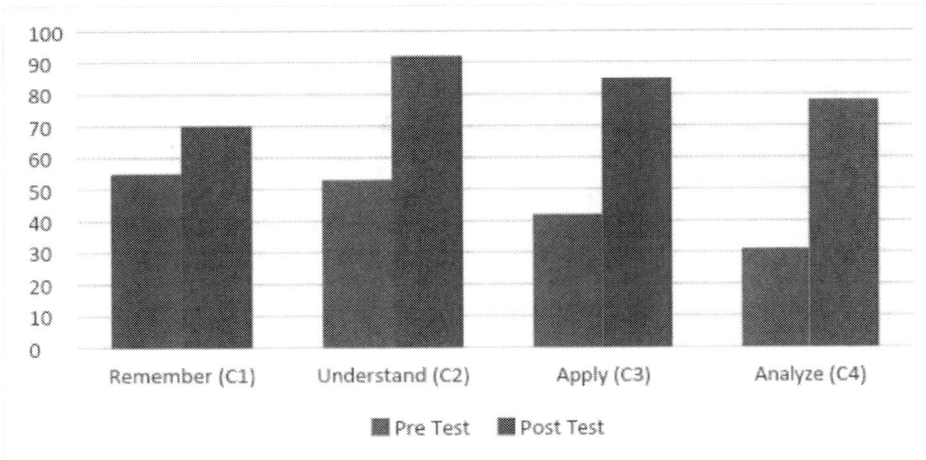

Figure 2. Improvement Result Learning Achievement of Knowledge

Based on Figure 2 could know that mean of student pre-test score learning achievement of knowledge is 45 and post-test score is 81 with n-gain is 0.64 that categorized middle. This indicates that improvement of learning achievement of knowledge was occurred after learning process by using guided inquiry model conducted and learning instrument were developed is used.

Based on research conducted by Sakdiah [12] shows that students who are taught with guided inquiry have the opportunity to obtain higher learning achievement of knowledge than those taught using conventional learning. Guided inquiry learning causes students to become more scientific and increase understanding of concepts. Guided inquiry learning causes students to actively participate in learning and improve scores better than traditional model.

Results above support and confirm the truth of theory by Jerome Bruner that suggest students to learn through participation actively with their own concepts and principles [13]. Based on this theory, student should able to involved actively within learning activity to find and construct their knowledge independently then appearing meaningful learning in the student mind.

3.2 Response

Learning effectiveness can also be shown based on student response after participating in all learning activities sequence. Student response is responses given by student about teaching and learning activities. Student response measured using instrument in form of questionnaire given after learning process is finished. On response questionnaire, it was giving some statements about anything along learning process as shown in Table 1.

Table 1. Student's Response Percentage

No.	Response	Percentage (%)	Category
1	Interest in learning activities	89,29	Very good
2	The renewal of the learning materials	88,78	Very good
3	The Ease to understand physics learning	88,10	Very good
4	Teacher guidance during learning	92,86	Very good
	Average	89,76	Very good

Based on Table 1 can know that student assessment toward some components within learning 89.29% of student stated enjoy along participating learning. This is because what implemented by teacher within learning is novel thing for them. It supported by Table 1 that 92.86% stated novelty on learning they participated. Positive response of student surely affected toward the improvement of student learning achievement which shown in Figure 1 and Figure 2. This is supported based on student responses where stated that science process skill can they understood easily, then it affected positively toward the improvement of learning achievement of science process skill. Positive response given by student also affected by learning model implemented. Guided inquiry may involve student to participate actively within learning through practice activity by trained science process skill so that student being more enthusiastic in every phases and steps within learning they participated.

This is supported based on research conducted by Azis [14] showing that student's responses to inquiry learning are positive, both in terms of subject matter or content, inquiry learning approaches used, science process skills being trained, learning tools used, learning management, as well as the test results of learning the understanding of knowledge and process skills undertaken.

4. Conclusion

Based on analysis of research result could be concluded that learning instrument of guided inquiry model is able to improve science process skill. This shows by the completeness of science process skill indicators and the learning outcomes with n-gain respectively 0.87 and 0.64 and categorized in high and middle. Besides that, learning effectiveness also supported with the positive result of student response 89.76. This result shows that learning instrument of guided inquiry model is highly effective to improve science process skill. Suggestion for the teacher is that the result of this research can be a reference for implementation of Physic learning in school. Based on those results indicates that guided inquiry learning model is effective to improve science process skill.

Some points are needed to be considered by teacher when implements guided inquiry model to improve science process skill, those are: (a) teacher who used guided inquiry model should prepare and manage time well in order to learn conducted effectively and efficient so that learning syntaxes were planned totally could be realized and learning goals can be achieved. (b) Teacher needs to evaluate every ends of learning in order to correct the disadvantages in the next learning.

Acknowledgments

The author thanked the Surabaya State University, principal and teacher of SMA Kemala Bhayangkari 1 Surabaya who has supported and assisted the research.

References

[1] Hosnan M 2014 *Pendekatan Saintifik dan Kontekstual dalam Pembelajaran abad 21* (Bogor: Ghalia Indonesia)
[2] Arifin Z 2012 *Evaluasi Pembelajaran* (Bandung: PT Remaja Rosdakarya)
[3] Provasnik S & Malley L 2016 *Highlights From TIMSS and TIMSS Advanced 2015* (Amerika: Institute of Education Science)
[4] Djamarah S 2000 *Guru dan Anak didik dalam Interaksi Edukatif* (Jakarta: Prestasi Pustaka)
[5] Mukhlis, Arifuddin M and Hartini S 2018 *Seminar Nasional Pendidikan* (Banjarmasin: FKIP Universitas Lambung Mangkurat) pp 102-109
[6] Trianto 2008 *Mendesain Pembelajaran Kontekstual di Kelas* (Surabaya: Cerdas Pustaka)
[7] Gulo W 2002 *Strategi Belajar Mengajar* (Jakarta: PT Grasindo)
[8] Arends 2007 *Learning to Teach* (Singapura: Mc.Graw Hill Higher Education)
[9] Rismawati, Iriwi L S, Yusuf I and Widyaningsih S W 2017 *Jurnal Pendidikan* **8** 12-24
[10] Stone R 2013 *Cara-Cara Terbaik Untuk Mengajar Sains* (Jakarta: PT. Indeks)
[11] Juhji 2016 *Jurnal Penelitian dan Pembelajaran IPA* **2** 58-70
[12] Sakdiah, Mursal and Syukri M 2018 *Jurnal IPA dan Pembelajaran IPA* **2** 41-48.
[13] Slavin E R 2011 *Educational Psychology Theory and Practice* (USA: Pearson)
[14] Azis M, Yuanita L and Rahayu Y S 2014 *Jurnal Penelitian Pendidikan Sains* **3** 410-419.

Development of Guided Inductive Inquiry Learning Support Materials Prototype on Energy for Junior High School Level to Improve Students' Conceptual Understanding: Validity and Practicality Study

M A Mahdiannur

Department of Natural Science, Universitas Negeri Surabaya, Indonesia.

e-mail: muhamadmahdiannur@unesa.ac.id

Abstract. This research purposed to describe the quality of learning support material prototype in energy topics for Junior High School. The quality-focused on this research were (1) validity and (2) practicality aspects. Guided inductive inquiry (GII) is a learning model to facilitate students to have a comprehensive understanding of concepts, develop complex thinking skills, and help internalize the concept of evidence and substantive concepts within students that they can use to solve problems, develop competencies, and improve learning outcomes. This research used Mafumiko's Model to develop a prototype of learning support materials based on design research. This study conducted at SMPN 1 Tanah Grogot, Paser Regency, East Kalimantan, and involved 3 (three) subject matter experts (SMEs) and 2 (two) observers during the classroom tryout. The SMEs examined the GII learning support materials valid and their appraisal reliable. The implementation of GII learning support materials during learning processes in the very good category and students' activities reflected the level of active participation. Based on these results, the quality of the prototype of GII learning support materials on energy topics for Junior High School level met the validity and practicality criteria and also improved the students' conceptual understanding.

1. Introduction

Science education emphasizes that students carry out activities according to the scientific method to improve their thinking skills and to apply it in solving problems. Thinking has a wide variety of which divided into scientific thinking (critical and creative), high order thinking, and logical thinking, although various types of thinking sometimes overlap among indicators. Thinking is a combination of knowledge, skills, and attitudes, and the thinking process generates a critical attitude [1,2]. The importance of the combination of three domains as a basis for being able to think also encourages the government to revise the curriculum according to the Indonesian National Qualifications Framework (INQF) to face the challenges of the 21st century characterized by the collaboration, disruption era, and internet of things. 2013 Curriculum characterizes by the three domains, namely attitudes (spiritual and social), knowledge, and skills.

Research in science education focuses on enhancing the 21st-century science curriculum itself through small-group learning [3–5]. Based on this definition, science education research focusing on the applicable curriculum. 2013 Curriculum itself is in line with the principles of science education globally, namely learning science as it was discovered and developed by scientists by emphasizing the incorporation of thinking skills, science process skills, and manipulative skills [1,6,7]. Imitation of the

Published under licence by IOP Publishing Ltd

behavior of scientists in conducting scientific investigations must be carried out by students so that students become competent in using inquiry skills that are accompanied by the ability to think scientifically. Adaptation of the behavior patterns of scientists can help shape scientific literacy that is useful for solving problems according to the scientific method. The formation of scientific behavioral habits in inquiry processes must begin at an early age to strengthen the ability to think scientifically [8,9]. The habit of scientific behavior is important because the concept itself formed through interaction in formal school settings [10]. The scientific behavior has implications for the importance of the science learning process in junior high schools because inquiry processes are the science itself [11,12].

Science has been introduced in Indonesia since early childhood education to the high school level. Early childhood and elementary school student have studied science content in thematic learning and integrated science. Science as a subject began to be applied at the junior high school level and then developed into subjects of physics, biology, and chemistry through a specialization system at the senior high school level. The awareness toward science education in Indonesia is fundamental, but the level of ability, conceptual understanding, and scientific literacy of Indonesian students is still low in every international assessment. Based on observations, it seems that science learning focused on traditional ways, which tend to be learning-memorizing and minimal scientific investigation activities. Traditional (conventional) learning is not always bad, but learning with inquiry processes is more successful in improving student understanding [13]. The natural science subject teachers are also less able to manage exciting and challenging science learning in identifying and exploring students' prior knowledge [11]. The inquiry process generally divided into two, namely inductive inquiry and deductive inquiry [1]. Inductive inquiry with a guided level [1], is very suitable for science learning for junior high school students [14]. On the other hand, the inductive approach to science learning based-on inquiry is very suitable because it starts from proving specific things and then moves towards general things, namely by matching the results of experimental activities with substantive concepts. The combination of laboratory activities with substantive concepts will improve students' conceptual understanding [15,16].

The factor of science knowledge content also has a significant influence in determining science learning itself. One concept that is important in science is energy. The concept of energy itself has long been studied by students and is cross-disciplinary [17]. The concept of energy itself is eclectic, which is divided into four domains and reinforced by various experimental results [18]. Taking into account the complex characteristics of energy material, curriculum objectives, and how science taught to students based on inquiry learning and various inquiry approaches need a learning support material that can mediate the learning process so that students can master scientific thinking skills and improve their conceptual understanding. Learning support material must be structured in such a way that it forms a unified with the interventions that will be designed according to the intended goals. The learning support material on the energy topic must meet the three criteria, namely validity, practicality, and effectiveness [19], to improve students' conceptual understanding. The three criteria must be fulfilled through an iterative process that involves designing, testing, and evaluation to obtain theory and design relevant to a situation [19]. The fulfillment of each of these criteria is not achieved simultaneously at one time but can be fulfilled gradually and separately [19]. Validity and practicality are two factors that must be fulfilled first because validity is related to fit of the learning devices with state-of-the-art of science itself and curriculum intention, while practicality indicates that learning support materials were successfully implemented by teachers and students in the learning process [19]. Therefore, this study focused on achieving the validity and practicality criteria for the prototype of GII learning support material on energy for Junior High School level to improve students' conceptual understanding.

2. Method

This research was part of the design-research with micro-phases criteria [19]. The developmental phase applied in this research described as follows:

2.1. Preparation
In this stage, the design specifications for the development of the prototype of learning support materials established based on intended goals.

Table 1. Development design specification.

No	Developmental Focus	Design Specification
1.	Teaching preparation	• Learning objectives • Learning times • Experimental material • Barriers and its solution list
2.	Subject knowledge	• Questions to feed the discussion • Important scientific concepts explanation • Students' questions during learning example
3.	Learning model	• GII model syntax • Students' grouping • Scaffolding distribution
4.	Assessment and feedback	• Rubrics • Assessment instruments

2.2. The Developmental Phase
This process adopted Mafumiko's model. The Mafumiko model is a development model that emphasizes the micro phases of each development process that are flexible, and each phase can be presented separately [19]. The process of designing a prototype of learning support materials adjusted to the guidelines and specifications that have been set and revised through a classroom tryout. The process of developing the learning support material prototype emphasized on improving quality. The first stage focused on achieving validity criteria and then tested its practicality in a classroom tryout. Before and after the classroom tryout, the students are given a pretest and posttest to evaluate their gain. To evaluate students' conceptual understanding, we used normalized change [20], which more sophisticated than Hake's normalized gain [21].

2.3. Validity Appraisal
The validity of the prototype of learning support materials examined by three SMEs consisting of science education experts, assessment experts, and science content experts. The evaluation of validity uses the screening from the SMEs. The prototype of learning support materials validity calculated using Aiken's V formula and the average reliability of all expert evaluations ($r_{xx'}$), calculated using Ebel formulas [22]. The reliability of the assessment instrument also determined based on trial on small groups of students using the Cronbach Alpha test.

2.4. Practicality Examination
After the SMEs appraisal the validity, the practicality of using learning support material prototype by the teachers and students in the actual classroom settings need to be tested. The testing location determined at SMPN 1 Tanah Grogot, Paser Regency, East Kalimantan Province in the First Semester, Academic Year 2014/2015. The research subjects were 30 students of Class VII A. During the practicality testing process in a classroom, and it involves two observers who snapshot student activities every five minutes. The observers' reliability is calculated using the Emmer and Millet formulas [23].

3. Result and Discussion
3.1. Result
Based on the screening method, the SMEs appraisal the validity of the learning support material prototype by filling out the validity appraisal form. Based on the results of the appraisal of the SMEs

in general, all components of the learning support material prototype for the Junior High School level on energy material topics have met the criteria of validity. The results of the appraisal of the SMEs also met the reliability criteria. The results of the validity appraisal shown in Table 2, Table 3, and Table 4.

Table 2. Lesson plans validity.

Lesson Plans	Validity Coefficient	Reliability	Explanation
I	0.97	0.81	Valid and the examination from the SMEs reliable
II	0.95	0.95	Valid and the examination from the SMEs reliable
III	0.95	0.87	Valid and the examination from the SMEs reliable
Average	0.96	0.88	Valid and the examination from the SMEs reliable

Table 3. Student worksheet validity.

Student Worksheet	Validity Coefficient	Reliability	Explanation
I	0.91	0.84	Valid and the examination from the SMEs reliable
II	0.95	0.84	Valid and the examination from the SMEs reliable
III	0.96	0.81	Valid and the examination from the SMEs reliable
Average	0.94	0.83	Valid and the examination from the SMEs reliable

Table 4. Student book and assessment instrument validity.

Type	Validity Coefficient	Reliability	Explanation
Student book	0.94	0.85	Valid and the examination from the SMEs reliable
Assessment instrument	0.96	0.85	Valid and the examination from the SMEs reliable

Based on the data in Table 2, Table 3, and Table 4, the results of validity for all components of the learning support materials prototype > 0.90 and the reliability of the SMEs' appraisal > 0.80. These results indicate that the SMEs agree that the learning support material prototype met the validity criteria. Specifically, for assessment instruments, before use, it tested for consistency with a small group of students besides the research subject—the instrument reliability test results using the Cronbach Alpha test. The test results showed that the assessment instrument has good consistency (0.71).

Testing the practicality criteria is based on the implementation of all learning scenarios and the use of all components of the learning support material prototype in a classroom tryout. Also, student learning activities used as parameters in determining the practicality criteria in this study. The learning process in the GII model begins by displaying phenomena packed in multimedia that can arouse curiosity and students' willingness to ask questions and identify problems. This process is then used by the teacher to conduct more intensive discussions and guide students to carry out inquiry activities in small groups. The experimental process assisted by administering the distribution of scaffolding. The process of carrying out laboratory activities was initially quite difficult for students and teachers to do, but over time could proceed according to the lesson plan. The students then interpret the data and develop and test conclusions through discussion among group members by utilizing student textbooks that developed according to the learning objectives and discussion at the time of presenting the data at the end of the learning process before the ending of the lesson. The implementation of each lesson plan according to observations results by observers is more than 95%. The results of the lesson plan implementation summarized in Table 5.

Table 5. Recapitulation of the lesson plan implementation.

Learning	Percentage of Implementation	Explanation
Lesson plan I	97.22	Very Good Criteria
Lesson plan II	98.15	Very Good Criteria
Lesson plan III	100.00	Very Good Criteria

Student activities generally dominated by inquiry processes. The inquiry activities indicated that students have successfully adapted and are willing to carry out various instructed activities according to the GII model. Various inquiry activities can reduce students' inappropriate behaviors and begin to express opinions on the material being studied. The students' activities summarized in Table 6.

Table 6. Recapitulation of the students' activities.

Students' Activities	Percentage of Implementation on Meeting			Explanation
	I	II	III	
Pay attention to the teacher's explanation and asking	32.36	31.96	32.49	Very Good Criteria
Work in a team	15.35	15.64	15.51	Very Good Criteria
Inquiry activities	36.52	36.05	35.45	Very Good Criteria
Communication	15.77	16.35	16.55	Very Good Criteria
Inappropriate behavior	0	0	0	Not Observed
Total	100	100	100	-
Reliability (observers' agreement)	95%	98%	98%	Reliable

The implementation of student-centered learning and supported by the prototypes of learning support materials also have a positive impact on improving students' understanding of energy concepts. The mean of pretest 33.11 with a range of 33.33, while the mean of posttest increased to 84.00 with a range of 40.00 (Table 7). The average value of normalized change (c) shows good results, which is 0.78 with a standard deviation of 0.172 and a range of 0.56, and the majority of students are in the intermediate and high c clusters (Figure 1).

Table 7. Result of pretest and posttest before and after the intervention.

Parameter(s)	Pretest	Posttest
Mean	33.11	84.00
Median	33.33	86.67
Mode	20.00	100.00
Maximum	53.33	100.00
Minimum	20.00	60.00
Standard Deviation	11.001	13.171
Standard Error	2.008	2.405

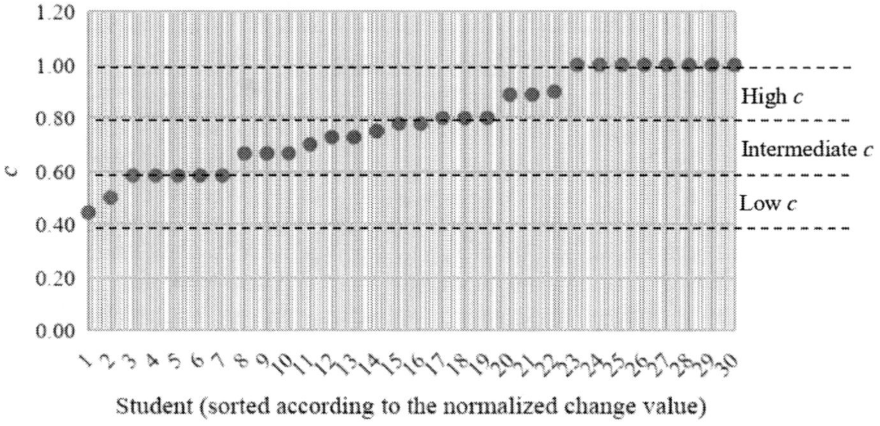

Figure 1. The plot of students' clusters indicates improvement in conceptual understanding.

Based on the data in Table 5, Table 6, Table 7, and Figure 1, all scenarios in each lesson plan can be implemented in a classroom tryout, and student activities are also following the inquiry processes and have an impact on the improvement of students' conceptual understanding according to the test results, before and after intervention. Based on these data, the learning support material prototype met the criteria of practicality.

3.2. Discussion

The results of the SMEs appraisal showed that all components of the prototype of learning support materials had met the criteria of validity. These results reflected that the development of material under the four principles of the organization of energy materials, all stages of learning could be carried out by teachers and students, and construction is coherent among the curricular components. All components of the prototype of learning support materials developed by combining scientific inquiry and social inquiry to improve students' thinking skills. The entire component integrated into the syntax of the GII learning model. These results are expected to improve the students' ability to think and develop a conceptual understanding that emphasizes the process of identifying previous experiences and knowledge of students on things studied like scientists because of the inquiry process is the science itself [11,12].

The learning support materials prototype has met the validity criteria, then proven by the results of the implementation of all the learning scenarios that have been previously designed (see Table 5). These results indicate that the SMEs have been able to appraisal whether the prototype learning support materials can be implemented or not in a classroom tryout setting. Student activities also tend to be dominated by inquiry activities, such as working in teams, conducting scientific inquiry activities, and simple scientific communication. Inductive inquiry processes assisted by scaffolding distribution according to the guided level tend to minimize the appearance of inappropriate behavior, and also improved the students' conceptual understanding. Inquiry processes attracted students to be active and gain meaningful concepts from the learning process [11,12]. Adaptation of the behavior patterns of scientists is beneficial for developing scientific thinking skills that are supported by the mastery of strong content knowledge [6,7] to solve problems. The active student participation levels also confirm that guided inductive inquiry-based learning is better than traditional learning [13,14]. It also shows that students do not need too long to be able to participate in inductive inquiry-based learning because it is following the development process of children who are always curious about new things [8,9]. Also, cooperation in teams during the experimental activities provides opportunities for students to interact with each other because concepts formed through interaction [16]. Data from the research showed how important the planning process and stages in the development of learning support materials. The SMEs have an important role in examining whether a prototype learning support material is by scientific developments and can implement in the learning process in a classroom. The validity and practicality criteria an inseparable unity. Fulfilments of these criteria (validity and practicality) gradually, shows that the Mafumiko's model is design research or design experiment design that is suitable for researchers who cannot conduct large-scale fieldwork [19]. Validity and practicality have an essential role in improving the learning process quality and learning achievement.

4. Conclusion

Based on the results of the appraisal of SMEs and the results of the classroom tryout indicated that the GII Learning Support Materials Prototype on Energy for Junior High School Level met the validity and practicality criteria and also improved the students' conceptual understanding. However, to evaluate the effectiveness of learning support materials prototypes requiring more classroom tryout processes to obtain the same trend of results continuously.

Acknowledgments

We would thank the anonymous reviewers who have provided their suggestions to improve the quality of this paper.

References

[1] Orlich D C, Harder R J, Callahan R C, Trevisan M S and Brown A H 2010 *Teaching Strategies: A Guide to Effective Instruction* (Boston: Wadsworth Publishing)

[2] González-García F J, Blanco-López Á, España-Ramos E and Franco-Mariscal A-J 2019 The Nature of science and citizenship: a delphi analysis *Res. Sci. Educ.*

[3] Bennett; J, Lubben F, Hogarth S and Campbell B 2005 Systematic reviews of research in science education: rigour or rigidity? *Int. J. Sci. Educ.* **27** 387–406

[4] Stokes A, King H and Libarkin J C 2007 Research in science education: Threshold concepts *J. Geosci. Educ.* **55** 434–8

[5] Bone E K and Ross P M 2019 Rational curriculum processes: revising learning outcomes is essential yet insufficient for a twenty-first century science curriculum *Stud. High. Educ.* 1–12

[6] Fadzil H M and Saat R M 2014 Exploring the influencing factors in students' acquisition of manipulative skills during transition from primary to secondary school *Asia-Pacific Forum Sci. Learn. Teach.* **15** Article 3, 1-18

[7] Abrahams I and Reiss M J 2015 The assessment of practical skills *Sch. Sci. Rev.* **96** 40–4

[8] McNerney K and Hall N 2017 Developing a framework of scientific inquiry in early childhood: An action research project to support staff development and improve science teaching *Early Child Dev. Care* **187** 206–20

[9] van der Graaf J, van de Sande E, Gijsel M, and Segers E 2019 A combined approach to strengthen children's scientific thinking: Direct instruction on scientific reasoning and training of teacher's verbal support *Int. J. Sci. Educ.* **41** 1119–38

[10] Sikder S and Fleer M 2015 Small science: Infants and toddlers experiencing science in everyday family life *Res. Sci. Educ.* **45** 445–64

[11] Kotuľáková K 2013 Teachers' focus on pupils prior conceptions in Inquiry-Based teaching *Rev. Sci. Math. ICT Educ.* **7** 53–71

[12] Wecker C, Rachel A, Heran-Dörr E, Waltner C, Wiesner H and Fischer F 2013 Presenting theoretical ideas prior to inquiry activities fosters theory-level knowledge *J. Res. Sci. Teach.* **50** 1180–206

[13] Fraser J M, Timan A L, Miller K, Dowd J E, Tucker L and Mazur E 2014 Teaching and physics education research: Bridging the gap *Reports Prog. Phys.* **77** 032401

[14] Wheeler L and Bell R 2012 Open-ended inquiry: Practical ways of implementing inquiry in the chemistry classroom *Sci. Teach.* **79** 33–9

[15] Nichols K, Burgh G and Kennedy C 2017 Comparing two inquiry professional development interventions in science on primary students' questioning and other inquiry behaviours *Res. Sci. Educ.* **47** 1–24

[16] Roberts R, Gott R, and Glaesser J 2010 Students' approaches to open-ended science investigation: the importance of substantive and procedural understanding *Res. Pap. Educ.* **25** 377–407
[17] Lee H-S and Liu O L 2009 Assessing learning progression of energy concepts across middle school grades: The knowledge integration perspective *Sci. Educ.* **94** 665–88
[18] Coelho R L 2014 On the concept of energy: Eclecticism and rationality *Sci. Educ.* **23** 1361–80
[19] Kennedy-Clark S 2013 Research by design: Design-based research and the higher degree research student *J. Learn. Des.* **6**
[20] Marx J D and Cummings K 2007 Normalized change *Am. J. Phys.* **75** 87–91
[21] Bao L 2006 Theoretical comparisons of average normalized gain calculations *Am. J. Phys.* **74** 917–22
[22] Azwar S 2012 *Reliabilitas dan Validitas* (Yogyakarta: Pustaka Pelajar)
[23] Good T L and Brophy J E 2008 *Looking in Classrooms* (Boston, MA: Allyn and Bacon)

Development of OrSAEv Model Learning Materials to Strengthen the Characters of Tsunami Disaster Mitigation for Middle School Students (Preliminary Study)

B U Isna[1], Madlazim[2,*], E Hariyono[3] and N Suprapto[4]

[1, 2, 3, 4]Postgraduate, State University of Surabaya, INDONESIA.

*The corresponding author: madlazim@unesa.ac.id

Abstract. Geologically, Indonesia is a disaster-prone country, especially tsunami. One strategic step that can be taken to minimize the impact of the tsunami is through the education sector as the fundamental sector forming student character. Therefore, this research is aimed to determine the process of strengthening students' character about earthquake and tsunami disaster mitigation, to find out the experiences of students facing earthquake and tsunami disasters and to find out the disaster learning that has been implemented in schools. This research belongs to a quantitative descriptive type of research. The subjects of this research were 30 students of ninth grade at Donomulyo 1 Public Middle School. The process of collecting the data was done by giving questionnaires and doing interviews. Questionnaires were given to students in the form of questionnaires about strengthening the character of earthquake and tsunami disaster mitigation, and questionnaires about students' experiences in dealing with tsunami disasters related to learning in class, then interviews were conducted with two science teachers at the school. Samples were taken through the process of purposive sampling which was based on input from science teachers at Donomulyo 1 Public Middle School. Based on the results of the data in related to the analysis and discussion that have been conducted, it can be concluded that there are contradictory results. Where based on the questionnaire is obtained that the character of tsunami disaster mitigation in Donomulyo 1 Public Middle School students is good. However based on the results of interviews with two science teachers and questionnaires how students facing the tsunami and earthquake disasters, the results show that facilities, experience of facing the disasters and disaster learning in classroom are still lack of it. Therefore, it is necessary to test its application to find out whether the character is truly embedded within the students' mind.

1. Introduction

Disaster is a series of events that threaten and disrupt people's lives and livelihood caused by natural and/or non-natural factors as well as the human factor, resulting in casualties, environmental damage, property loss, and psychological impacts [1]. Indonesia is ranked 7th as the most vulnerable to natural disaster risk [2].

Geologically, Indonesia is a meeting point of three active tectonic plates namely the Indo-Australian, the Eurasian, and the Pacific plates. Based on the perspective of geography, geology, climatology, and demography. Therefore, Indonesia has various disaster threats that can occur at any time.

Indonesia is referred to as an archipelago because it consists of thousands of islands stretching from Sabang to Merauke. Since Indonesia consists of islands, it is also prone to tsunamis. Tsunami is a very large ocean wave produced by changes in the vertical mass of water and is caused by a sudden disturbance of the water mass in the deep sea [3]. Tsunamis with large casualties were those that occurred in Aceh in 2004 and Palu in 2018. The earthquake that resulted in a tsunami caused a lot of casualties and losses in the economic field.

Besides in Aceh, the area prone to tsunamis in Indonesia is the southern coast of Java [4]. The area is directly adjacent to the Indian Ocean so that it is potentially affected by the tsunami. There are 5744 villages in Indonesia in the tsunami-prone areas, where Donomulyo sub-district, South Malang, is in the list of villages mentioned [5]. As a sub-district directly adjacent to the southern coast of Java island, Donomulyo has several beaches. Six out of ten villages in this sub-district have a coastline area. Starting from Jonggring Saloka Beach in Mentaraman Village, Ngliyep Beach and Nglurung Beach in Kedungsalam Village, and Modangan Beach in Sumberoto Village.

The most deaths from tsunami in Indonesia are due to people's unpreparedness towards the arrival of tsunami, the short distance between wave generation areas and settlements, the absence of early warnings, insufficient infrastructure for evacuation, and low residential areas to the sea level so that the tsunami travels far enough to the mainland [6]. The key to dealing with problems caused by tsunamis is providing knowledge to the public about the importance of tsunami disaster mitigation. Mitigation is an action taken to reduce the impact caused by a disaster [7]. In addition, at present, the government has implemented a program to improve the quality of education in accordance with that has been formulated in the national education objectives set out in the Constitution No. 31 of 1945, article 1 which states that "Every citizen has the right to education." One of the quality issues of Indonesian education is about the education of disaster mitigation.

Based on the cases of natural disasters that are prone to occur in Indonesia, the serious handling from various parties and pragmatic strategies in shaping Indonesian people to be aware of the dangers of natural disasters are very necessary. One strategic step that can be taken is through the education sector as the fundamental sector in forming student character [8]. Through the education sector, knowledge about disaster mitigation can be provided intensively by educators. Disaster education is one of the efforts to increase the students' knowledge capacity about disasters covering the definition of the disaster itself, types, signs and impacts of disaster, pre-during-post disaster efforts and efforts to reduce disaster risk and vulnerability, as well as disaster vulnerability in the area [9]. The 2013 curriculum developed by the government appears to have been based on awareness so that students in Indonesia recognize the disaster potential. This is especially evident in natural sciences subjects for Junior High level (SMP/MTs). This shows how important disaster mitigation education is in Indonesia. Disaster mitigation education taught in schools will shape the students' character who are prepared for every disaster that occurs. The ability of students to understand the potential of disasters that are located around their homes (local area) is very important, knowledge and skills become information that is very helpful when a disaster occurs in anticipation of early attitudes and values that encourage students to act prosaically, responsible and responsive when family and community are threatened [10]. The preparedness will be formed if students have a provision in terms of knowledge and disaster mitigation skills that can be embedded in the school environment in Indonesia in learning activities.

However, in reality, it is very rare for basic competencies provided into good, appropriate, interesting, enjoyable, and beneficial learning plans and practices for students. Most learning processes seem to rely more on text from books and student workbook which is practical to provide material summaries and practice questions. Not many take the initiative to develop real contextual learning. Therefore, innovative models and learning tools are needed specifically to strengthen the mitigation disaster character, especially the tsunami disaster for students. Learning model is a holistic approach in planning learning with its attributes including the theoretical framework, orientation towards the materials learned, and procedures [11]. A suitable learning model to strengthen the character of disaster mitigation in students is the OrSAEv (Orientasi, Siap Siaga, Aksi, dan Evaluasi--Orientation, Preparedness, Action, and Evaluation). The instructional impact of the OrSAEv learning model is to improve disaster preparedness, especially knowledge on the disaster, disaster response attitudes, and post-disaster evacuation skills. The accompanying impact of the OrSAEv learning model is to increase awareness and motivation towards the environment and nature in Indonesia in particular [12].

Based on the description above, this research is aimed to determine the process of strengthening students' character about earthquake and tsunami disaster mitigation, to find out the experiences of students facing earthquake and tsunami disasters and to find out the disaster learning that has been implemented in schools. The researchers will conduct research under the title of Development of OrSAEv Model Learning Materials to Strengthen the Characters of Tsunami Disaster Mitigation for Middle School Students.

2. Research Method

This quantitative research was conducted at Donomulyo 1 Public Middle School on July 17th, 2019. Donomulyo 1 Public Middle School is surrounded by several beaches, these are 12 km from Jonggring Saloka Beach in Mentaraman Village, 16 km from Ngliyep Beach, 12 km from Nglurung Beach in Kedungsalam Village, and 14 km from Modangan Beach in Sumberoto Village. This research is carried out development research. The development plan that will be carried out is to develop learning materials in the form of syllabus, lesson plan, handout, worksheet and disaster character assessment.

Methods of data collection in this research were questionnaire and interview methods. The instruments that used in this research are (1) questionnaire about strengthening the characteristics of earthquake and tsunami disaster mitigation (2) questionnaire about students' experiences in facing tsunami disasters and classroom learning refer to the standard earthquake procedures issued by FEMA (Federal Emergency Management Agency) and disaster guidebooks published by the volcanology center and Geological Disaster mitigation. (3) interview sheet for the teacher.

Purposive sampling technique was used during the assessment of Donomulyo 1 Public Middle School teachers in which the samples of this research were 30 graders from 9A class in the education year of 2019/2020. Data analysis carried out was univariate analysis. Univariate analysis was carried out on the variables used in this research. The purpose of this analysis is to see an overview of the frequency distribution of each variable studied. The results of the analysis will be presented in the form of a frequency distribution table. Interpretation of the level of student character regarding earthquake and tsunami disaster mitigation according to Arikunto, character is divided into 3 categories, namely (1) Good = If the subject is able to answer 76% -100% correctly of all questions (2) Enough = If the subject is able to answer correctly 56% -75% of all questions (3) Less = If the subject is able to answer correctly 40% -55% of all questions [13].

This research is conducted by two different activities continuously, first is preliminary research to determine the students' character in facing tsunami disaster in Donomulyo 1 Public Middle School. The next stage of research is to develop disaster-learning materials based on the 4D development method by Thiagarajan, Dorothy S. Semmel, and Melvyn I. Semmel in 1974. Advanced development research is carried out in four-phase, namely: namely define, design, develop, and disseminate [14]. The disseminate phase aims to implement the materials, find out the effectiveness of the materials, and obtain proven learning materials. Please follow these instructions as carefully as possible so all articles within a conference have the same style to the title page. This paragraph follows a section title so it should not be indented.

3. Result and Discussion

The result contains questionnaire percentage answers about strengthening the character of earthquake and tsunami disaster mitigation table, questionnaire percentage answers for students' experiences in facing earthquake and tsunami disasters and classroom learning table, and interviews' result with teachers.

3.1 Result of questionnaire about strengthening the characteristics of earthquake and tsunami disaster mitigation

This questionnaire contains eight questions in five indicators of characters with multiple choice answer type. The result is presented in **Table 1.**

Table 1. Questionnaire Percentage Answers About Strengthening The Character of Earthquake and Tsunami Disaster Mitigation

Character's Indicators	**Questions**	**Percentage Answers**	
		Correct (n=30) %	**Incorrect (n=30) %**
Caring	1. When in class and an earthquake occurs, what we should do with friends who are around us is ...	93.3	6.7
	2. After knowing what we have to do when a tsunami occurs, what we should do is …	93.3	6.7
Religious	3. Earthquake and tsunami disasters can occur at any time especially in areas close to the coast, as humans, we should...	100	0
Orderly	4. When the sirens sound a sign of an earthquake and tsunami, what must be done is ...	80	20
Alert	5. The action that must be taken before a tsunami occurs is ...	60	40
	6. When on the beach and an earthquake occurs, the action that should be taken is ...	70	30
	7. As a student, what needs to be prepared to deal with a tsunami disaster that can come at any time is ...	100	0
Love cleanliness and environment	8. After the tsunami was declared safe by the BMKG, what must be done with damaged objects is ...	80	20
Averages		81.7	18.3

It is known that 30 students were taken as the sample for tsunami mitigation character questionnaires. Tsunami mitigation character indicators include caring, religious, orderly, stay alert, and love cleanliness and environment. Based on the above data, the average percentage of strengthening the tsunami mitigation character of students is 81.7% in the strong or good category according to Arikunto. In detail, it can be explained that in the caring indicator, there are 28 students who answered correctly with the percentage of 93.3% on both questions. On the religious indicators, all students answered correctly with the percentage of 100%. On the orderly indicator, there were 24 students who answered correctly with the percentage of 80%. On the stay alert indicator, where there are three questions, the first question related to actions that must be taken before the tsunami is only 18 students who answered correctly with the percentage of 70%, namely determining the closest evacuation route from the house to the evacuation site. C's answer choice is to take refuge in the house. In the second question, related to the actions that should be taken while on the beach and an earthquake, there were 21 students answered correctly with the percentage of 70%, which was immediately heading to a high place, the rest mostly chose to wait until the tsunami incident was announced. In the third question, all students answered correctly about the things that are needed to be prepared in the face of the tsunami. In the last indicator, which is love of cleanliness and environment, there are 24 students correctly answered with the percentage of 80%. These results are theoretical results, while characters are derived from the results of learning directly or observing others. Direct learning can be in the form of lectures and discussions about character, while observations are obtained through daily experience what is seen in the environment and related to attitudes and values [15]. So the above theoretical results need to be tested for application to find out whether the character is truly embedded within the students' mind.

3.2 Result of questionnaire about students' experiences in facing earthquake and tsunami disasters and classroom learning

This questionnaire contains twenty questions with yes or no answers. The result is presented in **Table 2.**

Table 2. Questionnaire Percentage Answers for Students' Experiences in Facing Earthquake and Tsunami Disasters and Classroom Learning

ı.	Questions	Percentage of Answers	
		Yes (n=30) %	**No ı=30) %**
·	school close to the source of the disaster?	100	0
	Has it ever been felt an earthquake when a school?	100	0
	Have you ever been warned by the teacher to always get closer to God because a disaster can happen anytime?	80	20
	Does your teacher warn you to help and care for each other in the event of a disaster?	100	0
	Have you ever been explained about the character (caring, religious, orderly, stay alert, hygiene and environment) regarding things before, during and after a disaster occur?	40	60
	Can an earthquake not be predicted when it will occur?	100	0
	Have you ever learned about the earthquake and tsunami disaster at school?	46,7	53,3
	Are there school subjects' specifically discussing about disaster?	46,7	53,3
	Have disaster management skills been carried out during the teaching and learning process?	0	100
	ɔur school ever conducted a tsunami disaster mitigation simulation training conducted by the school?	0	100
	Do you have medicines and first aid kits at your school?	100	0
	ı know of the existence of equipment/equipment for first aid for disasters, such as first aid kits or medicines, other stretchers at school?	50	50
	Do you have a disaster evacuation route in your school?	0	100
	Are there in your school students who are members of the PMR (Red Cross volunteers) who are trained in disaster management?	80	20
	Do you have books, VCDs, or posters about disaster in the school especially about earthquakes and tsunamis? Does your school have a disaster warning system?	0	100

'our school have a disaster warning system?	0	100
you participated in activities/training/meetings about first aid?	50	50
Have you ever taken part in scouting activities and training, such as making rigging, installing tents, and making stretchers?	40	60
Have you ever told/shared knowledge and skills on making rigging, installed tents, and made stretchers for schoolmates?	50	50
Have you ever participated in disaster simulation activities and training or lectures on disaster?	0	100

Based on the data above especially from question number 1 and 2, it is shown that students realize if their school belongs to a disaster-prone school. From question number 3, 4, 5 and 17, it is shown that students theoretically have strong character of disaster mitigation. Question number 6 to 20 showed that the level of disaster preparedness at Donomulyo 1 Public Middle School is still very low. This can be seen from the answers of each student, which there are still students who feel they have never learned about the earthquake and tsunami disaster, and have never been done and provided facilities to face the disasters. There are some reasons why it can be happened, first because there is no proper learning material for disaster mitigation. Second, the lack of support from the government to conduct counselling related to disaster mitigation. Third, there are no facilities such as disaster mitigation book, poster, evacuation route that provide to support disaster mitigation education. In its implementation, disaster mitigation education can't be done only by one side. To make the people develop a conscience about disaster mitigation and gain a culture for disaster prevention and preparedness, what to be needed is to provide nation-wide policies and determinism to execute these policies. So, the government, the social institutions, the schools, and the people, in general, should work together collaboratively and incessantly, to be successful in the long run [16].

3.3 Result of Interview with the teachers

3.3.1 Result of Interview with the teacher 1

Teacher 1 has taught at Donomulyo Middle School 1 for 22 years and she is a science teacher. During the interview, it was explained that the disaster that often happened at Donomulyo 1 Public Middle School was earthquake and landslide. The most recent disaster occurred was an earthquake on July 15th, 2019 at around 8:00 a.m., when the learning took place. It was also said that there were no signs of the earthquake happened. Apart from that, there was no loss whatsoever other than the time of the lesson taken. Teacher 1 said that previously there had never been counselling or socialization related to any disaster mitigation as a whole. Socialization is only obtained by students who take PMR extracurricular activities. In addition, he explained that there were no disaster evacuation routes in the school, so when an earthquake occurred the teachers instructed students to move to a safer place. Regarding learning in class, it was explained that there was indeed Basic Competence related to disaster mitigation in seventh grade, but in the process of learning the material was only delivered by the lecturing method.

3.3.2 Result of Interview with the teacher 2

The second teacher interviewed was also a science subject teacher. As far as 26 years the teacher has been teaching at Donomulyo 1 Public Middle School, there had never been a tsunami and it was assumed that there would not be a tsunami in the school because the location of the school was on a hill. It was also explained that natural disaster that often occurs as earthquakes and landslides. The most recent occurrence was an earthquake on July 15th, 2019, at which time the learning process took place around 8:00 a.m. At that moment teachers were inviting all students to leave their respective classes and move to a field near the school. Because the strength of the earthquake at that time was not quite strong, he said that there was no damage whatsoever in the school. Previously, at Donomulyo 1 Public Middle

School there had never been any socialization related to disaster mitigation from the SAR Team or others. It was known that only students who took extracurricular activities from PMR and Scouts were given training related to disaster mitigation by their respective supervisors. Regarding disaster mitigation learning in the classroom, it was explained that there were indeed Basic Competencies related to disaster mitigation in grade 7, but in the process of learning the material was only delivered by the lecturing method and based on the questions in the workbook.

From the results of interviews conducted with two science teachers at Donomulyo 1 Public Middle School, it was found that there was a lack of awareness regarding the potential of the area there for the tsunami. This is quite reasonable because the location of the school is on a hill while the location of the nearest beach which was Ngliyep located below the hill. They only understand that the disasters that are likely to occur there are earthquakes and landslides, even though the location of the school is surrounded by four beaches. During the earthquake on July 15th, 2019, the students' condition was not conducive because the number of teachers was not comparable to the number of students. The students should first be given insight into disaster mitigation so that they can save themselves and can help their friends who have difficulty in the process of evacuation. In addition, there should be an evacuation route so students do not need to wait for teacher instructions when earthquake and tsunami disaster suddenly occur. In the learning process, it is better if the material on disaster mitigation is not only delivered by the lecture method and working on the questions, but also by showing them ways to deal with disasters when suddenly coming in the form of posters and simulating evacuation activities for earthquake and tsunami. It happens because there is no effective learning material to support learning process. However, learning material provides the much needed sensory experiences needed by the learners for an effective and meaningful behavioural change [17]. In addition, so far there has never been a government in providing counselling related to disaster mitigation in schools. So, the character of disaster mitigation and experience of students in dealing with earthquakes and tsunamis as well as tsunami disaster learning that has been implemented in schools is still very lacking.

4. Conclusion

Based on the results of the data in related to the analysis and discussion that have been conducted, it can be concluded that there are contradictory results. Where based on the questionnaire is obtained that the character of tsunami disaster mitigation in Donomulyo 1 Public Middle School students is good. However based on the results of interviews with two science teachers and questionnaires how students facing the tsunami and earthquake disasters, the results show that facilities, experience of facing the disasters and disaster learning in classroom are still lack of it. Therefore, it is necessary to test its application to find out whether the character is truly embedded within the students' mind.

Acknowledgments
I am deeply indebted to my adviser, Prof. Dr. Madlazim, M.Si for warm support, inspiration and thoughtful guidance. I express my gratitude to my second and third adviser, Dr. Eko Hariyono, M. Pd and Nadi Suprapto, S.Pd., M.Pd., Ph.D for the advice and help. I am especially grateful for the head master of Donomulyo 1 Public Middle School for the granted permission to do the research in Donomulyo 1 Public Middle School. I express my gratitude to the science teachers as the interviewees. And special thanks to 30 graders from 9A class in the education year of 2019/2020 as the sample of the research. I wish to express my deep thanks to my parents and Dwi Endah Elfiana for her help and cooperation during my research.

References

[1] Undang- Undang Nomor 24 Tahun 2007 Tentang Penanggulangan Bencana.
[2] UNESCO 2012 *UN System Task Team on the Post-2015 UN Development Agenda*
[3] Sunarto, Marfai A, Djati M, Sutikno, and Lavigne F 2018 *Penaksiran Multirisiko Bencana di Wilayah Kepesisiran Parangtritis* Yogyakarta Gajah Mada University Press 99
[4] Utomo K S, Muryani C, Nugraha S 2016 *J. Geo Eco* **4** 1

Journal of Physics: Conference Series **1417** (2019) 012082 doi:10.1088/1742-6596/1417/1/012082

[5] Badan Nasional Penanggulangan Bencan 2008 *Pedoman Penyusunan Rencana Penanggulangan Bencana* Jakarta 53
[6] Triatmadja R 2010 *Tsunami Kejadian, Penjalaran, Daya Rusak, Dan Mitigasinya* Yogyakarta Gadjah Mada University Press 134
[7] Dennis F N, Yaulie D Y, Stanley D S and Karouw 2016 *J. Teknik Elektro dan Komputer* **5** 2
[8] Rizaldy and David 2018 *Implementasi Pendidikan Mitigasi Bencana di Sekolah-Sekolah di Indonesia Sebagai Upaya Pembentukan Karakter Siswa Siap Siaga* Semarang Prosiding Pit Ke-5 Riset Kebencanaan Iabi Universitas Andalas
[9] Mardiyati and Sofi 2017 *Dasi Sigab (Dalang Siswa Siap Siaga Bencana): Model Pendidikan Kebencanaan Sebagai Ekstrakurikuler Berbasis Kearifan Lokal Di Daerah Rawan Bencana Di Indonesia* Semarang 23
[10] Selby and Kagawa 2012 *Disaster Risk Reduction in School Curricula* 67
[11] Arends R I 2012 *Learning to Teach ninth edition* New York McGraw-Hill 256
[12] Madlazim, Rahmadiarti F, Masriyah, Indana S and Sunarti 2019 *Development of the OrSAEv Learning Model to Improve Disaster Preparedness for Prospective Teacher.* Surabaya: Unesa
[13] Arikunto S 2006 *Prosedur Penelitian Suatu Pendekatan Praktik* Ed Revisi VI Jakarta: Penerbit PT Rineka Cipta.
[14] Thiagarajan D 1974 *Instructional Development for Training Teachers of Exceptional Children: A Sourcebook.* Washington D.C: National Center for Improvement of Educational Systems
[15] Djemari M 2016 *Penilaian Pendidikan Karakter* Yogyakarta Universitas Negeri Yogyakarta
[16] Ozmen F 2006 *J. of Emereld Insight* 383
[17] Ajoke A R 2012 *J. of Humanities and Social Science Invention* **6** 36

Development of The OrSAEv Learning Materials: Preliminary Study of Student's Knowledge about Earthquake and Tsunami Mitigation

D E Elfiana[1], Madlazim[1,*], E Hariyono[1] and Munasir[1]

[1] Postgraduate, Universitas Negeri Surabaya, Indonesia.

*The corresponding author: madlazim@unesa.ac.id

Abstract. Geologically, Indonesia is a disaster-prone country, especially tsunami. As a country has kind of disaster and so in the schools of Indonesia should be started to give education of hazard. Therefore, this study aims to determine the level of student knowledge about the earthquake and tsunami disasters, know the experience of students facing earthquake and tsunami disasters, and determine the conditions of counseling and disaster learning that have been applied in school. This type of research is a quantitative descriptive. The subjects of this study were 28 students of MTsN 7 Blitar. Data collection techniques carried out by observation, tests, and questionnaires. The sample of respondents was taken by purposive sampling based on advice from the science teacher at MTsN 7 Blitar. The results of the study are as follows: 1) In terms of the level of student knowledge about the earthquake and tsunami disasters can be categorized as less. 2) student's experiences in facing with disasters and actions to save themselves in situations and conditions are categorized as not ready. 3) Disaster counseling and learning activities have never been carried out. After adding up the average percentage of each indicator, it is found that the knowledge and preparedness of the community in dealing with the earthquake and tsunami disaster based on their knowledge is still in the category of less prepared with a percentage of 47.9%, so there is still a need for better knowledge and readiness for students reach the level of full preparedness in the face of disaster. Learning methods and materials are needed that can be used to increase student knowledge about earthquake and tsunami mitigations. Therefore, it can be concluded that disaster learning materials need to be developed.

1. Introduction

Disasters according to UU No 24 Tahun 2007 are "events or series of events that threaten and disrupt human life caused by natural and / or non-natural factors as well as human factors that cause loss of human life, environmental damage, property loss, and psychological impact "[1]. Disaster is a serious disruption to society that results in losses to humans, material, economic, or the environment that extends beyond the ability of affected communities to cope with their own resources [2]. Disasters can be in the form of earthquakes, tsunamis, volcanic eruptions, floods, droughts, hurricanes and landslides [3]. Based on these definitions it can be concluded that a disaster is an event that harms humans whether caused by natural factors or human factors themselves.

Indonesia is one of the countries with the highest threat of natural disasters in the world. Horrible disasters, such as earthquakes and tsunamis seem to be 'very familiar' with life in Indonesia in recent times [4]. This is due to Indonesia's geographical position located at the end of the movement of three world plates, namely Eurasia, Indo-Australia, and the Pacific. In this condition, Indonesia cannot escape from the disaster that has befallen [4]. A total of 8 regencies in East Java are prone to tsunami disasters that stretch in the southern lane of East Java. The 8 regencies include Banyuwangi, Jember, Lumajang, Malang, Blitar, Tulungagung, Trengalek, and Pacitan. Blitar Regency is located on the southern coast of Java and has a vast stretch of beach. These areas are located in the hills and prone to landslides.

One of the schools located in disaster-prone areas in Blitar District is MTsN 7 Blitar or commonly known as MTsN Mojorejo located in Wates sub-district which is an area close to Jolosutro beach and 3 other beaches. Disaster mitigation learning materials seem very urgent in Indonesia, and specifically in Blitar district. The reason is that there have been several natural disasters, such as earthquakes, landslides, tornadoes, floods, and tidal waves.

Mitigation includes protection activities and actions that can be started from the preparation before the disaster takes place, assessing the danger of disaster, disaster management in the form of rescue, rehabilitation, and relocation. Knowledge, understanding, and behavioral skills in preventing, detecting, anticipating disasters effectively can be transformed and socialized [5]. Early childhood socialization and types of lessons that touch directly on natural phenomena, such as science. Provision of knowledge and life skills needed by students, especially early classes so that when a disaster occurs students can make efforts to save themselves and can also help others [6]. But the problem that arises is that the Indonesian people still do not have a high enough level of awareness of the disaster. Lack of awareness can increase community risk against disaster threats [7]. The 2013 curriculum developed by the government appears to have been based on awareness so that students in Indonesia recognize the disaster potential. This is especially evident in natural sciences subjects for Junior High level (SMP/MTs). This shows how important disaster mitigation education is in Indonesia. Disaster mitigation education taught in schools will shape the students' character who are prepared for every disaster that occurs.

However, in reality, it is very rare for basic competencies provided into good, appropriate, interesting, enjoyable, and beneficial learning plans and practices for students. Most learning process seem to rely more on text from books and student workbook which is practical to provide material summaries and practice questions. Not many take the initiative to develop the real contextual learning. Therefore, innovative models and learning tools are needed specifically to strengthen the mitigation disaster character, especially the tsunami disaster for students.

Disaster mitigation learning model is learning that operates using the flow: (1) preparation before the disaster takes place, (2) assessing the danger of disaster, (3) disaster management, such as rescue, rehabilitation and relocation, (4) providing knowledge, understanding and behavioral skills in preventing, (5) detection and anticipation of disasters effectively can be transformed, and (6) socialization [8]. A suitable learning model to improve the knowledge of disaster mitigation in students is the OrSAEv (Orientasi, Siap Siaga, Aksi, dan Evaluasi/Orientation, Preparedness, Action, and Evaluation).The first stage, Orientation (Or) aims to attract students' interests, focus students' attention, and motivate them to play an active role in the learning process. The second stage, Disaster Preparedness (S), conceptual learning is carried out to provide basic knowledge about the disaster. This basic knowledge of disaster includes knowledge of potential disaster threats, vulnerabilities, capacities, and disaster risks. The third stage, Action (A) is practical learning carried out with the intention that students can have skills in disaster. The fourth stage, Evaluation (Ev) aims to evaluate the disaster preparedness of students and prepare follow-up [9]. This OrSAeV learning integrates ICT in the learning process. This is based on a multi-representation theory which shows that students who are supported by multi-representation learning will be easier to master and improve learning outcomes [9].

Based on the above considerations, learning materials that support disaster learning are needed so that students' knowledge in dealing with disasters increases. To determine the level of knowledge and preparedness of students, prior to the development of the device, an analysis of the level of knowledge and preparedness of students in the face of disasters is carried out first. Therefore, the research was conducted "Development of OrSAeV Learning materials: Preliminary Study of Student's knowledge about earthquake and tsunami mitigation".

2. Research Method

This type of research is quantitative descriptive research. Methods of data collection in this study were interview methods, test methods, and questionnaire methods. The instruments used in collecting data in this study include (1) Knowledge and preparedness test sheets to deal with earthquake and tsunami disasters. (2) Questionnaires of existing questionnaires refer to the standard earthquake procedures issued by FEMA (Federal Emergency Management Agency) and disaster guidebooks published by the volcanology center and Geological Disaster mitigation. (3). Interview sheet for the teacher.

The subjects of this study were 28 students in grade VII at MTsN 1 Blitar. Data analysis carried out was univariate analysis. Univariate analysis was carried out on the variables used in this study. the purpose of this analysis is to see an overview of the frequency distribution of each variable studied. the results of the analysis will be presented in the form of a frequency distribution table. Interpretation of the level of student knowledge regarding earthquake and tsunami disaster mitigation according to Arikunto, knowledge is divided into 3 categories, namely: (1) Good = If the subject is able to answer 76% -100% correctly of all questions. (2) Enough = If the subject is able to answer correctly 56% -75% of all questions. (3) Less = If the subject is able to answer correctly 40% -55% of all questions [10].

This research is conducted by two different activities continuously, first is a preliminary study to determine the level of knowledge about disaster mitigation in MTsN 7 Blitar. The next stage of research is to develop disaster learning materials based on the 4D development method by Thiagarajan, Dorothy S. Semmel, and Melvyn I. Semmel (1974). Advanced development research is carried out in four stages, namely: namely define (definition), design (planning), develope (development), and disseminate [11]. The disseminate stage aims to implement the materials, find out the effectiveness of the materials, and obtain proven learning materials.

3. Result and Discussion

The results of this study aim to obtain an overview of students' knowledge and preparedness in dealing with the earthquake and tsunami disasters. The test sheet consisted of 20 questions about tsunami knowledge and preparedness. The questionnaire sheet was divided into 2 parts, the first was a questionnaire about the experiences during a disaster which consisted of 8 points and the second was a questionnaire about the condition of counseling and learning disaster. Interview sheet to find out the history of disasters that have occurred in the Wates sub-district, the conditions of counseling and disaster learning in school.

3.1 Result of interviews with teachers

3.1.1 Result of interviews with teachers 1

Teacher 1 is a science teacher at MTsN 7 Blitar who has been teaching for 23 years. The teacher said that in Blitar MTsN 7 there were frequent minor earthquakes and there was a strong earthquake around 2005 at 1 pm and caused damage to people's homes. "minor earthquakes occur frequently during lesson, It is not to cause harm in school but students panic during class". According to teacher 1 in disaster learning is usually taught with powerpoint media "for a rescue simulation when a disaster has never been done". Disaster specific learning materials do not yet exist. Disaster counseling has never been done. There is no evacuation route at the school, the school building consists of 2 floors but there is only one staircase

3.1.2 Result of interviews with teachers 2

Teacher 2 is the deputy head of the curriculum at MTsN 7 Blitar. Disasters that often occur in the Wates area are landslides and earthquakes. The earthquake did not cause damage and heavy losses. According to teacher 2 regarding disaster counseling it was stated that "there has never been a disaster counseling, the discourse is there but it has never been carried out. There was no encouragement from the government and the principal to specifically teach disaster.

From interviews conducted with 2 teachers in MTsN 7 Blitar, it is known that disaster counseling has never been done so there is no awareness in being alert to the possibility of a disaster. The absence of encouragement from the government or the school causes disaster-based learning to be carried out only briefly without the direct action of a disaster simulation. School facilities do not support disaster learning specifically, besides there is also no evacuation route in the school and there is only one stair so that at any time there is a disaster it will be difficult to evacuate students on the upper floors.

3.2 Level of student knowledge about earthquake and tsunami disasters

Based on data obtained from a knowledge test with 20 questions given to 28 students of MTsN 7 Blitar. where junior high school student's knowledge about disaster preparedness is calculated based on the respondent's answer, if the wrong answer gets a value of 1 and if the correct answer gets a value of 2. The percentage of student's answers in answering earthquake and tsunami knowledge tests is presented in **Table 1.**

Table 1. Distribution of respondents answers according to knowledge test about the earthquake and tsunami in MTsN 7 Blitar

No	Questions of disaster knowledge	Student answers	
		True (n=28) %	False (n=28) %
1.	A series of events both caused by humans, nature and a combination of both that can cause casualties, property losses, and environmental damage are called ...	75	25
2.	Activities in the disaster management cycle in the pre-disaster phase are	57,1	42,9
3.	A series of efforts to reduce disaster risk, for example, the installation of a disaster warning system is a step ...	78,6	21,4
4.	Activities to find out the actions that are fast and appropriate in dealing with disasters in order to minimize losses due to disasters are...	64,3	35,7
5.	what causes earthquakes?	53,6	46,4
6.	What action is taken when an earthquake occurs?	64,3	35,7
7.	what salvation will be done when a tsunami strikes?	57,1	42,9
8.	When does a tsunami occur?	71,4	28,6
9.	Which are the main causes of the Tsunami	46,4	53,6
10	The meaning of the word "Tsunami"	7,1	92,9
11.	The deadliest tsunami ever recorded	10,7	89,3
12.	What are the warning signs of the coming of the Tsunami	46,4	53,6
13.	Where is the most active Tsunami area?	17,9	82,1
14.	The Tsunami Sound is approaching...	46,4	53,6
15.	If an earthquake produces a Tsunami, what should you do	50	50
16.	What do you do if the seawater suddenly drops	39,3	60,7
17.	A place to save yourself in the event of an earthquake and tsunami	57,1	42,9
18.	Does every earthquake cause a tsunami	57,1	42,9
19.	Characteristics of a tsunami resistant building / house	21,4	78,6
20.	What will you do if an earthquake occurs and your position is in the room?	46,4	53,6
Average		47,9	52,1

The level of student knowledge about earthquake and tsunami disaster mitigation after adding the percentage of each question then divided by the number of questions obtained 47.9% results or in intervals of 3 categories of knowledge according to Arikunto, then expressed in the less category [10].

3.3 Students' experiences in facing earthquake and tsunami disasters.

To find out student's experiences in facing disasters, 8 questionnaire questions are presented in **Table 2**.

Table 2. Distribution of respondents according to the disaster experience of MTsN 7 Blitar students

No	Questions about the Experience of Earthquakes and Tsunamis	Student answers	
		Yes (n=28) %	No (n=28) %
1.	Have you ever experienced an earthquake	100	0
2.	Is your school close to the source of the disaster (beach/mountain)	100	0
3.	An earthquake that is felt to have a direct impact on yourself	85.7	14.3
4.	The earthquake that you feel, makes you introspective wherever you are	53.6	46.4
5.	Do you experience panic when you feel the earthquake?	60.7	39.3
6.	Because the direct impact that you feel is not too large, your attitude towards the recent earthquake has become normal.	35.7	53.6
7.	The earthquake event that you feel makes you want to dig more information about earthquake events and how to react to them	57.1	42.9
8.	Besides earthquakes, have there been other disasters (landslides, tsunamis)	85.7	14.3

From table 2 it is known that all respondents had felt the earthquake that occurred in their area. Even though the felt earthquake event did not have a large direct impact, it did have the panic effect of some respondents, with a percentage of 60.7%. This is in accordance with UU No 7 Tahun 2007 that disasters have psychological impacts [1]. In addition to the earthquake, in the sub-district of Wates, there were also landslides and large waves. Therefore knowledge and readiness are needed if at any time a disaster occurs. Because the direct impact that they feel is not too large, the attitude towards the recent earthquake has become normal. To make the people develop a conscience about disaster mitigation and gain a knowledge for disaster prevention and preparedness, what to be needed is to provide nation-wide policies and determinism to execute these policies. So, the government, the social institutions, the schools, and the people in general should work together collaboratively and incessantly, to be successful in the long run [12].

3.4 The conditions of education and disaster learning in schools

In this questionnaire, six questions were given regarding counseling and learning about disaster mitigation in MTsN 7 Blitar which are presented in **Table 3.**

Table 3. Distribution of respondents' answers about counseling and student learning in MTsN 7 Blitar

No	Counseling and learning	Student answers	
		Yes (n=28) %	No (n=28) %
1.	Had received an earthquake counseling	0	100
2.	Had received a tsunami counseling	0	100
3.	Never been given learning about disaster mitigation	100	0
4.	Learning disaster mitigation is accompanied by simulation practices	100	0
5.	There is a specific book on disaster mitigation learning	78.6	21.4
6.	There is an evacuation route at school	100	0

Counseling about earthquake and tsunami disasters does not yet exist. According to 78.6% of students answered that there are no books and special learning of disaster mitigation learning, 21.4 students answered that the earthquake and tsunami disaster learning had been done, indicating that disaster learning had actually been carried out but because it was not accompanied by interesting learning so most did not leave an impression on the student's memory. Facilities that support disaster learning are also not yet in school. According to all school respondents, there was also no evacuation route, this was in accordance with what the teacher had stated in the interview.

The experience of an earthquake that does not cause a large loss makes the alert level is low. People tend to underestimate signs of disaster. In addition, there was no encouragement from the government and local agencies so that disaster preparedness education was considered unnecessary special learning. It has an impact on disaster mitigation knowledge of students who are still in the low category based on the disaster knowledge test. School community preparedness in reducing disaster risk is still at the stage of knowing rescue actions but not yet accompanied by skills so the action is needed in the form of simulations in dealing with disasters [13]. Simultaneous learning can increase student's enthusiasm for learning based on an interview with teachers.

Preparedness is strongly influenced by children's cognitive development, where children develop their thought processes so that initiatives arise in carrying out the skills taught and psychological development so that children are able to anticipate, identify and control themselves against actions that should be taken to be alert when disasters occur and increase awareness of others in facing disaster [14] [15]. Someone is said to be prepared to face a disaster if they have the knowledge in anticipating, responding to, and recovering effectively from the impacts of events or conditions of hazard that may be present, will soon or currently exist [16]. Seeing the reality of the importance of student knowledge in disaster mitigation, researchers increasingly believe that disaster mitigation education must be implemented as soon as possible. This is because MTsN 7 Blitar is at a radius of approximately 12 km from Jolosutro beach. The choice of learning model also greatly influences the condition of student's backgrounds and student's interest in learning [17]. While learning methods and learning resources also need to be developed to realize innovative learning. Students who receive disaster preparedness education have increased knowledge about disasters, mitigation actions, and emergency response. Disaster education in schools should be supported by appropriate learning materials [18]. With appropriate learning, methods can increase student enthusiasm so disaster preparedness can increase. Disaster preparedness increases with increasing knowledge of disasters, appropriate attitudes towards disasters and more appropriate actions in dealing with disasters [19]. The alignment of teacher interview answers and the results of student questionnaires about the conditions of disaster learning in school can formulate that the need for disaster learning is accompanied by appropriate learning models and accompanied by direct simulations in dealing with disasters, therefore the development of learning materials is needed to be developed.

4. Conclusion

Based on the results, analysis, and discussion it was found that knowledge about tsunami mitigation in MTsN 7 Blitar was still in the less category based on tests of knowledge about the tsunami and earthquake disasters that have been conducted. All of the respondents stated that they had felt an earthquake but had never felt a large impact. Disaster counseling at MTsN 7 Blitar has never been done. Learning methods and materials are needed that can be used to increase student knowledge about earthquakes and tsunamis and how to deal with them. Someone is said to be ready to face a disaster if they have the knowledge to anticipate, respond to, and recover effectively from the impacts of an event or condition of a hazard that may exist, will soon or currently exist. Therefore it can be concluded that the development of earthquake and tsunami disaster learning developed for science teachers and students needs to be developed.

Acknowledgments

The author would like to thank Prof. Dr. Madlazim. M.Si. for the support, advice, and inspiration in doing this research. Thanks to MTsN 7 Blitar (Blitar, Indonesia) for allowing us to conduct research, as well as and VII-B class students and teachers who have many roles in this research. Thanks to my sister Bethari Uswatul Isna for her help and support during the research.

References

[1] Undang-undang No 24 2007 *Penanggulangan Bencana Nasional* (Jakarta: Departemen Dalam Negeri Jakarta) p 2

[2] ISDR 2004 *Living with Risk A Hundred Positive Example of How People are Making The World Safer* (United Nation Publication Geneva Switzerland: Online) http://www.unisdr.org/eng/about_isdr/bd-lwr-2004-eng.htm

[3] Saptadi G and Djamal H 2012 *Jurnal Penanggulangan Bencana* **3**(2) 1-13

[4] Direktorat Vulkanologi dan Mitigasi Bencana, Departemen Energi dan Sumberdaya Mineral 2005 *Gempa Bumi dan Tsunami* (Bandung: Alfa Beta)

[5] Tanaka N K and Jinadasa B S N 2011 *Lands Eco Engineering*, **7**(1) 127-135.

[6] National Research Council 2007 *Earth Sci and App from Space: National Imperatives and for the Next Decade and Beyond* (Washington DC: Academy of Science)

[7] Paripurno E T, Munadi K, Koesuma S, Ismail N, and Mardiatmo D 2019 *Buku Panduan Pembelajaran Kebencanaan Untuk mahasiswa di Perauruan Tinggi* (Jakarta: Kemenristekdikti)

[8] Maryani N 2010 *J. Geo. Gea.* **10** (1) 17-21

[9] Madlazim, Rahmadiarti F, Masriyah, Indina S, Sunarti 2019 *Development of the OrSAEv Learning Model to Improve Disaster Prepareness for Prospective Teacher* (Surabaya: Unesa)

[10] Arikunto S 2006 *Prosedur Penelitian Suatu Pendekatan Praktik* Ed Revisi VI (Jakarta: PT Rineka Cipta)

[11] Thiagarajan dkk 1974 *Instructional Development for Training Teachers of Exceptional Children: A Sourcebook* (Washington DC: National Center for Improvement of Educational Systems)

[12] Ozmen F 2006 *J. of Emereld Insight* 383

[13] Pangesti A D H 2012 *Gambaran Tingkat Pengetahuan dan Aplikasi Kesiapan Bencana pada Mahasiswa Fakultas Ilmu Keperawatan Universitas Indonesia* (Universitas Indonesia: online)

[14] Brunner J and Lewis D 2006 *Planning for Emergencies, Principal leadership* **6**(8) p 65-66

[15] Trianto 2010 *Model Pembelajaran Inovatif Berorientasi Konstruktivis* (Surabaya: Pustaka Ilmu)

[16] Asian Disaster Reduction Response Network (ADRRN) 2009 *Terminology Pengurangan Resiko Bencana* (Malaysia: MERCY)

[17] Thomas J W 2000 *A Review of Research on Project-Based Learning* (California: The Autodesk Foundation Online) Http:/ /www.autodesk.com/foundation, accessed at 29 Juli 2019

[18] Pribadi K, and Yuliawati A K, 2009 *Jurnal Pendidikan* **9** 3

[19] Ramli D 2014 *Penerapan Pelatihan Siaga Bencana dalam Rangka Meningkatkan Pengetahuan, Sikap, dan Tindakan Komunitas SMA Negeri 5 Banda Aceh* **1**(1) 26-34

Journal of Physics: Conference Series **1417** (2019) 012084 doi:10.1088/1742-6596/1417/1/012084

Metacognitive Skills: A Solution in Chemistry Problem Solving

U Azizah, H Nasrudin, Mitarlis

Chemistry Department, Universitas Negeri Surabaya, Indonesia

E-mail: utiyaazizah@unesa.ac.id

Abstract. One of the skills needed in the 21st century is thinking skills including metacognitive skills. Many studies showed that undergraduate students who used metacognitive skills are more successful in learning. This study aims to illustrate using metacognitive skills as a solution in chemistry problem solving and undergraduate students' respond. The design used One Group Pre-test and Post-test Design. The subject of this study was the undergraduate students of chemistry class academic year 2019/2020 as many as 31 people. To obtain the profile data of metacognitive skills and undergraduate students' respond through an essay test, interview, and questionnaire. Research data were analysed by quantitative descriptive. The results showed that: (1) performance of metacognitive skills of undergraduate students has high category in chemistry problem solving on each indicator. The indicators were goal setting, identify the known knowledge, determining the learning strategies, monitoring the relevance of the known knowledge with learning strategies, monitoring the goal achievement in the making conclusions, and evaluating the process and thinking outcomes, and (2) most of the undergraduate students are pleasant to join to this learning activity. Based on the results of this study, metacognitive skills can be used as a solution in chemistry problem-solving.

1. Introduction

Metacognitive skills are mental abilities that consciously be arranged, controlled, and examined for their thinking processes. Moreover, metacognitive skills are higher-order thinking skills regarding ways of thinking that involve cognitive processes [1]. Metacognitive skills have three components, namely planning skills, monitoring skills, and evaluation skills [2]. Metacognitive skills are very important to be empowered in the education system because they can empower high-level thinking and increase academic success [3; 4]. Metacognitive skills and undergraduates' student achievement have a positive relationship [5]. Undergraduate students who have good metacognitive skills will have good cognitive learning outcomes too.

The development of metacognitive skills is important because they can develop their learning atmosphere to be more meaningful, develop thinking skills, engage in problem-solving processes, and choose strategies to improve cognitive performance in the future [6; 7]. Therefore, learning metacognitive skills needs to be empowered.

In fact, learning of colligative solutions has not been able to empower metacognitive skills and improve the cognitive learning outcomes of chemistry undergraduate students. The metacognitive skills were still low be shown by planning skills about 57.68% of undergraduate students. It means that they cannot set goals and plan strategies for problem-solving. Moreover, the monitoring skills were about 66.52% of undergraduate students and means they have not been able to relate known knowledge with strategies used to solve problems. The evaluation skills were about 72.39% of undergraduate students that means difficult to evaluate the process and the results of their thinking.

In fact, learning of colligative properties matter has not been able to empower metacognitive skills and improve the cognitive learning outcomes of chemistry undergraduate students. The metacognitive skills were still low be shown by 57.68 % of undergraduate students who could not set goals and plan strategies for problem-solving (planning skills). Moreover, 66.52% of undergraduate students have not been able to relate known knowledge with strategies used to problem-solving (monitoring skills) and 72.39% of undergraduate students difficult to evaluate the process and the results of their thinking (evaluation skills).

The low metacognitive skills of undergraduate students have an impact on learning difficulties and errors in understanding colligative properties matter. This error occurs in understanding the concept of vapour pressure related to the effect of the presence of solutes, sub-microscopic images of the process of evaporation of electrolytes and non-electrolytes associated with differences in their boiling points. Moreover, misunderstanding also occurs in the concept of osmotic pressure, and the relationship between the increasing boiling point and the decreasing in freezing point, molar mass and Van't Hoff factor. This is consistent with research findings that learning difficulties and misunderstandings occur in the concept of the colligative properties of solutions [8]. Based on this, the empowerment of metacognitive skills must be carried out on undergraduate students, so the undergraduate students can plan, monitor, and evaluate their learning outcomes to find out deficiencies and can make improvements to achieve better learning outcomes.

Empowerment of metacognitive skills in chemistry done by applying undergraduate students' ability in solving chemistry problems. The steps of problem-solving covering understanding the problem, make a plan, do the plan, and look back at the results obtained [9]. The steps in the chemistry problem-solving process are in accordance with metacognitive skills, namely planning skills done by undergraduate students with making a plan, monitoring skills did when undergraduate students do the plan, and evaluation skills by a look back at the results obtained. This meant the ability in solving chemistry problems important to mastering by the undergraduate student and being a provision in dealing with academic problems and daily life. Eggen and Kauchak also emphasized the importance of problem-solving ability who stated that problem-solving ability independently are long-term goals and undergraduate students need to have the ongoing experience to achieve this goal [10]. The statement reinforced by an opinion that ongoing experience, undergraduate students more developed, both in cognitive, affective as well as psychomotor aspects [11].

The importance of metacognitive in problem-solving is shown from the results of research that with metacognitive skills, university students can focus more on solving problems and can manage the learning process and solve scientific problems correctly so as to improve undergraduate student academic achievement [12]. Humans who have high metacognitive skills will do problem-solving better [13]. Through metacognitive skills, each undergraduate student can plan and manage time, choose the right strategy and provide understanding in learning, and monitor the progress of learning by reflecting the use of strategies, effective solutions, and self-efficacy of each undergraduate student in problem-solving [14].

Chemistry problem solving with metacognitive skills in this study is the ability of university students to develop cognitive strategies and involve awareness to regulate their own cognitive activities that include awareness in planning, monitoring, and evaluating, as well as being aware of their use in chemistry problem-solving. The problem presented in this study is a non-routine chemistry problem. The non-routine chemistry problem in question is a problem to find/solve questions related to chemistry problems that do not have specific procedures so that further thought needed so that they can immediately find a method/procedure to solve the problem.

Based on the description above, the objectives of this study are: (a) Describe the performance of undergraduate students' metacognitive skills in learning chemistry problem solving on each indicator. The indicators were setting goals, identifying known knowledge, determining learning strategies, monitoring the relevance of existing knowledge to the learning strategies used, monitoring the achievement of objectives in making conclusions, and evaluating the process and results of thought, and (b) Explain student responses in teaching and learning.

2. Method

This research used the One-Group Pre-test Post-test Design and the difference between the initial test and the final test is to be the effect of the treatment. The research was conducted in the undergraduate students of the chemistry department of Universitas Negeri Surabaya academic year 2019/2020 as many as 31. Samples taking using a purposive sampling technique by selecting undergraduate students who have low metacognitive skills based on pre-research data.

Metacognitive skills measured through essay tests and undergraduate students' respond measured using the questionnaire. All instruments validated with validity test and reliability test. The essay tests and the university students' response questionnaire sheet have been validated tests by the experts. Test instrument reliability was done by using the KR-21 formula. The question of metacognitive skills consists of 12 items for the colligative properties of solutions. The indicators of the metacognitive skills are goal-setting (P1), identification the known knowledge (P2), determining the learning strategies (P3), monitoring the relevance of knowledge which has been owned with learning strategies are used (M1), monitoring the achievement of the goal in the making conclusions (M2), and evaluating the process and outcomes of thinking (E) [15]. Scoring of metacognitive skills obtained from the scoring rubric of metacognitive skills integrated with essay form written tests adapted from Corebima [16]. The metacognitive skill scored changed to a scale of 100, and the value obtained categorized according to the level of attainment of its metacognitive skills. The levels of achievement and these categories are very high (85 - 100), high (70 - 84), medium (55 - 69), low (40 - 54), and very low (0 - 39). The analysis for metacognitive skills used percentage, while the interviews and questionnaire result descriptively.

3. Result and Discussion

Undergraduate Chemistry students at Universitas Negeri Surabaya carry out the process of chemistry problem-solving in the material of the colligative properties implemented the stages of planning skills, monitoring skills, and evaluation skills. The measurement of problem-solving process by 12 essays test items consisted of 6 items for planning skills, 4 items for monitoring skills, and 2 items for evaluation skills.

3.1 Planning Skills

Illustration of achieving undergraduate students' metacognitive skills for problem-solving in planning skills that includes goal setting (P1), identifying known knowledge (P2), and determining learning strategies (P3) indicators shown in **Figure 1**.

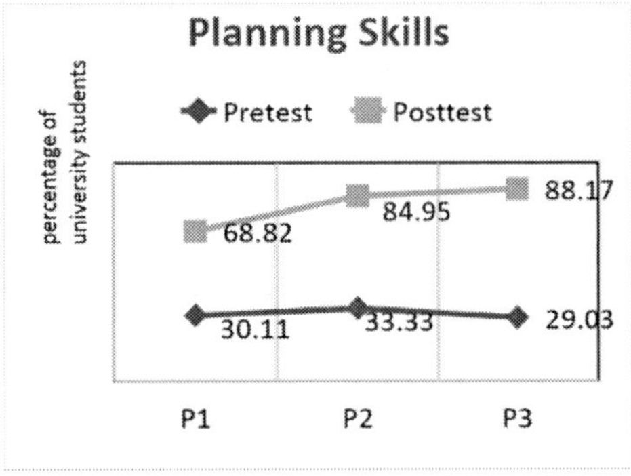

Figure 1. Percentage of Achievement of Each Indicator in Planning Skills

The results of planning skills for P2 and P3 indicators are high and very high because 70% of undergraduate students already have a value of ≥ 70, except for P1. The metacognitive skills of most undergraduate students' in problem-solving at the planning skills stage before learning is at a very low category level. These abilities increase to high and very high after learning, especially on indicators identify known knowledge and determine learning strategies in problem-solving. This fact shows that

Journal of Physics: Conference Series **1417** (2019) 012084 doi:10.1088/1742-6596/1417/1/012084

after learning, most of undergraduate students have been able to empower metacognitive skills at the planning skills stage, especially in indicators identifying known knowledge and determining learning strategies in problem-solving of colligative properties of solutions.

However, on the goal-setting indicator, some of the undergraduate students are less able to set goals according to the problem that solved. This can occur when undergraduate students determine the goal of problem-solving by analysis the experimental images measuring osmotic pressure at 300 K using 1M glucose solution (B) in a small tube that is put into pure water (A) and separated by a semipermeable membrane/osmosis membrane producing a pressure of 24.6 atmospheres (**Figure 2**)

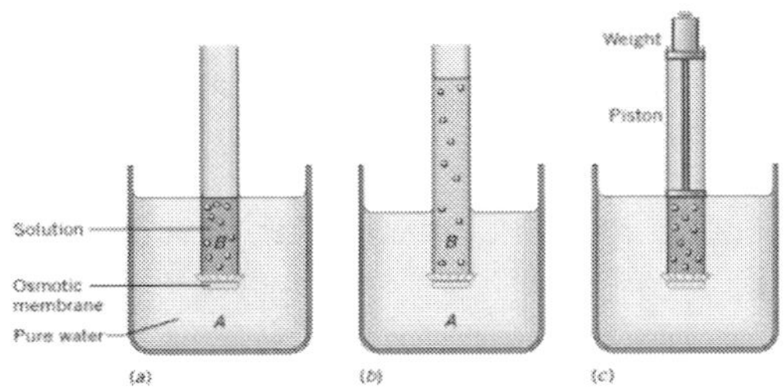

Figure 2. Measurement of osmosis pressure

Some of the undergraduate students still have errors in determining the goal of problem-solving based on the phenomenon in **Figure 2**. The goal of problem-solving should be to determine the relationship between osmotic pressure and the concentration of solution and temperature. The purpose of problem-solving will direct undergraduate students to evaluate the performance carried out in developing concepts, namely in measuring osmotic pressure using Van't Hoff's opinion where osmotic pressure is directly proportional to concentration and temperature, □ $V = n_2 R T$ with □ is osmotic pressure at T temperature from a solution containing n_2 mole of solute in a solution.

$$\frac{\pi \,.\, V}{T} = constant = 0{,}082\ L\ atm\ K^{-1}\ mole^{-1}$$

The undergraduate student's error in setting goals based on the results of the interview was that he could not understand the function of the semipermeable membrane and the theories proposed by Van't Hoff.

An initial indicator of metacognitive skills in planning skills is the ability to set learning goals based on phenomena (P1). This planning skill is in accordance with Jbeili's opinion, that the component of the cognitive arrangement of planning skills includes goal setting, activating relevant resources, and choosing the right strategy [17]. Another indicator is the skill of identifying known knowledge in observing phenomena (P2) and determining strategies to solve problems (P3). Both indicators in the planning skills stage have shown that all of undergraduate students have understood the knowledge and strategies needed to solve problems. This is in accordance with the positive response from undergraduate students stating that learning of the colligative properties of solutions can help in understanding material and practising metacognitive skills to identify the known knowledge through phenomena and determine strategies to problem-solving (Figure 1). One of them asserted:

"The phenomenon presented has been able to empower metacognitive skills and ease in understanding the material."

The need for planning skills in chemistry problem solving is in accordance with the opinion which states that undergraduate students must be able to solve problems by identifying knowledge already possessed and relevant to the phenomenon [18]. This confirmed by Rysz that when someone sees a problem in a phenomenon, it would indirectly have to identify relevant knowledge about the problem [19]. This is also consistent with the opinion that the selection of strategies to be used, learning resources that need to be collected, how to start, and which ones should be followed or implemented include planning skills in problem-solving [20].

3.2 Monitoring Skills

Problem-solving in monitoring skills in the pre-test and post-test shown in **Figure 3**. The overall results of monitoring skills have a very high category. The metacognitive skills of most undergraduate students in problem-solving at the monitoring skills stage before learning have very low categories. This ability shifts to a very high level after learning. This fact shows that after learning, most undergraduate students have been able to empower metacognitive skills in the monitoring skills stage in problem-solving of the colligative properties of solutions

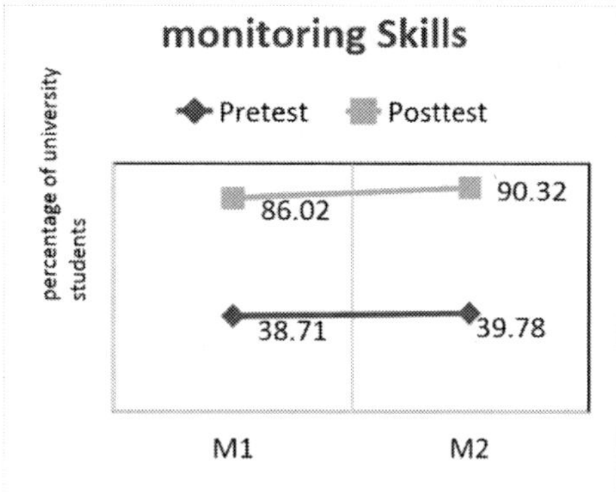

Figure 3. Percentage of Achievement of Each Indicator in Monitoring Skills

At the stage of problem-solving shows that undergraduate students monitor the relevance of knowledge already known and the learning strategies design used. The need for monitoring skills supported by statement Rickey and Stacy's that undergraduate student who applies metacognitive skills in good monitoring skills can increase success in problem-solving [21]. Likewise, the findings of the Fresenborg and Kaune study revealed that metacognitive skills in monitoring skills are the control of the problem-solving process [22]. Another statement that monitoring activities are direct awareness of how we carry out cognitive activities [20] and the results of the study found that a monitoring activity is a manifestation of the dimensions of goal development or to-do list [23]. This finding is in line with the opinion that metacognitive skills in internal undergraduate students are the key to controlling the success of learning situations [24].

Undergraduate students do problem-solving regarding the difference of freezing point of glucose solution and silver nitrate solution, as follows:

a. Analysis the phenomenon about glucose solutions (non-electrolyte substances) containing 1.80 g glucose (molar mass 180 g mole^{-1}) and a solution of silver nitrate (electrolyte substance) containing 1.70 g silver nitrate (molar mass 170 g mole^{-1}) each dissolved in 160 g of water. The silver nitrate solution ionizes as much as 83%. If Kf of water = 1.85 ^{0}C/m
b. Setting goals and identifying knowledge that is known based on phenomena
c. Determining the learning strategies problem solving
d. Implementing the task completion plan based on the planned problem-solving strategy

$$m_{glucose} = \frac{mole}{kg\ of\ solvent} = \frac{g}{molar\ mass\ x\ kg\ of\ solvent} = \frac{1{,}80\ g}{180\ g\ mol^{-1}\ x\ 0.160\ kg} = 0.0625\ m$$

$\Delta T_f = K_f.\ m = 1.85 \times 0.0625 = 0.117$

$T_f = 0 - 0.117 = -0.117^0C$ (freezing point of glucose)

$$m_{silver\ nitrate} = \frac{mole}{kg\ of\ solvent} = \frac{g}{molar\ mass\ x\ kg\ of\ solvent} = \frac{1.70\ g}{170\ g\ mol^{-1}\ x\ 0.160\ kg} = 0.0625\ m$$

$$\alpha = 0.83 = \frac{i-1}{2-1} \rightarrow\ i = 1.83$$

$\Delta T_f = i.\ K_f.\ M = 1.83 \times 1.85 \times 0.0625 = 0.212$

$T_f = 0 - 0.212 = -0.212^0C$ (freezing point of silver nitrate)

3.3 Evaluation Skills

Problem-solving in evaluation skills at the pre-test and post-test shown in **Figure 4**. The majority of undergraduate students have metacognitive skills in chemistry problem-solving at the evaluation skills stage before college learning is at a very low category level. This ability only shifted to quite high after college learning. This fact shows that after the learning process, the majority of undergraduate students still do not optimally empower metacognitive skills at the evaluation skills stage in problem-solving of the colligative properties of solutions.

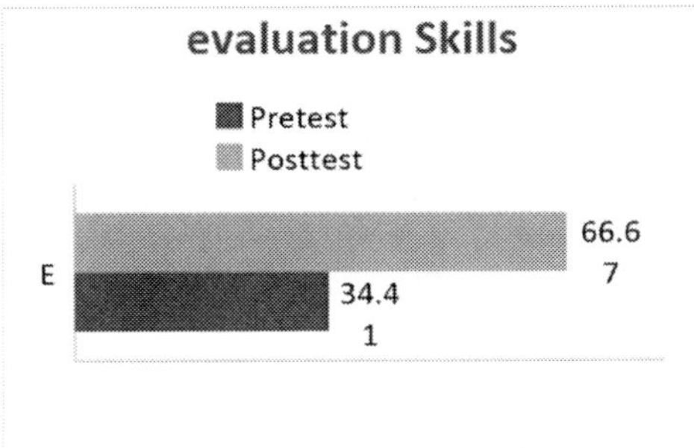

Figure 4. Percentage of Achievement Indicator in Evaluation Skills

According to the researchers' notes, some of the undergraduate students were less able to evaluate the known knowledge through the problem-solving process. This happens when undergraduate students have been able to determine the freezing point of silver nitrate and glucose solutions, but they cannot explain why to use that thinking because they do not understand the difference between the electrolyte and non-electrolyte solutions. Undergraduate students must be able to explain that in a solution of silver nitrate and glucose with the same molality and different freezing points because there are dissolved particles. The decrease in the freezing point of AgCl solution is greater than the decrease in the freezing point of glucose because AgCl is an electrolyte that has an ionization rate that causes a decrease in the freezing point of the solution is greater than glucose solution that is not ionized or no ionization rate. This is in accordance with Van't Hoff's law. According to the interview results, undergraduate students stated that they were not used to evaluating results or concepts that found using theoretical and empirical studies. This is a consistent response from undergraduate students as explained below.

"The implementation of learning from the colligative properties matter that has been followed is good and interesting, but it is difficult to evaluate the process and results of thinking using concepts that have been learned in problem-solving process."

In evaluation skills, undergraduate students must be able to evaluate their knowledge, learning objectives, learning strategies in building concepts and reflection assessment in the process of thinking/learning. This is consistent with the opinion that reflections carried out at the evaluation skills stage can be in the form of objectives that have been achieved, which learning strategies are more efficient, assess how learning strategies are applied in other contexts, and respect oneself after learning or completing assignments. In evaluation skills, undergraduate students train to think about what they had been learned [23]. So, the application of evaluation skills in learning can help to evaluate undergraduate students' way of thinking to be better in the future [25].

3.4 Undergraduate students Response

The interesting lectures obtained from the responses of undergraduate students using the response questionnaire, and the results presented in **Table 1**.

Table 1. Undergraduate Students' View in Response

Ease in practising metacognitive skills	
a.	Goal setting
b.	Identification the known knowledge through phenomena
c.	Determining the learning strategies or thinking
d.	Monitored the relevance of the known knowledge, goal achievement, learning strategies, and the implementation of strategies
e.	Evaluated the relevance of the known knowledge, goal achievement, and learning strategies that have been used

Based on **Table 1**, the tendency of undergraduate students to respond to lectures has a high category. Meanwhile, most of undergraduate students expressed interest in taking lectures on other material and other subjects by applying metacognitive skills. Nevertheless, the lowest response (63.64%) found in terms of evaluating the relevance of the known knowledge, goal achievement, and learning strategies that have been used (**Table 1**). This was consistent with college student comments as described below.

"The lectures implementation on colligative properties of solution that have been followed were good and interesting, but it was difficult to evaluate the relevance of the known knowledge, goal achievement, and learning strategies that have been used in problem-solving."

In line with this comment, another undergraduate student also comments as follows.

"The activity of planning skills, monitoring skills, and evaluation skills have not used to do before, so often forget to be done, for example in setting learning goals and communicating the results of self-evaluation."

4. Conclusion

Based on the objectives and results of the research can be concluded that: (1) performance of metacognitive skills of undergraduate students has high criteria in chemistry problem solving on each indicator. The indicators were goal setting, identify the known knowledge, determining the learning strategies, monitoring the relevance of the known knowledge with learning strategies, monitoring the goal achievement in the making conclusions, and evaluating the process and thinking outcome, and (2) most of undergraduate students are pleasant to join to this learning activity. Based on the results of this study, metacognitive skills used as a solution in chemistry problem-solving.

5. References

[1] Livingston J A 2003 *Metacognition: An Overview.* ERIC Dokümen No: ED474273, http://www.eric.ed.gov/ERICDocs/data/ericdocs2sql/

[2] Lai E R 2011 *Metacognition: Literature review,* (Online), https://moodle.elac.edu/pluginfile.php/111973/mod_resource/content/0/Metacognition_Literature_Review_Final.pdf

[3] Flavell JH 1979 *Metacognition and cognitive monitoring: A new area of cognitive-developmental inquiry.* American Psychology. **34(10)** 906-911

[4] Chisholm JM 1999 The Effect of Metacognition, Critical Thinking, Gender and Gender Role Identification on Academic Achievement in The Middle Years. *Thesis*. Halifax, Nova Scotia: Mount Saint Vincent University.

[5] Rahman UF 2010 Impact Of Metacognitive Awareness On Performance Of Students In Chemistry. *Contemporary Issues In Education Research,* **3 (10)**: 39—44.

[6] Lawson AE 2009 *Science Education*. 25 336

[7] Temmel S 2013 Problem of Education in the 21st Century. 511

[8] Pınarbaş T, Sozbilir M, and Canpolat N 2009 Prospective chemistry teachers' misconceptions about colligative properties: Boiling point elevation and freezing point depression. *Chemistry Education Research and Practice.* **10(4)** 273-280.

[9] Polya G 2002 *Model Problem Solving dalam Pembelajaran*. (Jakarta: Pustaka Buku)

[10] Eggen PD and Kauchak DP 2012. *Strategi dan model pembelajaran*. Edisi 6. (Jakarta: PT Indeks).

[11] Sanjaya W 2008 Perencanaan dan Desain Sistem Pembelajaran. (Jakarta: Kencana).

[12] Safari Y and Meskini H 2016 The Effect of Metacognitive Instruction on Problem Solving Skills in Iranian Students of Health Sciences. *Global Journal of Health Science*, **8(1)**, 150–156. https://doi.org/10.5539/gjhs.v8n1p150

[13] Kazemi F, Yektayar M, and Abad AMB 2012 Investigation The Impact of Chess Play on Developing Meta-Cognitive Ability and Math Problem-Solving Power of Students at Different Levels of Education. *Procedia - Social and Behavioral Sciences*, **32**, 372–379.

https://doi.org/10.1016/j.sbspro.2012.01.056
[14] Baten E Praet M and Desoete A 2017 The Relevance and Efficacy of Metacognition for Instructional Design in the Domain of Mathematics. ZDM, **49(4)**, 613–623. https://doi.org/10.1007/s11858-017-0851-y
[15] Azizah U and Nasrudin H 2018 Empowerment of Metacognitive Skills through Development of Instructional Materials on the Topic of Hydrolysis and Buffer Solutions. *J. Phys.: Conf. Ser. 953012199.* doi: 10.1088/1742-6596/953/1/012199.
[16] Corebima AD 2010 "Berdayakan keterampilan berpikir selama pembelajaran sains demi masa depan kita." Makalah disajikan dalam *Seminar Nasional Pendidikan sains di Prodi Pendidikan Sains PPs UNESA*, tanggal 16 Januari 2010.
[17] Jbeili IMA 2003 The Effect of metacognitive scaffolding & cooperative learning on mathematics performance and mathematical reasoning among fifth-grade students in Jordan.
[18] Gok T 2010 "A new approach: Computer-assisted problem-solving." *Asia-Pacific Forum on Science Learning and Teaching.* **11(2)**, 1-22.
[19] Rysz T 2004 *Metacognition in learning elementary probability and statistics* (Unpublished doctoral dissertation). University of Cincinnati.
[20] Woolfolk A 2008. *Psychology in education*, (New York: Pearson Education Limited)
[21] Rickey D and Stacy 2000 "The role of metacognition in learning chemistry." *Journal of Chemical Education,* **77(7)** 915–920.
[22] Fresenborg and Kaune 2007 Modelling Classroom Discussion and Categorizing Discursive and Metacognitive Activities. In *Proceeding of CERME 5, 1180 – 1189.*
[23] Pulmones R 2007 "Learning chemistry in a metacognitive environment." *The Asia Pacific – Education Researcher*, **16(2)** 165-183.
[24] Lin X 2001 *Designing Metacognitive Activities*. ETR&D, **49(2)**, 23–40. ISSN 1042–1629.
[25] Hacker DJ, Dunlosky J and Graesser AC 2009 *Handbook of metacognition*. (New York: Routledge Taylor and Francis).

Acknowledgements
The author thanked the first-year university students of Chemistry Education program who are directly involved in empowering metacognitive skills.

Development of A-SSI Learning Media (*Android Social Scientific Issues)* to Improve Science Literation in Earth Coating Subject for First Grade of Junior High School

R M Jalil[1], T Prastowo[2], W Widodo[1]

[1]Science Department, Faculty of Mathematics and Natural Sciences, State University of Surabaya, Kampus Unesa Ketintang, Surabaya 60231, Indonesia
[2]Physics Department, Faculty of Mathematics and Natural Sciences, State University of Surabaya, Kampus Unesa Ketintang, Surabaya 60231, Indonesia
(restu.17070795028@mhs.unesa.ac.id)

Abstract. This study aims to provide an alternative learning process that was supported by the use of A-SSI media (Android Social Scientific Issues) by including actual issues so that they could train and improve students' scientific literacy skills. The developing A-SSI media must be tested for feasibility based on three aspects, they were validity, effectiveness, and practicality. This study used a pretest-posttest design and tested on grade VII junior high school students. A-SSI media that would be used in learning process is validated by experts through a validation sheet. The results obtained based on the validators' assessment are 88.5% with the criteria "very good". In addition, effectiveness of A-SSI media is tested through literacy tests and student responses. The results obtained in the literacy test, literacy skills of students were increased with an average n-gain of 0.66 with "moderate" category, there was a response given by students in the learning process with A-SSI media, 88.9% students gave a positive response of the using of A-SSI media. Then in the last aspect, that was practicality of media was reviewed through student activities and implementation of learning. The results obtained in the implementation of learning, 89.8% of activities designed in RPP were carried out well, while the activities of students based on the results of observation were very good and showed activities that supported scientific literacy skills.

1. Introduction

Science is a way of thinking, science as a way of investigating, science as a body of knowledge, and interactions with technology and society, so it means that science learn about object and phenomena that occur in nature. Studying science is one way to develop thinking skills to investigate objects and phenomena that occur in daily life. Through the investigation activities, facts and concepts will be obtained then can be developed. Science's learning process should contain three dimensions, they are product, process, and development of scientific attitude. This statement shows that science's learning process is not only focused on facts and concepts, but also work processes, ways of thinking, problem-solving processes, and development science attitude. These three dimensions can be fulfilled with the help of the teacher as a facilitator in establishing learning strategies. One of the suitable strategies for creating right science learning process is by associating science learning with literacy.

Science literation is one of the important aspects for students to understand the environment, health, economy, and other problems faced by modern society. The problems around students can be used as a source of learning to understand them in terms of science. Unlike the case that happened in school shows that science learning in Indonesia still not supportive of science learning based on natural science itself. Based on PISA's research, shows literacy skills of students. Although Indonesia has increased literacy skill compared to 2012's result, Indonesia's literacy skills are still low compared to the OECD average. The suitability of scientific literacy with the purpose of science learning, it is expected that scientific literacy can improve students' skill in understanding and solving problems. Science learning will be better if it is supported by using suitable technology, such as smartphones. Smartphones are one of the important aspects of the life of digital generation like today. In this modern era, many things that were previously only can be done through computers but can be done by using mobile phone. Users of this smartphone are not only adults but also students who are sophisticated in operating a smartphone. The use of smartphones in science learning process will help students to package science concepts effectively that can be accessed anytime and anywhere.

In this case, the use of smartphones as a media for science's learning process will help students gain meaningful experience, and make the information more memorable. This is in line from [1] which states that using media especially in science materials can help students gain meaningful experiences through observation, exploration, and experimentation, and make it easier for students to understand abstract scientific concepts to be more concrete. The learning media developed in this study is APK (Android Application Package), an android package which is generally used to store applications and then operate it on android device. In addition, this APK media developed because researchers have not found any research that develops ap media to improve student's literacy skills. APK media developed in this study was made on powerpoint that has been edited in such a way as to add videos or photos if needed which are converted in form of WTPPT then can be changed to APK through an even application so it can be operated on android-based smartphones. Applications that need to be prepared in developing this media are WTPPT, Make APK Tool, Java, Flash Player, Any Video Converter and PowerPoint.

Apk media in this study consisted of concept maps, earth layer's material. There are pictures and videos that are expected to help students to understand the information. Researchers also included authentic problems to help students associate information that has been obtained with those problems or phenomena. Based on these problems, the researchers will do research with the title "*Development of A-SSI Learning Media (Android Social Scientific Issues) to Improve Student's Science Literation in Earth Coating Subject for First Grade of Junior High School*".

Media comes from Latin, namely medium. The term can be defined as an intermediary of information from the source to the recipient in [2]. According to [3], learning media can increase learning motivation and quality in learning process. This is also indicated because learning process contains five components, they are teacher as a communicator, learning material, students as communicants, and learning purpose. The cognitive function of visual media is shown by the use of visual symbols or images to facilitate goals to be achieved. Whereas the last, compensatory function of learning media can be seen from the results of research that show visual media provides content for understanding text, so students who are weak in reading can understand text and remember it again. Based on [4], states many benefits by using media in learning process, including learning process will be more interesting so that it will grow student learning motivation. In addition, by using media, the teaching and learning process will increase students' activity.

2. Methods

The type of research used in this study is Research and Development (R&D) with a 4D development model (define, design, develop, disseminate). Implementation of the development of this research is limited to 3 stages, namely define, design, and develop. This 4D model used in this study uses a model proposed by [5].

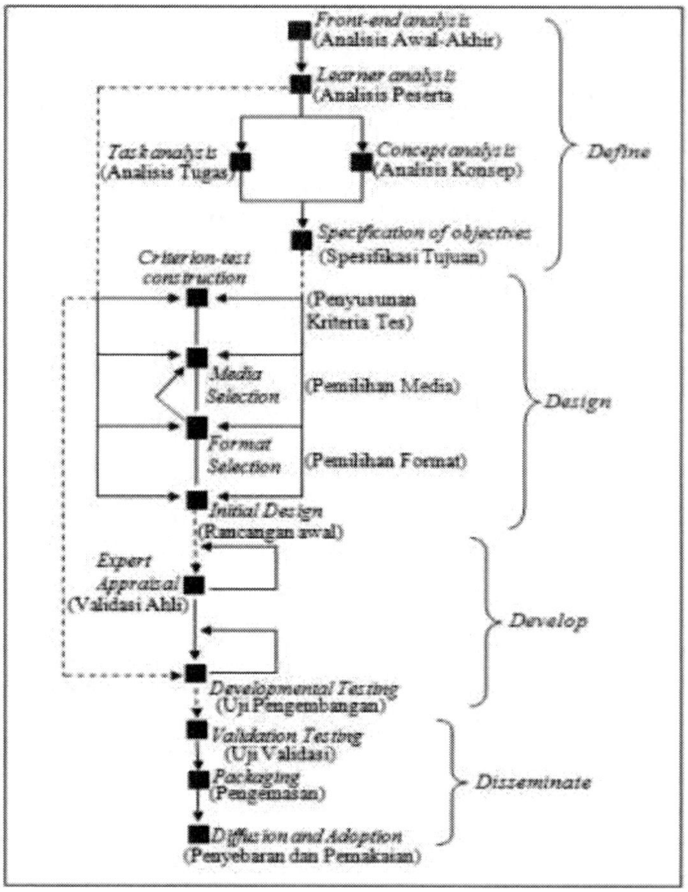

Figure 1. The development phase of 4D is divided into 4, namely define, design, develop, and disseminate. (available from http://conferences.uin-malang.ac.id for academic purpose only).

The research was conducted using a pre-experimental research design, one group pretest-posttest design to investigate the students' learning outcomes of Earth layer material. This research was tested in a limited way to 10 people of class VII junior high school students who were randomly selected by considering a number of groups and students. So the trial was carried out in 3 groups with each group consisting of 3-4 people. Instruments used in this study amounted to 5 instruments, namely: media validation sheet, teacher activity observation sheet, student activity observation sheet, student response questionnaire, and learning outcomes test.

Analysis technique used to assess each aspect related to the feasibility of A-SSI media is feasibility validity is obtained from average percentage of validators, where A-SSI is declared feasible and valid if the percentage of each aspect in instrument reach minimum 61%. Practicality is obtained based on observation sheet of teacher activities, observation sheet of students activities. The observation sheet of teacher activities is obtained from observer assessments as long as the teacher manages class based on the lesson plan. Media learning can be concluded practical if the percentage of each aspect reach minimum of 61%. The observation sheet of student activities is obtained from observer's assessment during learning process, in which students are assessed every 5 minutes. Effectiveness is obtained based on the complete competencies of student learning outcomes with different completeness value. Outcome test that is given to students is literacy question with level range from 1-5. Besides learning outcome, effectiveness of media is showed by questionnaire responses. Questionnaire responses of students were obtained from the average results of percentage of student assessment after using A-SSI media and said to be feasible if the average score percentage for each aspect of a positive or negative question scored minimum 61%.

3. Discussion

The results of the feasibility media include the results of feasibility in terms of validity, practicality, and effectiveness. This results can be seen in Table 1 below:

Table 1. Validation Result of A-SSI Media

No	Validated Aspects	Percentage(%)	Criteria
1.	General Aspects	89.3	Very good
2.	Material Aspects	89.5	Very good
3.	Technical Aspects	86.7	Very good
	Average	88.5	Very good

Based on [6] states in order to know the validity of developed media, the product must have been validated or assessed by an expert. Therefore, the feasibility based on this validity's aspect is reviewed from the validator's assessment of A-SSI media. Media validation consists of three aspects, namely general, material, and technical aspects. Based on [7], media can be categorized as good or very good if the percentage obtain minimum of 61%. Based on table 1, the results of media validation can be seen that A-SSI media is stated to be very feasible with an average percentage of 88,5% with "very good" criteria.

The practicality of a media seen from the ease of users in using A-SSI media. Practicality can be measured from three aspects, implementation of learning, student's response. Besides, this practicality also can be assessed from student's activities. The result of teacher activities is shown in Table 2 below:

Table 2. Observation of Teacher Activity

No.	Observed Aspects	First Meeting		Second Meeting		Third Meeting	
		Percentage (%)	Criteria	Percentage (%)	Criteria	Percentage (%)	Criteria
1.	Initial activity	89.3	Very good	91.1	Very good	89.6	Very good
2.	Main activity	94.2	Very good	92.5	Very good	92.1	Very good
3.	Closing activity	84.7	Very good	85.9	Very good	88.7	Very good
	Total Aspects	89.4	Very good	89.8	Very good	90.1	Very good
	Observation of teacher activity					**89.8**	Very good

Based on Table 2, results of observation of teacher activities can be seen that learning with science A-SSI media on earth layer material was carried out well in first, second, and third meetings with the percentage of each meeting is 89.4%, 89.8%, and 90.1%. Overall, the observation of teacher activities has been very good with an average percentage of 89.8%. The results of observation of students activities during learning using A-SSI media are presented in Table 3 below:

Table 3. Observation of Teacher Activity

Appeared Activity	Percentage of student's activities (%)			Average (%)
	Pert I	Pert II	Pert III	
Observing	30.00	24.99	24.99	26.66
Asking	15.00	10.62	11.25	12.29
Doing an experiment	-	13.12	-	4.37
Discussing	18.12	13.12	14.99	15.41
Doing the worksheet	15.00	15.62	18.12	16.25
Communicating	7.50	8.12	10.62	8.75
Other activities	13.75	14.37	17.50	15.21

Student activities observed in this study are student activities that support the learning process using A-SSI media. The observed activities consisted of 7 activities, namely observing, asking, conducting experiments, discussing, working on LKPD, communicating, and other activities. Observations were made by observers to observe 5 randomly selected students. Observations are carried out for 5 minutes in three meetings.
Based on Table 3 above, it can be seen that those activities of students as a whole start from observing to the teacher, asking question, answering, reading, working with tools, discussing, and doing the worksheet in a row to obtain an average percentage of 26.66%, 12.29%, 4.37%, 15.41%, 16.25%, 8.75%, and 15.21%. The most frequent student activity is observing. The last aspect of practicality is student response. Student responses show how student's responses after learning process using A-SSI media. Student response is presented in Table 4 below:

Table 4. Student Response

No	Rated Aspect	Student's Response	
		Yes (%)	No (%)
1.	Does the use of A-SSI (Android Social Scientific Issues) media make the learning process more enjoyable?	100	0
2.	Is the appearance of A-SSI (Android Social Scientific Issues) interesting?	100	0
3.	Is the A-SSI (Android Social Scientific Issues) media easy to use?	90	10
4.	Can A-SSI (Android Social Scientific Issues) media help you to understand the material of Earth Layer?	90	10
5.	Can the A-SSI (Android Social Scientific Issues) media function properly when used in science learning process?	90	10
6.	Are the components in A-SSI (Android Social Scientific Issues) media safe to use?	100	0
7.	Does the use of S-SSI (Android Social Scientific Issues) media can make you active in learning process?	70	30
8.	Can A-SSI (Android Social Scientific Issues) media train student's ability to analyze the phenomena happened around them?	80	20
9.	Does A-SSI (Android Social Scientific Issues) media help students to relate learning material to the phenomena that occur?	80	20
	Average	**88,9**	**11,1**

Based on Table 4, it can be observed that the student responses of A-SSI media were 88.9% with very good criteria. It means the media were easy to use. Practicality assessed based on product design that can be used realistically in [6]. In this research, a simple product can be states as realistic or not through the trial phase. Feasibility based on practical aspects in terms of observation of teacher activities as in Table 2, student activities as in Table 3, and student response as in Table 4. According to [7], observation of teacher activities can be organized as good media if the percentage obtained reaches minimum of 61%. Therefore, observations of teacher act as a whole are categorized very well with a percentage of 89.8%.

Effectiveness is assessed based on the results of the use of product development [6]. In this study, A-SSI media were declared feasible based on aspects of effectiveness in terms of students' learning outcomes. Researcher gave literacy questions with 1-5 level range. Comparison of student's literacy skills at pretest and posttest is shown in the graph below:

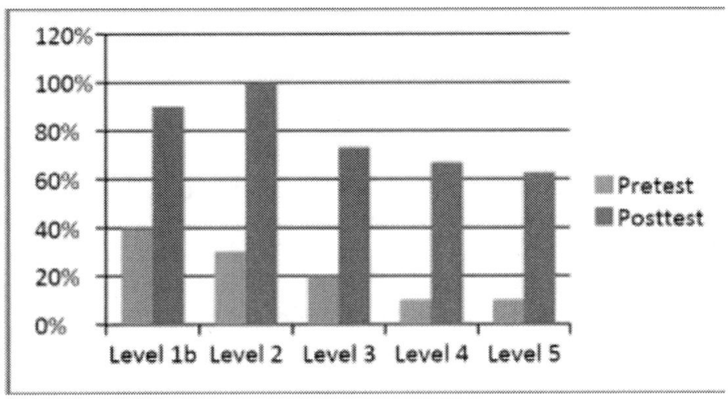

Figure 2. The graph shows significant differences in answered questions with different literacy levels at pretest and posttest.

The graph above shows a significant difference in scientific literacy skills at pretest and posttest of students. All literacy levels from level 1b-5 have increased with a range of 35%-70%. For pretest question, student's ability to answer the lowest literacy level question 1b, get 40%, then level 2 and 3 are 30% and 20%. As for high level of literacy, levels 4 and 5 get 10%. Students have increased in answering questions of posttest, question with level 1b are answered by 90% of students, then at level 2, 100% of students can answer the question. As for the levels 3,4, and 5, get a percentage of 72%, 66.7%, and 62.5%. The problem of literacy in this study is divided into three knowledge content, namely content, procedural, and epistemic knowledge. Based on these three aspects of knowledge, the increase in student's literacy skills is shown in the following graph:

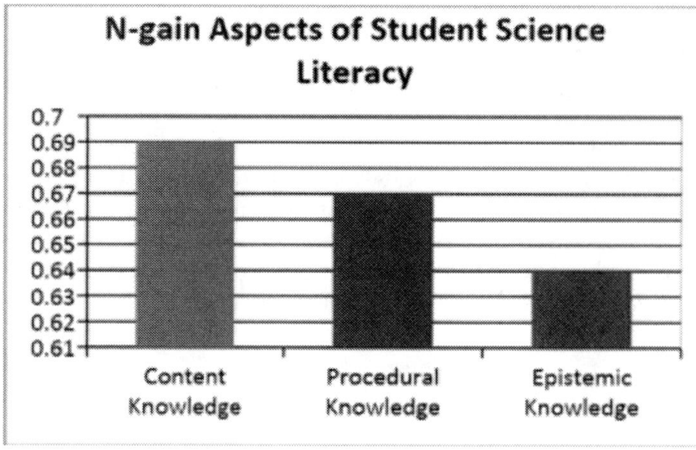

Figure 3. N-gain in each literacy knowledge content that is content knowledge, procedural content, epistemic content, get different scores.

Literacy that was tested on students, the average n-gain obtained in the aspects of content, procedural, and epistemic knowledge was 0.69 (medium), 0.67 (medium), and 0.64 (medium). The improvement of student literacy aspects is supported by the existence of A-SSI media which lists issues that occur around students by linking science concepts, so media can train students scientific literacy skills. However, in using this media, media must go through validation process that is assessed by 3 validators to describe whether the A-SSI media developed is suitable in science learning process.

Student's scientific literacy has increased supported by student's activities during the learning process. Based on data in Table 2, activity with the highest percentage is in observation activity. Observations made on learning with A-SSI media are observing scientific issues through videos and articles. This is suitable with statement which explains that scientific literacy refers to aspects of understanding and knowledge related to science issues, especially related to physical knowledge, life and nature so that observing videos on science issues and reading articles on socioscientific issues that can help improve student's scientific literacy skills. Then proceed with other activities, namely asking questions, conducting experiments, discussing, doing LKPD, communicating and other activities are activities that support student's scientific literacy abilities.

The student's activities who support learning process can not be separated from teacher's activities as a mediator who can provide learning process that supports student's activities. Through a good learning process with tools that support student activities, it can be seen whether the learning process can train student's scientific literacy skills. The implementation of learning assessed by two observers received a percentage of 87.9% and 93.2% at first meeting. The second meeting of the implementation of learning received percentage of 90% and 91.4%. Whereas in the third meeting, observers 1 and 2 gave assessments of 90.1% and 92.8% respectively. These results are very good results which indicate that the teacher has carried out the learning process according to the design that has been made. In addition, students are also actively involved in learning.

This is suitable with [7] which states that teachers' activity is good or very good if te percentage obtained minimum 61% so that the average activity that has been prepared in learning design has been carried out. These results are also in line with [8], which shows that the use of media supports activities of teachers, do the learning's implementation is also good.

4. Conclusion

Based on problem's formulation, research objectives, research results and discussion in this study, it can be concluded that A-SSI media is declared suitable for use based on aspects of validity, practicality, and effectiveness. This is supported by the results of study as follows: (1) The feasibility of A-SSI media to improve student's scientific literacy in first grade of junior high school in earth material is very good with an average percentage of general, material, and technical aspects of 88.9%, (2) The feasibility of A-SSI media to improve scientific literacy of students in earth layer material based on practical aspects was stated to be practical with the percentage of learning implementation 89.8% with very good criteria supported by activities shown by students in learning process and there was a response given by students in the learning process with A-SSI media, 88.9% students gave a positive response of the using of A-SSI media, (3) The feasibility of A-SSI media to improve scientific literacy skills in earth layer material based on effectiveness aspects was declared effective by increasing scores from pretest to posttest. Posttest's average score was 70.5 with an n-gain category. Nine students' n-gain has "medium" category and 1 student with "high" category.

Acknowledgements
Author sincerely thanks Tjipto Prastowo, Ph.D., Dr. Wahono Widodo, M.Si., Fery Sunarti, S.Pd., Faudina Permatasari, and all of the members of science postgraduate student at Science Department, Unesa for helping author in preparing the manuscript. The author also acknowledges the supports from Junior High School Lab of the State University of Surabaya for giving permission to complete the author's data. Also, special thanks to MISEIC 2019 for allowing author to be a speaker on the seminar.

References

[1] Putra Y S 2016 *Theoretical Review: Teori Perbedaan Generasi.* **9** 18
[2] Smaldino S E *et al* 2011 *Instructional Technology and Media for Learning. Teknologi Pembelajaran dan Media untuk Belajar* (Jakarta: Kencana Prenadamedia Group)
[3] Daryanto 2010 *Media Pembelajaran* (Yogyakarta : Gava Media)
[4] Sudjana N and Rivai A 2010 *Media Pengajaran* (Bandung: Sinar Baru Algesindo)
[5] Thiagaraja, Sivasailan A O *Instructional Development for Training Teachers of Exceptional Chidren* (Washington D.C.: Indiana University
[6] Nieveen N *et al* 2010 *An Introduction To Educational Design Research* (Netherlands: Netzodruk, Enschede)
[7] Riduwan 2010 *Skala Pengukuran Variabel-Variabel Penelitian.* (Bandung: Alfabeta)
[8] Aryani A K, Suwono H, and Parno 2016 *Prosiding Seminar Pendidikan IPA Pascasarjana UM*

Fostering Prospective Physics Teachers' Creativity in Analysing Education for Sustainable Development Based Curricula

M N R Jauhariyah*[1], E Hariyono[2], E N Abidin[3], and B K Prahani[4]
[1, 2, 3, 4] Physics Department, Universitas Negeri Surabaya, INDONESIA.

e-mail: mukhayyarotinjauhariyah@unesa.ac.id

Abstract. This study describes how to foster the creativity of prospective physics teachers in analysing ESD-based curriculum and describing the profile of prospective physics teachers' work as a form of ESD-based curriculum analysis. This study was a qualitative descriptive study using qualitative observation by observing behaviour consisting of independence, integration, motivations, judgment, flexibility, evaluation, questions, opportunity, and frustrations. Then, questionnaire gave to explore more about the attitude about analysis curriculum based on ESD. Besides the work of the prospective teacher in the form of the results of curriculum analysis was assessed based on the four basic components of the curriculum, namely objectives, content, the learning process designed, and the assessment design. To foster the creativity of a prospective teacher to produce an ESD-based curriculum analysis, all of them can make an analysis based on the basic component of the curriculum but still weak in integrating global issues of SDGs and provide solutions to the problems studied.

1. Introduction

In 2015, the United Nations General Assembly set the Sustainable Development Goals (SDGs) that collected 17 global goals for the year 2030 [1]. To achieve these global goals, a curriculum based on sustainable development is needed in quality education, in this case, education leads to issues for solving global problems in the SDGs. Efforts to encourage people to constructively and creatively face global challenges along with creating resilient and sustainable communities can be pursued through education for sustainable development or called ESD [2]. ESD contains main issues related to sustainable development in teaching and learning; for example, climate change, disaster risk reduction, biodiversity, poverty reduction, and sustainable consumption [3]. Through these main issues, ESD is believed to ensure that all people obtain knowledge, values, and skills for a better life and society [4-5] as well as helping students to reduce the gap between local and global problems [6-7].

ESD is more focused on developing the attitudes and skills owned by communities to deal with regional and global issues related to environmental aspects through education [8] and able to make easy life as well as finding relevant solutions upon the issues [9]. Although it has been introduced by UNESCO since 2005, this program is less concerned seriously related to the development of science education programs in higher education [10]. Therefore, educators must be able to apply ESD-based learning.

ESD does not rule out the possibility to be applied in learning physics, as one a branch of science. To implement ESD-based learning, it is necessary to analyse the curriculum so that ESD-based learning can be designed. Also, educators must have creativity in designing and managing the learning process so that they can infiltrate certain physical material with sustainable development main issues, this is one of the major problems for higher education educators [11]. Sustainable development in higher education involves a vibrant and holistic manner of learning [11]. Learning physics at the secondary level does not fully contain the ESD-based curriculum framework so that the hope of realizing a resilient and sustainable society still cannot be implemented. Physics learning still refers to textbooks and textual problem solving with case studies that have not oriented to global issues or attain experimental activities on a laboratory scale [12], not yet reaching the realm of finding solutions to global problems that are now happening on earth. Brundiers analyses how important sustainability competencies are, such as problem-solving skills and the ability to work with experts and stakeholders which need to be acquired in real-world learning opportunities [13].

Creative thinking is a rapidly growing worldwide community trend that promotes sustainable development projects, value chains, and supply chain relationship management [14]. Therefore, efforts need to be made to foster the creativity of prospective physics teachers in the analysis of ESD based curricula to support the SDGs program. To carry out this goal, a research subject with a high level of thinking ability is required and has a good academic background. In Mathematics and Natural Sciences Faculty of Surabaya State University (Universitas Negeri Surabaya), in which the research conducted, distinguished class program, which is selected through academic achievement indexes, English language proficiency and ability fields through interviews and written tests, is chosen to be the subject of the research to foster creativity in compiling an ESD based curriculum analysis.

Creativity education and environmental education are significant fields of studies but stay distant, with little efforts being made to bring them closer together [15]. The nature of the connection between organizational creativity and sustainability, however, is frequently ignored [16]. Creativity can be a critical element of addressing environmental sustainability and a widespread framework is provided [17]. It is a significant catalyst for both learning products and services to be innovative [16]. By applying new thoughts, it plays a significant role in enhancing learning processes and learning product designs [14]. Lozano claims that creativity and organizational learning can help defy traditional Newtonian and Cartesian mental models while promoting more sustainable societies [18].

The role of creative skills within the limits of green technology and sustainable development has so far received very little attention [14]. Although creativity as a problem-solving strategy and a main driver of sustainability has been improved [19], this skill has not been trained in physics learning in the form of ESD-based curriculum analysis. In this study, ESD is a new topic inserted for students as prospective physics teachers (PPTs) in our study program. So, the first thing we do is foster PPTs' creativity in presenting ideas based on global issues in the physics learning curriculum design. The aims of this study are described: (1) how to foster the creativity of PPTs in ESD-based curriculum analysis and (2) describing the profile of PPTs' work as a form of ESD-based curriculum analysis.

2. Research Methods

This study was a qualitative descriptive study by observing lectures behaviour consisting of independence, integration, motivations, judgment, flexibility, evaluation, questions, opportunity, and frustration [20]. The subjects of this study were 20 prospective physics teachers (PPTs) from the superior class of the 2016 Physics Education study program who conducted a school curriculum study course in semester 5 (odd semester of 2018/2019 academic year). Also, the work of prospective physics teachers in the form of the results of curriculum analysis was assessed based on the four basic components of the curriculum, namely objectives, content, the learning process designed, and the assessment design [21]. The content prepared is expected not only to textbook material but also to global issues. The product of the curriculum study was analysed the content for physics, the global issues, and the solution to the global problem discussed.

3. Results and Discussion

3.1. *How to foster the creativity of prospective physics in analysing ESD-based curricula*

The creativity of PPTs is grown in several ways, namely: (1) encourage PPTs to learn autonomously; (2) have a collaborative, socially inclusive learning style; (3) motivate the PPTs to master factual understanding so that they have a strong foundation for divergent thinking, (4) delay judging the thoughts of PPTs until they have been obviously formulated and developed, (5) foster flexible thinking, (6) promote self-assessment in PPTs, (7) seriously give the recommendations and issues of PPTs, (8) offer possibilities for PPTs to work with a broad range of materials and under various circumstances, and (9) help PPTs to learn how to deal with failure and frustration [20]. The entire process was carried out well by lecturers in the lectures on the school curriculum study for superior class students of 2016 by making a more operational checklist through 5 sub-items based on 9 items of behaviour carried out by lecturers in the classroom to foster student creativity.

The independence aspect can be carried out by encouraging PPTs to learn independently in the way: (a) encouraged the PPTs to show what they learned on their own, that is school curriculum study especially physics curriculum of national curriculum in Indonesia with inserted of ESD; (b) teach the PPTs about the basic knowledge of curriculum development and information about ESD framework, leave then to find out more about curriculum development; (c) renounce questions for PPTs to find out for themselves; (d) after teaching basics, leave space for individual learning; (e) leave open-ended question about the development of curriculum based on ESD to find the answers. This is difficult at the beginning because it requires students to be familiar with a learning strategy that expects PPTs to be fully involved and able to learn independently before fostering creativity.

The integration aspect can be carried out by (a) given opportunities to PPTs to share ideas and views about the development of curriculum, physics learning, ESD, and the others; (b) given opportunities to PPTs to do group work, in this case, they are grouped into 10 groups; (c) encouraged the PPTs to contribute the lesson in an active way in every learning activities; (d) encouraged the PPTs to ask questions and make suggestions; (e) expected the PPTs to wok in groups co-operatively. In this effort to foster creativity, lectures gave the same opportunities to all PPTs but not all of them provide active responses that are in line with expectations. They tend to be shy to ask questions so there are only a few of them who take advantage of the opportunities.

To provide motivation, the things that are have been done in order to foster creativity are: (a) learning the basic knowledge about curriculum and development and information about ESD as well as the skills of ESD with good emphasize; (b) highlight the significant of mastering the basics emphasize the importance of mastering the essentials, the basics of curriculum development and ESD curriculum framework exist; (c) expect the fundamental knowledge/skills of PPTs to learn well; (d) not a major concern to move rapidly to the next subject because it takes a good emphasize on basics knowledge/skills; (e) the PPTs learning is more important than only covering the syllabus of school curriculum study courses. In this aspect, the PPTs are expected to have a meaningful experience with basic knowledge/skills before carrying out an analysis of the school curriculum based on ESD.

In the framework of the decision making, it takes a process in aspects of judgment such as (a) before taking a stand, get the PPTs to explore their thoughts; (b) follow up on PPTs' questions to make them think; (c) did not give own views immediately on PPTs' ideas; (d) comment on PPTs' ideas only after more thorough exploration; (e) encouraged PPTs to do things differently, even if it takes time. Beside that, encourage flexible thinking is needed to foster creativity so the lectures must use the flexibility aspect that is: (a) probe PPTs ideas to encourage thinking; (b) encourage PPTs to ask question freely; (c) encourage PPTs to think in different direction; (d) like PPTs, it takes time to think differently; (e) allow PPTs to deviate from their instructions. But not all students take these opportunities so only a part of PPTs that can produce ESD-based curriculum analysis.

For the evaluation aspect, we should be: (a) expect PPTs to check their work; (b) offer PPTs the opportunities to share strengths and weaknesses; (c) provide PPTs inspect their work in advance of the lectures; (d) PPTs have the opportunities to assess themselves; (e) allow PPTs to demonstrate their work to each other prior to submission. But in the reality of learning, not all these evaluation items are carried out due to several obstacles including because students are not accustomed to doing self-assessments.

To foster creativity, we need the question strategies such as: (a) follow up on PPTs' questions; (b) listen to PPTs' questions carefully; (c) do not slightly reject the suggestion of PPTs; (d) listen to the recommendations of PPTs even if they are not practical; (e) listen patiently when PPTs ask stupid questions. Other than that the opportunity need to foster creativity through (a) encourage PPTs to experiment with what they learned; (b) appreciate placing PPTs into distinct uses what they have learned; (c) encourage PPTs to do various stuff with which they have learned; (d) don't worry that PPTs try their thoughts and deviate; (e) enable PPTs to go beyond what they are taught. Another thing that can foster creativity in frustrating. The lectures can foster the PPTs' creativity through (a) giving PPTs the opportunity to come when they are frustrated for emotional support, (b) helping PPTs who experience failure to regain confidence; (c) helping PPTs to draw lesson from their failures; (d) increase frustration among PPTs as part of the learning process; (e) encourage PPTs who have not found other alternatives.

Table 1. PPTs responses toward ESD-based curriculum analysis

Statement	Attitude Scale (%)				
	SDA	DA	N	A	SA
ESD is a new material in the School Curriculum Study	5	0	5	65	25
ESD is a very interesting material	0	0	0	35	65
ESD competencies can be integrated into the physics curriculum	0	0	0	65	35
ESD is very important for students	0	0	15	35	50
I find it difficult to relate global issues to curriculum analysis	5	35	35	25	0
I feel challenged and gain new experiences in analysing ESD based curricula	0	0	5	55	40
I was motivated to produce a good ESD based curriculum analysis	0	0	5	80	15
I was frustrated when looking for solutions to global problems as a characteristic of an ESD based curriculum	15	10	65	5	5

SDA=Strongly Disagree; DA=Disagree, N=Neutral; A=Agree; SA=Strongly Agree

After learning about the school curriculum study by inserting an ESD curriculum framework using a strategy to foster creativity in CFTIndex developed by Kaycheng Soh [20], students are given a questionnaire of responses in the form of an attitude scale. The response questionnaire (**Table 1**) contained statements about students' attitudes toward the ESD-based curriculum and their experience in analysing the ESD-based curriculum.

Table 1 shows most of the PPTs agree that ESD is an interesting material and can be integrated into the physics curriculum according to the PPTs' view. But none of them strongly agree if it is difficult to relate the global issue to physics curriculum analysis. Even though the results of the curriculum analysis made by these PPTs have not fully demonstrated the integration of global issues in the physics curriculum. The neutral attitude of 35% PPTs to the difficulty to relate global issues into curriculum analysis, Mirza gives the argument, *"In my opinion, the measure of the relevance of the concept of physics to global issues also depends on the context and the solution that can be resolved through physics or not."*

The 25% PPTs agree with the statement, Agista is one of the PPTs that agrees,

> *"I agree with the global issues applied to the physics curriculum. It's just that it is quite difficult to apply the physics curriculum. Because it requires analytical skills and critical thinking skills to understand it, and this tends to be subjective. If it is indeed going to be applied, it requires instruments that are consistent enough so that differences of thought between each student can be overcome and learning objectives can be achieved properly."*

Nikmatus is one of the 35% PPTs that disagree with the 5th statement because she thinks that ESD can be applied in physics learning so that physics learning can be varied and deeper for the ESD discussion. Elok gives her reason, *"I feel quite understand about ESD because it has been taught by lecturers during learning before"*.

Ilham in one of the 5% PPTs that strongly disagree with the statement that difficult to relate global issues to curriculum analysis.

> *"In my opinion, the issue of ESD is increasingly vigorous because in the 21st century we are increasingly aware of the limited available resources and so it must be wise to manage them. In the era of big data as it is now very easy to obtain information depending on how we filter it, starting from Amazon forest fires which are covered by international media for almost 2*

weeks, the problem of plastic waste in the sea, many people think that the issue of global warming is a hoax, to the socio-political problems between nations where the solution is the goal of education and the survival of humanity itself."

Most of the PPTs feel challenged and gain new experiences in analysing ESD based curricula and was motivated to produce a good ESD based curriculum analysis. But 5% of PPTs agree that they were frustrated when looking for solutions to global problems as a characteristic of an ESD based curriculum. Rohma said,

"That it is difficult to find solutions to global problems as characteristic of the ESD-based curriculum. That is because they are not literate enough on emerging issues at global, national, or local levels. As with the issue of climate change issues, what appears in the media is the language of a scientist so they do not know the relevance and the influence of these issues, and what the solution is."

Most of the PPTs have a neutral attitude about the frustration when looking for the solution to global problems as a characteristic of an ESD based curriculum. Elok thinks that whether or not it is easy to connect global issues into curriculum analysis depends on the material and its inspiration. On the other hand, 10% PPTs disagree with that statement. Nikmatus disagreed because ESD-based learning could be included in physics learning. ESD does not cause frustration, it can be easier because it is more emphasized in daily life that the use of ESD can be done in the current era. Mirza disagreed with the statement because according to her,

"Finding solutions to global issues related to ESD curriculum analysis could be communicated and involved students in related discussions. So, in my opinion, the teacher was not the only source of information and problem-solvers for solving the most problematic issues related to ESD issues."

Different from the others, Ilham and Agista gave an argument on why they chose strongly disagree with the last statement in the questionnaire. Ilham found it very easy to find ideas and solutions to these issues. Many international organisations and individuals who submit their ideas, for example, WWF, NatGeo, WHO, etc. who not only submit their ideas but also realise them in the form of social movements, journals, magazines, etc. Agista gives the reasons below:

"I chose strongly disagree because I feel happy when learning physics is directly connected with global issues, so the concept is no longer abstract. Maybe indeed for some people it is quite difficult, but for me, I am very interested and more into, 'Ow...so the application of the concept of physics is to solve this problem...' And it was very helpful and made it easier for me to understand the concepts of physics, it will not be frustrating."

Although there are quite a lot of PPTs who find it easy to connect global issues in analysing curriculum and do not feel frustrated by the search for solutions to the problems discussed, not many students can produce an integrated physics curriculum analysis with global issues and offer solutions to these problems. This can be seen in the have provided a new experience for students further explanation of the profile of the work of PPTs in the form of ESD-based curriculum analysis. Nevertheless, activities to integrate global issues in physics curriculum analysis so that they can be motivated to develop ESD-based physics learning.

3.2. *The profile of PPTs' work as a form of ESD-based curriculum analysis*

The result of PPTs' curriculum analysis based on ESD was analysed based on the components of the basic curriculum and ESD content with global issues and the offered solution of global problems.

Table 2. The analysis result on the PPTs' work as a form of ESD-based physics curriculum

Group	The objectives based on basic competence of national curriculum	The physics content	The learning model used	Design laboratory activities	Assessment of learning used	Global Issue	Offered Solution
1	Yes	Dynamic Fluid	Classroom Discussion	-	Affective, cognitive	-	-

2	Yes	Heat Transfer	Cooperative Learning, Problem-Based Learning, Discovery	Real laboratory	Affective, cognitive, literation skills, performance assessment authentic assessment	Climate Change	Cozy Tropical House
3	Yes	Mechanical Waves Characteristic	Guided Inquiry	Real laboratory	Affective, cognitive, psychomotor	Tsunami	Power Plants Use Ocean Wave Energy
4	Yes	Work and Energy	Cooperative Learning	Real laboratory	Affective, cognitive, psychomotor	Renewable Energy	-
5	Yes	Equilibrium	Problem-Based Learning	Real laboratory	Affective, cognitive, psychomotor	Unsafety Bridge	The bridge parabolic form for the buffer to make equilibrium and safety bridge
6	Yes	Refraction of Light	Project-Based Learning	Projects to make the simple telescope	Affective, cognitive, psychomotor	-	-
7	Yes	Static Electricity	Problem Solving	Virtual laboratory	Affective, cognitive	-	-
8	Yes	The Kinetic Theory of Gases	Cooperative Learning	Virtual and Real laboratory	Affective, cognitive, psychomotor	-	-
9	Yes	The Motion of the Planets and Satellites	Project-Based Learning	project to make props for the solar system	Affective, cognitive, psychomotor	-	-
10	Yes	Collision	Inquiry	Real laboratory and virtual laboratory	Affective, cognitive, psychomotor	Frequent car accidents	Collision application for car safety calculation

In an effort to foster the creativity of PPTs to produce an ESD-based curriculum analysis, all of them can make analysis based on the basic component of curriculum, 50% of them that can associate with global issues, and only 20% of them that provide content that pay attention to the global issues in the SDGs and provide solutions to the problems studied physically. These findings are similar to the previous study that the ability of students to integrate the ESD competences is relatively low [7]. From five global issues associated with physics material, the most interesting issue and the offered solution are given by group 2. **Figure 1** shows one of the results of an ESD-based curriculum study for heat transfer which is associated with the global issue on climate change created by PPTs.

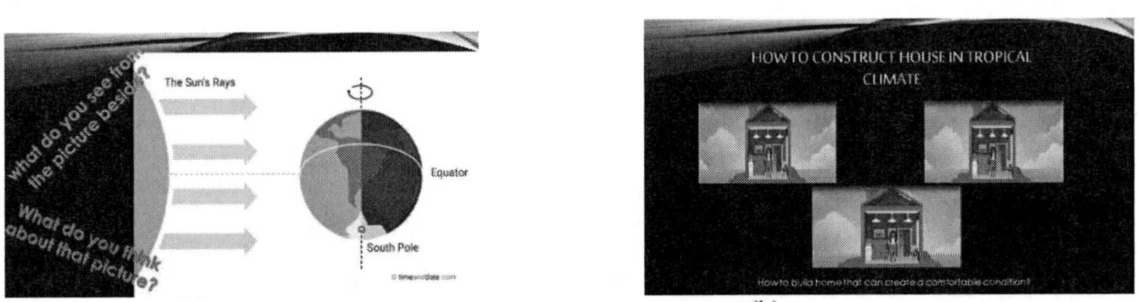

(a) (b)

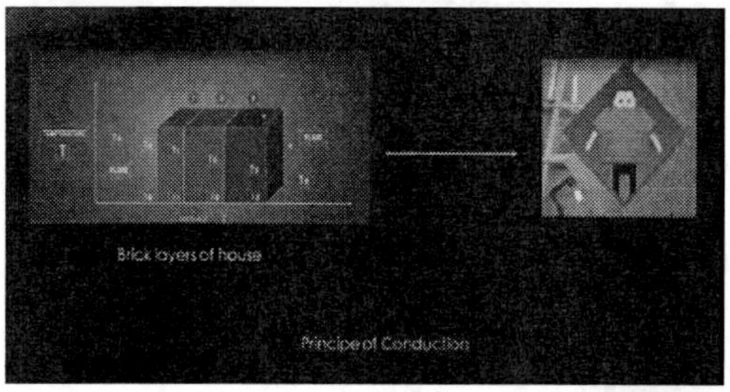

(c)

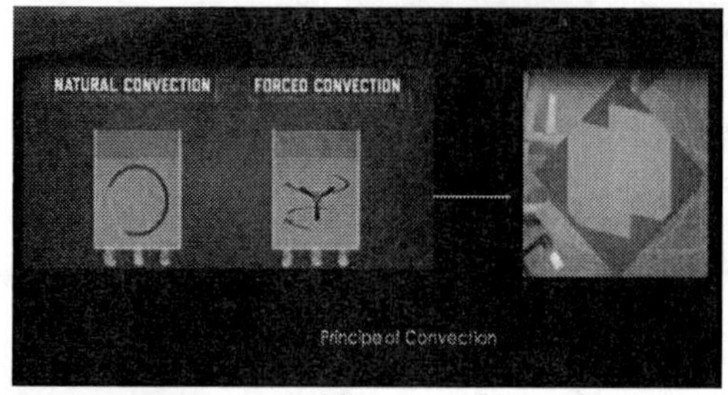

(d)

(e)

Figure 1. Content study of heat transfer associated with climate change: (a) equinox phenomenon, (b) problem solving in building home with comfortable condition in tropical climate, (c) bricklayers of cozy tropical house utilizing the principle of conduction, (d) glasses layers for windows utilizing the principle of convection, (e) green plants to reduce radiation.

Based on the equinox phenomenon presented by group 2, there is a moral message for humans to better protect the earth and better respect the surrounding natural environment. Group 2 advised that the community be wiser in responding to natural situations and conditions and be able to adapt well, be able to take advantage of the increase in the earth's temperature with positive things such as solar power, utilization in agriculture and marine affairs, and so on. In offering a solution in the form of a cozy tropical house, group 2 provides an environmental analysis so that the use of environmentally friendly materials is given more attention and reduces the use of air conditioners, socially required community organizations in building environmentally friendly homes, economically need to pay attention to spending on building materials that are more economical and utilizing natural materials that are easily found. The PPTs in group 2 can combine several aspects of ESD, namely environmental issues with social changes and economic growth [22,7]

3.3. Discussion

Based on the results of the implementation of learning using strategies to foster PPTs creativity [20] in compiling ESD-based curriculum analysis, it can be seen that there are only a few PPTs who have creativity in analysing and providing solutions to global problems in the main ESD issues. It because this is the first experience in pouring creativity into an ESD-based curriculum framework. PPTs are still fixated on the task of curriculum analysis without including ESD as other assignments in lectures programmed in the semester. Although all participating school leaders and educators are encouraged to create space (both physical and virtual spaces) and provide essential resources for teachers and students to demonstrate their creativity and imagination [23]. It is strongly suggested that interdisciplinary collaboration be used and that schools be encouraged to create partnerships with external industries and creative professionals and establish a meaningful relationship with other schools [23].

Problem-solving skills are needed to implement ESD-based learning [13]. It more attention needs to be given to the development of students' creativity [24]. In the previous study, some students did not understand the characteristics of an appropriate learning strategy in accommodating problem-solving skills [7]. Table 2 shows that three groups choose the learning models which practice social skills such as classroom discussion and cooperative learning, it is not suitable for the content of ESD and ESD competencies. Lectures can introduce the real-world experience and service-learning activities to support and create sustainability skills among PPTs [11]. Because sustainable development is not educational to teach PPTs that sustainability is a set event to be repeated, pre- and expertly determined [11].

Another finding is that learners who engage in service-learning acquire more advanced metacognitive skills, better skills in strategic planning and task assessment, better ability to discriminate against irrelevant data and better understand customer requirements and limitations [25]. Service-learning is a pedagogy that encourages instructional experiences involving learners, provides meaningful work, and reflects on structured operations that fulfil recognized community requirements [25-26]. Buhl et. al propose Design Thinking as an appropriate strategy to guarantee that sustainability elements are already taken into consideration in the original stage of the procedures of organizational innovation [27].

Kuo noted that creativity-relevant knowledge, expertise, and skills may not necessarily be acquired through internet-connected devices. In many circumstances, this kind of creativity-relevant resources can be obtained through reading, learning from the instructions of the educator, or even immersing the participant in an ethos full of freedom, respect, ambiguity tolerance, and open-mindedness [28]. While students in this study were not yet about the environment as requested by Kuo. Another interesting characteristic of Taiwan's creativity education that may not be implemented perfectly in creative learning programs in other nations is building a green building, using renewable energy and adopting innovative leadership [28].

The subject learners need to work on could also trigger frustrations [11], it can foster creativity if PPTs use this situation as apart of learning processes [20]. All the adverse images of the present and the future, such as like global warming, violations of human rights, terrorism often leads to adverse emotions such as powerlessness and apathy [11] make PPTs think how to reduce the negative images into positive images so they can create some solution for the global issues discussed.

4. Conclusion

There are several findings in effort to foster student creativity in analysis the ESD-based curriculum, namely: (a) ESD is interesting materials in school curriculum study and need the effort to foster PPTs creativity in analysis the ESD-based curriculum; (b) In general, PPTs feel challenged and motivated to integrate ESD in the physics curriculum even though they have some difficulties and frustration; (c) The ability of PPTs to analysis the ESD-based curriculum is good according to the component of basics curriculum development, but still weak in integrate the global issues and offered solution as well as in selecting appropriate learning strategies for problem solving skills; (d) Student that not able to integrate ESD in analysing physics curriculum are expected to give more attention and give them well service-learning.

Acknowledgements

Authors would like to thank the Faculty of Mathematics and Natural Sciences Surabaya State University (Universitas Negeri Surabaya) for funding this research in the faculty policy research scheme in 2019. This research is a preliminary study in 2018 which continues this year for the qualitative data exploration process. As for the assessment of the creativity of students in the next generation is done in the odd semester 2019/2020. This research is part of the faculty funding research with contract numbers B/29719/UN38.3/LT.02.05/2019 and the date of May 07th, 2019.

References

[1] General Assembly 21 October 2015 *Resolution adopted by the General Assembly on 25 September 2015* (United Nations: General Assembly)
[2] Nasibulina A 2015 *Procedia Soc. Behav. Sci.* **214** 1077
[3] Kankovskaya A R 2016 *Procedia CIRP* **48** 449
[4] Dannenberg S and Grapentin T 2016 *J. Futures Studies* **20** (3) 7

[5] Klarin T 2018 *Zagreb Int. Rev. Econ. Bus.* **21** (1) 67
[6] Gosselin DC, Manduca C, Bralower T J, and Mogk D 2013 *Eos, Transactions, American Geophysical Union* **94** (25) 221
[7] Hariyono E, Abadi, Liliasari, Wijaya A F C, Fujii H 2018 *J. Penelit. Fis. Apl.* **8** (2) 61
[8] Tanaka H 2009 *Education for Sustainability Development in Japan* (Tokyo: ESDRC Rikkyo University)
[9] Bakhati P 2015 *J. Train. Dev.* **1** (1) 21
[10] Filho W L, Raath S, Lazzarini B, Vargas VR, de Souza L, Anholon R, Quelhas OLG, Haddad R, Klavins M, and Orlovic VL 2018 *J. Clean. Prod.* **199** 286295
[11] Molderez I & Fonseca E 2017 *J. Clean. Prod.* (*Accepted Manuscript* DOI: 10.1016/j.jclepro.2017.04.062)
[12] Bimba A, Idris N, Mahmud R, Abdullah R, Abdul-Rahman S-S, Bong C H 2013 *Res. Notes Inf. Sci.* **14** 621
[13] Brundiers K & Wiek A 2010 *Innovation High. Educ.* **36** (2) 107
[14] Awan U, Sroufe R, Kraslawski A 2019 *J. Clean. Prod.* (*Accepted Manuscript* DOI: 10.1016/j.jclepro.2019.03.308)
[15] Cheng V M Y 2019 *Think. Sk. Creat.* **33** 100567
[16] Shrivastava P 2014 *J. Clean. Prod.* **79** 4
[17] Stables K 2009 *Int. J. Technol. Des. Educ.* **19** 199-219
[18] Lozano R 2014 *Sustain. Dev.* **22** 205-216
[19] Mitchell I K, Walinga J 2017 *J. Clean. Prod.* **140** 1872
[20] Soh K 2018 *Creativity Fostering Teacher Behaviour: Measurement and Research* (Singapore: World Scientific Publishing Co. Pte. Ltd.)
[21] Sukmadinata N S 2001 *Pengembangan Kurikulum: Teori dan Praktek (Curriculum Development: Theory and Practice)* (Bandung: PT Remaja Rosdakarya)
[22] Paw J B, Gericke N, Olsson D, and Berglund T 2015 *Sustainability* **7** (11) 15693
[23] Kagawa F 2007 *Int. J. Sustain. High. Educ.* **8** (3) 317
[24] Watson J 2018 *Teach. Teach. Educ.* 71 206
[25] Clevenger C M & Ozbek M E 2013 *J. Construction Eng. Manag.* **139** (12)
[26] Shepard K 2008 *Int. J. Sustain. High. Educ.* **9** (1) 87
[27] Buhl A, Schmidt-Keilich M, Muster V, Blazejewski S, Schrader U, Harrach C, Schäfer M, Süßbauer E 2019 *J. Clean. Prod.* **231** 1248
[28] Kuo H, Burnard P, McLellan R, Cheng Y, Wu J 2017 *Think. Sk. Creat.* (*Accepted Manuscript* http://dx.doi.org/10.1016/j.tsc.2017.02.005)

Implementation of KS-STEM Project: Bridging the STEM curriculum into Science education

N Suprapto[1] and C-H Ku[2]

[1]Department of Physics, Universitas Negeri Surabaya, Surabaya, Indonesia.
[2]Graduate Institute of Science Education, National Dong Hwa University, Hualien, Taiwan.

e-mail: nadisuprapto@unesa.ac.id

Abstract. Currently, interest in STEM education is growing in the world. In Indonesia, STEM practice is gradually developing. Through the partially "KS-STEM Project", this study was addressed. The project is a longitudinal cross-country project between Taiwan and Indonesia. In regards to the three philosophical foundations (epistemology-ontology-axiology), this study addresses three research questions, as follows: 1) To what extent does the Fuel Cell Car Science Kit facilitate junior high school students' understanding of the essence of STEM education; 2) To what extent do the students' outcomes deal with fuel Cell Car Science Kit; 3) What topics can students learn from the Fuel Cell Car Science Kit. The research used a descriptive qualitative approach by using instruments: "Fuel Cell Car Science Kit" as a STEM kit. Participants of the study included 32 junior high school students in East Java, Indonesia. The results of the study included: (1) the description of how STEM Kit facilitates the students' understanding of the essence of STEM education; (2) the students' outcomes dealing with the STEM Kit; and (3) several topics that students learn from the STEM Kit. The implication of the current study is to introduce and to involve Indonesian students, especially junior high school students to STEM education.

1. Introduction

According to the NARST strands, there are fifteen (15) issues majoring in science education. One of them is science learning in informal contexts. This strand is specified for teaching and learning in community programs, after-school programs, outdoor settings, museums, and communications media. The strand is also giving more opportunities among science educators and science researchers to integrate science, technology, and society. However, technology and society are not enough, we need mathematics and engineering. In other words, we need the integration of science-technology-engineering-mathematics (STEM).

The term "STEM" began as "SMET," standing for science, mathematics, engineering, and technology. In the 1990s, "the National Science Foundation (NSF) revised the term in order to emphasize the importance of these four separate disciplines" [1]. "The acronym was changed into STEM to help promote it, yet there is still a considerable number of Americans that associate STEM with stem cell research" [2]. Throughout the last decade, there has been developing of interest in STEM education [3]. Currently, the acronym "STEM" literally stands for science, technology, engineering, and mathematics. It can be defined in two ways: "(1) the more traditional way, it is the individual 'silo' fields of science, technology, engineering and mathematics education; and (2) the concept of integrated STEM education; it includes the teaching and learning practices when the subjects are purposefully integrated" [4]. This new strand has a strong philosophical foundation. Chesky and Wolfmeyer [2] addressed the epistemology, axiology, and ontology of STEM as a singular unit as shown in Figure 1.

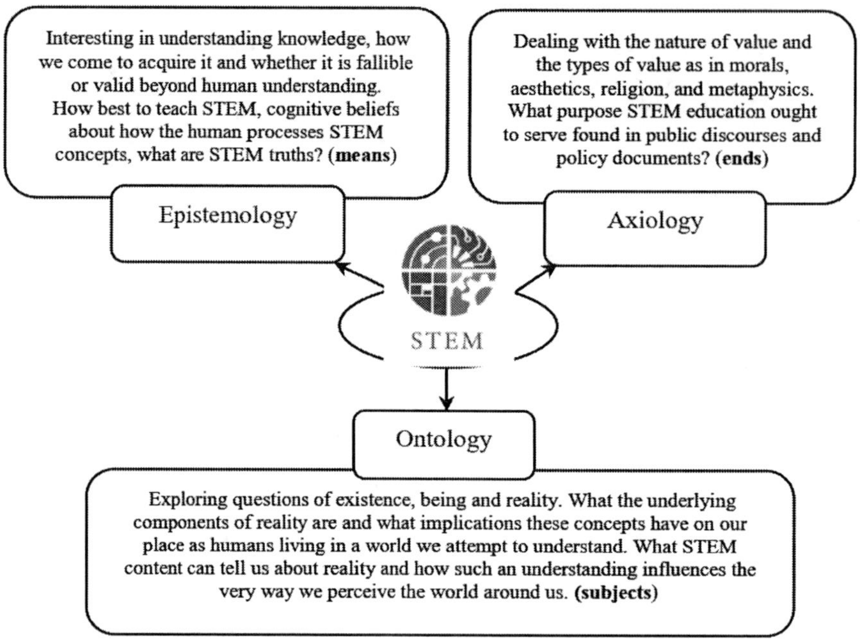

Figure 1. Epistemology, axiology, and ontology of STEM (reorganized from Chesky and Wolfmeyer [2])

The teaching and learning process in STEM realizes through blended unilaterally or one filed may be the dominant base field. Therefore, the elucidations of its exact meaning very broadly. These range from very specific (e.g., mathematics, physics, chemistry, biology, information sciences, and engineering) [5] to very extensive (e.g., "an across the discipline of practices and processes that transcend disciplinary lines and from which knowledge and learning of a particular kind emerges") [6]. Therefore, understanding across disciplines is a very important aspect of STEM education.

Recently, the concept of STEM became popular in Indonesia. The STEM practice is gradually developing in Indonesia [7], such as "a partnership program between local schools in Riau province and Honeywell about science and technology" [8]; "an ongoing project between Columbia University and the *Institut Pertanian Bogor* to improve the teaching of STEM in Indonesian high schools" [9]; and "an innovation strategy to build students' disaster literacy through STEM-D (Science, Technology, Engineering, Mathematics and Disaster) Education" [10]. However, "some of these programs were more emphasized on the secondary and higher education rather than lower levels of education, such as elementary and junior high school" [7]. For the young generation, Bottia et al. [1] endorsed that "the feasible approaches for inspiring, reinforcing and preparing more of the nation's youth to choose a STEM pathway for their futures". Therefore, this study dedicated to junior high school student in principal integrating STEM education.

Through the partially of "KS-STEM Project" (KS = Ku-Suprapto), this study was addressed. The project is a longitudinal cross-country project between Taiwan and Indonesia. The objectives of the project are to: "(1) provide all students with skills needed for success in the 21st century; (2) improve academic performance; (3) narrow achievement gaps; and (4) foster potential interest in STEM professions" [7]. For practicality, the Fuel Cell Car Science Kit was utilised to integrate between formal science curricula in junior high school and STEM education in this research. "The Kit guided students to invent their own clean energy applications using fuel cells and renewable hydrogen created using solar energy and water. The set was equipped with a complete curriculum on renewable energy with easy experiment, manual kits and background history on the technology. The sub-topics of the experiment include the effect of heat on solar panels, finding the solar panel's maximum power, electrolysis mode (generating Hydrogen and Oxygen from water) and fuel cell mode (generating electricity from Hydrogen and Oxygen)" [7].

Based on the rationale above, the aim of the current study is to introduce Indonesian students, especially junior high school students to STEM education. Finally, in regards to the three philosophical foundations above, this study addresses three research questions, as follows:
1) To what extent does the Fuel Cell Car Science Kit facilitate junior high school students' understanding of the essence of STEM education?
2) To what extent do the students' outcomes deal with Fuel Cell Car Science Kit?
3) What topics can students learn from the Fuel Cell Car Science Kit?

2. Method

Participants of the study included 32 students grade eight from East Java, Indonesia. The research used a descriptive qualitative approach by using instruments: STEM-Fuel Cell Car Science Kit. The apparatus was developed by Horizon Educational Company.

2.1. Fuel Cell Car Science Kit

The STEM- Fuel Cell Car Kit empowers students to ascertain the principles behind the real-scale fuel cell vehicles currently being rolled out across the globe. "A reversible Proton Exchange Membrane (PEM) fuel cell first electrolyzes water to separate oxygen and hydrogen molecules, this hydrogen is then converted into electrical energy to power the car" [11-12].

2.2. KIT contents

The Fuel Cell Car Science Kit consists of: "chassis with LED light & motor, reversible PEM fuel cell, Hydrogen tank, Oxygen tank, inner gas containers, 0.75 Watt solar cell, 2 x 2mm connecting leads, battery pack with connecting leads (2 X AA batteries), plastic plug pins for fuel cell, and transparent silicon tubing" [11]. Figure 2 shows the STEM: fuel cell car apparatus.

Figure 2. STEM - Fuel cell car apparatus

3. Result and Discussion

3.1. To what extent does the Fuel Cell Car Science Kit facilitate junior high school students' understanding of the essence of STEM education

The Fuel Cell Car Science Kit provides an activity that makes it possible for the students to experience and sense the STEM education. Through the following student' activities, they got the meaning of being a STEM member. Totally, we train 9 activities (see Table 1) to students in order to explore the physical variables based on the prototype of fuel cells car. By implementing the learning model of guided inquiry, the students got the meaning about the essence of STEM, the interconnection among science in a fuel cell, math in a fuel cell, technology and engineering in a fuel cell. The guided inquiry was chosen as a learning model because of the educational level of students in junior high with the main character is the transition from concrete to abstract thinking.

Table 1. Some activities from STEM kit.

Student activities
"Assembly and operation of the kit"
"The Effect of Heat on Solar Panels"
"The Effect of Shade on Solar Panels"
"The Effect of Tilt Angle on Solar Panels"
"Finding the Solar Panel's Maximum PowerPoint"
"Electrolysis Mode Generating Hydrogen and Oxygen from Water"
"Fuel Cell Mode Generating Electricity from Hydrogen and Oxygen"
"Determining the Minimum Voltage for Water Decomposition"
"Polarization States for Hydrogen Fuel Cells"

3.2. To what extent do the students' outcomes deal with fuel Cell Car Science Kit?

To answer the second research question, through student activities in class which clearly exhibits the science of electrolysis and the breakdown of water into oxygen and hydrogen. The following is the example of student outcomes that include a solar panel for powering electrolysis.

Objectives: "Build and test a hydrogen fuel cell car to explore the concepts of electrolysis and renewable energy" [11-12].

Materials: Fuel cell car science kit, distilled water, stopwatch

Procedure:

Procedure

These discussion questions will help you assemble your car. Read each of them carefully and discuss your responses with your group, then use your answers to put your car together. As you assemble your car, write down any interesting observations you have in the Observations section below.

1. The body of your car (attached to the motor) has many places where other pieces can attach to it. Discuss with your group where you think individual pieces should go.
2. Where would you attach the wheels? Does it matter which wheels go where?
3. How would you attach the H_2 and O_2 cylinders to the body of the car? Does it matter where they go?
4. Look at the fuel cell. It has two short tubes attached to it. Are there any other places on the fuel cell where you could attach the longer pieces of tubing?
5. When turned on with electric current, the fuel cell will produce hydrogen and oxygen gas from the water inside it. How will you trap the gas so that it doesn't float away?
6. What source of electricity will be better at separating the hydrogen and oxygen in the water: the solar cell or the battery pack? How should the electricity source be connected?
7. How do you know when the fuel cell is generating hydrogen? How can the hydrogen be used to power the car?

Observations:

Write down anything interesting you observe while building or running the car.

Once your car is assembled, see how long it will run on its hydrogen fuel. Try making more hydrogen and running the car again. How long can you get the car to run? Try at least four times and write your results in the Data Table 1.

Try and make some changes to the car to make it run faster. Could you increase its fuel capacity? Can you decrease friction in some way? Does it matter what kind of surface it runs on? Discuss some possible changes you could make to your car and run your experiment to see if your changes result in a longer run time. Put your results in the second Data Table 2.

Figure 3. Sample of procedure in a fuel cell car experiment in the student worksheet.

Data Table 1

TRIAL	TIME (SEC)
1	
2	
3	
4	
5	

Data Table 2

HOW CAR WAS CHANGED	TIME (SEC)

Figure 4. Sample of the data table for students in the student worksheet.

3.3 What topics can students learn from the Fuel Cell Car Science Kit?

There are two main topics students can learn from the Fuel Cell Car Science Kit, namely electrolysis and fuel cell itself. The following are the example of student handbook in STEM class project (see Vignette 1 and Vignette 2).

a. Electrolysis

Michael Faraday first formulated the principle of electrolysis in 1820. The word "lysis" means to dissolve or break apart, so the word "electrolysis" literally means to break substances apart by using electricity. The process occurs in an electrolyte, a watery solution or a salt melting that gives the ions a possibility to transfer between two electrodes. The electrolyte is the connection between the two electrodes, which are also connected to a direct current. If we apply an electrical current, the positive ions migrate to the cathode while the negative ions will migrate to the anode. At the electrodes, the cations will be reduced and the anions will be oxidized.

Electrolysis is a chemical reaction that uses an electric current to break the bonds that hold molecules together. An electrolyzer is a machine that can introduce an electric current into a liquid using two pieces of metal called electrodes. The electric current can cause ions to move within compounds, which allows the reaction of electrolysis to take place. Ion is a particle that is electrically-charged (it can be a positive or negative charge), or an atom or molecule that had gained or lost electrons. Anode is the negatively charged terminal of a fuel cell or battery that is supplying current. Cathode is the positively charged terminal of a fuel cell or battery that is supplying current.

Vignette 1. *Electrolysis*

The simple process of electrolysis is shown in Figure 5. Electrolysis is a process in which an electric current can decipher an electrolyte substance. That is, the electrolysis process changes in electrical energy into chemical energy (redox reaction). There are three main characteristics of an electrolysis process, which are as follows.

a. The presence of free ions in an electrolyte solution. These free ions will receive or give electrons which are flowed through the solution.
b. There is an external DC current source, for example, a battery.
c. It has two electrodes, namely the cathode and the anode, in the electrolysis cell.

In electrolysis cells, the cathode functions as a place for the reduction reaction and as a negative pole, while the anode is the place for the oxidation reaction and as a positive pole.

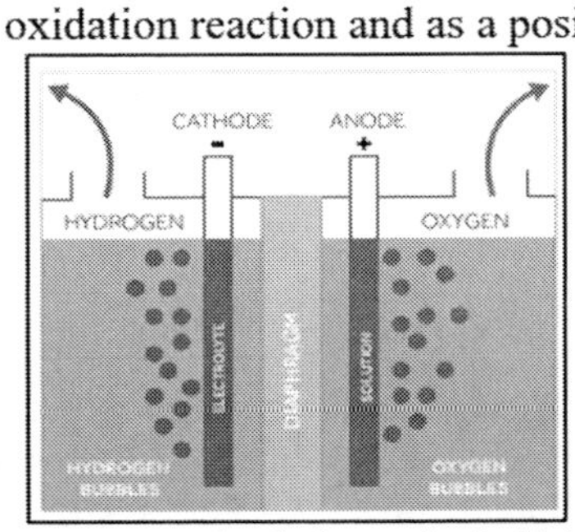

Figure 5. Standard electrolysis [11-12].

b. Fuel Cell

Fuel cells convert chemical energy directly into electricity and heat with high efficiency. These devices can be used anywhere at any time, for as long as necessary as long as fuel is supplied. Fuel cells are one of the few alternative energy devices that can be used for many applications. They can power portable electronics, automobiles, houses, buildings, and even space ships. The basic technology behind a fuel cell is simple: it consists of an electrolyte layer sandwiched between a porous anode and cathode on either side.

Hydrogen is broken into protons and electrons on the anode side, and is combined with oxygen to produce water and heat on the cathode side. Protons are transported from the anode to the cathode through the electrolyte, and the electrons are carried to the cathode over the external circuit [10]. Both the anode and cathode contain a catalyst to speed up the chemical reactions. Figure 6 shows an example of a typical proton exchange membrane (PEM) fuel cell with the following chemical reactions:

Anode: H_2 (g) $\rightarrow$ $2H^+$ (aq) + $2e^-$
Cathode: ½ O_2 (g) + $2H^+$ (aq) + $2e^-$ $\rightarrow$ H_2O (l)
Overall: H_2 (g) + ½ O_2 (g) $\rightarrow$ H_2O (l) + electric energy + waste heat

Fuel cells can utilize a variety of fuels to generate power in addition to hydrogen. Chemicals such as phosphoric acid, carbonate, methanol, solid oxides, and ethanol can all be used to power a fuel cell, but the chemical reactions involved is of course different from the one described above. Fuel for many of these types of fuel cell can be made from fossil fuels as well as many sources of biomass, including from municipal wastes, sewage sludge, forestry residues, land fill sites, agricultural and animal waste. Though fuel cells are a very efficient and clean source of energy, their power output is relatively small and can sometimes generate a lot of waste heat. Still, they are an important part of the future of energy.

Vignette 2. *Fuel Cell.*

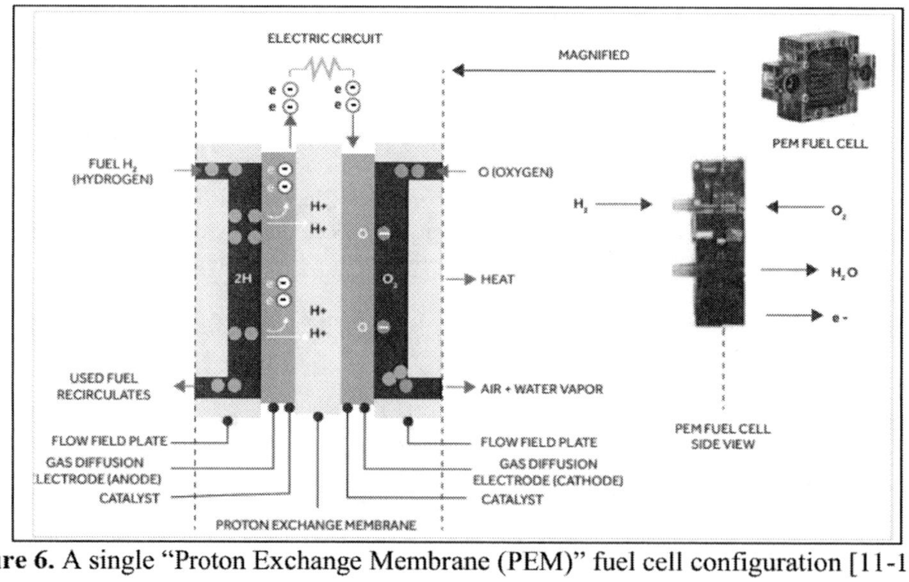

Figure 6. A single "Proton Exchange Membrane (PEM)" fuel cell configuration [11-12].

4. Conclusion

There were some conclusions of the study. Through the partially of "KS-STEM Project", this study was addressed. The project is a longitudinal cross-country project between Taiwan and Indonesia. The objectives of the project are to: "(1) provide all students with skills needed for success in the 21st century; (2) improve academic performance; (3) narrow achievement gaps; and (4) foster potential interest in STEM professions". For practicality, the Fuel Cell Car Science Kit was used to integrate between formal sciences curricula in junior high school and STEM education in this research. The main

Journal of Physics: Conference Series **1417** (2019) 012087 doi:10.1088/1742-6596/1417/1/012087

findings of the study included some description: (1) the STEM Kit facilitates the junior high school students' understanding of STEM education; (2) the students' outcomes dealing with the STEM Kit; and (3) several topics that students learn from the STEM Kit. The implication of the current study is the KS project engages Indonesian students, especially junior high school students to STEM education.

Acknowledgements
This project is a mutual collaboration between Physics Education Program, Universitas Negeri Surabaya and Graduate Institute of Science Education, National Dong Hwa University Taiwan.

References

[1] Bottia M C, Stearns E, Mickelson R A, Moller S and Parker A D 2015 The Relationships Among High School STEM Learning Experiences and Students Intent To Declare and Declaration of A STEM Major In College *Teachers College Record* **117** 3 1-46
[2] Chesky N Z and Wolfmeyer M R 2015 *Philosophy of STEM education: A critical investigation* (New York, NY: Palgrave Macmillan)
[3] Goodwin M, Cooper M, McCormick A, Patton C and Whitehair J 2014 *Implementing a whole-school STEM program: Successes, surprises, and lessons learned.* The 4th IEEE integrated STEM education conference, March 8, 2014, Princeton, NJ
[4] Yakman, G. 2010. *What is the point of STE@M? A brief overview.* Retrieved from http://www.academia.edu/8113832/What_is_the_Point_of_STEAM_A_Brief_Overview_of_STEAM_Education.
[5] Gonzalez H B and Kuenzi J J. *Science, Technology, Engineering, and Mathematics (STEM) Education: A Primer. (CRS Report No. R42530)* (Washington, DC: Library of Congress Congressional Research Service). 2012.
[6] Moon J and Singer S. 2014.*Education Week* **31** 32
[7] Suprapto N 2016 *J Turkish Sci. Educ.* **13** 75
[8] Honeywell 2014 *Honeywell promotes STEM in Indonesian schools.* Retrieved from http://www.honeywell.com/newsroom/news/2014/08/honeywell-promotes-stem-inindonesian-schools/ on June 12, 2015.
[9] Columbia Global 2014 *Improving STEM education in Indonesia* (Columbia University)
[10] Sampurno P J, Sari Y A and Wijaya A. D 2015 *Inter. J. of Learn. & Teach.* **1** 73
[11] Horizon Educational 2015 *STEM KITS* (Singapore: Horizon Educational)
[12] Spiegel C 2009 *Renewable Energy: Science Education E-book* (Prague: Horizon Educational)

Activities of students in using worksheet based on Contextual Teaching and Learning

A M Fauziah[1], T Nurita[1]

[1]Department of Science, Universitas Negeri Surabaya, Indonesia

e-mail: annurilfauziah@unesa.ac.id

Abstract. This study aims to describe the activities of students in using student worksheets based on Contextual Teaching and Learning (CTL). The method used is descriptive quantitative. Student worksheets based on Contextual Teaching and Learning that are used in learning consist of three sub-layers are atmosphere, lithosphere, and hydrosphere. In CTL-based student worksheets, there are 7 main components of CTL are constructivist, questioning, modelling, finding, community learning, reflection, and authentic assessment. Learning outcomes using CTL-based Student Worksheets are easier for students to understand learning material, this is shown by students in solving problems on student worksheets properly. Based on the results of the study it can be concluded that learning by using Student Worksheets based on Contextual Teaching and Learning helps students more easily understand the subject matter.

1. Introduction

The Indonesian government always strives to improve the quality and quality of education. One effort that has been made is through improving the curriculum. The curriculum applied in Indonesia today is the 2013 Curriculum. Learning in the 2013 Curriculum emphasizes learning that can develop student creativity and student-centred. In addition, the Ministry of Education and Culture [1] emphasizes that the 2013 Curriculum is also expected to encourage students to be better able to make observations, ask questions, reason, and communicate about what they have obtained or known after receiving learning materials [1].

The learning process as expected by Curriculum 2013 is certainly expected to be implemented in all subjects both Natural Sciences (Science) subjects and other subjects taught in Junior High School or equivalent. Science learning is currently demanded active learning which is implicated in physical or mental activities, not only includes hands-on activities but also minds-on [2]. The science learning process in the 2013 Curriculum emphasizes providing direct experience to develop competencies in order to explore and understand the natural environment scientifically [3].

Science is a systematic study of nature and how it affects life and the environment [4]. This definition is certainly very supportive of what is expected in the learning process in the 2013 Curriculum. Indeed science does not only consist of a collection of knowledge or various types of facts that must be memorized but Indeed science also consists of an active process of using thoughts in studying natural phenomena that have not been disclosed. Science (Natural Sciences) is a subject at the junior high school level related to how to systematically find out about nature.

Based on Piaget's Cognitive Development Theory, children of middle age (12-15 years) have not fully been able to think abstractly. During the learning process, the presence of concrete objects is still needed. Even so, semicircular objects are introduced. But at this junior high school level, children have begun to be able to apply thinking patterns that can lead them to understand and solve problems. The ability to understand certain material learned is included in the category of understanding. At the level of understanding, students are required to be able to answer questions in their own words and by giving examples related to principles or concepts. In science learning, there are many concepts that must be learned, but these concepts should not be a problem for students because science is very much found in everyday life. According to [5] understanding concepts is the ability of students who not only understand but also can apply the concepts given in solving problems, even to understand new concepts.

The results of the interview with the VII A grade science teacher at SMP 2 Menganti Gresik said that the level of conceptual understanding of students in class VII A was considered to be lacking. Middle exam of odd semester results shows that 70% of students in class VII A has not reached the completion limit. In addition to conducting observations in the form of tests of students' understanding, researchers also gave questionnaires to respondents regarding the process and learning media used during science lessons. Based on the results of the questionnaire it is known that students have never been given a learning device in the form of Student Activity Sheet (SAS) by the teacher during the science learning process in the classroom, students also never do lab work and make presentations. The teacher in class VII science teaching said that so far the science learning process only used the 2013 Curriculum Science package.

There must be an effort made by the teacher so that the concepts that students receive become better. Learning that can be used to foster student awareness of science as a science related to daily life and able to facilitate students in understanding the material is through a contextual approach or Contextual Teaching and Learning (CTL). The CTL approach is a learning approach that links the conditions of everyday life with the concept of learning. According to [6] with learning using the CTL approach, educators are expected to be able to associate the learning material that they teach with real-world situations of students to encourage students to make a connection between their knowledge and application in their daily lives involving seven main components of CTL learning, namely : (1) constructivism, (2) asking, (3) inquiry, (4) learning society, (5) modeling, (6) reflection, and (7) authentic assessment. [7] also revealed that contextual-oriented learning is expected by students to be able and accustomed to discovering their own concepts in the material being studied.

The CTL approach with 7 main components in the learning process can be trained to students with the help of SAS. According to the [2], the purpose of providing SAS is to provide opportunities for students to play an active role and help students to develop and find concepts through science process skills, and also as a guide for teachers and students in the learning process. SAS can be used as a guide or steps in the activities of experiments, demonstrations and discussions. In general, SAS is a translation or development that comes from student books. The results of the questionnaire distributed to respondents stated that 89% of students agreed that science learning uses SAS to improve their cognitive abilities.

Based on the review of learning topics in Curriculum 2013, one of the material that requires learning resources in the form of SAS to help understand students' concepts is the topic of the structure and layers of the earth. These topics are in class VII in Basic Competencies 3.10 Describe the earth layer, volcanoes, earthquakes, and risk reduction actions before, during, and post-disaster according to the threat of disasters in the area and KD 4.10 Communicate efforts to reduce disaster risk and impact nature and actions to save themselves in the event of a disaster in accordance with the type of disaster threat in their area. This topic was chosen because on this topic, students are required to understand the layers of the earth which consist of atmosphere, lithosphere and hydrosphere. concepts that students must understand in the earth layer material include Explaining the Earth's atmosphere, comparing the layers in the atmosphere, giving examples of activities that occur in the atmospheric layers, concluding the causes of holes in ozone, explaining the lithosphere's earth layer, summarizing tectonic theories plate, describes the process of recycling water, interpreting the evaporation and condensation process, classifying the water recycling cycle. This material can not be separated from the environment of students, through asking questions, learning communities, giving modelling, reflection, finding and constructivists as well as authentic assessments later students will be encouraged to find concepts on the topic themselves.

Some of the research that has been carried out regarding the use of the Contextual Teaching and Learning approach in learning shows a positive impact on students, including the Research by [8] showing that the use of CTL-based modules can improve students' reading skills. This can be seen from the calculation of the t-test in SPSS 17.0 for windows which shows that the experimental group's post-test score increased significantly than the control group's score. The results of the interviews also show that students get many benefits through the use of CTL in classroom activities, including helping students gain a better understanding of the material through various media used in the learning process that helps them understand the material being studied; motivate students to learn because they can share knowledge and work together in completing assignments.

2. Method

The method used in writing this article is a quantitative descriptive method. According to [9], descriptive methods are procedures for solving problems that are investigated by describing or describing the state of the subject or object of research (someone, institution, society, etc.) at present based on visible facts.

3. Results and Discussion

Learning by using the Contextual Teaching and Learning approach that is trained through the student activity sheet is able to facilitate students in understanding the subject matter. This is indicated by the completeness of students in completing tasks on the increasingly increasing worksheet. Increased student completeness in completing worksheets assignments is shown in Table 1.

Table 1 Increased Student Completeness

No	Indicator	*Worksheet 1*	*Worksheet 2*	*Worksheet 3*
1	Interpretation	70%	100%	100%
2	Summarize	83%	83%	100%
3	Example	67%	100%	100%
4	Compare	100%	100%	100%
5	Categorize	73%	100%	100%
6	Explain	87%	87%%	93%
7	Conclude	87%	97%	100%

The Student Activity Sheet used in learning uses the CTL approach. According to [10] in essence, CTL is a concept that helps teachers associate the material they teach with real-world situations and encourage students to make connections between the knowledge they have and their application in their lives as family members and society. CTL learning as an approach has 7 principles or components that underlie the implementation of the learning process, namely constructivism, asking questions, modelling, finding, learning society, reflection, and authentic assessment [6].

Constructivism emphasizes the building of one's own understanding actively, creatively, and productively based on prior knowledge and knowledge and from meaningful learning experiences [11]. The SAS developed has included examples and cases in real life as in worksheet 3 namely floods and droughts that sometimes occur in Gresik. This is to support the constructivist CTL principle in learning. Through the CTL constructivist principle, students can understand the material in the indicator explaining that is explaining the efforts that must be made during a flood (worksheet 3) and so on because students will learn more easily by finding their own knowledge so that they can more easily understand and explain the knowledge obtained. Students can also relate the knowledge gained with examples of application in daily life (exemplifying). As shown in the worksheet 3 students were asked to exemplify the waters on earth. Students will be facilitated in terms of exemplifying the existence of constructivist CTL principles because students have found a lot in real life. According to [7] Learning activities are packaged into a process of constructing knowledge so that learning starts from what students know. It is expected that students are able to find new ideas and knowledge (concepts, principles), apply ideas, then learners look for effective learning strategies to achieve competence and provide satisfaction with discovery.

In worksheets that have been developed, there is a column asking to be filled by students. After linking material to students' daily lives in the constructivist component, the teacher provokes students to ask questions regarding learning. In CTL learning that has been done, the teacher does not simply convey information but provokes students to find themselves. Through the principle of CTL, questions help students in understanding indicators of concepts summarize because these questions can help students to summarize the material they obtained. Worksheet 2 (atmosphere) students will be led to make questions related to the temperature of the earth, with these questions will help students to simplify summarizing the causes of the earth's temperature which increases as well as worksheet 1 and 3. Therefore, the asking role is very important as the teacher's way of guiding and directing students to find every material they learn [7].

Students conduct an investigation to find answers to the questions that exist. The CTL modelling component helps students to carry out the investigation process. The three CTL based worksheet used are equipped with CTL modelling components for students, namely in the form of an inquiry design. After carrying out the investigation plan of the earth convection process contained in worksheet 1 (lithosphere) students will be facilitated to understand the material being studied because students see and are directly involved in the process of finding the answer. The CTL component found is a core part of CTL-based learning activities. The knowledge and skills students acquire are not the result of remembering a set of facts, but the results of finding themselves. After conducting an investigation students will find data related to the material being studied. Examples of finding activities are on worksheet 2 (atmosphere) students do activities to find differences in the temperature of the earth through investigation. Through these activities, students can compare different temperatures of the earth.

Students will analyze the data obtained in groups after finding the results of the investigation. Through the principle of the learning community, students will conduct analysis by answering questions to help understand the material of the earth's layers. The concept of a learning community (Learning Comunity) in CTL learning outcomes is obtained through cooperation with other people, friends, between groups, other sources and not just teachers [6]. Learning outcomes can be obtained by discussions between friends, between groups, and between those who know who don't know, both inside and outside the classroom.

At the end of the worksheet based on Contextual Teaching and Learning, there are reflection activities. CTL reflection components as according to [6] is a process of deposition of experiences that have been learned by reordering the events or learning events that have been passed. In CTL learning, each end of the learning process, the teacher provides opportunities for students to reflect on or recall what they have obtained. Through reflection activities, students can improve their understanding of the material they have learned. Reflection also helps students strengthen students' memories of the material they have learned. Worksheet developed has been completed with reflection activities at the end of the activity, such as on worksheet 2 (atmosphere). In worksheet 2 students and teachers review the learning outcomes related to increasing the temperature of the earth so that students can deduce the causes of the temperature of the earth which is increasingly increasing.

The seventh component of the CTL approach is authentic assessment. Real assessment of authentic assessment is the process by which the teacher collects information about the development of student learning. This assessment is needed to find out whether students really learn or not, whether student learning experiences have a positive influence on students' intellectual or mental development. CTL learning emphasizes the learning process not just on learning outcomes [6]. During the learning process, each group will be assessed based on the learning assessment sheet as an authentic assessment. Authentic assessment is used to give rewards to students at the end of the teaching and learning process. The purpose of giving rewards or awards so students are more enthusiastic to improve their achievements. Ngalim Purwanto [12] explains the purpose of giving rewards is to educate children so they can feel happy because their deeds or work get an award. In addition, the purpose of giving rewards is also to increase students' willingness to improve or enhance their achievements.

4. Conclusion

The learning process using the CTL approach that is implemented in Student Activity Sheets can help students understand the subject matter more easily. This is indicated by an increase in students' completeness in solving questions on the worksheet.

5. Acknowledgement

The researchers acknowledge to the Science Department, Mathematics and Natural Sciences Faculty, Universitas Negeri Surabaya have provided the facilities for doing research and literature for constructing the article.

References

[1] Kemendikbud 2016 *Permendikbud Nomor 22 Tahun 2016 Tentang Standart Prroses Pendidikan dan Menengah* (Jakarta: Kemendikbud)
[2] Depdiknas 2008 *Panduan Pengembangan Bahan Ajar* (Jakarta: Departemen Pendidikan Nasional)
[3] Mulyasa 2013 *Pengembangan dan Implementasi Kurikulum 2013* (Bandung: PT Remaja Rosdakarya)
[4] Ibrahim R 2010 *Perencanaan Pengajaran* (Jakarta: Rineka Cipta)
[5] Woolfolk and Nicolich 2004 *Educational Psychology for Teacher* (Jakarta: Inisisasi Press)
[6] Sanjaya W 2016 *Strategi Pembelajaran Berorientasi Standart Proses Pendidikan* (Jakarta: Kencana Prenada Media Group)
[7] M Hosnan 2014 *Pendekatam Saintifik dan Kontekstual dalam Pembelajaran Abad 21* (Bogor: Ghalia Indonesia)
[8] Khaefatunnisa 2015 The Effectiveness of Contextual Teaching and Learning in Improving Students' Reading Skill In Procedural Text (A Quasi-Experimental Study of The Second Grade Students at One Vocational School in Bandung) *Journal of English and Education,* **3**(1) 80 95
[9] Hadari N 2005 *Penelitian Terapan* (Jakarta: Rineka Cipta)
[10] Suprijono Agus 2009 *Cooperative Learning: Teori dan Aplikasi PAIKEM* (Yogyakarta: Pustaka Pelajar)
[11] Muslich M 2011 *Melaksanakan PTK (Penelitian Tindakan Kelas) Itu Mudah* (Jakarta: PT. Bumi Aksara)
[12] Martha E D 2016 *Implementasi Pemberian Reward Kepada Siswa SD Muhammadiyah Bantul Kota.* (Skripsi FIP Universitas Negeri Yogyakarta)

The Implementation of Guided Discovery Learning Using Virtual Lab Simulation To Reduce Students' Misconception on Mechanical Wave

S Admoko, M Yantidewi, R Oktafia

Physics Department, Faculty of Mathematics and Natural Sciences, State University of Surabaya, Indonesia

E-mail: setyoadmoko@unesa.ac.id

Abstract. This study aims to determine the effectiveness of the implementation of guided discovery learning with virtual lab: Physics Education Technology (PhET) simulation in reducing students' misconceptions in mechanical wave physics. The design of this study was pre-experimental design with One Group Pretest-Posttest Design. This research was conducted in Mojokerto, East Java, Indonesia. Thirty five 11th grade high school students and one teacher participated in this study. This study applied the certainty of response index (CRI) with Two-tier Diagnostic Test to determine students' misconception. An analysis of differences in the average of misconception between pretest and posttest and also normalized gain analysis had been performed to find out the effectiveness of guided discovery learning implemetation towards students' misconception reduction. The treatment was declared effective if the average of misconception in pretest was greater than the postest and the normalized gain was negative, which meant there was a decrease in misconception minimum in the moderate category (**$\langle g \rangle \leq -0{,}3$**). The results showed that in the initial tests the 3 highest students' misconceptions occured on the concepts: wavelength, wave velocity and the nature of the wave. But the implementation of guided discovery learning was able to reduce students' misconceptions significantly. Based on the statistical analysis of T test, it was found that $t_{count} > t_{table}$ with $\alpha = 0.05$ with $1.96 > 1.71$. This meant that there were significant differences in average misconception. Through gain analysis, it is also known that $\langle g \rangle = -1{,}13$ where the average misconception at pretest is 53,33% to 19,55% for posttest. This decrease was included in the high category. According to these analysis results, the application of the Guided Discovery Learning model using Virtual Lab. PhET was able to reduce students' misconceptions.

1. Introduction

Understanding is to know everything and understand correctly. A person's ability to understand concepts leads to the ability possessed in mastering the signs of objects [1]. Understanding is one type of learning outcome that is much higher when compared to knowledge. But to reach understanding, it is necessary to have a process of knowing a material well before [2]. The lack of understanding certain concepts will influence the formation of scientific concepts in students' cognitive structures [3].

There are still many students who do not understand and do not master the concept of Physics. One of the difficulties in Physics is the occurrence of errors in concepts. In other words, they have different concepts from scientific knowledge. This is what we call misconception [4, 5, 6]. Misconception can affect students because it can obstruct students to receive new knowledge [7]. There are many ways to identify a misconception. One is using the certainty of response index (CRI) model. CRI is a confidence measurement in the given answer [8]. Using CRI will be quite easy and simple in distinguishing students who experience misconceptions and those who don't. It is a method for the honesty of students in answering a question by filling in the CRI level.

Learning that provides opportunities for active students in studying and do not cause many misconceptions is guided discovery. The findings referred to here are guided discoveries which involve the process of interaction between teacher and student, where students seek for themselves the desired concept through a sequence of questions set by the teacher. This learning gives students the freedom to find a concept [9, 10].

The discovery activities in learning physics cannot be separated from experimental and laboratory activities. From the results of observations at a senior high school in Mojokerto regarding the inadequate state of the Physics laboratory, where there is a lack of tools and materials for students to conduct experiments. Some tools are rusty or not suitable for use. One alternative method so that discovery activities can be carried out in learning is using virtual lab simulation. This method has also never been applied in school. By using virtual simulations it can provide a solution to the state of the laboratory that has equipment problems in the lab [11]. This simulation can provide opportunities for students to learn and apply the concepts they have in real-world skills. Using virtual lab simulation media can increase the effectiveness of understanding student concepts and minimize existing misconceptions [12].

There are a number of Physics concepts which have quite high concept weaknesses including wave propagation, wave superposition and particle movement in a mechanical wave system [13]. This concept allows misconception, it can be seen that many mechanical wave phenomena are found in everyday life. Based on the results of interviews, we find those still students who do not understand the physical meaning of wavelengths, wave velocities, and frequencies. Students only understand mathematically the meaning of wave propagation, which is when given a problem to snap the rope quickly, students will assume that the wave velocity will be large. In this case, the concept of students is different from scientific knowledge. Seeing this problem, the researcher conducted the guided discovery learning using virtual lab simulation in reducing the students' misconception of mechanical wave material.

2. Research Methods

This research is quantitative research with a pre-experimental design approach. The form of research used was a one-group pretest-posttest design. This research was conducted at a high school in Mojokerto. Thirty five 11th grade high school science students as a research sample. Analysis of student conception was carried out through a diagnostic test in the form of pretest and posttest using CRI as a measure of confidence in the answers given.

Table 1 How to determine the level of understanding of students' concepts [14]

Answer	Low CRI (<2,5)	High CRI (>2,5)
Correct answer	Correct answer and low CRI (CL) lack of knowledge (lucky guess)	Correct answer and high CRI (CH) knowledge of correct concepts
Wrong answer	Wrong answer and low CRI (WL) lack of knowledge	Wrong answer and high CRI (WH) misconception

Table 1 demonstrates the four potential mixes of the appropriate response (Correct or Wrong) and the CRI (High or Low) with respect to an individual understudy. For a student and for a given question, a correct answer with a low CRI (CL) indicates a lack of knowledge and a correct answer and a high CRI (CH) points to knowledge of correct concepts. A wrong answer with a low CRI (WL) suggests a lack of knowledge, while a wrong answer with a high CRI (WH) indicates the presence of misconceptions.

The results of the pretest and posttest were analyzed statistically. An analysis of differences in the average of misconception between pretest and posttest use paired t-test and also normalized gain analysis had been performed to find out the effectiveness of guided discovery learning implemetation towards students' misconception reduction. The treatment was declared effective if the average of misconception in pretest was greater than the postest and the normalized gain was negative, which meant there was a decrease in misconception minimun in the moderate category ($\langle g \rangle \leq -0{,}3$).

3. Results and Discussion

The profile of the initial concept understanding is based on the value of the pretest.

Table 2 Profile of Students' Understanding Concept

Absent Numbers	Percentage (%)				Category of Misconception
	Lucky Guess	Lack of Knowledge	Correct Concept	Misconception	
28	0,00	0,00	11,11	88,89	High
7	0,00	11,11	11,11	77,78	High
34	0,00	0,00	22,22	77,78	High
5	0,00	0,00	33,33	66,67	Medium
6	0,00	0,00	33,33	66,67	Medium
11	0,00	0,00	33,33	66,67	Medium
15	0,00	0,00	33,33	66,67	Medium
16	0,00	0,00	33,33	66,67	Medium
19	0,00	0,00	33,33	66,67	Medium
27	0,00	0,00	33,33	66,67	Medium
30	0,00	0,00	33,33	66,67	Medium
3	0,00	33,33	11,11	55,56	Medium
14	11,11	22,22	11,11	55,56	Medium
17	0,00	0,00	44,44	55,56	Medium
22	0,00	0,00	44,44	55,56	Medium
24	0,00	0,00	44,44	55,56	Medium
29	0,00	0,00	44,44	55,56	Medium
21	11,11	33,33	11,11	44,44	Low
31	0,00	0,00	55,56	44,44	Low
35	0,00	0,00	55,56	44,44	Low
12	22,22	33,33	11,11	33,33	Low
23	0,00	22,22	44,44	33,33	Low

32	33,33	33,33	22,22	11,11	Low
33	33,33	44,44	11,11	11,11	Low
26	44,44	55,56	0,00	0,00	Low

Based on the profile of conceptual understanding of class: 29% of students understood the concept. Students who had lack knowledge are 12% and those who answered by guessing are 6%. Misconceptions that occured are also still large at 53%. Misconceptions that occured could be said to be quite large because it exceeded the number of students in the class. Most of the misconceptions that occured made it possible for most students to be rarely given the problem of deepening the concept which prioritized mathematical elements. Reinforced again with the students' reasoning that was not in accordance with scientific knowledge.

Misconceptions that occur in students are divided into several levels of categories, namely low, medium and high. This is used to find out how many students experience irregularities. This level is obtained by determining the upper and lower limits of the percentage perception data. Students who are categorized as high in misconceptions are equal to 12% of the total students. Students who experience a high level of misconception are students who get a large percentage of misconceptions> 75.087%, moderate levels with a percentage of 75.087% $\leq x \leq$ 31.579% and low levels with a percentage of <31.579%.

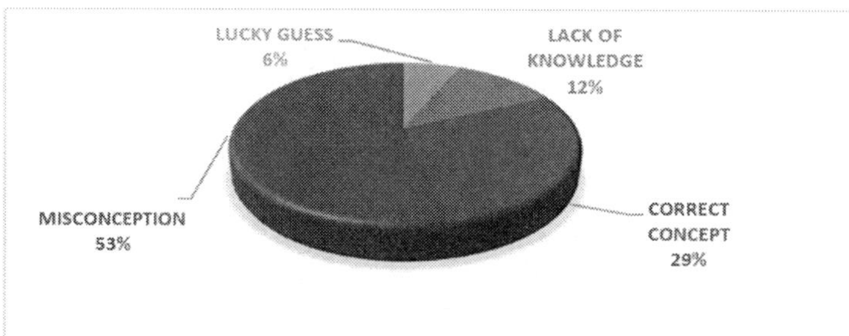

Figure 1 Profile of Understanding Concept

Understanding the concept of mechanical wave material is obtained from the pretest values that have been mapped into several concept categories that were asked in the CRI model diagnostic test questions. The following is the mapping.

Table 3 Profile Understanding Concept Of Mechanical Wave

Concept	Percentage (%)			
	Lucky Guess	Lack of Knowledge	Correct Concept	Misconception
wavelength	0,00	20,00	4,00	76,00
wave nature	4,00	13,33	28,00	54,66
fast propagation factors	8,00	14,67	25,33	52,00
period	4,00	8,00	44,00	44,00
amplitude	16,00	4,00	40,00	40,00

The concept of mechanical waves that experienced the highest misconception with a percentage of 76% is the wavelength. When students do diagnostic tests, they often experience errors between the concept of period and wavelength. Most students are less careful in understanding the components in the graph such as information located on the x and y axes. Thereof, when shown the closest distance between two peaks on the graph of the relationship between the size of the deviation and the distance traveled, students equate the graph of the deviation and time relationships frequently. As a result, students often answer with the same answer, namely the period.

The second highest misconception (54.67%) is understanding the wave nature. Where students assumed that by increasing the width of the gap that passed through the wave, the wavelength also increased. However, there was an error in understanding that what the students actually observed in the picture was that the diffraction was not the large wavelength produced.

The third highest misconception is the propagation factor with a percentage of 52%. Students assumed that the magnitude of the wave velocity would change in proportion to the change in amplitude. Students assumed that if the amplitude was large then it also had a large thrust so that the resulting wave propagation happened to be relatively large.

After being given the pretest, students conducted a treatment applying guided discovery learning models assisted by virtual lab simulation. To ensure guided discovery learning models assisted by virtual lab simulation is carried out well, the steps in preparation and implementation are as follows: (1) in the preparation process the development of learning tools is carried out consisting of: Syllabus, Learning Implementation Plan, Student Textbooks, Student Worksheet, learning evaluation questions in the form of Two-tier Diagnostic Test and also the observation sheet of the implementation of learning. The worksheets developed in the study are worksheets that provide experimental instructions with the PhET Laboratory as a source of data that students will get. To test the validity of all learning tools and instruments that have been developed, a validation process is carried out by two experts on the learning tools and instruments. The results of the validity of the learning tools and instruments conducted by two experts in the study stated that they were valid and proper to use. (2) to ensure the treatment goes well in the implementation process, an assessment of the implementation of learning is carried out by two observers. The results of the assessment of the delay in learning conducted in three meetings obtained an average value of 90.6%, which means that the learning process was carried out in the excellent category.

To find out the influence of the model applied in reducing misconceptions in students performed paired t-test. That is a test carried out on the subject for the conditions before and after the process. In this study, this test was used to calculate the difference between the pretest misconception score and posttest and to find out if there were differences in it.

Table 4 Misconception Score and Reduction of Misconception

Absent Numbers	Score		Percentage (%)		<g>	Category
	Pretest Misconception	Posttest Misconception	Pretest Misconception	Posttest Misconception		
28	8	4	88,89	44,44	-4	High
7	7	0	77,78	0	-3,5	High
34	7	0	77,78	0	-3,5	High
6	6	1	66,67	11,11	-1,67	High
15	6	1	66,67	11,11	-1,67	High
16	6	1	66,67	11,11	-1,67	High
27	6	1	66,67	11,11	-1,67	High
30	6	1	66,67	11,11	-1,67	High
19	6	2	66,67	22,22	-1,33	High
22	5	0	55,56	0	-1,25	High
14	5	1	55,56	11,11	-1	High
24	5	1	55,56	11,11	-1	High
29	5	1	55,56	11,11	-1	High

5	6	4	66,67	44,44	-0,67	Medium
11	6	4	66,67	44,44	-0,67	Medium
31	4	1	44,44	11,11	-0,6	Medium
3	5	3	55,56	33,33	-0,5	Medium
17	5	3	55,56	33,33	-0,5	Medium
23	3	1	33,33	11,11	-0,33	Medium
21	4	3	44,44	33,33	-0,2	Low
35	4	3	44,44	33,33	-0,2	Low
32	1	0	11,11	0	-0,13	Low
12	3	3	33,33	33,33	0	Low
33	1	2	11,11	22,22	0,13	Low
26	0	3	0	33,33	0,33	Low
Average	4,8	1,76	53,33	19,55	-1,13	High

From analysis of data on Table 3, we obtained $t_{count} > t_{table}$ with $\alpha = 0.05$ which was large with 1.96 > 1.71. Thus, using the guided discovery model assisted by virtual lab simulation can reduce students' misconceptions in mechanical wave material significantly. Through the gain test, the class was categorized as having a high decrease with value was -1,13 which was 88% of students who experienced a decrease. The increasing understanding of this concept was due to the improvement and strengthening of students' concepts in construction forming a new concept that had been done in guided discovery learning, namely the data collection phase. There were two students experiencing the greatest decrease in misconceptions, namely students numbers 7 and 34 with a decrease of 77.78%. As seen in the table above, the misconceptions of the two students are completely reduced, as is the case with students absent number 22. This could be seen from the posttest value of misconceptions obtained which was equal to 0%. In this class, only 4% of students did not decrease in misconception. This was because the understanding of the students' initial concepts in this class was still carried on their initial knowledge, which was called resistance [15]. Misconceptions were resistant or can be said as difficult to change and a tendency to persevere even though a treatment about a concept justification had been given. In addition to originating from resistant misconceptions, it was also caused by observation and reasoning which was considered inaccurate and the level of intelligence possessed by students in the low category.

Besides that, there were two students in the class who experienced an increase in misconceptions, namely students with absent numbers 26 and 33. The increasing misconception was on the concept of wave nature. The learning of wave nature carried out at the second meeting required students to have a high level of precision when understanding the phenomenon of wave nature in water ripples. This was due to the application of the wavelength concept in it, so it could be said that the learning taught was more complex than learning at the first meeting. When viewed from the results of the two students' pretest, the initial understanding of the concept possessed by the two students belonged to the category of insufficient knowledge. This could lead to difficulties in learning which had a high level of understanding in it. Still, the existence of students who experienced misconceptions showed that the design of the applied learning model needed improvement for the two students in order to reduce the misconceptions that they experienced maximally and thoroughly [16]. This could be one of the factors why in this study there were still students who experienced an increase in misconceptions.

4. Conclusion

Based on data analysis of the pre-test on the mechanical wave, it was found that students' misconceptions were very high. The students' biggest misconceptions occur in conceptualization: wavelength, wave properties and factors that influence the velocity of wave propagation. Three of the students' biggest misconceptions are: (1) in reading the graph, students find it difficult to distinguish between periods and wavelengths (2) in the diffraction wave process the greater the gap that is passed then the wavelength will change even greater. (3) the greater the amplitude of a wave, the greater the velocity of the wave propagation. Application of Guided Discovery Learning model using Virtual Lab. PhET was conducted to reduce students' misconceptions on the concepts of mechanical waves. From the analysis of data on the pre-test and post-test shows that: there is a significant difference in student understanding between before and after the application of Guided Discovery learning. Based on the facts obtained in this study it can be concluded that the application of the Guided Discovery Learning model using Virtual Lab. PhET was able to reduce students' misconceptions on all concepts in mechanical waves.

References

[1] Hermawanto 2013 *J. Pendidikan Fis.* **9** 67
[2] Audina 2017 *J. Ilmiah Pendidikan Fis.* **1 (1)**
[3] Maulidar 2016 *J. Pendidik Sains Indonesia* **4 (2)** 69
[4] Berg, Euwe Van Den. 1991. *Miskonsepsi Fisika dan Remediasi*. Salatiga: Universitas Kristen Satya Wacana
[5] F. Türker, December 2005, "Developing a Three-Tier Test to Assess High School Students' Misconceptions Concerning Force and Motion" Cankaya/Ankara: Thesis of Middle East Technical University.
[6] X. Vamvakoussi and S. Vosniadou," How A Review and Comparison of Diagnostic Instruments to Identify Students' Misconceptions in Science Many Decimals Are There Between Two Fractions? Aspects of Secondary School Students' Understanding of Rational Numbers and Their Notation,", Cognition and Instruction, 2010, vol. 28, pp 181
[7] Nawati 2017 *J. Penelitian Pemb Fisi* **8** 32
[8] Tayubi 2005 *Mimbar Pendidikan No. 3/XXIV/2005*
[9] Maulidar 2016 *J. Pendidik Sains Ind*, **4 (2)** 69
[10] Yuliani K and Saragih S 2015 *Journal of Education and Practice* **6 (24)** 116
[11] Arianti 2016 *J. Pendidikan Fisika dan Teknologi 2(4)*
[12] Suhandi 2009 *Jurnal Pengajaran MIPA* **13** (1)
[13] Kennedy 2011 *Can J Phys* **89** 1155-1161
[14] Hasan S, Bagayoko D and Kelley E L 1999 *Phys. Educ.* **34(5)** 294
[15] Chairunnisa 2016 *J. EduBio Tropik.* **4 (1)** 17
[16] Witanecahya (2014) *Jurnal Inovasi Pendidikan Fisika (JIPF)* **3 (3)**

Science teacher's conception about importance of Geoscience learning: A case study of junior high schools in Surabaya Indonesia

S Anggrayni[1], Madlazim[1] and E Hariyono[1]

[1]Postgraduate, State University of Surabaya, Indonesia.

e-mail: ekohariyono@unesa.ac.id

Abstract. The aim of this study is to identify the science teacher's conception of the importance of geoscience learning. This study was possessed the conception profile of science teacher about geoscience learning through questionnaire instrument and curriculum analyses to 17 science teachers at three Junior High Schools in Surabaya as the participants. A descriptive quantitative approach is used to analyse the data. Based on the results obtained, almost all the participant stated that geoscience learning is important to teach but more than 50% of science teachers do not teach it because of insufficient time constraints. Based on the curriculum analyses, can be known that the learning method which is done using textbooks only without involving issues which are happening in the present. Therefore, it can be concluded that the developed geoscience learning curriculum for pre-service science teachers needs to be developed.

1. Introduction

For the past several years, geoscience learning has been of considerable interest to science education researchers. The quality of learning depends on the teacher's understanding of the geoscience subject matter, which reflects the quality of education. Education is one of the main vehicles for the intellectual and professional development of human resources and plays a very important role in supporting a stronger and globally competitive Indonesia. Therefore, it is the most powerful weapon that can be used to change the world [1]. To realize good quality education, science teachers, curriculum developers, and resource researchers must be organized and have easy access to overcome geoscience misconceptions [2].

A conception about geoscience is very important for humanity to successfully respond to these challenges and develop in the coming decades. Several supporting factors are (1) global finance depends on access to reliable and safe energy sources, most of which depend on geoscience knowledge; (2) air, soil, and mineral resources become very scarce in some areas due to the increase in human population and global industrialization; and (3) there are many questions in terms of sea-level rise, drought, forest fires, desertification and intensification of storms as a result of global climate change widely accepted [3]. Geoscience is important to enhance the disaster mitigation of Indonesian society. The science curriculum for junior high school in Indonesia has some content to prepare students to understand the layers of the earth, volcanoes, earthquakes, and risk reduction actions before, during, and post-disaster according to the threat of disasters in the area. Thus, this understanding can increase his/her knowledge and concepts of science that are useful in everyday life as well as to develop skills, to investigate the environment, and solve problems [4].

The science curriculum that has been formulated is very suitable with the conditions of the earth in Indonesia because the events of the disaster and the associated social and economic impacts are currently increasing. It is undeniable that Indonesia is a country closed by natural disasters [5]. In the last decade, it showed the highest number and impact of disasters, while 2015 was declared the hottest year ever. The Asia Pacific region has become the most frequent place for this disaster, while Indonesia is one of the most at risk of disasters and the effects of climate change [6].

In fact, there were 350 natural disasters occurred in 2016 worldwide, killing 10.273 people and affecting over 204 million people. The estimated amount of economic damage came close to US$147.4 billion [7]. The impacts of natural disasters in four indices, occurrence, number of deaths, number of affected people and economic damage based on the region that were covered in Figure 1.

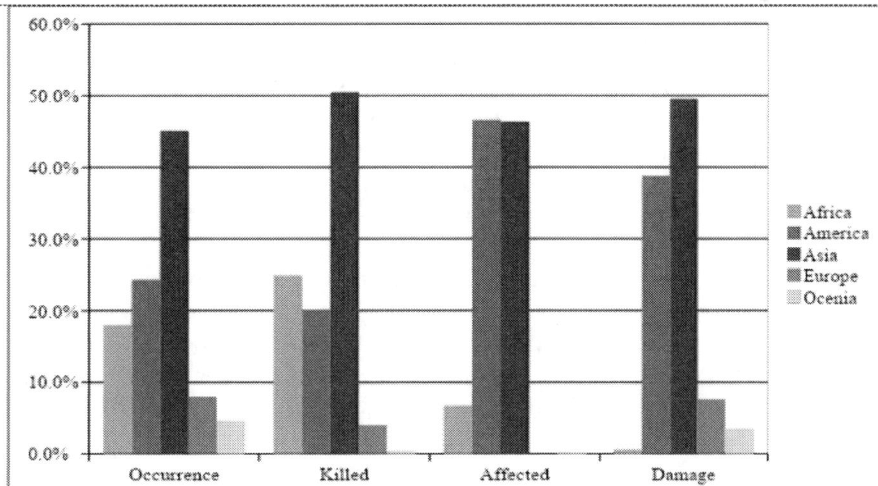

Figure 1. The natural disaster's impact by region, 2016 [7].

As shown in Figure 1, Asia ranks the first among all regions in the categories of disaster occurrence, the number of killed people, and economic damage, accounting for 45.1 per cent, 50.5 per cent, and 49.5 per cent, respectively. On the other hand, America is ranked the highest in the number of affected people, which is largely attributed to the storm in America.

While the impacts of natural disasters in Asia based on occurrence, number of deaths, number of affected people and economic damage that were covered in Figure 2 below.

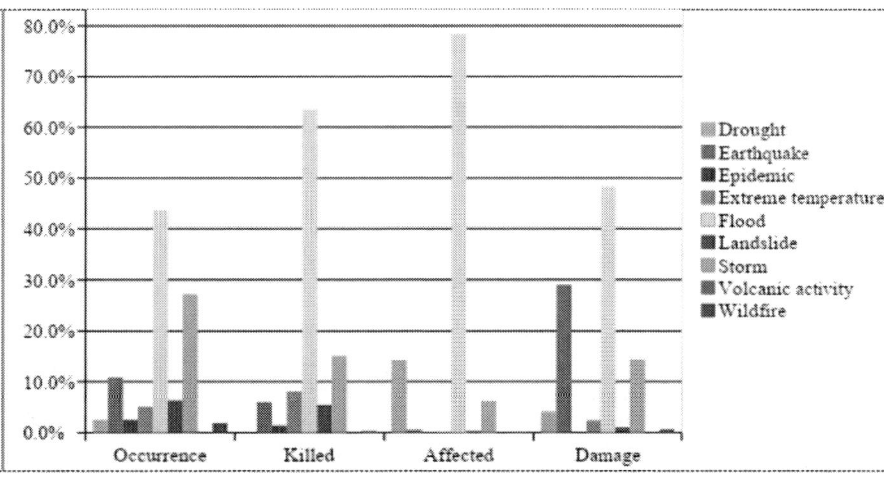

Figure 2. The natural disasters impact in Asia by disaster type, 2016 [7].

Based on Figure 2 above, we can be known that flood is the most dominant disaster in all categories. Flood tops at 43.7 per cent and 63.6 per cent while in the number of occurrences and killed people. In the number of affected people, the flood has the largest shares too, 78.3 per cent. Flood brings the heaviest economic damage by 48.3 per cent.

Based on EM-DAT (Emergency Event Database) data from the two figures above, it is known that more disasters that occur are from non-geophysical disasters (hydrology, meteorology, and climatology) than those originating from geophysical disasters. Therefore, the total damage and total losses resulting are more. However, the number of deaths is significantly caused by geophysics disasters, namely earthquakes and volcanic activity [6].

Volcanic phenomena become an inseparable part of the lives of Indonesian people [8]. Geographically, almost all Indonesia regions are prone to natural disasters, especially volcanic disasters. Indonesia has the most volcanoes in the world [9]. There are 147 volcanoes, and 76 of them are active volcanoes and spread along the islands of Sumatra, Java, Sulawesi and the Lesser Sunda [10]. This condition brings consequences or impacts both positive and negative for Indonesia society, especially those who live in the volcanic region. The positive impact is the potential of natural resources (fertile land, beautiful landscapes, many deposits of metals, non-metals, oil and gas) caused by the presence of volcanoes. Negative impacts are dangerous or vulnerable to geological disasters such as earthquakes, volcanic eruptions and landslides [9].

Another fact, many volcanic disasters show that the increasing number of deaths is caused by the weakness of the disaster preparedness system and the low understanding of disaster risk to the surrounding community. Indonesia is the fifth country with the largest population in the world where 60% of the population lives on stone slabs, but that number has not been balanced with a good level of understanding of disaster preparedness [11]. Beginning in 1815 and early 2014, Indonesia had 128 volcanic eruptions with deaths reaching 78647 people. The event was symbolized as a disaster with the second-largest eruption after the earthquake and tsunami. The National Disaster Management Agency (BNPB) also noted that from 2010 to 2014, there were 36 eruptions with a total of 432 people had died, and 2217 people had injured. These data illustrate the vulnerability of the society for the risk of disasters. If we cannot prevent and reduce significantly, the damage caused by extreme events is a major disaster for life [8].

One of the factors predicted as the cause of the number of disaster victims is the low knowledge and skills' the society in minimizing disaster risk. The ability to reduce disaster risk greatly helps them to protect people and economic assets [12]. Certainly, it cannot be done spontaneously to manage the disaster, but it must be planned in good long-term management [13]. This disaster management should be integrated into geoscience learning in school. Therefore, knowledge related to disaster mitigation and response techniques needs to be shared by every Indonesian society, so as to minimize the risk of disasters that occur [11].

This paper is intended to identify the science teacher's conception of the importance of geoscience learning. The function of this identified is to know how important this subject matter is taught, how the learning process is carried out at school, and what learning resources are used, so that the researchers can improve the appropriate learning strategy to create a generation that has excellent disaster mitigation.

2. Research Method

Participants of the study included 17 science teachers from three Junior High Schools in Surabaya, Indonesia, which five (5) teachers from Junior High School 21 Surabaya and six (6) teachers from each Junior High School 22 and 36 Surabaya. The research used a descriptive quantitative approach by using a questionnaire instrument created by the researchers. The data will be analysed using percentage descriptive analysis to determine the science teachers' conception about the importance of geoscience learning and how the learning method used. Basic competence (KD) that supporting the geoscience curriculum is KD 3.10 and KD 3.11 on the Junior High School science curriculum, each of them is KD 3.10 state "explain the layers of the earth, volcanoes, earthquakes, and actions risk reduction before, during, and after a disaster according to the threat of disaster in the area" and KD

Journal of Physics: Conference Series **1417** (2019) 012090 doi:10.1088/1742-6596/1417/1/012090

3.11 state "analyse the solar system, the rotation and revolution of the earth and the moon, and their impact on life on earth" [4].

3. Result and Discussion

The questionnaire has some question about geoscience learning. The first question is about the importance of geoscience learning based on science teachers' opinion. The following Figure 3 is the science teachers' response to the importance of geoscience learning.

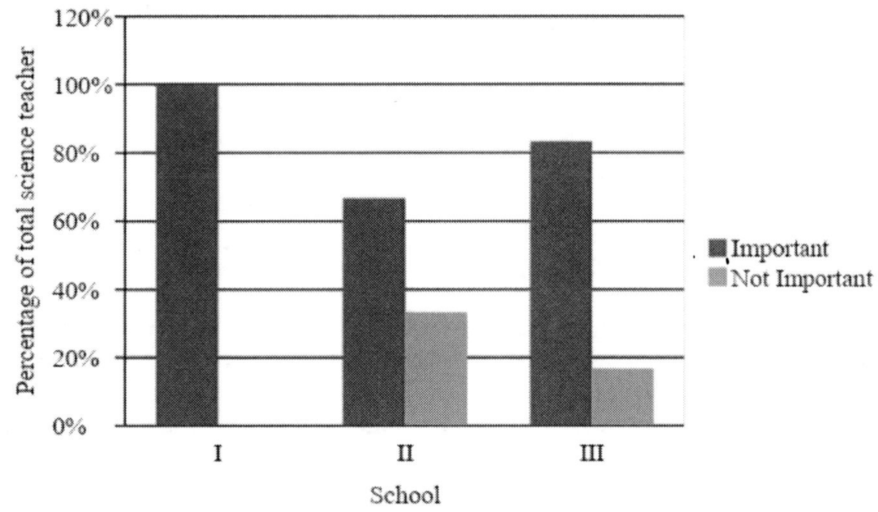

Figure 3. The importance of geoscience learning according to the science teachers from three Junior High Schools in Surabaya.

Based on Figure 3 above, it can be known that more than 50% of the participants think geoscience learning is important to teach. In school, I, all of the science teacher implemented the geoscience subject matter in the Science lesson. On the other hand, several science teachers in another school aren't teaching the subject matter. The following Figure 4 shows the implementation of geoscience learning in several junior high schools in Surabaya.

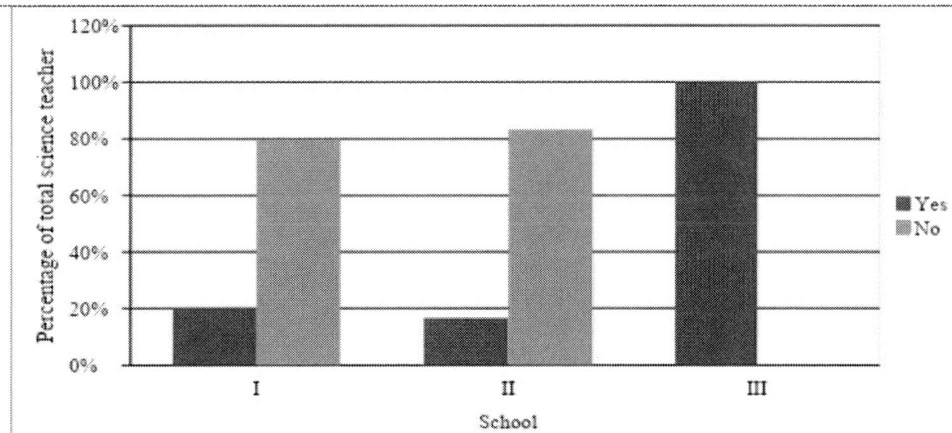

Figure 4. The implementation of geoscience learning from the three junior high schools in Surabaya.

As shown in Figure 6 above, science teachers that did not teach geoscience subject matter were more than 50%. Based on the interviews, his/her reasons are (1) the geoscience subject matter is located in the most last chapter, so the time is often not available due to explain it; (2) the students are only given individual assignments, such as making clippings and papers; and (3) the science teachers only display videos of related phenomena about geoscience.

In addition to that, science teachers' learning methods are adjusted to the characteristics of the students. For the students who have characteristics such as elementary school students, the suitable learning method is discourse. However, some classes were implementing another learning method, such as discussion.

Each school has different geoscience learning resources. The students of Junior High School 21 Surabaya just use their books in science learning. In Junior High School 22 Surabaya, the students use the internet as another learning resource. While in Junior High School 36 Surabaya, there are various kinds of learning resources, such as the science library and solar system model media. But based on the curriculum analyses, can be known that the learning method which is done using textbooks only without involving issues which are happening in the present.

Geoscience teachers have a very large contribution to make to universal education by offering place-based, locally relevant science education on topics of critical importance to society. These topics, such as water, natural hazards, and resource development, bring larger global concerns, such as energy and climate change, to a relevant, local level that can affect individual lives [14]. In this case, to achieving innovative learning, surely the teacher must be creative in applying geoscience learning by integrating various kinds of values that apply in society. For example, the developed geoscience learning showed that the implementation of the learning of Science, Environment, Technology and Society (SETS) was integrated with local wisdom that was able to reconstruct and increase the disaster management knowledge [15].

In addition, to improve the quality of Indonesian science education we suggested that science teachers should plan to teach strategic and systematic, implementation, evaluation systems, and continuous improvement of the education system in Indonesia [1]. So that, we need to do the reconstruction of the design of geoscience learning that prioritizes the issues of the earth through the subject of Earth and Space Sciences (IPBA) to the pre-service teachers, because the students' understanding of geoscience is very important to answer the earth issues in future.

As Laurie et al. (2016) suggested education of sustainable development (ESD) skills contributes in many ways to quality education in primary and secondary schools. Teaching and learning change in all contexts when the curriculum included sustainable content and the ESD pedagogy promotes the learning of the skills, perspectives, and values needed to grow a sustainable society [16]. This could suggest that ESD skills could effectively improve students' conceptual understanding and insightful problem-solving in geosciences phenomena through the subject of Earth and Space Sciences (IPBA).

4. Conclusion

Based on the results which are obtained of this research, it can be concluded that almost all of the science teachers stated that geoscience learning is important to teach. However, more than 50% of science teachers do not teach it due to several reasons, and the reasons are (1) there is no available time to explain the geoscience subject matter because it is located in the most last chapter; (2) the students are only given individual assignments, such as making clippings and papers; and (3) the science teachers only display videos of related phenomena about geoscience. While the learning method and learning resources also need to be developed to realize innovative learning. Therefore it can be concluded that the developed geoscience learning curriculum for pre-service science teachers needs to be developed.

References

[1] Sukasni A and Efendy H 2017 *Int. J. of Educ.* 9 183
[2] Francek M 2013 *Int. J. of Sci. Educ.* 35 31
[3] Wysession M E, LaDue N, Budd D A, Campbell K, Conklin M, Kappel E, Lewis G, Raynold R, Ridky R W, Ross R M, Taber J, Tewksbury B, and Tuddenham P 2012 *J. of Geo. Educ.* 60 95

[4] Kementerian Pendidikan Dan Kebudayaan 2017 *Model Silabus Mata Pelajaran Sekolah Menengah Pertama/Madrasah Tsanawiyah (SMP/MTs)* (Jakarta: Kemendikbud) pp 1-78
[5] OCHA 2014 *Dampak Bencana Alam Meningkat* (Buletin Kemanusiaan Indonesia: Online). http://indonesia.humanitarianresponse.info
[6] Djalante R 2016 Nat. Hazards Earth Syst. Sci. Discuss 112 1
[7] EM-DAT (Emergency Event Database) 2016 *Natural Disaster Data Book 2016 An Analytical Overview* (Asian Disaster Reduction Center)
[8] Hariyono E, Liliasari S, Tjasyono B, and Madlazim 2016 AIP Conf. Proc. 1708 080004
[9] Pujianto, Prabowo, and Wasis 2018 *J. of Phys. Conf. Series* 1006 012013
[10] Hariyono E and Liliasari S 2018 *InTech Open Access* ISBN: 978-953-51-5610-9
[11] Hidayati D 2012 *J. of Disaster Research* 7 1
[12] UNESCO 2012 *UN System Task Team on the Post-2015 UN Development Agenda*
[13] Situmorang T P 2017 *Int.Conf. on Public Policy, Social Computing and Development* 141 35
[14] Locke S, Libarkin J, and Chang C Y 2012 *J. of Geoscience Educ.* 60 199
[15] Atmojo S E, Rusilowati A, Dwiningrum S I A, Skotnicka M *Jurnal Pendidikan IPA Indonesia* 2 203
[16] Laurie R, Tarumi Y N, Mckeown R, Hopkins C *Journal for Environment Education* 10 226

Journal of Physics: Conference Series **1417** (2019) 012091 doi:10.1088/1742-6596/1417/1/012091

Practicality and effectiveness of Jigsaw-Modified Learning Models integrated on ARCS (Attention, Relevance, Confidence, and Satisfaction) motivation in Invertebrate material to complete student learning outcomes

Sheren R. Windy[1], Muslimin Ibrahim[2], Sunu Kuntjoro[2]

[1] Science Education, Post Graduate Program, Universitas Negeri Surabaya
[2] Universitas Negeri Surabaya

e-mail: raditasheren@gmail.com

Abstract. This study aims to produce a Jigsaw-modification learning model integrated with ARCS motivation that is valid, practical, and effective with learning tools that support learning. This research is a type of development research with a 4-D (Four-D Models) development model and experimental design of The One Group Pretest and Posttest Design. The preparatory activities were carried out at Postgraduate, UNESA while the trials on 30 students class X MIA-1 students have held at High School 1 Madiun in January 2019 until March 2019. The practicality of the model and the learning tools were reviewed from the classical implementation with an average of 3.95 with very good categories and student activities at 3.6 in the good category. The effectiveness of the model and learning device is based on the results of increasing student motivation gain score learning outcomes of 0.6 with moderate criteria, completeness of 100% learning outcomes with high gain scores. Based on the results of data analysis and discussion of research, it can be concluded that Jigsaw-modification learning models integrated the ARCS motivation that was developed properly, in terms of effectiveness and practicality. Learning models and devices developed can complete student learning outcomes seen from aspects of motivation and improvement in learning outcomes.

1. Introduction

Problems faced by students sometimes cannot be solved alone. Every student has learning problems, abilities, learning styles, types of thinking, different levels of motivation. Besides the demands of the curriculum that requires students to understand the material and connect it to daily life can also form a problem in students. Jigsaw cooperative learning can provide motivation for students to cultivate a good attitude. Jigsaw is learning that can improve metacognition skills and student learning outcomes [1]. In addition to solving learning problems, with Jigsaw students can also solve their social problems.

Classroom learning can build motivation among students. Students will be motivated in learning if there is value in the knowledge presented by the teacher through the learning model and students are optimistic about success [2]. Motivation in the classroom can lead to encouragement, security, and support for students [3]. According to Keller [4], teachers can develop a picture by creating strategies to stimulate and maintain student motivation. Many students have difficulty understanding the concepts that have been given by the teacher, especially in invertebrate material where students have difficulty distinguishing the characteristics of one invertebrate from one another and distinguishing invertebrates from vertebrates [5]. Studying invertebrates needs to use learning models that are in accordance with the material. The results of the study [6] concluded that using a modified Jigsaw learning model, Jigsaw I Aronson, and Jigsaw II Slavin could improve student learning outcomes by developing a modified Jigsaw learning model. Metacognition skills are measured based on the rubric that is integrated with the problem description, metacognitive awareness using MAI (Metacognitive Awareness Inventory), and cognitive learning outcomes with description test questions measured by the Hart scale. The modified Jigsaw cooperative learning model can increase the ability of students in the lower (AB) academic category to the upper category (AA).

Based on this background, the authors in this study need to study the Development of Integrated Jigsaw-Modification Models of ARCS Motivation in Invertebrate Materials to Improve Student Learning Outcomes.

2. Methods

This research used a quantitative descriptive type. The subject is in one class, that is X MIA-1. The design of the research used One group pretest-posttest which can be described as follows [5]:

$$O_1 \ X \ O_2$$

2.1 *Analysis of the implementation of problem-solving models*

The instrument that was used included observation sheets of implementation of jigsaw-modified models. Activities of the teacher were observed by two observers were sought the mode. Observation of the implementation of the problem-solving model was assessed by observers using a score of 0-4.

2.2 *Analysis of Jigsaw-modified*

The compilation of the Jigsaw-modified learning model syntax developed is to combine Jigsaw Aronson and Jigsaw Slavin. The modified Jigsaw model has more efficient learning steps or syntax compared to the Jigsaw syntax on Aronson and Slavin. Students need a transition time to move from the original group to the expert group and back to the original group with a very long time and can cause a decrease in student motivation [7]. Aronson and Slavin Jigsaw models need to be reduced and added an element of motivation so as to produce a new learning model that is Jigsaw-modification by applying learning strategies to students. The Jigsaw-modified learning model syntax is (1) Ministro Group *(expert group)*, (2) Colosseum group *(origin group)*, (3) Evaluation, (4) Reward.

Assessment analysis of the results of the feasibility of learning is done by calculating the score according to the reference on the learning feasibility sheet of students was determined using the following formula [5]:

Table 1. Interpretation Criteria for Workability Scores

Score Interval	Category
3,60 ≤ score ≤ 4,00	Very accomplished
2,60 ≤ score ≤ 3,59	Accomplished
1,60 ≤ score ≤ 2,59	Less accomplished
1,00 ≤ score ≤ 1,59	Not accomplished

Analysis of student learning outcomes in knowledge competencies was carried out using individual and classical completeness obtained from student's grade.

$$\text{Student score} = \frac{\Sigma\, value\ obtained}{\Sigma\, maximum\ value} x\ 100\%$$

According to Hake [8], the normalized gain score test is a method of analyzing the improvement of the results of the pretest and posttest. The normalized gain score test category is shown in the table below:

$$(g) = \frac{posttest\ score - pretest\ score}{100 - pretest\ score}$$

Furthermore, the N-gain criteria was used to determine the category of critical thinking skills described as: (1)learning outcome with "high gain" if (g) $\geq$0,7; (2) learning outcome with "medium gain" if 0,3$\leq$(g)<0,7; and (3) the learning outcome with "low gain" if (g)<0,3 [7].

3. Result and Discussions

The research developed aims to describe the feasibility of learning tools that are valid, practical, and effective. The validity of a device can be seen by the validity of Jigsaw-modification learning model and learning device. Effectiveness is measured by the results of student responses and student learning outcomes during learning using the Jigsaw-modification model integrated with ARCS motivation. Learning with the Jigsaw-modification learning model integrated with ARCS motivation (Attention, Relevance, Confidence, and Satisfaction) can be used as a reference as a source of biology learning, especially in invertebrate and vertebrate submerses in Kingdom Animalia material.

3.1. Jigsaw-modified model integrated with ARCS motivation learning models lead to learning, such as goals, evolution, environment, and management systems.

The integrated Jigsaw-modified learning model of ARCS motivation has learning objectives namely individual and group success. Learning outcomes maintain motivation, and students social skills will develop well. The preparation of the Jigsaw-modified learning model that was developed was to combine Jigsaw Aronson and Jigsaw Slavin. The modified Jigsaw model has more efficient learning steps or syntax compared to the Jigsaw syntax on Aronson and Slavin. The addition of the motivational element is supported [9] theory which compiles a set of motivational principles. The integrated Jigsaw-modified learning model of ARCS motivation is learning that includes activities that are used to carry out learning well. The steps on the modified Jigsaw have been modified. Modification of the steps or syntax of learning on Jigsaw has been reduced from the previous stage and added ARCS motivation so that student motivation is maintained. Students need time to transition from expert groups and to the original group is very long and resulted in decreased student motivation. The Jigsaw-modified syntax integrates ARCS motivation as well as the activities of the teacher and students in each phase. The chairs and tables are neatly arranged in the classroom and easily moved to make learning run well [7].

The learning model is a conceptual framework that describes a systematic procedure in organizing learning experiences to achieve certain learning goals, and serves as a guide for learning designers and instructors in planning teaching and learning activities. The term teaching model refers to a particular approach to instruction that includes its goals, syntax, environment, and management system [7]. The definition of a learning model leads to a specific learning approach including its goals, syntax, environment, and its management system. The model has characteristics such as: (1) logical theological rationales compiled by its creators or developers, (2) the rationale for what and how students learn (learning objectives to be achieved (3) the teaching behavior needed for the model to be successfully implemented and, (4) the learning environment needed for the learning objectives to be achieved [10].

3.2. The implementation of Jigsaw-modification model integrated with ARCS motivation

The implementation of teaching and learning activities and classroom atmosphere is based on the management of teaching and learning activities and classroom management. Student enthusiasm and teacher activities are related to classroom management. Based on the analysis of the implementation of the lesson plan on learning activities carried out in class X MIA 1 has a very good final average of 3.85. Learning resources have a high level of reliability that is $\geq$ 0.75 (75%) [11] with a very valid category so it can be stated that the learning resources developed are suitable for use in learning because they show consistent results and are presented with components coherent. Learning resources are sheets containing instructions, the guidance of statements, and understanding so that students can expand and deepen their understanding of the material to be studied [4].

Implementation of learning in the first trial was conducted by 4 observers. The material provided includes Mollusca, Arthropoda, and Echinoderms material of invertebrates. The results of observation of the implementation of the Jigsaw-modification model integrated with ARCS motivation in table 2.

Aspects	Average Meeting I	Average Meeting II	Average Meeting III	Average	Category
Phase 1 - Ministro Group (expert group)- *(Attention).*	4	4	4	4	Very accomplished
Phase 2 - Colosseum group (origin group)- *(Attention, Relevance).*	4	4	4	4	Very accomplished
Phase 3 – Evaluation- *(Confidence, Satisfaction).*	4	4	4	4	Very accomplished
Fase 4 – Reward - *(Satisfaction)*	3,25	3,5	3,75	3,5	Accomplished
Total Average				**3,75**	**Very accomplished**

The learning activities carried out had a very good final average of 3.75 with the very accomplished category. Each activity described in the three materials is adjusted to the syntax of the development of learning models. Each phase has a different average that is in phase 1, phase 2, and phase 3 of 4, and phase 4 of 3.5 with a very accomplished category.

3.3. The learning outcomes using Jigsaw-modification model integrated with ARCS motivation

Learning outcomes are obtained through tests before learning (pretest) and tests after learning (posttest) with the integrated Jigsaw-modification model of ARCS motivation. Before learning, students are given a pre-test to determine the ability of students initial knowledge and after learning given a post-test to find out and check or know the completeness of student learning outcomes. The description of the pre-test and post-test scores of jigsaw-modified will be illustrated in Figure 1.

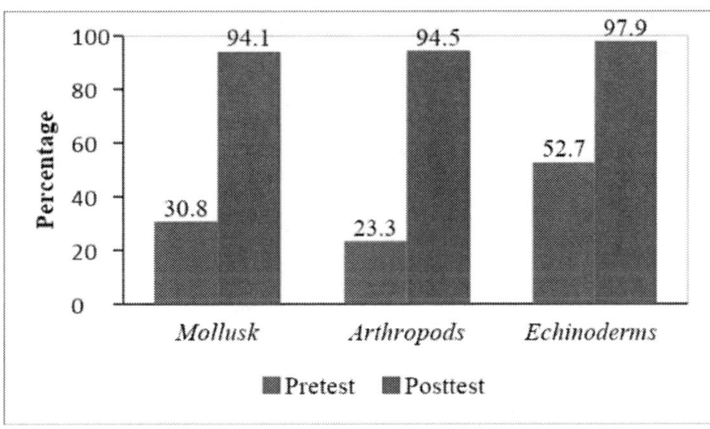

Figure 1 Graph of the pretest and posttest with a score of jigsaw-modified

In class X MIA 1 the average value of pretest and posttest on Mollusca material is 30.8 and 94.1 classically. The Arthropod material, the pretest and posttest were 23.3 and 94.5 respectively. Echinoderms material has pretest and posttest values of 52.9 and 97.9. All three materials have a percentage of completeness at pretest of 8% and posttest of 100% which indicates that each material does not have mastery achievement which is 8% at pretest where students have not been given learning with completeness criteria that are not complete and 100% complete at the hottest where students have been given learning with completeness criteria that are complete.

Teaching and learning activities are carried out based on the lesson plans that have been developed in the second meeting and the third meeting with higher achievement scores compared to the first meeting because the award is less apparent in the first and second meetings and then corrected in the third meeting. Giving appreciation can maintain student motivation at the end of learning. The awarding is a feedback from the teacher for students which can be in the form of praise to appreciate the hard work of students. In teaching and learning activities there is no real appreciation from researchers such as non-academic matters. However, an award is given to students not only from non-academic aspects but also praise [12]. Group rewards are based on the development of each individual to stay motivated. Giving awards with the title of a good team, great team, and super team in every learning is a motivation and self-development of students to be even better [13]. Two things in cooperative learning that are important are the first by using cooperative learning to improve student achievement while at the same time improving the ability of social relationships, fostering attitudes to accept self-deprivation and others and can increase self-esteem [13]. Second, learning to think, solve problems, and integrate knowledge with skills can be discussed in cooperative learning. Cooperative learning can encourage its members to make maximum efforts to achieve success.

Observing can occur in real objects or simulations that can be used to stimulate students to learn and ask questions. This shows the integrated Jigsaw-modified learning model of ARCS motivation can be used to apply to learn to invertebrate material or other material. Each stage is developed using a scientific approach that shows students can find concepts and communicate the results of the discussion. When students are involved in a group, there will be a scaffolding process, where students in one group work together and provide assistance to other group members who are less capable in the learning process [13]. Scaffolding is a learning process where students help other students in the closest zone so that students are able to actually reach their closest potential zone gradually according to the rhythm of learning.

The pretest and posttest data of students who have been obtained are then carried out to gain score test to find out an increase in student learning outcomes between before and after treatment is given [14]. The description of the pre-test and post-test scores of jigsaw-modified will be illustrated in Figure 2.

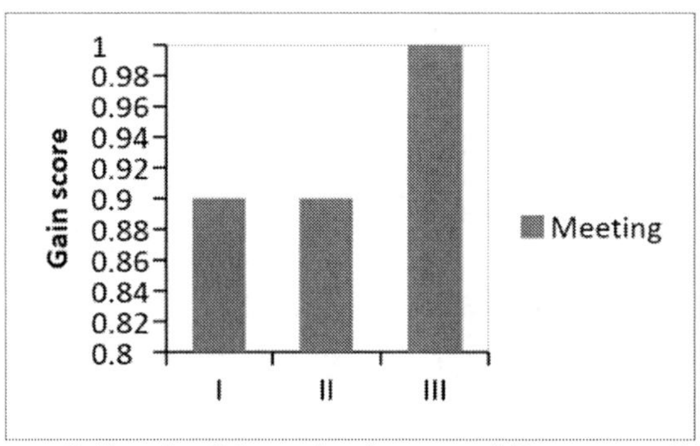

Figure 2 Graph of the gained score of jigsaw-modified

The results of the calculation of the gain score test for the 24 students who were given treatment showed that the material on Mollusca and Arthropoda was 0.9 and 0.9 with the criteria for the gain score that was high and the Echinodermata material was 1 with a high gain score criterion. The class that use integrated Jigsaw-modified learning models of ARCS motivation show good results for classroom learning and can improve student learning outcomes. Based on the recapitulation results of learning the three classes above learning using the integrated Jigsaw-modification model of ARCS motivation can be used properly in the classroom.

Group learning is learning that involves students to be active in groups. Learning in which there are students who are active in groups and communicating with their environment to seek knowledge is learning student centre. Learning with good student activities also supports the completeness of student learning outcomes well too. In addition to the completeness, student learning outcomes also improved which can be seen from the value of the pretest and posttest on each material provided. After learning to use the integrated Jigsaw-modified learning model ARCS, students' understanding of the concepts taught has increased with a high category based on the results of the gain score test from the pretest to the posttest score. Modification jigsaw is a model developed for learning activities that helps students easily achieve their knowledge with effective time and with integrated motivation ARCS can maintain student motivation from the beginning to the end of learning.

Students express their ideas better as a result of cooperation and an environment where they can discuss ideas easily created. It was revealed thanks to students' opinions that in a laboratory environment where teacher efficiency is increasing in class and teacher accessibility becomes easier for unsolved problems; any learning deficiencies resulting from the uproar disappear [12]. A study conducted to students shows that Jigsaw-modified learning can be used to practice the ability in metacognition skills for students in the aspects of knowledge and skills with efficient time and social interaction combined with ARCS motivation that can maintain student motivation from the beginning to the end of learning. These results show the superiority of the integrated Jigsaw-modified learning model of ARCS motivation in completing student learning outcomes [13].

4. Conclusions

Based on the results of data analysis and discussion of the research, it can be concluded that the Jigsaw-modification learning model integrates the ARCS (Attention, Relevance, Confidence, and Satisfaction) motivation which is feasible, in terms of effectiveness and practicality. Learning models and devices developed can complete student learning outcomes seen from aspects of attitude and knowledge. Future studies can use the Jigsaw-modification learning model integrated with ARCS motivation (Attention, Relevance, Confidence, and Satisfaction) on other materials so that student learning outcomes can increase. The time allocation used should be adjusted to the learning activities taking place. Skill assignments resulting from group drawings are given assignment limitations.

Acknowledgements

The authors would like to express an appreciation to Dr. Sifak Indana, M.Pd, Dr. Fida Rachmadiarti, M.Kes as a validator. The highest appreciation is present to two lecturers, Prof. Dr. Muslimin Ibrahim, M.Pd, Dr. Sunu Kuntjoro, M.Si, Dr. Eko Hariyono, M.Pd., and Dr. Yuni Sri Rahayu, M. Sc., that have guided research from the beginning until the article finish.

References

[1] Yuwono, Cornelius Sri Murdo 2012 *Pengembangan Model Pembelajaran Kooperatif Jigsaw modifikasi dari Aronson dan Slavin serta pengaruhnya terhadap keterampilan metakognisi dan hasil belajar Biologi siswa berkemampuan akademik berbeda di SMA Kota Denpasar* Disertasi. Program Studi Pendidikan Biologi: Program Pascasarjana, Universitas Negeri Malang

[2] Malik S 2014 Effectiveness of ARCS Model of motivational design to overcome non completion rate of students in distance education. *Turkish Online Journal of Distance Education-TOJDE* **15** (2) 14.

[3] Arends R I 2008 *Learning to Teach: Seven Edition* (New York: McGraw-Hill)

[4] Karsli F, Sahin C 2009 Developing worksheet based on science process skills: factors affecting aolubility *Asian-Pacific Forum of Learning and Teaching* **10**(1) 15.

[5] Sugiyono 2010 *Metode penelitian pendidikan pendekatan Kuantitatif, Kualitatif, dan R&D* (Bandung: Alfabeta)

[6] Tulasi G P 2007 *Methods of teaching elementary Science* (New Delhi: Arora Offset Press)

[7] Hake R R 1998 *American J. Physics* **66** 1 64-74

[8] Arends R I 2012 *Learning to Teach: Ninth Edition* (New York: McGraw-Hill)
[9] Keller J M 2010 *Motivational design for learning and performance: The ARCS Model Approach.* (New York: Springer)
[10] Kardi S, Mohammad Nur 2000 *Pengajaran Langsung* (Surabaya: University States of Surabaya)
[11] Borich G D 1994 *Observation skills for effective teaching* (New York: Mc. Millan Publishing Company)
[12] Aydin Abdullah, Biyikli Filiz 2017 The Effect of Jigsaw technique on the students laboratory material recognition and usage skills in General Physics Laboratory-I Course *Universal Journal of Educational Research* **5**(7) 1073-1082 (Turkey: Kastamonu University)
[13] Slavin R E 2010 *Cooperative Learning Teori, Riset, dan Praktik* (Bandung: Nusa Media)
[14] Bryce T G K, Mc.Call J, MacGregor R I J, Weston R A J 1990 *Techniques for assessing process skills in Practical Science* (Oxford: Heinemann Educational Books)

Pre-service Physics Teachers Perception about Concept Map, Mind map, and Link Map for Innovative Learning

Nurita Apridiana Lestari[1], Budi Jatmiko[2], Dwikoranto[3], Binar Kurnia Prahani[4], Utama Alan Deta[5], and Rayendra Wahyu Bachtiar[6]

[1-5]Physics Department, Universitas Negeri Surabaya, Indonesia.

[6]Physics Education Program, Universitas Jember, Indonesia.

E-mail: nuritalestari@unesa.ac.id

Abstract. This study aims to describe students' perceptions of the characteristics of assisted learning mapping methods such as concept maps (CM), mind maps (MM), and link maps (LM). The study was conducted with qualitative and quantitative data collection with a sample of 10 students as pre-service physics teachers. Quantitative data becomes a supporting part of the qualitative description. Each student was asked to make a concept map, mind map, and link maps with physics material. Furthermore, students' performance is analyzed in the description form of the differences and effectiveness of the mapping when applied in physics learning. These results indicated that mapping has a positive effect on the development of students' thinking abilities, especially in identifying the right concepts for physics material.

1. Introduction

Innovative learning is the main aspect to show the renewal of the educational process. The embodiment of innovative learning is to use effective learning systems and media. The effectiveness of learning can indicate the achievement of thinking skills to be measured [1,2]. Measurement of students' thinking skills done by using maps, which have shapes that are designed so that students can find concepts precisely and structured [3]. The maps that can indicate the construction of student knowledge can be in the form of concept maps (CM), mind maps (MM), and link maps (LM). All three have different characteristics, so they can be used as indicators to measure students' thinking skills.

Concept maps can be in the form of two concepts that are connected by conjunctions to form propositions. The proposition is the relationship of one concept (information) with another concept. A concept map is a type of graphic organizer used to help students organize and represent knowledge of a subject [4]. Concept maps begin with the main idea (or concept) and then branch out to show how that main idea can be broken down into specific topics. When created correctly and thoroughly, concept mapping is a powerful way for students to reach high levels of cognitive performance. A concept map is also not just a learning tool, but an ideal evaluation tool for educators measuring the growth of and assessing student learning. As students create concept maps, they reiterate ideas using their own words and help identify incorrect ideas and concepts; educators can see what students do not understand, providing an accurate, objective way to evaluate areas in which students do not yet grasp concepts fully [5].

Mind mapping is an alternative thought of the whole brain to linear thinking. Mind mapping can reach in all directions and capture various thoughts from various angles [6]. Mind mapping is a creative, effective way of recording and mapping thoughts so that it makes it easy to place information into the brain and take information out of the brain [7].

Link Maps was first suggested by Lindstrøm [8], that conceptual knowledge is structured in a web linkage where each symbol represents each concept, and where the relationship between symbols is a unity between the concepts themselves [9]. The three mappings have been quite commonly used as supporting methods in the learning process, especially physics learning. Several studies have previously tested the effectiveness of the use of mapping in learning. However, of course, its use must be appropriate and in line with the learning objectives to be achieved. Identification is needed to determine the characteristics of the maps so that they can be applied as well as possible.

This study purposes exploring students'perception of the differences between Concept Map, Mind Map, and Link Map and the usefulness in physics learning. It is expected that students will be able to choose and use mapping as a method to support the learning process.

2. Method

This study uses a descriptive, experimental model. Methods of collecting data using quantitative and qualitative methods, with quantitative data acting as supporting qualitative data. The schema of the research carried out is shown in Figure 1. This research was conducted in last year's student in the physics department consisting of 10 students. Data collection is done by asking students to make maps that are then analyzed both qualitatively and quantitatively to obtain results. The instrument has been validated by experts and empirical validation and has a reliability of 0.605 included in the high category.

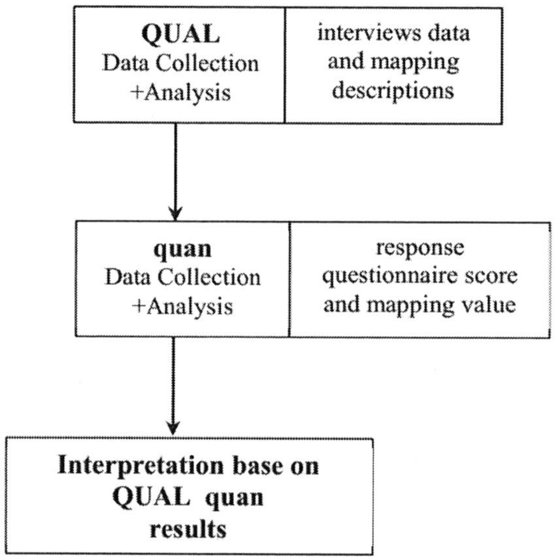

Figure 1: Step of Qualitative and Quantitative Data Collecting

3. Result and Discussion

The last year students who are studying as prospective physics teachers (class of 2016) are selected as participants because they are considered to have all the abilities needed in research. The skills are mastering physics material and pedagogy. A total of 10 students were asked to take part in the research process, which included interview stages, preparation of mapping (CM, MM, and LM), and filling in the questionnaire responses. The data obtained consisted of qualitative and quantitative data. The qualitative data are descriptions of student answers during the interview process and description of the mapping made. While the quantitative data are the result of the questionnaire response scores related to the effectiveness of using mapping in physics learning.

3.1. Interview
Qualitative data were obtained from interviews with students regarding the characteristics of concept maps, mind maps, and link maps. Examples of questions and answers during an interview are as follows.

Question:
How is the difference between the concept map, mind map, and link map in terms of the made process?
Answer:
From the made process, the concept map is the most severe form. Because it must be neat, sequential, and choose the correct concept.
Question:
What about the adequacy of physical concepts that can be learned in the form of mapping?
Answer:
The most physics concepts that can be written in the form of maps are on the mind map because we can not only write concepts but also other types of illustrations and are easier to understand.

Based on the results of responses from students who have made the mapping, the characteristics of each map obtained are as follow:
1. Concept Map, many rules must be fulfilled to produce a correct map, which must be sequential (hierarchy), not all concepts can be written down, need a more in-depth explanation of concept definitions (may not overlap with other terms, such as laws, formulas, objects, principle)
2. Mind Map, many colors, various materials can be written, each branch shows part of the central concept, interesting to apply to students in school
3. Link Map can use various colors, can write many things both concepts, laws, symbols, and images

The results obtained are categorized into several sub-sections, which can be summarized as follows.

Table 1. Conclusion about The Mapping

Categorized	Advantages	Weakness	Effectiveness
CM	Neatly arranged	Complicated, Lots of Rules	Enough
MM	Colourfull	Too abstract	Excellent
LM	Free style	difficult for others to read	Good

3.2. Mapping
Each student was asked to make a mapping with the provisions of 3 map forms for one material. The following is an example of an analysis of the results of the mapping made by students.
3.2.1. Concept Map (CM)
Descriptively, the sample concept map created by students indicates a collection of concepts with a fluid material that is interconnected with one another. The advantage of the concept map is that the concepts appear to be systematically arranged by conjunctions as a compliment. While the weakness is that the words or sentences contained in the concept box are not all concepts related to the fluid. For example, there are the words "Tegak Lurus Bidang" and "Diteruskan". These words are general words that can be used in any field and do not include concepts in the fluid matter. The correct concepts, for example, are "Zat yang Mengalir', "Gaya Apung", and so on.

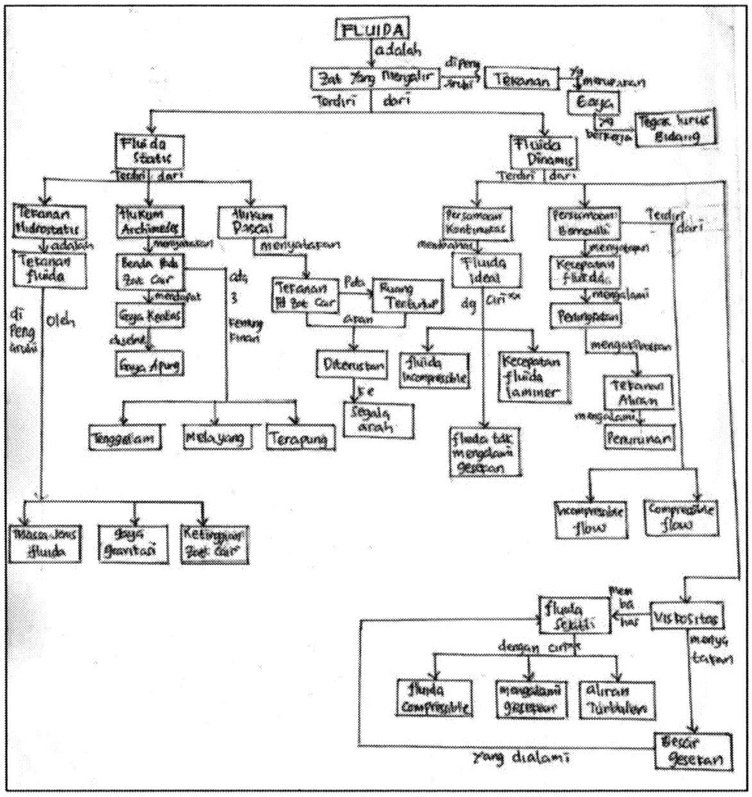

Figure 2. Example of the Mapping: Concept Map with Fluida's Material

3.2.2. Mind Map (MM)

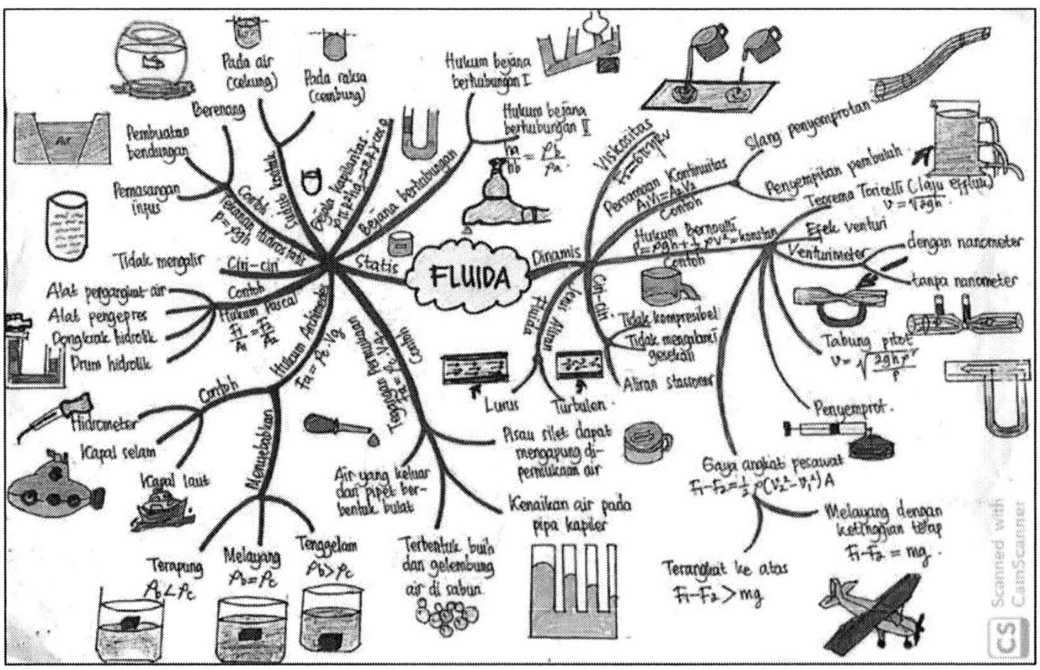

Figure 3. Example of the Mapping: Mind Map with Fluida's Material

Based on the mind map results made by students can be analyzed as follows, the concepts that written are connect with the fluid material. The image is one of the attractions in making this mind map. In addition to increasing student creativity, it also helps to understand the concept of fluid. The advantages of mind mapping include its ''free-form" and unconstrained structure. There are no limits on the ideas and links that can be made, and there is no necessity to retain an ideal structure or format. Mind mapping thus promotes creative thinking, and encourages ''brainstorming". A disadvantage of mind mapping is that the types of the made links are limited to simple associations. The absence of clear links between ideas is a constraint. Mind maps have been said to be idiosyncratic in terms of their design, often hard for others to read, representing only hierarchical relationships (in a radial form); inconsistent in terms of the level of detail; and often too complex and missing the ''big picture" [10].

3.2.3. Link Map (LM)

The use of link maps in the learning process is the rarest among the other maps. However, its effectiveness is no less good with CM and MM. Link Map is proven to be able to influence the conceptual changes that students have [11]. In theory, the preparation of LM cannot be arbitrary, although it does not have to write concepts regularly, they must be the right concepts and connected with other concepts. Following is an example of LM that has been made.

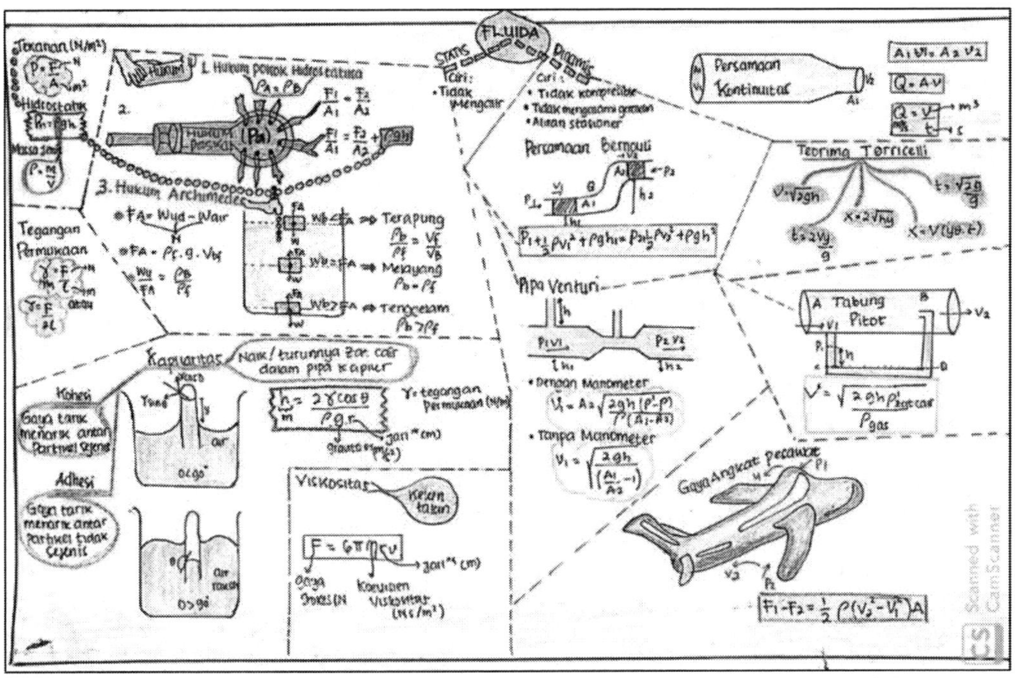

Figure 4. Example of the Mapping: Link Map with Fluida's Material

The link map is made quite well and attractive. The concepts listed are also appropriate. However, the drawback is that the interconnection between concepts is not yet apparent, so the advantages of using link maps cannot be identified

3.3. Response

At the end of the activity, students were asked to fill in the response questionnaire related to the effectiveness of the application of mapping (CM, MM, and LM) in the physics learning process. Based on the results of the questionnaire, it is known that students prefer to make Mind Maps compared to Concept Maps and Link Maps. They also believe that if Mind Maps are applied in classroom learning will increase student motivation to learn to understand physics concepts. The results of the recap and analysis of the response questionnaire are shown in Figure 5 below.

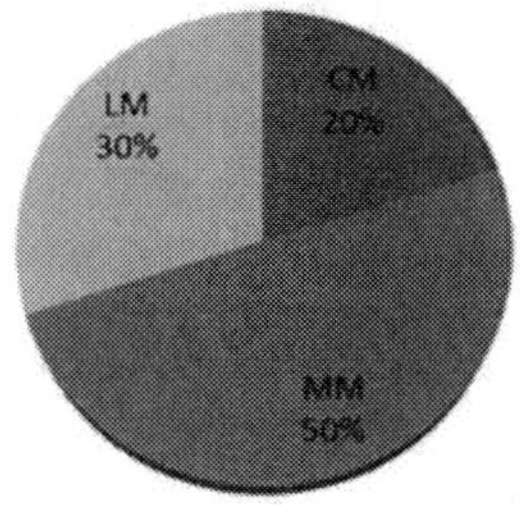

Figure 5. Percentage of the effectiveness of map methods for students

Conclusions

The results show a difference between the use of mapping methods, which consist of Concept Map, Mind Map, and Link Map. These differences are analyzed both in terms of content and form and effectiveness when used in learning. The results obtained are that a mind map is a form of mapping that is considered the most effective if used in learning. The reason is related to its easy preparation, interesting shape, and the number of concepts that can be listed quite a lot. This study is recommended for further researchers who want to explore and analyze mapping methods. s.

References

[1] Mazlina H, Halim A and Yusrizal Y 2018 Self-Description and Observers' Perspective Toward Science Teachers' Ability in Using Questioning Technique in Middle School *J. Penelit. Fis. dan Apl.* **8** 106–14

[2] Lestari N A, Suprapto N, Deta U A and Yantidewi M 2018 Implementation of Multimodel Active Learning to Improve Basic Teaching Skills of Pre-Service Physics Teachers *Journal of Physics: Conference Series* vol 1108 p 12119

[3] Forstadt L A and Doore B 2012 Program Planning with Problem Mapping to Better Understand Need *J. Ext.* **50** n1

[4] Mukhopadhyay K, Mukherjee S, Dhok A, Chatterjee C and Ghosh J 2019 Use of Concept Map as a reinforcement tool in Undergraduate Curriculum: An analytical study *J. Adv. Med. Educ. Prof.* **7** 118

[5] Hyater-Adams S, Fracchiolla C, Finkelstein N and Hinko K 2018 Critical look at physics identity: An operationalized framework for examining race and physics identity *Phys. Rev. Phys. Educ. Res.* **14** 10132

[6] Adodo S O 2013 Effect of mind-mapping as a self-regulated learning strategy on students' achievement in basic science and technology *Mediterr. J. Soc. Sci.* **4** 163

[7] Gagić Z Z, Skuban S J, Radulović B N, Stojanović M M and Gajić O 2019 The Implementation Of Mind Maps In Teaching Physics: Educational Efficiency And Students' Involvement *J. Balt. Sci. Educ.* **18**

[8] Lindstrøm C and Sharma M D 2009 Link maps and map meetings: Scaffolding student learning *Phys. Rev. Spec. Top. Educ. Res.* **5** 10102

[9] Ahda S 2017 Links Maps - Inquiry Learning (LMIL) Model to Improve Physics Thinking Ability *IOSR J. Res. Method Educ.* **07** 62–9

[10] Davies M 2011 Concept mapping, mind mapping and argument mapping: what are the differences and do they matter? *High. Educ.* **62** 279–301

[11] Futihat A 2014 Representasi Synthesis Problems Melalui Aktivitas Inquiry-Link Maps untuk Menganalisis Perubahan Konseptual Termodinamika *J. Pendidik. Sains* **2**

Science in A Black Box: Can Teachers Address Science from Socio-Scientific Issues?

E Erman[1], D A P Sari[1]

[1]Natural Science Department, Faculty of Mathematics and Natural Science, Universitas Negeri Surabaya

E-mail: erman@unesa.ac.id

Abstract. This study aims to determine teachers' ability to address science from socio-scientific issues (SSI) as predictors of scientific literacy that consider as a black box in schools. A total of 25 junior high school science teachers in some areas of East Java volunteered to be involved in this survey. Data collected by addressing macroscopic issues (AMI) tests and questionnaires were analyzed descriptively through content analysis and percentage. The results indicate that science teachers generally have difficulty in addressing science from SSI. The authentication worksheet shows that almost all AMI that can be addressed only reaches 50% maximum. Less than 50% of science teachers have heard of scientific literacy but have not been able to describe it. The implication, the student may not become a scientifically literate person if teachers are not scientifically literate, and SSI is not involved in the learning process.

1. Introduction

Nowadays, everyone cannot possibly avoid a connection between science and technology. Almost all aspects of human life are always related to science, such as food, clothing, various equipment in everyday life and natural phenomena (such as floods, earthquakes, volcanic eruptions), and so on. Therefore, understanding science issues in society or socio-scientific issues (SSI) becomes very important for students. Contextual learning that presents science issues in the classroom will significantly help students learn science and even achieve scientific literacy [1,2].

Contextual learning based on scientific issues, especially those issues that viral on social media or society, is still rarely done because these scientific issues are generally presented in a macroscopic manner so that they do not seem to contain scientific aspects [2]. At least, the conceptual level of scientific literacy is needed [3] to be able to address scientific issues in contextual learning [1,4]. The ability to understand the concept across contexts or transformation also determines the success of contextual learning [5]. Therefore, socio-scientific issues (SSI) consider as a science in a black box in this paper.

The 'Science for All' paradigm must be realized to respond to science in a black box. Science is essential for not only scientists but also all citizens who want to live or survive in an era which almost all aspects of human life depend on and determine by science and technology. Scientific literacy, which is the minimum ability that must be possessed by all citizens, is a manifestation of that paradigm. Therefore, textual-based learning should be replaced with contextual-based learning so that students achieve scientific literacy skills quickly [1,6,7]. Textual learning will only alienate students from scientific literacy because science will be difficult to learn, and its benefits become unreal [4].

Contextual-based learning will facilitate students' achievement of meaningful learning and knowledge construction [1,6]. However, these achievements could be obtained if the teachers master science concepts and science literacy, at least on the conceptual level.

This study reveals the extent of the extent to which science teachers are close to socio-scientific issues through AMI (addressing macroscopic issues). AMI measures the teacher's ability to define, describe, and explain SSI's scientific aspects and explain SSI [2].

1.1. Scientific Literacy

Science literacy is a minimum ability to master science and use it to explain science issues, even to overcome problems in life that related to science. [3] It is illustrated that science literacy as a scientific competency, which is the integration of science, context, and scientific attitude possessed by every citizen [8,9]. Science literacy will only be achieved if a person has minimal scientific knowledge (products and processes) involved in the context of his life. That is why science learning through SSI will train students to achieve scientific literacy [2].

Science issues in social media are generally macroscopic in order to be easily understood by the public. Nevertheless, SSI implicitly contains science [2], such as the use of hazardous materials in food, health, and sports. Understanding science with different contexts is not easy because it requires the ability to transform from one context to another, both terms, symbols, and their meanings [5]. As a result, SSI is still a black box that requires the role of the teacher to open it by addressing all aspects of science in it [1,4,6].

1.2. Addressing Socio-scientific Issues: Identify, Define, Describe, and Explain

Science in a black box of SSI can only be understood if students can identify, define, describe, and explain aspects of science contained in it, both explicitly and implicitly [2]. With this ability, students will be able to explain SSI, even overcome problems related to science. The more microscopic aspects that being addressed, the easier it is to explain these scientific issues in a meaningful, detailed, and comprehensive manner [2].

1.3. Teacher and Students' Scientific Literacy

Teachers have a strategic role in building students' scientific literacy. Characteristics of teachers who have scientific literacy can be detected from learning tools and the learning process [1,4], and evaluation oriented to scientific literacy [10]. Science learning using SSI and the ability to explain the benefits of science in the classroom shows that teachers have scientific literacy skills. Besides, teachers' scientific literacy can also be reviewed from the evaluation of the extent to which the learning is oriented towards scientific literacy.

2. Method

This preliminary study was conducted using a survey design involving 25 junior secondary science teachers in Kediri Region, East Java, Indonesia voluntarily based on their ability to address SSI in their area, to describe the concept of scientific literacy (context, competence, knowledge, and attitude), and to write examples of test questions multiple choice oriented scientific literacy. Two types of instruments, namely AMI (addressing macroscopic issues) and questionnaires, is used to measure the teacher's ability to identify, define, describe, explain scientific issues, apply science to explain socio-scientific issues (SSI) and describe teacher attitudes towards scientific literacy. AMI is applied through authenticated worksheets in groups (five people per group), while the questionnaire is individualized. Data were analyzed descriptively using content analysis and percentage. Content analysis is done using rubrics with an interval scale of values of 1 to 4 in 4 components of AMI [2]. In the first three rubrics, score 4 will be given to a group if all science aspects of scientific issues are identified, defined, described, and explained; score 3 if all science aspects that are explicitly written in the issues, but $\geq 50\%$ of implicitly science aspects that are not written in the issues are identified, defined, described, and explained; score 2 if less than 50% science aspects for both explicitly and implicitly science aspects; score 1 if less than 50% science aspects that are explicitly written and not for implicitly one. Lastly, in applying rubric, score 4 is given to the group if their description concerning SSI covered questions aspects: who, where, when, what, why, and how. Score 3 is given if only covered aspects what, why, and how. Score 2 is given if only covered aspects where, when, and who. The last, Score 1 is given if the description is not correct.

3. Result and Discussion

The research data and discussion are displayed separately. The data display is followed by an interpretation, while the discussion is displayed in another section.

3.1. Result

Research data are divided into two parts, namely the ability of AMI teachers (Table 1) and teacher responses to the scientific literacy paradigm (Figure 1).

Table 1 Teachers' Score in AMI test

No	AMI Component	Groups					Scores		
		A	B	C	D	E	X+	Xm	X-
1	Identifying scientific issues	3	2	2	1	2	3	2.0	1
2	Defining scientific issues	2	2	2	1	1	2	1.6	1
3	Describing scientific issues	2	3	2	2	2	3	2.2	2
4	Explaining scientific issues	1	1	2	2	1	2	1.4	1
5	Applying science to explain SSI	2	2	2	2	2	2	2.0	2
Total scores for each group		10	10	10	8	8			

Note: X+, Xm, and X- are the maximum score, mean, and minimum score, respectively. Score 4 means all science aspects of scientific issues are identified, defined, described, explained, and applied; score 3 if all science aspects that are explicitly wrote, but ≥ 50% of implicitly science aspects that are not wrote in the issues are identified, defined, described, explained, and applied; score 2 if < 50% science aspects for both explicitly and implicitly; score 1 if < 50% explicitly science aspects and 0% for implicitly one [2].

Table 1 shows that science teachers are generally only able to reach 50% in almost all AMI components (score 2 from maximum score 4). Each group is only reached a total score of 8 to 10 from the maximum score of 20. For component 4 (explaining scientific issues), which is the ability to master science material, the mean score is only 1.4, which is indicated that many teachers got difficulty to explain science aspects of SSI. Many teachers also experienced difficulties in component 2 (defining science aspects). Overall, there are two groups (D and E) still got a total score of 40% (score 8), while groups A, B, and C got 50% (score 10). Teachers' response to scientific literacy survey is as follow.

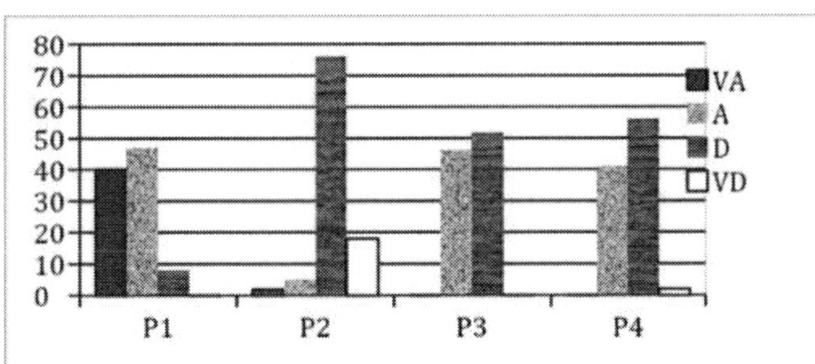

Note: P1 considers scientific literacy as a new term, P2 (already knows the concept of scientific literacy), P3 (beneficial scientific literacy), and P4 (wants to apply scientific literacy); VA (very agree, A (agree), D (disagree), and VD (very disagree)

Figure 1 Teachers' Response to Scientific Literacy Paradigm

Figure 1 shows that some teachers have heard about scientific literacy (P1, P2, P3, and P4). However, less than 50% have just heard about it through this study or sound as a new term, as indicated by respondents' disagreement in P2. Some teachers still thought that scientific literacy was useless (P3), and they did not even want to apply scientific literacy in their class.

3.2. Discussion

This study examines how close science teachers are to science issues in society. In addition to revealing the teacher's understanding of scientific literacy triggered by a number of questions as follow: 1) Have teachers been applying contextual learning-based principles? 2) Have teachers applied the principles of meaningful learning? and 3) Have teachers applied constructivist-based learning? and even 4) Has the teacher implemented real inquiry in their learning? One of the scientific literacy components is the ability to use science in a context according to needs [8,11]. The inquiry model used in science learning will only succeed if it is in accordance with students' experience (context) and capacity [12].

Based on the results of data analysis, it was found that science teachers still did not know about scientific literacy in general. Although some of the teachers have heard about it, they do not know in detail. Nearly all components in the AMI test showed that the teacher's ability was still carried by 50% (Table 1). Judging from the concept of scientific literacy [3], teachers' scientific literacy in explaining science issues in society is still at the functional level and even the nominal level. In this condition, the science teacher is still not connected with science issues in the community [2]. Generally, science learning has not been done contextually but still tends to be textual, which is conveying learning material under what is written in learning references or textbooks. Textual learning will only make students learning cognitively difficult to achieve deep learning [13]. Inquiry model that has been implemented in Indonesia since 2013 will be successful if it is done within the boundaries of students' thinking experiences and capacities [12,14]. The claim is supported by questionnaire data, which show that respondents still do not know what is meant by scientific literacy and do not even agree to apply the principles of scientific literacy in learning. The success of the application of the inquiry model must also be supported by scientific thinking skills, abstract thinking, and high-level thinking, both teachers and students [15].

This study shows that science learning has not been oriented to scientific literacy. This condition will have an impact on students' low scientific literacy, students' cognitive development so that it can have an impact on students' attitudes towards science that are less sympathetic to science. Therefore, efforts to socialize scientific literacy in schools and its application in learning should be of concern to all parties involved, the government, universities, including the education authorities in Indonesia.

4. Conclusion

The study concluded that the teachers' ability to address science issues in society is still quite low, which is still below 50%. Science teachers have heard the term of scientific literacy but have not known in detail about scientific literacy. So far, science learning in schools has not been oriented to scientific literacy, which shows that schools are still not close to scientific literacy. The low level of scientific literacy is also an indicator of the ability to construct science and enjoy learning science.

Acknowledgments

This research was supported by Faculty of Mathematics and Natural Sciences, the Universitas Negeri Surabaya. We thank our colleagues from Science Education Department who help us during this research.

References

[1] Erman E and Liliasari L 2012 Exercise to Analyze Sports Cases to Improve Mastery of the Biochemical Concepts of Sports Science Students *J. Pendidik dan Pembelajaran* **19** 94–101

[2] Erman E, Liliasari L, Ramdani M and Wakhidah N 2019 Addressing Macroscopic Issues: Helping Student Form Associations Between Biochemistry and Sports and Aiding Their Scientific Literacy *Int. J. Sci. Math. Educ.* 1–23

[3] Bybee R W 1997 *Achieving scientific literacy: From purposes to practices.* (ERIC)
[4] Parchmann I, Gräsel C, Baer A, Nentwig P, Demuth R and Ralle B 2006 "Chemie im Kontext": A symbiotic implementation of a context-based teaching and learning approach *Int. J. Sci. Educ.* **28** 1041–62
[5] Linn M C and Eylon B-S 2006 Science Education: Integrating views of learning and instruction.
[6] Potter N M and Overton T L 2006 Chemistry in sport: context-based e-learning in chemistry *Chem. Educ. Res. Pract.* **7** 195–202
[7] Passos R M, Sé A B, Wolff V L, Nobrega Y K M and Hermes-Lima M 2006 Pizza and pasta help students learn metabolism *Adv. Physiol. Educ.* **30** 89–93
[8] Bybee R, McCrae B and Laurie R 2009 PISA 2006: An assessment of scientific literacy *J. Res. Sci. Teach. Off. J. Natl. Assoc. Res. Sci. Teach.* **46** 865–83
[9] Osborne J, Simon S and Collins S 2003 Attitudes towards science: A review of the literature and its implications *Int. J. Sci. Educ.* **25** 1049–79
[10] OECD 2016 PISA 2015 Assessment and analytical framework
[11] OECD 2006 *Assessing scientific, reading and mathematical literacy: A framework for PISA 2006* (Paris: Publications de l'OCDE)
[12] Barrow L H 2006 A brief history of inquiry: From Dewey to standards *J. Sci. Teacher Educ.* **17** 265–78
[13] Wrenn J and Wrenn B 2009 Enhancing learning by integrating theory and practice. *Int. J. Teach. Learn. High. Educ.* **21** 258–65
[14] Dewey J 1986 Experience and education *The Educational Forum* vol 50 pp 241–52
[15] Erman E, Wasis W, Susantini E and Azizah U 2018 Scientific Thinking Skills: Why Junior High School Science Teachers Cannot Use Discovery and Inquiry Models In Classroom *International Conference on Science and Technology (ICST 2018)*

Improving Student's Creative Thinking Skill through Blended Learning using Schoology

F D Roqobih[1], Yuliani[2], and Y S Rahayu[2]

[1] Science Education Department, Postgraduate, Universitas Negeri Surabaya, Indonesia.

[2] Biology Department, Universitas Negeri Surabaya, Indonesia

E-mail: fikky911@gmail.com

Abstract. Creative thought is needed in the industrial revolution era 4.0 because learning can take place everywhere and every time. So, the role of technology also needed to support this learning. However, technology cannot fully replace face to face learning. One of the alternative efforts to overcome this problem is by implementing blended learning. Blended learning is learning that combines face-to-face learning in the classroom and online learning that can be done everywhere. This study aimed to evaluate the effectiveness of this learning to improve student's creative thinking skills in Bryophyte material through blended learning using Schoology. This research implemented to 22 students of Grade X by using one group pre-test and post-test design pattern. Data obtained were the student's activity and the level of student's creative thinking skills. The instrument used are observation sheets, pre-test, and post-test sheets. Data were analyzed descriptive-quantitatively. Data on creative thinking skills were analyzed based on gain score, student's creativity level, and completeness of creative learning indicators. Results of this study indicated that learning process combination of both face-to-face learning and online learning were very successful in improving student's creative thinking skill significantly, based on learning the outcome, percentage of student's activity and completeness each creative learning indicators

1. Introduction

Indonesian Ministry of Education in 2013 officially began implementing the 2013 Curriculum (K-13), which was expected to support the previous curriculum. The curriculum continued to be refined to answer the challenges in the era of industrial revolution 4.0. The 2013 curriculum invites students as active learners to be student-oriented or student-centered learning, that is, students as learning centers so that students are more active in building knowledge in their minds that develop skills in solving problems.

PISA in 2015 found that Indonesia was ranked 62nd out of 70 participant countries. Furthermore, based on the results of the TIMSS study in 2015, Indonesian students were weak in the content and cognitive aspects. Students better master routine and simple knowledge. Indonesian students need reinforcement in their ability to organize, integrate information, conclude and generalize knowledge, and use data to solve various problems [1]. The weaknesses of students must be overcome by increasing their thinking skills, so Indonesian students are well prepared to compete with the global community in the future. In the era of industrial revolution 4.0, students must have creative and innovative thinking.

Shwab states that industrial revolution 4.0 emphasizes the development of new technologies [2]. The 4.0 industrial revolution affected not only business and government but also education. In the era of industrial revolution 4.0, students must have creative and innovative thoughts so that they can think creatively [3]. So, the abilities needed in the future are related to professional engineering skills, integration, communication, ethics, independent learning, and cooperation, and the ability to think creatively [4]. There are four indicators of creative thinking, namely fluency (ability in composing a concept), flexibility (ability to change thoughts in response to a change), originality (ability to think or express different ideas), elaboration (ability to add details in an essential response. One way to create students to think creatively is by implementing Blended Learning using Schoology [5].

Blended learning is a teaching and learning activity that integrates technology in conventional learning or a combination of face-to-face learning and online learning [6]. Blended learning has various advantages such as flexibility in terms of location and time, sharing of online learning resources, the interaction between students, and interaction between students and teachers both inside and outside the classroom. Because of blended learning, students have more time to learn and explore the concepts of a material [7].

Blended learning is a learning method that can be applied to biological material, especially in material with a broad scope. In the case of Bryophyta topic, the tenth grader learns this extensive material because it includes the topic of three divisions, namely Hepatophyta (liverworts), Antherophyta (hornworts), and Bryophyta (leaf moss). They are grouped by their characteristics [8].

Some researches about blended learning have been conducted, but there is still little research which concerns about the implementation of blended learning, especially using Schoology software for Bryophyte materials to improve student's creative thinking skill. So, this study aimed to evaluate the effectiveness of this learning to improve student's creative thinking skills in Bryophyte material through blended learning using Schoology.

2. Method

2.1. Research Objects
This research applied to 22 students of grade X senior high school who have heterogenous of thinking skills. The instrument used were observation sheet of student's activity on both face to face and online learning and written test before and after learning to evaluate student's creative thinking skill.

2.2. Research Design
The type of this research was pre-experimental design with one group pre-test and post-test design [9]. Students will be given the test before learning with this method to determine their initial creative thinking skills. Then students will learn about Bryophyta topic with blended learning using Schoology software. After that, a test to determine their final creative thinking skill is conducted. The improvement also will be calculated.

2.3. Data collecting Method and Instrument
Data collection methods were through observations and written test. An individual student was stated to be creative if their test score was more than 81,6, and their improvement was evaluated by gain score. A student stated to be improved since their score of N-gain more than 70. Mastery of learning indicators was determined by the percentage of student concept mastery for each indicator. The instrument used were observation sheet of student's activity on both face to face and online learning and pre-test post-test sheet to evaluate student's creative thinking skill.

3. Result and Discussion
Effect of blended learning implementation using Schoology on Bryophyte materials for grade X was analyzed based on student's activity both on face to face learning and online learning and by pre-test and post-test scores. The observations sheet observed student activity. All the activities in the class ran well even though there were several problems. Some students did not comment on the content that had been

already uploaded by the teacher. This problem was overcome by giving more instruction repeatedly to remind students. This intervention was suitable because giving repeated instructions is very necessary for students to remind them that students go through all stages of learning desired by the teacher [10]. Student activity in face to face learning was excellent, with the percentage 98,4% (see Table 1).

Table 1. Student Activity on Face to Face Learning

Activity	Percentage (%)
Student read general instruction inside Student Worksheet	91
Student read objectives inside Student Worksheet	91
Student read procedures, materials, and equipment that necessary before observation.	100
Student did a practical activity based on worksheet	100
Student did observation in a group	100
Student made questions about Bryophyte (*fluency*).	100
Student made observations step of Bryophyte(*flexibility*).	100
Student analyzed the data about morphological characteristics of Bryophyta specimen(*originality*).	100
Students sketched Bryophyte specimen (*flexibility*).	100
Student could determine the division of each Bryophyte (*originality*).	100
Student could make a report about one of Bryophyte division in the form of a powerpoint slide (*elaboration*).	100
Total Average	98,4

Those student activities classified into the top category. Students were trained to read objectives and procedures before they did the observation. Then students analyzed, draw a sketch, grouped, and determined Bryophyte division of some plants based on their morphological characteristics. All those steps indicated that students had been already through creative learning. Such activities occur when they write some questions, they have to write some questions that related to the Bryophyte. In this case, they learned about fluency in making an idea. Then, they wrote steps of observation they learned about flexibility, which supported them to think from a different point of view. When they determined the division of each plant means they have their own opinion, and they must give some details for their decision. They already learned about originality and elaboration. Meanwhile, Student activity in online learning using Schoology also involved in outstanding category that showed by average percentage 93,6% (Table 2).

Table 2. Students Activity on *Online Learning*

Activity	Percentage (%)
Student sign in their own Schoology account	100
Student commented on information shared by the teacher in Group	86
Student participated in discussion within Schoology account	82
Student did Task 1 about reproduction, metagenesis, and roles of Bryophyte within *Schoology*.	100
Student did Quiz that had uploaded by the teacher in Schoology as a practice	100
Total Average	93,6

Based on Table 2. Student online learning activities were also categorized very well, with the percentage of student activity of 93,6%. The advantages of online learning using the Schoology application, which students can participate in discussions in the chat room with other students and teacher, work on online tasks as learning resources such as Task 1 about the reproduction, metagenesis, and roles of Bryophyte within *Schoology* and answer questions about Bryophyte in Quiz features as an exercise to make student get more profound understanding of that materials since they could share many pieces of information about that material with their friends wherever they are. It shows that blended learning had various advantages such as flexibility in terms of location and time, sharing of online learning resources, the interaction between students and teachers both inside the class and outside of class with online learning [7]. Students have more time to learn and understand more about Bryophyte concept. However, some online activities are not maximal, which were 86% of students commented on information, and 82% of students participate in a group discussion in Schoology. This condition was because researchers did not emphasize and repeat instruction on Schoology page to do those activities. Instructions are given only at the beginning of the course, whereas that instruction has a crucial role as a means of reminding students in learning. The teacher also must give more modeling of using Schoology through direct instruction [11].

Students after learned Bryophyte through both face to face learning and online learning. They had to make a product in the form of slides that contained their article of resume about a bryophyte division, then the metagenesis scheme, and the role of bryophyte based on their resources such as books and the internet. Their product was assessed with a product rubric. They used peer assessment so that a group will assess another group's product.

Student creative thinking skill was measured by a test of creative thinking ability that is given before (pre-test) and after the learning process (post-test). The results obtained were used as a reference in determining the category of creative thinking skills. The results of the pre-test and post-test then to be calculated to describe the improvement of student's creativity by using the N-gain score. The improvement of students could be categorized as high if their improvement was more than 70 (see Figure 1).

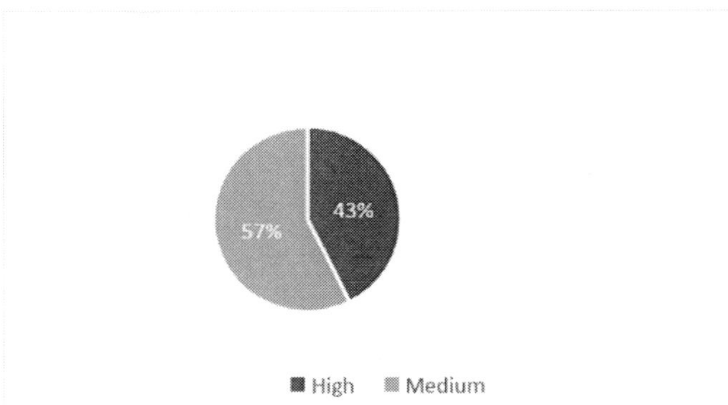

Figure 1. Improvement of Students Learning Outcome

Based on Figure 1, 57% students got significant improvement, and the rest, which was 47%, got a medium improvement of learning outcome. Besides, student's creative thinking skills can be described by Figure 2. that shows significant improvement of their creativity before and after learned Bryophyta using Blended Learning with Schoology. Students categorized as creative if the sore more than 61,2, and if the sore was more than 81,6, their creativity could be categorized as very creative.

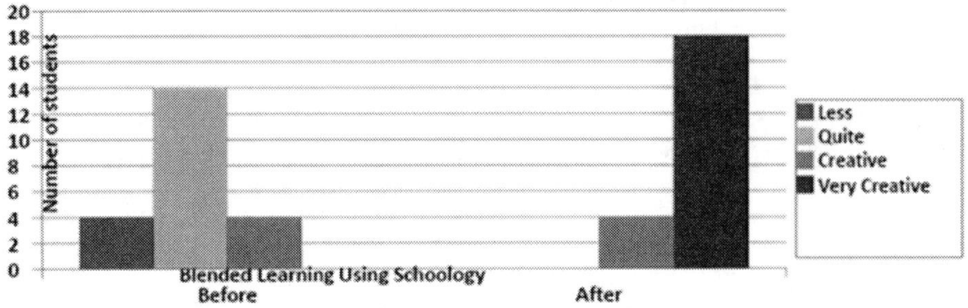

Figure 2. Level of Student's Creative Thinking Skill Before and After Blended Learning

Based on Figure 2, almost all students got to improve on their creative thinking skills in Bryophyte material because they had been learned with blended learning using Schoology. Blended learning is proven to improve successful creative thinking skills with some research that had been done by several experts. Blended Learning proved to have a positive effect on creative thinking, a significant effect on the creativity of learning outcomes, and a positive effect on the mastery of student concepts [4]. Schoology in Blended Learning also can increase student creativity in writing skills [12]. Then, another study stated that Blended Learning is proven to increase student creativity [13].

Improvement of student's creative thinking skills also can be shown by the completeness of creative learning indicators (see Table 3). This table shows the completeness of each indicator in face-to-face learning, online learning, and the whole learning process.

Table 3. Completeness of Each Creative Learning Indicators

Indicators	**Percentage (%)**
Face to face learning	
3.8.1	100
3.8.2	90,9
3.8.3	100
Online Learning	
3.8.4	81,8
3.8.5	100
3.8.6	90,9

Theoretically, Blended Learning is based on constructivist and collaborative learning theory [14]. The concept of collaboration is rooted in the social constructivist theory [15]. With social theory as a theoretical basis, collaborative learning is extended to online collaborative learning with the support of new technology. In blended learning, students learn in collaborative situations using online tools to support discovery-based learning. The basis of cooperative learning and collaborative learning in constructivism is that knowledge built and transformed by students. The environment and Blended Learning strategy have characteristics to facilitate the use of the principles and elements of the theory of constructivism in the learning process, which can improve student's creative thinking, analyzing, problem solving skills, knowledge construction, and collaborative work, through various learning strategies and supporting computer technology tools [16].

4. Conclusion and Suggestion

Based on analysis and discussion, it could be concluded as the results of this study indicated that: (a) learning process of both face-to-face learning and online learning were very successful based on student's activity and (b) succeed to improve student's creative thinking skill significantly, based on learning outcome and completeness each creative learning indicators.

The suggestion that could be given is (1) teachers must fully understand about managing their Schoology account before guiding the student in this learning so blended learning could take place well, (2) teachers should manage students so that the learning runs on time.

Acknowledgments
We are very grateful to Prof. Endang Susantini, M.Si., Dr. Tarzan Purnomo, M.Si as a validator of learning equipment and for their valuable comments for the improvement of this research. Then thank to all academic civitas of the State University of Surabaya for the support to finish this research.

References
[1] OECD 2018 *Survey International Program for International Students Assessment (PISA)* retrieved on 19th Mei 2018 from http: // www.oecd.org/pisa
[2] Shwab K 2016 *The Fourth Industrial Revolution* (New York: Crown Business)
[3] Fisk P 2017 *Education 4.0 the future of learning will be dramatically different, in school and throughout life* retrieved on 3 March 2019 from http://www.thegeniuswork.com
[4] Wu H Y and Lin H Y 2011 A Hybrid Approach to Develop an Analytical Model for Enhancing the Service Quality of E-Learning *Computers & Education* **54**(1) pp 1318-1338
[5] Sung H Y and Hwang G J 2013 A collaborative game-based learning approach to improving student's learning performance in science course *Computers & Education* **63** pp 43-51
[6] Poon J 2013 Blended Learning: An institutional approach for enhancing student's learning experiences *Merlot Journal of Online Learning and Teaching* **9**(2) pp 71-85
[7] Eryilmaz M 2015 The Effectiveness of *Blended Learning* Environments *Contemporary Issues In Education Research – 4th Quarter Journal* **8**(4) 251-256
[8] Glime J M 2017 Bryophyte Ecology Volume 1: Physiological Ecology Chapter 2 - Life Cycles and Morphology" retrieved on 5 April 2019 from https:// digitalcommons. mtu.edu/bryophyte-ecology1/1
[9] Arikunto S 2010 *Prosedur Penelitian Suatu Pendekatan Praktik* (Jakarta: Rineka Cipta)
[10] Geng G 2011 Investigation of Teachers' Verbal and Non-verbal Strategies for Managing Attention Deficit Hyperactivity Disorder (ADHD) students' Behaviours within a Classroom Environment *Australian Journal of Teacher Education* **36** pp 17-30
[11] Geraci L and Cabe D P 2006 Examining the basis for illusory recollection: The role of remember/know instructions *Psychonomic Bulletin& Review* **13** (4) pp 66-73
[12] Sicat A S 2015 Enhancing College Student's Proficiency in Business Writing Via *Schoology International Journal of Education and Research* **3** pp 160-178
[13] Bazhenov R I and Luchaninov D V 2014 Use of blended learning elements for the formation of a humanitarian student's creative initiative at learning modern information technologies *Life Science Journal* **11** (11s) pp 371-374
[14] Oliver M and Trigwell K 2005 Can 'Blended Learning' Be Redeemed? *E-Learning Journal* **2** pp 17-26
[15] Lam J 2015 Collaborative Learning Using Social Media Tools in a Blended Learning Course *Hybrid Learning: Innovation in Educational Practices* **9167** 187–98
[16] Al-Huneidi A and Schreurs J 2013 Constructivism Based Blended Learning in Higher Education *In World Summit on Knowledge Society* (Berlin, Heidelberg: Springer) pp 581–591

The Scientific Literacy Profile of Tsunami Disaster Mitigation of Non-Science Undergraduate Student in Universitas Negeri Surabaya

Utama Alan Deta[1], Putri Zulaiha Ria Agustina[1], Rizki Nur Fadillah[1], Irvan Prakoso[1], Nurita Apridiana Lestari[1], Meta Yantidewi[1], and Binar Kurnia Prahani[1]
[1]Department of Physics, Universitas Negeri Surabaya, Indonesia

E-mail: utamadeta@unesa.ac.id

Abstract. Scientific literacy needs to be developed in the school as the primary goal of science education. The purpose of the research is to describe the scientific literacy profile of a tsunami disaster for non-science undergraduate students in Universitas Negeri Surabaya. The research is a qualitative descriptive study using questionnaires and interviews for collecting the data. The sample of the study was 73 first-year non-science students from Universitas Negeri Surabaya. Based on the data analysis, it was found that the scientific literacy of non-science undergraduate students in the tsunami disaster is below average, with 40.67 scores. The fact that Indonesia had a Tsunami disaster in Aceh and Palu was not igniting student curiosity about this disaster. This phenomenon leads the student to have low scientific literacy. The student assumes that Surabaya has little risk of having a Tsunami, so they think they do not need to acknowledge the scientific facts related to Tsunami.

1. Introduction

Science literacy considered an essential skill in a developing country, especially Indonesia. Science literacy enables a person to make conclusions and the right decisions [1]. Moreover, science literacy trains critical thinking and science process skills [2]. These two skills are the basics skills for developing science and technology. Scientific literacy also considered as the root of the progressive change in science education [3].

One basic competency in Science Education in Indonesia is analyzing a kind of natural disaster and natural disaster management [4]. Natural Disaster Mitigation in Indonesia now become a hot research topic. Recently, Indonesia crushed by many natural disasters, especially tsunami. Many researchers try to investigate the tsunami in Aceh [5], the tsunami in Palu [6-7], and the potential tsunami in Lombok [8]. Furthermore, many science educators develop learning strategies and instruments to improve student awareness of natural disasters [9-11].

The research about an understanding of Disaster Mitigation has been held by Gusman et al., Rusilowati et al., Maryani, Pawestriana, and Maknun. [9-13]. However, there is no study about the scientific literacy profile of tsunami disaster mitigation. Most of the research analyses the disaster mitigation through problem-solving and non-structural learning [9-10]. For example, Rusilowati et al. study the integration of disaster mitigation and Science Environment Technology and Society (SETS) in the learning model for increasing the understanding and skills in the event of a natural disaster [9]. Also, Maryani and Yani explore the socialization of disaster mitigation is better integrated with culture or local wisdom in Sundanese society and culture [10]. Therefore, this research aims to describe scientific literacy profile of tsunami disaster mitigation, especially in non-science undergraduate student in Universitas Negeri Surabaya.

The understanding of disaster mitigation not only must be understood by scientists but also in the other society. Natural disasters may happen to everyone without exception, so it is crucial to understand disaster mitigation. This research was conducted on non-science students to find out how far the understanding of natural disaster mitigation, especially in non-science undergraduate students.

2. Research Method

The research used a qualitative descriptive method. The method used to describe and analyze the scientific literacy profile of tsunami disaster mitigation of Non-Science Undergraduate Student in Universitas Negeri Surabaya. The research was held in April 2019. The subject of the research is the undergraduate student of Management Study Program in Universitas Negeri Surabaya. The data obtained by questionnaire and interview. There is one text about tsunami and ten research questions related to the topic in the survey. The interview used to investigate the understanding and scientific literacy of students deeply. The data were analyzed quantitatively and qualitatively. Quantitative analysis was conducted on data of students' scientific literacy achievement scores, whereas the interview was analyzed qualitatively.

3. Results and Discussion

Based on the research questionnaire and data analysis, the number of correct answers can be seen in Figure 1.

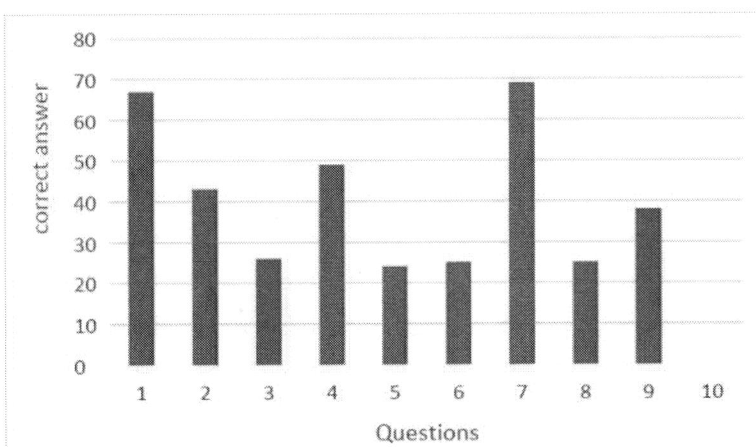

Figure 1. *The Results of the Disaster Mitigation Questionnaire*

The data in Figure 1 present the number of correct answers from 73 respondents of non-science students with ten questions. From the graph, it can be seen how the literacy ability in disaster mitigation is still low, with an average of 40.67. Many non-sciences students lack an understanding of the natural disasters that occur in Indonesia, especially those that arise in Aceh and Palu. Lack of student curiosity in reading information or news about disaster information causes students to assume that the impact of natural disasters is not too risky. They are not motivated to explore, even though disaster information is still a trending topic.

1. The following information, which can affect the potential disaster in Indonesia [14]? (A) the location between the four plate (B) have a large area **(C) Tropical climate** (D) The low land is smaller than the ocean, and many areas are a swamp

From the 1st question, students aware of the potential of natural disasters in Indonesia. Sixty-seven respondents answered correctly, according to geographical location, Indonesia, located in a tropical climate region and impact on high rainfall, strong winds, and high sea waves. It makes weathering on the ground, thus triggering natural disasters. Most of the students understand the potential causes of natural disasters. It is expected that respondents have prepared themselves if natural disasters come and may reduce the victims of natural disasters. However, this awareness is not following with the curiosity to explore the natural disaster mitigation.

8. Counted from the beginning of the earthquake in Lombok in August 2018, around 355 aftershocks occurred [15]. Why does this happen? **(A) Stabilization** (B) Normalization (C) Deformation (D) Earthquake

Stabilization is a process to return to normal conditions in an earthquake, and aftershocks are a form of stabilization after a massive earthquake occurs. Because of the large scale of the broken area, there are rocks from unstable folds, so stabilization occurs to return to normal conditions [16]. From the 8th question, from 73 of the first-year undergraduate non-science students, only 25 students answer the question correctly. It is about 2/3 students who are less motivated exploring the disaster topic; even the issue is still trending. Similarly, in the case of a tsunami (in Question 3), about sixty-six percent of students also did not understand the tsunami disaster risk.

3. Based on the history and data of natural disasters, which countries often occur Tsunami? (A) Singapore (B) United States **(C) Japan** (D) Indonesia

Japanese territory located in the Pacific Ring of Fire is the most active earthquake belt in the world. That is why Japan often experiences earthquakes and volcanic eruptions. The Pacific Ring of Fire is an imaginary horseshoe-shaped zone that follows the edge of the Pacific Ocean [17]. The natural disaster topic and mitigation are explained in junior high schools and some even at the elementary school. However, two-third of the undergraduate non-science student cannot answer the question. The fact that two disasters (earthquake and tsunami) were announced as a national disaster did not encourage students to explore more about natural disaster mitigation.

10. Pay attention to the following statement. I. The water will recede suddenly. II. Many fish in the sea die suddenly III. BMKG Information IV. The light stays on even though there is no electricity V. There was a thunderous roar. VI. An earthquake occurred VII. The weather changes instantly, and the clouds turn dark.

From this statement, which is the most appropriate sign of a tsunami?
(A) I - V - VI
(B) VII - III - IV
(C) I - II - III
(D) V - VI - VII

In question 10 that related to how a person understanding the signs of a tsunami, there were no respondents who answered correctly. Students do not understand the sign of a tsunami. If they know the sign of a tsunami, it will reduce the impact of the tsunami, especially in disaster victims.

The natural disaster topic and mitigation are explained in junior high schools and some even at the elementary school. However, two-third of the undergraduate non-science student cannot answer the question. The fact that two disasters (in Questions 3 and 8) were announced as a national disaster did not encourage students to explore more about natural disaster mitigation.

Based on the interview, the student assumes that Surabaya has little risk of having a Tsunami. They think they do not need to acknowledge the scientific facts related to Tsunami. Moreover, the student prefers to watch the information in the video format. However, even the information about Tsunami is provided online in video format, the viewer of the video is very low, compared with the entertainment or other non-educational content.

The lack of student literacy regarding tsunami disaster mitigation is in line with Maryani's research [11]. The public does not understand the knowledge of disaster mitigation, starting from knowledge about the disaster, disaster management, facilities, and infrastructure in disaster mitigation, as well as knowledge of the natural disaster mitigation law [11]. Therefore, the socialization of disaster mitigation is essential and must be packaged as attractive as possible so that the public can easily understand the concept of disaster mitigation.

Nowadays, disaster mitigation socialization by collaborating with the local culture and natural disasters has already a fundamental concept through shape buildings so that the community is sufficient to associate scientific knowledge [17-18]. This approximation expects that society will easily accept socialization. Each region has its own culture and local wisdom, reducing the risk of natural disasters, such as the shape buildings, location of the buildings, and many others [12-13]. Maknun found that local wisdom needs to be elaborated and distributed to the public [13]. In socialization, Abidin believes the need for the development of learning tools reducing the impact of natural disasters, for example, make learning tools in science subject that interest and make students learn them [19]. Finally, the community's understanding of the concept of disaster mitigation will be increased.

4. Conclusion

Based on the results and discussion, the literacy ability of non-science undergraduate students is quite low, especially in the concept of tsunami disaster mitigation. This phenomenon is caused by the lack of student motivation to literate and student attention to the impact of the tsunami disaster. The government needs to conduct more extensive socialization regarding disaster mitigation, especially the tsunami disaster.

Appendix (Research Question)

1. Pay attention to the following statement:
   (A) Located between the meeting of 3 plates
   (B) Has a wide area
   (C) **tropical climate**
   (D) Lowlands are smaller than oceans to many parts of a swamp
   Which affects the potential for disasters in Indonesia?

2. If we can assume the speed of the tsunami with the speed of the vehicle, which one suitable?
   (A) Train

(B) Motorcycle
(C) Airplane
(D) Car

3. Based on the history and Data of natural disasters, wich countries often occur Tsunami?
(A) Singapore
(B) United States
(C) Japan
(D) Indonesia

4. As reported by BNPB: Losses Due to the Palu Earthquake Reached Rp.18.4 Trillion that the Tsunami on the Palu destroyed hundreds of houses, damaged public facilities, and people died. These conditions are included in ...
(A) Disaster
(B) Threats
(C) Vulnerability
(D) Problems

5. Below the included in the principle of disaster management, except:
(A) Humanity
(B) Fast and Right
(C) Justice
(D) Togetherness

6. The following activities are included in the Post-Disaster phase
(A) Socialization and Training
(B) Preliminary Study
(C) Reconstruction and Rehabilitation
(D) Evacuation and Rescue

7. On April 12, 2019, an earthquake occurred in the Central Sulawesi region. With a magnitude of 6.9 at 18:40 WIB. Who has the authority and duty to issue warnings and the latest information about the earthquake until it is declared safe again?
(A) BNPB
(B) SAR TEAM
(C) BMKG
(D) RESIMENT

8. Counted from the beginning of the earthquake in Lombok in August 2018, around 355 aftershocks occurred [15]. Why does this happen?
(A) Stabilization
(B) Normalization
(C) Deformation
(D) Earthquake

9. Disaster mitigation in Indonesia is an activity that ....
(A) Continuous program
(B) Social
(C) Incidental
(D) Togetherness

10. Pay attention to the following statement.
    I. The water will recede suddenly.
    II. Many fish in the sea die suddenly
    III. BMKG Information
    IV. The light stays on even though there is no electricity
    V. There was a deafening roar.
    VI. An earthquake occurred
    VII. The weather changes instantly, and the clouds turn dark.
    From this statement, which is the most appropriate sign of a tsunami?
    (A) I - V - VI
    (B) VII - III - IV
    (C) I - II - III
    (D) V - VI - VII

References

[1] Martinez-hernandez K and Kimaru I 2015 Perspectives on Science Literacy: A Comparative Study of the United States and Kenya *Educational Research International* **4**(2) pp 25-34
[2] Hobson A 2008 The Surprising Effectiveness of College Scientific Literacy Courses *Physics Teacher* **46**(7) pp 404-406
[3] Sadler TD and Zeidler DL 2009 Scientific Literacy, PISA, and Socioscientific Discourse: Assessment for Progressive Aims of Science Education *Journal of Research in Science Teaching* **46**(8) pp 909-921
[4] Republic of Indonesia 2018 *Minister of Education and Culture Regulation Number 37 of 2018*
[5] Koshimura S, Oie T, Yanagisawa H, and Fumihiko I 2009 Developing Fragility Functions for Tsunami Damage Estimation using Numerical Model and Post-Tsunami Data from Banda Aceh Indonesia *Coastal Engineering Journal* **51**(3) pp 243-273
[6] Taro A 2018 Coastal Subsidence Induced Several Tsunamis During the 2018 Sulawesi Earthquake *Journal of Disaster Research Scientific Communication Online* **13** pp 1-3 sc20181204
[7] Wekke IS, Rajindra, Dinil P, Samad MA, Yani A, and Umam R 2019 Educational Institution on Responding Disasters in Palu of Indonesia *Proceedings of International Conference on Islam and Higher Education* https://osf.io/preprints/inarxiv/drc8q/
[8] The ASEAN Coordinating Centre for Humanitarian Assistance on Disaster Management (AHA Centre) 2018 *Earthquake in Lombok Indonesia Situation Update Number 8* https://ahacentre.org/situation-update/situation-update-no-8-m-6-4-and-7-0-lombok-earthquake/
[9] Gusman, Tanioka Y, Matsumoto H, and Iwasaki S 2009 Analysis of the Tsunami Generated by the Great 1977 Sumba Earthquake that Occurred in Indonesia *Bulletin of the Seismological Society of America* **99**(4) pp 2169-2179
[10] Rusilowati A, Supriyadi, Binadja A, and Mulyani SES 2012 Mitigasi Bencana Alam Berbasis Pembelajaran Bervisi Science Environment Technology and Society *Jurnal Pendidikan Fisika Indonesia* **8**(1) pp 51-60
[11] Maryani E 2010 Model Pembelajaran Mitigasi Bencana Dalam Ilmu Pengetahuan Sosial Di Sekolah Menengah Pertama *Journal of Geography* **10**(1) pp 42-58
[12] Pawestriana F 2016 Community Knowledge in Disaster Mitigation the Eruption of Slamet Volcano in Melung Village, Kedungbanteng District, Banyumas Regency *Geo Edukasi* **5**(2) pp 17-24
[13] Maknun J 2015 Pembelajaran Mitigasi Bencana Berorientasi Kearifan Lokal pada Pelajaran IPA di Sekolah Menengah Kejuruan *ATIKAN: Jurnal Kajian Pendidikan* **5**(2) pp 143-156
[14] Badan Nasional Penanggulangan Bencana 2017 https://bnpb.cloud/dibi/kondef
[15] Ramdhani J 2018 *Ada 564 Gempa Susulan, Tanggap Darurat di Lombok Ditambah 7 Hari* https://news.detik.com/berita/d-4150139/ada-564-gempa-susulan-tanggap-darurat-di-lombok-ditambah-7-hari
[16] Yuhananisa P, Yayan PN, Lailatul K, Rahil L, Wahyu SP, Ria O 2014 Sistem Telemetri Simulasi Bencana Kegempaan dan Deteksi Peningkatan Kadar Konduktivitas Belerang pada Gunung

Berapi dengan Sistem Wireless 802.15.4 *Jurnal Pendidikan Fisika dan Aplikasinya (JPFA)* **4**(2) pp 15-21

[17] Wifayanti E J and Santosa B J 2014 Estimasi Pola Bidang Sesar Dan Moment Tensor Gempa Bumi Jepang Pada Tahun 2003 Menggunakan Analisis Waveform 3 Komponen *Jurnal Sains dan Seni ITS* **3**(2) pp.B126-B131

[18] Zamzami L 2014 Kearifan Budaya Lokal Masyarakat Maritim untuk Upaya Mitigasi Bencana di Sumatera Barat *Jurnal Antropologi: Isu Isu Budaya Lokal* **16**(1) pp 37-48

[19] As'ari, H R and Nandang 2016 Kajian Nilai Kearifan Lokal Masyarakat Adat Kampung Naga dalam Pengelolaan Lingkungan Berbasis Mitigasi Bencana *Proceedings of Seminar Nasional Geografi UMS VII* https://publikasiilmiah.ums.ac.id/handle/11617/8546

Attitude, Knowledge, and Application of STEM Owned by Science Teachers

E Sujarwanto*[,1,2], Madlazim[1], and M Ibrahim[1]

[1] Science Education, Postgraduate of Universitas Negeri Surabaya, Surabaya, Indonesia.
[2] Physics Education, Universitas Siliwangi, Tasikmalaya, Indonesia.

e-mail: eko.sujarwanto@unsil.ac.id

Abstract. Providing STEM (Science Technology Engineering Mathematics) education and apply it in learning is influenced by teacher's knowledge and attitude towards The purpose of this study is to determine the teachers and lecturers' attitudes, knowledge, and application of STEM education. Data were collected from the responds of science lecturers and teacher toward a set of STEM instuments. The results point out that respondents strongly agreed to apply STEM education in learning and are sure that students would get added value when applying STEM education. Regarding aspect of knowledge, more than 50% of the respondents have adequate knowledge of the term STEM although more than 50% of the respondents do not grasp the idea of weaknesses and advantages of implementing integrated STEM. Furthermore, 76.2% of respondents tried to implement integrated STEM when teaching with a percentage between 26% -75%. The application of integrated STEM involving the discipline of engineering has a lower percentage of application than integration with other disciplines. The findings also report that obstacles to the implementation of integrated STEM are the ability to apply the mathematics concepts, technological tools available, the duration of learning time, and the process of connecting between disciplines in STEM. This research implies that there is a need to increase the understanding the interrelationships between aspects of STEM, and need to develop STEM-based learning models to support the application of integrated STEM in Science learning.

1. Introduction

Industrial Revolution 4.0 has a massive impact on the development of society. Developments that occur in the community in terms of science and technology require students to be better prepared to deal with it. Students need to be prepared because in the future human resources of school age do not become a burden of development but can be a major factor in national development. The preparation that students need to do is that students are allowed to experience learning in a multidisciplinary context [1]. Students can be prepared to face current developments in a multidisciplinary context through STEM Education.

STEM education begins with a learning program prepared by the United States Government to create high school and tertiary graduates to be able to compete globally [2]. STEM education program, which originated in the United States [2] [3], then spread to several other countries such as Australia, India, Malaysia, and Turkey [3][4][5]. Although STEM education initially developed in western countries, some developing countries experienced an increase in the use of STEM [4]. The application of STEM education as a form of preparation for facing global competition, especially in the era of industrial revolution 4.0, is carried out by both advanced countries and developing countries.

STEM is an integration of Science (S), Technology (T), Engineering (E), Mathematics (M). STEM education is a unique approach in the context of the interrelationships between Science, Technology, Engineering, and Mathematics [6][7]. STEM education as an attempt to combine all or two of the STEM disciplines in certain units (e.g. chapters or concepts) based on interdisciplinary relationships and real-world problems [8]. STEM education has the characteristics of student-centered learning. Thus, STEM education aims to prepare students to face the development of science and technology, be ready in the world of work, and be able to solve problems.

The success of STEM Education is influenced by how STEM Education is implemented within the framework of the learning model. Besides, Teachers' ability in STEM Education also determines the success of STEM education. Teachers are one of the main factors in Education, as well as STEM Education. Learning design that connects interdisciplinary in STEM Education is a complex activity. Those complex activities require teachers can make students understand how to implement STEM in real-world problems [9]. The teaching role in STEM learning can be viewed from aspects of knowledge, attitudes, and how to implement integrated STEM in learning.

There are several notions of attitude. The definition of attitude as an assessment of an object that can be stated with certain conditions such as pleasant or unpleasant, good or bad [10]. While another researcher state that attitudes are feelings towards a habit or object that is raised through negative, positive, or neutral expressions [11]. Attitudes also show mental and emotional entities that characterize one's actions and thoughts [12]. This study defines the attitude towards STEM as the teacher's level of curiosity towards STEM Education, the teacher's agreement to apply the integration of STEM during learning, and what the teacher thinks about STEM Education.

The next aspect surveyed in this study is knowledge about STEM. Knowledge is a testing situation that emphasizes the process of remembering, either by way of recognition or recalling an idea, object, or phenomenon [13]. Knowledge can be defined as one's familiarity and awareness of ideas or objects and the range of one's thoughts on certain information [14]. Based on information processing theory, knowledge comes from information obtained from the environment is then through storing and encoding processes stored in long term memory [15]. Knowledge stored in Long term memory can be divided into declarative knowledge, procedural knowledge, conditional knowledge [16]. In this study, STEM knowledge is all the information that the teacher has about the characteristics of STEM, the advantages, and the disadvantage of STEM and the integration between disciplines in STEM.

Application is an action involving an idea or procedure in an activity to achieve a certain goal. The application of STEM in this study means the efforts made by Teachers during learning to involve students in integrated STEM activities. Few teachers know how to apply STEM education. Although STEM education has been tried to be adapted in several countries, the percentage of teachers who know how to apply in learning is still small [9].

Science as an aspect of STEM is an important discipline for the development of society in science and technology. Science as a basic science includes physics, biology, and chemistry. Technological sophistication originates from Science which collaborates with mathematics and is applied through engineering principles. At the education level, Science teachers become one of the important aspects of the implementation of STEM education. Then, it is necessary to research attitudes, knowledge, and application of STEM in learning by the Science teacher. Research on integrated STEM can provide information to those interested in STEM education to identify shortcomings, barriers, and design the implementation of STEM Education that is suitable for its purpose.

2. Method

This is a survey research conducted in June 2019 and July 2019. The survey conducted focused on the attitudes, knowledge, and application of STEM by Science teachers. Teachers of Science here are teachers and lecturers of Science, especially Physics, Biology, and Science (junior high school level). The respondents are 21 people consisting of Lecturers and Teachers with teaching experience between 5 years and 10 years. Definitions of terms for STEM attitudes, STEM knowledge, and STEM applications have been mentioned in the introduction. The definition of the term research focus is adjusted to the instrument used. The instruments used are the results of adaptations of Attitude, Knowledge, and Application of STEM instruments [17]. The instrument adaptation is to adjust the subject matter of the science teacher alone to become a Physics, Biology, and Natural Sciences Lecturer and Teacher. The instrument consists of 30 statements, 2 closed-ended questions, and 2 open-ended questions. Attitude, Knowledge, and Application of STEM instruments are valid and reliable [17]. The data collection process is done online using Google Form. Data analysis was done by calculating the percentage of each response for each item statement and question.

3. Result and Discussion

This research produced data on the attitudes, knowledge, and application of STEM by Science teacher. Besides, there are data on the difficulties and obstacles that will be faced by the teacher when implementing the STEM Education approach in learning and the opinion of the teacher regarding the integrated STEM approach.

3.1. Attitude toward STEM

Attitudes toward STEM were measured using 3 statements and teachers were asked to determine the level of agreement/disagreement. Statement 1 about the agreement to apply STEM in learning, statement 2 about the teacher's belief that integrated STEM will provide added value to students, and statement 3 concerning the curiosity of the teacher how to apply STEM in learning well. The results of a survey of attitudes towards STEM are shown in Figure 1. Figure 1 shows more than 95% of respondents gave a positive response ("agree" and "strongly agree") for each statement. This result shows that the respondents strongly agreed to apply the STEM education in learning and sure that students would get added value when applying for STEM education. Furthermore, respondents are very interested in knowing more about STEM education and its application.

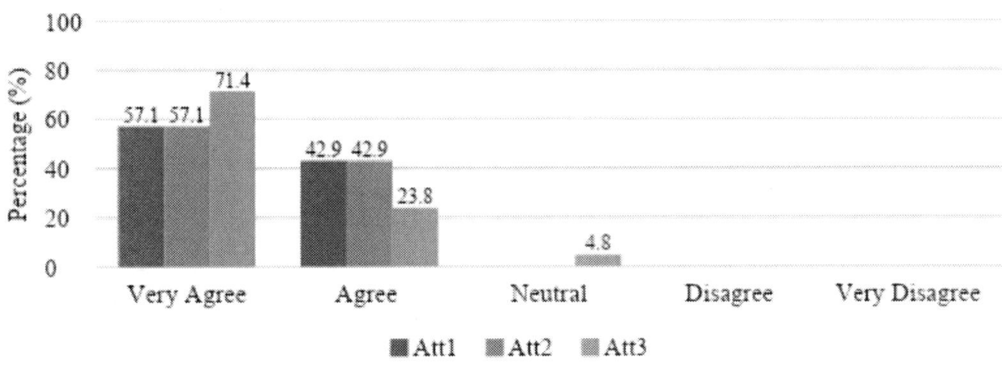

Figure 1. The response of Teacher about Attitude toward STEM (Att1: statement 1 about attitude; Att2: statement 2 about attitude; Att3: statement 3 about attitude)

The teacher's attitude towards STEM education influences the implementation of integrated STEM in learning. The teacher's attitude towards an object or idea can be transferred to students [18]. The teacher's negative attitude towards STEM education tends to make teachers reluctant to teach with the integrated STEM approach [19]. The teacher's positive attitude towards an idea will influence students' attitudes as well [20]. This study shows that teachers have a positive attitude towards STEM. A positive attitude towards STEM should be the initial capital to increase knowledge about STEM education and apply it in learning.

3.2. Knowledge about STEM

There are 4 statements to find out the STEM knowledge that the teacher has. Four statements about STEM knowledge consist of statement 1 about the familiarity of the term STEM, statement 2 concerning the nature of STEM in education, statement 3 is about awareness of the weaknesses and advantages of integrated STEM, and statement 4 about interdisciplinary integration in STEM. The survey results regarding STEM knowledge are shown in Figure 2. As many as 90.4% of respondents knew the term STEM. As many as 52.4% of respondents are aware that STEM is not a teaching method. The majority of respondents (76.2%) also believe that integrated STEM integrates at least 2 disciplines. However, only 23.8% of respondents had enough knowledge about the weaknesses and advantages of integrated STEM. These results indicate that respondents have enough STEM knowledge and need to increase knowledge about the disadvantages and advantages of implementing integrated STEM in learning.

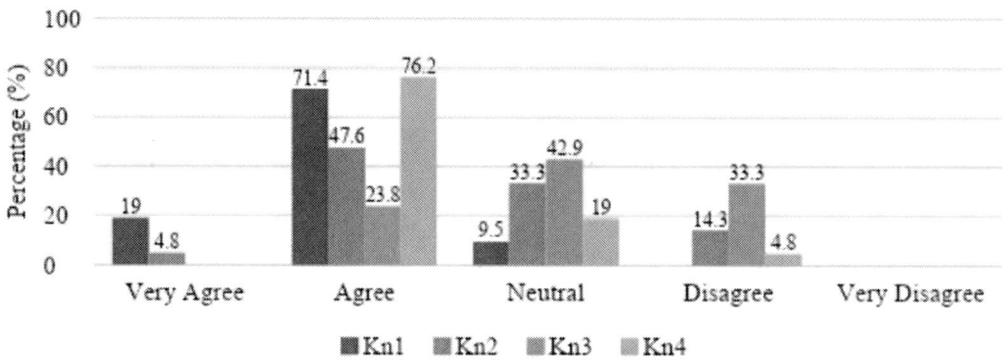

Figure 2. The response of Teacher about Knowledge of STEM (Kn1: statement 1 about knowledge; Kn2: statement 2 about knowledge; Kn3: statement 3 about knowledge; Kn4: statement 4 about knowledge)

Teachers have an important role to play in preparing learners to face advances in technology and science that occur in society. Teacher preparation in STEM education is an important factor [21]. One of the preparations for teachers in STEM education is to increase understanding of how interdisciplinary is integrated in STEM. Integrated STEM involves at least two disciplines in STEM [6]. The same idea was proposed, that is integrated STEM integrates some or all of the disciplines in STEM [8]. Teachers also need to have pedagogical content knowledge besides content knowledge to support the professional competence of the teacher [16]. Knowledge is needed to create a strong basis for educational decisions and policies [22].

Misunderstanding about integrated STEM can occur when applying it in learning. Misunderstanding about integrated STEM for example only focuses on the development of science and mathematics [2], integration with technology and engineering is still minimal [23], does not teach high-order thinking skills [23], and focuses only in one discipline that is teachers' background [6]. Misunderstanding about STEM experienced by the teacher is likely to occur since the teacher is still a prospective teacher. The prospective teacher has a misunderstanding about STEM [16][24]. Thus, in the context of STEM education, lecturers play a role in preparing a prospective teacher to face the development of technology and science in society and provide appropriate knowledge about STEM.

3.3. Application of STEM

Integrated STEM integrates at least 2 STEM disciplines in learning. The items in Application STEM must accommodate the number of possible STEM integration, namely the integration of 2 disciplines, 3 disciplines and 4 disciplines. The number of possible STEM integration is 7 items. Each item can have 3 to 4 statements. The survey results on STEM implementation are shown in Table 1.

Table 1. The response of teacher about the application of STEM

Item	No.	Very Agree (%)	Agree (%)	Neutral (%)	Disagree (%)	Very Disagree (%)
Application of Science-Technology (Ap-ST)	Ap-ST1	23.8	47.6	23.8	4.8	0
	Ap-ST2	4.8	66.7	19	9.5	0
	Ap-ST3	9.5	66.7	4.8	19	0
	Ap-ST4	4.8	76.2	14.3	4.8	0
Application of Science-Engineering (Ap-SE)	Ap-SE1	28.6	61.9	9.5	0	0
	Ap-SE2	4.8	61.9	28.6	4.8	0
	Ap-SE3	4.8	71.4	23.8	0	0
Application of Science-Mathematic (Ap-SM)	Ap-SM1	4.8	76.2	19	0	0
	Ap-SM2	9.5	71.4	19	0	0
	Ap-SM3	14.3	71.4	14.3	0	0
Application of Science-Technology-Engineering (Ap-STE)	Ap-STE1	9.5	52.4	28.6	9.5	0
	Ap-STE2	0	66.7	23.8	9.5	0
	Ap-STE3	4.8	61.9	33.3	0	0
Application of Science-Technology-Mathematic (Ap-STM)	Ap-STM1	14.3	71.4	9.5	4.8	0
	Ap-STM2	9.5	66.7	23.8	0	0
	Ap-STM3	19	57.1	14.3	9.5	0
Application of Science-Engineering-Mathematic (Ap-SEM)	Ap-SEM1	19	52.4	19	9.5	0
	Ap-SEM2	0	42.9	42.9	14.3	0
	Ap-SEM3	4.8	57.1	38.1	0	0
Application of Science-Technology-Engineering-Mathematic (Ap-STEM)	Ap-STEM1	4.8	52.4	42.9	0	0
	Ap-STEM2	4.8	71.4	23.8	0	0
	Ap-STEM3	9.5	52.4	33.3	4.8	0
	Ap-STEM4	4.8	57.1	33.3	4.8	0

The application of Science-Technology integration presents a statement about the use of electronic equipment and ready-to-use technology when teaching science by teachers or lecturers, the activity of students using simple technology or using certain procedures to make products, and use the internet by students. More than 60% of respondents gave a positive response ("agree" and "strongly agree") for each statement in item science-technology applications. Average of 9.5% of respondents gave a negative response ("disagree") for items of science-technology applications. The highest percentage of negative responses to statement 3 is the use of the internet by students. Based on open-ended questions, teachers confirm that there is no internet at some school and school policy for restrictions on carrying mobile phones. As many as 9.5% of respondents gave negative responses to statements related to the activity of students making a product. This shows that there are teachers who have not been oriented to the application of concepts to create a product.

The application of the Science-Mathematic integration presents a statement about the presentation of data on science practicum in the form of quantitative data, the use of simple statistics for data analysis, and statements about the use way-of-thinking mathematically to decision-making. The application of the integrated STEM type of Science-Mathematic integration has zero negative responses. Except for the type of Science-Mathematic integration, there must always be a negative response to other item's statement. This shows that teachers and lecturers in the natural sciences are familiar with the use of mathematics in their teaching. As a science teacher and lecturer, especially Physics, don't be too stressed on mathematical operations but forget the essence of the physics concept that is represented mathematically. This effect on students focusing more on mathematical operations than physics concept [25]. Focus can mean that students give more time to understand mathematical operations because they have difficulties in mathematical aspects. This was revealed from the results of the open-ended questions at the end of the instrument.

The application of STEM integration involving Technology and Engineering disciplines has a negative response and more neutral responses than the neutral response to the integration of Science-Mathematics. The integration of Engineering and Technology is still minimal [24]. However, in recent years there have been efforts to try to integrate Engineering and Technology in STEM learning for example through engineering design [26][27], utilizing learning software [28], and implementing integrated project-based and STEM Teaching [29].

3.4. Application of STEM and the Difficulties Faced

Based on the survey results on the application of integrated STEM, the majority of respondents gave a positive response. The aspect of application integrated STEM was followed up by asking questions about how much teacher and lecturer teach using integrated STEM. The results are shown in Table 2. Based on Table 2, there are 76.2% of respondents tried to implement integrated STEM when teaching with a percentage between 26% -75% and 76.2% of respondents tried to implement integrated STEM when teaching with a percentage between 26% -75% at each semester. Although teachers who have sufficient knowledge about the weaknesses and benefits of integrated STEM only 23.8%, most of the respondents have implemented integrated STEM in learning. Teachers and lecturers have to know that not all content can be related to technology and engineering that exist today, for example, the content of mathematics and physics that more focus on theoretical [6]. The application of learning approaches can vary depending on the material. Variations in learning approaches are needed and useful so students are not bored.

Table 2. Response about estimated time to teach using integrated STEM

Questions	<25%	26%-50%	51%-75%	76%-100%
In general, what percentage of you teach using technology tools, designing technology or designing certain concepts with students and analyzing mathematically? on science learning in the classroom.	9.5	23.8	52.4	14.3
In general, what percentage of you integrate the mathematical, technological and engineering approaches in teaching science in the classroom each semester?	9.5	28.6	47.6	14.3

In the final section of the survey, respondents were asked to answer 2 open-ended questions about opinions on the application of integrated STEM and the difficulties that will be faced when implementing integrated STEM. Opinions about the implementation of integrated STEM can be categorized into 3 categories, those are STEM education strengthens science in knowledge point of view and usefulness point of view, STEM education prepares students to face the 21st century and industrial revolution 4.0, and STEM education is difficult to implement and requires more ability to understand STEM. Meanwhile, opinions about the difficulties that will be faced when applicating integrated STEM can be categorized into 4 categories, those are the problem of learning time, facilities that support STEM education, the ability of teachers to integrate STEM, and the ability of students in mathematics.

The application of integrated STEM not only teaches each STEM discipline together. The focus in the integration of S-T-E-M in learning is on the core contents of each discipline and interdisciplinary processes [7, 30]. The implementation of integrated STEM must make the link between S-T-E-M clear and relate to real-world problems [6, 31]. The application of STEM education increased in various countries but not many knew about the method or approach to the application of STEM education instruction [9].

4. Conclusion

This research was conducted to know the attitudes, knowledge, and application of STEM in learning Science by teachers and lecturers. Based on data analysis and discussion, this study concludes that the majority of teachers and lecturers of Science have a positive attitude towards STEM education, have sufficient knowledge regarding the term STEM but do not have sufficient knowledge about the weaknesses and advantages of applicating STEM education. Conclusions related to the application of STEM education are teachers and lecturers of Science applying STEM through the integration of Mathematics-Science, but integration involving engineering and technology is minimal. Science teachers and lecturers give the opinion that learning time, facilities that support STEM education, the ability of teachers to integrate STEM, and the ability of students in terms of mathematics are the difficulties that will be faced when applicating integrated STEM. Teachers and lecturers of Science have the opinion that STEM education can strengthen science from a knowledge point of view and usefulness point of view, prepares students to compete and face technological and scientific developments in society, and need more abilities to integrate STEM because it will be more difficult to implement. The implication of this research is that there is a need to increase the understanding of teachers and lecturers about STEM education, especially the weaknesses and advantages of using STEM, increasing understanding of the interrelationships between aspects of discipline in STEM, and need to develop STEM-based learning models to support the application of integrated STEM in Science learning.

Reference

[1] Dare E A, Ellis J A, and Roehrig G H 2018 Understanding Science Teachers' Implementation of Integrated STEM Curricular Units Through A Phenomenological Multiple Case Study *International Journal of STEM Education* **5**(4)
[2] Breiner J M, Harkness S S, Johnson C C, and Koehler C M 2012 What is STEM? A Discussion About Conceptions of STEM in Education and Partnership *School Science and Mathematics* **112**(1) pp 3-11
[3] Gonzalez H B and Kuenzi J J 2012 *Science, Technology, Engineering, and Mathematics (STEM) Education: A Primer* Congressional Research Service
[4] Thomas B and Watters J J 2015 Perspectives on Australian, Indian and Malaysian approaches to STEM education *International Journal of Educational Development* **45** pp 42-53.
[5] Ceylan S and Ozdilek Z 2015 Improving a Sample Lesson Plan for Secondary Science Courses within the STEM Education *Proceedia – Social and Behavioral Sciences* **177**
[6] Sanders M 2009 STEM, STEM education, STEMmania *The Technology Teacher* **68**(4) pp 20–26
[7] English L D 2016 STEM Education K-12: Perspectives on Integration *International Journal of STEM Education* **3**(1) p 2-8
[8] Moore T, Stohlmann M, Wang H., Tank K, Glancy A, and Roehrig G 2014 Implementation and integration of engineering in K-12 STEM education. In S Purzer, J Strobel, & M Cardella (Eds.) *Engineering in Pre-College Settings: Synthesizing Research, Policy, and Practices* (pp. 35–60). West Lafayette: Purdue University Press
[9] Kelley T R and Knowles J G.2016 A Conceptual Framework for Integrated STEM Education *International Journal of STEM Education* **3**(11) pp 2–11
[10] Maio G and Haddock G 2014 *The Psychology of Attitudes and Attitude Change* London England: Sage Publications

[11] Pryor B W, Pryor C R, and Kang R 2016 Teachers' thoughts on Integrating STEM into Social Studies Instruction: Beliefs, Attitudes, and Behavioral Decisions *The Journal of Social Studies Research* **40** pp 123–136
[12] Perloff R M 2016 *The Dynamics of Persuasion: Communication and Attitudes in the Twenty-First Century* (New York:Routledge)
[13] Bloom B, Englehart M, Furst E, Hill W, and Krathwohl O 1956 *Taxonomy of educational objectives: The classification of educational goals: Handbook 1. The cognitive domain* (White Plains NY: Longman)
[14] Thomson D 1998 *The concise Oxford dictionary: 9th edn* (England: Oxford University Press)
[15] Moreno R 2010 *Educational Psychology* (New Jersey:John Wiley & Sons)
[16] Eggen P D and Kauchak D P 2010 *Educational psychology: windows on classrooms* (New Jersey: Pearson Education)
[17] Wahono B and Chang C-Y 2019 Development and Validation of a Survey Instrument (AKA) Towards Attitude, Knowledge and Application of STEM *Journal of Baltic Science Education* **18**(1) pp 63-76
[18] Deemer S 2004 Classroom goal orientation in high school classrooms: revealing links between teacher beliefs and classroom environments *Educational Research* **46** pp 73–90
[19] Appleton K 2003 How do beginning primary school teachers cope with science? Toward an understanding of science teaching practice *Journal for Research in Science Teaching* **33** pp 1–25
[20] Arends R I 2009 *Learning to Teach* (New York: McGraw-Hill).
[21] Rinke C R, Gladstone-Brown W, Kinlaw C R, and Cappiello J 2016 Characterizing STEM teacher education: Affordances and constraints of explicit STEM preparation for elementary teachers *School Science & Mathematics* **116**(6) pp 300-309
[22] Salkind N J (ed) 2008 *Encyclopedia of Educational Psychology* (California:Sage Publication)
[23] Hoachlander G and Yanofsky D 2011 Making STEM real: by infusing core academics with rigorous real-world work, linked learning pathways prepare students for both college and career *Educational Leadership* **68**(3) pp 60–65
[24] Radloff J and Guzey S 2016 Investigating Preservice STEM Teacher Conceptions of STEM Education *Journal of Science Education and Technology* **25**(5) pp 759-774
[25] Kuo E, Hull M M, Gupta A, and Elby A 2013 How students blend conceptual and formal mathematical reasoning in solving physics problems *Science Education* **97**(1) pp 32–57
[26] English L D and King D T 2015 STEM learning through engineering design: fourth-grade students' investigations in aerospace *International Journal of STEM Education* **2**(14)
[27] T Tati, Firman H, and Riandi R 2017 The Effect of STEM Learning through the Project of Designing Boat Model toward Student STEM Literacy *Journal Physics: Conference Series* **895** 012157
[28] Alan B, Zengin F K, and Kececi G 2019 Using stem applications for supporting integrated teaching knowledge of pre-service science teachers *Journal of Baltic Science Education* **18**(2) pp 158-170
[29] Siew N M and Ambo N 2018 Development and Evaluation of An Integrated Project-Based and STEM Teaching and Learning Module on Enhancing Scientific Creativity Among Fifth Graders *Journal of Baltic Science Education* **17**(6) pp 1017-33
[30] Urban M J and Falvo, D A 2016 Preface. In M J Urban & D A Falvo (Eds.) *Improving K-12 STEM education outcomes through technological integration* (Hershey, PA: IGI Global)
[31] Pearson G 2017 National academies piece on integrated STEM *The Journal of Educational Research* **110**(3) pp 224-226

AUTHOR INDEX

AUTHOR INDEX

AUTHOR INDEX

Institute of Physics
Dirac House, Temple Back
Bristol BS1 6BE UK

ISSN: 1742-6588
ISBN 978-1-7138-1322-4

CURRAN ASSOCIATES INC.
proceedings.com